T0381316

# Fundamental of Mathematical Tools for Thermal Modeling of Solar Thermal and Photo-voltaic Systems-Part-I

Gopal Nath Tiwari

# Fundamental of Mathematical Tools for Thermal Modeling of Solar Thermal and Photo-voltaic Systems-Part-I

 Springer

Gopal Nath Tiwari
BERS Public School (BPS)
Ballia, Uttar Pradesh, India

ISBN 978-981-99-7087-2      ISBN 978-981-99-7085-8  (eBook)
https://doi.org/10.1007/978-981-99-7085-8

This Springer imprint is published by the registered company Springer Nature Singapore Pte Ltd.
The registered company address is: 152 Beach Road, #21-01/04 Gateway East, Singapore 189721, Singapore

Paper in this product is recyclable.

# Dedication

**I have been always blessed unconditionally by**

**Guru Ji Padma Shri Prof. M. S. Sodha**

**Guru Ji: Professor M. S. Sodha** was born in Ajmer on February 08, 1932, Rajasthan, and pursued his all education from Uttar Pradesh, including his higher education from Allahabad University, Allahabad (UP), India. His father was a teacher in a government school in Uttar Pradesh. After his Ph.D. from Allahabad University, he got job in Defense Research and Development Organization (DRDO) in New Delhi. Before his joining, he sought the blessing from his father and his father advised was as follows:

(a) If someone comes to you for any kind of help, then you think that God blessed you in a position to help others.
(b) Before taking any decision, you place yourself in his position and expect the results from your boss.

Our Guru Ji has followed his father's advice throughout his life, and we are celebrating his 92nd birthday in the year 2024.

**Be happy with good health and help nature for longer life.**

and

**I also present this book while remembering my dearest entire UG, PG and Ph.D. students at IIT Delhi to whom I have taught**

# Preface

There are many books on solar energy thermal and photo-voltaic (PV) system for undergraduates (UG), postgraduates (PG), researchers and practitioner in the market. Few books are based on numerical examples, problems and exercises for a better understanding of the subject covered in each book. Basically, one needs basic knowledge of solar radiation and heat transfer to analyze the solar thermal and photo-voltaic system in terms of the useful thermal/electrical energy and overall and instantaneous thermal and electrical efficiency. After gaining basic knowledge of solar radiation and heat transfer, there is a requirement of writing basic energy balances of solar energy thermal, electrical (photo-voltaic) as well as photo-voltaic thermal (PVT) system to be solved for various unknown parameters under consideration for a given design and climatic parameters.

The purpose of writing this book is based on many queries by M.Tech students, research scholars and new faculty who desired to work in solar energy in my academic career of 45 years. To clear the basic mathematical concept among the learners, this book will help in starting work in the area of solar thermal and photo-voltaic (PV) system. This book will also provide a platform for many to understand the optimization of sensitive parameters which matters in terms of thermal energy (first law of thermodynamics) and exergy (second law of thermodynamics) output. Further, I suggest to the reader of this book to undertake some basic solar energy application-based courses at undergraduate (UG)/postgraduates (PG) level for gaining the knowledge of working principle of solar thermal and photo-voltaic (PV) devices, namely liquid/air flat plate collector, law of thermodynamics, solar water heating system, solar dryers, evacuated tubular collector, solar distillation system, solar houses, biogas, type of PV module with their basic parameters [fill factor (FF), short circuit current (Ioc) and open circuit voltage (Voc)] and other renewable energy system. The suggested books are as follows:

(a) Handbook of Solar Energy by G.N.Tiwari, A. Tiwari and Shyam. Springer, 2016
(b) Solar Engineering of Thermal Process by J.A. Duffie and W.A. Beckman, John and Wiley, 2002

(c)  Solar Energy by G.N.Tiwari Narosa, New Delhi and Alpha Science (UK), 2002, etc.

(d)  Developments in Environmental Durability for Photovoltaics by Basant Agrawal and G. N. Tiwari, Pira International Ltd., UK, 2008.

(e)  Fundamentals of Photovoltaic Modules and Their Applications by G. N. Tiwari and Swapnil Dubey, Royal Society of Chemistry (RSC), (UK), 2010

(f)  Advanced Renewable Energy Sources by G. N. Tiwari and R.K. Mishra, Royal Society of Chemistry (RSC), (UK), 2011.

(g)  Building Integrated Photovoltaic Thermal Systems by Basant Agrawal and G.N.Tiwari, Royal Society of Chemistry (RSC), (UK), 2010

This book has been written in nine chapters along with appendix at the end of book and some important useful constants used in examples in the beginning. There are total of more than five hundred examples covering most areas of solar energy thermal (SET), photo-voltaic (PV) and photo-voltaic thermal (PVT). The first chapter deals with many examples about the availability of hourly and monthly solar radiation in terms of sun-earth angles by using various models on any inclined with respect to horizontal with any orientation surface. The special case for horizontal surface has also been discussed. Fifty-one examples have been given based on solar radiation. The basic heat transfer including conduction, convection, radiation and mass transfer/evaporation and their corresponding heat transfer coefficient with examples have been addressed in Chapter II. The second chapter also described a total heat and an overall heat transfer coefficient with the help of thermal circuit analysis. About seventy-nine examples based on various basic heat transfer, energy and exergy have been briefly reported in the second chapter. With the help of various heat transfer coefficients, the energy balance equation for single and double glass/glazed, wetted solid surfaces, opaque and semi-transparent photo-voltaic (PV) modules along with packing factor with examples in each case has been addressed in terms of sol-air temperature in Chapter III. Effect of wind velocity on their performance has also been discussed in brief. Thermoelectric cooler integrated PV module has been also considered with examples. There are fifty-nine examples based on various energy balances which will better understand of thermal modeling. Chapter IV has analyzed the performance of single- and double-glazed liquid flat plate collectors (FPC) along with the outlet fluid temperature, the rate of thermal energy, mass flow rate/heat removal factor and an instantaneous thermal efficiency. The four combinations, namely parallel, series, parallel and series and series and parallel combination of liquid flat plate combination, have also been discussed with examples. In this case, the working fluid as water has been considered. Fifty-four examples have been given in Chapter IV. In similar pattern of Chapter IV, the flat plate air collector also referred as solar air collector has been developed by using various expressions for top loss coefficient (Ut) and bottom loss coefficient (Ub), an overall loss (UL) coefficient and solar air collector efficiency factor (F′) in Chapter V with fifty-two examples. Flow rate factor has also been discussed with single (FR) as well N-solar air connected in series (FRN) in the same chapter. Sustainable photo-voltaic thermal (PVT) liquid collectors have been discussed in terms of an overall thermal (first law of thermodynamics) and an overall exergy (second law of thermodynamics) in Chapter VI. The various types of PV modules, namely opaque (tedlar and flexible base) and semi-transparent, have been considered for analysis of PVT liquid collector performance in terms of thermal energy, thermal exergy and overall exergy. The cases, namely constant electrical efficiency and variable electrical efficiency in basic energy balances, have also been considered. There are fifty examples based on PVT liquid collectors. The concept of integration of compound parabolic concentrator with PVT liquid, PVT air and PVT-TEC collector has been analyzed in Chapter VII. The comparison in the performance between single and N-collectors connected in series in terms of thermal energy, thermal exergy and an overall exergy has been done with more than fifty examples. In Chapter VIII, a comparison in optical and thermal efficiency of solar energy systems has been carried out in terms of fractional efficiency. Further, the analysis of Fresnel lens integration with evacuated tubular collector, semi-transparent PVT collector and PVT-TEC collector has been done with many examples. There are sixty-two examples in Chapter VIII. Further, thermal energy and exergy matrices based on embodied energy for solar energy system discussed in previous chapters have been presented with some remarkable conclusions in the last chapter with fifty-one examples. A brief glossary has been given at the end of book along with appendix.

This book has been written for interdisciplinary students and research scholars with basic mathematical background belonging to all branches of basic science and engineering courses. The outcome of this book is based on my forty years of teaching experience for undergraduate (UG) and postgraduate (PG) students at IIT Delhi. During my stay at IIT Delhi, I have learned a lot from my beloved Guru ji Padma Shree Prof. M.S. Sodha. He has guided me in all aspects of life and for this I am not able to return too much to guru ji for making our peaceful life.

I am also thankful to my entire UG, PG and research scholar who inspired me to write this book. I believe that there may be some mistakes in many examples so I request all readers to point out that mistake if any, and I will assure you that it will be incorporated in the revised version.

The book will be released by Dr. Alok Srivastava, Chief Guest, USA, SOLARIS 2024 on February 08, 2024.

I am extremely thankful to Ms. Swati Meherishi, Editorial Director, Engineering, Materials Science, Energy, Water, Climate, Springer, New Delhi, India, for her support and encouragement during the preparation of manuscript.

*Further, last but not least, thanks to Mrs. Kamalawati Tiwari, Mr. Ghanshyam Nath Tiwari, Ms. Ritu Mishra, Mr. Deepak. Ms. Gopika, Shrivats, Shri Ganeshu, and Ms. Shrivani (little lovely princess) for keeping patience during this period and helping in so many ways.*

Ballia, India                                                                                          Gopal Nath Tiwari

# Important Constants

**Approximate values of some constants in renewable energy sources which is multiple of three**

| Constants | Actual value | Approximate value |
|---|---|---|
| **A** | **A** | **A** |
| Absorptivity of bare surface | | 0.3 |
| Absorptivity of blackened surface | | >0.9 |
| Altitude of ozone ($O_3$) layer present in stratosphere | 12–25 Km | 12–24 Km |
| Average heat flux from center of Earth to Earth's surface | 0.06 W/m$^2$ | 0.06 W/m$^2$ |
| Average temperature of the Earth ($\approx$ 25 °C) | 298 K | 300 K |
| **B** | **B** | **B** |
| Band gap for silicon | 1.16 eV | 1.2 eV |
| Black body temperature of the Sun's surface | 5777 K | 6000 K |
| Basic convective heat transfer coefficient from outer bare surface with zero wind velocity | 2.8 | 3 |
| Boltzmann constant | $1.38 \times 10^{-23}$ J/K | $12 \times 10^{-24}$ J/K |
| Broad classification of thermal comfort (Physical, physiological, intermediate) parameters | 3 | 3 |
| **C** | **C** | **C** |
| Central core (0–0.23R) temperature of the Sun | $8–40 \times 10^6$ K | $9–30 \times 10^9$ K |

(continued)

(continued)

|   | Constants | Actual value | Approximate value |
|---|-----------|--------------|-------------------|
|   | Convective heat transfer coefficient for air with V as a wind velocity | $2.8 + 3$ V | $3 + 3$ V |
|   | Convective and radiative heat transfer coefficient from bare surface to flowing air | $(5.7 + 3.8$ V$)$ W/m$^2$K | $(6 + 3$ V$)$ W/m$^2$K |
|   | Cooking time by solar cooker | 2–3 h | 3 h |
|   | Climatic zone in India | 6 | 6 |
|   | **D** | **D** | **D** |
| 1 | Diameter of the Sun ($2R_S$) | $1.39 \times 10^9$ m | $1.5 \times 10^9$ m |
| 2 | Distance of the Sun from the Earth | $1.5 \times 10^{11}$ m | $150 \times 10^9$ m |
| 3 | Diameter of the earth ($D = 2R$) | 13,000 km | $1.5 \times 10^6$ m |
|   | Density of air | 1.2 kg/m$^3$ | 1.2 kg/m$^3$ |
| 4 | Density of water | 997 kg/m$^3$ | 990 kg/m$^3$ |
|   | Dry biomass in biosphere | $250 \times 10^9$ ton/year | $240 \times 10^9$ ton/year |
|   | **E** | **E** | **E** |
| 5 | Energy generated at center core of the Sun | 90% | 90% |
|   | Effect of climatic parameters on yield |  | 9–12% |
|   | Effective density of states in conduction bands | $2.82 \times 10^{19}$ cm$^3$ | $27 \times 10^{18}$ cm$^3$ |
|   | Emissivity of surface | 0.9 | 0.9 |
|   | Efficiency of solar cells in standard conditions | 15% | 15% |
|   | Efficiency of PV module with Si solar cell | 12% | 12% |
|   | Energy contained in visible region | 47% (502.6 W/m$^2$) | 48% |
|   | Energy contained in infrared region | 51.02% (697.4 W/m$^2$) | 51% |
|   | Energy contained in ultraviolet (UV) region |  |  |
|   | Energy produced in one fusion reaction inside sun | 26.7 MeV | 24 MeV |
|   | **F** | **F** | **F** |
|   | Fermentation temperature of slurry for biogas production | 37 °C | 36 °C |
|   | Fin efficiency |  | 0.9 |

(continued)

(continued)

| Constants | Actual value | Approximate value |
|---|---|---|
| Flat plate collector (FPC) efficiency factor ($F'$) | 0.7 | 0.6 |
| Flow rate factor ($F_R$) | <1.0 | <0.9 |
| FPC connected in series | | $\leq 3$ |
| **G** | **G** | **G** |
| Gas-turbine operates | 600–1200 °C | 600–1200 °C |
| Geothermal energy from the Earth | $300 \times 10^{12}$ W | $300 \times 10^{12}$ W |
| **H** | **H** | **H** |
| Heating concepts (Direct, indirect, and isolated) | 3 | 3 |
| Heating value of coal | 29,000 kJ/kg | 30,000 kJ/kg |
| Heating value of biogas | 20,000 kJ/kg | 20,000 kJ/kg |
| Heating value of wood/straw | 15,000 kJ/kg | 15,000 kJ/kg |
| Heating value of gasolene/ kerosene | 42,000 kJ/kg | 42,000 kJ/kg |
| Heating value of methane | 50,000 kJ/kg | 51,000 kJ/kg |
| High-temperature geothermal well | $\geq$150 °C | $\geq$150 °C |
| Hydropower system electrical efficiency (Pelton wheel turbine base) | 90% | 90% |
| **I** | **I** | **I** |
| Ideal efficiency of solar still | 60% | 60% |
| Intermediate comfort parameters | 6 | 6 |
| Insulation thickness | 0.10 m | 0.09 m |
| **J** | **J** | **J** |
| Junction thickness near n-type semiconductor in Si | 0.15 $\mu$m | 0.15 $\mu$m |
| **L** | **L** | **L** |
| Latent heat of vaporization | $2.3 \times 10^6$ J/kg | $3 \times 10^6$ J/kg |
| Long wavelength radiation from Earth | 10 $\mu$m | 9 $\mu$m |
| Long wavelength radiation exchange ($\Delta R$) between ambient and sky | 60 W/m$^2$ | 60 W/m$^2$ |
| Low-temperature geothermal well | $\leq$150 °C | $\leq$150 °C |
| **M** | **M** | **M** |
| Mature tree consumed, $CO_2$ | 12 Kg of $CO_2$ | 12 Kg of $CO_2$ |

(continued)

(continued)

| Constants | Actual value | Approximate value |
|---|---|---|
| Maximum temperature in concentrating collector | | 3000 °C |
| Maximum wind power extraction | a = 1/3 | a = 1/3 |
| Maximum hydropower mechanical factor | $u_t = u_j/3$ | $u_t = u_j/3$ |
| Maximum power coefficient (WECS) | 59% | 60% |
| Maximum efficiency of WECS with ideal Pelton wheel curved turbine | 100% | 90% |
| Mean sun–earth angles (three w.r.t. center of earth and three w.r.t. observer on earth) | 6 | 6 |
| Methane presence in biogas | 60% | 60% |
| **N** | **N** | **N** |
| Number of sunshine hours (average) | 5 | 6 |
| Number of solar cell in standard PV module for 18 V | | 36 |
| **O** | **O** | **O** |
| Optimum till angle for maximum solar radiation, degree | Φ±15 | Φ±15 |
| Order of radiation heat transfer coefficient | 6 W/m²K | 6 W/m²K |
| Order of convective heat transfer coefficient between hot plate and water | 100 W/m²K | (90–300 )W/m²K |
| Overall heat transfer coefficient for glazed FPC single glazed | | 6 W/m²K |
| Optimum temperature for body of human | 35–37 °C | 36 °C |
| Optimum depth of water in basin of solar still | 0.02–0.03 m | 0.03 m |
| Optimum wind velocity for wind turbine | 10 m/s | 9 m/s |
| One photon has energy | 3.2883 eV | 3 eV |
| OTEC power generation, ΔT | ≥15 °C | ≥15 °C |
| **P** | **P** | **P** |
| Physical comfort parameters | 9 | 9 |
| Physiological comfort parameters | 6 | 6 |

(continued)

(continued)

| | Constants | Actual value | Approximate value |
|---|---|---|---|
| | Propane in liquefied petroleum gas (LPG) | 90% | 90% |
| | **R** | **R** | **R** |
| | Radius of Earth (R) | 6500 km | $6.5 \times 10^6$ m |
| | Rate of evaporation from free water surface ($\dot{q}_{ew}$) | $0.016 \times h_{cw} \times (Pw{-}rPa)$ | $0.015 \times h_{cw} \times (Pw{-}rPa)$ |
| | **S** | **S** | **S** |
| | Saturation current in reverse bias ($I_o$) | $10^{-8}$ A/m$^2$ | $0.1 \times 10^{-9}$ A/m$^2$ |
| | Sky temperature (°C) | ($T_a$-12) | ($T_a$-12) |
| 7 | Solar constant | 1367 W/m$^2$ | 1500 W/m$^2$ |
| | Solar intensity in terrestrial region | | 900 W/m$^2$ |
| | Solar energy for photosynthesis | $30 \times 10^{12}$ W | $30 \times 10^{12}$ W |
| | Solar energy for wind and wave conversion | $300 \times 10^{12}$ W | $300 \times 10^{12}$ W |
| | Solar energy for hydropower | $40 \times 10^{15}$ W | $30 \times 10^{15}$ W |
| | Solar energy for sensible heating | $80 \times 10^{15}$ W | $90 \times 10^{15}$ W |
| | Specific of water | 4190 J/kg °C | 4200 J/kg °C |
| | Specific heat of air | 1 kJ/kg K | 1 kJ/kg K |
| 8 | Short wavelength radiation | 0.23–2.6 μm | 0.3–3.0 μm |
| 9 | Sunshine hour at equator | 12 h | 12 h |
| | Standard test condition | 1000 W/m$^2$ and 25 °C | 900 W/m$^2$ and 24 °C |
| 10 | Stefan–Boltzmann constant | $5.67 \times 10^{-8}$ W/m$^2$ K$^4$ | $60 \times 10^{-9}$ W/m$^2$ K$^4$ |
| | Sunshine hour at north pole | 24 h | 24 h |
| | **T** | **T** | **T** |
| | Thickness of n-type semiconductor in silicon solar cell | 0.2 μm | 0.3 μm |
| | Thickness of p-type semiconductor in silicon solar cell | 0.50 mm | 0.60 mm |
| | Threshold intensity<br>(a) Winter<br>(b) Summer | | >150 W/m$^2$<br>>150 W/m$^2$ |
| | The rate of heat generated by healthy person during sleeping | 60 W | 60 W |
| | The rate of heat generated by healthy person during hard work | 600 W/m$^2$ | 600 W/m$^2$ |
| | The body temperature of human being | 37 °C | 36 °C |

(continued)

(continued)

|    | Constants | Actual value | Approximate value |
|----|-----------|--------------|-------------------|
|    | Thermal conductivity of insulating material (K) | 0.03–0.04 W/m K | 0.03 W/m K |
|    | Transmittivity of window glass | 0.9 | 0.9 |
|    | The rate of ventilation/ infiltration | $0.33\, NV\, (T_r - T_a)$ | $0.33\, NV\, (T_r - T_a)$ |
|    | Tidal power from planetary motion | $3 \times 10^{12}$ W | $3 \times 10^{12}$ W |
|    | Thermal solar energy absorbed by earth | $120 \times 10^{15}$ W | $120 \times 10^{15}$ W |
|    | Type of radiation (beam, diffuse, and reflector) | 3 | 3 |
|    | **U** | **U** | **U** |
|    | UV solar radiation | 0–0.30 $\mu$m | 0–0.30 $\mu$m |
|    | **V** | **V** | **V** |
|    | Vehicle using gasoline produces | 2.5 Kg of $CO_2$/litre | 2.4 Kg of $CO_2$/litre |
|    | V group impurity concentration | $10^{15}$ cm$^3$ | $10^{15}$ cm$^3$ |
|    | Velocity of light, c | 299,792,458 m/s | $3 \times 10^8$ m/s |
|    | **W** | **W** | **W** |
| 11 | Wein's displacement law | $\lambda T = 2897.6\ \mu$m K | 3000 $\mu$m K |
| 13 | Wavelength radiation from the sun | 0–30 $\mu$m | 0–30 $\mu$m |

# Contents

# About the Author

**Prof. Gopal Nath Tiwari** (G. N. Tiwari) hails from Sagarpali, Ballia, in the state of Uttar Pradesh, India. After completing his basic primary education from Ballia, he moved to Varanasi for higher education and completed his studies from Banaras Hindu University (BHU). Later, he joined IIT Delhi in the year 1977 under the mentorship of Padma Shri Prof. Mahendra Singh Sodha, who introduced Prof. Tiwari to solar energy research. He was superannuated in June 2016. To his credit, he has published more than 700 research papers with h-index of 100 and supervised more than 100 Ph.D. students. After superannuation, he moved back to his hometown Ballia to implement the work done at IIT Delhi and provide basic education to nearby children through BERS PUBLIC SCHOOL, BALLIA (UP). He has also established a 'Sodha Energy Research Park' at Margupur in Ballia (UP), to impart energy education to locals.

# Chapter 1
# Solar Radiation

## 1.1 Introduction

In this chapter, we will discuss some basic laws, formula and numerical examples for evaluating hourly, daily and monthly average variation of solar radiation on horizontal and inclined surface in extra-terrestrial and terrestrial region. Solar radiation, coming from the Sun at temperature of 5777 K ($\sim$ 6000 K) reaches on earth through atmosphere as shown in Fig. 1.1. The monthly average value of extra-terrestrial radiation ($I_{\text{ext}}$) is known as solar constant ($I_{\text{sc}}$), and its value is 1367 W/m$^2$. The mathematical expression for $I_{\text{ext}}$ and $I_{\text{sc}}$ will be discussed in coming sections. This will help to all reader to evaluate solar radiation on horizontal as well as inclined surface to analyze the performance of most of solar thermal (ST), photo-voltaic (PV) and photo-voltaic thermal (PVT) systems in terrestrial region. Appendix A–I provides important data for conversion of units, solar radiation, wavelength of radiation, embodied energy/ coefficient, steam table, solar cell properties, physical parameters of various working fluid and other values of turbidity of atmosphere, etc., to be used in the thermal modeling [1]. Generally, most of the solar system is analyzed either on hourly or monthly basis. For this, one needs either hourly or monthly average solar radiation as input parameter.

In order to evaluate hourly and monthly average solar radiation, the declination angle plays an important role. For this, one should consider the value of '$n$' as follows:

(a) For hourly variation of solar radiation irrespective of extra-terrestrial and terrestrial region the value of should be $n$th day of year which varies from $n = 1$ to 365/366.

(b) For average monthly variation of solar radiation irrespective of extra-terrestrial and terrestrial region, the value of '$n$' should be considered from each month given in Table 1.6.

© Bag Energy Research Society 2024
G. N. Tiwari, *Fundamental of Mathematical Tools for Thermal Modeling of Solar Thermal and Photo-voltaic Systems-Part-I*,
https://doi.org/10.1007/978-981-99-7085-8_1

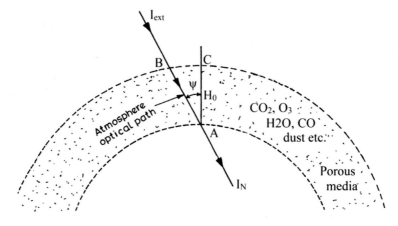

**Fig. 1.1** Propagation of solar radiation from extra-terrestrial ($I_{ext}$) region to terrestrial region ($I_N$) here $\psi = \theta_z$

## 1.2  Some Basic Laws and Definitions

The most of basic laws, definitions and examples have been considered from Tiwari [2, 3].

### 1.2.1  Stefan–Boltzmann Law

It describes the heat power radiated ($E$) from a back body in terms of its temperature ($T$ in Kelvin), and it is expressed by

$$E = \sigma T^4 \tag{1.1a}$$

where, $\sigma = 8.703 \times 10^{-7} \ W/m^2 \ K^4$

### 1.2.2  Planck's Law Radiation

The solar radiation/irradiance ($E_{\lambda b}$) from any blackbody at temperature $T$ in Kelvin as a function of wavelength ($\mu m$) is governed by Planck's law radiation. It is expressed as

$$E_{\lambda b} = \frac{C_1}{\lambda^5 [\exp\{C_2/(\lambda T)\} - 1]} \tag{1.1b}$$

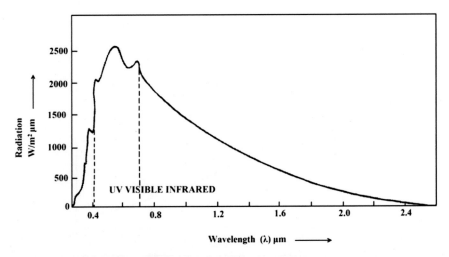

**Fig. 1.2** Solar spectral with wavelength in the absence of atmosphere variation of $I_{ext}$ with month of the year (from Tiwari and Mishra [1])

where, $E_{\lambda b}$ represents the energy emitted per unit area per unit time per unit wavelength ($\mu$m) interval at a given wavelength. Here, $C_1 = 3.742 \times 10^8$ W $\mu$m$^4$/ m$^2$ ($3.7405 \times 10^{-16}$ m$^2$ W) and $C_2 = 14{,}387.9$ $\mu$m K ($0.0143879$ m K). The variation of $E_{\lambda b}$ with wavelength is shown in Fig. 1.2. In the case of solar radiation emitted from the sun at 6000 K and earth at 300 K, Eqs. 1.1a and 1.1b are also applicable, and it is shown in Fig. 1.3.

The total emitted radiation from zero to any wavelength ($\lambda$) from the sun can be obtained from Eqs. 1.1a and 1.1b as

$$E_{0-\lambda,b} = \int_0^\lambda E_{\lambda b}d\lambda \tag{1.2}$$

If Eq. 1.2 is divided by Eq. 1.1a ($\sigma T^4$), then integral can be rearranged as follows:

$$f_{0-\lambda T} = \frac{E_{0-\lambda T}}{\sigma T^4} = \int_0^{\lambda T} \frac{C_1 d(\lambda T)}{\sigma (\lambda T)^5 \left[\exp(C_2/\lambda T) - 1\right]} \tag{1.3}$$

Equation 1.3 is a function of $\lambda T$. The value of $f_{0-\lambda T}$ for different $\lambda T$ in $\mu$m K is given in Table 1.1.

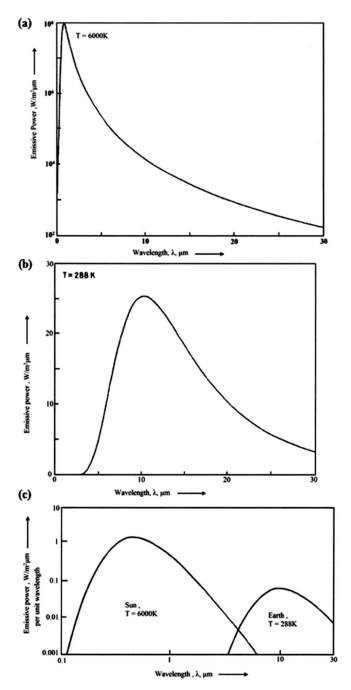

**Fig. 1.3** Effect of temperature of blackbody on emissive power, **a** $T = 6000$ K, **b** $T = 288$ K and **c** comparison

**Table 1.1**  Value of $f_{0-\lambda T}$ for different $\lambda T$ in $\mu$m K for even increment of $\lambda T$

| $\lambda T$, $\mu$m K | $f_{0-\lambda T}$ | $\lambda T$, $\mu$m K | $f_{0-\lambda T}$ | $\lambda T$, $\mu$m K | $f_{0-\lambda T}$ |
|---|---|---|---|---|---|
| 1000 | 0.0003 | 4500 | 0.5643 | 8000 | 0.8562 |
| 1100 | 0.0009 | 4600 | 0.5793 | 8100 | 0.8601 |
| 1200 | 0.0021 | 4700 | 0.5937 | 8200 | 0.8639 |
| 1300 | 0.0043 | 4800 | 0.6075 | 8300 | 0.8676 |
| 1400 | 0.0077 | 4900 | 0.6209 | 8400 | 0.8711 |
| 1500 | 0.0128 | 5000 | 0.6337 | 8500 | 0.8745 |
| 1600 | 0.0197 | 5100 | 0.6461 | 8600 | 0.8778 |
| 1700 | 0.0285 | 5200 | 0.6579 | 8700 | 0.8810 |
| 1800 | 0.0393 | 5300 | 0.6693 | 8800 | 0.8841 |
| 1900 | 0.0521 | 5400 | 0.6803 | 8900 | 0.8871 |
| 2000 | 0.0667 | 5500 | 0.6909 | 9000 | 0.8899 |
| 2100 | 0.0830 | 5600 | 0.7010 | 9100 | 0.8927 |
| 2200 | 0.1009 | 5700 | 0.7107 | 9200 | 0.8954 |
| 2300 | 0.1200 | 5800 | 0.7201 | 9300 | 0.8980 |
| 2400 | 0.1402 | 5900 | 0.7291 | 9400 | 0.9005 |
| 2500 | 0.1613 | 6000 | 0.7378 | 9500 | 0.9030 |
| 2600 | 0.1831 | 6100 | 0.7461 | 9600 | 0.9054 |
| 2700 | 0.2053 | 6200 | 0.7451 | 9700 | 0.9076 |
| 2800 | 0.2279 | 6300 | 0.7618 | 9800 | 0.9099 |
| 2900 | 0.2506 | 6400 | 0.7692 | 9900 | 0.9120 |
| 3000 | 0.2730 | 6500 | 0.7763 | 10,000 | 0.9141 |
| 3100 | 0.2958 | 6600 | 0.7831 | 11,000 | 0.9318 |
| 3200 | 0.3181 | 6700 | 0.7897 | 12,000 | 0.9450 |
| 3300 | 0.3401 | 6800 | 0.7961 | 13,000 | 0.9550 |
| 3400 | 0.3617 | 6900 | 0.8022 | 14,000 | 0.9628 |
| 3500 | 0.3829 | 7000 | 0.8080 | 15,000 | 0.9689 |
| 3600 | 0.4036 | 7100 | 0.8137 | 16,000 | 0.9737 |
| 3700 | 0.4238 | 7200 | 0.8191 | 17,000 | 0.9776 |
| 3800 | 0.4434 | 7300 | 0.8244 | 18,000 | 0.9807 |
| 3900 | 0.4624 | 7400 | 0.8295 | 19,000 | 0.9833 |
| 4000 | 0.4829 | 7500 | 0.8343 | 20,000 | 0.9855 |
| 4100 | 0.4987 | 7600 | 0.8390 | 30,000 | 0.9952 |
| 4200 | 0.5160 | 7700 | 0.8436 | 40,000 | 0.9978 |
| 4300 | 0.5327 | 7800 | 0.8479 | 50,000 | 0.9988 |
| 4400 | 0.5488 | 7900 | 0.8521 | $\infty$ | 1 |

**Example 1.1**

Evaluate the total energy contained in (a) ultraviolet (UV) region ($0 < \lambda < 0.40$ μm), (b) visible region ($0.40$ μm $< \lambda < 0.70$ μm) and (c) infrared region ($0.70$ μm $< \lambda < \infty$) as shown in Fig. 1.2 by using either Eq. 1.3 or Table 1.1. The value of solar constant ($I_{sc}$) is 1367 W/m².

**Solution**

By using the value of Table 1.1,

(a)  The value of $\lambda T$ at 5777 K for ultraviolet region $= 0.4 \times 5777 = 2310.8$ μm K. From Table 1.1,

    The energy contained in fraction at $\lambda T = 2300$ is 0.1200,

    The energy contained in fraction at $\lambda T = 2400$ is 0.1402.

    So, the energy contained in fraction per unit $\lambda T = [(0.1402 - 0.1200)/100] = 0.000202$.

    Then, the energy contained in fraction for 10.8 μm K $= 0.000202 \times 10.8 = 0.00218$ (approximately 0.0004).

    The energy contained in fraction between 0 and 2301.8 μm K $= 0.1200 + 0.00218 = 0.1222$ (12.22%).

    The energy contained between 0 and 2301.8 μm K $= 0.1222 \times 1367 = 167.05$ W/m².

(b)  The value of $\lambda T$ at 0.70 μm for the visible region $= 0.7 \times 5777 = 4043.9$ μm K. From Table 1.1,

    The energy contained in fraction at $\lambda T = 4000$ is 0.4829,

    The energy contained in fraction at $\lambda T = 4100$ is 0.4987.

    So, the energy contained in fraction per unit $\lambda T = [(0.4987 - 0.4829)/100] = 0.000158$.

    Then, the energy contained in fraction for 43.9 μm K $= 0.000158 \times 43.9 = 0.0069$.

    The energy contained in fraction between 0 and 2301.8 μm K $= 0.4829 + 0.0069 = 0.4898$ (48.98%).

    The energy contained between 2301.8 and 4043.9 μm K $= (0.4898 - 0.1222) \times 1367 = 502.51$ W/m².

(c)  The remaining $100\text{–}48.98 = 51.02\%$ (697.44 W/m²) is contained in the infrared region.

It is clear from above calculation that there is significant change in energy contained from the range of ultraviolet (UV) to infrared regions (Fig. 1.1).

## 1.2.3  Wien's Displacement Law

According to Wien's displacement law, the product of the wavelength corresponding to the maximum of solar radiation/irradiance from the sun at temperature of $T = 5777$ K (5504 °C) Kelvin and temperature of sun is constant, and it can written as

$$\lambda_{\max} T = 2897.8 \, \mu m \, K \qquad (1.4)$$

**Example 1.2**

Prove Wien's displacement law by using Planck's law radiation given by Eqs. 1.1a and 1.1b.

**Solution**

Differentiate Eq. 1.1 with respect to $\lambda$ and equate it to zero to get maximum wavelength ($\lambda_{\max}$). After differentiation of Eqs. 1.1a and 1.1b, one gets

$$\frac{dE_{\lambda b}}{d\lambda} = \lambda^{-5}\left(e^{C_2/\lambda T} - 1\right)^2 e^{\frac{C_2}{\lambda T}} \cdot \frac{C_2}{T}\lambda^{-2}d\lambda + \left(e^{\frac{C_2}{\lambda T}} - 1\right)(-5\lambda^{-6})d\lambda = 0$$

Since $\dfrac{C_2}{\lambda T} \le 1$, then, $e^{\frac{C_2}{\lambda T}} = 1 + e^{\frac{C_2}{\lambda T}}$ and, $e^{\frac{C_2}{\lambda T}} \to 1$, then

$$\frac{C_2}{\lambda T} = 5 \Rightarrow \lambda_{\max} T = 2897.8 \, \mu m \, K$$

This value is very close to the value given by Eq. 1.4.

**Example 1.3**

Calculate the maximum monochromatic emissive power at 288 K.

**Solution**

From the above example, $\lambda_{\max} = 2897.8/288 = 10.06 \, \mu m$.

By using Eq. 1.2, the maximum monochromatic power will be

$$E_{b\lambda} = \frac{3.742 \times 10^8}{(10.06)^5[\exp(14387/2897.6) - 1]} = 25.53 \quad W/m^2 \, \mu m$$

The value of $E_{\lambda b}$ is the same as reported in Fig. 1.3b at $\lambda_{\max} = 10.06 \, \mu m$.

### 1.2.4 Solar Intensity in Extra-terrestrial Region

The monthly variation of solar radiation/intensity in extra-terrestrial region which is measured on a plane normal to the solar radiation on the $n$th day of the year, Fig. 1.1, is given in terms of solar constant $I_{sc}$ as:

$$I_{ext} = I_{sc}\left[1 + 0.033 \cos\left(\frac{360 \times n}{365}\right)\right] \qquad (1.5)$$

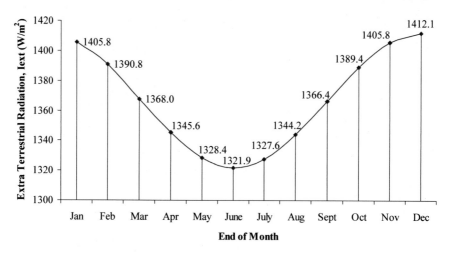

**Fig. 1.4**  Variation of $I_{ext}$ with month of the year

where, solar constant, $I_{sc}$, is an average value of monthly variation of extra-terrestrial region.

The monthly variation of extra-terrestrial solar radiation ($I_{ext}$) is also shown in Fig. 1.4. It is to be noted that for leap year, February month is having 29 days and accordingly value of $n$ is increased by 1 in Eq. 1.5 after February 28. For normal year, the variation of extra-terrestrial radiation on earth's surface for $n$ given in Table 1.6 for each month using Eqs. 1.6a and 1.6b can be evaluated as follows:

$$\text{For January 2022, } n = 17, \ I_{ext} = 1410 \ \text{W/m}^2 \tag{1.5a}$$

$$\text{For June 2022, } n = 162, \ I_{ext} = 1324 \ \text{W/m}^2 \tag{1.5b}$$

## 1.2.5  Air Mass

Referring to Fig. 1.1, the atmospheric attenuation is characterized by a term called air mass. It is defined as the ratio of the optical thickness of the atmosphere through which beam/direct solar radiation passes to the vertical depth/thickness of atmosphere if the sun is at zenith.

An expression for air mass referring to Fig. 1.1 is defined as follows:

$$\text{Air mass}(m) = \frac{\text{path length traveled in the atmosphere}}{\text{vertical depth of atmosphere}} = \frac{AB}{AC}$$

$$= \frac{m_o}{H_o} = \sec\theta_z = \frac{1}{\cos\theta_z} \text{ for } \theta_z \leq 0 \quad [\text{here } \psi = \theta_z] \quad (1.6a)$$

At noon, $\theta_z = 0$, $m = 1$ for $\theta_z = 60°$, $m = 2$ and $m = 0$ for outside the earth atmosphere.

Further, Kasten [4] has also proposed the modified expression for air mass as follows:

$$m = \left[\cos\theta_Z + 0.15 \times (93.885 - \theta_Z)^{-1.253}\right]^{-1} \quad (1.6b)$$

## Example 1.5

Derive an expression for air mass in terms of zenith angle ($\theta_z$) and altitude angle ($\alpha$).

**Solution**

In Eq. 1.6a,

$$\text{Air mass} = \frac{1}{\cos\theta_z} \quad (\text{Eq. 1.6a})$$

Here, $\cos\theta_z$ is given by Eq. 1.12. Since $\alpha = 90 - \theta_z$, hence

$$\text{Air mass} = \text{cosec}\alpha$$

## Example 1.6

Evaluate air mass ($m$) of normal direct irradiance coming from sun at New Delhi at 2.30 p.m. on February 20, 2023.

**Solution**

Here, $\cos\theta_Z = 0.587$ and $\theta_Z = 54.03°$ at 2.30 p.m. on February 20, 2023, New Delhi.

Now, substituting the above values in Eq. 1.6b, we have

$$m = \left[0.587 + 0.15((93.885 - 54.03))^{-1.253}\right]^{-1} = 1.699$$

The variation of air mass with time of the day for the latitude of Delhi for different number of days of the year is shown in Fig. 1.5. It is observed that the sunshine hours are shorter and air mass is higher for the month of December 21, in comparison with other days as expected.

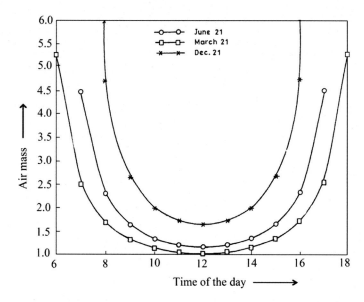

**Fig. 1.5** Variation of air mass with hour of day

### 1.2.6 Solar Time

Time based on the apparent angular motion of the sun across the sky, with solar noon denoting the time the sun crosses the meridian of the observer.

The difference in minutes between solar time and standard time is

$$\text{Solar time} - \text{Standard time} = 4(L_{st} - L_{loc}) + E \tag{1.7}$$

where, $L_{st}$ is the standard meridian for the local time zone. The $L_{st}$ for India has the value of $81°44'$. $L_{loc}$ is the longitude of the location in question (in degrees west) given in Table 1.2, $E$ is the equation of time (in minutes), and it is expressed as

$$E = 229.2(0.000075 + 0.001868 \cos B - 0.032077 \sin B$$
$$- 0.014615 \cos 2B - 0.04089 \sin 2B) \tag{1.8}$$

where, $B = (n - 1)360/365$, $n =$ day of the year.

The equation of time (minutes: seconds) for typical days for different months for Delhi (Longitude $77°12'E$) is given in Table 1.3.

**Table 1.2** Latitude, longitude and elevation for different places in India

| Place | Latitude ($\phi$) | Longitude ($L_{loc}$) | Elevation ($E_0$) |
|---|---|---|---|
| New Delhi | 28°35′ N (≈30°) | 77°12′ E | 216 m above msl |
| Mumbai | 18°54′ N | 72°49′ E | 11 m above msl |
| Chennai | 13°00′ N | 80°11′ E | 16 m above msl |
| Kolkata | 22°32′ N | 88°20′ E | 6 m above msl |
| London | 51°30′ N | 00°07′ W | 35 m above msl |
| New York | 40°42′ N | 74°00′ W | 10 m above msl |
| Paris | 48°51′ N | 02°21′ E | 35 m above msl |
| Moscow | 55°45′ N | 37°37′ E | 156 m above msl |
| Singapore | 01°17′ N | 103°50′ E | 6 m above msl |
| Beijing | 39°54′ N | 116°23′ E | 50 m above msl |
| Berlin | 52°31′ N | 13°23′ E | 34 m above msl |

**Table 1.3** Sun's equation of time ($E$) (minutes: second)

| Month | 1 | 8 | 15 | 22 |
|---|---|---|---|---|
| January | − (3:16) | − (6:26) | − (9:12) | − (11:27) |
| February | − (13:34) | − (14:14) | − (14:15) | − (13:41) |
| March | − (12:36) | − (11:04) | − (9:14) | − (7:12) |
| April | − (4:11) | − (2:07) | − (0:15) | (1:19) |
| May | 2:50 | 3:31 | 3:44 | 3:30 |
| June | 2:25 | 1:15 | − (0:09) | − (1:40) |
| July | − (3:33) | − (4:48) | − (5:45) | − (6:19) |
| August | − (6:17) | − (5:40) | − (4:35) | − (3:04) |
| September | − (0:15) | 2:03 | 4:29 | 6:58 |
| October | 10:02 | 12:11 | 13:59 | 15:20 |
| November | 16:20 | 16:16 | 15:29 | 14:02 |
| December | 11:14 | 8:26 | 5:13 | 1:47 |

## Example 1.7

Determine the solar time (ST) corresponding to 12.00 noon Indian Standard Time (IST) (Longitude 81°54′E on May 8, 1995, for Delhi.

## Solution

Equation of time for May 8 is 3 min 31 s (Table 1.3). The longitude correction, in this case, would be negative as Delhi is west of standard meridian. Using Eq. 1.7, we have

$$ST = 12 \text{ h } 0 \min 0 \text{ s} + 3 \min 31 \text{ s} - 4 \min \left(81°54' - 77°12'\right)$$
$$= 11\text{h } 45 \min 23 \text{ s}.$$

Here, $1° = 60'$ or 4 min (Appendix A).

### 1.2.7  Declination ($\delta$)

It is defined as the angle between the line joining the centers of the sun and the earth and its projection on the equatorial plane. It varies with number of days in year. For a given $n$th day of year, Cooper's empirical relation for calculating the solar declination angle (in degrees) has been used, and it is given by

$$\delta = 23.45 \sin\left[(284 + n) \times \frac{360}{365}\right] \tag{1.9}$$

where, $n = n^{\text{th}}$ day of the year ($1 \leq n \leq 365/366$).

This solar declination angle ($\delta$) changes due to rotation of earth about an axis. Its maximum value is 23.45° on December 21 and minimum is $- 23.45°$ on June 21 as shown in Fig. 1.6.

If $n$ is considered as monthly average day as given in Table 1.6 along with declination value, then all calculation for solar radiation will be monthly average value, and it is used to evaluate monthly performance of any solar system.

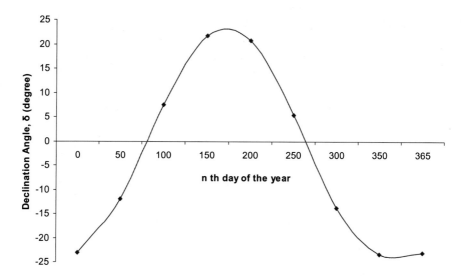

**Fig. 1.6** Variation of declination angle with $n^{\text{th}}$ day of year

**Example 1.8a**

Calculate the declination ($\delta$) on September 23, 2023.

**Solution**

For September 23, 2023, $n = 266$. From Eq. 1.9, we have

$$\delta = 23.45 \sin\left[\frac{360}{365}(284 + 266)\right] = -1.008$$

**Example 1.8b**

Calculate the declination ($\delta$) for monthly average day of September 2023.

**Solution**

For September 2023 the monthly average day ($n$) = 258. From Eq. 1.9, we have

$$\delta = 23.45 \sin\left[\frac{360}{365}(284 + 258)\right] = 2.22$$

So, one can observed that for daily variation, the declination angle ($\delta$), Example 1.8a, is different than the average monthly variation, Example 1.8b.

### 1.2.8 Hour Angle ($\omega$)

An expression for hour angle, $\omega$ (in degree), is given by

$$\omega = (ST - 12) \times 15 \tag{1.10}$$

where, $ST$ is local solar time. The value of hour angle for different local time is given in Table 1.4.

**Table 1.4** Value of hour angle with time of the day (for northern hemisphere)

| Time of the day (h) | 6 | 7 | 8 | 9 | 10 | 11 | 12 |
|---|---|---|---|---|---|---|---|
| Hour angle (°) | − 90 | − 75 | − 60 | − 45 | − 30 | − 15 | 0 |
| Time of the day (h) | 12 | 13 | 14 | 15 | 16 | 17 | 18 |
| Hour angle (°) | 0 | + 15 | + 30 | + 45 | + 60 | + 75 | + 90 |

**Table 1.5**  Surface azimuth angle ($\gamma$) for various orientations in northern hemisphere

| Surface orientation | $\gamma$ |
|---|---|
| Sloped toward south | 0° |
| Sloped toward north | − 180° |
| Sloped toward east | − 90° |
| Sloped toward west | + 90° |
| Sloped toward southeast | − 45° |
| Sloped toward southwest | + 45° |

**Table 1.6**  Recommended average days for months and values of $n$ by months [2]

| Month | $n$ for $i$th day of month | For the average day of the month | | |
|---|---|---|---|---|
| | | Date | Day of year ($n$) | Declination ($\delta$) |
| January | $i$ | 17 | 17 | − 20.9 |
| February | $31 + i$ | 16 | 47 | − 13.0 |
| March | $59 + i$ | 16 | 75 | − 2.4 |
| April | $90 + i$ | 15 | 105 | 9.4 |
| May | $120 + i$ | 15 | 135 | 18.8 |
| June | $151 + i$ | 11 | 162 | 23.1 |
| July | $181 + i$ | 17 | 198 | 21.2 |
| August | $212 + i$ | 16 | 228 | 13.5 |
| September | $243 + i$ | 15 | 258 | 2.2 |
| October | $273 + i$ | 15 | 288 | − 9.6 |
| November | $304 + i$ | 14 | 318 | − 18.9 |
| December | $334 + i$ | 10 | 344 | − 23.0 |

### 1.2.9  Surface Azimuth Angle ($\gamma$)

It is an angle between line due to south and projection of normal to inclined surface in a horizontal plane. The value of surface azimuth angle ($\gamma$) for different orientation is given in Table 1.5.

### 1.2.10  Angle of Incidence ($\theta_i$)

It depends on latitude (Table 1.2), inclination of plane ($\beta$) with respect to horizontal plane, surface azimuth angle ($\gamma$), declination ($\delta$) (Fig. 1.3) and hour angle ($\omega$) (Table 1.4) and its expression for $\cos\theta_i$ is given by

$$\cos \theta_i = (\cos \Phi \cos \beta + \sin \Phi \sin \beta \cos \gamma) \cos \delta \cos \omega + \cos \delta \sin \omega \sin \beta \sin \gamma$$
$$+ (\sin \Phi \cos \beta - \cos \Phi \sin \beta \cos \gamma) \times \sin \delta \qquad (1.11a)$$

For south facing surface, surface azimuth angle $(\gamma) = 0$, then Eq. 1.11a becomes

$$\cos \theta_i = [\cos(\Phi - \beta)] \cos \delta \cos \omega + [\sin(\Phi - \beta)] \times \sin \delta \qquad (1.11b)$$

For a horizontal plane facing due south, $\gamma = 0$, $\beta = 0$, then the angle of incidence $\theta_i$ becomes equal to zenith angle, i.e., $\theta = \theta_z$ (Zenith angle). Equations 1.11a and 1.11b reduces for $\cos \theta_z$ as

$$\cos \theta_z = \cos \phi \cos \delta \cos \omega + \sin \delta \sin \phi \qquad (1.12)$$

The zenith angle $(\theta_Z)$ will be 90° in the morning as well as in the evening with respect to noon time before sunrise and after sunset, and then $\omega = \omega_s$. So, Eq. 1.12 can be written as follows:

$$\cos 90 = 0 = \cos \varphi \cos \delta \cos \omega_s + \sin \varphi \sin \delta$$

or,

$$\cos \omega_s = - \tan \varphi \tan \delta$$

or,

$$\omega_s = \cos^{-1}(- \tan \varphi \tan \delta) \qquad (1.13a)$$

The total hour angle from sunrise to sunset is given by

$$2\omega_s = 2 \cos^{-1}(- \tan \varphi \tan \delta) \qquad (1.13b)$$

Since, $15° = 1$ h, the number of daylight (sunshine) hours $(N)$ is given by

$$\text{Day length} = N = 2 \cdot \frac{\omega_s}{15} = \frac{2}{15} \cdot \cos^{-1}[- \tan \phi \cdot \tan \delta] \qquad (1.14)$$

Since, the sum of altitude $(\alpha)$ and zenith angle $(\theta_z)$ is 90° and hence

$$\alpha + \theta_z = 90 \quad \text{or} \quad \theta_z = 90 - \alpha$$

Then, Eq. 1.12 can be as

$$\cos\theta_z = \cos\alpha = \cos\phi\cos\delta\cos\omega + \sin\delta\sin\phi \qquad (1.15)$$

The value of sunshine hour, $N$, obtained from Eq. 1.14 will be always higher than actual observed bright sunshine hour, and hence $\frac{n}{N}$ will be always less than one. The $\frac{n}{N}$ is used to evaluate diffuse component from known total either daily or monthly solar radiation.

**Example 1.9a**

Calculate the angle of incidence of beam radiation on a surface located at New Delhi at 1:30 (solar time) on February 1, 2023 if the surface is tilted $45°$ from the horizontal and pointed $30°$ west of south ($1° = 60'$).

**Solution**

In the given problem, the value of $n$th day is $31 + 1 = 32$.

$\delta = -17.52°$ (Eq. 1.9); $\omega = +22.5°$ (Table 1.4).

$\gamma = 30°$; $\beta = 45°$; $\varphi = +28° 35' = 28.58° = $ (Table 1.2).

Now, the angle of incidence can be calculated by using Eq. 1.11a as follows:

$$\cos\theta_i = (\cos\Phi\cos\beta + \sin\Phi\sin\beta\ \cos\gamma)\cos\delta\cos\omega + \cos\delta\sin\omega\sin\beta\sin\gamma +$$
$$(\sin\Phi\cos\beta - \cos\Phi\sin\beta\ \cos\gamma)\times\sin\delta$$

$$\cos\theta_i = \left[\cos(28.58)\cos 45° + \sin(28.58)\sin 45°\cos 30°\right]\cos(-17.52°)\cos(22.5°)$$
$$+ \cos\left(-17.52°\right)\sin\left(22.5°\right)\sin 45°\sin 30°$$
$$+ \sin(28.58)\cos 45° - \cos(28.58)\sin 45°\cos 30°\right] \times \sin(-17.52°)$$
$$= [0.6209 + 0.2929] \times 0.8810 + 0.1290 + [0.3382 - 0.5377] \times (-0.3010)$$
$$= 0.8050 + 0.1290 + 0.0600 = 0.934$$

$$\theta_i = \cos^{-1}(0.934) = 20.9°$$

**Example 1.9b**

Calculate the angle of incidence of beam radiation on a surface located at New Delhi at 1:30 (solar time) for an average monthly day of February 2023 if the surface is tilted $45°$ from the horizontal and pointed $30°$ west of south ($1° = 60'$).

**Solution**

In the given problem, the value of an average day of month ($n$) is 47 (Table 1.6).

$\delta = -13.0°$ (Eq. 1.9); $\omega = +22.5°$ (Table 1.4)

$\gamma = 30°$; $\beta = 45°$; $\varphi = +28° \, 35' = 28.58°$ (Table 1.2).

Now, the angle of incidence can be calculated by using Eq. 1.11a as follows:

$$\cos\theta_i = (\cos\Phi\cos\beta + \sin\Phi\sin\beta\cos\gamma)\cos\delta\cos\omega + \cos\delta\sin\omega\sin\beta\sin\gamma$$
$$+ (\sin\Phi\cos\beta - \cos\Phi\sin\beta\cos\gamma) \times \sin\delta$$

$$\cos\theta_i = \big[\cos(28.58)\cos 45° + \sin(28.58)\sin 45°\cos 30°\big]\cos(-13°)\cos(22.5°)$$
$$+ \cos(-13°)\sin(22.5°)\sin 45°\sin 30°$$
$$+ \sin(-13°)\big[\sin(28.58)\cos 45° - \cos(28.58)\sin 45°\cos 30°\big]$$
$$= [0.6209 + 0.2929] \times 0.9002 + 0.1318 + [0.3382 - 0.5377] \times (-0.2249)$$
$$= 0.8226 + 0.1318 + 0.0448 = 0.999$$

$$\theta_i = \cos^{-1}(0.999) = 2.56°$$

In this case too, one can observe that incidence angle $\theta_i$ for $n$th day (February 1) is significantly higher than monthly average day of February.

**Example 1.10a**

Evaluate the number of sunshine hours for New Delhi on December 22 and June 22, 2013.

**Solution**

For the present example, we have

$\varphi = 28.58°$ (New Delhi); for December 22, 2013 $n = 356$, and $\delta = -23.44°$. From Eq. 1.14, one gets

$$N = \frac{2}{15}\cos^{-1}\big[-\tan(-23.44°)\tan(28.58°)\big]$$
$$= \frac{2}{15}\cos^{-1}[(0.434)(0.545)]$$
$$= \frac{2}{15}\cos^{-1}[0.237] = 10.18\,\text{h}$$

Similarly, for June 22, 2013, $n = 173$; $\delta = 23.45°$ (Eq. 1.9).

From Eq. 1.12, we have

$$N = \frac{2}{15}\cos^{-1}(-\tan 23.45°\tan 28.58°) = 13.82\,\text{h}$$

**Example 1.10b**

Evaluate the number of sunshine hours for New Delhi for an average day for December and June 2023.

**Solution**

For the present case of average day of month, we have from Table 1.6 as

$\varphi = 28.58°$ (New Delhi); for December 2023 $n = 344$, and $\delta = -23$. From Eq. 1.14, one gets

$$
\begin{aligned}
N &= \frac{2}{15} \cos^{-1}\left[-\tan(-23°)\tan(28.58°)\right] \\
&= \frac{2}{15} \cos^{-1}[(0.4245) \times (0.545)] \\
&= \frac{2}{15} \cos^{-1}[0.2313] = 10.22\,\text{h}
\end{aligned}
$$

Similarly, for June 2023, $n = 162$; $\delta = 23.1°$ (Eq. 1.9)
From Eq. 1.14, we have

$$
N = \frac{2}{15} \cos^{-1}(-\tan 23° \times \tan 28.58°) = 13.78\,\text{h}
$$

One can see that there is not much difference in sunshine hours for a given variation of date in a particular day of month.

**Example 1.11a**

Calculate zenith angle of the sun at New Delhi at 2.30 PM on February 20, 2023 (particular $n$th day in month).

**Solution**

We have $n = 31 + 20 = 51$; $\Phi = 28° 35' = 28.58°$ (Table 1.2) and $\delta = -11.58°$ (Eq. 1.9); $\omega = 37.5°$ (Table 1.4).
From Eq. 1.12, we have

$$
\begin{aligned}
\cos\theta_z &= \cos(28.58°)\cos(-11.58°)\cos(37.5°) \\
&\quad + \sin(-11.58°)\sin(28.58°) = 0.6825 - 0.0960 = 0.587 \\
\theta_z &= \cos^{-1}(0.587) = 54.09°.
\end{aligned}
$$

**Example 1.11b**

Calculate zenith angle of the sun at New Delhi at 2.30 PM for monthly average day of February 2023.

**Solution**

We have for monthly average day of February 2023 $n = 47$; $\Phi = 28°35' = 28.58°$ (Table 1.2) and $\delta = -13°$ (Eq. 1.9); $\omega = 37.5°$ (Table 1.4).
    From Eq. 1.12, we have

$$\cos\theta_z = \cos(28.58°)\cos(-13°)\cos(37.5°) + \sin(-13°)\sin(28.58°)$$
$$= 0.6788 - 0.1076 = 0.5711$$

$$\theta_z = \cos^{-1}(0.587) = 55.16°$$

In this case too, there is not much difference between zenith angle for a particular day and average day of the month.

### 1.2.11  Solar Radiation/Irradiation in Terrestrial Region for a Given Value of Solar Radiation in Extra-terrestrial Region

An expression for direct solar radiation emitted from the sun entering into terrestrial region, $I_N$, Fig. 1.1 is given by

$$I_N = I_{ext} \times \tau_{atm} \tag{1.16a}$$

where, $\tau_{atm} = \exp[-(m.\varepsilon.T_R + \alpha)]$ is atmospheric transmittance. The '$m$' is the air mass, it is given by Eq. 1.6b and $T_R$, turbidity factor, and it is defined as cloudiness/haziness factor for the lumped atmosphere. The values of $T_R$ and $\alpha$ for different weather and flat land conditions (cases a, b, c and d as defined earlier) are given in Appendix C.1. The expression for $I_{ext}$ is given by Eq. 1.5.
    The expression for $\varepsilon$, known as integrated optical thickness of the terrestrial clear and dry atmosphere/Rayleigh atmosphere (dimensionless), is given by

$$\varepsilon = 4.529 \times 10^{-4} \times m^2 - 9.66865 \times 10^{-3} \times m + 0.108014$$

Further, more simple expression for the direct solar radiation entering into terrestrial region, $I_N$, Fig. 1.1 through atmosphere from extra-terrestrial region is also given by

$$I_N = I_{ext}\exp[-T_R/(0.9 + 9.4\sin\alpha)] \tag{1.16b}$$

where, an expression of $\sin\alpha = \cos\theta_z$ is given by Eq. 1.15 and the values of turbidity ($T_R$) for different region and month are given in Appendix C.2.

**Further, there are many expressions for $I_N$ developed by various researchers for different weather condition given in Appendix C.3.**

**Example 1.12a**

Calculate atmospheric transmittance $(\tau_{atm})$ for New Delhi $(\varphi = 28°35')$ for blue sky condition for the average monthly day of January and June at 12 noon.

**Solution**

Known parameters: $\omega = o$, $\varphi = 28°35' \approx 30$ for New Delhi (Tables 1.4 and 1.6),

$$\tau_{atm} = \exp[-(m.\varepsilon.T_R + \alpha)] \text{ (Eq. 1.16a)}$$

(a) **For January:** $n = 17(\delta = -20.9°)$; $T_R = 2.25$ and $\alpha = 0.07$ (Appendix C.1).

$$\text{Air mass (m)} = \frac{1}{\cos\theta_z} \quad \text{(Eq.1.6a)}$$

Here,

$$\cos\theta_z = [\cos\varphi\cos\delta\cos\omega + \sin\delta\sin\varphi]\left[Eq.1.12\right]$$
$$= [0.866 \times 0.9342 + (-0.3567) \times 0.5]$$
$$= [0.8090 - 0.17835] = 0.6307$$

$$\text{Air mass}(m) = \frac{1}{0.6307} = 1.5855$$
$$\varepsilon = 4.529 \times 10^{-4} \times m^2 - 9.66865 \times 10^{-3} \times m + 0.108014$$

or,

$$\varepsilon = 4.529 \times 10^{-4} \times (1.5855)^2 - 9.66865 \times 10^{-3} \times 1.5855 + 0.108014$$
$$= 0.001138 - 0.01533 + 0.108014 = 0.0938$$

The atmospheric transmittance $(\tau_{atm})$ for January is

$$\tau_{atm} = \exp[-(1.5855 \times 0.0938 \times 2.25 + 0.07)] = \exp(-0.4046) = 0.6674$$

(b) **For June:** June and 162 $(\delta = 23.1°)$; here $T_R = 2.47$ and $\alpha = 0.28$ (Appendix C.1)

$$\cos\theta_z = [0.866 \times 0.9198 + (0.3923) \times 0.5] = [0.7965 + 0.1962] = 0.9927$$

$$\text{Air mass (m)} = \frac{1}{0.9927} = 1.0074$$

$$\varepsilon = 4.529 \times 10^{-4} \times (1.0074)^2 - 9.66865 \times 10^{-3} \times 1.0074 + 0.108014$$
$$= 0.0004596 - 0.0094 + 0.108014 = 0.0991$$

**The atmospheric transmittance ($\tau_{atm}$) for June is**

$$\tau_{atm} = \exp[-(1.0074 \times 0.0991 \times 2.47 + 0.28)] = \exp(-0.5266) = 0.5906$$

**Example 1.12b**

Evaluate direct solar radiation emitted from the sun entering into terrestrial region, $I_N$, for the average monthly day of January and June at 12 noon (January, $n = 17$ and June, $n = 162$, Table 1.6).

**Solution**

For known values from Eqs. 1.5a and 1.5b, we have

For January 2022, $n = 17$, $I_{ext} = 1410\,\text{W/m}^2$.

For June 2022, $n = 162$, $I_{ext} = 1324\,\text{W/m}^2$.

From Example 1.12a, we have $\tau_{atm} = 0.6674$ for $n = 17$ and $0.5906$ for $n = 162$, then

$$I_N = 1410 \times 0.6674 = 941\,\text{W/m}^{-2}\text{ for January } n = 17$$

and,

$$I_N = 1324 \times 0.5906 = 781\,\text{W/m}^{-2}\text{ for January } n = 162.$$

**Example 1.13**

Determine the temperature of sun in the month of June for the following data: $I_{ext} = 1320\,\text{W/m}^2$,

Sun's diameter $(2R_s) = 1.39 \times 10^9$ m,

Mean sun-earth's distance $(L_{se}) = 1.5 \times 10^{11}$ m and $\sigma = 5.67 \times 10^{-8}\,\text{W/m}^2\,\text{K}^4$.

**Solution**

The amount of solar radiation emitted by the surface of the sun $= \sigma T_s^4 (4\pi R_s^2)$.

The total emitted radiation from the sun will be received by a sphere having a radius equal to mean sun-earth distance $= I_{ext} \times (4\pi L_{se}^2)$. Hence,

$$\sigma T_s^4 (4\pi R_s^2) = I_{ext} \times (4\pi L_{se}^2)$$

or,

$$T_s = \left[ \frac{I_{ext}\left(4\pi L_{se}^2\right)}{\sigma\left(4\pi R_s^2\right)} \right]^{\frac{1}{4}} = 5738.5\,\text{K}.$$

**Example 1.14**

Calculate the average earth's temperature in the absence of atmosphere.

**Solution**

The rate of solar radiation per m² in extra-terrestrial region ($I_{sc}$) is 1367 W/m². The diameter of the earth ($D_e$) is $12.75 \times 10^6$ m.

The total heat flow in from the sun to the earth = projected area of the earth × solar flux = $\pi(D_e/2)^2 I_{sc} = (\pi/4)(12.75 \times 10^6\,\text{m})^2\,1367\,\text{W/m}^2 = 1.75 \times 10^{14}$ kW.

Hence, it is assumed that all the incoming solar radiation is absorbed by the earth. Thus,

outward radiation = the surface area of the earth × $\varepsilon\sigma T_e^4 = \pi D_e^2 \varepsilon\sigma T_e^4$

$$= \pi\left(12.75 \times 10^6\,\text{m}\right)^2 \cdot 5.672 \times 10^{-11}\,(\text{kW/m}^2\,\text{K}^4) \cdot T_e^4$$

If the radiation due to nuclear decay and tidal friction with the moon on the earth, which is about 0.1% of the solar energy inflow, is neglected, then

All the incoming solar radiation = Outward radiation

or,

$$\pi\left(12.75 \times 10^6\,\text{m}\right)^2 \times 5.672 \times 10^{-11}\left(\text{kW/m}^2\,\text{K}^4\right).\left(T_e^4\right) = 1.75 \times 10^{14}\,\text{kW}$$

or,

$$T_e = 278.7\,K = 5.7°C$$

where, $\varepsilon = 1$.

If the total heat flow in from the sun to the earth is multiplied by 0.7 due to the earth's albedo, then $T_e = 255\,\text{K} = -18\,°\text{C}$ (Appendix A) a frozen world. This indicates that existing atmosphere between the sun and the earth blocks maximum outgoing radiation so that the average earth's temperature has been raised to about 15 °C.

**Example 1.15**

What is the fraction of the outgoing radiation from the earth which is blocked by the atmosphere?

**Solution**

Assume that 30% of incoming radiation from the sun is reflected back to extra-terrestrial region through atmosphere. This means that 70% of incoming radiation is absorbed by the earth. If we assume average earth's surface about 15 °C = 288 K, then

$$\text{Fraction emitted} = \frac{0.7(\text{total solar radiation inflow from the sun})}{\pi D^2 \sigma T^4}$$

$$= \frac{0.7 \times (1.75 \times 10^{14})}{3.14 \times (12.75 \times 10^6)2 \times 5.672 \times 10^{-11} \times (288)^4}$$

$$= 0.606.$$

The atmospheric outward transmission of radiant energy = 0.606/0.7 = 0.86.
Eighty-six percent of the inward transmission of solar energy is blocked by atmosphere.

## 1.3  Solar Radiation on Horizontal and Inclined Surfaces in Terrestrial Region

### 1.3.1  Hourly Solar Radiation on Horizontal

Basically, there are two types of radiation on the horizontal surface, namely beam/direct radiation and diffuse radiation. These are as follows

#### 1.3.1.1  Beam Radiation ($I_b$)

It is normal component of solar radiation in terrestrial region, $I_N$ (Eqs. 1.6a and 1.6b) in W/m$^2$ propagating along the line joining the receiving horizontal surface and the sun. It has a direction and also termed as direct radiation.

Mathematically, it is expressed as follows:

$$I_b = I_N \cos\theta_z = I_{ext} \exp[-T_R/(0.9 + 9.4\sin\alpha)]\cos\theta_z \tag{1.17}$$

where, an expression for $\cos\theta_z$ and $I_N$ are given by Eqs. 15, 1.6a and 1.6b, respectively.

**Example 1.16a**
Evaluate beam/direct radiation on horizontal surface for the average day of month of January, $n = 17$ and June $n = 162$ for Examples 1.12a and 1.12b at 12 noon $\omega = o$.

**Solution**

**From Examples 1.12a and 1.12b, we have**

$$I_N = 1410 \times 0.6674 = 941 \text{ W/m}^2 \text{ for January } n = 17$$

and,

$$I_N = 1324 \times 0.5906 = 781 \text{ W/m}^2 \text{ for January } n = 162$$

**For January 2022, $n = 17$, $I_N = 941 \text{W/m}^2$, $\cos \theta_z = 0.6307$**

**For June 2022, $n = 162$, $I_N = 781 \text{W/m}^2$, $\cos \theta_z = 0.9927$**

From Eq. 1.17, one gets

**For January 2022, $n = 17$, $I_b = I_N \cos \theta_z = 941 \times 0.6307 = 593.48 \text{ W/m}^2$**

**For June 2022, $n = 162$, $I_b = I_N \cos \theta_z = 781 \times 0.9927 = 775.30 \text{ W/m}^2$**

This shows that beam radiation in June (775.30 W/m$^2$) is more than the beam radiation in the month of January (593.48 W/m$^2$) as expected.

### 1.3.1.2  Diffuse Radiation ($I_d$)

It is the solar radiation (in W/m$^2$) scattered by aerosols and other particulates present in atmosphere from bottom layer of atmosphere. It does not have any definite and will depend on difference between extra-terrestrial, Eq. 1.15 and terrestrial, Eq. 1.15 solar radiation, i.e. ($I_{ext} - I_N$). Therefore, the expression for diffuse radiation in W/m$^2$ on horizontal surface has been proposed by Singh and Tiwari [4] as follows:

$$\begin{aligned} I_d &= K_1(I_{ext} - I_N) \cos \theta_z + K_2 \\ &= K_1 I_{ext}\left[1 - \exp[-T_R/(0.9 + 9.4 \sin \alpha)]\right] \cos \theta_z + K_2 \end{aligned} \tag{1.18}$$

where, the numerical values for $K_1$ and $K_2$ for different weather are given in Appendix C.1.

**Example 1.16b**
Evaluate diffuse radiation on horizontal surface for the month of January $n = 17$ and June $n = 162$ for Examples 1.12a, 1.12b and 1.16a.
**Solution**

**From Example 1.12a and 1.12b, Eq. 1.15 and Appendix C.1, we have**

**For January 2022, $n = 17$, $I_N = 941 \text{ W/m}^2$, $I_{ext} = 1410 \text{ W/m}^2$, $\cos \theta_z = 0.6307$, $K_1 = 0.47$ and $K_2 = -13.17$, so from Eq. 1.18, we have**

$$I_d = K_1(I_{ext} - I_N)\cos\theta_Z + K_2 = 0.47(1410 - 941) \times 0.6307 + (-13.17)$$
$$= 126.15\ \text{W/m}^2$$

**For June 2022**, $n = 162$, $I_N = 781$ W/m$^2$, $I_{ext} = 1324$ W/m$^2$, $\cos\theta_z = 0.9927$, $K_1 = 0.27$ and $K_2 = 31.86$, so from Eq. 1.18, we have

$$I_d = K_1(I_{ext} - I_N)\cos\theta_Z + K_2 = 0.27(1324 - 781)$$
$$\times 0.9927 + 31.86 = 177.40\ \text{W/m}^2.$$

### 1.3.1.3   The Total Radiation (*I*)

It is the sum of the beam (direct) and diffuse radiation in w/m$^2$. It is also known as global radiation, and it can be mathematically expressed as follows:

$$I(t) = I_b(t) + I_d(t) \tag{1.19}$$

However, $I_b(t)$ and $I_d(t)$ are both time depended, but it will be noted as $I_b$ and $I_d$, respectively.

**Irradiation (radiant exposure)**: The incident solar radiation/irradiance on surface in J/m$^2$ per hour or per day is known as irradiation. It is obtained by integrating beam/diffuse/total radiation over an hour or day.

**Example 1.16c**
Evaluate total radiation on horizontal surface for the month of January $n = 17$ and June $n = 162$ for Examples 1.16a and 1.16b.

**Solution**

**From Examples 1.16a and 1.16b, we have**

**For January 2022**, $n = 17$, $I(t) = I_b(t) + I_d(t) = 593.48 + 126.15 = 719.63$ W/m$^2$

**For June 2022**, $n = 162$, $I(t) = I_b(t) + I_d(t) = 775.30 + 177.40 = 952.78$ W/m$^2$.

## *1.3.2   Solar Radiation on Inclined Surfaces*

There are three type of solar radiation on inclined surface having inclination ($\beta$). The third type of radiation is reflected radiation in addition to beam and diffuse radiation on inclined surface.

### 1.3.2.1  Beam Radiation

If surface/plane has inclination of $\beta$ and solar radiation incident makes an angle $\theta_i$ (Eqs. 1.11a and 1.11b) with respect to normal to inclined surface/plane, then the rate of beam/direct radiation on inclined plane can be evaluated as follows:

$$I_{b\beta} = I_N \cos \theta_i \tag{1.20a}$$

or,

$$I_{b\beta} = I_N \cos \theta_i = I_N \cos \theta_i \frac{\cos \theta_i}{\cos \theta_z} = I_N \cos \theta_i R_b = I_b R_b \tag{1.20b}$$

where, $R_b = \frac{\cos \theta_i}{\cos \theta_z}$ is conversion factor for beam radiation from horizontal to inclined surface/plane.

**Example 1.17a**

Evaluate (i) beam/direct radiation on south oriented ($\gamma = 0$) inclined plane/surface at angle of 45° and (ii) conversion factor for beam radiation ($R_b$) for the month of January $n = 17$ and June $n = 162$ for Examples 1.12a and 1.12b at 12 noon ($\omega = o$).
**Solution**

**Case (i):** From Eq. 1.17, we have

$$\cos \theta_i = (\cos \Phi \cos \beta + \sin \Phi \sin \beta \cos \gamma) \cos \delta \cos \omega + \cos \delta \sin \omega \sin \beta \sin \gamma$$
$$+ (\sin \Phi \cos \beta - \cos \Phi \sin \beta \cos \gamma) \times \sin \delta \text{ with } \omega = \gamma = 0$$

(a)  **For January:** $n = 17$ ($\delta = -20.9°$).

   For $\omega = \gamma = 0$, we have

$$\cos \theta_i = (\cos 30 \cos 45 + \sin 30 \sin 45) \cos \delta \cos 0$$
$$+ (\sin 30 \cos 45 - \cos \Phi 30 \sin 45) \times \sin \delta$$
$$= (0.866 \times 0.7071 + 0.5 \times 0.7071) \times 0.9342$$
$$+ (0.5 \times 0.7071 - 0.8661 \times 0.7071) \times -0.3567$$
$$= 0.9023 + 0.0924 = 0.9947$$

Now, $I_{b\beta} = I_N \cos \theta_i = 941 \times 0.9947 = 936.02 \text{ W/m}^2$ (Eq. 1.12)

(b)  **For June:** June and 162 ($\delta = 23.1°$);

   Also,

$$\cos \theta_i = (\cos 30 \cos 45 + \sin 30 \sin 45) \cos \delta \cos 0$$
$$+ (\sin 30 \cos 45 - \cos 30 \sin 45) \times \sin 23.1$$

$$= (0.866 \times 0.7071 + 0.5 \times 0.7071) \times 0.9198$$
$$+ (0.5 \times 0.7071 - 0.8661 \times 0.7071) \times 0.3923$$
$$= 0.884 - 0.2588 = 0.2299$$

Now, $I_{b\beta} = I_N \cos \theta_i = 781 \times 0.2299 = 179.55$ W/m$^2$ (Eq. 1.12)

**Case (ii):**

(a) **For January:** $n = 17$ ($\delta = -20.9°$), $\cos \theta_z = 0.6307$ (Example 1.12a).

The conversion factor $= R_b = \frac{\cos \theta_i}{\cos \theta_z} = 0.9947/0.6307 = 1.5771$.

(b) **For June:** June and 162 ($\delta = 23.1°$);

The conversion factor $= R_b = \frac{\cos \theta_i}{\cos \theta_z} = 0.2299/0.9927 = 0.2315$ (Example 1.12a).

### 1.3.2.2 Diffuse Radiation

In this case, the diffuse radiation will be different, and it will depend on inclination of the surface. It can be expressed as

$$I_{d\beta} = I_d \cdot R_d \tag{1.21a}$$

where,

$$R_d = \text{Conversion factor for diffuse radiation}$$
$$= (1 + \cos \beta)/2; \ R_d = 1 \text{ for } \beta = 0 \text{ (horizontal surface)} \tag{1.21b}$$

**Example 1.17b**
Evaluate (i) diffuse radiation and (ii) conversion factor ($R_d$) on south oriented ($\gamma = 0$) inclined plane/surface at angle of 45° for the month of January $n = 17$ and June $n = 162$ for Example 1.16b at 12 noon ($\omega = o$).

**Solution**

**For January 2022, n = 17**

(i) $I_{d\beta} = I_d \times (1 + \cos \beta)/2 = 126.15 \times 0.8536 = 107.68$ W/m$^2$ and (ii) $R_d = (1 + \cos \beta)/2 = 0.8536$.

**For June 2022, n = 162**

(i) $I_{d\beta} = I_d \times (1 + \cos \beta)/2 = 177.40 \times 0.8536 = 151.42$ W/m$^2$ and (ii) $R_d = (1 + \cos \beta)/2 = 0.8536$.

### 1.3.2.3    Reflected Radiation

In this case, inclined surface will receive the reflected beam and diffuse both radiations. In this case it will be determined by the following expression.

$$I_r = (I_b + I_d)\, R_r = I(t) R_r \tag{1.22a}$$

where, $\rho$ is the reflectance of ground surface surrounding the inclined plane and

$$R_r = \text{Conversion factor for reflected radiation}$$
$$= \rho(1 - \cos\beta)/2;\; R_r = 0\,\text{for}\,\beta = 0\,\text{(horizontal surface)} \tag{1.22b}$$

**Example 1.17c**

Evaluate reflected solar radiation and conversion factor $(R_r)$ on south oriented $(\gamma = 0)$ inclined plane/surface at angle of $45°$ for the month of January $n = 17$ and June $n = 162$ for Example 1.16c at 12 noon $(\omega = o)$ with ground reflectivity of 0.2.

**Solution**

**For January 2022, n = 17**

$$I_r = I(t)\rho(1 - \cos\beta)/2 = 719.63 \times 0.2 \times 0.1464 = 21.07\,\text{W/m}^2\;\text{and}$$
$$R_r = \rho(1 - \cos\beta)/2 = 0.2 \times 0.1464 = 0.0293$$

**For June 2022, n = 162**

$$I_r = I(t)\rho(1 - \cos\beta)/2 = 952.78 \times 0.2 \times 0.1464 = 27.89\,\text{W/m}^2\;\text{and}$$
$$R_r = \rho(1 - \cos\beta)/2 = 0.2 \times 0.1464 = 0.0293$$

### 1.3.2.4    The Total Radiation ($I_\beta$)

It is the sum of the beam (direct), diffuse radiation and reflected radiation in W/m$^2$.

$$I_\beta = I_{b\beta} + I_{d\beta} + I_r = I_b R_b + I_d R_d + I(t) R_r \tag{1.23}$$

The effective ratio of solar energy incident on a surface to that on a horizontal surface, $R'$, is

$$R' = \frac{I_\beta}{I(t)} = \frac{(I_b R_b + I_d R_d)}{I_b + I_d} + R_r \tag{1.24}$$

Equation 1.23 can be used for hourly/daily as well average monthly solar radiation variation for converting beam and diffuse radiation on any inclined and orientation surface from known data on horizontal surface.

**Example 1.17d**
Evaluate total solar radiation on south oriented ($\gamma = 0$) inclined plane/surface at angle of 45° for the month of January $n = 17$ and June $n = 162$ for Examples 1.16c at 12 noon ($\omega = o$) with ground reflectivity of 0.2.

**Solution**

**For January 2022, $n = 17$**

$$I_\beta = I_{b\beta} + I_{d\beta} + I_r = 936.02 + 107.68 + 21.07 = 1064.77\,\text{W/m}^2$$

**For June 2022, $n = 162$**

$$I_\beta = I_{b\beta} + I_{d\beta} + I_r = 179.55 + 151.42 + 27.89 = 358.86\,\text{W/m}^2$$

In Examples 1.17a–1.17d, this has been indicated that total solar radiation on inclined surface in winter (1064.77 W/m²) is significantly higher in comparison with summer condition (358.86 W/m²).

# 1.4   Solar Radiation on Horizontal Surface in Extra-terrestrial Region

The solar radiation in extra-terrestrial region may be in hourly, for a given time interval $I_0(\text{W/m}^2) = I_{ext}\cos\theta_Z$, daily in Joule per m² (J/m²) and daily average hourly in W/m². These can be evaluated for an average day of each month from January to December as given in Table 1.6.

An expression for each case has been derived as follows:

## 1.4.1   Hourly Solar Radiation in W/m² and J/m² on Horizontal Surface in Extra-terrestrial Region

For a given time, $t$, the solar radiation incident on a horizontal surface/plane in W/m² in extra-terrestrial region (outside the atmosphere) ($I_0$) W/m² is given by

$$I_0(\text{W/m}^2) = I_{ext}\cos\theta_z$$

Substitute $I_{ext}$ and $\cos\theta z$ from Eqs. 1.5–1.15, one gets

$$I_o = I_{sc}\left[1.0 + 0.033\cos\left(\frac{360n}{365}\right)\right](\cos\varphi\cos\delta\cos\omega + \sin\delta\sin\varphi)\ (W/m^2),$$

$$(1.25a)$$

where, $I_{sc}$ is the solar constant.

The numerical value of $n$ can be considered for two cases, namely.

**Case (i):** The $n$ is $n$th day of the year for hourly variation for a given day in year and

**Case (ii):** The $n$ is an average day of each month for monthly variation of solar intensity for year. Here, $I_o$ is in $W/m^2$. The recommended average days and the values of $n$ for each month are given in Table 1.6.

$$\text{Since } \frac{W}{m^2} = \frac{J}{s\,m^2} = \frac{3600\,s}{1\,h} \times \frac{J}{s\,m^2}$$

$$= \frac{3600}{1\,h} \times \frac{J}{m^2} = \frac{3600}{\frac{2\pi}{24}} \times \frac{J}{m^2} = \frac{24 \times 3600}{2\pi} \times \frac{J}{m^2}$$

$$= \frac{12 \times 3600}{\pi} \times \frac{J}{m^2} \text{ (Appendix A)}$$

Equation 1.25a in $J/m^2$ becomes

$$I_o = \frac{12 \times 3600}{\pi} \times I_{sc}\left[1.0 + 0.033\cos\left(\frac{360n}{365}\right)\right]$$

$$(\cos\varphi\cos\delta\cos\omega + \sin\delta\sin\varphi)$$

$$(1.25b)$$

### 1.4.2  Solar Radiation Between Time $t_1$ ($\omega_1$) and $t_2$ ($\omega_2$) in J/ m² on Horizontal Surface in Extra-terrestrial Region

It can be obtained by integrating Eq. 1.25b between $\omega_1$ and $\omega_2$ (where $\omega_2$ is higher than $\omega_1$) in $J/m^2$ as

$$\Delta I_0 = \int_{\omega_1}^{\omega_2} I_0 d\omega = \frac{12 \times 3600}{\pi} \times I_{sc}\left[1.0 + 0.033\cos\left(\frac{360n}{365}\right)\right]$$

$$\left[\cos\varphi\cos\delta\int_{\omega_1}^{\omega_2}\cos\omega d\omega + \sin\delta\sin\varphi\int_{\omega_1}^{\omega_2}d\omega\right]$$

or,

$$\Delta I_0 = \frac{12 \times 3600}{\pi} \times I_{sc}\left[1.0 + 0.033\cos\left(\frac{360n}{365}\right)\right]$$

$$\left[\cos\varphi\cos\delta\{\sin\omega_2 - \sin\omega_1\} + \sin\delta\sin\varphi(\omega_2 - \omega_1)\frac{2\pi}{360°}\right]. \tag{1.26}$$

We know that $360° = 2\pi \approx 1° = \frac{2\pi}{360}$.

### 1.4.3 Daily Monthly Average Solar Radiation on Horizontal Surface in J/m² in Extra-terrestrial Region

For calculations of daily solar radiation on a horizontal surface in extra-terrestrial region $(H_o)$ n J/m², Eq. 1.25b should be integrated over the period from sunrise $(-\omega_s)$ to sunset $(+\omega_s)$ and one gets

$$H_o = \frac{12 \times 3600}{\pi} \int_{-\omega_s}^{-\omega_s} I_{sc}\{[1.0 + 0.033\cos(360n/365)]$$

$$[\cos\varphi\cos\delta\cos\omega + \sin\delta\sin\varphi]\}d\omega$$

or,

$$H_o = \frac{2 \times 12 \times 3600}{\pi} \int_{0}^{-\omega_s} I_{sc}\{[1.0 + 0.033\cos(360n/365)]$$

$$[\cos\varphi\cos\delta\cos\omega + \sin\delta\sin\varphi]\}d\omega$$

or,

$$H_o = \frac{24 \times 3600}{\pi} I_{sc}$$

$$\{[1.0 + 0.033\cos(360n/365)]$$

$$[\cos\varphi\cos\delta \int_{0}^{\omega_s}\cos\omega d\omega + \sin\delta\sin\varphi\int_{0}^{\omega_s}d\omega]\}$$

or,

$$H_o = \frac{24 \times 3600}{\pi} I_{sc}$$

$$\{[1.0 + 0.033\cos(360n/365)]$$

$$[\cos\varphi\cos\delta\cos\omega_s + \frac{2\pi\omega_s}{360}\sin\delta\sin\varphi]\}\text{MJ/m}^2 \tag{1.27}$$

Equation 1.27 will be used to determine.

**Case (i):** The daily extra-terrestrial radiation, $\overline{H}_0$, in MJ/m$^2$ and J/m$^2$ (1 MJ $= 10^6$ J) can be obtained; if $n$ is considered for that day, Eq. 1.9 for evaluation of declination angle other than the value given in Table 1.6, then $H_0$ will be daily solar radiation in extra-terrestrial region.

**Case (ii):** The average monthly daily extra-terrestrial radiation, $\overline{H}_0$, also in MJ/m$^2$ and J/m$^2$ (1 MJ $= 10^6$ J) can be obtained by using average day numbers ($n$) and declination angle ($\delta$) for other month from Table 1.6.

### 1.4.4 The Monthly Average Solar Radiation on Horizontal Surface in W/m² in Extra-terrestrial Region

Further, the daily hourly average solar radiation in W/m$^2$ in extra-terrestrial region ($H_{0\text{cx}}$) can be evaluated as

$$H_{oex}(\text{W/m}^2) = \frac{H_o(\text{J/m}^2)}{N \times 3600} = \frac{H_o}{N \times 3600} \text{ W/m}^2 \tag{1.28}$$

where, $N$ = sunshine hours for a given day ($n$), Table 1.6 $= \frac{2\omega_s}{15} = \omega_s = \cos^{-1}(-\tan\varphi\tan\delta)$ (Eq. 1.14).

**Example 1.18a**
Calculate the daily solar radiation in J/m$^2$ and MJ/m$^2$ on a horizontal surface ($H_o$) in extra-terrestrial region at latitude 30° N on May 31, 2022.

**Solution**

In the given problem, $n = 151$ and $\delta$ (from Eq. 1.9) $= 21.90°$, $\varphi = 30°$.

From Eq. 1.15, $\omega_s = \cos^{-1}(-\tan\varphi\tan\delta) = 103.41°$.
   Now, the daily solar radiation on a horizontal surface in extra-terrestrial region can be obtained from Eq. 1.19 as follows:

$$H_o = \frac{24 \times 3600 \times 1367}{\pi}\left[1.0 + 0.033\cos\left(\frac{360 \times 151}{365}\right)\right]\left(\cos 30° \cos 21.90° \sin 103.41°\right.$$
$$\left. + \frac{\pi\,103.41}{180}\sin 30° \sin 21.90°\right)$$

or,

$$H_o = 37.614 \times 10^6[1.0 + 0.033(-0.85717)]$$
$$(0.866 \times 0.9278 \times .972735 + 1.8 \times 0.50 \times 0.373)$$

$$H_o = 36.55 \times 10^6 \times (0.74139 + 0.3357) = 39.37 \times 10^6 \text{ J/m}^2$$

or,

$$H_o = 39.37 \, \text{MJ/m}^2.$$

In this case, number of sunshine hour $(N) = \frac{2\omega_s}{15} = \frac{2 \times 103.41}{15} = 13.788 \, \text{h}.$

So, solar radiation in W/m$^2$ in extra-terrestrial region $= \frac{39.37 \times 10^6}{13.788 \times 3600} =$ 793.16 W/m$^2$.

**Example 1.18b**

Calculate the average monthly solar radiation in J/m$^2$ and MJ/m$^2$ and monthly daily average solar radiation on a horizontal surface in W m$^2$ in extra-terrestrial region at latitude 30 N on for May 2022.

**Solution**

In the given problem, for $n = 135$ and $\delta$ (Eq. 1.9) $= 18.8°$ $\varphi = 30°$ (Case (ii) of Sect. 1.4.3).

From Eq. 1.15, $\omega_s = \cos^{-1}(-\tan\varphi \tan\delta) = \cos^{-1}[-\tan 30° \tan 18.8°] = 101.33°.$

Now, the daily solar radiation on a horizontal surface in extra-terrestrial region can be obtained from Eq. 1.27 as follows:

$$H_o = \frac{24 \times 3600 \times 1367}{\pi} \left[ 1.0 + 0.033 \cos\left(\frac{360 \times 135}{365}\right) \right]$$
$$\left( \cos 30° \cos 18.8° \sin 101.33° + \frac{\pi \, 103.41}{180} \sin 30° \sin 18.8° \right).$$

or,

$$H_o = 37.614 \times 10^6 [1.0 + 0.033(-0.8639)]$$
$$(0.866 \times 0.9466 \times 0.9805 + 1.8 \times 0.50 \times 0.3222) \, \text{J/m}^2$$

or,

$$H_o = 26.82 \times 10^6 \times (0.8038 + 0.2899) = 29.6 \times 10^6 \, \text{J/m}^2$$
$$\text{or } H_o = 29.6 \, \text{MJ/m}^2$$

In this case, the number of sunshine hour $(N) = \frac{2\omega_s}{15} = \frac{2 \times 101.33}{15} = 13.51 \, \text{h}.$

So, from Eq. 1.27, monthly average solar radiation for May 2022, in W/m$^2$ in extra-terrestrial region $(H_{0ex})$ can be evaluated as

$$H_{oex} = \frac{29.6 \times 10^6}{13.51 \times 3600} = 608.60 \, \text{W/m}^2.$$

**Example 1.19**

Repeat Examples 1.18a and 1.18b for the average day of the June month ($n = 162$).

**Solution**

For $n = 162$, $\delta = 23.45 \sin\left[\frac{360}{365}(284 + 162)\right] = 23.45 \times \sin 439.89 = 23.45 \times 0.9845 = 23.08$ (Eq. 1.9) and $\varphi = 30°$.

From Eq. 1.15, $\omega_s = \cos^{-1}(-\tan\varphi\tan\delta) = \cos^{-1}(-0.5773 \times 0.4262) = \cos^{-1}(-0.2461) = 75.75$.

Now, the daily solar radiation on a horizontal surface in extra-terrestrial region can be obtained from Eq. 1.19 as follows:

$$H_o = \frac{24 \times 3600 \times 1367}{\pi}\left[1.0 + 0.033\cos\left(\frac{360 \times 162}{365}\right)\right]$$
$$\left(\cos 30° \cos 23.08° \sin 75.75° + \frac{\pi \times 75.75}{180}\sin 30° \sin 23.08\right)$$

or,

$$H_o = 37.614 \times 10^6[1.0 + 0.033(-0.9383)]$$
$$(0.866 \times 0.9196 \times 0.9692 + 1.32 \times 0.50 \times 0.3920)$$

or,

$$H_o = 37.57\,\text{MJ/m}^2.$$

In this case, the number of sunshine hour ($N$) $= \frac{2\omega_s}{15} = \frac{2 \times 75.75}{15} = 10.10\,\text{h}$

From Eq. 1.27, the daily average solar radiation on month June ($n = 162$), in W/m$^2$ in extra-terrestrial region

$$H_{oex} = \frac{37.57 \times 10^6}{10.01 \times 3600} = 1033.23\,\text{W/m}^2.$$

**Example 1.20**

Calculate the solar radiation in MJ/m$^2$ on the horizontal surface for Examples 1.18a and 1.18b between 10 and 11 a.m.

**Solution**

Given: $\omega_1 = -30°$ and $\omega_2 = -15°$ (Eq. 1.9).

There are two methods to evaluate solar radiation in MJ/m$^2$ as

(a)  **First Method by Using Eq. 1.26**:

From Eq. 1.18, one gets,

$$
I_0 = \frac{12 \times 3600}{\pi} \times 1367 \left[ 1 + 0.033 \times \cos\left(\frac{360 \times 151}{365}\right) \right]
$$

$$
\left[ \cos 30° \times \cos 21.9° \times \{\sin(-15°) - \sin(-30°)\} \right.
$$

$$
\left. + \left\{ \frac{\pi(-15 - (-30))}{180} \right\} \sin 30° \times \sin 21.9° \right]
$$

$$
= 4.43 \times 10^6 \, \text{J/m}^2 = 4.43 \, \text{MJ/m}^2
$$

(b)  **Second method by using Eq. 1.25a**:

The hourly extra-terrestrial radiation in Wh/m$^2$ can also be approximated by using the relation Eq. 1.25a as

$$
I_0 = 3600 \times I_{sc} \left[ 1 + 0.033 \cos\left(\frac{360n}{365}\right) \right] [\cos\varphi \times \cos\delta \times \cos\omega + \sin\varphi \sin\delta]
$$

where, $\omega$ is evaluated at the midpoint of the hour, i.e., $\omega = (\omega_1 + \omega_2)/2 = -22.5°$.

Now,

$$
I_0 = 3600 \times 1367 \times 0.972 \times (0.743 + 0.187) = 4.44 \times 10^6 \, \text{J/m}^2
$$

$$
= 4.44 \, \text{MJ/m}^2
$$

The differences between the hourly radiations evaluated by both methods will be slightly higher at times near sunrise and sunset.

## 1.5  Average Solar Radiation on Horizontal Surface in Terrestrial Region

### 1.5.1  Monthly Average Beam Radiation $\left(\overline{I}_b\right)$ on Horizontal Surface

The hourly average beam radiation $\left(\overline{I}_b\right)$ on horizontal surface in W/m$^2$ in terrestrial region for a given month can be obtained by multiplying the hourly average solar radiation in extra-terrestrial region Eq. 1.28 by atmospheric transmittance, Eqs. 1.16a and 1.16b, as follows:

$$\overline{I}_b = H_{oex} \times \tau_{atm} = H_{oex} \times \exp[-(m.\varepsilon.T_R + \alpha)] \qquad (1.29)$$

Equation 1.29 can be used to evaluate monthly average beam radiation for each month by using different $n$ and $\delta$ for each month (Table 1.6).

### 1.5.2 Monthly Average Diffuse Radiation on Horizontal Surface

Following Eq. 1.18, monthly average diffuse radiation $\left(\overline{I}_d\right)$ can be evaluated as

$$\overline{I}_d = K_1\left(H_{oex} - \overline{I}_b\right) + K_2 \qquad (1.30)$$

The values of $K_1$ and $K_2$ for each month and weather conditions are given in Appendix C.1.

### 1.5.3 Monthly Average Total Radiation on Horizontal Surface

The monthly average total radiation on horizontal surface is sum of Eqs. 1.29 and 1.30, and it can be expressed as follows:

$$\overline{I}(t) = \overline{I}_b + \overline{I}_d \qquad (1.31)$$

**Example 1.21**
Calculate monthly average beam radiation $\left(\overline{I}_b\right)$ on month June ($n = 162$), in W/m² in terrestrial region for Example 1.19.

**Solution**

From Eq. 1.29, we have

$$\overline{I}_b = H_{oex} \times \exp[-(m.\varepsilon.T_R + \alpha)]$$

Here, from Eqs. 1.16a and 1.16b, for $n = 162'$

$$\tau_{atm} = \exp[-(m.\varepsilon.T_R + \alpha)] = 0.5906$$

Then, an average beam radiation is

$$\overline{I}_b = 1033.23 \times 0.5906 = 610.23 \text{ W/m}^2 \text{(Examples 1.12a and 1.19a)}$$

**Example 1.22**
Calculate monthly average diffuse radiation $\left(\overline{I}_d\right)$ on month June ($n = 162$), in W/m² in terrestrial region for Example 1.21.

**Solution**

**From Eq. 1.30, we have**

$$\overline{I}_d = K_1\left(H_{oex} - \overline{I}_b\right) + K_2$$

Here, $H_{0ex} = 1033.23$ W/m², $\overline{I}_b = 610.23$ W/m² (Example 1.30), $K_1 = 0.27$ and $K_2 = 31.86$ (Example 1.16b and Appendix C.1 for June, $n = 162$).
 Now, the monthly average diffuse radiation $\left(\overline{I}_d\right)$ is

$$\overline{I}_d = 0.27(1033.23 - 610.23) + 31.86 = 146.07 \text{ W/m}^2.$$

**Example 1.23**
Calculate monthly average total radiation $\left[\overline{I}(t)\right]$ on month June ($n = 162$) in W/m² in terrestrial region.
 From Eqs. 1.13a and 1.13b we have the following

$$\overline{I}(t) = \overline{I}_b + \overline{I}_d$$

Here, $\overline{I}_b = 610.23$ W/m², $\overline{I}_d = 146.07$ W/m² (Examples 1.21 and 1.22).
 So, the monthly average total radiation $\left(\overline{I}_d\right)$ is

$$\overline{I}(t) = 610.23 + 146.07 = 756.3 \text{ W/m}^2.$$

## 1.5.4 Daily Solar Radiation on Horizontal Surface in J/m² in Terrestrial Region

By using the concept of Eqs. 1.16a and 1.26, the daily solar radiation on horizontal surface in J/m² in terrestrial region can be written as

$$H = H_o \times \tau_{atm} \tag{1.32}$$

where, $\tau_{atm} = \exp[-(m.\varepsilon.T_R + \alpha)]$ is atmospheric transmittance (Eq. 1.16a) and an expression for $H_0$ is given by Eq. 1.27.

**Example 1.24**
Calculate the daily solar radiation on horizontal surface in MJ/m² in terrestrial region for Example 1.19.

**Solution**

From Example 1.19, we have

$$H_o = 37.57 \, \text{MJ/m}^2$$

From Example 1.12a for the average day of the June month ($n = 162$), we have atmospheric transmittance as

$$\tau_{\text{atm}} = \exp[-(1.0074 \times 0.0991 \times 2.47 + 0.28)] = \exp(-0.5266) = 0.5906$$

The daily solar radiation on horizontal surface in MJ/m$^2$ in terrestrial region can be obtained from Eq. 1.32 as

$$H = H_o \times \tau_{\text{atm}} = 37.52 \times 0.5906 = 22.16 \, \text{MJ/m}^2.$$

## 1.6   Hourly Solar Radiation for Known Solar Radiation in Extra-terrestrial Region

### 1.6.1   Hourly Solar Radiation in W/m² from Daily Solar Radiation (H₀) in J/m² on Horizontal Surface in Extra-terrestrial Region

Collares Pereira and Rabl [5] have proposed the following relation to evaluate the hourly total solar radiation from known daily solar radiation data in J/m$^2$ in extra-terrestrial region:

$$r_t = \frac{I(t)_{\text{ext}}}{H_o} \quad \text{with } H_o \text{ is in MJ/m}^2 \text{ (Eq. 1.27)}$$

or,

$$I(t)_{\text{ext}} = r_t \times H_o \frac{\text{MJ}}{\text{m}^2} = \frac{r_t \times H_o}{N \times 3600} \, \text{MW/m}^2 \qquad (1.32a)$$

where, $N$ is sunshine hours and

$$r_t = \frac{\pi}{24}(a + b \times \cos \omega) \frac{\cos \omega - \cos \omega_s}{\sin \omega_s - (2\pi \omega_s/360) \cos \omega_s} \qquad (1.32b)$$

The constants '$a$' and '$b$' are given by

$$\left.\begin{array}{c} a = 0.409 + 0.5016 \sin(\omega_s - 60) \\ \text{and} \\ b = 0.6609 - 0.4767 \sin(\omega_s - 60) \end{array}\right\}$$

where, $\omega$ is the hour angle in degrees, Table 1.4 for the time in question and $\omega_s$ is the sunset hour angle, and it is given by Eq. 1.13a as

$$\omega_s = \cos^{-1}(-\tan \varphi \tan \delta)$$

Gueymard model [6] has modified Collares Pereira and Rabl model [5] divided by a factor '$f$' which is given by

$$f = \frac{a + 0.5b(\omega_0 - \sin \omega_0 \cos \omega_0)}{\sin \omega_0 - \frac{2\pi \omega_s}{360} \cos \omega_0} \text{(Appendix C.3)} \qquad (1.33)$$

Now, after dividing Eq. 1.32b by '$f$' one gets

$$r_t = (a + b \cos \omega) \frac{\frac{\pi}{24}(\cos \omega - \cos \omega_0)}{a + 0.5b\left(\frac{2\pi \omega_s}{360} - \sin \omega_0 \cos \omega_0\right)} \qquad (1.34a)$$

with $\omega_0 = \frac{2\pi \omega_s}{360}$ (in radian) $= \omega_s$ (in degree).
So, Eqs. 1.34a and 1.34b becomes as

$$r_t = (a + b \cos \omega) \frac{\frac{\pi}{24}(\cos \omega - \cos \omega_s)}{a + 0.5b\left(\frac{2\pi \omega_s}{360} - \sin \omega_s \cos \omega_s\right)} \text{ with}$$

$$\frac{r_0}{f} = \frac{\frac{\pi}{24}(\cos \omega - \cos \omega_s)}{a + 0.5b\left(\frac{2\pi \omega_s}{360} - \sin \omega_s \cos \omega_s\right)} \qquad (1.34b)$$

with

$$\frac{r_0}{f} = \frac{\frac{\pi}{24}(\cos \omega - \cos \omega_0)}{a + 0.5b\left(\frac{2\pi \omega_s}{360} - \sin \omega_0 \cos \omega_0\right)}$$

**Example 1.25**
Calculate the constants '$a$' and '$b$' of Eq. 1.32b for the average day of the June month ($n = 162$ and $\delta = 23.1°$) along with sunshine hours.

**Solution**

**For June 2022, n = 162**, $\varphi = 30°$ (Table 1.2) and $\delta = 23.1$ (Table 1.6).
    Now, $\omega_s = \cos^{-1}(-\tan \varphi \tan \delta) = \cos^{-1}(-0.5773 \times 0.4265) = 104.25°$
    Sunshine hour $= \frac{2\omega_s}{15} = 13.9$ and hence

$$a = 0.409 + 0.5016 \sin(\omega_s - 60) = 0.409 + 0.5016 \times 0.6978 = 0.7590$$
$$b = 0.6609 - 0.4767 \sin(\omega_s - 60) = 0.6609 - 0.4767 \times 0.6978) = 0.3282.$$

### Example 1.26a

Calculate hourly solar intensity for Example 1.25 for the average day of the June month ($n = 162$ and $\delta = 23.1°$) at 6 a.m. (early morning).

### Solution

Here, $\omega_s = 104.25°$, $N = 13.9$ h, $a = 0.7682$ and $b = 0.3195$ (Example 1.25); $H_o = 37.57 \times 10^6$ J/m$^2$ (Example 1.19).

Equation 1.32 is the hourly variation of total intensity, and it is given by

$$I(t)_{ext} = \left[ \frac{\pi}{24} (0.7590 + 0.3282 \times \cos \omega) \right.$$
$$\left. \frac{\cos \omega - \cos 104.25}{\sin 104.25 - (2\pi \times 104.25/360) \cos 104.25} \right] \frac{37.57 \times 10^6}{13.9 \times 3600} \text{ W/m}^2$$

or,

$$I(t)_{ext} = \left[ \frac{\pi}{24} (0.7590 + 0.3282 \times \cos \omega) \right.$$
$$\left. \frac{\cos \omega - \cos 104.25}{\sin 104.25 - (2\pi \times 104.25/360) \cos 104.25} \right] \frac{37.57 \times 10^6}{13.9 \times 3600} \text{ W/m}^2$$

or,

$$I(t)_{ext} = \left[ \frac{\pi}{24} (0.7590 + 0.3282 \times \cos \omega) \right.$$
$$\left. \frac{\cos \omega - \cos 104.25}{\sin 104.25 - (2\pi \, 104.25/360) \cos 104.25} \right] \frac{37.57 \times 10^6}{13.9 \times 3600} \text{ W/m}^2$$

Consider the different value of $\omega$ at different time from Table 1.4. For example at 6 a.m. $\omega = -90$ (Table 1.4), then

$$I(t)_{ext} = 982 \left[ (0.7590 + 0.3282 \times \cos \omega) \frac{\cos \omega - (-0.2462)}{0.9692 - 1.8185 \times (-0.2462)} \right] \text{ W/m}^2$$

or,

$$I(t)_{ext} = 982 \left[ (0.7590 + 0.3282 \times \cos(-90)) \frac{\cos(-90) - (-0.2462)}{0.9692 - 1.8185 \times (-0.2462)} \right]$$

or,

$$I(t)_{ext} = 982 \left[ 0.7590 \times \frac{\cos(-90) - (-0.2462)}{0.9692 - 1.8185 \times (-0.2462)} \right]$$

$$= 982\left[0.07590 \times \frac{0.2462}{1.4169}\right] = 129.5\,\text{W/m}^2$$

**Example 1.26b**

Calculate hourly solar intensity for Example 1.25 for the average day of the June month ($n = 162$ and $\delta = 23.1°$) at 12 noon (at noon).

**Solution**

At noon, $\omega = 0$, noon time.

From Example 1.26a, we have

$$I(t)_{\text{ext}} = 982\left[(0.7682 + 0.3282 \times \cos(0))\frac{\cos(0) - (-0.2462)}{0.9692 - 1.8185 \times (-0.2462)}\right]\text{W/m}^2$$

$$I(t)_{\text{ext}} = 982\left[(0.7682 + 0.3282 \times 1)\frac{1 - (-0.2462)}{0.9692 - 1.8185 \times (-0.2462)}\right]\text{W/m}^2$$

$$I(t)_{\text{ext}} = 982\left[1.0964 \times \frac{1.2462}{1.4165}\right] = 947.22\,\text{W/m}^2.$$

**Example 1.27**

Calculate hourly solar intensity for Example 1.25 for the average day of the June month ($n = 162$ and $\delta = 23.1°$) at morning 6 a.m. and 12 noon.

**Solution**

**In this case, $\omega = 90°$ (morning) and $0°$ (noon).**

From Example 1.25, we have $\omega_s = 104.250$; $a = 0.7682$ and $b = 0.3195$.

In this case, the hourly variation of total intensity will be modified, and it is given by

$$I(t)_{\text{ext}} = \left[(0.7682 + 0.3282 \times \cos\omega)\frac{\frac{\pi}{24}(\cos\omega - \cos 104.25)}{a + 0.5b\left(\frac{2\pi\omega_s}{360} - \sin\omega_s\cos\omega_s\right)}\right]$$
$$\frac{37.57 \times 10^6}{3600}\,\text{W/m}^2$$

or,

$$I(t)_{\text{ext}} = [(0.7682 + 0.3282 \times \cos\omega)$$
$$\frac{\frac{\pi}{24}(\cos\omega - \cos 104.25)}{0.7682 + 0.5b\left(\frac{2\pi \times 104.25}{360} - \sin 104.25\cos 104.25\right)}]$$
$$\frac{37.57 \times 10^6}{13.9 \times 3600}\,\text{W/m}^2$$

or,

$$I(t)_{ext} = 982[(0.7682 + 0.3282 \times \cos \omega)$$
$$\frac{(\cos \omega - \cos 104.25)}{0.7682 + 0.5 \times 0.3195[1.8185 - 0.9692 \times (-0.2461)]}\bigg] W/m^2.$$

(a)  **For morning,** $\omega = 90°$

$$I(t)_{ext} = 982[(0.7682 + 0.3282 \times \cos 90)$$
$$\frac{(\cos 90 - \cos 104.25)}{0.7682 + 0.5 \times 0.3195[1.8185 - 0.9692 \times (-0.2461)]}\bigg] W/m^2$$

or,

$$I(t)_{ext} = 982[(0.7682 + 0.3282 \times 0)$$
$$\frac{(0 - \cos 104.25)}{0.7682 + 0.5 \times 0.3282[1.8185 - 0.9692 \times (-0.2461)]}\bigg] W/m^2$$

or,

$$I(t)_{ext} = 982\bigg[0.7682 \times \frac{[0 - (-0.2461)]}{0.7682 + 0.5 \times 0.3282[1.8185 - 0.9692 \times (-0.2461)]}\bigg] W/m^2$$

$$I(t)_{ext} = 982\bigg[0.7682 \times \frac{0.2461}{1.1058}\bigg] = 167.88 \ W/m^2$$

(b)  **For morning,** $\omega = 0°$

$$I(t)_{ext} = 982[(0.7682 + 0.3282 \times \cos 0)$$
$$\frac{(\cos 0 - \cos 104.25)}{0.7682 + 0.5 \times 0.3195[1.8185 - 0.9692 \times (-0.2461)]}\bigg] W/m^2$$

or,

$$I(t)_{ext} = 982[(0.7682 + 0.3282 \times 1)$$
$$\frac{(1 - \cos 104.25)}{0.7682 + 0.5 \times 0.3195[1.8185 - 0.9692 \times (-0.2461)]}\bigg] W/m^2$$

or,

$$I(t)_{ext} = 982\bigg[1.0877 \times \frac{[1 - (-0.2461)]}{0.7682 + 0.5 \times 0.3195[1.8185 - 0.9692 \times (-0.2461)]}\bigg] W/m^2$$

$$I(t)_{ext} = 982\left[1.0964 \times \frac{1.2461}{1.1058}\right] = 1207.22 \, \text{W/m}^2$$

## 1.6.2 Hourly Total Solar Radiation [I(t)] in W/m² on Horizontal Surface in Terrestrial Region from Hourly Solar Radiation $\left[I(t)_{ext}\right]$ in W/m² on Horizontal Surface in Extra-terrestrial Region

### 1.6.2.1 Beam Radiation on Horizontal Surface from Known Extra-terrestrial Radiation $I(t)_{ext}$

From Eq. 1.32a, we have hourly variation of solar radiation (only direct/beam) in extra-terrestrial region as follows:

$$I(t)_{ext} = r_t \times H_o \frac{MJ}{m^2} = \frac{r_t \times H_o}{N \times 3600} \, \text{MW/m}^2$$

If this solar radiation passes from extra-terrestrial region to terrestrial region through atmosphere, it will be attenuated by a factor known as atmospheric transmittance given by Eq. 1.16a as

$$\tau_{atm} = \exp[-(m.\varepsilon.T_R + \alpha)]$$

Now, the hourly beam radiation in terrestrial region can be obtained by multiplying Eq. 1.32a with Eq. 1.16a as

$$I_b = I(t)_{ext} \times \tau_{atm} \tag{1.35a}$$

where, $\tau_{atm} = \exp[-(m.\varepsilon.T_R + \alpha)]$

### 1.6.2.2 Diffuse Radiation from Known Extra-terrestrial Radiation $\left[I(t)_{ext}\right]$

Following Eq. 1.30, one can write an expression for diffuse radiation as

$$I_d = K_1\left[I(t)_{ext} - I_b\right] + K_2 \tag{1.35b}$$

### 1.6.2.3   Total Radiation from Known Extra-terrestrial Radiation $\left[I(t)_{ext}\right]$

Following Eq. 1.31, one can write an expression for total radiation in terrestrial region as

$$I(t) = I_b + I_d \tag{1.35c}$$

Equation 1.35c can be rewritten as

$$I(t) = I(t)_{ext} \times \tau_{atm} + K_1\left[I(t)_{ext} - I(t)_{ext} \times \tau_{atm}\right] + K_2 \tag{1.36}$$

**Example 1.28**

Evaluate total hourly radiation $[I(t)]$ in terrestrial region from known extra-terrestrial region $[I(t)_{ext}]$ for Example 1.27.

**Solution**

**From Example 1.27 , we have**

(a)   **For morning, $\omega = 90°$**

$$I(t)_{ext} = 982\left[0.7682 \times \frac{0.2461}{1.1058}\right] = 167.88 \text{ W/m}^2$$

(b)   **For morning, $\omega = 0°$**

$$I(t)_{ext} = 982\left[1.0964 \times \frac{1.2461}{1.1058}\right] = 1207.22 \text{ W/m}^2$$

From Example 1.24, we have atmospheric transmittance for the month of June as

$$\tau_{atm} = \exp[-(1.0074 \times 0.0991 \times 2.47 + 0.28)]$$
$$= \exp(-0.5266) = 0.5906$$

From Appendix C.1 for month of June, we have

$$K_1 = 0.27 \text{ and } K_2 = 31.86$$

From Eq. 1.36, we can evaluate total solar radiation on horizontal surface in terrestrial region as

(a)   **For morning, $\omega = 90°$**

$$I(t) = 167.88 \times 0.5906 + 0.27[167.88 - 167.88 \times 0.5906]$$
$$+ 31.86 = 99.15 + 50.41 = 149.56 \text{ W/m}^2$$

(b) **For morning, $\omega = 0°$**

$$I(t) = 1207.22 \times 0.5906 + 0.27[1207.22 - 1207.22 \times 0.5906]$$
$$+ 31.86 = 712.98 + 165.32 = 878.30 \, \text{W/m}^2$$

### 1.6.3 Hourly Total Solar Radiation on Inclined Surface, $[I(t)_\beta]$ in Terrestrial Region for Known Daily Solar Radiation on Horizontal Surface, $I(t)_{ext}$ in Extra-terrestrial Region

Following Eq. 1.23 and Sect. 1.6.2, the total radiation on inclined surface, $[I(t)_\beta]$ in terrestrial region can be written as follows:

$$I(t)_\beta = [I(t)_{ext} \times \tau_{atm}] \times R_b$$
$$+ \{K_1[I(t)_{ext} - I(t)_{ext} \times \tau_{atm}] + K_2\}R_d + I(t) \times R_r \quad (1.37)$$

where expressions for $R_b$, $R_d$ and $R_r$ are given by Eqs. 1.21a and 1.21b–1.23 with Examples 1.17a–1.17d.

**Example 1.29**
Estimate total solar radiation on inclined surface, $[I(t)_\beta]$ in terrestrial region for Example 1.28 at inclination of $\beta = 45°$ n the noon time, $\omega = 0°$ in the month of June ($n = 162$).

**Solution**

From Example 1.28, we have
   **For noon, $\omega = 0°$:** $I_b = 712.98 \, \text{W/m}^2$, $I_d = 165.32 \, \text{W/m}^2$ and $I(t) = 878.30 \, \text{W/m}^2$
   From Examples 1.17a–1.17d for June ($n = 162$): $R_b = 0.2315$, $R_d = 0.8536$, and $R_r = 0.0293$.
   Now, the total solar radiation for inclined surface is determined from Eq. 1.37 as

$$I(t)_\beta = 712.98 \times 0.2315 + 165.32 \times 0.8536 + 878.30 \times 0.0293$$
$$= 164.83 + 141.12 + 25.73 = 331.68 \, \text{W/m}^2$$

## 1.7 Evaluation of Solar Radiation for Known Solar Radiation Data and Bright Sunshine Hour in Terrestrial Region for a Given Solar Radiation in Extra-terrestrial Region

In earlier section, we have computed solar radiation on horizontal and inclined in terrestrial region without knowing any solar radiation data's in terrestrial region. However, nowadays some solar radiation data are available by Metrological Department of most countries, and hence we will discuss how solar radiation can be evaluated on horizontal/inclined surface in terrestrial region.

### 1.7.1 Evaluation of the Monthly Average of Daily Total Radiation $(\overline{H})$ in J/m² on a Horizontal Surface in Terrestrial Region for Known Solar Radiation in Extra-terrestrial Region

For many applications, one needs the knowledge of the monthly average of daily solar radiation $(\overline{H})$ in J/m² available on a horizontal surface. It can be calculated with 10% accuracy using following correlation:

$$\frac{\overline{H}}{\overline{H}_o} = a + b\left(\frac{\overline{n}}{\overline{N}}\right) \tag{1.37a}$$

where, $\overline{n}$ and $\overline{N}$ are the observed/measured monthly average of daily bright sunshine hours (known) and total length of average day of month, Eq. 1.14. $\overline{H}$ and $\overline{H}_0$ are the monthly average of daily total solar radiation (unknown) on a horizontal surface in terrestrial and extra-terrestrial region (known) (case (ii) of Eq. 1.27) respectively.

The regression coefficients $a$ and $b$ are given by

$$a = -0.309 + 0.539 \cos \varphi - 0.0693 E_o + 0.290(\overline{n}/\overline{N}) \text{ and,}$$
$$b = 1.527 - 1.027 \cos \varphi + 0.0926 E_o - 0.359(\overline{n}/\overline{N}) \tag{1.37b}$$

where, $\varphi$ is latitude of the place and $E_0$ is the height of the place (altitude/elevation in kilometers) above sea level (Table 1.2).

The regression coefficients can also be evaluated as following:

$$a = -0.110 + 0.235 \cos \varphi + 0.323(\overline{n}/\overline{N}) \text{ and}$$
$$b = 1.449 - 0.553 \cos \varphi - 0.694(\overline{n}/\overline{N}) \tag{1.37c}$$

For different period of year, the rang of $\bar{n}/N$ for validity of above correlations is given below

$$0.2 \le (\bar{n}/\bar{N}) \le 0.6 \quad \text{for monsoon period}$$
$$0.4 \le (\bar{n}/\bar{N}) \le 0.9 \quad \text{for pre-monsoon and}$$
$$\text{post-monsoon period}$$

**Example 1.30**

Evaluate the monthly average solar radiation in J/m² and W/m² on a horizontal surface for 12.1 h of the observed monthly average daily hours of bright sunshine ($n$) observed at New Delhi in May 2013.

**Solution**

Given $\delta = 18.79°$ (Eq. 1.9, Table 1.6), $\varphi = 28°35' = 28.58°$ (New Delhi) and $\omega_s = \cos^{-1}(-\tan 28.58° \tan 18.79°) = 100.68°$ (Eq. 1.13a).

The average monthly extra-terrestrial radiation is obtained as (Eq. 1.27),

$$H_o = \frac{24 \times 3600 \times 1367}{\pi} \left[ 1.0 + 0.033 \cos\left( \frac{360 \times 135}{365} \right) \right]$$
$$(\cos(28.58)° \cos(18.79)° \sin(100.68)°$$
$$+ \frac{\pi \times 100.68}{180} \sin(28.58)° \sin(18.79)°\right) = 39.96 \, \text{MJ/m}^2$$

The number of daylight hours, Eq. 1.14 is obtained as

$$N = \frac{2}{15} \cos^{-1}(-\tan 28.58° \tan 18.79°) = 13.42$$

The values of '$a$' and '$b$' can be obtained from Eq. 1.37c as follows:

$$a = -0.110 + 0.235 \cos(28.58°)$$
$$+ 0.323(12.1/13.42) = 0.388 \quad \text{and}$$
$$b = 1.449 - 0.553 \cos(28.58°)$$
$$- 0.694(12.1/13.42) = 0.338$$

The above values of '$a$' and '$b$' are used in Eq. 1.37a to obtain the monthly average daily solar radiation incident on the horizontal surface as follows

$$\frac{\bar{H}}{\bar{H}_o} = a + b\left(\frac{\bar{n}}{\bar{N}}\right) = 0.388 + 0.338\left(\frac{12.1}{13.42}\right) = 0.693$$

For $H_o = 39.96 \, \text{MJ/m}^2$, we get $\bar{H} = 27.66 \, \text{MJ/m}^2$, Eq. 1.37a.

Now, an average monthly solar radiation $\left[\overline{I(t)}_{monthly}\right]$ in W/m$^2$ in terrestrial region is evaluated as

$$\overline{I(t)}_{monthly} = \frac{\overline{H}(\text{MJ/m}^2)}{n \times 3600\,\text{s}} = \frac{27.66 \times 10^6\,\text{J/m}^2}{12.1 \times 3600\,\text{s}} = 634\,\text{W/m}^2$$

Similarly, an average monthly solar radiation $\left[I(t)_{monthly}\right]$ in W/m$^2$ in terrestrial region for other months can be determined. Such monthly data are useful to determine the monthly performance of solar energy system.

## 1.7.2  Clearness Index

**The clearness index is required to compute diffuse radiation for known data of solar radiation $I(t)$, hourly in W/m²; $H$, daily in J/m² and $\overline{H}$, monthly in J/m² in terrestrial region.**

### 1.7.2.1  Hourly Clearness Index ($k_T$)

It is ratio of hourly data of solar radiation $[I(t)]$, Eq. 1.19, in terrestrial region to hourly data solar radiation, Eq. 1.25a, in extra-terrestrial region which is given by

$$k_T = \frac{I(t)}{I_o} \tag{1.38a}$$

where, $I(t)$ is measured hourly solar radiation and $I_0$ is obtained from Eq. 1.25a.

For estimation of hourly total solar radiation on any inclined surface and orientation, see Sect. 1.3.2, hourly values of beam and diffuse radiation on horizontal surface should be known. For this one should know various clearness indexes.

If $I_d/I(t)$, the ratio of hourly diffuse radiation in W/m$^2$ to the hourly total solar radiation on a horizontal surface in W/m$^2$ at a given time of hour of a day, then there was correlation developed by Orgill and Hollands (1977), and it is given by

$$\begin{aligned}
I_d/I(t) &= 1.0 - 0.249 k_T \quad \text{for } k_T < 0.35 \\
I_d/I(t) &= 1.557 - 1.84 k_T \quad 0.35 < k_T < 0.75 \\
I_d/I(t) &= 0.177 \quad \text{for } k_T > 0.35
\end{aligned} \tag{1.38b}$$

### 1.7.2.2  Daily Clearness Index ($K_T$)

For this one should know daily measured solar radiation in terrestrial region which can be obtained as

$$H = \sum_{i=1}^{n} I_i(t) \times 3600, \ \text{J/m}^2 \tag{1.38c}$$

The $n$ is measured bright sunshine hours.

A daily clearness index, $K_T$, is the ratio of daily solar radiation Eq. 1.38c, on horizontal surface in terrestrial region to daily solar radiation on horizontal surface, Eq. 1.27, in J/m$^2$ in the extra-terrestrial radiation for that day.

$$K_T = \frac{H}{H_o} \tag{1.38d}$$

where, $H$ is measured daily total solar radiation, Eq. 1.38c, and $H_o$ is obtained from case (i) of Eq. 1.27 in J/m$^2$.

For known $K_T$ from above Eq. 1.38d, an expression for ration of diffuse radiation to daily total solar tradition, $H_d/H$, has been proposed by Collares-Pereira and Rable [5] as follows:

$$\left. \begin{array}{ll} H_d/H = 0.99 & \text{for } K_T \leq 0.17 \\ H_d/H = 1.188 - 2.72K_T + 9.473K_T^2 & \text{for } 0.17 < K_T < 0.75 \\ \quad - 21.865K_T^3 + 14648K_T^4 & \\ H_d/H = -0.54K_T + 0.632 & \text{for } 0.75 < K_T < 0.80 \\ H_d/H = 0.2 & \text{for } \geq 0.80 \end{array} \right\}$$

## 1.7.3  Monthly Clearness Index $\left(\overline{K}_T\right)$

Monthly measured solar radiation can be evaluated as follows:

$$\overline{H} = \frac{\sum_{j=1}^{n_0}[\sum_{i=1}^{n} I_i(t) \times 3600], \ \text{J/m}^2}{n_0} \tag{1.38e}$$

where, $n_0$ is number of days in the month.

The monthly average clearness index, $\overline{K}_T$, is the ratio of monthly average solar radiation on a horizontal surface, Eq. 1.38e in terrestrial region to the monthly average extra-terrestrial solar radiation, Case (ii) of Eq. 1.27, i.e.,

$$\overline{K}_T = \frac{\overline{H}}{\overline{H}_o} \qquad (1.38f)$$

where, $\overline{H}$ is measured monthly solar radiation and $\overline{H}_o$ is given by Eq. 1.38e.

For known $\overline{K}_T$, sunshine hour, $n$, based on the measured data for Chennai, New Delhi and Pune, Gopinathan (1988) has proposed the following correlation

$$\frac{\overline{H}_d}{\overline{H}} = 1.194 - 0.838\overline{K}_T - 0.0446\frac{\overline{n}}{\overline{N}} \qquad (1.38g)$$

The prediction based on the above equation has been compared with that obtained using following correlations:

$$\frac{\overline{H}_d}{\overline{H}} = 1.403 - 1.672\overline{K}_T \qquad (1.38h)$$

and,

$$\frac{\overline{H}_d}{\overline{H}} = 0.931 - 0.814\frac{\overline{n}}{\overline{N}} \qquad (1.38i)$$

## Example 1.31

Evaluate beam $\left(I_{b\beta}\right)$, diffuse $\left(I_{d\beta}\right)$, reflected $(I_r)$ and total solar radiation $\left(I_\beta\right)$ on a south oriented ($\gamma = 0$) tilted surface inclined at $\beta = 45°$ at a place having latitude $40°$ N for 2–3 p.m. on May 31, 2015. Given that $I = 1.04\,\mathrm{MJ/m^2}$ and $\rho = 0.60$.

## Solution

For present case: $n = 151$; $\delta = -11.6°$ (Eq. 1.9); $\Phi = 40°$ N; $\omega_2 = 45°$, $\omega_1 = 30°$ (Eq. 1.10) and $\gamma = 0$.

The hourly extra-terrestrial solar radiation in J/m² (Eq. 1.18) is given by

$$I_0 = \frac{12 \times 3600 \times 1367}{\pi}\left[1 + 0.033\cos\left(\frac{360 \times 151}{365}\right)\right]$$
$$\times \left[\cos 40° \times \cos(-11.6°) \times (\sin 45° - \sin 30°)\right.$$
$$\left. + \frac{\pi \times (45 - 30)}{180}\sin 40° \times \sin(-11.6°)\right]$$

or,

$$I_0 = \frac{12 \times 3600 \times 1367}{\pi} \times 0.972 \times 0.121 = 2.21\,\mathrm{MJ/m^2}.$$

The hourly clearness index is obtained by Eq. 1.22a as

$$k_T = \frac{I}{I_o} = \frac{1.04}{2.21} = 0.471$$

For the above value of $k_T$, the diffuse component can be evaluated from Eq. 1.22b and is given by

$$\frac{I_d}{I} = 1.557 - 1.84k_T = 0.690$$

For a given value of $I$ $(1.04\,\text{MJ/m}^2)$

$$I_d = 0.690 \times 1.04 = 0.718\,\text{MJ/m}^2 \text{ and}$$
$$I_b = (1.04 - 0.718) = 0.322\,\text{MJ/m}^2$$

The conversion factor for beam radiation at midpoint of the hour ($\omega = 37.5°$) and $\gamma = 0$ and it is given by

$$R_b = \frac{\cos(40° - 45°)\cos(-11.6°)\cos 37.5° + \sin(-11.6°)\sin(40°)}{\cos 40° \cos(-11.6°)\cos 37.5° + \sin(-11.6°)\sin 40°},$$
$$[\text{Eqs. 1.11, 1.12, and 1.20b}]$$
$$= 1.38$$

The total solar radiation, Eq. 1.23, is given as

$$I_\beta = 0.322 \times 1.38 + 0.718 \times \frac{(1 + \cos 45°)}{2} + 1.04 \times 0.60 \times \frac{(1 - \cos 45°)}{2}$$

or,

$$I_\beta = 0.425 + 0.613 + 0.091 = 1.129\,\text{MJ/m}^2$$

Contributions of beam, diffuse and reflected components are 0.425, 0.613, and 0.091 $\text{MJ/m}^2$ respectively.

The solar radiation in $\text{W/m}^2 = \frac{I_\beta}{3600} = \frac{1.129 \times 10^6}{3600}\,\text{W/m}^2 = 313.61\,\text{W/m}^2$.

## 1.8 Additional Examples

**Example 1.32**
**Determine the temperature of sun ($T_S$), in the month of January, for the following given data:**

**Solar constant = 1367 W/m²**

**Sun diameter ($2R_S$) = 1.39 × 10⁹ m**

**Sun-earth distance $(L_{se}) = 1.5 \times 10^{11}$ m**

**Solution**

The amount of solar radiation emitted by the surface of the sun $= \sigma\, T_s^4 \times (4\pi\, R_s^2)$, Eq. 1.1a.

The total emitted radiation from the sun will be received by a sphere having a radius equal to mean sun-earth distance $(L_{se}) = I_{ext} \times (4\pi L_{se}^2)$, Eq. 1.5.

Since, the solar radiation emitted by the sun = solar radiation received by a sphere having radius of $L_{se}$

or,

$$\sigma\, T_s^4 \left(4\pi\, R_s^2\right) = l_{ext} \times \left(4\pi\, L_{se}^2\right).$$

From above equation, one gets

$$T_s = \left[\frac{I_{ext}\left(4\pi\, L_{se}^2\right)}{\sigma\left(4\pi\, R_s^2\right)}\right]^{\frac{1}{4}}$$

For the month of January $I_{ext} = 1410$ W/m$^2$, Eq. 1.5 and 1.5a for $n = 17$ (Table 1.6). After substituting the value of constant, $R_s = 0.695 \times 10^9$ m, $\sigma = 5.67 \times 10^{-8}$ W/m$^2$ K$^4$, one gets

$$T_s = \left[\frac{1410 \times 4\pi (1.5 \times 10^{11})^2}{5.67 \times 10^{-8} \times 4\pi (0.695 \times 10^9)^2}\right]^{\frac{1}{4}} = 5833\,K$$

This indicates that there is little variation in the temperature of the sun ($T_S = 5833$ K).

**Example 1.33**
**Calculate the declination angle ($\delta$) for March 31 in a leap year.**
**Solution**

From Eq. 1.9, the expression for declination angle ($\delta$) is given by

$$\delta = 23.45 \sin\left[(284 + n) \times \frac{360}{365}\right]$$

For March 31 in a leap year, $n = 31$(January) $+ 29$ (February) $+ 31$ (March) $= 91$, then

$$\delta = 23.45 \sin\left[(284 + 91) \times \frac{360}{365}\right] = 4.016°$$

**Example 1.34**
**Calculate the hour angle (ω) at 2.30 p.m.**

**Solution**

An expression for hour angle, $\omega$ (in degree), Eq. 1.10, is given by

$$\omega = (ST - 12) \times 15°$$

where, ST is local solar time $= 2.30$ p.m. $= 12 + 2.5 = 14.5$ h.
Then,

$$\omega = (14.5 - 12) \times 15° = 37.5°.$$

**Example 1.35**
**Find out the daily variation of the extra-terrestrial solar intensity ($I_{ext}$) and the declination angle (δ) for the month of June 2020 (leap year).**

**Solution**

The extra-terrestrial solar radiation measured on a plane normal to the radiation from the sun on the $n$th day of the year is given in terms of solar constant $I_{sc}$ can be calculated from Eq. 1.5 as

$$I_{ext} = I_{sc}\left[1 + 0.033\cos\left(\frac{360 \times n}{365}\right)\right]$$

Similarly, declination (δ) for the $n$th day of year can be calculated from Eq. 1.9 as

$$\delta = 23.45\sin\left[(284 + n) \times \frac{360}{365}\right]$$

For the month of June, the values of $n$ can be calculated by conventional way for year having 365/366 days. For June 1, $n = 31 + 29 + 31 + 30 + 31 + 1 = 153$. The computed values of extra-terrestrial solar intensity and declination angle for each day of June for leap year are given in Table 1.8.

**Example 1.36**
**Calculate the hourly direct radiation on horizontal surface ($I_b$) and on inclined surface with an inclination of $45°\left[I_{b\beta}\right]$ due south on January 15, 1995, for city condition in terrestrial region.**

**Solution**

For Delhi $\varphi = 28.58°$; $n = 15$ for January 15; $\omega = 0$ at noon and $\gamma = 0$.
Substituting the value of $n = 15$ in Eq. 1.9, then

$$\delta = 23.45 \sin\left[(284 + 15) \times \frac{360}{365}\right] = -21.26.$$

From Eqs. 1.11a, 1.11b and 1.12, one can evaluate

$$\cos \theta_i = \cos(\Phi - \beta) \cos \delta \cos \omega + \sin \delta \sin(\Phi - \beta)$$
$$= \cos(28.56 - 45) \cos(-21.26) \cos 0$$
$$+ \sin(-21.26) \sin(28.56 - 45) = 0.$$
$$9591 \times 0.9319 + (-0.3626) \times (-0.2830) = 0.9964$$

and,

$$\cos \theta_z = \cos 28.56 \cos(-21.26) \cos 0 + \sin(-21.26) \sin 28.56 = 0.8783$$
$$\theta_z = \cos^{-1}(0.8783) = 28.56 \text{ and } \alpha + \theta_z = 90, \text{ then}$$
$$\alpha = 90 - \theta_z = 90 - 28.56 = 61.43$$

Further, from Eq. 1.6a for $n = 15$, one gets

$$I_{ext} = 1367\left[1 + 0.033 \cos\left(\frac{360 \times 15}{365}\right)\right] = 1410 \text{ W/m}^2$$

Now, $T_R = -3.1$ for city condition, Table 1.7/Appendix C.2, we can get

$$I_N = 1410 \exp[-3.1/(0.9 + 9.4 \sin 61.43)]$$
$$= 1410 \times 0.7127 = 1005 \text{ W/m}^2, \text{ Eq. 1.16b.}$$

For horizontal surface

$$I_b = I_N \cos \theta_z = 1005 \times 0.8783 = 882.89 \text{ W/m}^2, \text{ Eq.1.17}$$

For inclined surface at 45°

$$I_{b\beta} = 1005 \times 0.9964 = 1001.38 \text{ W/m}^2, \text{ Eq.1.20a}$$

**Table 1.7**  Turbidity factor for different months (Appendix C.2)

| Month | 1 | 2 | 3 | 4 | 5 | 6 | 7 | 8 | 9 | 10 | 11 | 12 |
|---|---|---|---|---|---|---|---|---|---|---|---|---|
| Region | | | | | | | | | | | | |
| Mountain | 1.8 | 1.9 | 2.1 | 2.2 | 2.4 | 2.7 | 2.7 | 2.7 | 2.5 | 2.1 | 1.9 | 1.8 |
| Flat land | 2.2 | 2.2 | 2.5 | 2.9 | 3.2 | 3.4 | 3.5 | 3.3 | 2.9 | 2.6 | 2.3 | 2.2 |
| City | 3.1 | 3.2 | 3.5 | 3.9 | 4.1 | 4.2 | 4.3 | 4.2 | 3.9 | 3.6 | 3.3 | 3.1 |

**Table 1.8** Daily variation of $I_{ext}$ and declination ($\delta$) for the month of June 2020

| Day of June | Value of $n$ | $I_{ext}$ (extra-terrestrial radiation) (W/m$^2$) | Declination ($\delta$) (W/m$^2$) |
|---|---|---|---|
| 1 | 153 | 1327.58 | 22.17 |
| 2 | 154 | 1327.20 | 22.3 |
| 3 | 155 | 1326.84 | 22.42 |
| 5 | 156 | 1326.5 | 22.53 |
| 6 | 157 | 1326.16 | 22.6 |
| 7 | 158 | 1325.84 | 22.74 |
| 8 | 159 | 1325.53 | 22.84 |
| 9 | 160 | 1325.23 | 22.93 |
| 10 | 161 | 1324.94 | 23.01 |
| 11 | 162 | 1324.66 | 23.08 |
| 12 | 163 | 1324.40 | 23.15 |
| 13 | 164 | 1324.15 | 23.2 |
| 14 | 165 | 1323.92 | 23.26 |
| 15 | 166 | 1323.69 | 23.31 |
| 16 | 167 | 1323.48 | 23.35 |
| 17 | 168 | 1323.28 | 23.38 |
| 18 | 169 | 1323.10 | 23.4 |
| 19 | 170 | 1323.92 | 23.43 |
| 20 | 171 | 1322.77 | 23.44 |
| 21 | 172 | 1322.62 | 23.44 |
| 22 | 173 | 1322.49 | 23.44 |
| 23 | 174 | 1322.37 | 23.45 |
| 24 | 175 | 1322.26 | 23.42 |
| 25 | 176 | 1322.17 | 23.4 |
| 26 | 177 | 1322.09 | 23.37 |
| 27 | 178 | 1322.02 | 23.33 |
| 28 | 179 | 1321.97 | 23.29 |
| 29 | 180 | 1321.93 | 23.24 |
| 30 | 181 | 1321.90 | 23.18 |

Beam radiation on horizontal surface $I_b = 882.89$ W/m$^2$ has lower value in comparison with inclined surface 1001.38 W/m$^2$ for city condition in terrestrial region.

**Example 1.37**
**Prove that $\cos \theta_z = \sin \alpha$.**

**Solution**

Since, we know $\theta_z = 90 - \alpha$

Taking cosine of both sides, we get

$$\cos \theta_z = \cos(90 - \alpha) = \sin \alpha$$

**Example 1.38**

Calculate the solar altitude angle ($\alpha$) at 1 a.m. for March 31 in leap year.

**Solution**

From Eq. 1.10, one has

$$\omega = (ST - 12) \times 15 = (13 - 12) = 15°$$

and $\delta = 4.016°$ for $n = 31 + 29 = 31 = 91$.
From Eq. 1.12, we have

$$\cos \theta_z = \cos \phi \cos \delta \cos \omega + \sin \delta \sin \phi$$

After substituting the following values, we get

$$\cos \theta_z = 0.88, \, \alpha + \theta_z = 90 \text{ and hence, } \alpha = 90 - 28.35 = 61.64°.$$

The solar altitude angle ($\alpha$) for March 31 in leap year at 1 a.m. is 61.64°.

**Example 1.39**

**Calculate $\overline{N}$ for each month for Delhi location by using Table 1.6 for average n for each month and declination angle.**

**Solution**

The number of daylight hours (sunshine hours) (N) is given by 1.14

$$\text{Daylight hours} = N = 2 \cdot \frac{\omega_s}{15} = \frac{2}{15} \cdot \cos^{-1}[-\tan \phi \cdot \tan \delta].$$

Delhi, $\varphi = 30°$, the value of $\delta$ taken from Table 1.6.

For New Delhi

For January month

$$\text{Daylight hours} = N = 2 \cdot \frac{\omega_s}{15} = \frac{2}{15} \cdot \cos^{-1}\left[-\tan(30°) \cdot \tan(-20.9°)\right]$$
$$\overline{N} = 10.30 \, \text{h}$$

For February month

$$\text{Daylight hours} = N = 2 \cdot \frac{\omega_s}{15} = \frac{2}{15} \cdot \cos^{-1}\left[-\tan(30°) \cdot \tan(-13°)\right]$$
$$\overline{N} = 10.98\,\text{h}$$

For March month

$$\text{Daylight hours} = N = 2 \cdot \frac{\omega_s}{15} = \frac{2}{15} \cdot \cos^{-1}\left[-\tan(30°) \cdot \tan(-2.4°)\right]$$
$$\overline{N} = 11.81\,\text{h}$$

For April

$$\text{Daylight hours} = N = 2 \cdot \frac{\omega_s}{15} = \frac{2}{15} \cdot \cos^{-1}\left[-\tan(30°) \cdot \tan(9.4°)\right]$$
$$\overline{N} = 12.73\,\text{h}$$

For May

$$\text{Daylight hours} = N = 2 \cdot \frac{\omega_s}{15} = \frac{2}{15} \cdot \cos^{-1}\left[-\tan(30°) \cdot \tan(18.8°)\right]$$
$$\overline{N} = 13.51\,\text{h}$$

For June

$$\text{Daylight hours} = N = 2 \cdot \frac{\omega_s}{15} = \frac{2}{15} \cdot \cos^{-1}\left[-\tan(30°) \cdot \tan(23.1°)\right]$$
$$\overline{N} = 13.81\,\text{h}$$

For July

$$\text{Daylight hours} = N = 2 \cdot \frac{\omega_s}{15} = \frac{2}{15} \cdot \cos^{-1}\left[-\tan(30°) \cdot \tan(21.2°)\right]$$
$$\overline{N} = 13.71\,\text{h}$$

For August

$$\text{Daylight hours} = N = 2 \cdot \frac{\omega_s}{15} = \frac{2}{15} \cdot \cos^{-1}\left[-\tan(30°) \cdot \tan(13.5°)\right]$$
$$\overline{N} = 13.06\,\text{h}$$

For September

$$\text{Daylight hours} = N = 2 \cdot \frac{\omega_s}{15} = \frac{2}{15} \cdot \cos^{-1}\left[-\tan(30) \cdot \tan(2.2°)\right]$$

$$\overline{N} = 12.16 \text{ hours}$$

For October

$$\text{Daylight hours} = N = 2 \cdot \frac{\omega_s}{15} = \frac{2}{15} \cdot \cos^{-1}\left[-\tan(30°) \cdot \tan(-9.6°)\right]$$

$$\overline{N} = 11.25 \text{ h}$$

For November

$$\text{Daylight hours} = N = 2 \cdot \frac{\omega_s}{15} = \frac{2}{15} \cdot \cos^{-1}\left[-\tan(30°) \cdot \tan(-18.9°)\right]$$

$$\overline{N} = 10.48 \text{ h}$$

For December

$$\text{Daylight hours} = N = 2 \cdot \frac{\omega_s}{15} = \frac{2}{15} \cdot \cos^{-1}\left[-\tan(30°) \cdot \tan(-23.0)\right]$$

$$\overline{N} = 10.10 \text{ h}$$

The $\overline{N}$ for each month for Delhi location are 10.30, 10.98, 11.81, 12.73, 13.81, 13.51, 13.71, 13.06, 12.16, 12.16, 10.48a and 10.10 h. This shows that the maximum sunshine occurs in the month of July (seventh month of year).

**Example 1.40**
**Plot the variation of monochromatic emissive power with $\lambda$ in $\mu$m for different temperatures of blackbody (6000, 3000, 1000 and 288 K).**

**Solution**

Using Eq. 1.1b, the variation of monochromatic emissive power with $\lambda$ in $\mu$m for different temperatures of blackbody (6000, 3000, 1000 and 288 K) is shown in Fig. 1.7. From Fig. 1.7, one can conclude that emissive power decreases with increase of surface temperature. In this case, one can draw the following conclusions:

(i)   At 6000 K which the temperature of sun, emits maximum power. This is short wavelength radiation due to Wien's displacement law.
(ii)  At 288 K which is the temperature of sun, emits minimum power. This is long wavelength radiation due to Wien's displacement law.

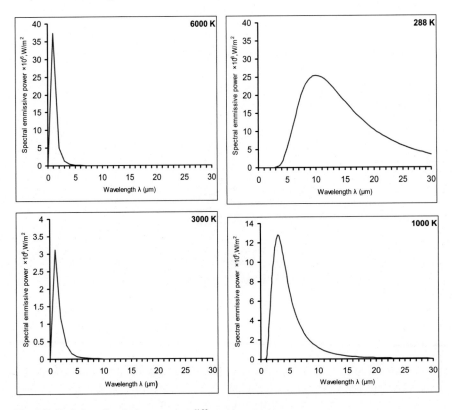

**Fig. 1.7** Variation of emissive power at different temperature

The comparison of both is shown in Fig. 1.7.

**Example 1.41**
**Plot the Conversion Factor for Beam Radiation ($R_b$) with Latitude for Different days from January 1, i.e., $N = 0, 40, 80, 120$ and $160$ for a Given Hour Angle ($\omega$).**

**Solution**

The variation of $R_b$, Eq. 1.20b with $n$th day of the year for different latitude is shown in Fig. 1.8. Figure 1.8 shows that conversion factor for beam radiation has significant effect at higher latitude and minimum effect at lower latitude for a given hour angle. However, this effect reduces as $n$th day of the year increases.

**Example 1.42**
**Plot the Conversion Factor for Beam Radiation ($R_b$) with Hour Angle ($\omega$) for Different days from January 1, i.e., $N = 0, 40, 80, 120$ and $160$ for a Given Latitude ($\Phi = 30°$).**

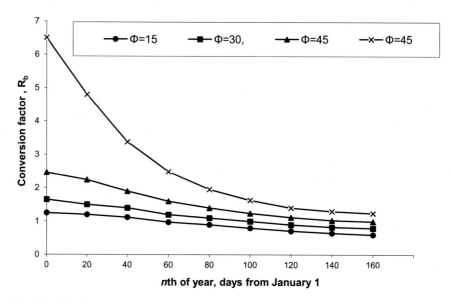

Fig. 1.8 Variation of conversion factor for inclined surface, $R_b$, with $n$th day of the year

## Solution

The variation of $R_b$, Eq. 1.20b, with $n$th day of the year for different hour angle is shown in Fig. 1.9. One can see that conversion factor for beam, $R_b$, decreases with increase of $n$th day of year but little effect due to hour angle.

## Example 1.43

**Plot the variation of $R_b$ with $n$th day of year from January 1 for a given $\beta$, $\Phi$ and $\omega$.**

## Solution

The variation of $R_b$ with $n$th day of year from January 1 for a given $\beta = 45°$, $\Phi = 30°$ and $\omega = 0°$ can be obtained by Eq. 1.20b. The results are shown in Fig. 1.9b. The similar results are expected as reported in Fig. 1.9a.

## Example 1.44
**Calculate and plot the variation of air mass ($m$) with hour angle for different $n$th day of the year for given latitude.**

## Solution

An expression of air mass ($m$) is given by Eq. 1.6a as

$$\text{Air mass } (m) = 1/\cos\theta_z$$

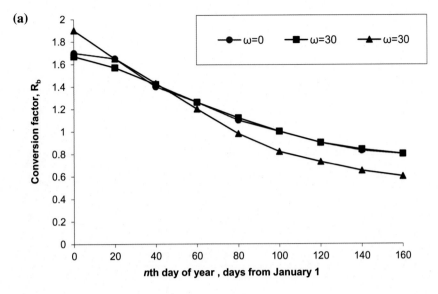

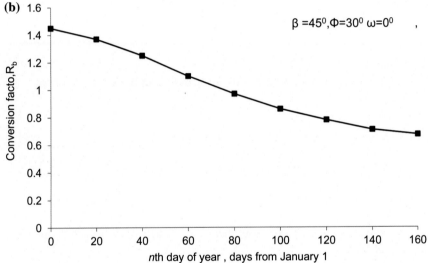

**Fig. 1.9 a** Variation of beam radiation conversion factor with $n$th day of year for different hour angle (triangle value should be 60) **b** Variation of conversion factor for beam radiation with $n$th day of year

The variation of air mass with time of the day for latitude of Delhi for different number of days (January 1, June 30 and March 31) is shown in Fig. 1.10. It is clear from this figure that air mass is higher in January month as compared to March and June. Further, it is to be noted that air mass ($m$) is higher in early morning and evening time. There is not much significant variation about noon time.

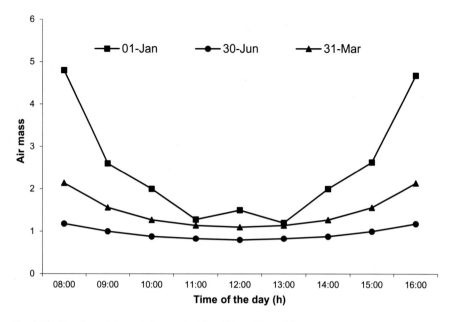

**Fig. 1.10** Hourly variation of air mass ($m$) for different day of the year

**Example 1.45**
**Plot the hourly variation of air mass for different latitude for a given $n$.**

**Solution**

An expression of air mass ($m$) is given by Eq. 1.6a as

$$\text{Air mass } (m) = 1/\cos\theta_z$$

The variation of air mass with time of the day for different latitude of a day (March 31) is shown in Fig. 1.11. It is clear from Fig. 1.11 that air mass is higher for higher latitude (Srinagar 34.08°). However, there is not much effect of latitude for a given $n$th day of year.

**Example 1.46**
**Plot the variation of sunshine hours (N) for different latitude.**

**Solution**

Equation 1.14 has been computed, and the results are shown in Fig. 1.12a for different month. There is no variation in sunshine hour for the month of March and September 21 as expected.

**Example 1.47**
Plot the variation of $R_b$ with $n$th day of year from January 1 for a given $\beta$, $\Phi$ and $\omega$.

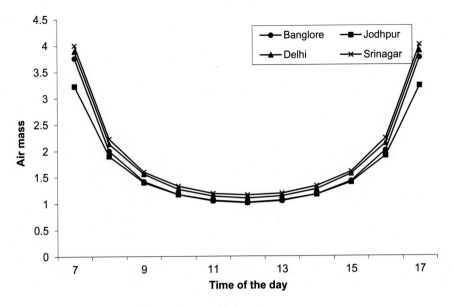

**Fig. 1.11** Hourly variation of air mass ($m$) at different cities/latitude

**Solution**

The variation of $R_b$ with $n$th day of year from January 1 for a given $\beta = 45°$, $\Phi = 30°$ and $\omega = 0°$ can be obtained by Eq. 1.20b. The results are shown in Fig. 1.12 b.

From Eq. 1.14, one gets an expression for sunshine hours as

$$\text{Daylight hours} = N = 2 \cdot \frac{\omega_s}{15} = \frac{2}{15} \cdot \cos^{-1}[-\tan\phi \cdot \tan\delta]$$

$$\text{Day} = 15\,\text{June}\ \delta = 18.8°$$

The variation of sunshine hours with latitude is shown in Fig. 1.12. It is clear from figure that the sunshine hours decrease as latitude increases as per our expectation. It plays an important role in determining the value of hourly, daily and monthly average value of solar radiation in terrestrial region. The maximum value of sunshine hour is about 11.4 h for given parameters.

**Example 1.48**
**Evaluate solar altitude for zenith angle of zero.**

**Solution**

We know $\theta_z = 90 - \alpha$ as per definition of zenith angle, Fig. 1.1.
Then,

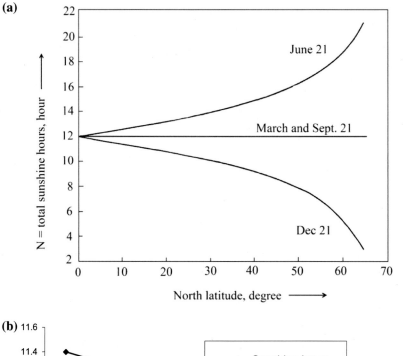

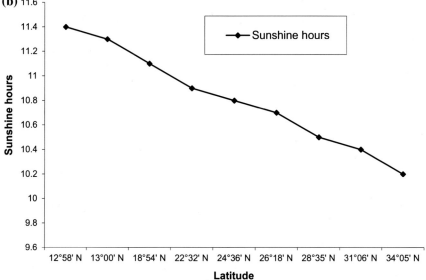

**Fig. 1.12  a** Variation of sunshine hours with north latitude in different month **b** Variation of sunshine hour with latitude

$\alpha = 90 - \theta_z = 90^\circ$ in the early morning condition.

**Example 1.49**
**Find out nth day for zero declination angles.**

**Solution**

**We know that**

$$\delta = 23.45 \sin\left[(284 + n) \times \frac{360}{365}\right]$$

For declination zero, one should have

$$\sin\left[(284 + n) \times \frac{360}{365}\right] = 0$$

or,

$$(284 + n) \times \frac{360}{365} = 0,\ 180,\ 360,\ 540\ \text{etc.}$$

First value of $n$ will be as follows

$$(284 + n) \times \frac{360}{365} = 360 \cong n = 81$$

and,
second value of $n$ will be as follows

$$(284 + n) \times \frac{360}{365} = 540 \cong n = 263.5$$

**Example 1.50**
**Find out nth day for maximum declination angles.**
**Solution**

For maximum declination angle, one should have

$$\sin\left[(284 + n) \times \frac{360}{365}\right] = 1 = \sin 90,\ \sin 270,\ \sin 450$$

This gives

$$\left[(284 + n) \times \frac{360}{365}\right] == 450 \cong n = 172.25$$

**Example 1.51**

Calculate hour angle at ST = 12 h.

**Solution**

As we know that the hour angle is given by

$$\omega = (ST - 12) \times 15 = (12 - 12) = 0°.$$

# References

1. Tiwari, G. N., & Mishra, R. K. (2012). *Advance renewable energy sources.* RSC Publishing.
2. Tiwari, G. N. (2002). *Solar energy: Fundamentals, design, modelling and applications.* CRC Press (USA) and Alpha Science International (UK) also published by Narosa Publishing House
3. Tiwari, G. N. (2016). *Tiwari Arvind and Shyam.* Springer.
4. Singh, H. N., & Tiwari, G. N. (2005). *Energy, 30*, 1589.
5. Collares-Pereira, M., & Rabl, A. (1979). *Solar Energy, 22*, 155.
6. Gueymard, C. (1986). Journal of solar energy engineering. *Transactions on ASME, 108*, 320.
7. Kasten, F., & Young, A. T. (1989). *Applied Optics, 28*, 4735.

# Chapter 2
# Basic Element of Heat Transfer

## 2.1 Introduction

As we have seen in Chap. 1 that solar radiation is emitted from the sun at temperature of 5777 K ($\sim$6000 K) having wavelength range between 0 (zero) and $\infty$ (infinity). The solar radiation passes through atmosphere and finally received by planet earth in the wavelength range of 0.26–2.6 μm needed by all living organism. The wavelength range of 0.26–2.6 μm is known as short wavelength due to Wien's displacement law. The variable solar radiation reaching on earth creates temperature range due to heating and emits long wavelength radiation which is blocked by atmosphere. So, one can concludes that there is strong relation between temperature and wavelength. Further, heat transfer depends on temperature difference with respect to surrounding and hence there will temperature domain and wavelength domain for heat transfer. For a known temperature domain, the wavelength domain can be determined by using Wien's displacement law. Finally, heat transfer mainly depends on temperature.

Energy coming from the sun is in the form of electromagnetic wave (*e/m*) (thermal energy) as well as photon (*hυ*). Thermal energy can be transferred generally from higher to lower temperature as a result of temperature difference ($\Delta T$). No net heart transfer takes place between two mediums at the same temperature. The rate of heat transfer in a particular direction is directly proportion to the temperature gradient ($\Delta T$) in that direction. The thermal energy transfer is basically a heat transfer phenomena. It is a well-established topic. It does not require detailed deliberation due to small operating temperature difference ($\Delta T$) in most of the solar thermal applications. Very few application of solar energy is at high operating temperature difference ($\Delta T$), i.e., solar power generation.

To deal with heat transfer, one must know law of thermodynamics. There are four laws of thermodynamics, namely the zeroth law, first law, second law and third law. However, only first law and second law were important to analyze solar energy system by applying thermal modeling based on heat transfer.

© Bag Energy Research Society 2024                                                67
G. N. Tiwari, *Fundamental of Mathematical Tools for Thermal Modeling of Solar Thermal and Photo-voltaic Systems-Part-I*,
https://doi.org/10.1007/978-981-99-7085-8_2

## 2.2   First Law of Thermodynamics [1]

For thermal modeling of solar energy system which operates at medium temperature range, i.e., $15\,°C < T < 100\,°C$, the **first law of thermodynamics** is applicable. It is also termed as **energy conservation** law for thermodynamic systems. The energy available is low grade energy, and the process is also reversible. In this case, the lost energy can be recovered.

According to the law of **conservation of energy**,

The available solar energy = the useful energy + the lost energy

or,

The rate of available solar energy, $I(t)$, $\dfrac{W}{m^2}$

$$= \text{the rate of useful energy, } \dot{q}_u, \frac{W}{m^2}$$

$$+ \text{the rate of lost energy, } \dot{q}_L, \frac{W}{m^2}$$

or,

$$\dot{q}_u\left(\frac{W}{m^2}\right) = I(t)\left(\frac{W}{m^2}\right) - \dot{q}_L\left(\frac{W}{m^2}\right) \tag{2.1a}$$

The rate of lost thermal energy is in the form of thermal/heat loss. It may be in different modes, namely conduction, convection and radiation.

Equation 2.1a can also be written as

$$\dot{q}_u = I(t) - U_L(T - T_a). \tag{2.1b}$$

**The following points to be noted that in Eq. 2.1**

(i)   The rate of useful energy, $\dot{q}_u$, is unknown.
(ii)  The rate of incident solar energy is known.
(iii) The overall heat transfer coefficient/heat transfer coefficient ($U_L$) is design parameter depends on surface treatment, is also known and
(iv)  Temperature, $T$, is unknown which depend on the rate of solar intensity and surface treatment.

Further, it is necessary to understand heat/thermal energy transfer by different modes, namely **(i) conduction, (ii) convection** and **(iii) radiation** as mentioned earlier. It is also important to note the following:

(i)   **Each heat transfer mode acts independently and,**
(ii)  **Only one heat transfer mode behaves strongly in comparison with other two modes**

Generally, all three modes of heat transfer are involved in an overall heat transfer problems [6]. Corresponding to each mode, there is heat transfer coefficients. All heat transfer depends on temperature difference.

All three heat transfer modes will be discussed in the next sections.

## 2.3 Heat Transfer by Conduction

Conduction is the transfer of heat/thermal energy in solid from the hot surface at temperature $T_2$ to cold surface $T_1$. The thermal vibration and collision between free electrons in solid are responsible for thermal conduction in solids. There are mainly two types of solids used in thermal modeling of solar thermal system which are conductor and insulator. The conductor is used for fast transfer of thermal energy/ heat obtained due to solar energy while insulator is used to minimize thermal/heat loss.

### Fourier's Law of Heat Conduction

Basically, heat transfer is a three-dimensional problem. In solar energy system, most of units are more than 1 m² and hence heat conduction from edges/corner is neglected and once consider only one-dimensional heat flow across thickness of solid.

Let us consider steady-state heat transfer through a large vertical/horizontal plane wall of (a) thickness $\Delta x = L$, (b) surface area $A$ and temperature difference between two surface, i.e., $\Delta T = T_2 - T_1$. On the basis of experiments, it can be shown that the rate of heat transfer $\dot{Q}$ in $W$ through the vertical/ horizontal surface wall can be expressed as

$$\dot{Q} = -KA\frac{\Delta T}{\Delta x},\qquad(2.2a)$$

where, the constant $K$ is the thermal conductivity of the materials (Appendix Table E.3), which has ability to transport thermal energy heat from higher temperature to lower temperature through conduction. For limiting case, $\Delta x \to 0$, Eq. 2.2a reduces to differential form

$$\dot{Q} = -KA\frac{dT}{dx}.\qquad(2.2b)$$

Equation 2.2b is known as Fourier's law of one-dimensional heat conduction. Here $\frac{dT}{dx}$ is slope (temperature gradient) of the temperature curve on a $T-x$ diagram at point $x$. The negative sign indicates the temperature decreases as one move from higher temperature point toward the lower temperature. Positive direction of $x$ is taken from higher temperature to lower temperature.

The thermal conductivity of a material $(K)$ is defined as the rate of heat transfer per unit surface area $(A = 1\,\text{m}^2)$ per unit thickness $(\Delta x = 1\,\text{m})$ between two points having unit temperature difference ($\Delta T = 1\,°C$). Higher value of thermal conductivity gives higher rate of heat transfer rate through the material. In fact, thermal conductivity depends on temperature but in temperature range of solar thermal system, there is not significant variation and hence it is considered as constant, Appendix Table E.3.

Equation 2.2b in $W$ can be rearranged as follows:

$$\dot{Q}_k = A\frac{K}{L}(T_2 - T_1). \tag{2.2c}$$

Here, −ve sign has been removed because the length has no negative sign. Equation 2.2c can also be written for the rate of heat transfer per unit area in $W/m^2$ due to conduction as

$$\dot{q}_k = \frac{\dot{Q}}{A} = h_k(T_2 - T_1) = \frac{K}{L}(T_2 - T_1) = h_k \Delta T \text{ or } h_k = \frac{\frac{\dot{Q}}{A}}{(T_2 - T_1)}, \tag{2.2d}$$

where, $h_k$ is known as either heat transfer coefficient due to conduction or conductive heat transfer coefficient. It is defined as the ration of the rate of heat transfer due to conduction per unit area to unit temperature difference in $W/m^2\,°C$, and it is expressed by

$$h_k = \frac{K}{L}. \tag{2.2e}$$

Here, it is important to mention that the unit of heat transfer coefficient ($h_k$) can be expressed either in $W/m^2\,°C$ or in $W/m^2\,K$ due to cancelation of 273 in temperature difference.

Further, it is important to note that the according to Ohm's law, the current flow in wire can be expressed as follows:

$$I = \frac{V}{R} \tag{2.3a}$$

where, $R = \rho\frac{L}{A}$, then

$$I = \frac{A}{\rho L} \times V \text{ or } \frac{I}{A} = \frac{1}{\rho L} \times V. \tag{2.3b}$$

There is similarity between Eqs. 2.2d and 2.3b with respect to the rate of heat flow (thermal power) and the rate of charge flow (current) through the conducting material is directly proportional to temperature different and potential difference respectively. So proportionality constant will also be similar. Hence, it is said that

The heat transfer coefficient ($h_k$) is inversely proportional to thermal resistance ($r_k$) and vice versa and, it can be expressed as

$$h_k = \frac{1}{r_k} \quad \text{or,} \quad r_k = \frac{L}{K} = \frac{1}{h_k}, \frac{m^2 K}{W} \quad \text{or} \quad \frac{m^2\, ^\circ C}{W}, \quad \text{with,} \quad r_k = \rho L \qquad (2.3c)$$

From Eq. 2.3c, one can infer that

(a)  For faster heat transfer, one has to choose the materials having low thermal resistance ($r_k$) and large value of heat transfer coefficient ($h_k$) and

(b)  For least heat transfer, one has to choose the materials having large thermal resistance ($r_k$) and low value of heat transfer coefficient ($h_k$).

***Example 2.1*** Evaluate the thermal resistance ($r_k$) and heat transfer coefficient ($h_k$) of some metals and non-metals material given in Appendix Tables E.3 and E.4.

**Solution**

The thermal resistance ($r_k$) and heat transfer coefficient ($h_k$) of some metals' and non-metals' materials based on their physical properties given in Appendix Tables E.3 and E.4 are calculated by using Eq. 2.3c which have been given in Tables 2.1 and 2.2, respectively. From these tables, it is clear that

(a)  The materials with high thermal conductivity should be used for fast heat transfer as mentioned above, Table 2.1. For example, it is used as absorber in solar system and heat exchanger.

(b)  The material with high thermal conductivity should be used for minimum heat transfer as mentioned above, Table 2.2. For example, it is used as insulating material cork board/glass wool) in solar system to reduce heat losses from hot surface and

(c)  There is no choice for other non-metal which is used for construction of building (brick/stone), Table 2.2.

***Example 2.2*** Evaluate the rate of heat transfer ($\dot{q}_k$) per $m^2$ through a rectangular plane wall of 0.10 m thick and thermal conductivity ($K$) of (a) 0.84 W/m K (brick) and (b) 0.04 W/m K (cork board/insulation) in a steady-state uniform having surface temperatures of $T_1 = 21.1\,^\circ C$ (cold) and $T_2 = 71.1\,^\circ C$ (hot).

**Table 2.1**  Thermal resistance ($r_k$) and heat transfer coefficient ($h_k$) for some metals

| S. No. | Metals | Thermal conductivity, $K$, (W/m K) | Thickness, $L$, (m) | Thermal resistance, $r_k$, ($m^2$ K/W) | Heat transfer coefficient, $h_k$ (W/$m^2$ K) |
|---|---|---|---|---|---|
| 1 | Aluminium | 137 | 0.003 | $2.1898 \times 10^{-5}$ | $0.4566 \times 10^5$ |
| 2 | Iron | 73 | 0.003 | $4.1096 \times 10^{-5}$ | $0.2433 \times 10^5$ |
| 3 | Copper | 383 | 0.003 | $0.07833 \times 10^{-5}$ | $12.766 \times 10^5$ |
| 4 | Bronze | 326 | 0.003 | $0.09202 \times 10^{-5}$ | $10.867 \times 10^5$ |

**Table 2.2** Thermal resistance ($r_k$) and heat transfer coefficient ($h_k$) for some non-metals

| S. No. | Non-metals | Thermal conductivity, $K$, (W/m K) | Thickness, $L$, (m) | Thermal resistance, $r_k$, (m² K/W) | Heat transfer coefficient, (W/m² K) |
|---|---|---|---|---|---|
| 1 | Building brick | 0.69 | 0.30 | 0.4347 | 2.3004 |
| 2 | Stone 1-2-4 mix | 1.37 | 0.30 | 0.2189 | 4.5682 |
| 3 | Marble | 2.07–2.94 | 0.30 | 0.2803–0.1020 | 3.5676–9.804 |
| 4 | Glass window | 0.78 | 0.003 | 0.003846 | 260.0104 |
| 5 | Cark board/ground | 0.043 | 0.10 | 2.32558 | 0.4300 |

**Solution**

From Eq. 2.3b, we have the following expression for $\dot{q}_k$

$$\dot{q}_k = \frac{\dot{Q}_k}{A} = \frac{K(T_2 - T_1)}{L} = \frac{0.84 \times 50}{0.10} = 42 \text{ W/m}^2 \text{ (brick) and}$$

$$\dot{q}_k = \frac{\dot{Q}_k}{A} = \frac{K(T_2 - T_1)}{L} = \frac{0.04 \times 50}{0.10} = 2.0 \text{ W/m}^2 \text{ (cork board/insulation)}.$$

The value of conductive heat transfer coefficient ($h_k$) is given by

$$h_k = \frac{K}{L} = \frac{0.84}{0.10} = 8.4 \text{ W/m}^2 \text{ K (brick) and}$$

$$h_k = \frac{K}{L} = \frac{0.04}{0.10} = 0.40 \text{ W/m}^2 \text{ K (cork board/insulation)}.$$

It is important to note that the rate of heat transfer per unit area in brick is 21 times faster than cork board and heat transfer coefficient is five times more than cork board due to insulating properties of cork board for same thickness. Hence, the either cork board or any equivalent insulating material of thickness of 0.10 m is preferred for reducing the heat losses from back of flat plate collector.

***Example 2.3*** Evaluate the rate of heat transfer ($\dot{q}_k$) per m² through a joint two rectangular plane wall/surface of each 0.10 m thick and thermal conductivity ($K$) of (a) 0.84 W/m K (brick, $K_b$) and (b) 0.04 W/m K (cork board/insulation, $K_i$) in a steady-state uniform having surface temperatures of $T_1 = 21.1$ °C (cold) at back of cork board and $T_2 = 71.1$ °C (hot) of brick.

**Solution**

In this case, we have to first calculate the total thermal resistance of combined wall/surface by using Eqs. 2.3a, 2.3b and 2.3c as

$$r_{kT} = \frac{L}{K_b} + \frac{L}{K_i} = \frac{0.10}{0.84} + \frac{0.10}{0.04} = 0.1190 + 2.5 = 2.619 \frac{m^2 \, K}{W}.$$

Further, an inverse of the total thermal resistance is nothing but an effective conductive heat transfer coefficient, $h_{keff}$, and it becomes as $h_{keff} = \frac{1}{r_{kT}} = \frac{1}{2619} = 0.3818 \, W/m^2 \, K$.

This total thermal resistance will be referred as an overall heat transfer coefficient ($U$) which will be discussed later on.

So, the rate of heat transfer ($\dot{q}_k$) per m$^2$ through a joint two rectangular plane wall/ surface can be determined as

$$\dot{q}_k = h_{keff}(T_2 - T_1) = 0.3818 \times 50 = 19.09 \, W/m^2.$$

From the present example, one should note that

(i)  If one desires to reduce the heat flow from one surface to other surface of solid, locally available insulating material with proper thickness should be used such in building walls/roof due to higher thermal resistance and,

(ii)  If one desires to increase the heat flow from one surface to other surface of solid, locally available conducting material with proper thickness should be used such as brick and glass in building walls/roof due to low thermal resistance.

## 2.4 Heat Transfer by Convection [1, 2]

By using the definition of heat transfer coefficient due to conduction, Eq. 2.3b, it becomes simpler to write an expression for the rate of heat transfer between the fluid and the boundary surface in W/m$^2$ due to convection as follows:

The rate of heat transfer per unit area in W/m$^2$ due to convection is given by

$$\dot{q}_c = \frac{\dot{Q}}{A} = h_c(T_2 - T_1), \tag{2.4a}$$

where, $h_c$ is the local convective heat transfer coefficient in W/m$^2$ K $\left(W/m^2 \, {}^\circ C\right)$.

Also, the expression for $h_c$ can be expressed as

$$h_c = \frac{\dot{q}_c}{(T_2 - T_1)}. \tag{2.4b}$$

It is well-established that the convective heat transfer generally occurs at boundary layer of solid and liquid/fluid which is a very complex problem. However, researchers have developed some empirical relation to find out numerical value of convective heat transfer coefficient which depends on physical properties of fluid/liquid, Appendix E as well as operating temperature. For this, one needs some dimensional number to start to find out convective heat transfer coefficient. These dimensional numbers are as follows:

## 2.4.1   *Dimensionless Heat Convection Parameters*

**Nusselt number (Nu):** It is defined as the ratio of convective heat transfer coefficient, $h_c$, to the conductive heat transfer coefficient for fluids, $h_k = \frac{K}{L}$. Mathematically it is expressed as follows:

$$\text{Nusselt number, Nu} = \frac{h_c}{\frac{K}{L}} = \frac{h_c L}{K}. \tag{2.5a}$$

In the above equation, $K$ is the thermal conductivity of a fluid above solid surface and $h_c$ is convective heat transfer coefficient between solid and fluid/liquid.

**Reynold's number (Re):** It is defined for the heat transfer problem in force mode of operation. It is the ratio of the fluid dynamic force $\left(\rho u_0^2\right)$ to the viscous drag force $\left(\mu u_0/X\right)$, and it is given by

$$\text{Reynolds number, Re} = \frac{\rho u_0^2}{\mu u_0/L} = \frac{\rho u_0 L}{\mu} = \frac{u_0 L}{\nu}, \tag{2.5b}$$

where, $\rho$ is the density, $\mu$ is the dynamic viscosity and $\nu = $ kinematic viscosity $= \frac{\mu}{\rho}$ and $L$ is characteristic length for the system of interest. Reynold's number signifies the flow behavior in forced convection. Value of Reynold's number indicates the flow type (laminar flow/turbulent flow) of fluid.

In this case, the fluid can flow in the following configuration:

(a) **Through the cylindrical pipes:** In this case, the characteristic dimension is diameter of pipe **(L = D)**.
(b) **Over flat plate surface:** In this case, the characteristic dimension is length of flat surface, i.e., **L**.
(c) **Through parallel plate:** In this case, the characteristic dimension is distance between two plates, i.e., **d**.

**From Eq. 2.5b, one can see Reynold's number significantly depends on the flow velocity ($u_0$) and of fluid and the characteristic dimension. It plays an important role in determining the connective heat transfer coefficient in forced mode of operation unlike natural convective heat transfer coefficient.**

**Prandtl number (Pr):** Prandtl number relates fluid motion and heat transfer to the fluid. It is defined as the ratio of momentum diffusivity ($\mu/\rho$) to the thermal diffusivity $\left(K/\rho C_p\right)$, and it is given by

$$\text{Prandtl number, Pr} = \frac{\mu/\rho}{K/\rho C_p} = \frac{\mu C_p}{K}, \tag{2.5c}$$

where, $\mu$ is dynamic viscosity, $C_p$ is the specific heat at constant pressure, and $K$ is thermal conductivity of the fluid.

It is important to note that Reynold and Prandtl's number are not directly depends on temperature like Grashof's number, Eq. 2.5d. It depends through physical properties of fluid, Appendix E.

**Grashof number (Gr)**: It is the ratio of the buoyancy force to the viscous force. It is given by

$$\text{Grashof number, Gr} = \frac{g\beta'\rho^2 X^3 \Delta T}{\mu^2} = \frac{g\beta' X^3 \Delta T}{\nu^2}, \tag{2.5d}$$

where, $\beta'$ is the coefficient of volumetric thermal expansion, $g$ is the acceleration due to gravitational and $\Delta T$ is the operating temperature difference between the surface and the fluid.

**Rayleigh number (Ra)**: It is the ratio of the thermal buoyancy to viscous inertia. It is expressed as

$$\text{Rayleigh number, Ra} = \text{GrPr} = \frac{g\beta'\rho^2 L^3 C_p \Delta T}{\mu K} \tag{2.5e}$$

All the numbers defined in this section can be obtained using the properties of air and water tabulated in Appendix E. The properties are calculated at average temperature $T_f$ given as follows:

$$T_f = \frac{T_1 + T_2}{2}, \tag{2.6a}$$

where, $T_1$ is hot surface temperature and $T_2$ is fluid temperature.

The thermal expansion coefficient $(\beta')$ at hot surface temperature and fluid temperature is calculated using Eqs. 2.6b and 2.6c, respectively,

$$\beta' = \frac{1}{(T_2 + 273)} \quad \text{and} \tag{2.6b}$$

$$\beta' = \frac{1}{(T_f + 273)}. \tag{2.6c}$$

The characteristic dimension $(L)$ for different shapes is calculated as follows:

$$L = \frac{A(\text{Area})}{P(\text{Perimeter})}. \tag{2.6d}$$

In some specific cases, the characteristic dimension for rectangular horizontal surface is determined using relation given below:

$$L = \left(\frac{L_0 + B_0}{2}\right). \tag{2.6e}$$

The characteristic dimension $(L)$ appearing in denominator of **Nusselt number (Nu)** in Eq. 2.5a plays an important role. Based on examples given below, one should prefer characteristic dimension $(L)$ given by Eq. 2.6d. The temperature-dependent physical properties of air and water have been given in Appendix Tables E.1 and E.2 involved in dimensional number given in Eqs. 2.5a–2.5e.

Now, the heat transfer by convection is further classified as free and forced convection which will be discussed as follows:

## 2.4.2  Free Convection

In solar system analysis, inclination of solar system to receive maximum solar radiation is an important parameter to provide maximum solar energy. In this section, we will consider the various configurations including horizontal, inclined and vertical surface. An orientation does not play any role in heat transfer as far as analysis of solar system is concerned. We will consider two fluids, namely water and air, as a working fluid for evaluating heat transfer coefficient along with the rate of heat transfer. In the case of natural convection, the working fluid will be is stationary position.

In this case, the following configuration with respect to Table 2.3 for stagnant fluid will be discussed:

(a)  **Horizontal surface**

There is relation between **Nusselt number (Nu)**, **Grashof number (Gr)** and **Prandtl number (Pr)** to determine the convective heat transfer coefficient in limner flow/natural flow of fluid for horizontal surface/plane/plate as

$$\text{Nu} = \frac{hL}{K} = C'(\text{GrPr})^n K'. \tag{2.7}$$

The above relation is obtained by using dimensional analysis at boundary layer. Values of $C'$ and $n$ are estimated by using experimental data for systems with same geometrical shapes and size. For some geometrical shape/inclination used in solar thermal technology are given in Table 2.3. The $K'$ governs the entire physical behavior of the problem [7].

Some empirical relations used for free convention are also given in Table 2.4.

(b)  **For incline surface/plane/plate**

In this case, Eq. 2.7 will be modified by multiplying Gr by $\cos \beta$ as

$$\text{Nu} = \frac{hL}{K} = C'(\text{Gr} \times \cos \theta \times \text{Pr})^n K', \tag{2.8}$$

where, $\theta$ is inclination of hot plate with respect to vertical line, case (v) of Table 2.3. Its value is equal to latitude $\pm$ inclination $(\theta)$, depending on seasonal variation. The

**Table 2.3** Free convective heat transfer of various systems [1, 2]

| Case number | System | Schematic | $C'$ | $n$ | $k'$ | Operating conditions |
|---|---|---|---|---|---|---|
| Case (i) a | Horizontal cylinder | | 0.47 | 0.25 | 1 | Laminar flow condition |
| b | | | 0.1 | – | 1 | Turbulent flow condition |
| c | | | 0.0246 | 0.4 | $[Pr^{1/6}/(1+0.496Pr^{-2/3})]^{2/3}$ | Turbulent flow condition; to obtain local Nu use $C' = 0.0296$, $X = x$ |
| Case (ii) | Vertical cylinder with small diameter | | 0.686 | 0.25 | $[Pr/(1+1.05Pr)]^{1/4}$ | Laminar flow condition $\overline{Nu}_{local} = \overline{Nu} + 0.52(L/D)$ |
| Case (iii) a | Heated horizontal plate facing upward | | 0.54 | 0.25 | 1 | Laminar flow condition $(10^5 < GrPr < 2\times10^7)$ $X = (L_0 + B_0)/2$ Laminar flow condition $(10^7 < GrPr < 10^{11})$, $X = A/P$ For circular disc of diameter $D$, use $X = 0.9D$ |
| b | | | 0.14 | 0.33 | 1 | Turbulent flow condition $(2\times10^7 < GrPr < 3\times10^7)$, $X = (L_0 + B_0)/2$ |
| c | | | 0.15 | 0.33 | 1 | Turbulent flow condition $(10^7 < GrPr < 10^{11})$, $X = A/P$ |

(continued)

**Table 2.3** (continued)

| Case number | System | Schematic | $C'$ | $n$ | $k'$ | Operating conditions |
|---|---|---|---|---|---|---|
| Case (iv) | Heated horizontal plate facing downward | | 0.27 | 0.25 | 1 | Laminar flow condition only |
| Case (v) | Moderately inclined plane | | 0.8 | 0.25 | $\left[\dfrac{\cos\theta}{1+\left(1+\frac{1}{\sqrt{Pr}}\right)^2}\right]^{1/4}$ | Laminar flow condition (multiply Gr by cos θ in the formula for vertical plate) |
| Case (vi) | Two vertical parallel plates at the same temperature | | 0.04 | 1 | $(d/L)^3$ | Air layer |
| Case (vii) | Hollow vertical cylinder with open ends | | 0.01 | 1 | $(d/L)^3$ | Air column |
| Case (viii) | Two horizontal parallel plates hot plate uppermost | | 0.27 | 0.25 | 1 | Pure conduction $\dot{q} = K(T_h - T_c)/d$ Laminar flow condition (air) $(3 \times 10^5 < GrPr < 3 \times 10^{10})$ |

(continued)

**Table 2.3**  (continued)

| Case number | System | Schematic | $C'$ | $n$ | $k'$ | Operating conditions |
|---|---|---|---|---|---|---|
| Case (ix) | Two concentric cylinders | $X=\frac{1}{2}(d_o-d_i)$  $A=2\pi XL$ | 0.317 | 0.25 | $\left[X^3\left(\frac{1}{d_i^{3/5}}+\frac{1}{d_o^{3/5}}\right)^5\right]^{-1/4}$ | Laminar flow condition |
| Case (x) a | Two vertical parallel plates of different temperatures ($h$ for both surfaces) | for $\frac{L}{d}>3$  $X=d$  $\Delta T=T_h-T_c$ | 0.18 | 0.25 | $\left(\frac{L}{d}\right)^{-1/9}(Pr)^{-1/4}$ | Laminar flow condition (air) $(2\times10^4 < Gr < 2\times10^5)$ |
| b |  |  | 0.065 | 1/3 | $\left(\frac{L}{d}\right)^{-1/9}(Pr)^{-1/3}$ | Turbulent flow condition (air) $(2\times10^5 < Gr < 2\times10^7)$ |
| Case (xi) | Two inclined parallel plates | $\Delta T=T_h-T_c$  $X=d$ |  |  | $\overline{Nu}=\dfrac{\overline{Nu}_{vert}\cos\theta+\overline{Nu}_{horz}\sin\theta}{2}$ |  |
| Case (xii) a | Two horizontal parallel plates cold plate uppermost | $\Delta T=T_h-T_c$  $X=d$ | 0.195 | 0.25 | $Pr^{1/4}$ | Laminar flow condition (air) $(10^4 < Gr < 4\times10^5)$ |
| b |  |  | 0.068 | 1/3 | $Pr^{1/3}$ | Turbulent flow condition (air) $Gr > 4\times10^5$ |

**Table 2.4** Simplified equations for free convection from various surfaces to air at atmospheric pressure [1, 2]

| Cases | Surface | Laminar $10^4 < \mathrm{Gr}_f\mathrm{Pr}_f < 10^9$ | Turbulent $\mathrm{Gr}_f\mathrm{Pr}_f > 10^9$ |
|---|---|---|---|
| 1 | Horizontal hot plate facing upward | $h = 1.32(\Delta T/L)^{1/4}$ | $h = 1.52(\Delta T)^{1/3}$ |
| 2 | Hot plate facing downward | $h = 0.59(\Delta T/L)^{1/4}$ | |
| 3 | Vertical plane and cylinder | $h = 1.42(\Delta T/L)^{1/4}$ | $h = 1.31(\Delta T)^{1/3}$ |
| 4 | Horizontal cylinder | $h = 1.32(\Delta T/d)^{1/4}$ | $h = 1.24(\Delta T)^{1/3}$ |

'+' refers for winter condition and '−' refers for summer condition, and for annual it is equal to latitude. For horizontal surface $\cos\theta = 1$.

(c)  **For two vertical surface/wall**

In this case, the convective heat transfer coefficient, $h_c$ will be determined by using constants from case (v) of Table 2.3 with air as working fluid as

$$\mathrm{Nu} = \frac{hd}{K} = 0.04(\mathrm{Gr.Pr})^1 (d/L)^3 \text{ for same temperature} \qquad (2.9a)$$

and, case (x) of Table 2.3.

$$\mathrm{Nu} = \frac{hd}{K} = 0.065(\mathrm{Gr.Pr})^{1/3}\left(\frac{L}{d}\right)^{-1/9}(\mathrm{Pr})^{-1/3} \text{ for turbulent flow} \qquad (2.9b)$$

Here, $l$ is the length of vertical surface and $d$ is the spacing, i.e., characteristic dimension.

It is important to mention that above three configurations namely horizontal, inclined and vertical along with correlation between dimensional numbers is applicable for both water and air as working fluid. However, there is some other useful correlation suggested by Fuji and Imura (1972) for only air as a working fluid. These are based on extensive experimental studies. In this case, $\cos\theta$ has been considered as 1.

There are as follows:

(i)  Horizontal hot surface facing downward

$$\overline{\mathrm{Nu}}_e = 0.56[\mathrm{Gr}_e\mathrm{Pr}_e]^{1/4} \quad 10^5 < \mathrm{Gr}_e\mathrm{Pr}_e < 10^{11} \qquad (2.10a)$$

(ii)  Horizontal hot surface facing upward

$$\overline{\mathrm{Nu}}_e = 0.58(\mathrm{Gr}_e\mathrm{Pr}_e)^{1/5} \quad 10^6 < \mathrm{Gr}_e\mathrm{Pr}_e < 10^{11} \qquad (2.10b)$$

The thermal properties of air should be evaluated at $T_e = T_p - 0.25(T_p - T_s)$ with plate temperature of $T_p$ and surrounding temperature of $T_s$. This change in temperature will not affect the numerical value of dimensional number due to little change in thermal properties of air, Appendix Table E.1 in comparison with an average value of $T_p$ and $T_s$.

### (d) Inclined cylindrical tube

With the help of case (i) and case (ii) of Table 2.3, one can write dimensional number as follows:

$$\overline{Nu} = \frac{Nu_{vert} \cos\theta + Nu_{hori} \sin\theta}{2}. \qquad (2.10c)$$

The above equation can be used to evaluate convective heat transfer coefficient for glass tube used in evacuated tubular collector working under natural mode of operation and having an inclination of $\theta$ with respect to vertical line, case (v) of Table 2.3.

### (A) Water as a working fluid

***Example 2.4*** Estimate convective heat transfer coefficient for a horizontal rectangular hot surface (1.0 m × 0.8 m) facing upward, and it is maintained at 134 °C. The hot surface is exposed to water at 20 °C.

### Solution

For present exercise in both cases, the average temperature, Eq. 2.6a, $T_f = (134 + 20)/2 = 77\,°C$. The computation will be done for both characteristic dimension either for water or air as a fluid.

(i) **For the characteristic dimension** $(L = X) = (1.0 + 0.8)/2 = 0.90\,m$, **Eq. 2.6e**

From Appendix E, the water thermal properties at $T_f = 77\,°C$ will be

$$\mu = 3.72 \times 10^{-4}\,\text{kg/ms};\ K = 0.668\,\text{W/m K},$$
$$\rho = 973.7\,\text{kg/m}^3,\ Pr = 2.33\ \text{and}$$
$$\beta' = \frac{1}{77 + 273} = 2.857 \times 10^{-3}\,\text{K}^{-1}.$$

From Eq. 2.5d, the Grashof number can be calculated as

$$Gr = \frac{g\beta'\rho^2(\Delta T)X^3}{\mu^2} = \frac{9.8 \times 2.857 \times 10^{-3}(973.7)^2 \times 114 \times (0.9)^3}{(3.72 \times 10^{-4})^2} = 1.594 \times 10^{13}.$$

This is a turbulent flow and for heated plate facing upward, the values of $C = 0.14$ and $n = 1/3$ [case (iiib) in Table 2.3]. Now, a convective heat transfer coefficient from Eq. 2.7 can be calculated as

$$h_c = \frac{K}{L}(0.14)(Gr.Pr)^{1/3} = \frac{0.668}{0.9}(0.14)(1.594 \times 10^{13} \times 2.33)^{1/3}$$

$$= 3467 \text{ W/m}^2 \text{ K}$$

(ii)  **For the characteristic dimension** $L = \frac{A(\text{Area})}{P(\text{Perimeter})} = \frac{1 \times 0.8}{3.6} = 0.22$, **Eq. 2.6d**

The Grashof number and convective heat transfer coefficient become as

$$Gr = \frac{g\beta' \rho^2 (\Delta T) X^3}{\mu^2} = \frac{9.8 \times 2.857 \times 10^{-3}(973.7)^2 \times 114 \times (0.22)^3}{(3.72 \times 10^{-4})^2}$$

$$= 2.32 \times 10^{11}$$

*and,*

$$h_c = \frac{K}{L}(0.14)(GrPr)^{1/3} = \frac{0.668}{0.22}(0.14)(2.32 \times 10^{11} \times 2.33)^{1/3} = 3316 \text{ W/m}^2 \text{ K}.$$

So, there is marginal change (4%) in convective heat transfer coefficient ($h_c$) due to variation in characteristic dimension given by Eqs. 2.6d and 2.6e from solid surface to water as a working fluid.

***Example 2.5*** Repeat the Example 2.4 with same average water temperature of $T_f = 77\,°C$ with different difference between hot plate and surrounding water at 79 °C ($\Delta T = 2\,°C$) for actual operating condition in solar thermal system for characteristic dimension of $L = 0.22$.

**Solution**

In the present case, the Grashof number and convective heat transfer coefficient become as

$$Gr = \frac{g\beta' \rho^2 (\Delta T) X^3}{\mu^2} = \frac{9.8 \times 2.857 \times 10^{-3}(973.7)^2 \times 2 \times (0.22)^3}{(3.72 \times 10^{-4})^2} = 4.07 \times 10^9$$

*and,*

$h_c = \frac{K}{L}(0.14)(GrPr)^{1/3} = \frac{0.668}{0.22}(0.14)(4.07 \times 10^9 \times 2.33)^{1/3} = 899.77 \text{ W/m}^2 \text{ K}$,
Eq. 2.7.

From Examples 2.3 and 2.4 one can observe that the convective heat transfer coefficient depends significantly on temperature difference between hot plate and surrounding water temperature over hot plate.

***Example 2.6*** Estimate convective heat transfer coefficient for a 45° inclined rectangular surface/plate (1.0 m × 0.8 m) with respect to vertical direction is maintained at 134 °C. The hot surface is exposed to water at 20 °C.

**Solution**

For present exercise in both cases, the average temperature, Eq. 2.6a, $T_f = (134 + 20)/2 = 77\,°C$. The computation will be done for both characteristic dimension either for water or air as a fluid.

(j) **For the characteristic dimension $(L = X) = (1.0 + 0.8)/2 = 0.90\,\text{m}$, Eq. 2.6e**

From Appendix E, the water thermal properties at $T_f = 77\,°C$ will be

$$\mu = 3.72 \times 10^{-4}\,\text{kg/ms}; \quad K = 0.668\,\text{W/m K},$$
$$\rho = 973.7\,\text{kg/m}^3, \quad \text{Pr} = 2.33 \text{ and}$$
$$\beta' = 1/(77 + 273) = 2.857 \times 10^{-3}\,\text{K}^{-1}.$$

From Eq. 2.5d, the Grashof number can be calculated as

$$\text{Gr}_L = \frac{g\beta'\rho^2(\Delta T)X^3}{\mu^2} = \frac{9.8 \times 2.857 \times 10^{-3}(973.7)^2 \times 114 \times (0.9)^3}{(3.72 \times 10^{-4})^2}$$
$$= 1.594 \times 10^{13}.$$

This is a turbulent flow and for heated plate facing upward, the values of $C = 0.14$ and $n = 1/3$ [case (iiib) in Table 2.3]. Now, a convective heat transfer coefficient for inclined surface, Eq. 2.8 can be calculated as

$$h_c = \frac{K}{L}(0.14)(\text{Gr} \times \cos\theta \times \text{Pr})^{1/3}$$
$$= \frac{0.668}{0.9}(0.14)(1.594 \times 10^{13} \times 0.707 \times 2.33)^{1/3}$$
$$= 3050.96\,\text{W/m}^2\text{ K}.$$

(ii) **For the characteristic dimension $L = \frac{A\,(\text{Area})}{P\,(\text{Perimeter})} = \frac{1\times0.8}{3.6} = 0.22$, Eq. 2.6d**

**The Grashof number and convective heat transfer coefficient become as**

$$\text{Gr}_L = \frac{g\beta'\rho^2(\Delta T)X^3}{\mu^2} = \frac{9.8 \times 2.857 \times 10^{-3}(973.7)^2 \times 114 \times (0.22)^3}{(3.72 \times 10^{-4})^2}$$
$$= 2.32 \times 10^{11}$$

*and,*

$$h_c = \frac{K}{L}(0.14)(\text{Gr} \times \cos 45 \times \text{Pr})^{1/3} = \frac{0.668}{0.22}(0.14)(2.32 \times 10^{11} \times 0.707 \times 2.33)^{1/3},$$
$$= 2918.08\,\text{W/m}^2\text{ K}$$
Eq. 2.8.

So, there is 11–12% reduction in convective heat transfer coefficient ($h_c$) due to effect of inclination in the case of water as a fluid.

***Example 2.7*** Estimate convective heat transfer coefficient for a horizontal rectangular hot surface (1.0 m × 0.8 m) facing downwards, and it is maintained at 134 °C. The hot surface is exposed to water at 20 °C below hot plate.

**Solution**

For present exercise in both cases, the average temperature, Eq. 2.6a, $T_f = (134 + 20)/2 = 77$ °C. The computation will be done for both characteristic dimension either for water or air as a fluid.

(i)  **For the characteristic dimension $(L = X) = (1.0 + 0.8)/2 = 0.90$ m, Eq. 2.6e**

From Appendix E, the water thermal properties at $T_f = 77$ °C will be

$$\mu = 3.72 \times 10^{-4} \text{ kg/ms}; K = 0.668 \text{ W/m K},$$
$$\rho = 973.7 \text{ kg/m}^3, \text{Pr} = 2.33 \text{ and}$$
$$\beta' = 1/(77 + 273) = 2.857 \times 10^{-3}\text{K}^{-1}$$

From Eq. 2.5d, the Grashof number can be calculated as

$$\text{Gr} = \frac{g\beta'\rho^2(\Delta T)X^3}{\mu^2} = \frac{9.8 \times 2.857 \times 10^{-3}(973.7)^2 \times 114 \times (0.9)^3}{\left(3.72 \times 10^{-4}\right)^2}$$
$$= 1.594 \times 10^{13}.$$

This is a turbulent flow and for heated plate facing upward, the values of $C = 0.14$ and $n = 1/3$ [case (iiib) in Table 2.3]. Now, a convective heat transfer coefficient from Eq. 2.7 can be calculated as

$$h_c = \frac{K}{L}(0.14)(\text{Gr.Pr})^{1/3} = \frac{0.668}{0.9}(0.14)\left(1.594 \times 10^{13} \times 2.33\right)^{1/3} = 3467 \text{ W/m}^2 \text{ K}.$$

(ii)  **For the characteristic dimension $L = \frac{A(\text{Area})}{P(\text{Perimeter})} = \frac{1 \times 0.8}{3.6} = 0.22$, Eq. 2.6d.**

The Grashof number and convective heat transfer coefficient become as

$$\text{Gr} = \frac{g\beta'\rho^2(\Delta T)X^3}{\mu^2} = \frac{9.8 \times 2.857 \times 10^{-3}(973.7)^2 \times 114 \times (0.22)^3}{\left(3.72 \times 10^{-4}\right)^2}$$
$$= 2.32 \times 10^{11}$$

*and,*

$$h_c = \frac{K}{L}(0.14)(GrPr)^{1/3} = \frac{0.668}{0.22}(0.14)\left(2.32 \times 10^{11} \times 2.33\right)^{1/3} = 3316 \text{ W/m}^2 \text{ K}$$

So, there is marginal change (4%) in convective heat transfer coefficient ($h_c$) due to variation in characteristic dimension given by Eqs. 2.6d and 2.6e from solid surface to water as a working fluid.

***Example 2.8*** Evaluate convective heat transfer coefficient from top surface of a glass pipe to stagnant water inside for (a) horizontal tube, (b) vertical tube and (c) inclined tube for the following parameters:

Length of glass pipe $= 1.8$ m

Diameter of outer tube $= 0.058$ m

Diameter of inner tube $= 0.056$ m

Tube surface temperature $= 80\,^\circ$C

Stagnant water temperature $= 78\,^\circ$C

An average temperature $= 79\,^\circ$C

Characteristic dimension ($D$) $= 0.056$ m

Inclination of tube $= 45^\circ$.

**Solution**

In this case, the physical properties of water at $77 - 79\,^\circ$C will be approximately same, Appendix Table E.2, and hence from Example 2.4 we have

From Appendix Table E.2, the water thermal properties at $T_f = 77 - 79\,^\circ$C will be

$$\mu = 3.72 \times 10^{-4} \text{ kg/ms}; \quad K = 0.668 \text{ W/m K},$$
$$\rho = 973.7 \text{ kg/m}^3, \text{ Pr} = 2.33 \text{ and}$$
$$\beta' = 1/(79 + 273) = 2.841 \times 10^{-3} \text{ K}^{-1}$$

From Eq. 2.5d, the Grashof number can be calculated as

$$Gr = \frac{g\beta'\rho^2(\Delta T)X^3}{\mu^2} = \frac{9.8 \times 2.841 \times 10^{-3}(973.7)^2 \times 2 \times (0.056)^3}{(3.72 \times 10^{-4})^2} = 1.194 \times 10^9.$$

In this case, there will be laminar flow of water within tube filled with water. Now, Nusselt number.

(a)  Horizontal tube, case (i) of Table 2.3

$$\text{Nu}_{\text{hori}} = 0.47(\text{Gr.Pr})^{0.25} = 0.47(1.194 \times 10^9 \times 2.33)^{0.25} = 107.94.$$

Now, the inner convective heat transfer for horizontal tube will be evaluated as

$$h_c = \frac{K}{L}\text{Nu}_{\text{hori}} = \frac{0.668}{0.056} \times 107.94 = 1287.58 \text{ W/m}^2 \text{ K.}$$

(b)  Vertical tube, case (ii) of Table 2.3

$$\text{Nu}_{\text{vert}} = 0.686(\text{Gr.Pr})^{0.25}[\text{Pr}/(1+1.05\text{Pr})]^{1/4}$$

$$= 0.686(1.194 \times 10^9 \times 2.33)^{0.25} \times \left[\frac{2.33}{3.4465}\right]^{0.25} = 139.11$$

$$h_c = \frac{K}{L}\text{Nu}_{\text{hori}} = \frac{0.668}{0.056} \times 139.11 = 1659.37 \text{ W/m}^2 \text{ K.}$$

(c)  Inclined tube at 45°.

From Eq. 2.10b, we have

$$\overline{\text{Nu}} = \frac{\text{Nu}_{\text{vert}}\cos\theta + \text{Nu}_{\text{hori}}\sin\theta}{2}.$$

Substitute the values of $\text{Nu}_{\text{hori}}$ and $\text{Nu}_{\text{vert}}$ in the above equation, one gets

$$\overline{\text{Nu}} = \frac{139.11 \times 0.707 + 107.94 \times 0.707}{2} = 87.33.$$

Now, the convective heat transfer coefficient of inclined tube at 45° will be as follows:

$$h_c = \frac{K}{L}\overline{\text{Nu}} = \frac{0.668}{0.056} \times 87.33 = 1041.72 \text{ W/m}^2 \text{ K.}$$

(B)  **Air as a working fluid**

*Example 2.9* Estimate convective heat transfer coefficient for a horizontal rectangular hot surface (1.0 m × 0.8 m) facing upward and it is maintained at 134 °C. The hot surface is exposed to ambient air at 20 °C.

**Solution**

For present exercise, the average temperature, Eq. 2.6a, $T_f = (134 + 20)/2 = 77$ °C. The computation will be done for both characteristic dimensions for air as a fluid.

(i) **For the characteristic dimension** $(L = X) = (1.0 + 0.8)/2 = 0.90 \, \text{m}$, **Eq. 2.6e**

By using the physical properties of air at $T_f = 77\,°C$ (Appendix E) and $L = 0.90 \, \text{m}$

$$Gr.Pr = \frac{(9.8) \times (134 - 20) \times (0.90)^3 \times (0.697)}{(293) \times (2.08 \times 10^{-5})^2} = 4.51 \times 10^9.$$

Using case (iiib) in Table 2.3 for hot surface facing upward and turbulent flow condition, the heat transfer coefficient from Eq. 2.7 can be calculated as

$$h_c = \left(\frac{K}{L}\right) \times 0.14 \times (Gr \times Pr)^{0.333} = \left(\frac{0.03}{0.9}\right) \times (0.14) \times \left(4.51 \times 10^9\right)^{0.333}$$
$$= 7.74 \, \text{W/m}^2\,°C.$$

(ii) **For the characteristic dimension** $L = \frac{A(\text{Area})}{P(\text{Perimeter})} = \frac{1 \times 0.8}{3.6} = 0.22$, Eq. 2.6d

$$Gr.Pr = \frac{(9.8) \times (134 - 20) \times (0.22)^3 \times (0.697)}{(293) \times (2.08 \times 10^{-5})^2} = 0.658 \times 10^9$$

The convective heat transfer coefficient becomes

$$h_c = \left(\frac{K}{L}\right) \times 0.14 \times (Gr \times Pr)^{0.333} = \left(\frac{0.03}{0.22}\right) \times (0.14)\left(0.658 \times 10^9\right)^{0.333}$$
$$= 16.6 \, \text{W/m}^2\,°C.$$

**Example 2.10** Repeat Example 2.8 by using Eq. 2.10a and compare the results for hot plate facing upwards.

**Solution**

From Example 2.8, we have

$$Gr.Pr = \frac{(9.8) \times (134 - 20) \times (0.90)^3 \times (0.697)}{(293) \times (2.08 \times 10^{-5})^2} = 4.51 \times 10^9 \text{ for } L = 0.9 \text{ and}$$

$$Gr.Pr = \frac{(9.8) \times (134 - 20) \times (0.22)^3 \times (0.697)}{(293) \times (2.08 \times 10^{-5})^2} = 0.658 \times 10^9 \text{ for } L = 0.22.$$

From Eq. 2.10a, we have

$$\overline{Nu}_e = 0.56[Gr_e Pr_e]^{1/4} \quad 10^5 < Gr_e Pr_e < 10^{11}.$$

Now, the convective heat transfer coefficient, $h_c$, can be obtained as.

(i) **For the characteristic dimension** $(L = X) = (1.0 + 0.8)/2 = 0.90\,\text{m}$, **Eq. 2.6e**

$$h_c = \left(\frac{K}{L}\right) \times 0.56 \times (\text{Gr} \times \text{Pr})^{1/4} = \left(\frac{0.03}{0.9}\right) \times (0.56) \times \left(4.51 \times 10^9\right)^{1/4}$$

$$= 5.32\,\text{W/m}^2\,{}^\circ\text{C}.$$

(ii) **For the characteristic dimension** $L = \frac{A\,(\text{Area})}{P\,(\text{Perimeter})} = \frac{1 \times 0.8}{3.6} = 0.22$, Eq. 2.6d

$$h_c = \left(\frac{K}{L}\right) \times 0.56 \times (\text{Gr} \times \text{Pr})^{1/4} = \left(\frac{0.03}{0.22}\right) \times (0.56)\left(0.658 \times 10^9\right)^{1/4}$$

$$= 12.23\,\text{W/m}^2\,{}^\circ\text{C}$$

There is reduction of convective heat transfer coefficient by 26% but the order remains same.

**Example 2.11** Repeat Example 2.8 by using Eq. 2.10b and compare the results for hot plate facing downwards.

**Solution**

From Example 2.8, we have

$$\text{Gr.Pr} = \frac{(9.8) \times (134 - 20) \times (0.90)^3 \times (0.697)}{(293) \times \left(2.08 \times 10^{-5}\right)^2} = 4.51 \times 10^9 \text{ for } L = 0.9 \text{ and}$$

$$\text{Gr.Pr} = \frac{(9.8) \times (134 - 20) \times (0.22)^3 \times (0.697)}{(293) \times \left(2.08 \times 10^{-5}\right)^2} = 0.658 \times 10^9 \text{ for } L = 0.22.$$

From Eq. 2.10b, we have

$$\overline{\text{Nu}}_e = 0.58(\text{Gr}_e\text{Pre})^{1/5} \quad 10^6 < \text{Gr}_e\text{Pr}_e < 10^{11}.$$

Now, the convective heat transfer coefficient, $h_c$, can be obtained as

(i) **For the characteristic dimension** $(L = X) = (1.0 + 0.8)/2 = 0.90\,\text{m}$, **Eq. 2.6e**

$$h_c = \left(\frac{K}{L}\right) \times 0.58 \times (\text{Gr} \times \text{Pr})^{1/5} = \left(\frac{0.03}{0.9}\right) \times (0.58) \times \left(4.51 \times 10^9\right)^{1/5}$$

$$= 1.65\,\text{W/m}^2\,{}^\circ\text{C}.$$

(ii) **For the characteristic dimension** $L = \frac{A\,(\text{Area})}{P\,(\text{Perimeter})} = \frac{1 \times 0.8}{3.6} = 0.22$, Eq. 2.6d

$$h_c = \left(\frac{K}{L}\right) \times 0.58 \times (\text{Gr} \times \text{Pr})^{1/5} = \left(\frac{0.03}{0.22}\right) \times (0.58)(0.658 \times 10^9)^{1/5}$$
$$= 4.59 \text{ W/m}^2 \text{ °C.}$$

From Examples 2.9 and 2.10, one can offer that when hot plate is changed from upward direction to downward direction, the convective eat transfer coefficient is reduced to 1/3.

**Example 2.12** Compute convective heat transfer coefficient by using Eq. 2.10b for Example 2.6 for characteristic dimension of 0.22 m.

**Solution**

From Eq. 2.10b, we have an expression for convective heat transfer coefficient as

$$h_c = \frac{K}{L} \times \overline{\text{Nu}}_e = \frac{0.03}{0.22} \times 0.58(\text{Gr}_e \, \text{Pr}_e)^{1/5} = \frac{0.03}{0.22} \times 0.58 \times (0.658 \times 10^9)^{1/5}$$
$$= 4.5 \text{ W/m}^2 \text{ K.}$$

**Example 2.13** Compute convective heat transfer coefficient by using empirical expression for air, Table 2.4 for Example 2.9 in turbulent flow.

**Solution**

The convective heat transfer coefficient for air by using the empirical relation of case (1) in Table 2.4, one gets

$$h_c = 1.52(\Delta T)^{1/3} = 1.52(114)^{1/3} = 1.5 \times 4.85 = 7.28 \text{ W/m}^2 \text{ °C}$$

**Example 2.14** Repeat Example for $\Delta T = 2\,°\text{C}$ for hot plate facing upwards direction.

**Solution**

Following Example 2.8, the product of Grashof and Prandtl is obtained as

$$\text{Gr.Pr} = \frac{(9.8) \times 2 \times (0.22)^3 \times (0.697)}{(293) \times (2.08 \times 10^{-5})^2} = 1.15 \times 10^7.$$

Referring to case (iiia) of Table 2.3 with $\text{Gr.Pr} = 1.15 \times 10^7$, $C = 0.54$ and $n = 0.25$, the convective heat transfer coefficient becomes

$$h_c = \left(\frac{K}{L}\right) \times 0.54 \times (\text{Gr} \times \text{Pr})^{0.25} = \left(\frac{0.03}{0.22}\right) \times (0.54)(1.15 \times 10^7)^{0.25} =$$
4.58 W/m$^2$ °C, Eq. 2.7

Following conclusions based on Examples 2.3–2.7 can be derived:

(i) It is important to note that the convective heat transfer coefficient ($h_c$) changes from 3467 W/m$^2$ °C (Example 2.3) to 4–8 W/m$^2$ °C with change of fluid from water to air for given same other parameters because it depends on physical properties of fluid.

(ii) The characteristic dimension ($L$) has insignificant effect on convective heat transfer from solid surface to water/air as a working fluid.

(iii) There is similarity in convective heat transfer coefficient obtained either by dimensionless formula or empirical formula. The values are approximately same around 7–8 W/m$^2$ °C.

(iv) The temperature difference between hot plate and working fluid plays an important role in determining convective heat transfer coefficient.

**Example 2.15** Repeat Example for $\Delta T = 2$ °C for hot plate facing downward direction.

**Solution**

In this case, referring to case (iv) of Table 2.3 for Gr.Pr $= 1.15 \times 10^7$, $C = 0.27$ and $n = 0.25$, the convective heat transfer coefficient becomes

$$h_c = \left(\frac{K}{L}\right) \times 0.27 \times (Gr \times Pr)^{0.25} = \left(\frac{0.03}{0.22}\right) \times (0.27)(1.15 \times 10^7)^{0.25}$$

$$= 2.29 \text{ W/m}^2 \text{ °C}.$$

This indicated that for air as working fluid, if hot plate is reversed, then convective heat transfer coefficient becomes half. It is only true for air as working fluid due to low range of Gr.Pr. under laminar flow.

**Example 2.16** Estimate the rate of convective heat loss and coefficient from a horizontal rectangular surface (1.0 m × 0.8 m) which is maintained at 134 °C. The hot surface is exposed to a plate placed at a distance of 0.10 m above it with air as working media and temperature of plate are maintained at 20 °C.

**Solution**

The average air temperature, $T_f = (134 + 20)/2 = 77$ °C.

From Appendix Table C.4, at $T_f = 77$ °C $+ 273 = 350$ K °C; $\rho = 1$ kg/m$^3$, $v = 20.8 \times 10^{-6}$ m$^2$/s; $K = 0.030$ W/m K $Pr = 0.697$ and $\beta = 1/(77 + 273) = 2.857 \times 10^{-3}$ K$^{-1}$ due to exposure to another plate and the characteristic dimension $L = 0.10$ m

$$Gr = \frac{(9.8)\left[2.857 \times 10^{-3}\right] \times 1^2(134 - 20)(0.1)^3}{\left(20.8 \times 10^{-6}\right)^2} = 7.377 \times 10^6.$$

Using case (x)-a of Table 2.3 for turbulent flow condition for two horizontal parallel plates (cold plate uppermost), the heat transfer coefficient can be calculated as

$$h_c = \frac{K}{L}(0.068)(GrPr)^{1/3}Pr^{-1/3}$$

$$= \frac{0.030}{0.10}(0.068)(7.377 \times 10^6)^{1/3} = 3.97 \text{ W/m}^2\text{K}$$

Hence, the rate of heat loss from a vertical wall is,

$$q = h_c A(T_s - T_\infty) = (3.97)(1.0 \times 0.8)(134 - 20) = 362.14 \text{ W}.$$

From Examples 2.7 and 2.8 for air, one can observe that there is a change in numerical value of convective heat transfer coefficient from 7.47 to 3.97 W/m² K due change in condition from open to closed. However, their numerical value of convective heat transfer coefficient is of the order one digit.

***Example 2.17*** Estimate convective heat transfer coefficient for a 45° inclined rectangular surface/plate (1.0 m × 0.8 m) which is maintained at 134 °C. The hot surface is exposed to air at 20 °C.

**Solution**

For present exercise in both cases, the average temperature, Eq. 2.6a, $T_f = (134 + 20)/2 = 77$ °C. The computation will be done for both characteristic dimension for air as a fluid.

(i) **For the characteristic dimension** $(L = X) = (1.0 + 0.8)/2 = 0.90$ **m, Eq. 2.6e**

By using the physical properties of air at $T_f = 77$ °C (Appendix E) and $L = 0.90$ m

$$Gr.Pr = \frac{(9.8) \times (134 - 20) \times (0.90)^3 \times (0.697)}{(293) \times (2.08 \times 10^{-5})^2} = 4.51 \times 10^9.$$

Using case (iiib) of Table 2.3 for hot surface facing upward and turbulent flow condition, the heat transfer coefficient, $h_c$, from Eq. 2.8 can be calculated as

$$h_c = \left(\frac{K}{L}\right) \times 0.14 \times (Gr \times Pr \times \cos\theta)^{0.333} = \left(\frac{0.03}{0.9}\right)$$

$$\times (0.14) \times (4.91 \times 10^9 \times 0.707)^{0.333} = 2.49 \text{ W/m}^2 \text{ °C},$$

(ii) **For the characteristic dimension** $L = \frac{A(\text{Area})}{P(\text{Perimeter})} = \frac{1 \times 0.8}{3.6} = 0.22$, Eq. 2.6d

The product of Gr.Pr will be as follows:

$$Gr.Pr = \frac{(9.8) \times (134 - 20) \times (0.22)^3 \times (0.697)}{(293) \times (2.08 \times 10^{-5})^2} = 0.0717 \times 10^9.$$

The convective heat transfer coefficient, $h_c$, from Eq. 2.8 for inclined surface, Eq. 2.7 becomes

$$h_c = \left(\frac{K}{L}\right) \times 0.14 \times (\text{Gr.Pr} \times \cos\theta)^{0.333} = \left(\frac{0.03}{0.22}\right)$$
$$\times (0.14)(0.0717 \times 10^9 \times 0.707)^{0.333} = 7.03 \text{ W/m}^2 \text{ °C}.$$

**Example 2.18** Estimate the rate of convective heat loss and coefficient for inclined rectangular surface (1.0 m × 0.8 m) which having inclination of $\theta = 45°$ is maintained at 134 °C. The hot surface is exposed to a plate placed at a distance of 0.10 m above it with air as working fluid and temperature of plate are maintained at 20 °C.

**Solution**

With the help of Example 2.4 and Eq. 2.8, for inclined two horizontal parallel plates (cold plate uppermost), the convective heat transfer coefficient can be calculated as

$$h_c = \frac{K}{L}(0.068)(\text{Gr} \times \cos\theta \times \text{Pr})^{1/3}\text{Pr}^{-1/3}$$
$$= \frac{0.030}{0.10}(0.068)(7.377 \times 10^6 \times 0.7)^{1/3} = 3.49 \text{ W/m}^2 \text{ K}.$$

So, one can see that there is reduction of 12% variation in convective heat transfer coefficient which has marginal effect on the performance of solar energy system.

**Example 2.19** Repeat Example 2.5 for moderately incline plate.

**Solution**

In this case, from case (v) of Table 2.2, we have convective heat transfer coefficient for moderately incline plate as

$$h_c = \frac{K}{L} \times 0.8(\text{Gr Pr})^{0.25}\left[\frac{\cos\beta}{1+\left(1+\frac{1}{\sqrt{\text{Pr}}}\right)^2}\right]$$
$$= \frac{0.030}{0.10} \times 0.8(7.377 \times 10^6 \times 0.697)^{1/3}\left[\frac{0.707}{1+\left(1+\frac{1}{\sqrt{0.697}}\right)^2}\right]$$
$$= \frac{0.03}{0.10} \times 0.0968 \times 172.5 = 5.00 \text{ W/m}^2 \text{ K}.$$

**Example 2.20** Find out the rate of convective heat loss and coefficient from a parallel vertical rectangular surfaces (1.0 m × 0.8 m) which are maintained at same temperature of 77 °C with air as working fluid and exposed to surrounding temperature of 50 °C.

**Solution**

The known air temperature, $T_f = 77\,°C$. The temperature of both surface/plate and air will be same and $\Delta T = 77 - 50 = 27\,°C$.

From Appendix Table C.4, at $T_f = 77\,°C + 273 = 350\,K\,°C$; $\rho = 1\,kg/m^3$, $v = 20.8 \times 10^{-6}\,m^2/s$; $K = 0.030\,W/m\,K\,Pr = 0.697$ and $\beta = 1/(77 + 273) = 2.857 \times 10^{-3}\,K^{-1}$ due to exposure to another plate and the characteristic dimension $L = 0.10\,m$

$$Gr = \frac{(9.8)\left[2.857 \times 10^{-3}\right] \times 1^2 \times (27) \times (0.1)^3}{\left(20.8 \times 10^{-6}\right)^2} = 1.747 \times 10^6$$

Using case (vi) of Table 2.3 for turbulent flow condition for two vertical parallel plates at same temperature, the convective heat transfer coefficient from Eq. 2.9a can be calculated as

$$h_c = \frac{K}{L}(0.04)(Gr \times Pr)^1 \left(\frac{d}{L}\right)^3$$

$$= \frac{0.030}{0.10}(0.04)\left(7.377 \times 10^6 \times 0.697\right)^1 \left(\frac{0.10}{1}\right)^3$$

$$= 1.423 \times 10^{-5} \times 10^6 = 14.2\,W/m^2\,K.$$

**Example 2.21** Find out the rate of convective heat loss and coefficient from a parallel vertical rectangular surface (1.0 m × 0.8 m) which is maintained at 134 °C. The hot surface is exposed to another vertical plate placed at a distance of 0.10 m in front of it with air as working fluid and temperature of plate is maintained at 20 °C.

**Solution**

The average air temperature, $T_f = (134 + 20)/2 = 77\,°C$.

From Appendix Table C.4, at $T_f = 77°C + 273 = 350\,K\,°C$; $\rho = 1\,kg/m^3$, $v = 20.8 \times 10^{-6}\,m^2/s$; $K = 0.030\,W/m\,K$, $Pr = 0.697$ and $\beta = 1/(77 + 273) = 2.857 \times 10^{-3}\,K^{-1}$ due to exposure to another plate and the characteristic dimension $L = 0.10\,m$

$$Gr = \frac{(9.8)\left[2.857 \times 10^{-3}\right] \times 1^2(134 - 20)(0.1)^3}{\left(20.8 \times 10^{-6}\right)^2} = 7.377 \times 10^6$$

Using case (x)-b of Table 2.3 for turbulent flow condition for two horizontal parallel plates (cold plate uppermost), the heat transfer coefficient Eq. 2.9b can be calculated as

$$h_c = \frac{K}{d}0.065(Gr.Pr)^{1/3}\left(\frac{L}{d}\right)^{-1/9}(Pr)^{-1/3}$$

$$= \frac{0.03}{0.10} \times 0.065\left(7.377 \times 10^6\right)^{1/3} \left(\frac{1}{0.1}\right)^{-1/9} = 2.939 \text{ W/m}^2\text{K}.$$

Hence, the rate of heat loss from a vertical wall is,

$$q = h_c A(T_s - T_\infty) = (3.07)(1.0 \times 0.8)(134 - 20) = 279.984 \text{ W}.$$

From this example and Example 2.3 for air, one can observe that there is a change in numerical value of convective heat transfer from horizontal to vertical position.

***Example 2.22*** Estimate convective heat transfer coefficient for a 45 ° inclined rectangular surface/plate (1.0 m × 0.8 m) which is maintained at 134 °C. The hot surface is exposed to air at 20 °C. For case (x) of Table 2.1

**Solution**

For the case (xi) of Table 2.3, an average dimensional number for inclined surface/plate is given by

$$\overline{\text{Nu}} = \frac{\left[\overline{\text{Nu}}_{\text{vert}} \cos\theta + \overline{\text{Nu}}_{\text{horz}} \sin\theta\right]}{2}.$$

Here, an average Nusselt number $\left(\overline{\text{Nu}}\right)$ for horizontal, Eq. 2.9a and vertical, Eq. 2.9b surface can be obtained from Examples 2.4 and 2.9, respectively, as

$$\overline{\text{Nu}}_{\text{horizontal}} = \frac{h_c L}{K}(0.068)(\text{GrPr})^{1/3}\text{Pr}^{-1/3} = 13.23$$

and,

$$\overline{\text{Nu}}_{\text{vertical}} = \frac{h_c d}{K} = 0.065(\text{Gr} \cdot \text{Pr})^{1/3}\left(\frac{L}{d}\right)^{-1/9}(\text{Pr})^{-1/3}$$

$$= 0.065\left(7.377 \times 10^6\right)^{1/3}\left(\frac{1}{0.1}\right)^{-1/9} = 9.796$$

Now, an average dimensional number Nusselt number $\left(\overline{\text{Nu}}\right)$ for case (x) of Table 2.1 for inclined surface/plate at 45° can be evaluated as follows:

$$\overline{\text{Nu}} = \frac{[9.796 \times 0.707 + 13.23 \times 0.707]}{2} = 8.15$$

$$h_c = \overline{\text{Nu}} \times \frac{d}{K} = 8.15 \times \frac{0.10}{0.03} = 27.166.$$

From Examples 2.14 and 2.18, there is significant change in value of convective heat transfer coefficient for the same inclination of 45°.

### 2.4.3 Forced Convection [3]

For forced convection, the rate of heat transfer is enhanced by moving the fluid over the hot surface using external source of energy such as pump (liquid) or fan (air/gas) unlike natural convection. Externally supplied energy overcomes the pressure drop due to fluid motion and viscous force (frictional force) due to viscosity of fluid. The heat transfer is dominantly affected by flow behavior (laminar/turbulent) and fluid motion (momentum diffusivity) which in turn is directly related to **Reynold's number and Prandtl number**, respectively.

**Case (i): Flat surface**

Therefore the Nusselt number in forced convection depends on Reynold's number and Prandtl number. It is expresses for **horizontal flat surface** by following correlation:

$$Nu = C(RePr)^n K,  \qquad (2.11a)$$

where, $C$ and $n$ are constants for a given type of flow and geometry as given in Table 2.3 for natural circulation. The $K$ is a correction factor (shape factor) added, to obtain a greater accuracy.

For the horizontal hot plate facing upward and heated over its entire length, average Nusselt number is given by

$$\overline{Nu}_L = 0.664\,Re_L^{1/2}Pr^{1/3}.  \qquad (2.11b)$$

For the horizontal hot plate facing downward and heated over its entire length, average Nusselt number is given by

$$\overline{Nu}_L = 0.332 Re_L^{1/2}Pr^{1/3}.  \qquad (2.11c)$$

Similar to Eqs. 2.8, 2.11a and 2.11b can also be used for **inclined hot surface/plate** facing upward with modification given below:

$$Nu = C(Re_L Pr \times \cos\theta)^n K  \qquad (2.11d)$$

$$\overline{Nu}_L = 0.664(Re_L \times \cos\theta)^{1/2}Pr^{1/3},  \qquad (2.11e)$$

where, $\theta$ is the inclination of horizontal surface/plate with respect to vertical line, case (v) of Table 2.3.

**We will see later on in Examples 2.25 and 2.26, that Eq. 2.11e is most appropriate to evaluate convective heat transfer coefficient for water flowing over inclined surface.**

**Case (ii): Cylindrical tube**

The empirical relation for forced convective heat transfer through **horizontal cylindrical tubes** may be represented as,

$$\overline{Nu} = \frac{hD}{K} = C\mathrm{Re}_L^m \mathrm{Pr}^n K, \tag{2.12a}$$

where, $D = 4A/P$ is the hydraulic diameter (m); $P$ is the perimeter of the section (m) and $K$ is the thermal conductivity (W/mK) of the fluid passing through cylindrical pipe. The values of $C$, $m$, $n$ and $K$ for various conditions are given in Table 2.5.

For turbulent flow in smooth cylindrical tubes, the following correlation is recommended by Dithus and Boelter (1930), case (v) of Table 2.5,

$$Nu = 0.023\mathrm{Re}^{0.8}\mathrm{Pr}^n, \tag{2.12b}$$

where, $n = 0.4$ for heating of the fluid and $n = 0.3$ for cooling of the fluid.

In the entrance region of the tube the flow is not developed; for this region Nusselt (1931) recommended the relation:

$$Nu = 0.036\mathrm{Re}^{0.8}\mathrm{Pr}^{1/3}\left(\frac{D}{L}\right)^{0.055} \quad \text{for } 10 < \frac{L}{D} < 400. \tag{2.12c}$$

where, $L$ is the length and $D$, the diameter of the cylindrical tube.

For fully developed laminar flow in tubes at constant wall temperature, the correlation can be expressed as

$$Nu_d = 3.66 + \frac{0.0668(d/L)\mathrm{Re}\mathrm{Pr}}{1 + 0.04[(d/L)\mathrm{Re}\mathrm{Pr}]^{2/3}}. \tag{2.12d}$$

**Table 2.5** Value of constants for forced convection [1, 2]

| S. No. | Cross-section | $D$ | $C$ | $m$ | $n$ | $K$ | Operating conditions |
|---|---|---|---|---|---|---|---|
| Case (i) | | $d$ | 3.66 | 0 | 0 | 1 | Laminar flow case long tube Re < 2000, Gz < 10 |
| Case (ii) | | $d$ | 1.86 | 0.33 | 0.33 | $(D/L)^{1/3}(\mu/\mu_w)^{0.14}$ | Laminar flow case short tube for Re < 2000, Gz > 10 |
| Case (iii) | | $d$ | 0.027 | 0.8 | 0.33 | $(\mu/\mu_w)^{0.14}$ | Highly turbulent flow case of highly viscous liquids for 0.6 < Pr < 100 |
| Case (iv) | | $d$ | 0.023 | 0.8 | 0.4 | 1 | Turbulent flow case of gases Re > 2000 |

The heat transfer coefficient calculated from this relation is the average value over the entire length of the tube. When the tube is sufficiently long the Nusselt's number approaches a constant value of 3.66.

**Case (iii) Parallel flat plate**

In the case of solar air collector, we consider the parallel flat plate. Kays and Grawford proposed following dimensionless Nusselt number to evaluate convective heat transfer coefficient in forced mode of operation

$$Nu = 0.0158 \, Re^{0.8} \text{ for } Re > 2100,$$

$$\text{fully developed turbulent flow and } \frac{L}{D} \text{ is large.} \qquad (2.13)$$

The characteristic dimension $(D)$ is twice of two place spacing.

For an **inclined cylindrical tube, Reynold's number, Re,** is replaced by **Re.cos$\theta$**, where **$\theta$ is inclination of cylindrical tube with respect to vertical line**.

The thermophysical properties of water (or any base fluid) can be improved for higher values of heat transfer coefficient by mixing nanoparticles in water (or base fluid). The correlations for thermophysical properties of nanofluid are given in Appendix Tables E.7 and E.8.

First, we will calculate convective heat transfer coefficient in forced mode of operation by considering fluid as water for different configuration.

**(A)  Working fluid as water.**

**(i)  Over flat and inclined surface.**

***Example 2.23*** Evaluate the convective heat transfer coefficient and an average rate of heat transfer for water flows over a smooth hot flat plate at 32 °C and 1.0 m long above the free stream fluid temperature with velocity of 0.20 m/s. The water properties at the fluid film temperature are: $v = 7.66 \times 10^{-7}$ m²/s, $K = 0.621$ W/m K, Pr = 5.13, Appendix Table E.2.

**Solution**

The flow is laminar; therefore, from Eq. 2.5b, we have,

$$\text{Reynolds number, Re} = \frac{u_0 L}{v} = \frac{0.20 \times 1.0}{7.66 \times 10^{-7}} = 261,096$$

$$\overline{Nu}_L = 0.664 Re_L^{1/2} Pr^{1/3}.$$

The convective heat transfer coefficient from hot plate to water can be evaluated from Eq. 2.11b as

$$\overline{h}_c = \frac{K}{L} \overline{Nu}_L = \frac{K}{L} \times 0.664 Re_L^{1/2} Pr^{1/3}$$

$$= \frac{0.621}{1} \times 0.664 \times (261{,}096)^{1/2} \times (5.13)^{1/3}$$

$$= 360 \, \text{W/m}^2 \, \text{K}$$

The average rate of heat transfer per $m^2$ to the water is given by

$$\frac{\dot{Q}}{A} = \overline{h} \Delta T = (360)(32) = 11.53 \, \text{kW/m}^2.$$

**Example 2.24** Repeat Example 2.23 for hot surface facing downward.

**Solution**

In this case Eqs. 2.1a, 2.1b will be used without change in Reynold and Prandtl number.

The convective heat transfer coefficient from hot plate to water can be evaluated from Eq. 2.11b as

$$\overline{h}_c = \frac{K}{L} \overline{\text{Nu}}_L = \frac{K}{L} \times 0.332 \, \text{Re}_L^{1/2} \text{Pr}^{1/3}$$

$$= \frac{0.621}{1} \times 0.332 \times (261096)^{1/2} \times (5.13)^{1/3}$$

$$= 180 \, \text{W/m}^2 \, \text{K}.$$

The average rate of heat transfer per $m^2$ to the water is given by

$$\frac{\dot{Q}}{A} = \overline{h} \Delta T = (180)(32) = 2.76 \, \text{kW/m}^2.$$

**Example 2.25** Repeat Examples 2.23 and 2.24 by using Eq. 2.11d for an inclination of 45°.

**Solution**

From Eq. 2.11d, one has the following relation

$$\text{Nu} = C(\text{RePr} \cos \theta)^n K.$$

Referring to Table 2.3 case (iiia) for laminar flow, $C = 0.54$ and $n = 1/4 = 0.25$, $K = 1$, then

$$h_c = \frac{K}{L} \text{Nu} = \frac{K}{L} C(\text{RePr} \cos \theta)^n$$

$$K = \frac{0.621}{1} \times 0.54 \times (261096 \times 5.13 \times 0.707)^{0.25}$$

$$= 10.46 \, \text{W/m}^2 \text{K}$$

The average rate of heat transfer per m$^2$ to the water is given by

$$\frac{\dot{Q}}{A} = \bar{h}\Delta T = (10.46)(32) = 334.72 \text{ W/m}^2.$$

For inclined and downward surface, the convective heat transfer coefficient become half due to C = 0.27 = 5.23 W/m$^2$ K

***Example 2.26*** Repeat Examples 2.23 by using Eq. 2.11d for inclination of 45° by using Eq. 2.11e.

**Solution**

From Example 2.23, the convective heat transfer coefficient from hot plate to water can be evaluated from Eq. 2.11e as

$$\bar{h}_c = \frac{K}{L}\overline{Nu}_L = \frac{K}{L} \times 0.664(Re_L \cos\theta)^{1/2}Pr^{1/3}$$

$$= \frac{0.621}{1} \times 0.664 \times (261{,}096 \times 0.707)^{1/2} \times (5.13)^{1/3} = 302.94 \text{ W/m}^2 \text{ K}$$

The average rate of heat transfer per m$^2$ to the water is given by

$$\frac{\dot{Q}}{A} = \bar{h}\Delta T = (360)(32) = 9694.28\frac{\text{W}}{\text{m}^2} = 9.69 \text{ kW/m}^2.$$

**From Examples 2.25 and 2.26, one can easily conclude that 2.11e is most appropriate correlation to evaluate convective heat transfer coefficient for water flowing over inclined horizontal surface.**

(ii) **Through cylindrical horizontal and inclined tubes.**

***Example 2.27*** Evaluate the average convective heat transfer coefficient and the rate of heat transfer in W/m$^2$ in the fully developed velocity at 0.61 m/s in a 25.4 mm ID tube in a cylindrical tube with water at $T_w = 11$ °C flows with the tube wall temperature of $T_s = 20$ °C.

**Solution**

At average water and wall temperature of $\overline{T} = 15.5$ °C, the physical properties of water: $v_b = \frac{\mu}{\rho} = 1.12 \times 10^{-6}$ m$^2$/s, $\overline{K}_w = 0.585$ W/m K, $Pr_b = 7.88$ (Appendix Table E.2).

The Reynold number, Eq. 2.5b, for the present case $L = D$, characteristic length for cylindrical tube is given by

$$Re = \frac{Du_0}{v} = \frac{(0.0254)(0.61)}{1.12 \times 10^{-6}} = 1.38 \times 10^4,$$

This is turbulent flow condition and water heating, and hence, the Nusselt Number from Eq. 2.12b and case (iv) of Table 2.5 is given by,

$Nu_{D,\infty} = (0.023)(1.38 \times 10^4)^{0.8} \times (7.88)^{0.4} = 107.71$ (from Table 2.5)

The convective heat transfer coefficient from tube surface to working fluid (water) will be determined as

$$h_\infty = \frac{K_w}{D} \times Nu = \frac{0.585}{0.0254} \times 107.71 = 2480.6 \text{ W/m}^2 \text{ °C}$$

The rate of heat transfer in W/m² from cylindrical wall to the flowing water as fluid can be obtained as

$$\dot{q} = h_\infty(T_b - T_c) = 2480.6(20 - 11) = 22{,}325.4 = 22.325 \frac{\text{kW}}{\text{m}^2}.$$

**Example 2.28** Repeat the Example 2.27 for $\frac{L}{D} \geq 30$.

**Solution**

For $\frac{L}{D} > 30$, an expression for convective heat transfer coefficient can be written by using Eq. 2.12c as

$$h_c = \frac{K}{D} \times Nu = \frac{0.585}{0.0254} \times 0.036(1.38 \times 10^4)^{0.8}(7.88)^{1/3}$$
$$\left(\frac{1}{30}\right)^{0.055} \text{ for } 10 < \frac{L}{D} < 400$$

or,

$$h_c = 0.829 \times 2050.71 \times 1.989 \times 0.829 = 2804.48 \text{ W/m}^2 \text{ °C}.$$

The rate of heat transfer in W/m² from cylindrical wall to the flowing water as fluid can be obtained as

$$\dot{q} = h_c(T_b - T_c) = 2804.48(20 - 11) = 25{,}240.32 \text{ W/m}^2 = 25.240 \frac{\text{kW}}{\text{m}^2}.$$

From Examples 2.27 and 2.28, it can be seen that the length and diameter of cylindrical tube plays an important role in heat transfer under similar condition because there is not much change in physical properties of water during our temperature range of operation.

**Example 2.29** Evaluate the average convective heat transfer coefficient in the fully developed velocity at 0.61 m/s in a 25.4 mm ID tube in a cylindrical tube with water at $T_w = 20$ °C flows with the tube wall temperature of $T_s = 11$ °C.

**Solution**

Referring to Example 2.27, at average water and wall temperature of $\overline{T} = 15.5\,^\circ\text{C}$, the $\text{Pr}_b = 7.88$ (Appendix Table E.2) and the Reynold number, Eq. 2.5b, $\text{Re} = 1.38 \times 10^4$.

This is turbulent flow condition and water cooling, and hence, the Nusselt number from Eq. 2.12b and case (iv) of Table 2.5 is given by,

$$\text{Nu}_{D,\infty} = (0.023)\left(1.38 \times 10^4\right)^{0.8} \times (7.88)^{0.3} = 87.62.$$

The convective heat transfer coefficient from tube surface to working fluid (water) will be determined as

$$h_\infty = \frac{K_w}{D} \times \text{Nu} = \frac{0.585}{0.0254} \times 87.62 = 2017.9 \text{ W/m}^2\,^\circ\text{C}$$

From Examples 2.27 and 2.29, one can see that there is reduction of heat transfer coefficient by 16% from heating to cooling fluid (water).

***Example 2.30*** Evaluate the average convective heat transfer coefficient water flowing with velocity $(u_0)$ at 0.061 m/s in a 25.4 mm ID of cylindrical tube with water at $T_w = 11\,^\circ\text{C}$ flows with the tube wall temperature of $T_s = 20\,^\circ\text{C}$.

**Solution**

Referring to Example 2.27, at average water and wall temperature of $\overline{T} = 15.5\,^\circ\text{C}$, the $\text{Pr}_b = 7.88$ (Appendix Table E.2) and the Reynold number, Eq. 2.5b, is given by

$$\text{Re} = \frac{Du_0}{v} = \frac{(0.0254)(0.061)}{1.12 \times 10^{-6}} = 1.38 \times 10^3 = 1389 \le 2000,$$

This is laminar flow. From Table 2.5, the Nusselt number $\text{Nu} = 3.66$, and hence, the convective heat transfer coefficient from tube surface to working fluid (water) will be determined as

$$h_\infty = \frac{K_w}{D} \times \text{Nu} = \frac{0.585}{0.0254} \times 3.66 = 84.29 \text{ W/m}^2\,^\circ\text{C}.$$

So, one can conclude that the convective heat transfer coefficient from tube surface to working fluid (water) under turbulent condition, Examples 2.27, 2.28 and 2.29 is much-much larger value in comparison with laminar flow condition, Example 2.30. There will not be any effect of inclination on convective heat transfer coefficient for laminar flow.

***Example 2.31*** Evaluate the average convective heat transfer coefficient and the rate of heat transfer in W/m$^2$ in the fully developed velocity at 0.61 m/s in a 25.4 mm ID tube in a inclined at 45° cylindrical tube with respect to vertical line with water at $T_w = 11\,^\circ\text{C}$ flows with the tube wall temperature of $T_s = 20\,^\circ\text{C}$.

**Solution**

In this case, Reynolds number, Re, will be replaced by Re.cos $\theta$ in Eq. 2.12b as

$$\mathrm{Nu} = 0.023(\mathrm{Re}.\cos\vartheta)^{0.8}\mathrm{Pr}^n.$$

From Example 2.27, we have $\mathrm{Re} = 1.38 \times 10^4$ and $\mathrm{Pr} = 7.88$.

This is turbulent flow condition and water heating, and hence, the Nusselt number from Eq. 2.12b and Table 2.5 is given by,

$$\mathrm{Nu}_{D,\infty} = (0.023)\left(1.38 \times 10^4 \cos 45°\right)^{0.8} \times (7.88)^{0.4} = 81.62.$$

The convective heat transfer coefficient from tube surface to working fluid (water) will be determined as

$$h_\infty = \frac{K_w}{D} \times \mathrm{Nu} = \frac{0.585}{0.0254} \times 81.62 = 1.879.75 \ \mathrm{W/m^2\,°C}.$$

The rate of heat transfer in $\mathrm{W/m^2}$ from cylindrical wall to the flowing water as fluid can be obtained as

$$\dot{q} = h_\infty(T_b - T_c) = 1.879.75 \times (20 - 11) = 16917.72 \ \mathrm{W/m^2} = 16.92\frac{\mathrm{kW}}{\mathrm{m^2}}.$$

Due to inclination of cylindrical tube, the convective heat transfer coefficient from cylindrical tube surface to working fluid (water) is reduced by 24%.

***Example 2.32*** Evaluate convective heat transfer coefficient and the rate of heat transfer in W. The water at average bulk temperature ($T_b$) 82 °C is flowing through inside cylindrical tube having 0.020 m inner diameter with length of 4 m with average of velocity 2.0 m/s. The cylindrical tube wall temperature is at 43 °C below the value of $T_b$.

**Solution**

The fluid properties at $T_b = 82$ °C and $T_s = 43$ °C are,

$$\mu_b = 3.47 \times 10^{-4}\mathrm{kg} \times \frac{\mathrm{m}}{\mathrm{s}}, \ \rho_b = 970.2 \ \mathrm{kg/m^3},$$

$$v_b = 3.58 \times 10^{-7} \ \mathrm{m^2/s}, \ \mathrm{Pr}_b = 2.16, \ K_b = 0.673 \ \mathrm{W/m\,K}, \ \mathrm{and}$$

$$\mu_s = 6.16 \times 10^{-4}\mathrm{kg} \times \frac{\mathrm{m}}{\mathrm{s}}, \ \rho_s = 990.6 \ \mathrm{kg/m^3},$$

$$v_s = 3.58 \times 10^{-7}\mathrm{m^2/s}, \ \mathrm{Pr}_s = 4.04, \ K_s = 0.637 \ \mathrm{W/mK} \ (\text{Appendix Table E.2}).$$

The Reynold number at flowing water base temperature from Eq. 2.5b and using $L = D$, i.e., characteristic dimension for tube can be obtained as

$$\text{Re} = \frac{Du_0}{v_b} = \frac{(0.020)(2.0)}{3.58 \times 10^{-7}} = 1.12 \times 10^5,$$

The flow is highly turbulent. Since $T_s - T_b = 40\,°C$, and noting that, $\frac{L}{D} = \frac{4}{0.02} = 200 > 60$ we have $m = 0.8$ and $n = 1/3$, case (iii) of Table 2.5. Hence, the average convective heat transfer coefficient becomes

$$\bar{h}_c = \frac{K_b}{D}(0.027)(\text{Re})^{0.8}(\text{Pr})^{1/3}\left(\frac{\mu_b}{\mu_s}\right)^{0.14}$$

$$= (0.673/0.02)(0.027)\left(1.12 \times 10^5\right)^{0.8}(2.16)^{1/3}[(3.47/6.16)]^{0.14}$$

$$= 12{,}855.72 \times 0.9228 = 11.863.18 \text{ W/m}^2$$

$$= 116.2221 \text{ kW}$$

Hence, the rate of heat transfer from hot fluid to tube wall is given by

$$Q = h_c \times 3.14 \times \text{DL}(T_b - T_s) = 11{,}863, \, 18 \times 3.14 \times 0.02 \times 4 \times 39$$

$$= 116{,}221.20 \text{ W} = 116.221 \text{ kW}.$$

### (B) Working fluid as air

**Example 2.33** The ambient air at 27 °C is flowing with velocity of 4.57 m/s (a) above and (b) below through hot plate with effective area of 0.91 m × 0.61 m at temperature of 127°. Determine the convective heat transfer coefficient and the rate of heat transfer from hot plate to the air for both cases.

### Solution

The average temperature of hot plate and air, $T_f = (27 + 127)/2 = 77\,°C = 350\text{ K}$. The values of physical properties of air from Appendix Table E.1 are as follows:

$$v = 2.079 \times 10^{-5} \text{ m}^2/\text{s}, \text{ Pr} = 0.697, K = 3.001 \times 10^{-2} \text{ W/mK}.$$

The value of Reynolds number for above physical properties of air is given by:

$$\text{Re}_L = \frac{u_0 L}{v} = \frac{4.57 \times 0.91}{2.079 \times 10^{-5}} = 2 \times 10^5.$$

### Case (a): Airflowing above the hot plate

The flow is laminar, therefore, with the help of Eq. 2.11b; one can have an expression for an average convective heat transfer coefficient as

$$\bar{h} = \frac{K}{L}\overline{\text{Nu}} = \frac{K}{L}(0.664)\text{Re}_L^{1/2}\text{Pr}^{1/3}$$

$$= \frac{3.001 \times 10^{-2}}{0.91}(0.664)\left(2 \times 10^{5}\right)^{1/2}(0.697)^{1/3}$$
$$= 8.68 \ \text{W/m}^2 \ \text{K}$$

Here, the characteristic dimension becomes the length of hot plate unlike natural convection (Example 2.9).

Then, the rate of heat transfer from the hot plate to flowing air becomes

$$\dot{q} = \overline{h}A\Delta T = 8.68 \times 0.91 \times 0.61 \times (127 - 27) = 482 \ \text{W}.$$

**Case (b) Airflowing below the hot plate**

With the help of Eq. 2.11c, one can have an expression for an average convective heat transfer coefficient as

$$\overline{h} = \frac{K}{L}\overline{\text{Nu}} = \frac{K}{L}(0.332)\text{Re}_L^{1/2}\text{Pr}^{1/3}$$
$$= \frac{3.001 \times 10^{-2}}{0.91}(0.332)\left(2 \times 10^{5}\right)^{1/2}(0.697)^{1/3} = 4.34 \ \text{W/m}^2 \ \text{K}$$

Then, the rate of heat transfer from the hot plate to flowing air becomes

$$\dot{q} = \overline{h}A\Delta T = 4.34 \times 0.91 \times 0.61 \times (127 - 27) = 241 \ \text{W}.$$

The rate of heat transfer becomes half due to change of flow from the above hot plate to below hot plate due to tendency of hot air moving upwards direction.

***Example 2.34*** Repeat Example 2.33 for the hot plate facing upward and inclined at $45°$ from vertical direction.

**Solution**

Here, $\cos 45 = 0.707$.

**According to Eqs. 2.11d and 2.11e, the average dimensional Nusselt number becomes**

$$\overline{\text{Nu}}_L = 0.664(\text{Re}_L \times \cos\theta)^{1/2}\text{Pr}^{1/3}$$
$$\overline{\text{Nu}}_L = 0.664\left(2 \times 10^{5} \times \cos\theta\right)^{1/2}(0.697)^{1/3} = 221.41.$$

Substitute the values of an average Nusselt number in determining the average convective heat transfer coefficient as follows:

$$\overline{h} = \frac{K}{L}\overline{\text{Nu}} = \frac{3.001 \times 10^{-2}}{0.91} \times 211.41 = 7.3 \ \text{W/m}^2 \ \text{K}$$

One can observe that due to inclination of hot plate convective heat transfer coefficient is reduced as per expectation. In the case of inclined hot plate, its value is further reduced to half.

**Example 2.35** Determine the convective heat transfer coefficient and the rate of heat transfer from hot plate with effective area of 0.91 m × 0.61 m at temperature of 127° and another plate at 27 °C placed over at 0.10 m (characteristic dimension). The air is moving air with velocity of 4.57 m/s between two plates.

**Solution**

The average temperature of hot plate and air, $T_f = (27 + 127)/2 = 77\ °C = 350\ K$. The values of physical properties of air from Appendix Table E.1 are as follows:

$$v = 2.079 \times 10^{-5}\ m^2/s,\ Pr = 0.697,\ K = 3.001 \times 10^{-2}\ W/m\,K.$$

The value of Reynolds number for above physical properties of air with characteristic dimension of $L - 0.10$ m (Eq. 2.5b) unlike Example 2.33 is given by:

$$Re = \frac{u_0 L}{v} = \frac{4.57 \times 0.10}{2.079 \times 10^{-5}} = 2.198 \times 10^4.$$

This is laminar flow. Referring to case (xiia) of Table 2.3, the dimensional Nusselt number becomes

$$Nu = 0.195(Re \times Pr)^{0.25}Pr^{1/4}.$$

From above correlation, one can conclude that the Nusselt number in this case is independent of Prandtl number, Pr unlike Example 2.33, i.e.,

$$Nu = 0.195(Re)^{0.25}.$$

After substitution of $Re = 2.198 \times 10^4$ and $Pr = 0.697$, one gets

$$Nu = 0.195\left(2.198 \times 10^4 \times 0.697\right)^{0.25}(0.697)^{1/4}$$
$$= 0.195\left(2.198 \times 10^4\right)^{0.25} = 2.37.$$

Now, the convective heat transfer coefficient can be calculated as follows:

$$\overline{h} = \frac{K}{L}\overline{Nu} = \frac{3.001 \times 10^{-2}}{0.1} \times 2.37 = 0.711\ W/m^2\,K.$$

So, one can notice that there is significant drop in convective heat transfer coefficient due to flowing of air between two plates unlike open flowing (Example 2.34). It is only because of significant change in characteristic dimension from open to closed condition.

***Example 2.36*** Repeat Example 2.35 for parallel hot plates inclined at 45° from vertical line.

**Solution**

In this case, Reynolds number is, Re, is replaced by Re.$\cos\theta$ similar to Example 2.34. Hence, the Nusselt number becomes

$$\text{Nu} = 0.195(\text{Re. } \cos\theta)^{0.25} = 0.195\left(2.198 \times 10^4 \times 0.707\right)^{0.25} = 2.18 \text{ W/m}^2 \text{ K.}$$

In this case, there is marginal decrease in Nusselt number from 2.37 to 2.18 W/m² K.

Now, the convective heat transfer coefficient can be calculated as follows:

$$\bar{h} = \frac{K}{L}\overline{\text{Nu}} = \frac{3.001 \times 10^{-2}}{0.1} \times 2.18 = 0.654 \text{ W/m}^2 \text{ K.}$$

In this case, there is marginal change in heat transfer coefficient due to inclination.

***Example 2.37*** Repeat the Example 2.35 with change in characteristic dimension from 0.10 to 0.05 m with same operating condition.

**Solution**

In this case, Reynolds's number will be determined as

$$\text{Re} = \frac{u_0 L}{\nu} = \frac{4.57 \times 0.05}{2.079 \times 10^{-5}} = 1.099 \times 10^4.$$

From above correlation, one can conclude that the Nusselt number in this case is independent of Prandtl number, Pr unlike Example 2.33, i.e.,

$$\text{Nu} = 0.195(\text{Re})^{0.25}.$$

After substitution of Re $= 1.099 \times 10^4$ and Pr $= 0.697$, one gets

$$\text{Nu} = 0.195\left(1.099 \times 10^4\right)^{0.25} = 1996.$$

Now, the convective heat transfer coefficient can be calculated as follows:

$$\bar{h} = \frac{K}{L}\overline{\text{Nu}} = \frac{3.001 \times 10^{-2}}{0.05} \times 2.027 = 1.2 \text{ W/m}^2 \text{ K.}$$

There is 57% increase in convective heat transfer coefficient due to change in characteristic dimension from 0.1 to 0.05 m.

*Example 2.38* Repeat the Example 2.35 with change in flow velocity from 4.57 m/s to 2.28 m with same operating condition.

**Solution**

The value of Reynolds number for same physical properties of air with characteristic dimension of $L - 0.10$ m (Eq. 2.5b) is given by:

$$Re = \frac{u_0 L}{\nu} = \frac{2.28 \times 0.10}{2.079 \times 10^{-5}} = 1.099 \times 10^4.$$

This is laminar flow.

After substitution of $Re = 1.099 \times 10^4$ and $Pr = 0.697$, one gets

$$Nu = 0.195 \left(1.099 \times 10^4 \times 0.697\right)^{0.25} (0.697)^{1/4}$$
$$= 0.195 \left(1.099 \times 10^4\right)^{0.25} = 1.996.$$

Now, the convective heat transfer coefficient can be calculated as follows:

$$\bar{h} = \frac{K}{L} \overline{Nu} = \frac{3.001 \times 10^{-2}}{0.1} \times 1.99 = 0.359 \text{ W/m}^2 \text{ K}.$$

So, one can notice that there is drop by about 50% in convective heat transfer coefficient due to reduction of flowing of air by 50%. The heat transfer is reduced due to reduction of flow rate as per our expectation.

*Example 2.39* Repeat the Example 2.35 by using Eq. 2.13 for same physical and specification parameters.

**Solution**

From Example 2.35, we have

$$Re = \frac{u_0 L}{\nu} = \frac{4.57 \times 0.10}{2.079 \times 10^{-5}} = 2.198 \times 10^4 > 2100.$$

From Eq. 2.13, we have

$$Nu = 0.0158 \, Re^{0.8}.$$

Substitute $Re = 2.198 \times 10^4$ in the above equation, one gets

$$Nu = 0.0158 \left(2.198 \times 10^4\right)^{0.8} = 47.02.$$

Now, the convective heat transfer coefficient can be calculated as follows:

$$h = \frac{K}{D} \times \text{Nu} = \frac{0.03}{0.20} \times 47.02 = 7.05 \text{ W/m}^2 \text{ K}.$$

Here, characteristic dimension ($D$) is twice the spacing between two plates.

**Example 2.40** Evaluate convective heat transfer coefficient for two parallel plates inclined at 45° for Example 2.39.

**Solution**

**In this case**

From Eq. 2.13, we have

$$\text{Nu} = 0.0158(\text{Re} \times \cos\theta)^{0.8} = 0.0158(2.198 \times 10^4 \times 0.707)^{0.8} = 35.63.$$

Now, the convective heat transfer coefficient can be calculated as follows:

$$h = \frac{K}{D} \times \text{Nu} = \frac{0.03}{0.20} \times 35.63 = 5.34 \text{ W/m}^2 \text{ K}.$$

### 2.4.4 Effect of Nanofluid on Convective Heat Transfer

Most of dimensional numbers, namely Reynolds, Grashof, Prandtl numbers, depends on physical properties of working fluid which has been used to determine convective heat transfer coefficient. The solar energy system generally operates in medium temperature range between 15 and 100 °C; there is marginal change in physical properties of water as a working fluid, and hence there is marginal variation in convective heat transfer coefficient. If physical properties of water are changed significantly, then there is possibility of significant change in convective heat transfer coefficient. It has been found that if nanoparticles, Table 2.6 is mixed with water in certain percentage then it becomes nanofluid with much improved with physical properties. The physical properties of nanofluid can be determined from the various formulas given in Table 2.7.

(a)  **Natural convection on horizontal plate**

**Example 2.41** Calculate the convective heat transfer coefficient from horizontal surface having dimension of 1 m × 1 m to (a) water (base fluid) and (b) nanofluid ($Al_2O_3$ nanoparticles) placed over horizontal surface.

**Given**: Properties of $Al_2O_3$ nanoparticles (Table 2.6): $C_p = 880 (\text{J/kg K})$, $\varphi_p = 0.4\%(0.004)$, $\rho_p = 3.89 \times 10^3 (\text{kg/m}^3)$,

**Table 2.6** Thermal properties of nano-particles

| S. No. | Name of nanoparticles | Chemical formula | Density, $\rho$ (kg/m$^3$) | Specific heat, $C_p$ (J/ kg K) | Thermal conductivity, $K$ (W/m K) |
|---|---|---|---|---|---|
| 1 | Aluminum oxide | $Al_2O_3$ | 3890 | 880 | 39.5 |
| 2 | Titanium dioxide | $TiO_2$ | 4230 | 697 | 11.8 |
| 3 | Copper oxide | CuO | 6310 | 550 | 17.6 |

**Table 2.7** Physical properties of nanofluid

| S. No. | Name of physical properties | Expression |
|---|---|---|
| 1 | Density, kg/ m$^3$ | $\rho_{nf} = \varphi_p \rho_p + (1 - \varphi_p)\rho_w$ |
| 2 | Thermal conductivity, W/m K | $k_{nf} =$ $k_w \left[ 1 + (1.0112)\varphi_p + (2.4375)\varphi_p \left( \frac{47}{d_p(nm)} \right) - (0.0248)\varphi_p \left( \frac{k_p}{0.613} \right) \right]$ with $d_p = 21$ nm and $\varphi_p = \frac{V_p}{V_{bf} + V_p}$; $V = V_{bf} + V_p$ |
| 3 | Specific heat, J/kg K | $C_{nf} = \frac{[(\varphi_p \rho_p C_p + (1 - \varphi_p)\rho_w C_w)]}{\rho_w}$ |
| 4 | Dynamic viscosity, kg/ ms | $\mu_{nf} = \mu_w \left[ 1 + 2.5(\varphi_p) + 6.2\varphi_p^2 \right]$ |
| 5 | Thermal expansion coefficient, K$^{-1}$ | $\beta_{nf} = (1 - \varphi_p)\beta_w + \varphi_p \beta_p$ |

Thermal conductivity of particle $K_p = 39.5(W/m\,°C)$, diameter of nanoparticle, $(d_p) = 21$ nm.

Temperature of horizontal surface: $(T_h) = 72.25\,°C$, temperature of nanofluid $(T_{nf}) = 74.45\,°C$.

**Solution**

(a) **Base fluid only water**

From Appendix Table E.2, all the thermophysical properties of water can be obtained as

$$\rho_w = 977.16(kg/m^3), \mu_w = 3.984 \times 10^{-4}(kg/ms),$$
$$k_w = 0.661(W/m\,°C), C_w = 4.190 \times 10^3(J/kg\,K),$$

$$\beta_w = \frac{1}{273 + 74.45} = 0.0029 \text{ K}^{-1}.$$

Now, characteristic length $L = \dfrac{A}{P} = \dfrac{1}{4} = 0.25.$

$$Gr = \frac{g\beta L^3 \rho^2 \Delta T}{\mu^2} = \frac{9.8 \times 0.0029 \times (0.25)^3 \times (977.16)^2 \times 2.2}{(3.984 \times 10^{-4})^2} = 0.585 \times$$

$10^{10}$ and $Pr = \dfrac{\mu C_p}{k} = \dfrac{3.984 \times 10^{-4} \times 4.190 \times 10^3}{0.661} = 2.25$

$$Gr.Pr = 0.585 \times 10^{10} \times 2.25 = 1.31 \times 10^{10}.$$

This is laminar flow, and hence $C = 0.54$. $n = 0.25$, and $K' = 1$ case (iiia) of Table 2.2.

Further, from Eq. 2.7, we have

$$Nu = C(Gr\ Pr)^n = 0.54 \times \left(1.31 \times 10^{10}\right)^{0.25} = 1.83 \times 10^2.$$

Now,

$$\text{Convective heat transfer coefficient}(h_c) = Nu \times \left(\frac{k_w}{d}\right)$$

$$= 1.83 \times 10^2 \times \frac{0.661}{0.25} = 483\ \left(\text{W/m}^2\ {}^\circ\text{C}\right)$$

### (b) Nanofluid (40:60; $\varphi_p = 0.40$)

From Tables 2.6 and 2.7, the physical properties of nanofluid with aluminum oxide ($Al_2O_3$) particles are

$$C_{nf} = \frac{\left[\left(\varphi_p \rho_p C_p + \left(1 - \varphi_p\right)\rho_w C_w\right)\right]}{\rho_w}$$

$$= \frac{\left[0.4 \times 880 \times 3.89 \times 10^3 + 0.6 \times \left(977.16 \times 4.190 \times 10^3\right)\right]}{977.16}$$

$= 3915.2(\text{J/kg K}); \ \rho_{nf} = \varphi_p \rho_p + \left(1 - \varphi_p\right)\rho_w$

$= 0.4 \times 3.89 \times 10^3 + 0.6 \times 977.16 = 2.14 \times 10^3\left(\text{kg/m}^3\right);$

$\mu_{nf} = \mu_w\left[1 + 2.5(\varphi_p) + 6.2\varphi_p^2\right] = 3.984 \times 10^{-4}\left[1 + 2.5 \times 0.4 + 6.2 \times (0.4)^2\right]$

$= 11.92 \times 10^{-4}(\text{kg/ms})$

$$k_{nf} = k_w\left[1 + (1.0112)\varphi_p + (2.4375)\varphi_p\left(\frac{47}{d_p(\text{nm})}\right) - (0.0248)\varphi_p\left(\frac{k_p}{0.613}\right)\right]$$

$$= 0.661\left[1 + 2.4375 \times 0.4 \times \frac{47}{21} - 0.0248 \times 0.4 \times \frac{39.5}{0.613}\right]$$

$$= 0.661[3.182 - 0.6392] = 1.68(\text{W/m °C});$$

$$\beta_{nf} = (1 - \varphi_p)\beta_w + \varphi_p\beta_p = 0.0028 \text{ K}^{-1}.$$

In this case,

$$Gr = \frac{g\beta d^3 \rho^2 \Delta T}{\mu^2} = \frac{9.8 \times 0.0028 \times (0.25)^3 \times (2.14 \times 10^3)^2 \times 2.2}{(11.92 \times 10^{-4})^2} = 30.37 \times 10^8 \text{ and } Pr = \frac{\mu C_p}{k} =$$

$$\frac{11.92 \times 10^{-4} \times 3915.2}{1.68} = 2.778.$$

Now,

$$Nu = C(GrPr)^n = 0.54 \times (30.37 \times 10^8 \times 2.778)^{0.25} = 1.27 \times 10^2.$$

Further,

Convective heat transfer coefficient($h_{cw}$)

$$= Nu \times \left(\frac{k_{nf}}{L}\right) = 1.27 \times 10^2 \times \frac{1.68}{0.25} = 853.44 \left(\text{W/m}^2 \text{ °C}\right).$$

One can see that for 40:60 ratio of nanofluid and base fluid water, the convective heat transfer coefficient is increased by 100%.

***Example 2.42*** Repeat Example 2.41 for 60:40 ratios ($\varphi_p = 0.60$) between nanofluid and base fluid water.

**Solution**

From Tables 2.6 and 2.7, the physical properties of nanofluid with aluminum oxide ($Al_2O_3$) particles are

$$C_{nf} = \frac{\left[(\varphi_p\rho_pC_p + (1 - \varphi_p)\rho_wC_w)\right]}{\rho_w}$$

$$= \frac{\left[0.6 \times 880 \times 3.89 \times 10^3 + 0.4 \times (977.16 \times 4.190 \times 10^3)\right]}{977.16}$$

$$= 3777.9(\text{J/kg K}); \rho_{nf} = \varphi_p\rho_p + (1 - \varphi_p)$$

$$\rho_w = 0.6 \times 3.89 \times 10^3 + 0.4 \times 977.16 = 2.72 \times 10^3 (\text{kg/m}^3);$$

$$\mu_{nf} = \mu_w\left[1 + 2.5(\varphi_p) + 6.2\varphi_p^2\right]$$

$$= 3.984 \times 10^{-4}\left[1 + 2.5 \times 0.6 + 6.2 \times (0.6)^2\right] = 18.85 \times 10^{-4}(\text{kg/ms})$$

$$k_{nf} = k_w\left[1 + (1.0112)\varphi_p + (2.4375)\varphi_p\left(\frac{47}{d_p(\text{nm})}\right) - (0.0248)\varphi_p\left(\frac{k_p}{0.613}\right)\right]$$

$$= 0.661\left[1 + 2.4375 \times 0.6 \times \frac{47}{21} - 0.0248 \times 0.6 \times \frac{39.5}{0.613}\right]$$

$$= 0.661[4.2732 - 0.9588] = 2.19 \text{ (W/m °C)};$$

$$\beta_{nf} = \left(1 - \varphi_p\right)\beta_w + \varphi_p\beta_p = 0.0028 \text{ K}^{-1}.$$

In this case,

$$Gr = \frac{g\beta d^3\rho^2\Delta T}{\mu^2} = \frac{9.8\times0.0028\times(0.25)^3\times\left(2.72\times10^3\right)^2\times2.2}{(18.85\times10^{-4})^2} = 19.64 \times 10^8 \text{ and } Pr = \frac{\mu C_p}{k} =$$

$$\frac{18.85\times10^{-4}\times3777.9}{2.19} = 3.25.$$

Now,

$$Nu = C(GrPr)^n = 0.54 \times \left(19.64 \times 10^8 \times 3.25\right)^{0.25} = 1.52 \times 10^2;$$

Further,

$$\text{Convective heat transfer coefficient}(h_{cw})$$

$$= Nu \times \left(\frac{k_{nf}}{L}\right) = 1.52 \times 10^2 \times \frac{2.19}{0.25} = 1332 \left(\text{W/m}^2 \text{ °C}\right)$$

This shows that convective heat transfer coefficient increases with increase of fraction of nanofluid ($\varphi_p$).

***Example 2.43*** Repeat Example 2.41 for $\varphi_p = 1$ (100% nanofluid, $Al_2O_3$).

**Solution**

**Pure nanofluid (100%; $\varphi_p = 1, 0$).**
From Tables 2.6 and 2.7, the physical properties of nanofluid with aluminum oxide ($Al_2O_3$) particles are

$$C_{nf} = \frac{\left[\left(\varphi_p\rho_pC_p + \left(1 - \varphi_p\right)\rho_wC_w\right)\right]}{\rho_w}$$

$$= \frac{\left[1.0 \times 880 \times 3.89 \times 10^3\right]}{977.16} = 3503.78 \text{ (J/kg K)};$$

$$\rho_{nf} = \varphi_p\rho_p + \left(1 - \varphi_p\right)\rho_w = 1 \times 3.89 \times 10^3 = 3.89 \times 10^3 \left(\text{kg/m}^3\right);$$

$$\mu_{nf} = \mu_w\left[1 + 2.5\left(\varphi_p\right) + 6.2\varphi_p^2\right]$$

$$= 3.984 \times 10^{-4}\left[1 + 2.5 \times 1 + 6.2 \times (1)^2\right] = 38.64 \times 10^{-4} \text{ (kg/ms)}$$

$$k_{nf} = k_w\left[1 + (1.0112)\varphi_p + (2.4375)\varphi_p\left(\frac{47}{d_p(\text{nm})}\right) - (0.0248)\varphi_p\left(\frac{k_p}{0.613}\right)\right]$$

$$= 0.661\left[1 + 2.4375 \times 1 \times \frac{47}{21} - 0.0248 \times 1 \times \frac{39.5}{0.613}\right]$$

$$= 0.661[6.4553 - 1.980] = 3.21(\text{W/m °C});$$

$$\beta_{nf} = \left(1 - \varphi_p\right)\beta_w + \varphi_p\beta_p = \beta_p \approx 0.0028 \text{ K}^{-1}.$$

In this case,

$$\text{Gr} = \frac{g\beta d^3 \rho^2 \Delta T}{\mu^2} = \frac{9.8 \times 0.0028 \times (0.25)^3 \times (3.89 \times 10^3)^2 \times 2.2}{(38.64 \times 10^{-4})^2} = 9.56 \times 10^8 \text{ and } \text{Pr} = \frac{\mu C_p}{k} =$$

$$\frac{38.64 \times 10^{-4} \times 3503.78}{3.21} = 4.22.$$

Now, the Nusselt number is

$$\text{Nu} = C(\text{Gr Pr})^n = 0.54 \times (9.56 \times 10^8 \times 4.22)^{0.25} = 1.36 \times 10^2.$$

Further,

$$\text{Convective heat transfer coefficient}(h_{cw})$$

$$= \text{Nu} \times \left(\frac{k_{\text{nf}}}{L}\right) = 1.36 \times 10^2 \times \frac{3.21}{0.25} = 1747(\text{W/m}^2\,{}^\circ\text{C})$$

One can see that for 100% of nanofluid the convective heat transfer coefficient is maximum as per expectation.

***Example 2.44*** Calculate the convective heat transfer coefficient from horizontal surface having dimension of 1 m × 1 m to (a) water (base fluid) and (b) nanofluid (CuO nanoparticles) placed over horizontal surface.

**Given**: Properties of CuO nanoparticles (Table 2.6): $C_p = 550\,(\text{J/kg K})$, $\varphi_p = 0.4(40\%)$, $\rho_p = 6.31 \times 10^3\,(\text{kg/m}^3)$,

Thermal conductivity of particle $K_p = 17.6\,(\text{W/m}\,{}^\circ\text{C})$, diameter of nanoparticle, $(d_p) = 21$ nm.

Temperature of horizontal surface: $(T_h) = 72.25\,{}^\circ\text{C}$, Temperature of nanofluid $(T_{\text{nf}}) = 74.45\,{}^\circ\text{C}$.

The physical properties of base water is same as given in Example 2.40.

**Solution**

With the help of Tables 2.6–2.7 and Example 2.41, the physical properties of nanofluid with copper oxide (CuO) particles are

$$C_{\text{nf}} = \frac{\left[(\varphi_p \rho_p C_p + (1 - \varphi_p)\rho_w C_w)\right]}{\rho_w}$$

$$= \frac{\left[0.4 \times 550 \times 6.31 \times 10^3 + 0.6 \times (977.16 \times 4.190 \times 10^3)\right]}{977.16} = 3935\,(\text{J/kg K});$$

$$\rho_{\text{nf}} = \varphi_p \rho_p + (1 - \varphi_p)\rho_w = 0.4 \times 6.31 \times 10^3 + 0.6 \times 977.16$$

$$= 3.11 \times 10^3\,(\text{kg/m}^3);$$

$$\mu_{\text{nf}} = \mu_w\left[1 + 2.5(\varphi_p) + 6.2\varphi_p^2\right]$$

$$= 3.984 \times 10^{-4}\left[1 + 2.5 \times 0.4 + 6.2 \times (0.4)^2\right] = 11.92 \times 10^{-4}\,(\text{kg/ms})$$

$$k_{nf} = k_w \left[1 + (1.0112)\varphi_p + (2.4375)\varphi_p\left(\frac{47}{d_p \text{ (nm)}}\right) - (0.0248)\varphi_p\left(\frac{k_p}{0.613}\right)\right]$$

$$= 0.661\left[1 + 2.4375 \times 0.4 \times \frac{47}{21} - 0.0248 \times 0.4 \times \frac{17.6}{0.613}\right]$$

$$= 0.661[3.182 - 0.2848] = 1.92 \text{ (W/m}°\text{C)};$$

$$\beta_{nf} = (1 - \varphi_p)\beta_w + \varphi_p\beta_p = 0.0028 \text{ K}^{-1}.$$

In this case,

$$\text{Gr} = \frac{g\beta d^3 \rho^2 \Delta T}{\mu^2} = \frac{9.8 \times 0.0028 \times (0.25)^3 \times (3.11 \times 10^3)^2 \times 2.2}{(11.92 \times 10^{-4})^2} = 64.26 \times$$

$10^8$ and $\text{Pr} = \frac{\mu C_p}{k} = \frac{11.92 \times 10^{-4} \times 3935}{1.92} = 2.4429.$

Now,

$$\text{Nu} = C(\text{Gr Pr})^n = 0.54 \times (64.26 \times 10^8 \times 2.4429)^{0.25} = 1.91 \times 10^2.$$

Further,

Convective heat transfer coefficient $(h_{cw})$

$$= \text{Nu} \times \left(\frac{k_{nf}}{L}\right) = 1.91 \times 10^2 \times \frac{1.92}{0.25} = 1468 \text{ (W/m}^2°\text{C)}$$

One can see that, the convective heat transfer coefficient for CuO base nanofluid is significantly higher than both $Al_2O_3$ base nanofluid and water base fluid (Example 2.41) due to its higher thermal conductivity.

**Example 2.45** Repeat Example 2.44 for nanoparticle size of 15 nm.

**Solution**

In this case, only thermal conductivity of nanofluid is changed as follows:

$$k_{nf} = k_w \left[1 + (1.0112)\varphi_p + (2.4375)\varphi_p\left(\frac{47}{d_p \text{ (nm)}}\right) - (0.0248)\varphi_p\left(\frac{k_p}{0.613}\right)\right]$$

$$= 0.661\left[1 + 2.4375 \times 0.4 \times \frac{47}{15} - 0.0248 \times 0.4 \times \frac{17.6}{0.613}\right]$$

$$= 0.661[4.005 - 0.2848] = 2.49 \text{ (W/m}°\text{C)}$$

The Nusselt number will remain same, i.e., $\text{Nu} = 1.91 \times 10^2$.
From Example 2.55, we have

Convective heat transfer coefficient$(h_{cw})$

$$= \text{Nu} \times \left(\frac{k_{nf}}{L}\right) = 1.91 \times 10^2 \times \frac{2.49}{0.25} = 8404 \text{ (W/m}^2°\text{C)}$$

So, one can conclude that particle size of nanoparticles plays an important role in determining the convective heat transfer coefficient of nanofluid.

### (b) Forced convection on horizontal surface

In forced convection, the characteristic dimension becomes the length of horizontal surface, i.e., $L$ unlike natural convection.

***Example 2.46*** Evaluate the convective heat transfer coefficient for (a) water (base fluid) and (b) nanofluid ($Al_2O_3$ nanoparticles) flows over a smooth hot flat plate at 74.45 °C and 1.0 m long above with velocity of 0.20 m/s. The water properties at the fluid film temperature are: $v = 7.66 \times 10^{-7}$ m$^2$/s, $K = 0.621$ W/m K, Pr $= 5.13$, Appendix Table E.2.

**Given**: Properties of $Al_2O_3$ nanoparticles (Table 2.6): $C_p = 880$ (J/kg K), $\varphi_p = 0.4\%(0.004)$, $\rho_p = 3.89 \times 10^3$ (kg/m$^3$).

Thermal conductivity of particle $K_p = 39.5$ (W/m °C), diameter of nanoparticle, $(d_p) = 21$ nm.

### Solution

### (a) Base fluid only water

From Appendix Table E.2, all the thermophysical properties of water can be obtained as

$$\rho_w = 977.16 \, (\text{kg/m}^3), \, \mu_w = 3.984 \times 10^{-4} \, (\text{kg/m s}),$$
$$k_w = 0.661 \, (\text{W/m °C}), \, C_w = 4.190 \times 10^3 \, (\text{J/kg K}),$$

and, Prandtl number, $\text{Pr} = \frac{\mu C_p}{k} = \frac{3.984 \times 10^{-4} \times 4.190 \times 10^3}{0.661} = 2.25$ and

$$\text{Reynolds number, Re} = \frac{u_0 L}{v} = \frac{0.20 \times 1.0}{3.99 \times 10^{-7}} = 501,253;$$

$$v = \frac{\mu_w}{\rho_w} = \frac{3.984 \times 10^{-4}}{977.16} = 3.99 \times 10^{-7}.$$

Now, the Nusselt number is given by

$$\overline{\text{Nu}}_L = 0.664 \text{Re}_L^{1/2} \text{Pr}^{1/3} = 0.664 \times (501,253)^{1/2} \times (2.25)^{1/3} = 615.$$

Now,

Convective heat transfer coefficient$(h_c)$

$$= \text{Nu} \times \left(\frac{k_w}{d}\right) = 615 \times \frac{0.661}{1} = 407 \, (\text{W/m}^2 \, \text{°C})$$

## (b)  Nanofluid (40:60; $\varphi_p = 0.40$)

From Tables 2.6 and 2.7, the physical properties of nanofluid with aluminum oxide ($Al_2O_3$) particles are

$$C_{nf} = \frac{[(\varphi_p \rho_p C_p + (1 - \varphi_p)\rho_w C_w)]}{\rho_w}$$

$$= \frac{[0.4 \times 880 \times 3.89 \times 10^3 + 0.6 \times (977.16 \times 4.190 \times 10^3)]}{977.16}$$

$$= 3915.2 \ (J/kg \ K);$$

$$\rho_{nf} = \varphi_p \rho_p + (1 - \varphi_p)$$

$$\rho_w = 0.4 \times 3.89 \times 10^3 + 0.6 \times 977.16 = 2.14 \times 10^3 \ (kg/m^3);$$

$$\mu_{nf} = \mu_w[1 + 2.5(\varphi_p) + 6.2\varphi_p^2]$$

$$= 3.984 \times 10^{-4}[1 + 2.5 \times 0.4 + 6.2 \times (0.4)^2] = 11.92 \times 10^{-4} \ (kg/ms)$$

$$k_{nf} = k_w\left[1 + (1.0112)\varphi_p + (2.4375)\varphi_p\left(\frac{47}{d_p \ (nm)}\right) - (0.0248)\varphi_p\left(\frac{k_p}{0.613}\right)\right]$$

$$= 0.661\left[1 + 2.4375 \times 0.4 \times \frac{47}{21} - 0.0248 \times 0.4 \times \frac{39.5}{0.613}\right]$$

$$= 0.661[3.182 - 0.6392] = 1.68 \ (W/m \ ^\circ C)$$

In this case,
Reynolds number, $Re = \frac{u_0 L}{\nu_{nf}} = \frac{0.20 \times 1.0}{5.57 \times 10^{-7}} = 359{,}066; \ \nu = \frac{\mu_{nf}}{\rho_{nf}} = \frac{11.92 \times 10^{-4}}{2.14 \times 10^3 6} = $
$5.57 \times 10^{-7}$ and Prandtl number, $Pr = \frac{\mu_{nf} C_{nf}}{k_{nf}} = \frac{11.92 \times 10^{-4} \times 3915.2}{1.68} = 2.881$.
Now, from Eq. 2.11b, one has the following relation

$$\overline{Nu}_L = 0.664 Re_L^{1/2} Pr^{1/3} = 0.664 \times (359{,}066)^{1/2} \times (2.881)^{1/3} = 565.96$$

Further,

$$\text{Convective heat transfer coefficient}(h_{cw})$$

$$= Nu \times \left(\frac{k_{nf}}{L}\right) = 565.96 \times \frac{1.68}{1} = 960.81 \ (W/m^2 \ ^\circ C)$$

One can see that for 40:60 ratio of nanofluid and base fluid water, the convective heat transfer coefficient for forced mode is significantly increased like natural mode of operation.

**Example 2.47** Repeat Example 2.46 for inclined surface at angle of 45° with vertical plane.

**Solution**

From Eq. 2.11e one has an expression for Nusselt number as

$$\overline{Nu}_L = 0.664(Re_L \times \cos\theta)^{1/2}Pr^{1/3}$$

From Example 2.46, we have for nanofluid.
Reynolds number, $Re = \frac{u_0 L}{v_{nf}} = \frac{0.20 \times 1.0}{5.57 \times 10^{-7}} = 359{,}066$ and Prandtl number, $Pr = \frac{\mu_{nf}C_{nf}}{k_{nf}} = \frac{11.92 \times 10^{-4} \times 3915.2}{1.68} = 2.881$.

Now, the Nusselt number for inclined surface of 45° becomes as

$$\overline{Nu}_L = 0.664(Re_L \times \cos\theta)^{1/2}Pr^{1/3}$$
$$= 0.664 \times (359{,}066 \times 0.707)^{1/2} \times (2.881)^{1/3} = 47587.$$

Further, the convective heat transfer coefficient for inclined surface with nanofluid can be evaluated as follows:

Convective heat transfer coefficient$(h_{cw})$

$$= Nu \times \left(\frac{k_{nf}}{L}\right) = 475.87 \times \frac{1.68}{1} = 799.47 \left(W/m^2 \, ^\circ C\right).$$

This means the convective heat transfer coefficient for inclined surface is reduced.

***Example 2.48*** Repeat Example 2.46 for two parallel horizontal surfaces having spacing of 0.1 m.

**Solution**

In this case, only characteristic dimension $L = 0.1$ m.
Then, the convective heat transfer coefficient can be evaluated as follows:

$$\text{Convective heat transfer coefficient}(h_{cw}) = Nu \times \left(\frac{k_{nf}}{L}\right)$$

$$= 565.96 \times \frac{1.68}{0.10} = 9608.1 \left(W/m^2 \, ^\circ C\right).$$

In such configuration, the spacing between two parallel horizontal surface plays an important role.

***Example 2.49*** Evaluate the convective heat transfer coefficient for nanofluid ($Al_2O_3$ nanoparticles) flows over a smooth hot flat plate at 74.45 °C and 5.0 m long above with velocity of 0.20 m/s. The water properties at the fluid film temperature are: $v = 7.66 \times 10^{-7}$ m²/s, $K = 0.621$ W/m K, $Pr = 5.13$, Appendix Table E.2.

**Given**: Properties of $Al_2O_3$ nanoparticles (Table 2.6): $C_p = 880$ (J/kg K), $\varphi_p = 0.4\%(0.004)$, $\rho_p = 3.89 \times 10^3$ (kg/m³).

Thermal conductivity of particle $K_p = 39.5$ (W/m °C), diameter of nanoparticle, $(d_p) = 21$ nm.

**Solution**

**Nanofluid (40:60; $\varphi_p = 0.40$).**
From Tables 2.6 and 2.7, the physical properties of nanofluid with aluminum oxide $(Al_2O_3)$ particles are

$$C_{nf} = \frac{\left[(\varphi_p \rho_p C_p + (1-\varphi_p)\rho_w C_w)\right]}{\rho_w}$$

$$= \frac{\left[0.4 \times 880 \times 3.89 \times 10^3 + 0.6 \times (977.16 \times 4.190 \times 10^3)\right]}{977.16}$$

$$= 3915.2 \text{ (J/kg K)}; \rho_{nf} = \varphi_p \rho_p + (1-\varphi_p)\rho_w$$

$$= 0.4 \times 3.89 \times 10^3 + 0.6 \times 977.16 = 2.14 \times 10^3 \text{ (kg/m}^3\text{)};$$

$$\mu_{nf} = \mu_w\left[1 + 2.5(\varphi_p) + 6.2\varphi_p^2\right]$$

$$= 3.984 \times 10^{-4}\left[1 + 2.5 \times 0.4 + 6.2 \times (0.4)^2\right] = 11.92 \times 10^{-4} \text{ (kg/ms)}$$

$$k_{nf} = k_w\left[1 + (1.0112)\varphi_p + (2.4375)\varphi_p\left(\frac{47}{d_p \text{ (nm)}}\right) - (0.0248)\varphi_p\left(\frac{k_p}{0.613}\right)\right]$$

$$= 0.661\left[1 + 2.4375 \times 0.4 \times \frac{47}{21} - 0.0248 \times 0.4 \times \frac{39.5}{0.613}\right]$$

$$= 0.661[3.182 - 0.6392] = 1.68 \text{ (W/m °C)}.$$

In this case,
Reynolds number, $Re = \frac{u_0 L}{v_{nf}} = \frac{0.20 \times 5.0}{5.57 \times 10^{-7}} = 1.6 \times 10^6$; $v = \frac{\mu_{nf}}{\rho_{nf}} = \frac{11.92 \times 10^{-4}}{2.14 \times 10^3 6} = 5.57 \times 10^{-7}$ and Prandtl number, $Pr = \frac{\mu_{nf}C_{nf}}{k_{nf}} = \frac{11.92 \times 10^{-4} \times 3915.2}{1.68} = 2.881$.
Now, from Eq. 2.11b, one has the following relation

$$\overline{Nu}_L = 0.664 Re_L^{1/2} Pr^{1/3} = 0.664 \times (1.6 \times 10^6)^{1/2} \times (2.881)^{1/3} = 1211.50.$$

Further,

$$\text{Convective heat transfer coefficient}(h_{cw})$$

$$= Nu \times \left(\frac{k_{nf}}{L}\right) = 1211.50 \times \frac{1.68}{5} = 407.06 \text{ (W/m}^2 \text{ °C)}$$

One can see that for 40:60 ratio of nanofluid the convective heat transfer coefficient for forced mode is reduced by approximately half with increase of length of horizontal surface, Example 2.46.

### (c) Forced convection in a horizontal tube/pipe

In the case of horizontal tube/pipe, the diameter and length of tube/pipe plays an important role as mentioned in Eqs. 2.12a–2.12d.

***Example 2.50*** Evaluate the average convective heat transfer coefficient and the rate of heat transfer in $W/m^2$ in the fully developed velocity at 0.61 m/s in a 25.4 mm ID tube in a cylindrical tube with nanofluid (CuO), at $T_w = 11\,°C$ flows with the tube wall temperature of $T_s = 20\,°C$.

**Given**: Properties of CuO nanoparticles (Table 2.6): $C_p = 550$ (J/kg K), $\varphi_p = 0.4(40\%)$, $\rho_p = 6.31 \times 10^3$ $(kg/m^3)$,

Thermal conductivity of particle $K_p = 17.6$ $(W/m\,°C)$, diameter of nanoparticle, $(d_p) = 21$ nm.

Temperature of horizontal tube: $(T_h) = 74.45\,°C$, Temperature of nanofluid $(T_{nf}) = 72.25\,°C$.

**Solution**

With the help of Tables 2.6 and 2.7 and Example 2.41, the physical properties of nanofluid with copper oxide (CuO) particles are

$$C_{nf} = \frac{[(\varphi_p \rho_p C_p + (1 - \varphi_p)\rho_w C_w)]}{\rho_w}$$

$$= \frac{[0.4 \times 550 \times 6.31 \times 10^3 + 0.6 \times (977.16 \times 4.190 \times 10^3)]}{977.16} = 3935 \text{ (J/kg K)};$$

$$\rho_{nf} = \varphi_p \rho_p + (1 - \varphi_p)\rho_w = 0.4 \times 6.31 \times 10^3 + 0.6 \times 977.16$$
$$= 3.11 \times 10^3 \ (kg/m^3);$$

$$\mu_{nf} = \mu_w [1 + 2.5(\varphi_p) + 6.2\varphi_p^2]$$
$$= 3.984 \times 10^{-4}[1 + 2.5 \times 0.4 + 6.2 \times (0.4)^2] = 11.92 \times 10^{-4} \text{ (kg/ms)}$$

$$k_{nf} = k_w \left[1 + (1.0112)\varphi_p + (2.4375)\varphi_p \left(\frac{47}{d_p \text{ (nm)}}\right) - (0.0248)\varphi_p \left(\frac{k_p}{0.613}\right)\right]$$

$$= 0.661 \left[1 + 2.4375 \times 0.4 \times \frac{47}{21} - 0.0248 \times 0.4 \times \frac{17.6}{0.613}\right]$$

$$= 0.661[3.182 - 0.2848] = 1.92 \text{ (W/m\,°C)}$$

The Reynold number, Eq. 2.5b, for the present case $L = D$, characteristic length for cylindrical tube is given by

$$Re = \frac{Du_0}{v} = \frac{(0.0254)(0.61)}{3.83 \times 10^{-7}} = 4.045 \times 10^4,$$

Here,

$$v = \frac{\mu}{\rho} = \frac{11.92 \times 10^{-4}}{3.11 \times 10^3} = 3.83 \times 10^{-7}$$

$$\mathrm{Pr} = \frac{\mu C_p}{k} = \frac{11.92 \times 10^{-4} \times 3935}{1.92} = 2.4429.$$

This is turbulent flow condition for thermal heating, and hence, the Nusselt Number from Eq. 2.12b and case (iv) of Table 2.5 is given by,

$$\mathrm{Nu}_{D,\infty} = (0.023)\left(4.045 \times 10^4\right)^{0.8} \times (2.4429)^{0.4} = 159.35.$$

The convective heat transfer coefficient from tube surface to working fluid (water) will be determined as

$$h_\infty = \frac{K_{bf}}{D} \times \mathrm{Nu} = \frac{1.92}{0.0254} \times 159.35 = 12,045 \; \mathrm{W/m^2\,^\circ C}$$

The rate of heat transfer in $W/m^2$ from cylindrical wall to the flowing water as fluid can be obtained as

$$\dot{q} = h_\infty(T_b - T_c) = 12,045(74.45 - 72.25)$$
$$= 26,499.8 \; \frac{W}{m^2} = 26.50 \; \mathrm{kW/m^2}.$$

**Example 2.51**  Repeat the Example 2.50 for thermal cooling from nanofluid to tube wall.

**Solution**

Referring Eq. 2.12b, an expression for Nusselt number with $n = 1/3$ becomes

$$\mathrm{Nu}_{D,\infty} = (0.023)\left(4.045 \times 10^4\right)^{0.8} \times (2.4429)^{0.3} = 150.12$$

The convective heat transfer coefficient from tube surface to working fluid (water) will be determined as

$$h_\infty = \frac{K_{bf}}{D} \times \mathrm{Nu} = \frac{1.92}{0.0254} \times 150.12 = 11,347.56 \; \mathrm{W/m^2\,^\circ C}$$

The rate of heat transfer in $W/m^2$ from cylindrical wall to the flowing water as fluid can be obtained as

$$\dot{q} = h_\infty(T_b - T_c) = 11,347.56(74.45 - 72.25)$$
$$= 21,560 \; \frac{W}{m^2} = 21.56 \; \mathrm{kW/m^2}.$$

Therefore, the convective heat transfer coefficient and the rate of heat transfer is reduced if direction of heat flow is reversed.

### 2.4.5 Convective Heat Transfer Due to Wind [1, 2]

In previous section, we have discussed free and forced convective heat transfer coefficient and the rate of heat transfer from horizontal flat surface and cylindrical tube/pipe with and without inclination from vertical surface. Effect of various nanofluids, namely $Al_2O_3$, CuO and $TiO_2$, has also been discussed. Most of heat transfer mainly depends on physical properties of working fluid, namely air and water, characteristic dimension and concentration of nanofluid.

The correlation for the convective heat transfer coefficient for the flat plates exposed to the ambient air/wind has also been studied by many researchers. The correlations for this case depend on the wind speed ($V$). The typical correlation for the exposed surface to flowing wind over it has been given by different researchers as follows:

$$h_c = 5.7 + 3.8V \quad \text{for } 0 \le V \le 5 \, \text{ms}^{-1}. \tag{2.13a}$$

For zero wind speed Eq. 2.13a will give the heat transfer coefficient for natural convection. This correlation is an approximate correlation as it is valid for the wind blowing parallel to the exposed surface.

The correlation given in Eq. 2.13a includes the effect of convection and radiation both. The correlation for heat transfer coefficient due to only convection is given as:

$$h_c = 2.8 + 3.0V \quad \text{for } 0 \le V \le 7 \, \text{ms}^{-1}; \quad \text{Laminar flow} \tag{2.13b}$$

The sensibility of these parameters is also demonstrated through a comparison, and an another expression for convective heat transfer coefficient is given by,

$$h_c = 7.2 + 3.8V. \tag{2.13c}$$

Several other correlations are also available in the literature and generally, $h_c$ is determined from an expression in the formed expressed as

$$h_c = a + bV_a^n; \quad \text{Laminar flow} \tag{2.13d}$$

where $a = 2.8$, $b = 3$ and $n = 1$ for $V_a < 5$ m/s

and $a = 0$, $b = 6.15$ and $n = 0.8$ for $V_a > 5$ m/s.

[The source and reference of Eq. 2.13 can be obtained from Tiwari (2002)] [1, 2]. In all above equation, $h_c = 2.8$ W/m$^2$ K for $V = 0$ in case of Eqs. 2.13b and 2.13d, respectively. Therefore, the convective heat transfer coefficient ($h_c$) for closed loop system like inside building and solar distillation can be considered as 2.8 W/m$^2$ K.

However, it is important to note that Eq. 2.13 are independent of temperature of the surface and physical properties of air blowing over the surface. So, these are generally used for exposed surface to ambient air as well as enclosed room air.

**Example 2.52** Evaluate convective heat transfer coefficient for horizontal surface exposed to wind speed of I m/s by Using Eq. 2.13.

**Solution**

From Eq. 2.13a one has convective heat transfer coefficient as

$$h_c = 5.7 + 3.8V = 5.7 + 3.8 = 9.5 \text{ W/m}^2 \text{ K}.$$

From Eq. 2.13b one has convective heat transfer coefficient as

$$h_c = 2.8 + 3.0V = 5.8 \text{ W/m}^2 \text{ K}.$$

From Eq. 2.13c one has convective heat transfer coefficient as

$$h_c = 7.2 + 3.8V = 11 \text{ W/m}^2 \text{ K}.$$

From Eq. 2.13c one has convective heat transfer coefficient as

$$h_c = a + bV_a^n = 2.8 + 3 = 5.8 \text{ W/m}^2 \text{ K}.$$

**Example 2.53** Evaluate thermal resistance for each case of Example 2.52.

**Solution**

Following Eq. 2.3c and Example 2.1, thermal resistance is inversely proportional to heat transfer coefficient and hence

Thermal resistance of Eq. 2.13a is

$$r_c = \frac{1}{h_c} = \frac{1}{9.5} = 0.1052 \text{ m}^2 \text{ K/W}.$$

Thermal resistance of Eq. 2.13b is

$$r_c = \frac{1}{h_c} = \frac{1}{5.8} = 0.172 \text{ m}^2 \text{ K/W}.$$

Thermal resistance of Eq. 2.13c is

$$r_c = \frac{1}{h_c} = \frac{1}{11} = 0.0909 \text{ m}^2 \text{ K/W}.$$

Thermal resistance of Eq. 2.13d is

$$r_c = \frac{1}{h_c} = \frac{1}{5.8} = 0.172 \text{ m}^2 \text{ K/W}.$$

**Example 2.54** Evaluate total thermal resistance of metallic surface exposed to wind speed of 1 m/s by using the data's of Example 2.53 and Table 2.1.

**Solution**

Let us consider the thermal resistance due to wind which varies between 0.0909 and 0.172 m$^2$ K/W (Eqs. 2.13a–2.13d). Further, the thermal resistance of all metals is of the order of $10^{-5}$ (Table 2.2) which is negligible is comparison with thermal resistance due to wind flowing over it. Hence in such condition, thermal resistance due to metal can be neglected during analysis of solar thermal systems involving as absorbing surface.

**Example 2.55** Evaluate total an overall heat transfer coefficient of metallic surface exposed to wind speed of 1 m/s by using the data's of Example 2.54 and Table 1.1.

**Solution**

Since total thermal resistance of metal surface exposed to flowing wind over, it is mainly depends on thermal resistance due to wind, Example 2.54, and hence, an overall heat transfer coefficient of such metallic surface will be heat transfer coefficient only due to wind, Example 2.54.

**Example 2.56** Evaluate total thermal resistance and an overall heat transfer coefficient of non-metallic surface exposed to wind speed of 1 m/s by using the data's of Example 2.53 and Table 2.2.

**Solution**

Let us consider the maximum thermal resistance due to wind which is 0.172 m$^2$ K/W (Example 2.53). The total thermal resistance including non-metal and wind is given in Table 2.8
    From Table 2.8, one can infer the following important points which will be helpful to analyze solar thermal systems:

(a) The glass cover material also behaves as metallic materials because only heat transfer due to wind is significant due to low thermal resistance of glass material, case (4).
(b) In the case of insulating material, case (5), the role of wind is insignificant (0.172 $\frac{\text{m}^2 \text{ K}}{\text{W}}$) due to highest value of thermal resistance (2.32558 $\frac{\text{m}^2 \text{ K}}{\text{W}}$) due to insulating material.
(c) In other non-material case, one has to consider the thermal resistance of non-material as well wind effect in determining the overall heat transfer coefficient, cases (1–3).

**Table 2.8** Thermal resistance ($r_k$) and heat transfer coefficient ($h_k$) for some non-metals

| Case | Non-metals | Thermal conductivity, K, (W/m K) | Thickness, L, (m) | Thermal resistance, $r_k$, $\left(\frac{m^2K}{W}\right)$ | Total thermal resistance | Overall heat transfer coefficient, $(W/m^2\ K)$ |
|------|-----------|------------------|-----------|----------------|------------------|------------------|
| 1 | Building brick | 0.69 | 0.30 | 0.4347 | 0.6067 | 1.648 |
| 2 | Stone 1–2–4 mix | 1.37 | 0.30 | 0.2189 | 0.3909 | 2.558 |
| 3 | Marble | 2.07–2.94 | 0.30 | 0.2803–0.1020 | 0.4523–0.2740 | 2.211–3.6496 |
| 4 | Glass window | 0.78 | 0.003 | 0.003846 | 0.1758 | 5.688 |
| 5 | Cark board/ ground | 0.043 | 0.10 | 2.32558 | 2.49758 | 0.4000 |

## 2.5  Heat Transfer by Radiation

The radiation is a transmission/propagation through space/vacuum in the form of electromagnetic waves (e/m wave) from any surface to surrounding by Plank's law radiation, Sect. 1.2.2. Thermal radiation is the electromagnetic waves of wavelength between 4–100 μm in terrestrial region. All the rules applicable to light are also obeyed by thermal radiation. Thermal radiation when strikes the solid surface it can be absorbed, reflected, or refracted according to the same rule as that of light.

**Radiation Involving Real Surfaces**

When solar radiation ($I_T$) incident on a solid surface, a part of it is reflected ($I_r$), another is absorbed ($I_a$) and the rest is transmitted ($I_t$) through it if it is transparent solid body. According to first law of thermodynamics, the conservation of energy states that the total sum ($I_r + I_a + I_t$) must be equal to the incident solar radiation ($I_T$); thus,

$$I_r + I_a + I_t = I_T. \tag{2.14a}$$

Divide the above equation by $I_T$, one gets

$$\rho + \alpha + \tau = 1, \tag{2.14b}$$

where, $\rho = \frac{I_r}{I_T}$, the ratio of the solar energy reflected to the incident solar energy is called reflectivity; $\alpha = \frac{I_a}{I_T}$, the ratio of the solar energy absorbed to the incident solar energy is known as the absorptivity and $\tau = \frac{I_t}{I_T}$, the ratio of the solar energy transmitted to the incident solar energy is referred as the transmittance of the intercepting body.

For an opaque surface, $\tau = 0$; therefore $\rho + \alpha = 1$. However, when $\rho = \tau = 0$; $\alpha = 1$, that is, the substance absorbs the whole of the solar energy incident on it. Such a substance is called the blackbody. Similarly, for a white body which reflects the whole of the radiation falling on it, $\alpha = \tau = 0$, $\rho = 1$.

The energy which is absorbed is converted into heat and this heated body, by virtue of its temperature, emits radiation according to Stefan–Boltzmann's law ($\sigma T^4$, $T$ in Kelvin, Eq. 1.1a). The radiant energy emitted per unit area of a surface in unit time is referred to as the emissive power ($E_{\lambda b}$) given by Eq. 1.1b. However, it is defined as the amount of energy emitted per second per unit area perpendicular to the radiating surface in a cone formed by a unit solid angle between the wavelengths lying in the range $d\lambda$, it is called spectral emissive power $\left(E_{0-\lambda,b}\right)$ as given in Eq. 1.2. Further, emissivity, defined as the ratio of the emissive power of a surface to the emissive power of a blackbody of the some temperature, is the fundamental property of a surface.

## 2.5.1 Laws of Thermal Radiation

Some of other thermal radiation law's has also been defined as

### Sky Radiation

In order to evaluate radiation exchange from a horizontal body directly exposed to the sky, certain equivalent blackbody sky temperature is defined. This accounts for the fact that the atmosphere is not at a uniform temperature and that it radiates only in certain wavelength regions. Thus, the net radiation exchange between horizontal surface ($T_1$) with emittance ($\varepsilon$) and area ($A$) and sky temperature, $T_{sky}$, is given by

$$\dot{Q} = A\varepsilon\sigma\left(T_1^4 - T_{sky}^4\right). \tag{2.15}$$

In order to express the equivalent sky temperature $T_{sky}$, in terms of ambient air temperature, various expressions have been given by different people. These relations, although simple to use, are only approximations. The sky temperature, which is always less than ambient air temperature, can be expressed in terms of local air temperature as follows:

$$T_{sky} = 0.0552 T_a^{1.5}. \tag{2.16a}$$

where, $T_{sky}$, $T_1$ and $T_a$ are both in degrees Kelvin.
Another commonly used relation is given as

$$T_{sky} = T_a - 6 \tag{2.16b}$$

or,

$$T_{sky} = T_a - 12. \tag{2.16c}$$

***Example 2.57*** Evaluate sky temperature by using Eqs. 2.16a–2.16c for an ambient air temperature of 25 °C.

**Solution**

From Eq. 2.16a, we have

$$T_{sky} = 0.0552T_a^{1.5} = 0.0552 \times (273 + 25)^{1.5} = 283.96 \text{ K} = 10.96\,°C.$$

From Eq. 2.16b, we have

$$T_{sky} = T_a - 6 = 25 - 6 = 19\,°C.$$

From Eq. 2.16c, we have

$$T_{sky} = T_a - 12 = 25 - 12 = 13\,°C.$$

Equations 2.16a and 2.16c are very close to each other, and hence Eq. 2.16c is very simple to be used in thermal modeling of any solar systems.

### 2.5.2 Heat Transfer Coefficient Due to Radiation

The radiant heat exchange between two infinite parallel surfaces per $m^2$ ($A = 1 \text{ m}^2$) at temperatures $T_1$ and $T_{sky}$ may be given from Eq. 2.15 as,

$$\dot{q}_r = \varepsilon\sigma\left[T_{x=0}^4 - T_{sky}^4\right], \text{W/m}^2. \tag{2.17a}$$

The above equation may be rewritten as,

$$\dot{q}_r = \varepsilon\sigma\left(T_1^4 - T_a^4\right) + \varepsilon\sigma\left(T_a^4 - T_{sky}^4\right) \tag{2.17b}$$

or,

$$\dot{q}_r = \varepsilon\sigma\left(T_1^4 - T_a^4\right) + \varepsilon\Delta R \tag{2.17c}$$

where, $\Delta R = \sigma\left[(T_a + 273)^4 - (T_{sky} + 273)^4\right]$ is the difference between the long wavelength radiation exchange between ambient air temperature at $T_a$ to the sky temperature at $T_{sky}$. Since $T_a$ and $T_{sky}$ are at low temperature and hence according to Wein's displacement law, Eq. 1.4, the emitted radiation will be long wavelength radiation which is blocked by atmosphere.

Further, after linearization of first term of Eq. 2.17c, one can have

$$\dot{q}_r = \frac{\varepsilon\sigma\left(T_1^4 - T_a^4\right)}{(T_1 - T_a)}(T_1 - T_a) + \varepsilon\Delta R = h_r(T_1 - T_a) + \varepsilon\Delta R \qquad (2.17d)$$

where,

$$h_r = \frac{\varepsilon\sigma\left(T_1^4 - T_a^4\right)}{(T_1 - T_a)} = \varepsilon\sigma\left(T_1^2 + T_2^2\right)(T_1 + T_2) = \varepsilon \times 4\sigma\left(\overline{T}\right)^3 \quad \text{for } \overline{T} = \frac{T_1 + T_2}{2}.$$
$$(2.17e)$$

Here, the temperature is in Kelvin $T = 273 +$ temperature in °C. Even there is small difference of temperature between two surfaces, it will not affect much the value of $h_r$.

It is necessary to discuss here that the numerical value of $\Delta R$ becomes zero for the surfaces not directly exposed to sky condition for example as glazed solar system.

Equation 2.17a can also be linearized in terms of radiative heat transfer coefficient $(h_r)$ as

$$\dot{q}_r = \left\{\frac{\varepsilon\sigma\left[T_{x=0}^4 - T_{sky}^4\right]}{(T_{x=0} - T_a)}\right\}(T_{x=0} - T_a) = h_r(T_{x=0} - T_a), \text{ W/m}^2 \qquad (2.17f)$$

where, $h_r = \left\{\frac{\varepsilon\sigma\left[T_{x=0}^4 - T_{sky}^4\right]}{(T_{x=0} - T_a)}\right\}.$

**Example 2.58** Determine the rate of long wavelength radiation exchange $(\Delta R)$ between the ambient air $(T_a = 25\,°\text{C})$ and sky temperature (Eq. 2.17c) for hot climatic condition.

**Solution**

From Eq. 2.17c, we can have an expression for $\Delta R$ as follows:

$$\Delta R = \sigma\left[(T_a + 273)^4 - \left(T_{sky} + 273\right)^4\right]$$

From Eq. 2.18, we have

$$T_{sky} = 0.0552(25)^{1.5} = 10.96 \approx 11\,°\text{C}$$
$$= 25 - 6 = 19\,°\text{C}$$
$$= 25 - 12 = 13\,°\text{C}$$

Now,

$$\Delta R = 5.67 \times 10^{-8}\left[(25 + 273)^4 - (11 + 273)^4\right] = 78.28 \text{ W/m}^2$$
$$= 5.67 \times 10^{-8}\left[(25 + 273)^4 - (19.0 + 273)^4\right] = 34.93 \text{ W/m}^2$$
$$= 5.67 \times 10^{-8}\left[(25 + 273)^4 - (13.0 + 273)^4\right] = 67.78 \text{ W/m}^2$$

In this case too, $\Delta R = 78.28$ and $67.78$ W/m$^2$ are very close to each other in comparison with $\Delta R = 34.93$ W/m$^2$.

Here, it is important to mention that the value of $T_{sky}$ is nearly same for two cases hence the numerical value of average $\Delta R$ should be considered as $73.03 \approx 70$ W/m$^2$ for hot climatic condition.

***Example 2.59*** Determine the rate of long wavelength radiation exchange ($\Delta R$) between the ambient air ($T_a = 15\,°C$) and sky temperature (Eq. 2.17c) for cold climatic condition.

**Solution**

From Eq. 2.17c, we can have an expression for $\Delta R$ as follows:

$$\Delta R = \sigma\left[(T_a + 273)^4 - \left(T_{sky} + 273\right)^4\right].$$

From Eq. 2.18, we have

$$T_{sky} = 0.0552(15)^{1.5} = 3.2\,°C$$
$$= 15 - 6 = 9\,°C$$
$$= 15 - 12 = 3\,°C$$

Now,
$$\Delta R = 5.67 \times 10^{-8}\left[(15 + 273)^4 - (3.2 + 273)^4\right] = 60.10 \text{ W/m}^2$$
$$= 5.67 \times 10^{-8}\left[(15 + 273)^4 - (9.0 + 273)^4\right] = 35.85 \text{ W/m}^2.$$
$$= 5.67 \times 10^{-8}\left[(15 + 273)^4 - (3.0 + 273)^4\right] = 61.06 \text{ W/m}^2$$

In this case too, $\Delta R = 60.10$ and $61.06$ W/m$^2$ are very close to each other in comparison with $\Delta R = 35.85$ W/m$^2$ as per Example 2.57.

Here, it is also important to mention that the value of $T_{sky}$ is nearly same for two cases; hence the numerical value of average $\Delta R$ should be considered as $60.55 \approx 60$ W/m$^2$ for cold climatic condition.

***Example 2.60*** Find out the radiative heat transfer coefficient between the surface of a wall at $25\,°C$ and room air temperature at $T_a = 24\,°C$.

**Solution**

From Eq. 2.17e, we have

$$h_r = \varepsilon\sigma\left(T_1^2 + T_2^2\right)(T_1 + T_2) = 0.9 \times 5.67 \times 10^{-8}$$
$$\times\left[(273 + 25)^2 + (273 + 24)^2\right](595) = 5.37\,\text{W/m}^2\,°\text{C}$$

**Example 2.61** Find out the radiative heat transfer coefficient between the surface of a wall at $25\,°\text{C}$ and room air temperature at $T_a = 24\,°\text{C}$.

**Solution:**

Since, the temperatures are approximately same, hence from Eq. 2.17e, we have,

$$h_r = 4\varepsilon\sigma T^3 = 4 \times 0.9 \times 5.67 \times 10^{-8} \times (24.5 + 273)^3 = 5.37\,\text{W/m}^2\,°\text{C}$$

So, the radiative heat transfer coefficient is same as 5.37 by both method in Examples 2.60 and 2.61, respectively.

**Example 2.62** Find out the radiative heat transfer coefficient between the blackened hot surface at $80\,°\text{C}$ and exposed to ambient air temperature at $T_a = 50\,°\text{C}$.

**Solution**

From Eq. 2.17e, we have to use the following expression

$$h_r = \varepsilon\sigma\left(T_1^2 + T_2^2\right)(T_1 + T_2) = 0.9 \times 5.67 \times 10^{-8}$$
$$\times\left[(273 + 80)^2 + (273 + 50)^2\right](676) = 6.35\,\text{W/m}^2\,°\text{C}.$$

One can see in Examples 2.60–2.61 that there is not much difference between radiative heat transfer coefficient by changing operating temperature of the surface.

**Example 2.63** Estimate the rate of heat transfer in $\text{W/m}^2$ for Examples 2.60–2.62.

**Solution**

The rate of heat transfer in $\text{W/m}^2$ is given by

$$\dot{q}_r = h_r(T_1 - T_2).$$

For Examples 2.60–2.61,

$$\dot{q}_r = 5.37(25 - 24) = 5.37\,\text{W/m}^2.$$

For Example 2.62,

$$\dot{q}_r = 6.35(80 - 50) = 190.5\,\text{W/m}^2.$$

So, there is big rise in the rate of heat transfer in $\text{W/m}^2$.

## 2.6  Heat Transfer by Evaporation (Mass Transfer)

In addition to basic three heat transfer by conduction (Sect. 2.3), convection (Sect. 2.4) and radiation (Sect. 2.5) from solid surface, there is another fourth heat transfer from free water surface/free water molecules in the plant which is known as heat transfer due to evaporation/respiration. In this case, the one surface will be wetted surface for mass transfer to the either surrounding or any cooler surface unlike other cases discussed earlier.

For free water/wetted surface too, the convective heat transfer will always be present. The rate of heat transferred in this case is given as:

$$\dot{Q} = h_{cw}(T_w - T_1), \tag{2.18}$$

where, $T_w$ is free water/wetted surface (fluid) temperature and $T_1$ is another surface temperature kept above free water surface.

According to Lewis relation at interface of water and moist air (water vapor mixture), there is a relation between convective heat transfer coefficient and mass transfer as follows:

$$\frac{h_{cw}}{h_D} = \rho^0 C_{pa}. \tag{2.19}$$

By assuming Eq. 2.19, Malik et al. [4] have derived the following expression for heat transfer by evaporation in the case of solar distillation for 100% relative humidity above the water free surface

$$\dot{q}_{ew} = 16.276 \times 10^{-3} \times h_{cw} \times (P_w - P_1). \tag{2.20}$$

In Eq. 2.20, $P_w$ and $P_1$ are partial water vapor pressure at free water ($T_w$) and another surface ($T_1$) temperature, respectively. The values of partial vapor pressure at temperature $T$ for the ranges of temperature (10–90°C) can be obtained from the following expression [43]

$$P(T) = \exp\left[25.317 - \frac{5114}{T + 273}\right].$$

If the surface is exposed to atmosphere, then the Eq. 2.20 becomes

$$\dot{q}_{ew} = 16.276 \times 10^{-3} \times h_{cw} \times (P_w - \gamma P_a), \tag{2.21}$$

where, $\gamma$ is the relative humidity of air in fraction and Pa is partial vapor pressure at ambient air temperature.

Further, after linearization of Eqs. 2.20 and 2.21, one can have

$$\dot{q}_{ew} = h_{ew}(T_w - T_1) \quad \text{for closed water cycle} \tag{2.22a}$$

and,

$$\dot{q}_{ew} = h_{ew}(T_w - T_a) \quad \text{for open water cycle} \tag{2.22b}$$

where, the evaporation heat transfer coefficient for two cases, namely closed and open water cycle, will be as follows:

$$h_{ew} = \frac{16.276 \times 10^{-3} \times h_{cw} \times \left(\overline{P}_w - \overline{P}_1\right)}{(T_w - T_a)} \tag{2.23a}$$

and,

$$h_{ew} = \frac{16.276 \times 10^{-3} \times h_{cw} \times \left(\overline{P}_w - \gamma \overline{P}_a\right)}{(T_w - T_a)}. \tag{2.23b}$$

For wetted surface which is exposed to ambient moving air with velocity '$V$' then the numerical value of $h_{cw}$ can be considered as given by Eq. 2.13.

***Example 2.64*** Evaluate the evaporative heat transfer coefficient in $W/m^2\,°C$ and the rate of evaporative heat transfer in $W/m^2$ from free water surface $(T_w)$ at 35 °C to another condensing plate $(T_1)$ at 15 °C in a closed loop condition.

**Solution**

In a closed loop condition, the evaporation took place in solar distillation system with 100% relative humidity $(\gamma)$. The convective heat transfer coefficient $h_{cw} = 2.8\,W/m^2\,K$ for $V = 0$ from Eqs. 2.13b and 2.13d can be considered.

So, the evaporative heat transfer coefficient in $W/m^2\,°C$ can be evaluated from Eq. 2.23a as

$$\begin{aligned} h_{ew} &= \frac{16.276 \times 10^{-3} \times h_{cw} \times \left(\overline{P}_w - \overline{P}_1\right)}{(T_w - T_a)} \\ &= \frac{0.016276 \times 2.8 \times [5517.62 - 1730.02]}{35 - 15} = 8.63\,W/m^2\,°C. \end{aligned}$$

with $\overline{P}_w = \exp\left[25.317 - \frac{5114}{35+273}\right] = 5517.62$ and $\overline{P}_w = \exp\left[25.317 - \frac{5114}{15+273}\right] = 1730.02$.

So, the rate of evaporative heat transfer in $W/m^2$ for closed loop becomes

$$\dot{q}_{ew} = h_{ew}(T_w - T_1) = 8.63 \times (35 - 15) = 172.6\,W/m^2.$$

***Example 2.65*** Evaluate the evaporative heat transfer coefficient in $W/m^2\,°C$ and the rate of evaporative heat transfer in $W/m^2$ from free water surface $(T_w)$ at 35 °C to an ambient air temperature $(T_a)$ at 15 °C with a relative humidity of 50% in open condition.

**Solution**

The vapor pressure $P(T)$ in N/m$^2$ at any temperature $T$ (in °C) can be calculated from Eq. 3.48.

Thus, the vapor pressures at wetted and ambient air temperatures can be calculated as,

$$\overline{P}_w = \exp\left[25.317 - \frac{5114}{35 + 273}\right] = 5517.62 \text{ and}$$

$$\overline{P}_w = \exp\left[25.317 - \frac{5114}{15 + 273}\right] = 1730.02.$$

Using $h_c = 2.8$ W/m$^2$ °C (Eqs. 2.13b and 2.13d for $V = 0$) and substituting the values in Eq. 2.23b, we have the rate of evaporation in W/m$^2$ as:

$$\dot{q}_{ew} = 16.273 \times 10^{-3} \times 2.8 \times (5517.6 - 0.5 \times 1730) = 211.99 \text{ W/m}^2.$$

The evaporative heat transfer coefficient can be calculated as

$$h_{ew} = \frac{\dot{q}_{ew}}{(T_w - T_a)} = \frac{211.99}{35 - 15} = 10.60 \text{ W/m}^2 \text{ °C}.$$

***Example 2.66*** Repeat Example 2.65 for $\gamma = 0.9$

**Solution**

The rate of evaporation in W/m$^2$ as:

$$\dot{q}_{ew} = 16.273 \times 10^{-3} \times 2.8 \times (5517.6 - 0.9 \times 1730) = 180.46 \text{ W/m}^2.$$

The evaporative heat transfer coefficient can be calculated as

$$h_{ew} = \frac{\dot{q}_{ew}}{(T_w - T_a)} = \frac{180.46}{35 - 15} = 9.02 \text{ W/m}^2 \text{ °C}.$$

From Examples 2.64 to 2.66, one can observe the importance of relative humidity on the rate of evaporation.

***Example 2.67*** Evaluate the evaporative heat transfer coefficient in W/m$^2$ °C and the rate of evaporative heat transfer in W/m$^2$ from free moving water surface ($T_w$) at 35 °C with velocity at 0.5 m/s over the surface area of 1 m × 1 m to an ambient air temperature ($T_a$) at 15 °C with a relative humidity of 50% in open condition.

**Solution**

The physical properties of air at 25 °C (Appendix Table E.1) are $\rho = 1.1774$ kg/m$^3$, $C_p = 1.0050$ J/kg °C, $\mu = 1.983 \times 10^{-5}$ kg/m-s, $v = 15.68 \times 10^{-6}$ (m$^2$/s), $K \times 10^{-3} = 26.22 \times 10^{-3}$ W/m$^2$ K, Pr $= 0.708$.

Now, Reynolds number is obtained as

$$\text{Reynolds number, Re} = \frac{u_0 L}{\nu} = \frac{0.5 \times 1}{15.68 \times 10^{-6}} = 0.0316 \times 10^6 = 31{,}600$$

This is laminar flow and hence case (1) of Table 2.4 convective heat transfer coefficient can be calculated as follows:

$$h_c = 1.32(\Delta T/L)^{1/4} = 1.32\left[\frac{20}{1}\right]^{1/4} = 2.79 \, \text{W/m}^2\,{}^\circ\text{C}.$$

From Example 2.65, we have the rate of evaporation in W/m² as:

$$\dot{q}_{ew} = 16.273 \times 10^{-3} \times 2.79 \times (5517.6 - 0.5 \times 1730) = 211.23 \, \text{W/m}^2.$$

The evaporative heat transfer coefficient can be calculated as

$$h_{ew} = \frac{\dot{q}_{ew}}{(T_w - T_a)} = \frac{211.23}{35 - 15} = 10.56 \, \text{W/m}^2\,{}^\circ\text{C}.$$

Example 2.68 Repeat Example 2.68 with water flowing with velocity of 5 ms.

## 2.7  Total Heat Transfer Coefficient

The **total heat transfer per m²** $(\dot{q}_T)$ from any solid surface having either stationary/ flowing water to surrounding moist will be sum of heat transfer by conduction, Eq. 2.2e; natural (Eq. 2.7)/forced (2.11b) convection; radiation, Eq. 2.17e and evap- oration, Eqs. 2.23a and 2.23b. Since thermal conductivity of air is very small (Appendix Table E.1, $K \approx 0.033$ W/m °C), conductive heat transfer from any wetted solid surface to air can be neglected. So, the total rate of heat transfer from any solid surface can be expressed as follows:

$$\dot{q}_T = \dot{q}_c + \dot{q}_r + \dot{q}_{ew}. \tag{2.24}$$

The above equation can also be written in terms of respective heat transfer coefficient as

$$\dot{q}_T = (h_c + h_r + h_{ew})[T_s - T_a] \tag{2.25a}$$

or,

$$\dot{q}_T = h_1[T_s - T_a] \tag{2.25b}$$

where,

$$h_1 = h_c + h_r + h_{ew}. \tag{2.25c}$$

An expression for $h_c$, $h_r$ and $h_{ew}$ are given by Eqs. 2.17e/2.11b, 2.17e, 2.23a and 2.23b, respectively. It is important to mention here that $T_s$ represents the temperature of wetted solid surface as required in the analysis.

**Example 2.68** Evaluate the total heat transfer coefficient in W/m² °C and the rate of evaporative heat transfer in W/m² from free moving water surface ($T_w$) at 35 °C with velocity at 0.5 m/s over the surface area of 1 m × 1 m to an ambient air temperature ($T_a$) at 15 °C with a relative humidity of 10, 50, and 90% in open condition.

**Solution**:

For radiative heat transfer coefficient, we have to use the following expression from Eq. 2.17e as

$$h_r = \varepsilon\sigma\left(T_1^2 + T_2^2\right)(T_1 + T_2) = 0.9 \times 5.67 \times 10^{-8}$$
$$\times \left[(273 + 35)^2 + (273 + 15)^2\right](596) = 5.41 \, \text{W/m}^2 \, ^{\circ}\text{C}.$$

From Example 2.67, we have convective heat transfer coefficient

$$h_c = 2.79 \, \text{W/m}^2 \, ^{\circ}\text{C}.$$

From Example 2.65, we have the rate of evaporation in W/m² and total heat transfer coefficient in W/m² °C as:

(a) **For relative humidity of 10%**

$$\dot{q}_{ew} = 16.273 \times 10^{-3} \times 2.79 \times (5517.6 - 0.10 \times 1730)242.65 \, \text{W/m}^2.$$

The evaporative heat transfer coefficient can be calculated as

$$h_{ew} = \frac{\dot{q}_{ew}}{(T_w - T_a)} = \frac{242.65}{35 - 15} = 12.13 \, \text{W/m}^2 \, ^{\circ}\text{C}.$$

So, the total heat transfer coefficient $h_1$ from Eqs. 2.22a and 2.22b is

$$h_1 = h_r + h_c + h_e = 5.41 + 2.79 + 12.13 = 20.33 \, \text{W/m}^2 \, ^{\circ}\text{C}.$$

(b) **For relative humidity of 50%**

$$\dot{q}_{ew} = 16.273 \times 10^{-3} \times 2.79 \times (5517.6 - 0.5 \times 1730) = 211.23 \, \text{W/m}^2$$

The evaporative heat transfer coefficient can be calculated as

$$h_{ew} = \frac{\dot{q}_{ew}}{(T_w - T_a)} = \frac{211.23}{35 - 15} = 10.56\,\text{W/m}^2\,{}^\circ\text{C}.$$

So, the total heat transfer coefficient $h_1$ from Eqs. 2.22a and 2.22b is

$$h_1 = h_r + h_c + h_e = 5.41 + 2.79 + 10.56 = 18.76\,\text{W/m}^2\,{}^\circ\text{C}.$$

(c)  **For relative humidity of 90%**

The rate of evaporation in $\text{W/m}^2$ as:

$$\dot{q}_{ew} = 16.273 \times 10^{-3} \times 2.79 \times (5517.6 - 0.9 \times 1730) = 180.46\,\text{W/m}^2.$$

The evaporative heat transfer coefficient can be calculated as

$$h_{ew} = \frac{\dot{q}_{ew}}{(T_w - T_a)} = \frac{180.46}{35 - 15} = 9.02\,\text{W/m}^2\,{}^\circ\text{C}.$$

So, the total heat transfer coefficient $h_1$ from Eqs. 2.22a and 2.22b is

$$h_1 = h_r + h_c + h_e = 5.41 + 2.79 + 9.02 = 17.22\,\text{W/m}^2\,{}^\circ\text{C}.$$

In this example, one can notice that the total rate of heat transfer from water surface to moving ambient air is maximum for lowest relative humidity which generally happens in summer condition. Only due to this reason, maximum water evaporation occurs during summer period and then precipitation (rainfall) occurs after that in monsoon season.

## 2.8  Overall Heat Transfer Coefficient

The rate of heat transfer in $\text{W/m}^2$ from a medium generally fluid (air/water) at higher temperature to another media may be either air or fluid at lower temperature may occur through many solid with and without air cavity layers with different thermo-physical properties and different thicknesses. Also there may be more than one mode of heat transfer at interfaces namely conduction, convection, radiation and evaporation involved in the process. In such cases, the concept of overall heat transfer coefficient ($U$) is adopted for the evaluation of heat transfer coefficient from one medium to second medium via a third medium.

In such cases, the thermal resistance ($r$) of each medium, which is inverse of heat transfer coefficient ($\frac{1}{h}$), Eq. 2.2c, should be evaluated . Further, the total thermal resistance of whole configuration should be calculated. The inverse of total thermal resistance is known as an overall heat transfer coefficient. It is represented by 'U'.

### (A)  An overall heat transfer coefficient for N-parallel slabs [1, 2]

Consider a composite wall consists of $N$-layers of solid having thermal conductivity of $K_j$ with thickness of $L_j$, The $j$ varies from 1 to $N$. The thermal resistance of each layer is $r_j = \frac{L_j}{K_j}$. The total thermal resistance can be mathematically expressed as

$$R = \sum_{j=1}^{j=N} r_j = \left[\sum_{j=1}^{j=N} \frac{L_j}{K_j}\right] = \left[\frac{L_1}{K_1} + \frac{L_2}{K_2} + \frac{L_3}{K_3} + \cdots + \frac{L_N}{K_N}\right]. \qquad (2.26a)$$

If one surface of $N$-parallel slab is in contact of outside air and another last layer is in contact of inside air of building, then the total thermal resistance of $N$-parallel slab will be as follows:

$$R = \left[\frac{1}{h_o} + \frac{L_1}{K_1} + \frac{L_2}{K_2} + \frac{L_3}{K_3} + \cdots + \frac{L_N}{K_N} + \frac{1}{h_i}\right], \qquad (2.26b)$$

where, $h_o$ and $h_i$ are outside and inside heat transfer coefficient of a room of building. It can be sum of both convective and radiative heat transfer coefficient. If outside surface is wetted, then $h_o$ will be total heat transfer coefficient $h_1$, given by Eq. 2.25c.

The inverse of Eqs. 2.26a and 2.26b are known as conductive overall heat transfer coefficient $(U)$, and it can be written as follows:

$$U = \frac{1}{R} = \left[\frac{L_1}{K_1} + \frac{L_2}{K_2} + \frac{L_3}{K_3} + \cdots + \frac{L_N}{K_N}\right]^{-1} \qquad (2.27a)$$

and,

$$U = \frac{1}{R} = \left[\frac{1}{h_o} + \frac{L_1}{K_1} + \frac{L_2}{K_2} + \frac{L_3}{K_3} + \cdots + \frac{L_N}{K_N} + \frac{1}{h_i}\right]^{-1}. \qquad (2.27b)$$

Equation 2.27 is used to design of multilayer roof/walls of a building.

***Example 2.69*** Find out an overall heat transfer coefficient in W/m$^2$ °C of a slab shown in Fig. 2.1, having outside air temperature $(T_A)$ at 35 °C and the surface area of 1 m × 1 m and other surface is exposed to room air temperature $(T_B)$ of 20 °C. The thickness and thermal conductivity of three layer of slab, Appendix Table E.5, are as follows:

(a)  Brick [$K_1 = 0.84$ W/m K and $L_1 = 0.10$ m]
(b)  Lime stone [$K_1 = 1.5$ W/m K and $L_1 = 0.30$ m]
(c)  Cork board brick [$K_1 = 0.04$ W/m K and $L_1 = 0.05$ m]

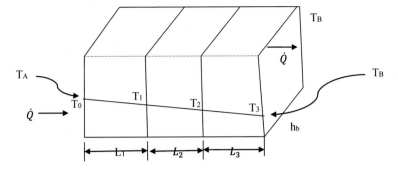

**Fig. 2.1** One-dimensional heat flow through parallel perfect contact slabs

**Solution**

From Example 2.68: $h_r = 5.41 \, \text{W/m}^2 \, ^\circ\text{C}$ and $h_c = 2.79 \, \text{W/m}^2 \, ^\circ\text{C}$ then

$$h_o = h_r + h_c = 5.41 + 2.79 = 8.2 \, \text{W/m}^2 \, ^\circ\text{C}$$

and,

$h_i = h_c = 2.79 \, \text{W/m}^2 \, ^\circ\text{C}$ due to inside surface exposed to $20 \, ^\circ\text{C}$ and convective heat transfer coefficient at $20 \, ^\circ\text{C}$ is $2.79 \, \text{W/m}^2 \, ^\circ\text{C}$, Example 2.68.

Now, an overall heat transfer coefficient ($U$) is given by

$$U = \frac{1}{R} = \left[ \frac{1}{h_o} + \frac{L_1}{K_1} + \frac{L_2}{K_2} + \frac{L_3}{K_3} + \frac{1}{h_i} \right]^{-1}$$

$$= \left[ \frac{1}{8.2} + \frac{0.10}{0.84} + \frac{0.30}{1.5} + \frac{0.05}{0.04} + \frac{1}{2.79_i} \right]^{-1}.$$

$$= [0.1219 + 0.1190 + 0.20 + 1.25 + 0.3584]^{-1}$$

$$= [2.049]^{-1} = 0.499 \, \text{W/m}^2 \, ^\circ\text{C}.$$

If there is no cork board, then

$$U = \frac{1}{R} = \left[ \frac{1}{h_o} + \frac{L_1}{K_1} + \frac{L_2}{K_2} + \frac{1}{h_i} \right]^{-1} = \left[ \frac{1}{8.2} + \frac{0.10}{0.84} + \frac{0.30}{1.5} + \frac{1}{2.79_i} \right]^{-1}$$

$$= [0.1219 + 0.1190 + 0.20 + 0.3584]^{-1} = [0.799]^{-1} = 1.25 \, \text{W/m}^2 \, ^\circ\text{C}.$$

So, one can observe that the use of cork board of thickness of 5 cm reduces an overall heat transfer coefficient from 1.25 to 0.499 W/m$^2$ °C by a factor of $\frac{1}{4}$. The use of cork board is used for reducing the heat loss from the inside room of a building to outside.

***Example 2.70*** Evaluate the rate of heat transfer in $W$ for Example 2.60 with and without cork board.

**Solution**

The rate of heat transfer in $W(\dot{Q}) = U A(T_A - T_B)$

$$= 1.25 \times 1 \times (35 - 20) = 18.75 \text{ W, without cork board}$$
$$= 0.499 \times 1 \times (35 - 20) = 7.485 \text{ W, with cork board.}$$

***Example 2.71*** Evaluate an overall heat transfer coefficient in $W/m^2 \, ^{\circ}C$ and the rate of heat transfer in $W$ for Example 2.70 for outside wetted surface with 10% relative humidity.

**Solution**

From Example 2.69, $h_1 = h_o = h_r + h_c + h_e = 5.41 + 2.79 + 12.13 = 20.33 \, W/m^2 \, ^{\circ}C$.

From Example 2.70, we have

$$U = \frac{1}{R} = \left[ \frac{1}{h_o} + \frac{L_1}{K_1} + \frac{L_2}{K_2} + \frac{L_3}{K_3} + \frac{1}{h_i} \right]^{-1}$$
$$= \left[ \frac{1}{20.33} + \frac{0.10}{0.84} + \frac{0.30}{1.5} + \frac{0.05}{0.04} + \frac{1}{2.79_i} \right]^{-1}$$

or,

$$[0.049 + 0.1190 + 0.20 + 1.25 + 0.3584]^{-1} = [1.976]^{-1} = 0.506 \, W/m^2 \, ^{\circ}C.$$

The rate of heat transfer in $W \, (\dot{Q}) = U A(T_A - T_B)$

$$= 0.506 \times 1 \times (35 - 20) = 7.59 \, W.$$

**(B)  An overall heat transfer coefficient for *N*-parallel slabs with $(N - 1)$ air gap/cavity [1, 2]**

The roof/wall of a building can have air cavity as shown in Fig. 2.2. It reduces the rate of heat gain/loss from inside room air to outside either for heating or cooling. In this case, the expression for an overall heat transfer coefficient $(U)$ includes the effect of air conductance $(C, \, W/m^2 \, ^{\circ}C)$. The air conductance, $C$, is equivalent to heat transfer coefficient used in determining the rate of heat flow through solid. The value of air conductance $(C)$ varies nonlinearly with the thickness of the air gap, and it becomes constant at larger air gaps as shown in Fig. 2.3. The inverse of air conductance is thermal resistance.

If one considers *N*-solid layers of different materials having air gaps between two solid layers, then the total thermal resistance can be written as

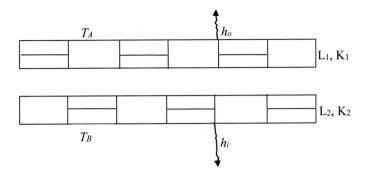

**Fig. 2.2** Configuration of parallel slabs with air cavity

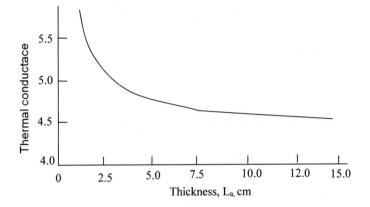

**Fig. 2.3** Variation of thermal air conductance in W/m$^2$ °C with air gap thickness in cm

$$R = \left[ \frac{1}{h_o} + \frac{L_1}{K_1} + \frac{1}{C_1} + \frac{L_2}{K_2} + \frac{1}{C_n} + \frac{L_3}{K_3} + \frac{1}{C_3} + \cdots + \frac{1}{C_{N-1}} + \frac{L_N}{K_N} + \frac{1}{h_i} \right].$$
(2.28a)

The inverse of Eq. 2.28b will be an overall heat transfer coefficient for $N$-layers of solid materials with $(N - 1)$ air gap

$$U = \left[ \frac{1}{h_o} + \frac{L_1}{K_1} + \frac{1}{C_1} + \frac{L_2}{K_2} + \frac{1}{C_n} + \frac{L_3}{K_3} + \frac{1}{C_3} + \cdots + \frac{1}{C_{N-1}} + \frac{L_N}{K_N} + \frac{1}{h_i} \right]^{-1}$$
(2.28b)

***Example 2.72*** Find out an overall heat transfer coefficient in W/m$^2$ °C of a two brick slabs of thickness 0.10 m and thermal conductivity of 0.84 W/m K with air cavity of thickness 5 cm shown in Fig. 2.2, having outside air temperature ($T_A$) at 35 °C and the surface area of 1 m × 1 m to an ambient air temperature ($T_A$) at 15 °C and other surface is exposed to room air temperature ($T_B$) of 20 °C.

**Solution**

From Example 2.69: $h_o = 8.2 \, \text{W/m}^2 \, °\text{C}$ and $h_i = h_c = 2.79 \, \text{W/m}^2 \, °\text{C}$.
    Here $K_1 = K_2 = 0.84 \, \text{W/m K}$; $L_1 = L_2 = 0.10 \, \text{m}$ and $C = 4.75 \, \text{W/m}^2 \, °\text{C}$, Fig. 2.3.
    Now, an overall heat transfer coefficient $(U)$ for two slabs with air cavity can be written from Eq. 2.28b as

$$U = \left[ \frac{1}{h_o} + \frac{L_1}{K_1} + \frac{1}{C} + \frac{L_2}{K_2} + \frac{1}{h_i} \right]^{-1} = \left[ \frac{1}{8.2} + \frac{0.10}{0.84} + \frac{1}{4.75} + \frac{0.10}{0.84} + \frac{1}{2.79_i} \right]^{-1}$$

$$= [0.1219 + 0.1190 + 0.21 + 0.1190 + 0.3584]^{-1}$$

$$= [0.9283]^{-1} = 1.077 \, \text{W/m}^2 \, °\text{C}.$$

Further, an overall heat transfer coefficient $(U)$ for two slabs without air cavity can be written from Eq. 2.28b as

$$U = \left[ \frac{1}{h_o} + \frac{L_1}{K_1} + \frac{L_2}{K_2} + \frac{1}{h_i} \right]^{-1} = \left[ \frac{1}{8.2} + \frac{0.10}{0.84} + \frac{0.10}{0.84} + \frac{1}{2.79_i} \right]^{-1}$$

$$= [0.1219 + 0.1190 + 0.21 + 0.1190 + 0.3584]^{-1}$$

$$= [0.7183]^{-1} = 1.39 \, \text{W/m}^2 \, °\text{C}.$$

One can see that introduction of air cavity reduces an overall heat transfer coefficient.

***Example 2.73*** Calculate the overall heat transfer coefficient $(U)$ for

(a)  Single concrete $(K = 0.72 \, \text{W/m} \, °\text{C})$ slab with thickness $(L)$ of $0.10 \, \text{m}$
(b)  Two-layered horizontal slab with same material and thickness
(c)  Two-layered horizontal slab with air cavity $(0.05 \, \text{m})$
(d)  Two-layered horizontal slab with two air cavity (each $0.05 \, \text{m}$ air gap) and

for the following parameters:

$h_o = 9.5 \, \text{W/m}^2 \, °\text{C}$, $L_1 = L_2 = 0.05$, $K_1 = K_2 = 0.72 \, \text{W/m} \, °\text{C}$; $C_1 = C_2 = 4.75 \, \text{W/m}^2 \, °\text{C}$ for $0.05 \, \text{m}$ air cavity and $h_i = 5.7 \, \text{W/m}^2 \, °\text{C}$.

**Solution**

From Eq. 2.27b, one can evaluate an overall heat transfer coefficient in $\text{W/m}^2 \, °\text{C}$ as follows:

(a)  For single concrete slab,

$$U = \left[ \frac{1}{9.5} + \frac{0.10}{0.72} + \frac{1}{5.7} \right]^{-1} = 2.38 \, \text{W/m}^2 \, °\text{C}.$$

(b) For two-layered horizontal slab,

$$U = \left[ \frac{1}{9.5} + \frac{0.05}{0.72} + \frac{0.05}{0.72} + \frac{1}{5.7} \right]^{-1} = 2.38 \, \text{W/m}^2 \, {}^{\circ}\text{C}.$$

(c) For two-layered horizontal slab with single air cavity, Eq. 2.28b,

$$U = \left[ \frac{1}{9.5} + \frac{0.05}{0.72} + \frac{1}{4.75} + \frac{0.05}{0.72} + \frac{1}{5.7} \right]^{-1} = 1.59 \, \text{W/m}^2 \, {}^{\circ}\text{C}.$$

(d) For two-layered horizontal slab with two air cavity gaps separated by metal foil,

$$U = \left[ \frac{1}{9.5} + \frac{0.05}{0.72} + \frac{1}{4.75} + \frac{1}{4.75} + \frac{0.05}{0.72} + \frac{1}{5.7} \right]^{-1} = 1.19 \, \text{W/m}^2 \, {}^{\circ}\text{C}.$$

It is clear from above calculation that an increase of the number of air cavity from one to two reduces $U$ from to 1.59 W/m$^2$ °C to 1.19 W/m$^2$ °C, respectively.

## 2.9 The Second Law of Thermodynamics

The second law of thermodynamics gives a relation between heat supplied to the system and work extracted on application of the heat as shown in Fig. 2.4.

Further, the thermal efficiency of a heat machine/engine working between two energy levels is defined in terms of absolute temperature in Kelvin as:

$$\eta = \left[ 1 - \frac{T_c}{T_h} \right], \tag{2.29}$$

where, $\eta$ = thermal efficiency of heat engine; $T_h$ = temperature hot reservoir (K); $T_c$ = temperature cold reservoir (surrounding) (K).

**Fig. 2.4** Heat engine cycle

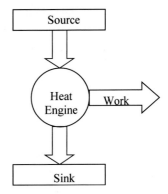

Here, it is also important to mention that the second law of thermodynamics is mostly used for higher operating temperature range ($\gg 400\,°C$).

Second law of thermodynamics also leads to the concept of **exergy** which is defined as the maximum possible work which can be extracted from the system as it reached in equilibrium with the cold reservoir/surroundings. For system in equilibrium with surroundings, exergy will be zero.

First law ensures that energy can never be destroyed which can be achieved only for reversible processes. On the contrary, exergy of the system depends on the irreversibility of the systems which is directly proportional to the increase in entropy of system in surrounding. When a process takes between two different temperatures, energy is destroyed. The destroyed energy is known as **anergy**.

### 2.9.1  Exergy Analysis

The quality of energy directly relates the exergy available from the system. The exergy analysis of any thermal process takes into account the irreversibility at every point of system and inefficiency of each components of the system. The prime objective of exergy analysis is improvement if efficiency and reduction in losses as maximum as possible. The optimization of solar thermal systems following exergy analysis will give insights on the quality of energy attainable from the system and exergy-based optimization will provide more efficient system which in turn give more return and capital saving.

**"Exergy is the property of system, which gives the maximum amount of useful work obtained from the system when it comes in equilibrium with a reference to the environment."**

Hence one can have:

Heat supplied (energy) = Available energy (exergy)
$$+ \text{ unavailable energy (anergy)} \qquad (2.30)$$

For the process taking place between two temperatures (source $T_1$ (in $K$)) and (sink $T_2$ (in $K$)), the maximum work ($W_{max}$) drawn during this process is given by:

$$W_{max} = \text{exergy} = \left(1 - \frac{T_0}{T_1}\right) \times Q_1, \qquad (2.31)$$

where, $Q_1$ is the thermal energy available at $T_1$.

***Example 2.74*** Calculate maximum work available $W_{max}$ from the heat source at $T_1 = 40, 60, 80\,°C$ and ambient temperature $= 20\,°C$ when $Q_1 = 150\,kWh$.

**Solution**

Use Eq. 2.31 for 40 °C, we have

$$W_{max} = \left(1 - \frac{20 + 273}{40 + 273}\right) \times 150 = 9.58\,\text{kWh}.$$

Similarly, for 60 and 80 °C,

$$W_{max} = 18\,\text{kWh and } 25.5\ \text{kWh, respectively.}$$

It is concluded that the maximum work is available at higher source temperature when sink temperature is constant.

### 2.9.2 Exergy Efficiency

Exergy efficiency is defined as the ratio of energy output to the exergy input. Performance of the system evaluated by incorporating exergy efficiency is based on the second law of thermodynamics unlike energy analysis which is based on the first law of thermodynamics. Exergy efficiency of the system is obtained by writing exergy balance of the system which includes exergy input, exergy output and exergy destroyed from the system.

The exergy balance for any system can be given following:

$$\sum \dot{E}x_{in} - \sum \dot{E}x_{out} = \sum \dot{E}x_{dest}. \tag{2.32a}$$

The exergy efficiency is improved by minimizing the inefficiency and irreversibility which can be achieved by minimizing the term $\sum \dot{E}x_{in} - \sum \dot{E}x_{out}$. The concept of 'improved potential (IP)' due to the improvement of exergy efficiency is directly related to the exergy efficiency and term $\sum \dot{E}x_{in} - \sum \dot{E}x_{out}$. It is defined as follows:

$$IP = (1 - \varepsilon)\left(\sum \dot{E}x_{in} - \sum \dot{E}x_{out}\right), \tag{2.32b}$$

where, $\varepsilon$ is the exergy efficiency of the system defined as: [1]

$$\varepsilon = \frac{\text{Rate of useful product energy}}{\text{rate of exergy input}} = \frac{\dot{E}x_{out}}{\dot{E}x_{in}} = 1 - \frac{\dot{E}x_{dest}}{\dot{E}x_{in}}, \tag{2.32c}$$

where, $\dot{E}x_{dest}$ is the rate of exergy destruction.

### 2.9.3  Solar Radiation Exergy

Exergy of any system is obtained due to the matter (substance) or field (radiation field, magnetic field, gravitational field, etc.) involved in the system. Exergy depends on the temperature of source if sink temperature is taken constant throughout out the process. Therefore the term exergy of heat should 'change in exergy of the heat source' [5]. Solar radiation exergy is defined as the input exergy from the sun which is available to solar thermal systems.

The solar radiations received from the sun are utilized by human being, wildlife habitat, plants and in almost all the processes on planet earth. It can be converted into work, heat or other necessary process for life on earth. The energy efficiency and exergy efficiency can be calculated from the work or heat equivalent of thermal radiations using the relations given in Table 2.9.

The unified efficiency expression ($U_{ee}$) expresses the capacity of utilization of thermal radiation. Unified efficiency expression ($U_{ee}$) for different input and output is given in Table 2.10. The solar radiation exergy can be estimated using following relation:

$$\dot{E}x_{sun} = \text{Thermal energy} \times U_{ee}. \tag{2.33}$$

If $I(t)$ is solar radiation, Chap. 1 falling on unit surface area; the energy contained

**Table 2.9**  Conversion efficiency of thermal radiation

| S. No. | Efficiency | Radiation to work conversion | Radiation to heat conversion |
|---|---|---|---|
| 1 | Energetic, $\eta_e$ | $\eta_e = W/e$ $\eta_{e\,max} = W_{max}/e$ | $\eta_e = 1 - (T_a/T)^4$ |
| 2 | Exergetic, $\eta_{ex}$* | $\eta_{ex} = W/W_{max}$ | $\eta_{ex} = W_q/W_{max}$ |

* $\eta_{ex}$ is exergetic efficiency
where, $W$ is the work performed due to utilization of the solar radiation and $W_{max}$ is the exergy of radiation

**Table 2.10**  Input, output and unified efficiency expression ($U_{ee}$) of utilization of thermal radiation

| S. No. | Researcher | Input | Output | $U_{ee}$ |
|---|---|---|---|---|
| 1 | Petela [5] | Radiation Energy | Absolute work | $1 - \frac{4}{3} \times \left(\frac{T_0}{T_s}\right) + \frac{1}{3} \times \left(\frac{T_0}{T_s}\right)^4$ |
| 2 | Battisti [7] | Radiation Energy | Useful work radiation exergy | $1 - \frac{4}{3} \times \left(\frac{T_0}{T_s}\right)$ |
| 3 | Lewis and Keoleian [8] | Heat | Net work of a heat engine | $1 - \left(\frac{T_0}{T_s}\right)$ |

where, $T_s$ and $T_0$ are surface temperature of sun and environment temperature in Kelvin at earth, respectively.

in solar radiation falling on any surface of area $A$ will be $\{I(t) \times A\}$. The equivalent exergy from the incident solar radiation is then given as [3]:

$$\dot{E}x_{sun} = \{A \times I(t)\} \times U_{ee} = \{A \times I(t)\} \times \left[1 - \frac{4}{3} \times \left(\frac{T_0}{T_s}\right) + \frac{1}{3} \times \left(\frac{T_0}{T_s}\right)^4\right]$$

(2.34)

$T_0 = $ Surrounding or ambient temperature $(K) = T_a$;
$T_s = $ Sun surface temperature $= T_{sun} = 6000\,K$.

***Example 2.75*** Calculate the unified efficiency $(U_{ee})$ and exergy by using the expression of Petela model and others when surrounding temperature $= 20\,°C$, $A = 2\,m^2$, $I(t) = 750\,W/m^2$.

**Solution**

Using Table 2.10 and Eq. 2.34, we have

(a) **By Petela model**

$$U_{ee} = \left[1 - \frac{4}{3} \times \left(\frac{20 + 273}{6000}\right) + \frac{1}{3} \times \left(\frac{20 + 273}{6000}\right)^4\right]$$
$$= [1 - 0.065 + 0.01627] = 0.9513.$$

(b) **By Battisti model**

$$U_{ee} = 1 - \frac{4}{3} \times \left(\frac{T_0}{T_s}\right) = \left[1 - \frac{4}{3} \times \left(\frac{20 + 273}{6000}\right)\right] = 1 - 0.065 = 0.935.$$

(c) **By Lewis model**

$$U_{ee} = \left[1 - \left(\frac{T_0}{T_s}\right)\right] = \left[1 - \frac{20 + 273}{6000}\right] = [1 - 0.0488] = 0.9511.$$

Now, the exergy of solar radiation.

(a) **By Petela model**

$$\dot{E}x_{sun} = 1500 \times 0.9513 = 1426\,W/m^2 \times 1.4\,kW.$$

(b) **By Battisti model**

$$\dot{E}x_{sun} = 1500 \times 0.935 = 1402.5\,W/m^2 \times 1.4\,kW.$$

(c)  **By Lewis model**

$$\dot{Ex}_{\text{sun}} = 1500 \times 0.9511 = 1426.65 \text{ W/m}^2 \times 1.4 \text{ kW}.$$

One can observe that the unified efficiency ($U_{\text{ee}}$) by Patela and Lewis model are approximately same as 0.95. This indicates that there are only 5% destruction of exergy of solar radiation. However, the exergy of solar radiation in kW is exactly same.

## 2.9.4  Thermal Exergy

It is known that the thermal energy available at medium ($\approx 100 - 150 \,°\text{C}$) and high operating temperature range ($> 300 \,°\text{C}$) is governed by first and second laws of thermodynamics, respectively. Hence, thermal energy obtained at medium ($\approx 100 - 150 \,°\text{C}$) and high operating temperature ranges ($> 300 \,°\text{C}$) is known as low and high grade energies. So, electrical energy is high grade energy due to high operating temperature range.

In the case of solar energy system, low grade energy can be converted into high grade energy by using the concept of entropy, Singh and Tiwari [9]. In this case, one can use the following expression for hourly and daily thermal exergy as follow:

(a)  **Hourly exergy**

If $\dot{m}_f$ and $C_f$ are the mass flow rate of fluid passing through solar system and its specific heat of fluid. The inlet fluid temperature at $T_{\text{fi}}$ of flowing fluid is heated up to $T_{\text{fo}}$, the outlet fluid temperature, then exergy of heated fluid will be determined, (Appendix I), from the following expression

$$\dot{Ex}_u = \dot{m}_f C_f \left[ (T_{\text{fo}} - T_{\text{fi}}) - (T_a + 273) \ln \frac{T_{\text{fo}} + 273}{T_{\text{fi}} + 273} \right]. \tag{2.35a}$$

(b)  **Daily exergy**

The daily exergy for flowing fluid through solar system will be sum of hourly exergy over sunshine hour as follows:

$$Ex_{\text{daily}} = \sum_{j=1}^{j=N} \dot{Ex}_{uj}. \tag{2.35b}$$

If $M_w$ and $C_f$ are the mass of stagnant fluid in solar system and its specific heat. Further, the stagnant fluid temperature at $T_{\text{fi}}$ is heated up to $T_{\text{fo}}$, the final temperature after some time say sunshine hours, then the daily exergy of solar system can be determined as follows:

$$Ex_{\text{daily}} = M_w C_f \left[ (T_{\text{fo}} - T_{\text{fi}}) - (T_a + 273) \ln \frac{T_{\text{fo}} + 273}{T_{\text{fi}} + 273} \right]. \qquad (2.36)$$

**Example 2.76** Evaluate hourly exergy for water mass flow rate of 0.20 kg s through flat plate collector heated from 20 to 35 °C by using solar radiation of 750 W/m$^2$.

**Solution**

**Here, ambient air temperature can be assumed to be** 20 °C and specific heat of water, $C_f$, is 4179 j/kg °C (Appendix Table E.2). Them from Eq. 2.35a, one gets

$$\dot{Ex}_u = \dot{m}_f C_f \left[ (T_{\text{fo}} - T_{\text{fi}}) - (T_a + 273) \ln \frac{T_{\text{fo}} + 273}{T_{\text{fi}} + 273} \right]$$

$$= 0.20 \times 4179 \left[ 15 - (293) \ln \frac{308}{293} \right] = 310.36 \text{ W}$$

**Example 2.77** Evaluate instantaneous exergy efficiency of Example 2.76.

**Solution**

From Example 2.76, $\dot{Ex}_u = 310.36 \text{ W/m}^2$ and $U_{\text{ee}} = 0.95$.
   Referring to Table 2.9, one has

$$\eta_{iex} = \frac{\dot{Ex}_u}{I_{ex}} = \frac{310.36}{750 \times 0.95} = \mathbf{0.43} \approx \mathbf{43\%}.$$

**Example 2.78** If water mass of 100 kg is heated from 20–70 °C during six sunshine hour during winter period with average solar radiation of 350 W/m$^2$ by flat plate collector of area 2 m$^2$, then evaluate the daily exergy of solar system.

**Solution**

Assuming $T_{fi} = T_a = 20$ °C and $C_f = 4179$ J/kg °C (Appendix Table E.2).
   From Eq. 2.36, we have an expression for daily thermal exergy as

$$Ex_{\text{daily}} = M_w C_f \left[ (T_{\text{fo}} - T_{\text{fi}}) - (T_a + 273) \ln \frac{T_{\text{fo}} + 273}{T_{\text{fi}} + 273} \right]$$

$$= 100 \times 4179 [50 - 293 \times 0.1576]$$

$$= 2,332,383 \text{ J} = 2,332.383 \text{ kJ}.$$

**Example 2.79** Calculate daily exergy efficiency of Example 2.78.

**Solution**

The daily exergy efficiency of solar collector will be determined as follows:

$\eta_{ex,daily}$

$$= \frac{Ex_{daily}}{Area\ of\ collector\ \times\ Average\ solar\ intensity\ \times\ Sunshine\ hour\ \times\ 0.95}$$

After substitution of daily exergy, area of collector, average solar intensity and sunshine hour, one gets

$$\eta_{ex,daily} = \frac{2,332,383\ J}{2\ m^2 \times 3506\ J/m^2s \times 6 \times 3600\ s} = 0.16.$$

### Additional Examples

*Example 2.80* Evaluate Carnot efficiency for machine working at temperature of 60 °C for ambient air temperature of 15 °C (winter) and 45 °C (summer), respectively.

### Solution

Given: $T_c = 15 + 273 = 288\ ^\circ C$ and $T_h = 15 + 273 = 288\ ^\circ C$; $T_h = 45 + 273 = 318\ ^\circ C$;

From Eq. 2.29, one has

$$\eta = \left[1 - \frac{T_c}{T_h}\right]$$

At $T_h = 288\ ^\circ C$, $\eta = \left[1 - \frac{288}{288}\right] = 0$, no work and no destruction.

At $T_h = 318\ ^\circ C$, $\eta = \left[1 - \frac{288}{318}\right] = 0.9056$.

If operating temperature of machine increases for a given ambient air temperature, $\eta$ will reduce to increase in destruction losses.

*Example 2.81* Repeat Example 2.76 with same parameters with fluid as an air.

### Solution

In this case too, an ambient air temperature can be assumed to be 20 °C and specific heat of air, $C_f$, is 1005 j/kg °C (Appendix Table E.2). Them from Eq. 2.35a, one gets

$$\dot{E}x_u = \dot{m}_f C_f \left[(T_{fo} - T_{fi}) - (T_a + 273)\ln\frac{T_{fo} + 273}{T_{fi} + 273}\right]$$

$$= 0.20 \times 1005\left[15 - (293)\ln\frac{308}{293}\right] = 74.64\ W$$

In the case of air as a fluid, the rate of exergy has been reduced from 310.36 W. (Example 2.76) to 74.64 W due to poor holding capacity of air.

*Example 2.82* Evaluate the instantaneous thermal efficiency for the rate of exergy obtained in Examples 2.76 and 2.81 for using solar radiation of 750 W/m².

**Solution**

The rate of exergy of solar radiation $(I_{ex}) = 750 \times 0.95 = 712.25$ (Example 2.76).
Following Example 2.77 for water, we have

$$\eta_{iex} = \frac{\dot{E}x_u}{I_{ex}} = \frac{310.36}{750 \times 0.95} = 0.43 \approx 43\%$$

For air, it is computed as

$$\eta_{iex} = \frac{\dot{E}x_u}{I_{ex}} = \frac{74.64}{750 \times 0.95} = 0.1048 \approx 10.48\%.$$

This indicates that an instantaneous exergy efficiency is reduced in same proposition as specific heat is reduced from 4179 to 1005 j/kg °C.

***Example 2.83*** Evaluate overall heat transfer coefficient, $U$, for Example 2.73 for two air gap of 0.025 m each for case (c).

**Solution**

Here, $C_1 = C_2 = 5.2 \, \text{W/m}^2 \, ^\circ\text{C}$ for 0.025 m air cavity (Fig. 2.3).
    **From Example 2.73, for two-layered horizontal slab with single air cavity of 0.05 m**: An overall heat transfer coefficient, U, can be evaluated as follows (Eq. 2.28b):

$$U = \left[ \frac{1}{9.5} + \frac{0.05}{0.72} + \frac{1}{4.75} + \frac{0.05}{0.72} + \frac{1}{5.7} \right]^{-1} = 1.59 \, \text{W/m}^2 \, ^\circ\text{C}.$$

Now, for two-layered horizontal slab with two air cavity of 0.025 m separated by aluminum foil, an overall heat transfer coefficient, U, can be evaluated as follows (Eq. 2.28b):

$$U = \left[ \frac{1}{9.5} + \frac{0.05}{0.72} + \frac{1}{5.2} + \frac{1}{5.2} + \frac{0.05}{0.72} + \frac{1}{5.7} \right]^{-1}$$
$$= [0.1063 + 0.0694 + 0.1923 + 0.1923 + 0.0694 + 0.1754]^{-1}$$
$$= 0.8051^{-1} = 1.24 \, \text{W/m}^2 \, ^\circ\text{C}$$

Thus, an overall heat transfer coefficient is reduced from 1.59 to 1.24 W/m² °C just by dividing same gap into two equal gap. This is most important in building design.

# References

1. Tiwari, G. N. (2004). *Solar energy: Fundamental, design, modelling and applications.* Narosa Publishing House, New Delhi and CRC Press.
2. Tiwari, G. N. (2016). *Arvind Tiwari and Shyam, hand book of solar energy.* Springer.
3. Holman, J. P. (1992). *Heat transfer.* McGraw Hill International Ltd.
4. Malik, M. A. S., Tiwari, G. N., Kumat, A., & Sodha, M. S. (1982). *Solar distillation.* Pergamon Press.
5. Petela, R. (2003). *Solar Energy, 74*(6), 469.
6. Szargut, J. T. (2003). *Energy, 28*(11), 1047.
7. Battisti, R., & Corrado, A. (2005). *Energy, 30*, 952.
8. Lewis, G., & Keoleian, G. (1996). *National Pollution Prevention Center.* University of Michigan.
9. Singh, R. G., & Tiwari, G. N. (2020). Daily exergy analysis of passive solar still for maximum yield. *Desalination and Water Treatment, 204*, 1–9. https://doi.org/10.5004/dwt.2020.26276

# Chapter 3
# Energy Balance

## 3.1 Introduction

In order to write the energy balance of solar thermal, photo-voltaic (PV) and photo-voltaic thermal (PVT) systems, it is mandatory to first understand the input energy (solar energy) and basic heat and mass transfer which we have discussed in Chaps. 1 and 2, respectively. Now, the most of solar system is glazed with either window glass or photo-voltaic (PV) module. So, we are going to discuss in brief basic energy balance of each component of solar system which is used to design various configurations of solar thermal, photo-voltaic (PV) and photo-voltaic thermal (PVT) systems.

## 3.2 Solar Thermal System

### 3.2.1 Single Window Glass Covers (Fig. 3.1)

The main component of any solar thermal system is glass cover which transmits short wavelength solar radiation but does not allow long wavelength radiation to escape again through glass cover after absorption by blacked surface. It can only allow to again passing through glass cover by various heat transfer coefficients known as an overall heat transfer coefficient.

Referring to Eq. 2.1a, the energy balance in W/m$^2$ of single glass window cover based on the first law of thermodynamics can be written as

$$\dot{q}_u = \tau I(t) - U_L(T_r - T_a) \tag{3.1}$$

© Bag Energy Research Society 2024
G. N. Tiwari, *Fundamental of Mathematical Tools for Thermal Modeling of Solar Thermal and Photo-voltaic Systems-Part-I*,
https://doi.org/10.1007/978-981-99-7085-8_3

**Fig. 3.1**  View of single
glass window

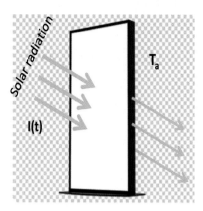

[The net rate of useful thermal energy gain to room air

= The rate of solar radiation transmitted through the glass window

− The rate thermal energy lost through the glass cover]

where, $\dot{q}_u$, the rate of useful thermal energy in W/m$^2$; $\tau$, the transmittivity of the glass cover; $I(t)$, the solar radiation incident on the glass cover (W/m$^2$); $U_L$, the overall heat transfer coefficient from room air to ambient air through glass cover (W/m$^2$ °C); $T_r$ (°C), the room air temperature and $T_a$, the ambient air temperature (°C).

**Example 3.1**  Evaluate an overall heat transfer coefficient ($U$) in W/m$^2$ °C and the rate of heat transfer in W/m$^2$ in to room for single glazed window for the following parameters.

$$I(t) = 500 \, \text{W/m}^2, \, \tau = 0.9, \, T_r = 20\,^\circ\text{C and } T_a = 10\,^\circ\text{C},$$
$$L_g = 5 \text{ mm and } K_g = 0.78 \text{ W/m}\,^\circ\text{C}.$$

**Solution**

Assuming there is no wind then from Eqs. 2.13a and 2.13b, we can have $h_o = h_c = 5.7$ W/m °C and $h_i = h_c = 2.8$ W/m °C.

Now, an overall heat transfer coefficient from room air to ambient air through the glass window, Eq. 2.27b, can be written as

$$U_L = \left[ \frac{1}{h_o} + \frac{L_g}{K_g} + \frac{1}{h_i} \right]^{-1} = \left[ \frac{1}{5.7} + \frac{0.005}{0.78} + \frac{1}{2.8} \right]^{-1}$$
$$= [0.175 + 0.0064 + 0.357]^{-1} = (0.5385)^{-1} = 1.857 \, \text{W/m}^2\,^\circ\text{C}$$

**The above calculation without glass window will be as follows:**

$$U_L = \left[\frac{1}{h_o} + \frac{1}{h_i}\right]^{-1} = \left[\frac{1}{5.7} + \frac{1}{2.8}\right]^{-1} = \left[\frac{1}{5.7} + \frac{1}{2.8}\right]^{-1}$$
$$= [0.175 + 0.357]^{-1} = (0.5332)^{-1} = 1.8796 \text{ W/m}^2\,^\circ\text{C}$$

One can observe that there is an increase of 1.2% in the numerical value of an overall heat transfer coefficient ($U$); hence in any glazed surface, effect of glass cover is generally not considered.

From Eq. 3.1, one has the rate of useful thermal energy (direct gain) as

$$\dot{q}_u = \tau I(t) - U_L(T_r - T_a) = 0.9 \times 500 - 1.8796(20 - 10) = 431.2 \text{ W/m}^2$$

**Example 3.2** Evaluate an overall heat transfer coefficient ($U$) in W/m$^2$ °C and the rate of heat transfer in W/m$^2$ in to room for double glazed window for the following parameters.

$I(t) = 500$ W/m$^2$, $\tau = 0.9$, $T_r = 20\,^\circ$C and $T_a = 10\,^\circ$C, $L_{g1} = L_{g2} = 5$ mm and $K_{g1} = K_{g2} = 0.78$ W/m °C, air conductance between two glass window, $C = 5$ W/m$^2$ °C for 2.5 cm air gap.

**Solution**

Assuming there is no wind then from Eqs. 2.13a and 2.13b, we can have $h_o = h_c = 5.7$ W/m$^2$ °C and $h_i = h_c = 2.8$ W/m$^2$ °C.

Now, an overall heat transfer coefficient from room air to ambient air through the glass window, Eq. 2.28b, can be written as

$$U_L = \left[\frac{1}{h_o} + \frac{L_{g1}}{K_{g1}} + \frac{1}{C} + \frac{L_{g2}}{K_{g2}} + \frac{1}{h_i}\right]^{-1} = \left[\frac{1}{5.7} + \frac{0.005}{0.78} + \frac{1}{5} + \frac{0.005}{0.78} + \frac{1}{2.8}\right]^{-1}$$
$$= [0.175 + 0.0064 + 0.20 + 0.0064 + 0.357]^{-1} = (0.7488)^{-1} = 1.3426 \text{ W/m}^2\,^\circ\text{C}$$

The above calculation without glass window will be as follows:

$$U_L = \left[\frac{1}{h_o} + \frac{1}{C} + \frac{1}{h_i}\right]^{-1} = \left[\frac{1}{5.7} + \frac{1}{5} + \frac{1}{2.8}\right]^{-1}$$
$$= [0.175 + 0.2 + 0.357]^{-1} = (0.732)^{-1} = 1.366 \text{ W/m}^2\,^\circ\text{C}$$

One can observe that there is an increase of 1.7% in the numerical value of an overall heat transfer coefficient ($U$); hence in any glazed surface, effect of glass cover is generally not considered.

From Eq. 3.1, one has the rate of useful thermal energy for double glazed wall as

$$\dot{q}_u = \tau^2 I(t) - U_L(T_r - T_a) = 0.9 \times 0.9 \times 500 - 1.366(20 - 10)$$
$$= 391.34 \ \text{W/m}^2$$

From Example 3.1, one can infer that

(i)   The numerical value of an overall heat transfer coefficient is reduced from single glazed wall to double glazed wall by 28.92% as per expectation and
(ii)  The net gain of thermal energy for double glazed wall is also reduced from 431.2 W/m$^2$ for single glazed to 391.34 W/m$^2$ for double glazed wall by 9.2%.

**Example 3.3** Evaluate the rate of heat loss from room air to ambient air for single, Fig. 3.1, and double glazed, Fig. 3.2, window during off-sunshine hours for Example 3.1.

**Solution**

From Examples 3.1 and 3.2, one have
$U_{L,\text{single}} = 1.8796 \ \text{W/m}^2 \, ^\circ\text{C}$ and $U_{L,\text{double}} = 1.366 \ \text{W/m}^2 \, ^\circ\text{C}$
Now, the rate of heat loss from room air to ambient air for single and double glazed window can be calculated with help of Eq. 3.1 with $I(t) = 0$ as follows:

$$\dot{q}_{u,\text{single}} = -U_{L,\text{single}}(T_r - T_a) = -1.8796(20 - 10) = -18.796 \ \text{W/m}^2 \text{ and}$$
$$\dot{q}_{u,\text{double}} = -U_{L,\text{double}}(T_r - T_a) = -1.366(20 - 10) = -13.66 \ \text{W/m}^2$$

**Fig. 3.2** Double glazed window

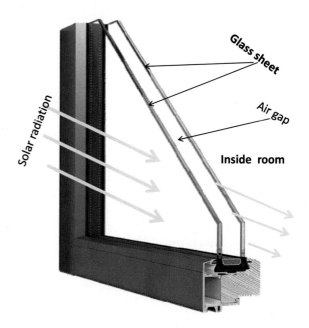

The negative sign shows the rate of heat loss during off-sunshine/night hours. During off-sunshine, the rate of heat loss from room air to ambient air from single to double glazed window during off-sunshine hours is reduced by 28.89% which is very useful for cold climatic condition to conserve the fossil fuel for thermal heating.

### 3.2.2  Single Bare Solid Surface (Fig. 3.3)

The energy balance equation for single bare solid surface exposed to solar radiation, $I(t)$, in W/m$^2$ can be written as follows:

$$-K\frac{\partial T}{\partial T}\bigg|_{x=0} = \alpha I(t) - h_c(T_{x=0} - T_a) - \varepsilon\sigma\left[T_{x=0}^4 - T_{sky}^4\right] \qquad (3.2)$$

$\left[\text{The rate of thermal energy (heat) is conducted at exposed surface at } x = 0\right]$
$= \left[\text{The rate of solar energy absorbed by surface at } x = 0\right]$
$- \left[\text{The rate of thermal energy lost to ambient from } x = 0 \text{ by convection}\right]$
$- \left[\text{The rate of thermal energy lost to sky from } x = 0 \text{ by radiation}\right].$

Equations 2.17, and 3.2 can be written as

$$-K\frac{\partial T}{\partial T}\bigg|_{x=0} = \alpha I(t) - h_c(T_{x=0} - T_a) - \varepsilon\sigma\left(T_{x=0}^4 - T_a^4\right) - \varepsilon\sigma\left(T_a^4 - T_{sky}^4\right)$$

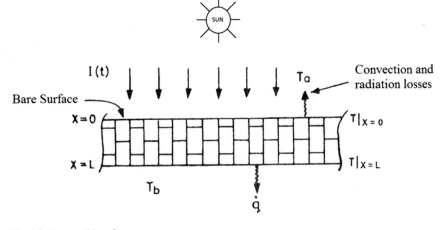

Fig. 3.3  Bare solid surface

or,

$$-K\frac{\partial T}{\partial T}\bigg|_{x=0} = \alpha I(t) - h_c(T_{x=0} - T_a) - \varepsilon\sigma\left(T_{x=0}^4 - T_a^4\right) - \varepsilon\Delta R \qquad (3.3)$$

where, $\Delta R = \sigma\left[(T_a + 273)^4 - \left(T_{sky} + 273\right)^4\right] = 70\,\text{W/m}^2$ for summer condition and $60\,\text{W/m}^2$ for winter condition is the difference between the long wavelength radiation exchange between ambient air temperature at $T_a$ to the sky temperature at $T_{sky}$.

Equation 3.3 can also be rewritten as

$$-K\frac{\partial T}{\partial T}\bigg|_{x=0} = \alpha I(t) - h_c(T_{x=0} - T_a) - h_r(T_{x=0} - T_a) - \varepsilon\Delta R$$

$$= \alpha I(t) - h_o(T_{x=0} - T_a) - \varepsilon\Delta R \qquad (3.4)$$

where, $h_o = h_c + h_r$, combined heat transfer coefficient by convection and radiation. The expression of $h_o = 5.7 + 3.8V$ with $V$ as wind velocity and $h_o$ is independent of temperature of the surface, Eq. 2.13a.

Further, Eq. 3.4 can be rewritten as

$$-K\frac{\partial T}{\partial T}\bigg|_{x=0} = h_o\left[\left\{\frac{\alpha I(t)}{h_o} + T_a - \frac{\varepsilon\Delta R}{h_o}\right\} - T_{x=0}\right] = h_o[T_{sa} - T_{x=0}] \qquad (3.5)$$

where,

$$T_{sa} = \frac{\alpha I(t)}{h_o} + T_a - \frac{\varepsilon\Delta R}{h_o} \qquad (3.6)$$

The $T_{sa}$ is a sol–air temperature. It can be defined as effective ambient air temperature including the effect of solar intensity and long wavelength radiation between ambient air and sky temperature.

One can also infer the following from Eq. 3.6

(a)  The sol–air temperature $(T_{sa})$ is higher than ambient air temperature during day time, i.e., $I(t) \neq 0$ and

(b)  The sol–air temperature $(T_{sa})$ is less than ambient air temperature either during off-sunshine hours, i.e., $T(t) = 0$ or low values of intensity, i.e., $\alpha I(t) < \varepsilon\Delta R$

**Example 3.4**  Find out threshold intensity when sol–air temperature becomes equal to ambient air temperature.

**Solution**

As per requirement $I(t) = I_{thresold}$ if $T_{sa} = T_a$.

Substitute above condition in Eq. 3.6 and one gets following as Eq. 3.7

$$T_s = \frac{\alpha I(t)_{threesold}}{h_o} + T_a - \frac{\varepsilon \Delta R}{h_o}. \tag{3.7}$$

From above, one has

$$I(t)_{threesold} = \frac{\varepsilon \Delta R}{\alpha}. \tag{3.8}$$

So, the threshold intensity depends on absorptivity of solid surface.

**Example 3.5** Evaluate sol–air temperature for winter condition for the following parameters:

$\alpha = 0.20$(bare) and $0.90$ (blackened; $\varepsilon = 0.9$ $I(t) = 350$ W/m²; $T_a = 10\,°C$; $\Delta R = 60$ W/m²; (winter condition, Example 2.58); $h_o = 5.7$ W/m² °C (Eq. 2.13a for $V = 0$).

**Solution**

From Eq. 3.6, we have an expression for sol–air temperature as

$$T_{sa} = \frac{\alpha I(t)}{h_o} + T_a - \frac{\varepsilon \Delta R}{h_o}.$$

Substitute an appropriate value of given parameters for $\alpha = 0.20$, one has

$$T_{sa} = \frac{0.20 \times 350}{5.7} + 10 - \frac{0.9 \times 60}{5.7} = 12.2 + 10 - 9.47 = 12.81\,°C.$$

Further, substitute an appropriate value of given parameters for $\alpha = 0.90$, one has

$$T_{sa} = \frac{0.90 \times 350}{5.7} + 10 - \frac{0.9 \times 60}{5.7} = 55.26 + 10 - 9.47 = 55.79\,°C$$

One can see that effect of $\Delta R$, the long wavelength radiation exchange between ambient air and sky cannot be neglected for any surface weather it is bare or blackened.

**Example 3.6** Evaluate sol–air temperature for summer condition for the following parameters:

$\alpha = 0.20$ (bare) and $0.90$ (blackened); $\varepsilon = 0.9$; $I(t) = 500$ W/m²; $T_a = 25\,°C$; $\Delta R = 70$ W/m²; (summer condition, Example 2.59); $h_o = 5.7$ W/m² °C (Eq. 2.13a for $V = 0$).

**Solution**

From Eq. 3.6, we have an expression for sol–air temperature as

$$T_{\text{sa}} = \frac{\alpha I(t)}{h_o} + T_a - \frac{\varepsilon \Delta R}{h_o}.$$

Substitute an appropriate value of given parameters for $\alpha = 0.20$, one has

$$T_{\text{sa}} = \frac{0.20 \times 500}{5.7} + 25 - \frac{0.9 \times 70}{5.7} = 17.5 + 25 - 11.05 = 31.45\,^{\circ}\text{C}$$

Further, substitute an appropriate value of given parameters for $\alpha = 0.90$, one has

$$T_{\text{sa}} = \frac{0.90 \times 500}{5.7} + 25 - \frac{0.9 \times 70}{5.7} = 78.94 + 25 - 11.05 = 92.89\,^{\circ}\text{C}$$

From Examples 3.5 and 3.6, one can observe that effect of $\Delta R$, the long wave-length radiation exchange between ambient air and sky on sol–air temperature varies from 10 to 40%, and hence it cannot be neglected for any surface weather it is bare or blackened.

### 3.2.3  Wetted Bare Surface (Fig. 3.4)

Following Eq. 3.4, the energy balance equation in W/m² for wetted surface can be written as follows:

$$-K\frac{\partial T}{\partial T}\bigg|_{x=0} = \alpha I(t) - h_1(T_{x=0} - T_a) - \varepsilon \Delta R \tag{3.9}$$

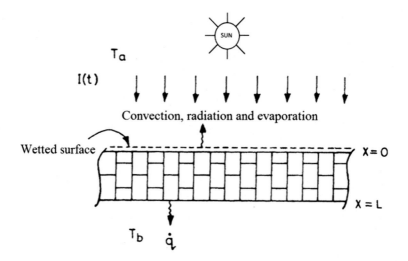

**Fig. 3.4**  Wetted solid surface

where, $h_1 = h_c + h_r + h_{ew}$ is total heat transfer coefficient from wetted surface to ambient air including convection, radiation and evaporation.

Equation 3.7 can be rearranged in terms of sol–air temperature as

$$-K \frac{\partial T}{\partial T}\bigg|_{x=0} = h_1[T_{sa} - T_{x=0}] \qquad (3.10)$$

where, $T_{sa} = \frac{\alpha I(t)}{h_1} + T_a - \frac{\varepsilon \Delta R}{h_1}$, sol–air temperature for wetted surface of solid layer.

**Example 3.7** Calculate sol–air temperature for wetted surface (10% relative humidity) of Example 3.6 for summer condition with absorptivity of 0.20.

**Solution**

For wetted surface with relative humidity of 10%: $h_1 = 20.33 \, \text{W/m}^2 \, °C$, Example 2.68.

From Eq. 3.10, one has

$$T_{sa} = \frac{\alpha I(t)}{h_1} + T_a - \frac{\varepsilon \Delta R}{h_1} = \frac{0.20 \times 500}{20.33} + 25$$
$$- \frac{0.9 \times 70}{20.33} = 29.92 - 3.099 = 26.82 \, °C$$

One can observe that sol–air temperature for bare surface with $\alpha = 0.20$ is reduced from 31.45 to 26.82 °C by a factor of 14.72%.

### 3.2.4 Single Blackened and Glazed Solid Surface (Fig. 3.5)

In this case, the blackened surface of solid surface is covered by window glass to reduce the thermal losses from blackened surface to ambient air. Such configuration is referred as blackened and glazed surface used for thermal heating of a building. There is no effect of long wavelength radiation exchange between ambient air and sky on sol–air temperature due to glass window between blackened surface and sky. That means that the blackened surface is not directly exposed to sky.

For single blackened and glazed surface of solid, Eq. 3.4 can be modified as follows:

$$-K \frac{\partial T}{\partial T}\bigg|_{x=0} = \alpha \tau I(t) - U_t(T_{x=0} - T_a) \qquad (3.11)$$

where, $\tau$ is the transmittivity of window glass cover and $U_t$ is an overall top heat transfer coefficient from blackened surface to ambient air through window glass cover, and its expression is given by

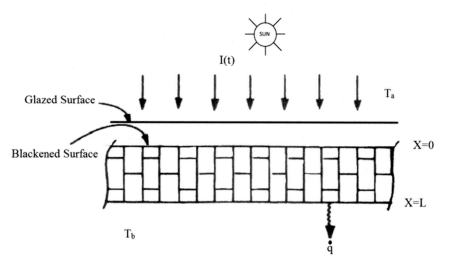

**Fig. 3.5** Single glazed solid surface

$$U_t = \left[ \frac{1}{h_o} + \frac{L_g}{K_k} + \frac{1}{h_i} \right]^{-1} \approx \left[ \frac{1}{h_o} + \frac{1}{h_i} \right]^{-1} , \text{ from Example 3.1.} \qquad (3.12a)$$

If there is a gap of 10 cm between blackened surface and window glass cover, then an expression for $U_t$ becomes as

$$U_t = \left[ \frac{1}{h_o} + \frac{L_g}{K_k} + \frac{1}{C} \right]^{-1} \qquad (3.12b)$$

Equation 3.8 can also be written in terms of sol–air temperature as

$$-K \frac{\partial T}{\partial T}\bigg|_{x=0} = \alpha \tau I(t) - U_t(T_{x=0} - T_a) = U_t \left[ \frac{\alpha \tau I(t)}{U_t} + T_a - T_{x=0} \right]$$

$$= U_t(T_{sa} - T_{x=0}) \qquad (3.13)$$

where, $T_{sa}$ is sol–air temperature for blackened and glazed surface and it is given by

$$T_{sa} = \frac{\alpha \tau I(t)}{U_t} + T_a \qquad (3.14)$$

**Example 3.8** Evaluate the top overall heat transfer coefficient, $U_t$, and sol–air temperature for blackened and single glazed solid surface, $T_{sa}$, for the following parameters:

$\alpha = 0.90$; $\tau = 0.9$; $I(t) = 500 \text{ W/m}^2$; $T_a = 25 \,^\circ\text{C}$; $h_o = 5.7 \text{ W/m}^2\,^\circ\text{C}$ (Eq. 2.13a for $V = 0$); $L_g = 5$ mm and $K_g = 0.78$ W/m °C, Appendix E; $C = 4.75$ W/m$^2$ °C (for 5 cm air gap, Fig. 2.3), $h_o = 2.8$ W/m$^2$ °C.

**Solution**

From Eq. 3.12b, one has an expression for the top overall heat transfer coefficient, $U_t$, as

$$U_t = \left[\frac{1}{h_o} + \frac{L_g}{K_k} + \frac{1}{C}\right]^{-1} = \left[\frac{1}{5.7} + \frac{0.005}{0.78} + \frac{1}{4.75}\right]^{-1}$$
$$= [0.175 + 0.0064 + 0.21]^{-1} = [0.3919]^{-1} = 2.55 \text{ W/m}^2 \text{ °C}$$

If Eq. 3.12a is used for $U_t$, then

$$U_t = \left[\frac{1}{h_o} + \frac{L_g}{K_k} + \frac{1}{h_i}\right]^{-1} = \left[\frac{1}{5.7} + \frac{0.005}{0.78} + \frac{1}{2.8}\right]^{-1} = 1.8796 \text{ W/m}^2 \text{ °C, Example 3.1.}$$

It is due to lower value of heat transfer coefficient between solid surface and glass cover.

The sol–air temperature for single blackened and glazed surface, $T_{sa}$, can be obtained from Eq. 3.9 as

$$T_{sa} = \frac{\alpha \tau I(t)}{U_t} + T_a = \frac{0.9 \times 0.9}{2.55} \times 500 + 25 = 183.82 \text{ °C}$$

From Examples 3.6 and 3.7 one can observe that the solar air temperature is increased from 92.89 to 183.82 °C almost double due to glazed surface, and hence glazed surface should be only used for thermal heating of a building.

### 3.2.5  Bottom Loss Coefficient of Single Blackened and Glazed Solid Surface

In Sect. 3.2.3, the energy balance in terms of W/m$^2$ for top blackened and glazed surface has been written, Eq. 3.7, in a steady-state condition. In a steady-state condition, there is no variation of temperature with time. In the case of bottom heat loss coefficient for solid surface, the energy balance can be written in terms of an overall bottom heat transfer coefficient, $U_b$, which can be defined as follows:

$$U_b = \left[\frac{L_i}{K_i} + \frac{1}{h_i}\right]^{-1} \tag{3.15}$$

where, $L_i$ and $K_i$ are thickness and thermal conductivity of solid surface below glazed surface. Its value may be for conducting material as well as insulating materials depending upon applications. If thermal heating of a building is needed by solar energy, it should be conducting material (Appendix Tables E.3 and E.4), and it should be insulating material if surface heating is needed just like flat plate collector (Appendix Table E.4).

**Example 3.9** Calculate the overall bottom loss coefficient for conducting (brick: $K_i = 0.69$ W/m °C) and insulating (cord board: $K_i = 0.043$ W/m °C) material for thickness of 10 cm thickness.

**Solution**

From Eq. 3.15, we can evaluate the overall bottom loss coefficient for conducting material as

$$U_b = \left[ \frac{L_i}{K_i} + \frac{1}{h_i} \right]^{-1} = \left[ \frac{0.10}{0.69} + \frac{1}{2.8} \right]^{-1} = [0.1449 + 0.3571]^{-1}[0.5020]^{-1}$$
$$= 1.992 \text{ W/m}^2 \text{ °C}.$$

From Eq. 3.10, we can evaluate the overall bottom loss coefficient for insulating material as

$$U_b = \left[ \frac{L_i}{K_i} + \frac{1}{h_i} \right]^{-1} = \left[ \frac{0.10}{0.043} + \frac{1}{2.8} \right]^{-1} = [2.3256 + 0.3571]^{-1} = [2.6827]^{-1}$$
$$= 0.3727 \text{ W/m}^2 \text{ °C}$$

Further, the overall bottom loss coefficient for insulating material can be approximated as

$$U_b \approx \frac{K_i}{L_i} = 0.43 \text{ W/m}^2 \text{ °C}$$

We can conclude the following from Example 3.8:

(a) The conducting materials should be used for transferring the thermal energy/heat from exposed solar energy to inside room due to higher thermal conductivity (0.69 W/m °C) in comparison with insulating materials (0.043 W/m °C).
(b) The insulating materials should be used to collect the thermal energy/heat from exposed solar energy of blackened surface to have minimum heat loss from blackened surface to ambient air though insulating solid surface.
(c) The overall bottom loss coefficient, $U_b$ can also be approximated as $\frac{K_i}{L_i}$ because of its dominance in comparison with convective heat transfer coefficient.

**Example 3.10** Estimate the total overall loss coefficient, $U_L$ for blackened and glazed surface of solid material for Examples 3.8 and 3.9.

**Solution**

From Examples 3.8 and 3.9, one has the following

$U_t = 2.55 \ \text{W/m}^2 \, ^\circ\text{C}; \ U_b = 1.992 \ \text{W/m}^2 \, ^\circ\text{C}$ (for conducting solid) and $U_b = 0.3727 \ \text{W/m}^2 \, ^\circ\text{C}$ for insulating solid.

So, the total overall loss coefficient, $U_L$ for conducting material can be calculated as follows:

$$U_L = U_t + U_b = 2.55 + 1.992 = 4.542 \ \text{W/m}^2 \, ^\circ\text{C}$$

The total overall loss coefficient, $U_L$ for insulating material can be calculated as follows:

$$U_L = U_t + U_b = 2.55 + 0.3727 = 2.92 \ \text{W/m}^2 \, ^\circ\text{C}$$

This indicates that the total thermal loss coefficient from blackened and glazed surface to ambient air is more in conducting solid than insulating solid.

**Example 3.11** Estimate the total overall loss coefficient, $U_L$ for bare and wetted surface (10% relative humidity) of solid material for Examples 3.6, 2.68 and 3.8.

**Solution**

(a) **For bare surface**

From Examples 3.6 and 3.9: $U_t = h_o = 5.7 \ \text{W/m}^2 \, ^\circ\text{C}$ and $U_b = 1.992 \ \text{W/m}^2 \, ^\circ\text{C}$ (for conducting solid) and $U_b = 0.3727 \ \text{W/m}^2 \, ^\circ\text{C}$ for insulating solid.

So, the total overall loss coefficient, $U_L$ for conducting material can be calculated as follows:

$$U_L = U_t + U_b = 5.7 + 1.992 = 7.692 \ \text{W/m}^2 \, ^\circ\text{C}.$$

The total overall loss coefficient, $U_L$, for insulating material can be calculated as follows:

$$U_L = U_t + U_b = 5.7 + 0.3727 = 6.0727 \ \text{W/m}^2 \, ^\circ\text{C}$$

(b) **For wetted surface**

From Examples 2.68 and 3.8: $U_t = h_1 = 20.33 \ \text{W/m}^2 \, ^\circ\text{C}$ and $U_b = 1.992 \ \text{W/m}^2 \, ^\circ\text{C}$ (for conducting solid) and $U_b = 0.3727 \ \text{W/m}^2 \, ^\circ\text{C}$ for insulating solid.

So, the total overall loss coefficient, $U_L$, for conducting material can be calculated as follows:

$$U_L = U_t + U_b = 20.33 + 1.992 = 22.332 \ \text{W/m}^2 \, ^\circ\text{C}.$$

The total overall loss coefficient, $U_L$, for insulating material can be calculated as follows:

$$U_L = U_t + U_b = 20.33 + 0.3727 = 20.7027 \ \text{W/m}^2 \, ^\circ\text{C}$$

From Example 3.10 and 3.11, one can see that the total thermal loss coefficient for wetted surface is maximum (22.332 $\text{W/m}^2 \, ^\circ\text{C}$) and minimum (4.542 $\text{W/m}^2 \, ^\circ\text{C}$) for blackened and glazed surface to ambient air for conducting solid. The same results are for insulating materials.

### 3.2.6  The Rate of Heat Transfer

The rate of heat transfer from solar energy exposed surface to ambient air through solid layer can be written as

$$\dot{q}_u = U_L(T_{\text{sa}} - T_a) \tag{3.16}$$

where, an expression for $U_L$ and $T_{\text{sa}}$ for different treatment of solid surface are given in Examples 3.10 and 3.11 and by Eqs. 3.6, 3.10 and 3.14, respectively.

**Example 3.12** Estimate the rate of heat transferred per m$^2$ from solar energy exposed surface to ambient air through bare, wetter and blackened and glazed conducting solid for Examples 3.9 and 3.10.

**Solution**

From Eq. 3.16, the rate of heat transfer in W/m$^2$ can be evaluated from the following expression as

$$\dot{q}_u = U_L(T_{\text{sa}} - T_a)$$

For bare surface ($\alpha = 0.20$): $T_{\text{sa}} = 31.45\,^\circ\text{C}$ and $U_L = 7.692 \ \text{W/m}^2 \, ^\circ\text{C}$.

(a) **For bare surface ($\alpha = 0.20$)**: $T_{\text{sa}} = 31.45\,^\circ\text{C}$, Example 3.6, and $U_L = 7.692 \ \text{W/m}^2 \, ^\circ\text{C}$, Example 3.11

$$\dot{q}_{u,\text{Ba}} = U_L(T_{\text{sa}} - T_a) = 7.692(31.45 - 25) = 49.99 \ \text{W/m}^2$$

(b) **For blackened surface ($\alpha = 0.90$)**: $T_{\text{sa}} = 92.89\,^\circ\text{C}$, Example 3.6, and $U_L = 7.692 \ \text{W/m}^2 \, ^\circ\text{C}$, Example 3.11

$$\dot{q}_{u,B} = U_L(T_{\text{sa}} - T_a) = 7.692(92.89 - 25) = 522.20 \ \text{W/m}^2$$

(c) **For wetted surface ($\alpha = 0.20$)**: $T_{\text{sa}} = 26.82\,^\circ\text{C}$, Example 3.7, and $U_L = 22.332 \ \text{W/m}^2 \, ^\circ\text{C}$, Example 3.11

$$\dot{q}_{u,W} = U_L(T_{\text{sa}} - T_a) = 22.332(26.82 - 25) = 29.75 \ \text{W/m}^2$$

(d) **For blackened surface and glazed surface ($\alpha$ = 90 and $\tau$ = 0.9):**
$T_{sa}$ = 183.82 °C, Example 3.8 and $U_L$ = 4.542 W/m² °C, Example 3.8

$$\dot{q}_{u,B\&G} = U_L(T_{sa} - T_a) = 4.542(183.82 - 25) = 721.36 \text{ W/m}^2$$

From the above calculation one have the following

$$\dot{q}_{u,B\&G} > \dot{q}_{u,B}[\text{Thermal heating}] > \dot{q}_{u,Ba} > \dot{q}_{u,W}[\text{Thermal cooling}]$$

The results are as per expectation to be used for building design for passive heating and cooling.

## 3.3 Photo-Voltaic (PV) Module

The working principle of solar cell is based on semiconductor devices. The PV module consists of many solar cells connected in series which is sandwiched between transparent toughen glass and rigid/flexible materials with the help of 100% transparent ethyl vinyl acetate (EVA). The PV module generates direct current (DC) by using photon of solar energy incident on top of n-p junction (For details see book by Tiwari [1] and Tiwari et al. [2]). Broadly, PV module is classified as

(a) **Opaque PV Module**: In this case, the incident solar radiation cannot be transmitted to other side of PV module, Figs. 3.6a and 3.7a. The base of opaque PV module is generally insulating tedlar material which is rigid.

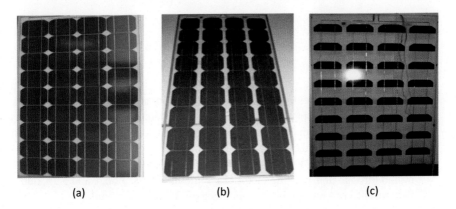

(a)          (b)          (c)

**Fig. 3.6** Photograph of single crystalline photo-voltaic module, **a** 75 $W_p$ opaque PV module; **b** 75 $W_p$ semi-transparent PV module and **c** 37 $W_p$ semi-transparent PV module [Area of each PV module = 0.605 m²]

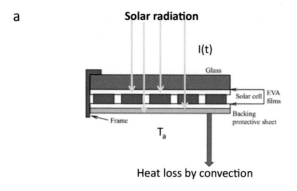

a

Solar radiation

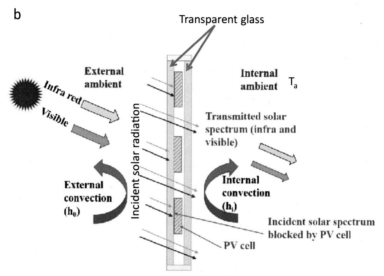

b

Fig. 3.7 **a** Cross-sectional view of horizontal opaque PV module, **b** Cross-sectional view of vertical semi-transparent PV module

(b) **Flexible PV Module**: In this case too, incident solar radiation cannot be transmitted to other side of PV module due to opaque nature of flexible sheet, namely copper and aluminum. The base of PV module is either copper or aluminum which is flexible unlike tedlar. It is also opaque.

(c) **Semi-transparent PV Module**: In this case, incident solar radiation is transmitted through non-packing factor of PV module, Figs. 3.6b, c and 3.7b, due to the presence of transparent glass in place of opaque PV module.

The area of each solar cell and PV module is 0.015 and 0.605 m², respectively, and number of solar cell in PV module (Fig. 3.6a, b) is thirty-six (36). The specification for a typical Si-PV module is summarized in Table 3.1.

The electrical efficiency ($\eta_{ec}$) of solar cell as a function of temperature in fraction is given by [3]

**Table 3.1** Specification of Si-based PV module under STC

| S. No. | Parameters | Numerical value |
|--------|-----------|-----------------|
| 1 | Current | 4.4 A |
| 2 | Voltage | 18 |
| 3 | Intensity | 1000 W/m$^2$ |
| 4 | Area | 0.606 m$^2$ |
| 5 | Packing factor | 0.89 |
| 6 | Efficiency in fraction | 0.12 |
| 7 | Number of solar cell | 36 |

$$\eta_{sc} = \eta_0[1 - \beta_0(T_c - 298)] \qquad (3.17)$$

where, $\eta_0$ is electrical efficiency of solar cell at standard test condition (STC) (solar flux of 1000 W/m$^2$ and surrounding temperature in 298 K); $\beta_0$ is temperature coefficient and $T_c$ is solar cell temperature $(K)$. The temperature coefficient $(\beta_0)$ for different solar cell material is given in Table 3.2.

If $T_c$ is solar cell temperature (°C), then Eq. 3.17 becomes

$$\eta_{sc} = \eta_0[1 - \beta_0(T_c - 25)] \qquad (3.18a)$$

**Table 3.2** Specifications for various silicon and non-silicon-based PV modules including temperature coefficient (Durisch et al. [5] and Virtuani et al. [6], Tiwari and Mishra [7])

| S. No. | Different solar cell materials | PV module efficiency $\eta_{m0}$ (%) | Expected life $n_{PV}$ (years) | Specific energy density/ coefficient $E_{in}$ (kWh m$^{-2}$) | $(E_{in})$ of PV module, $A_m =$ 0.71 m$^2$ (kWh) | Average temp. coefficient $\beta$ (°C$^{-1}$) |
|--------|-------------------------------|------|------|------|------|------|
| 1 | c-Si (single crystalline) | 16 | 30 | 1190 | 844.9 | 0.00535 |
| 2 | mc-Si (multi-crystalline silicon) | 14 | 30 | 910 | 646.1 | 0.00425 |
| 3 | nc-Si (nanocrystalline silicon) | 12 | 25 | 610 | 433.1 | 0.0036 |
| 4 | a-Si (amorphous silicon) | 6 | 20 | 378 | 268.38 | 0.00115 |
| 5 | CdTe (cadmium telluride) | 8 | 15 | 266 | 188.86 | 0.00205 |
| 6 | CIGS (copper indium gallium selenide) | 10 | 5 | 24.5 | 17.395 | 0.00335 |
| 7 | Average | 11 | 20 | 563.08 | 399.79 | 0.003292 |

Here, the surrounding temperature is in °C.

Further, it is important to mention here that Eq. 3.18a is valid only if $T_c \geq 25$, if $T_c < 25$ for harsh cold climatic condition then Eq. 3.18a reduces to

$$\eta_{sc} = \eta_0 \qquad (3.18b)$$

It is due to fact that the maximum electrical efficiency of solar cell may be $\eta_0$.

### 3.3.1 Packing Factor ($\beta_m$) of PV Module

It is defined as the ratio of total solar cell area to the total module area, and it can be expressed as:

$$\beta_m = \frac{\text{The area of one solar cell} \times \text{number of solar cell}}{\text{The area of PV module}} \qquad (3.19)$$

It is clear that $\beta_m$ is less than unity (pseudo solar cell), and it has maximum value of one when all area is covered by solar cell (rectangular mono-Si solar cell).

**Example 3.13** Calculate packing factor of single crystalline PV module of Fig. 3.6 b and c for the following parameters.

Area of PV module $= 0.605$ m$^2$; Area of solar cell in Fig. 3.6a and $b = 0.015$ m$^2$; Number of solar cell in module $= 36$.

**Solution**

From Eq. 3.19, one has

$$\beta_m = \frac{\text{The area of one solar cell} \times \text{number of solar cell}}{\text{The area of PV module}} = \frac{0.015 \times 36}{0.605} = 0.89$$

So, the packing factor of PV module shown in Fig. 3.6a and $b = 0.89$.

**Example 3.14** Repeat the Example 3.13 if area of each solar cell is 0.0075 (Fig. 3.6c).

**Solution**

From Eq. 3.19, one has

$$\beta_m = \frac{\text{The area of one solar cell} \times \text{number of solar cell}}{\text{The area of PV module}} = \frac{0.0075 \times 36}{0.605} = 0.445$$

So, the packing factor of PV module shown in Fig. 3.6c $= 0.445$. The packing factor plays a very important role in determining thermal as well as electrical power of solar thermal system.

**Example 3.15** Calculate the number of crystalline solar cell required for the following parameters given at standard test condition (STC):

Open circuit voltage ($V_{oc}$) at 25 °C = 500 mV (0.5 V).

Voltage drop at higher than 25 °C = 0.08 V.

Required voltage = 15 V.

**Solution**

The voltage available across the terminal of each crystalline solar cell at STC = 0.5–0.08 = 0.42 V.

The required number of crystalline solar cell to construct PV module = $\frac{15}{0.42}$ = 35.7 ≈ 36.

It is clear that the number of required solar cell depends on voltage of PV module which plays an important role to charge the batteries of given voltage. The voltage of PV module should be always higher than the voltage of batteries in real open condition.

### 3.3.2  Efficiency of PV Module

The electrical efficiency of PV module in fraction can be expressed as

$$\eta_m = \tau_g \times \eta_{ec} = \tau_g \times \eta_0[1 - \beta_0(T_c - 25)]$$
$$= \eta_{m0}[1 - \beta_0(T_c - 25)], \text{ in fraction} \tag{3.20}$$

where,

$$\eta_{m0} = \tau_g \times \eta_0, \text{ an electrical efficiency of PV module} \tag{3.21a}$$

Following Eq. 7.18b for harsh cold climatic condition, Eq. 3.21a will be reduced to be

$$\eta_m = \eta_{m0} \tag{3.21b}$$

The values of $\eta_{m0}$ and $\eta_0$ for different solar cell materials are given in Table 3.3. It is important to mention here that the values of thermal expansion ($\beta_0$) for solar cell and PV module are same. This shows that the electrical efficiency of PV module ($\eta_{m0}$) is less than electrical efficiency of solar cell due to the presence of glass over solar cell ($\eta_{sc}$).

It is also expressed as:

**Table 3.3** Electrical efficiency of various solar cells

| S. No. | Different solar cell materials | $\eta_{m0}$ | $\eta_0$ $[\tau_g = 0.9]$ clean glass |
|--------|-------------------------------|-------------|----------------------------------------|
| 1 | c-Si (single-crystalline) | 16 | 0.1777 |
| 2 | mc-Si (multi-crystalline silicon) | 14 | 0.1555 |
| 3 | nc-Si (nanocrystalline silicon) | 12 | 0.1333 |
| 4 | a-Si (amorphous silicon) | 6 | 0.0666 |
| 5 | CdTe (cadmium telluride) | 8 | 0.8888 |
| 6 | CIGS (copper indium gallium selenide) | 10 | 0.1111 |

$$\eta_{m0} = \frac{FF \times I_{sc} \times V_{oc}}{A_m \times I} \text{ in fraction} = \left(\frac{FF \times I_{sc} \times V_{oc}}{A_m \times I}\right) \times 100 \text{ in percentage}$$

(3.22a)

where, $A_m$ = area of PV module; $I = 1000\,\text{W/m}^2$ incident solar intensity on PV module and FF, $I_{sc}$ and $V_{oc}$ are fill factor, short circuit current and open circuit voltage of PV module at STC. The maximum value of fill factor (FF) of mc-crystalline PV module based on Si is 0.89 which is always less than one. The FF factor is provided by manufacturer of PV module. Equations 3.22a and 3.22b are independent of outdoor variable operating solar cell temperature like Eq. 3.20.

If $I_{sc}$ and $V_{oc}$ are measured in open sky condition for given variable solar intensity $[I(t)]$ for known FF, the electrical efficiency ($\eta_m$). PV module can be determined as

$$\eta_m = \frac{FF \times I_{sc} \times V_{oc}}{A_m \times I(t)} \text{ in fraction} = \left(\frac{FF \times I_{sc} \times V_{oc}}{A_m \times I(t)}\right) \times 100 \text{ in percentage}$$

(3.22b)

The value of $\eta_m$ in Eq. 3.22b is variable, and it depends on solar intensity and indirectly temperature of solar cell unlike $\eta_{m0}$ in Eq. 3.22a.

**Example 3.16** Evaluate electrical efficiency of multi-crystalline silicon (mc-Si) and nanocrystalline silicon (nc-Si) PV module in fraction at 50 and 70 °C.

**Solution**

From Eq. 3.20, we have an expression for electrical efficiency of PV module l as

$$\eta_m = \eta_{m0}[1 - \beta_0(T_c - 25)]$$

**For multi-crystalline silicon (mc-Si) solar cell:** $\eta_{m0} = 0.14$ and $\beta_0 = 0.00425$ (Table 3.2)

**At 50 °C.**

$$\eta_m = 0.14[1 - 0.00425(50 - 25)] = 0.125, \text{ and}$$

**At 70 °C.**

$$\eta_m = 0.14[1 - 0.00425(70 - 25)] = 0.11025$$

**For nanocrystalline silicon (nc-Si) solar cell:** $\eta_{m0} = 0.12$ and $\beta_0 = 0.0036$ (Table 3.2).

**At 50 °C.**

$$\eta_m = 0.12[1 - 0.0036(50 - 25)] = 0.1092, \text{ and}$$

**At 70 °C.**

$$\eta_m = 0.12[1 - 0.0036(70 - 25)] = 0.0984$$

One can see that an electrical efficiency decreases with increase of operating temperature.

**Example 3.17** Evaluate an electrical efficiency of solar cell for each PV module given in Table 3.1 under standard test condition (STC) for $\tau_g = 0.90$.

**Solution**

From Eq. 3.21, one gets  an electrical efficiency of PV module as

$$\eta_{m0} = \tau_g \times \eta_0,$$

or,

$$\eta_0 = \frac{\eta_{m0}}{\tau_g}$$

The results are given in Table 3.2. It indicates that solar cell electrical efficiency is more than PV module electrical efficiency.

**Example 3.18** Find out effect of dust deposition on PV module which reduces the transitivity of toughen glass from 0.9 to 0.7 and 0.5 for Example 3.17 (Table 3.3).

**Solution**

From Eq. 3.21, one has an expression for electrical efficiency of PV module as

$$\eta_{m0} = \tau_g \times \eta_0.$$

By using the above equation, the effect of $\tau_g$ for 0.7 and 0.5 is given in Table 3.4. One can see that an electrical efficiency of PV module is decreased by 49.97%.

**Table 3.4** Effect of dust deposition on electrical efficiency of PV module

| S. No. | Different solar cell materials | $\eta_{m0} = \tau_g \times \eta_0$ | $\eta_0$ [$\tau_g = 0.9$] clean glass | $\eta_{m0}$ [$\tau_g = 0.7$] | $\eta_{m0}$ [$\tau_g = 0.5$] |
|--------|-------------------------------|------------------------------------|---------------------------------------|------------------------------|------------------------------|
| 1 | c-Si (single-crystalline) | 16 | 0.1777 | 0.1244 | 0.0889 |
| 2 | mc-Si (multi-crystalline silicon) | 14 | 0.1555 | 0.1089 | 0.0778 |
| 3 | nc-Si (nanocrystalline silicon) | 12 | 0.1333 | 0.0933 | 0.0067 |
| 4 | a-Si (amorphous silicon) | 6 | 0.0666 | 0.0466 | 0.0333 |
| 5 | CdTe (cadmium telluride) | 8 | 0.8888 | 0.6222 | 0.4444 |
| 6 | CIGS (copper indium gallium selenide) | 10 | 0.1111 | 0.7778 | 0.5556 |

**Example 3.19** Evaluate the electrical efficiency of Si-based PV module at STC in fraction and percentage by using fill factor (FF) for data of Table 3.1.

**Solution**

From Eqs. 3.22a and 3.22b, we have an expression for electrical efficiency as

$$\eta_{m0} = \frac{\text{FF} \times I_{sc} \times V_{oc}}{A_m \times I} \text{ in fraction} = \left( \frac{\text{FF} \times I_{sc} \times V_{oc}}{A_m \times I} \right) \times 100 \text{ in percentage.}$$

From Table 3.1: FF $= 0.89$; $I_{sc} = 4.4$ Am; $V_{oc} = 18$ V; $A_m = 0.605$ m$^2$ and $I = 1000$ W/m$^2$ at STC.

Now, electrical efficiency of PV module ($\eta_{m0}$) in fraction $= \frac{\text{FF} \times I_{sc} \times V_{oc}}{A_m \times I} = \frac{0.89 \times 4.4 \times 18}{0.605 \times 1000} = 0.1165$.

Electrical efficiency of PV module ($\eta_{m0}$) in percentage $= \frac{\text{FF} \times I_{sc} \times V_{oc}}{A_m \times I} \times 100 = \frac{0.89 \times 4.4 \times 18}{0.605 \times 1000} \times 100 = 11.65\% \approx 12\%$.

### 3.3.3  Temperature Coefficient of PV Module

From Eq. 3.20, we have

$$\eta_m = \eta_{m0}[1 - \beta_0(T_c - 25)]$$

The above equation can be rearranged as follows:

$$\beta_0 = \frac{\eta_{m0} - \eta_m}{\eta_{m0}(T_c - 25)} \ll 1 \text{ in fraction} \qquad (3.23)$$

The values of $\eta_{m0}$ and $\eta_m$ can be obtained from Eqs. 3.22a and 3.22b, respectively.

**Example 3.20** Evaluate temperature coefficient for multi-crystalline silicon (mc-Si) and nanocrystalline silicon (nc-Si) PV module in fraction for the parameters at 50 °C for Example 3.16.

**Solution**

From Eq. 3.23, we have an expression for electrical efficiency of PV module l as

$$\beta_0 = \frac{\eta_{m0} - \eta_m}{\eta_{m0}(T_c - 25)}$$

**For multi-crystalline silicon (mc-Si) solar cell at 50 °C**

: $\eta_{m0} = 0.14$ and $\eta_m = 0.125$ Example 3.16 and $\beta_0 = 0.00425$ (Table 3.2).
Then,

$$\beta_0 = \frac{\eta_{m0} - \eta_m}{\eta_{m0}(T_c - 25)} = \frac{0.14 - 0.125}{0.14 \times (50 - 25)} = 0.00426$$

**For nanocrystalline silicon (nc-Si) solar cell at 50 °C**

: $\eta_{m0} = 0.12$ and $\eta_m = 0.1092$, Example 3.16 $\beta_0 = 0.0036$ (Table 3.2).
Then,

$$\beta_0 = \frac{\eta_{m0} - \eta_m}{\eta_{m0}(T_c - 25)} = \frac{0.12 - 0.1092}{0.12 \times (50 - 25)} = 0.0036$$

### 3.3.4 Electrical Output of PV Module

If solar radiation, $I(t)$, is incident on PV module having packing factor $\beta_m$, the direct current (DC) power obtained in W can be calculated as follows:

$$\dot{E}_{el} = \beta_m, \times \eta_m \times I(t) \times A_m \tag{3.24}$$

Since, solar radiation varies with time (Chap. 1), so electrical power will also vary. Hence, the daily electrical energy in Wh will be evaluated as

$$E_{el} = \sum_{i=1}^{i=N} \eta_{mi} \times I_i(t) \times A_m \tag{3.25}$$

where, $i$ varies from 1 to $N$. The $N$ is sunshine hours given by Eq. 1.14. Further, the module electrical efficiency also varies with temperature due to variation of $I(t)$, Eq. 3.20.

**Example 3.21** Evaluate electrical output of a mono-crystalline and nanocrystalline PV module for area of PV module ($A_m$) of 0.605 m$^2$ and solar intensity of 500 W/m$^2$ and $T_c = 70\,°C$ for Example 3.16.

**Solution**

From Eq. 3.24, we have an expression for electrical output of PV module in $W$ as

$$\dot{E}_{el} = \eta_m \times I(t) \times A_m$$

**For multi-crystalline silicon (mc-Si) solar cell:** $\eta_{m0} = 0.14$ (Table 3.2) and at 70 °C

$$\eta_m = 0.14[1 - 0.00425(70 - 25)] = 0.11025$$

The DC electrical power in $W$ can be evaluated as

$$\dot{E}_{el} = \eta_m \times I(t) \times A_m = 0.11025 \times 500 \times 0.605 = 33.35\ \text{W}$$

**For nanocrystalline silicon (nc-Si) solar cell:** $\eta_{m0} = 0.12$ (Table 3.2) and at 70 °C

$$\eta_m = 0.12[1 - 0.0036(70 - 25)] = 0.0984$$

The DC electrical power in $W$ can be evaluated as

$$\dot{E}_{el} = \eta_m \times I(t) \times A_m = 0.0984 \times 500 \times 0.605 = 29.76\ \text{W}$$

One can see that an electrical output is more for multi-crystalline silicon (mc-Si) solar cell multi-crystalline silicon (mc-Si) solar cell.

### 3.3.5  Energy Balance Equations for Opaque PV Modules, Figs. 3.6a and 3.7a

The energy balance equations in $W$ for opaque PV modules have been written with following assumptions:

- One-dimensional heat conduction.
- The system is in quasi-steady state.
- The ohmic losses between two solar cells connected in series of PV module are negligible.
- Heat capacity of tedlar, solar cell material and toughen glass are also neglected due to small temperature gradient across its thickness.

The energy balance in $W$ for any PV module, Fig. 3.7a, can be written as follows:

$$\tau_g[\alpha_c\beta_c I(t) + (1 - \beta_c)\alpha_T I(t)]A_m = [U_{tc,a}(T_c - T_a) + U_{bc,a}(T_c - T_a)]A_m$$
$$+ \eta_m\beta_c I(t)A_m \qquad (3.26)$$

where,

The first term of LHS $= [\tau_g\alpha_c\beta_c I(t)A_m] =$ the rate of incident solar radiation, $[I(t)]$ absorbed by all solar cell, $[\alpha_c]$ in packing area $[\beta_c A_m]$ of PV module after transmission, $[\tau_g]$ in W.

The second term in LHS $= [\tau_g(1 - \beta_c)\alpha_T I(t)A_m] =$ the rate of incident solar radiation, $I(t)$ absorbed by tedlar, $\alpha_T$ in non-packing area $[(1 - \beta_c)A_m]$ of PV module after transmission, $\tau_g$ in W.

The first term of RHS $= [U_{tc,a}(T_c - T_a)A_m] =$ the rate of thermal energy lost from solar cell to ambient air through top toughen glass cover in W.

The second term of RHS $= [U_{bc,a}(T_c - T_a)A_m] =$ the rate of thermal energy lost from solar cell to ambient air through tedlar at bottom of PV module in W.

The third term $= [\eta_m\beta_c I(t)A_m] =$ the rate of DC power produced by PV module in W.

For packing factor of one $[\beta_c = 1]$ in any PV module, Eq. 3.26 reduces to

$$[\tau_g\alpha_c I(t)A_m] = [U_{tc,a}(T_c - T_a) + U_{bc,a}(T_c - T_a)]A_m + \eta_m I(t)A_m \qquad (3.27)$$

**Example 3.22** Evaluate the top, $U_{tc,a}$ with $(V = 1 \text{ m/s})$ and without wind condition and bottom, $U_{bc,a}$ loss coefficient of opaque PV module for the following parameters:

Thickness of toughen glass, $L_g = 0.003$ m; thermal conductivity of toughened glass, $K_g = 0.688$ W/m °C (Appendix Table E.4 [8]) and outside heat transfer coefficient from top glass to ambient $h_0 = 5.7 + 3.8$ V (Eq. 2.13a).

Thickness of tedlar, $L_t = 0.0005$ m; thermal conductivity of toughened glass $K_t = 0.033$ W/m °C (Appendix Table E.4 [8]).

Bottom heat transfer coefficient from back of tedlar to ambient $h_i = 2.8 + 3$ V (Eq. 2.13b).

**Solution**

Following Eq. 2.27b and Example 3.1, we can write an expression for an overall top heat transfer coefficient from solar cell $(T_c)$ to ambient air over top toughened glass as

(a) **With wind velocity**

$$U_{t,ca} = \left[\frac{L_g}{K_g} + \frac{1}{h_o}\right]^{-1} = \left[\frac{0.003}{0.688} + \frac{1}{5.7 + 3.8 \times 1}\right]^{-1}$$
$$= [0.00436 + 0.10416]^{-1} = [0.1096]^{-1} = 9.12 \text{ W/m}^2 \text{ °C}$$

(b) **Without wind velocity**

$$U_{t,ca} = \left[ \frac{L_g}{K_g} + \frac{1}{h_o} \right]^{-1} = \left[ \frac{0.003}{0.688} + \frac{1}{5.7} \right]^{-1} = [0.00436 + 0.1754]^{-1}$$
$$= [0.1797]^{-1} = 5.56 \, \text{W/m}^2 \, {}^\circ\text{C}.$$

The above calculations show that the contribution of toughened glass is marginal; however, wind velocity has major role to play in energy balance.

Following Eq. 2.27b and Example 3.1, we can write an expression for an overall bottom loss transfer coefficient from solar cell ($T_c$) to ambient air through tedlar as

(a) **With wind velocity**

$$U_{b,ca} = \left[ \frac{L_t}{K_t} + \frac{1}{h_i} \right]^{-1} = \left[ \frac{0.0005}{0.033} + \frac{1}{2.8 + 3 \times 1} \right]^{-1}$$
$$= [0.01515 + 0.1724]^{-1} = [0.1876]^{-1} = 5.3304 \, \text{W/m}^2 \, {}^\circ\text{C}$$

(b) **Without wind velocity**

$$U_{b,ca} = \left[ \frac{L_t}{K_t} + \frac{1}{h_i} \right]^{-1} = \left[ \frac{0.0005}{0.033} + \frac{1}{2.8} \right]^{-1}$$
$$= [0.01515 + 0.3571]^{-1} = [0.3722]^{-1} = 2.686 \, \text{W/m}^2 \, {}^\circ\text{C}.$$

In this case too, effect of tedlar on $U_{b,ca}$ is marginal due to its smallest thickness.

**Example 3.23** Evaluate the overall heat transfer coefficient, $U_{bt}$, from bottom ambient air to top ambient air thorough tedlar, solar cell and toughened glass for Example 3.22 for following solar cell parameters with and without wind velocity:

Thickness of solar cell, $L_{sc} = 0.0003$ m; Thermal conductivity of solar cell, $K_{sc} = 0.036$ W/m °C (Appendix Table E.4 [8]).

**Solution**

**Following Eq. 2.27b, one has an expression for $U_{bt}$ as**

(a) **With wind velocity**

$$U_{bt} = \left[ \frac{1}{h_o} + \frac{L_g}{K_g} + \frac{L_{sc}}{K_{sc}} + \frac{L_t}{K_t} + \frac{1}{h_i} \right]^{-1}$$
$$= \left[ \frac{1}{5.7 + 3.8 \times 1} + \frac{0.003}{0.688} + \frac{0.0003}{0.036} + \frac{0.0005}{0.033} + \frac{1}{2.8 + 3 \times 1} \right]^{-1}$$
$$= [0.1053 + 0.00436 + 0.00909 + 0.0152 + 0.1754]^{-1}$$
$$= [0.3093]^{-1} = 3.233 \, \text{W/m}^2 \, {}^\circ\text{C}$$

## (b) Without wind velocity

$$U_{bt} = \left[ \frac{1}{h_o} + \frac{L_g}{K_g} + \frac{L_{sc}}{K_{sc}} + \frac{L_t}{K_t} + \frac{1}{h_i} \right]^{-1}$$

$$= \left[ \frac{1}{5.7} + \frac{0.003}{0.688} + \frac{0.0003}{0.036} + \frac{0.0005}{0.033} + \frac{1}{2.8} \right]^{-1}$$

$$= [0.1754 + 0.00436 + 0.00909 + 0.0152 + 0.3572]^{-1}$$

$$= [0.5612]^{-1} = 1.78 \, \text{W/m}^2 \, {}^\circ\text{C}$$

This indicates that wind velocity has significant role in determining the overall heat transfer coefficient.

**Example 3.24** Repeat Example 3.23 without toughened glass, solar cell and tedlar materials with wind velocity.

**Solution**

An expression for an overall heat transfer coefficient, $U_{bt}$, from bottom ambient air to top ambient air without tedlar, solar cell and toughened glass can be written as follows:

$$U_{bt} = \left[ \frac{1}{h_o} + \frac{1}{h_i} \right]^{-1} = \left[ \frac{1}{5.7 + 3.8 \times 1} + \frac{1}{2.8 + 3 \times 1} \right]^{-1}$$

$$= [0.1053 + 0.1754]^{-1} = [0.2807]^{-1} = 3.56 \, \text{W/m}^2 \, {}^\circ\text{C}$$

This shows that toughened glass, solar cell material and tedlar have marginal effect on an overall heat transfer coefficient.

**Example 3.25** Calculate the total heat transfer coefficient from solar cell to ambient air through toughened glass and tedlar, $U_L$, for Example 3.22 with and without wind velocity.

**Solution**

The total heat transfer coefficient from solar cell to ambient air through toughened glass and tedlar, $U_L$, is sum of top $(U_{t,ca})$ and bottom $(U_{b,ca})$ overall heat transfer coefficient.

It is expressed as

$$U_L = U_{t,ca} + U_{b,ca}$$

(a)  **With wind velocity**

We have

$U_{t,ca} = 9.14 \, \text{W/m}^2 \, {}^\circ\text{C}$ and $U_{b,ca} = 5.3304 \, \text{W/m}^2 \, {}^\circ\text{C}$, Example 3.22.

So, the total overall heat transfer coefficient from solar cell to ambient air through top toughened glass and tedlar becomes

$$U_L = U_{t,ca} + U_{b,ca} = 9.12 + 5.33 = 14.45 \, \text{W/m}^2 \, {}^\circ\text{C}$$

(b)  **Without wind velocity**

We have

$U_{t,ca} = 5.56 \, \text{W/m}^2 \, {}^\circ\text{C}$, and $U_{b,ca} = 2.686 \, \text{W/m}^2 \, {}^\circ\text{C}$, Example 3.22.

So, the total overall heat transfer coefficient from solar cell to ambient air through top toughened glass and tedlar without wind becomes

$$U_L = U_{t,ca} + U_{b,ca} = 5.56 + 2.686 = 8.246 \, \text{W/m}^2 \, {}^\circ\text{C}.$$

### 3.3.5.1  Analytical Expression for Solar Cell Temperature $(T_c)$

From Eq. 3.26, we have

$$\tau_g[\alpha_c\beta_c I(t) + (1 - \beta_c)\alpha_T I(t)]A_m$$
$$= \left[U_{tc,a}(T_c - T_a) + U_{bc,a}(T_c - T_a)\right]A_m + \eta_m\beta_c I(t)A_m$$

or,

$$\tau_g[\alpha_c\beta_c I(t) + (1 - \beta_c)\alpha_T I(t)] = U_L(T_c - T_a) + \eta_m\beta_c I(t) \qquad (3.28)$$

where, $U_L = U_{tc,a} + U_{bc,a}$, Example 3.25.

From Eqs. 3.22a and 3.22b, we also have

$$\eta_m = \eta_{m0}[1 - \beta_0(T_c - 25)]$$

After substitution of $\eta_m$ in Eq. 3.28, one gets an expression for solar cell temperature as

$$T_c = \frac{\tau_g[\alpha_c\beta_c + (1 - \beta_c)\alpha_T]I(t) + U_LT_a - [\eta_{m0}\beta_cI(t)](1 + 25\beta_0)}{[U_L - [\eta_{m0}\beta_cI(t)\beta_0]]} \tag{3.29}$$

For $\beta_c = 1$, the above equation reduces to

$$T_c = \frac{\tau_g\alpha_cI(t) + U_LT_a - [\eta_{m0}I(t)](1 + 25\beta_0)}{[U_L - [\eta_{m0}I(t)\beta_0]]} \tag{3.30}$$

For $\beta_c = 0$, the solar cell temperature becomes the tedlar temperature, i.e., $T_c = T_t$, the above equation reduces to

$$T_c = T_t = \frac{\tau_g\alpha_TI(t) + U_LT_a}{U_L} = \frac{\tau_g\alpha_TI(t)}{U_L} + T_a \tag{3.31}$$

Equation 3.31 is the temperature of glazed tedlar.

**Example 3.26** Calculate solar cell temperature $(T_c)$ with wind velocity of 1 m/s for multi-crystalline (mc-Si) solar cell for packing factor $(\beta_c)$ of 0.89, absorptivity of 0.9 and solar intensity of 500 W/m².

Other given parameters: $\tau_g$ = 0.95, $\alpha_T$ = 0.20, $U_L$ = 14.45 W/m² °C with wind velocity of 1 m/s and $U_L$ = 8.246 W/m² °C without wind velocity (Example 3.25) and $T_a = 25$ °C, $\eta_{m0} = 0.14$ and $\beta_0 = 0.00425$ (Table 3.2).

**Solution**

Here,

$$\tau_g[\alpha_c\beta_c + (1 - \beta_c)\alpha_T]I(t) = 0.95[0.9 \times 0.89 + 0.11 \times 0.20]$$
$$\times 500 = 390.93 \text{ W/m}^2 \text{ and } [\eta_{m0}\beta_cI(t)](1 + 25\beta_0)$$
$$= 0.14 \times 0.89 \times 500(1 + 25 \times 0.00425) = 68.92 \text{ W/m}^2$$

(a) **With wind velocity**

$$[U_L - [\eta_{m0}\beta_cI(t)\beta_0]] = 14.45 - (0.14 \times 0.89 \times 500 \times 0.00425)$$
$$= 14.18 \text{ W/m}^2 \text{ °C}$$

From Eq. 3.29, we have

$$T_c = \frac{\tau_g[\alpha_c\beta_c + (1 - \beta_c)\alpha_T]I(t) + U_LT_a - [\eta_{m0}\beta_cI(t)](1 + 25\beta_0)}{[U_L - [\eta_{m0}\beta_cI(t)\beta_0]]}$$
$$= \frac{390.93 + 361.25 - 68.92}{14.18} = 48.19 \text{ °C}$$

(b) **Without wind velocity**

$$[U_L - [\eta_{m0}\beta_c I(t)\beta_0]] = 8.246 - (0.14 \times 0.89 \times 500 \times 0.00425)$$
$$= 7.98 \, \text{W/m}^2 \, ^\circ\text{C}$$

From Eq. 3.29, we have

$$T_c = \frac{\tau_g[\alpha_c\beta_c + (1 - \beta_c)\alpha_T]I(t) + U_L T_a - [\eta_{m0}\beta_c I(t)](1 + 25\beta_0)}{[U_L - [\eta_{m0}\beta_c I(t)\beta_0]]}$$
$$= \frac{390.93 + 206.15 - 68.92}{7.98} = 66.18 \, ^\circ\text{C}$$

One can observe that solar cell temperature with wind velocity is lower by 18 °C due to cooling of PV module due to wind.

**Example 3.27** Repeat Example 3.26 for packing factor of $(\beta_c)$ of 0.5.

**Solution**

**For $\beta_c = 0.5$, we have**

$$\tau_g[\alpha_c\beta_c + (1 - \beta_c)\alpha_T]I(t) = 0.95[0.9 \times 0.50 + 0.50 \times 0.20] \times 500$$
$$= 261.125 \, \text{W/m}^2 \text{ and } [\eta_{m0}\beta_c I(t)](1 + 25\beta_0)$$
$$= 0.14 \times 0.5 \times 500(1 + 25 \times 0.00425) = 38.72 \, \text{W/m}^2$$

(a) **With wind velocity**

$$[U_L - [\eta_{m0}\beta_c I(t)\beta_0]] = 14.45 - (0.14 \times 0.5 \times 500 \times 0.00425)$$
$$= 14.30 \, \text{W/m}^2 \, ^\circ\text{C}$$

From Eq. 3.29, we have

$$T_c = \frac{\tau_g[\alpha_c\beta_c + (1 - \beta_c)\alpha_T]I(t) + U_L T_a - [\eta_{m0}\beta_c I(t)](1 + 25\beta_0)}{[U_L - [\eta_{m0}\beta_c I(t)\beta_0]]}$$
$$= \frac{261.125 + 361.25 - 38.72}{14.30} = 40.82 \, ^\circ\text{C}$$

(b) **Without wind velocity**

$$[U_L - [\eta_{m0}\beta_c I(t)\beta_0]] = 8.246 - (0.14 \times 0.5 \times 500 \times 0.00425)$$
$$= 8.097 \, \text{W/m}^2 \, ^\circ\text{C}$$

From Eq. 3.29, we have

$$T_c = \frac{\tau_g[\alpha_c\beta_c + (1 - \beta_c)\alpha_T]I(t) + U_LT_a - [\eta_{m0}\beta_cI(t)](1 + 25\beta_0)}{[U_L - [\eta_{m0}\beta_cI(t)\beta_0]]}$$

$$= \frac{261.125 + 206.15 - 38.72}{8.097} = 52.93\,°C$$

Here, one can conclude that if packing factor is reduced, then solar cell temperature is also reduced.

**Example 3.28** Repeat Example 3.26 for packing factor of $(\beta_c)$ of zero.

**Solution**

(a)  For $\beta_c = 0$, we have Eq. 3.31 as

$$T_c = T_t = \frac{\tau_g\alpha_TI(t) + U_LT_a}{U_L} = \frac{\tau_g\alpha_TI(t)}{U_L} + T_a$$

(i)  With wind velocity of 1 m/s

$$T_c = T_t = \frac{\tau_g\alpha_TI(t)}{U_L} + T_a = \frac{0.95 \times 0.20 \times 500}{14.45} + 25 = 31.71\,°C$$

(ii)  Without wind velocity $(V = 0)$

$$T_c = T_t = \frac{\tau_g\alpha_TI(t)}{U_L} + T_a = \frac{0.95 \times 0.20 \times 500}{8.246} + 25 = 36.52\,°C$$

The results are as per expectation. This condition acts as a glazed tedlar.

**Example 3.29** Repeat Example 3.26 for packing factor of $(\beta_c)$ of one.

**Solution**

**For $\beta_c = 1$ we have**

$$\tau_g[\alpha_c\beta_c + (1 - \beta_c)\alpha_T]I(t) = 0.95 \times 0.9 \times 500 = 427.5 \text{ W/m}^2 \text{ and}$$
$$[\eta_{m0}(t)](1 + 25\beta_0) = 0.14 \times 500(1 + 25 \times 0.00425) = 77.43 \text{ W/m}^2$$

(c)  **With wind velocity**

$$[U_L - [\eta_{m0}I(t)\beta_0]] = 14.45 - (0.14 \times 500 \times 0.00425) = 14.15 \text{ W/m}^2\,°C$$

From Eq. 3.30, we have

$$T_c = T_c = \frac{\tau_g \alpha_c I(t) + U_L T_a - [\eta_{m0} I(t)](1 + 25\beta_0)}{[U_L - [\eta_{m0} I(t) \beta_0]]}$$

$$= \frac{427.5 + 361.25 - 77.43}{14.15} = 50.63 \,^\circ C$$

### (d) Without wind velocity

$$[U_L - [\eta_{m0} I(t) \beta_0]] = 8.246 - (0.14 \times 500 \times 0.00425)$$
$$= 7.948 \, W/m^2 \,^\circ C$$

From Eq. 3.30, we have

$$T_c = \frac{\tau_g \alpha_c I(t) + U_L T_a - [\eta_{m0} I(t)](1 + 25\beta_0)}{[U_L - [\eta_{m0} I(t) \beta_0]]}$$

$$= \frac{427.5 + 206.15 - 77.43}{7.948} = 69.98 \,^\circ C$$

In this case of $\beta_c = 1$, one can observe that solar cell temperature is maximum due to maximum utilization of incident solar radiation.

### 3.3.5.2 Analytical Expression for PV Module Electrical Efficiency ($\eta_m$)

From Eqs. 3.22a and 3.22b, we also have an expression for PV module electrical efficiency ($\eta_m$)

$$\eta_m = \eta_{m0}[1 - \beta_0(T_c - 25)]$$

From Eq. 3.29, we have an expression for solar cell temperature ($T_c$)

$$T_c = \frac{\tau_g[\alpha_c \beta_c + (1 - \beta_c)\alpha_T]I(t) + U_L T_a - [\eta_{m0}\beta_c I(t)](1 + 25\beta_0)}{[U_L - [\eta_{m0}\beta_c I(t)\beta_0]]}$$

$$T_c - 25 = \frac{\tau_g[\alpha_c \beta_c + (1 - \beta_c)\alpha_T]I(t) + U_L T_a - [\eta_{m0}\beta_c I(t)](1 + 25\beta_0)}{[U_L - [\eta_{m0}\beta_c I(t)\beta_0]]} \\ \frac{- 25[U_L - [\eta_{m0}\beta_c I(t)\beta_0]]}{[U_L - [\eta_{m0}\beta_c I(t)\beta_0]]}$$

$$T_c - 25 = \frac{\tau_g[\alpha_c \beta_c + (1 - \beta_c)\alpha_T]I(t) + U_L T_a - [\eta_{m0}\beta_c I(t)] - 25U_L}{[U_L - [\eta_{m0}\beta_c I(t)\beta_0]]}$$

$$[1 - \beta_0(T_c - 25)]$$
$$= 1 - \beta_0 \left\{ \frac{\tau_g[\alpha_c \beta_c + (1 - \beta_c)\alpha_T]I(t) + U_L T_a - [\eta_{m0}\beta_c I(t)] - 25U_L}{[U_L - [\eta_{m0}\beta_c I(t)\beta_0]]} \right\}$$

$$[1 - \beta_0(T_c - 25)] = \frac{] - \beta_0\{\tau_g[\alpha_c \beta_c + (1 - \beta_c)\alpha_T]I(t)}{+ U_L T_a - [\eta_{m0}\beta_c I(t)] - 25U_L\}}{[U_L - [\eta_{m0}\beta_c I(t)\beta_0]]}$$

$$\eta_m = \eta_{m0} \frac{] - \beta_0\{\tau_g[\alpha_c\beta_c + (1 - \beta_c)\alpha_T]I(t) + U_L T_a - [\eta_{m0}\beta_c I(t)] - 25U_L\}}{[U_L - [\eta_{m0}\beta_c I(t)\beta_0]]} \qquad (3.32)$$

The threshold intensity $I(t)_{th}$ can be obtained by having $\eta_m$ of Eq. 3.32 less than one and hence

$$\frac{[U_L - [\eta_{m0}\beta_c I(t)\beta_0]] - \beta_0\{\tau_g[\alpha_c\beta_c + (1 - \beta_c)\alpha_T]I(t) + U_L T_a - [\eta_{m0}\beta_c I(t)] - 25U_L\}}{[U_L - [\eta_{m0}\beta_c I(t)\beta_0]]} < 1$$

or,

$$I(t)_{ths} > \frac{U_L[25 - T_a]}{\{\tau_g[\alpha_c\beta_c + (1 - \beta_c)\alpha_T] - [\eta_{m0}\beta_c]\}} \qquad (3.33)$$

To get DC power from PV module, the required condition should be achieved from the above equation. The required conditions are as follows:

(i)  $I(t)_{ths} > U_L[25 - T_a]$.
(ii)  The denominator, i.e., $\{\tau_g[\alpha_c\beta_c + (1 - \beta_c)\alpha_T] - [\eta_{m0}\beta_c]\}$ should be always positive.

In summer/winter condition, the term $[25 - T_a]$ may be positive/negative, sometime even PV produces the DC power which has been verified by us.

**Example 3.30** Evaluate the electrical efficiency and electrical power of multi-crystalline (mc-Si) PV module with an effective area of 1 m$^2$ for Example 3.26 with and without wind velocity of 1 m/s.

**Solution**

Here,

$$\tau_g[\alpha_c\beta_c + (1 - \beta_c)\alpha_T]I(t) = 0.95[0.9 \times 0.89 + 0.11 \times 0.20] \times 500$$
$$= 390.93 \text{ W/m}^2 \text{ and}$$
$$[\eta_{m0}\beta_c I(t)](1 + 25\beta_0) = 0.14 \times 0.89 \times 500(1 + 25 \times 0.00425)$$
$$= 68.92 \text{ W/m}^2$$

(a)  **With wind velocity**

$$[U_L - [\eta_{m0}\beta_c I(t)\beta_0]] = 14.45 - (0.14 \times 0.89 \times 500 \times 0.00425)$$
$$= 14.18 \text{ W/m}^2 \,^\circ\text{C}$$

From Eq. 3.29, we have

$$\eta_m = \eta_{m0} \frac{\begin{aligned}[U_L - [\eta_{m0}\beta_c I(t)\beta_0]] - \beta_0\{\tau_g[\alpha_c\beta_c + (1 - \beta_c)\alpha_T]I(t) \\ + U_L T_a - [\eta_{m0}\beta_c I(t)] - 25U_L\}\end{aligned}}{[U_L - [\eta_{m0}\beta_c I(t)\beta_0]]}$$

$$= 0.14 \frac{14.18 - 0.00425\{390.93 + 14.45 \times 25 - 0.14 \times 0.89 \times 500 - 25 \times 14.45\}}{14.18}$$

$$= 0.14 \frac{14.14 - 0.00425 \times (752.18 - 423.55)}{14.18}$$

$$= 0.123$$

Electrical power $(\dot{E}_{pv}) = \eta_m \times I(t) \times \beta_c \times A_m = 0.123 \times 500 \times 0.89 \times 1 = 54.735$ W.

(b) **Without wind velocity**

$$[U_L - [\eta_{m0}\beta_c I(t)\beta_0]] = 8.246 - (0.14 \times 0.89 \times 500 \times 0.00425)$$
$$= 7.98 \text{ W/m}^2 \,^\circ\text{C}$$

From Eq. 3.29, we have

$$\eta_m = \eta_{m0} \frac{\begin{aligned}[U_L - [\eta_{m0}\beta_c I(t)\beta_0]] - \beta_0\{\tau_g[\alpha_c\beta_c + (1 - \beta_c)\alpha_T]I(t) \\ + U_L T_a - [\eta_{m0}\beta_c I(t)] - 25U_L\}\end{aligned}}{[U_L - [\eta_{m0}\beta_c I(t)\beta_0]]}$$

$$= 0.14 \frac{7.98 - 0.00425\{390.93 + 8.246 \times 25 - 0.14 \times 0.89 \times 500 - 25 \times 8.246\}}{7.98}$$

$$= 0.14 \frac{7.987.98 - 0.00425 \times (597.08 - 268.45)}{7.98}$$

$$= 0.1155$$

Electrical power $(\dot{E}_{pv}) = \eta_m \times I(t) \times \beta_c \times A_m = 0.1155 \times 500 \times 0.89 \times 1 = 51.40$ W.

One can infer that the electrical efficiency of multi-crystalline (mc-Si) PV module with wind velocity is higher by 18 6.5% due to cooling of PV module due to wind.

**Example 3.31** Verify the results of Example 3.30 for PV module efficiency by using the expression given by Eqs. 3.22a and 3.22b as

$$\eta_m = \eta_{m0}[1 - \beta_0(T_c - 25)]$$

**Solution**

From Example 3.26, we have

(a) **With wind velocity**

$$T_c = 48.19\,°C$$

Then,

$$\eta_m = \eta_{m0}[1 - \beta_0(T_c - 25)] = 0.14$$
$$\times [1 - 0.00425(48.19 - 25)] = 0.126$$

(b) **Without wind velocity**

From Eq. 3.29, we have

$$T_c = 66.18\,°C$$

Then,

$$\eta_m = \eta_{m0}[1 - \beta_0(T_c - 25)] = 0.14 \times [1 - 0.00425(66.18 - 25)] = 0.1155$$

The results are similar to those obtained in Example 3.30.

**Example 3.32** Evaluate the electrical efficiency and power of opaque PV module for packing factor of one for solar cell temperature of Example 3.29.

**Solution**

From Example 3.31 and Eqs. 3.22a and 3.22b, we have

$$\eta_m = \eta_{m0}[1 - \beta_0(T_c - 25)]$$

(a) **With wind velocity**

From Example 3.29: $T_c = 50.63\,°C$

$$\eta_m = 0.14[1 - 0.00425(50.63 - 25)] = 0.1247 \text{in fraction}$$

Electrical power $\left(\dot{E}_{pv}\right) = \eta_m \times I(t) \times \beta_c \times A_m = 0.1247 \times 500 \times 1 \times 1 = 62.35$ W.

(b) **Without wind velocity**

From Example 3.29: $T_c = 69.98\,°C$

$$\eta_m = \eta_{m0}[1 - \beta_0(69.98 - 25)] = 0.1132 \text{ in fraction}$$

Electrical power $\left(\dot{E}_{pv}\right) = \eta_m \times I(t) \times \beta_c \times A_m = 0.1132 \times 500 \times 1 \times 1 = 55.60$ W.

**Example 3.33** Determine threshold intensity for Example 3.26 for summer condition.

**Solution**

From Eq. 3.33, we have an expression for threshold solar intensity, $I(t)_{ths}$, to get DC power from PV module as

$$I(t)_{ths} > \frac{U_L[25 - T_a]}{\{\tau_g[\alpha_c\beta_c + (1 - \beta_c)\alpha_T] - [\eta_{mo}\beta_c]\}}$$

From Example 3.30 for mc-Si-PV module, we have the following parameters to evaluate threshold solar intensity with wind velocity of 1 m/s as

$U_L = 14.45 \, \text{W/m}^2\,{}^\circ\text{C}$; $\eta_{mo} = 0.14$; $\beta_c = 0.89$ and $\beta_0 = 0.00425 C^{-1}$ (Table 3.2) $\alpha_c = 0.9$, $\alpha_T = 0.2$, $\tau_g = 0.95$ and $T_a = 25\,{}^\circ\text{C}$.

Now,

$$I(t)_{ths} > \frac{14.45[25 - T_a]}{\{0.95[0.9 \times 0.89 + (1 - 0.89)0.2] - [0.14 \times 0.89]\}}$$

$$= \frac{14.45[25 - 25]}{0.78185 - 0.1246} = \frac{0}{0.6574} = 0$$

This shows that for summer condition, threshold intensity should be more than zero for DC power production for any design parameters of PV module. We have seen that PV produces DC power at the time of sun starts shining early in the morning.

**Example 3.34** Determine threshold intensity for Example 3.26 for winter condition of 10 °C.

**Solution**

For winter condition with same design parameters of PV module, we have from Eq. 3.33,

$$I(t)_{ths} > \frac{U_L[25 - T_a]}{\{\tau_g[\alpha_c\beta_c + (1 - \beta_c)\alpha_T] - [\eta_{mo}\beta_c]\}}$$

$$= \frac{14.45[25 - 10]}{0.78185 - 0.1246} = \frac{216}{0.6574} = 329.71 \, \text{W/m}^2$$

One can see that the PV module takes some time to produce DC power after sunshine hours unlike summer condition, Example 3.31. It mainly depends on the total overall heat transfer coefficient, design parameters and ambient air temperature.

### 3.3.6 Energy Balance Equations for Semi-Transparent PV Module, Figs. 3.6b and 3.7b

In this case, the back sheet of opaque PV module, i.e., tedlar, is replaced by same toughened glass as shown in Fig. 3.7b. Due to this, incident solar radiation on non-packing area is transmitted to other side of PV module unlike opaque PV module. The assumptions made to write the energy balance of semi-transmitted PV module are same as made in Sect. 3.3.5.

The energy balance in W for any semi-transparent transparent PV module, Fig. 3.7b, can be written as follows:

$$\tau_g[\alpha_c \beta_c I(t)] A_m = \left[ U_{tc,a}(T_c - T_a) + U_{bc,a}(T_c - T_a) \right] A_m + \eta_m \beta_c I(t) A_m \quad (3.34)$$

where,

The term of LHS $= \left[ \tau_g \alpha_c \beta_c I(t) A_m \right] =$ the rate of incident solar radiation, $[I(t)]$ absorbed by all solar cell, $[\alpha_c]$ in packing area $[\beta_c A_m]$ of PV module after transmission, $\left[ \tau_g \right]$ in W.

The $\left[ \tau_g^2 (1 - \beta_c) \alpha_T I(t) A_m \right] =$ the rate of incident solar radiation, $[I(t)]$ on semi-transparent PV module is transmitted after two times, $\tau_g^2$ in W to back of PV module.

The first term of RHS $= \left[ U_{tc,a}(T_c - T_a) A_m \right] =$ the rate of thermal energy lost from solar cell to ambient air through top toughen glass cover in W.

The second term of RHS $= \left[ U_{bc,a}(T_c - T_a) A_m \right] =$ the rate of thermal energy lost from solar cell to ambient air through tedlar at bottom of PV module in W.

The third term $= [\eta_m \beta_c I(t) A_m] =$ the rate of DC power produced by PV module in W.

If packing factor, $[\beta_c]$, becomes one, then Eq. 3.34 reduces to

$$\tau_g[\alpha_c I(t)] A_m = \left[ U_{tc,a}(T_c - T_a) + U_{bc,a}(T_c - T_a) \right] A_m + \eta_m I(t) A_m \quad (3.35)$$

This is exactly same as Eq. 3.27.

**Example 3.35** Evaluate the top, $U_{tc,a}$ with ($V = 1$ m/s) and without wind condition and bottom, $U_{bc,a}$ loss coefficient of semi-transparent PV module for the following parameters:

Thickness of top and bottom toughen glass, $L_g = 0.003$ m; thermal conductivity of toughened glass, $K_g = 0.688$ W/m °C (Appendix Table E.4 [8]) and outside heat transfer coefficient from top glass to ambient $h_0 = 5.7 + 3.8\ V$ (Eq. 2.13a).

Bottom heat transfer coefficient from back of semi-transparent PV module to ambient $h_i = 2.8 + 3\ V$ (Eq. 2.13b).

**Solution**

Following Eq. 2.27b and Example 3.1, we can write an expression for an overall top heat transfer coefficient from solar cell ($T_c$) to ambient air over top toughened glass as

(a) **With wind velocity**

$$U_{t,ca} = \left[\frac{L_g}{K_g} + \frac{1}{h_o}\right]^{-1} = \left[\frac{0.003}{0.688} + \frac{1}{5.7 + 3.8 \times 1}\right]^{-1}$$

$$= [0.00436 + 0.10416]^{-1} = [0.1096]^{-1} = 9.12\ \text{W/m}^2\,^\circ\text{C}$$

(b) **Without wind velocity**

$$U_{t,ca} = \left[\frac{L_g}{K_g} + \frac{1}{h_o}\right]^{-1} = \left[\frac{0.003}{0.688} + \frac{1}{5.7}\right]^{-1}$$

$$= [0.00436 + 0.1754]^{-1} = [0.1797]^{-1} = 5.56\ \text{W/m}^2\,^\circ\text{C}$$

In this case of semi-transparent PV module too, the above calculations show that

(i) The contribution of toughened glass is marginal; however wind velocity has major role to play in energy balance.
(ii) The top overall heat transfer coefficient for opaque and semi-transparent PV module is exactly same.

Following Eq. 2.27b and Example 3.1, we can write an expression for an overall bottom loss transfer coefficient from tedlar $(T_t)$ to ambient air through tedlar as

(a) **With wind velocity**

$$U_{b,ca} = \left[\frac{L_g}{K_g} + \frac{1}{h_i}\right]^{-1} = \left[\frac{0.003}{0.688} + \frac{1}{2.8 + 3 \times 1}\right]^{-1}$$

$$= [0.00436 + 0.1724]^{-1} = [0.17676]^{-1}$$

$$= 5.6574\ (5.3304)\ \text{W/m}^2\,^\circ\text{C}$$

(b) **Without wind velocity**

$$U_{b,ca} = \left[\frac{L_g}{K_g} + \frac{1}{h_i}\right]^{-1} = \left[\frac{0.003}{0.688} + \frac{1}{2.8}\right]^{-1}$$

$$= [0.00436 + 0.3571]^{-1} = [0.3615]^{-1}$$

$$= 2.7663\ (2.686)\ \text{W/m}^2\,^\circ\text{C}$$

In this case, it is to be noted that the overall bottom heat transfer coefficient in case of semi-transparent PV module is more in comparison with opaque PV module (bracket value) and hence more cooling of semi-transparent PV module.

**Example 3.36** Evaluate the overall heat transfer coefficient, $U_{bt}$, from bottom ambient air to top ambient air thorough tedlar, solar cell and toughened glass for Example 3.22 for following solar cell parameters with and without wind velocity:

Thickness of solar cell, $L_{sc} = 0.0003$ m; Thermal conductivity of solar cell, $K_{sc} = 0.036$ W/m °C (Appendix Table E.4 [8]).

**Solution**

Following Eq. 2.27b, one has an expression for $U_{bt}$ as

(a) **With wind velocity**

$$U_{bt} = \left[ \frac{1}{h_o} + \frac{L_g}{K_g} + \frac{L_{sc}}{K_{sc}} + \frac{L_g}{K_g} + \frac{1}{h_i} \right]^{-1}$$

$$= \left[ \frac{1}{5.7 + 3.8 \times 1} + \frac{0.003}{0.688} + \frac{0.0003}{0.036} + \frac{0.003}{0.688} + \frac{1}{2.8 + 3 \times 1} \right]^{-1}$$

$$= [0.1053 + 0.00436 + 0.00909 + 0.00436 + 0.1754]^{-1}$$

$$= [0.2985]^{-1} = 3.35 \text{ W/m}^2 \text{ °C}$$

(b) **Without wind velocity**

$$U_{bt} = \left[ \frac{1}{h_o} + \frac{L_g}{K_g} + \frac{L_{sc}}{K_{sc}} + \frac{L_g}{K_g} + \frac{1}{h_i} \right]^{-1}$$

$$= \left[ \frac{1}{5.7} + \frac{0.003}{0.688} + \frac{0.0003}{0.036} + \frac{0.003}{0.688} + \frac{1}{2.8} \right]^{-1}$$

$$= [0.1754 + 0.00436 + 0.00909 + 0.0152 + 0.3572]^{-1}$$

$$= [0.5595]^{-1} = 1.79 \text{ W/m}^2 \text{ °C}$$

This indicates that wind velocity has significant role in determining the overall heat transfer coefficient, $U_{bt}$, even in semi-transparent PV module with

**Example 3.37** Repeat Example 3.23 without toughened glass, solar cell and tedlar materials with wind velocity.

**Solution**

An expression for an overall heat transfer coefficient, $U_{bt}$, from bottom ambient air to top ambient air without tedlar, solar cell and toughened glass can be written as follows:

$$U_{bt} = \left[ \frac{1}{h_o} + \frac{1}{h_i} \right]^{-1} = \left[ \frac{1}{5.7 + 3.8 \times 1} + \frac{1}{2.8 + 3 \times 1} \right]^{-1}$$

$$= [0.1053 + 0.1754]^{-1} = [0.2807]^{-1} = 3.56 \text{ W/m}^2 \text{ °C}$$

This shows that toughened glass, solar cell material and tedlar have no effect on an overall heat transfer coefficient of semi-transparent PV module; see Example 3.33.

**Example 3.38** Calculate the total heat transfer coefficient from solar cell to ambient air through semi-transparent PV module, $U_L$, for Example 3.32 with and without wind velocity.

**Solution**

The total heat transfer coefficient from solar cell to ambient air through semi-transparent PV module, $U_L$, is sum of top $(U_{t,ca})$ and bottom $(U_{b,ca})$ overall heat transfer coefficient.

It is expressed as

$$U_L = U_{t,ca} + U_{b,ca}$$

### (c)  With wind velocity

We have

$U_{t,ca} = 9.12\ \text{W/m}^2\,^\circ\text{C}$ and $U_{b,ca} = 5.6574\ \text{W/m}^2\,^\circ\text{C}$, Examples 3.22 and 3.35.

So, the total overall heat transfer coefficient from solar cell to ambient air through top toughened glass and tedlar becomes

$$U_L = U_{t,ca} + U_{b,ca} = 9.12 + 5.6574 = 14.78\ \text{W/m}^2\,^\circ\text{C}$$

### (d)  Without wind velocity

We have

$U_{t,ca} = 5.56\ \text{W/m}^2\,^\circ\text{C}$, and $U_{b,ca} = 2.7663\ \text{W/m}^2\,^\circ\text{C}$, Example 3.35.

So, the total overall heat transfer coefficient from solar cell to ambient air through top toughened glass and tedlar without wind becomes

$$U_L = U_{t,ca} + U_{b,ca} = 5.56 + 2.7663 = 8.33\ \text{W/m}^2\,^\circ\text{C}$$

#### 3.3.6.1   Analytical Expression for Solar Cell Temperature $(T_c)$

From Eq. 3.34, we have

$$\tau_g[\alpha_c\beta_c I(t)]A_m = \left[U_{tc,a}(T_c - T_a) + U_{bc,a}(T_c - T_a)\right]A_m + \eta_m\beta_c I(t)A_m \quad (3.36)$$

or,

$$\tau_g[\alpha_c\beta_c I(t)] = U_L(T_c - T_a) + \eta_m\beta_c I(t) \quad (3.37)$$

where, $U_L = U_{tc,a} + U_{bc,a}$, Example 3.25.

From Eqs. 3.22a and 3.22b, we also have

$$\eta_m = \eta_{m0}[1 - \beta_0(T_c - 25)]$$

After substitution of $\eta_m$ in Eq. 3.37, one gets an expression for solar cell temperature as

$$\tau_g[\alpha_c \beta_c I(t)] = U_L(T_c - T_a) + \eta_{m0}[1 - \beta_0(T_c - 25)]\beta_c I(t)$$

or,

$$T_c = \frac{[\tau_g \alpha_c - \eta_{m0}]\beta_c I(t) + U_L T_a - \eta_{m0}\beta_0\beta_c I(t) \times 25}{[U_L - \eta_{m0}\beta_0\beta_c I(t)]} \qquad (3.38)$$

For $\beta_c = 1$, the above equation reduces to

$$T_c = \frac{[\tau_g \alpha_c - \eta_{m0}]I(t) + U_L T_a - \eta_{m0}\beta_0 I(t) \times 25}{[U_L - \eta_{m0}\beta_0 I(t)]} \qquad (3.39)$$

For $\beta_c = 0$, the solar cell temperature becomes the toughened glass temperature, i.e., $T_c = T_g$, the above equation reduces to

$$T_c = T_g = T_a \qquad (3.40)$$

Equation 3.40 is the temperature of toughened glass which is equal to ambient air because there is no absorption of solar intensity in the glass cover.

**Example 3.39** Calculate solar cell temperature ($T_c$), electrical efficiency and electrical output of semi-transparent PV module with wind velocity of 1 m/s for multi-crystalline (mc-Si) solar cell for packing factor ($\beta_c$) of 0.89, absorptivity of 0.9 and solar intensity of 500 W/m$^2$ for semi-transparent PV module.

Other given parameters: $\tau_g$ = 0.95, $\alpha_T$ = 0.20, $U_L$ = 14.78 W/m$^2$ °C with wind velocity of 1 m/s and $U_L$ = 8.33 W/m$^2$ °C without wind velocity (Example 3.35) and $T_a = 25$ °C, $\eta_{m0} = 0.14$ and $\beta_0 = 0.00425$ (Table 3.2).

**Solution**

Here,
$$[\tau_g \alpha_c - \eta_{m0}]\beta_c I(t) + U_L T_a - \eta_{m0}\beta_c\beta_0 I(t) \times 25$$
$$= \{[0.95 \times 0.9 - 0.14]0.89 \times 500 + 14.78 \times 25$$
$$- 0.14 \times 0.89 \times 0.00425 \times 500 \times 25\}$$
$$= [318.18 + 369.5 - 66.19] = 687.68 \text{ W/m}^2 \text{ and}$$
$$[U_L - \eta_{m0}\beta_c\beta_0 I(t)] = 14.78 - 0.14 \times 0.89 \times 0.00425 \times 500$$
$$= 14.78 - 0.2648 = 14.515 \text{ W/m}^2 \text{ °C.}$$

## (a)  With wind velocity

From Eq. 3.38, we have an expression for solar cell

$$T_c = \frac{\left[\tau_g \alpha_c - \eta_{m0}\right] \beta_c I(t) + U_L T_a - \eta_{m0} \beta_0 \beta_c I(t) \times 25}{\left[U_L - \eta_{m0} \beta_0 \beta_c I(t)\right]}$$

$$= \frac{687.68}{14.515} = 47.38\,°C$$

Electrical efficiency of PV module $(\eta_m)$ $=$ $\eta_{m0}[1 - \beta_0(T_c - 25)]$ $=$ $0.14[1 - 0.00425(47.38 - 25)] = 0.1267$.

Electrical power $(\dot{E}_{pv}) = \eta_m \times I(t) \times \beta_c \times A_m = 0.1267 \times 500 \times 0.89 \times 1 = 56.37\,W$.

## (b)  Without wind velocity

Now,

$$[\tau_g \alpha_c - \eta_{m0}] \beta_c I(t) + U_L T_a - \eta_{m0} \beta_c \beta_0 I(t) \times 25$$
$$= \{[0.95 \times 0.9 - 0.14]0.89 \times 500 + 8.23 \times 25$$
$$- 0.14 \times 0.89 \times 0.00425 \times 500 \times 25\}$$
$$= [318.18 + 208.25 - 66.19] = 460.24\,W/m^2 \text{ and}$$

$$[U_L - \eta_{m0} \beta_c \beta_0 I(t)] = 8.33 - 0.14 \times 0.89 \times 0.00425 \times 500 = 8.33 - 0.2648$$
$$= 8.07\,W/m^2\,°C$$

From Eq. 3.38, we have

$$T_c = \frac{\left[\tau_g \alpha_c - \eta_{m0}\right] \beta_c I(t) + U_L T_a - \eta_{m0} \beta_0 \beta_c I(t) \times 25}{\left[U_L - \eta_{m0} \beta_0 \beta_c I(t)\right]}$$

$$= \frac{460.24}{8.07} = 57.03\,°C$$

Electrical efficiency of PV module $(\eta_m)$ $=$ $\eta_{m0}[1 - \beta_0(T_c - 25)]$ $=$ $0.14[1 - 0.00425(57.03 - 25)] = 0.1209$.

Electrical power $(\dot{E}_{pv}) = \eta_m \times I(t) \times \beta_c \times A_m = 0.1209 \times 500 \times 0.89 \times 1 = 53.81\,W$.

One can observe that solar cell temperature with wind velocity is lower by 9.65 °C due to cooling of PV module due to wind. However, the solar cell temperature of semi-transparent PV module is always less than the solar cell temperature of opaque PV module (Example 3.26).

**Example 3.40** Repeat Example 3.39 for packing factor of $(\beta_c)$ of 0.5.

**Solution**

**For $\beta_c = 0.5$, we have**

(a) **With wind velocity**

$$[\tau_g \alpha_c - \eta_{m0}]\beta_c I(t) + U_L T_a - \eta_{m0}\beta_c \beta_0 I(t) \times 25$$
$$= \{[0.95 \times 0.9 - 0.14]0.5 \times 500 + 14.78 \times 25$$
$$- 0.14 \times 0.5 \times 0.00425 \times 500 \times 25\}$$
$$= [178.75 + 369.5 - 37.18] = 511.07 \text{ W/m}^2 \text{ and}$$
$$[U_L - \eta_{m0}\beta_c \beta_0 I(t)] = 14.78 - 0.14 \times 0.5 \times 0.00425 \times 500$$
$$= 14.78 - 0.1655 = 14.61 \text{ W/m}^2 \,°C$$

From Eq. 3.38, we have an expression for solar cell

$$T_c = \frac{[\tau_g \alpha_c - \eta_{m0}]\beta_c I(t) + U_L T_a - \eta_{m0}\beta_0 \beta_c I(t) \times 25}{[U_L - \eta_{m0}\beta_0 \beta_c I(t)]}$$

$$= \frac{511.07}{14.61} = 35.00 \,°C$$

(b) **Without wind velocity**

Now,

$$[\tau_g \alpha_c - \eta_{m0}]\beta_c I(t) + U_L T_a - \eta_{m0}\beta_c \beta_0 I(t) \times 25$$
$$= \{[0.95 \times 0.9 - 0.14]0.5 \times 500 + 8.33 \times 25$$
$$- 0.14 \times 0.5 \times 0.00425 \times 500 \times 25\}$$
$$= [178.75 + 208.25 - 37.18] = 349.82 \text{ W/m}^2 \text{ and } [U_L - \eta_{m0}\beta_c \beta_0 I(t)]$$
$$= 8.33 - 0.14 \times 0.5 \times 0.00425 \times 500 = 8.33 - 0.1487$$
$$= 8.18 \text{ W/m}^2 \,°C$$

From Eq. 3.38, we have

$$T_c = \frac{[\tau_g \alpha_c - \eta_{m0}]\beta_c I(t) + U_L T_a - \eta_{m0}\beta_0 \beta_c I(t) \times 25}{[U_L - \eta_{m0}\beta_0 \beta_c I(t)]}$$

$$= \frac{349.82}{8.18} = 42.76 \,°C$$

Here, one can conclude that if packing factor is reduced, then solar cell temperature is also reduced. In this case too, the solar cell temperature of semi-transparent PV module is always less than the solar cell temperature of opaque PV module (Example 3.27).

**Example 3.41** Repeat Example 3.39 for packing factor of ($\beta_c$) of zero.

**Solution**

Here, for $\beta_c = 0$.

$$\left[\tau_g \alpha_c - \eta_{m0}\right] \beta_c I(t) + U_L T_a - \eta_{m0} \beta_c \beta_0 I(t) \times 25$$
$$= \{0 + 14.78 \times 25 - 0\} = 369.5 \, \text{W/m}^2 \text{ and}$$
$$[U_L - \eta_{m0} \beta_c \beta_0 I(t)] = 14.78 - 0 = 14.78 \, \text{W/m}^2 \, {}^\circ\text{C}$$

From Eq. 3.38, we have an expression for solar cell

$$T_c = \frac{\left[\tau_g \alpha_c - \eta_{m0}\right] \beta_c I(t) + U_L T_a - \eta_{m0} \beta_0 \beta_c I(t) \times 25}{[U_L - \eta_{m0} \beta_0 \beta_c I(t)]}$$
$$= \frac{369.5}{14.78} = 25\,{}^\circ\text{C} = T_a$$

In this case, it is independent of wind velocity and it is same as Eq. 3.40.

**Example 3.42** Repeat Example 3.39 for packing factor of ($\beta_c$) of one.

**Solution**

**For $\beta_c = 1$, we have**

(a) **With wind velocity**

$$\left[\tau_g \alpha_c - \eta_{m0}\right] \beta_c I(t) + U_L T_a - \eta_{m0} \beta_c \beta_0 I(t) \times 25$$
$$= \{[0.95 \times 0.9 - 0.14] \times 1 \times 500 + 14.78 \times 25 - 0.14 \times 1 \times 0.00425 \times 500 \times 25\}$$
$$= [357.5 + 369.5 - 74.36] = 652.64 \, \text{W/m}^2 \text{ and}$$
$$[U_L - \eta_{m0} \beta_c \beta_0 I(t)] = 14.78 - 0.14 \times 1 \times 0.00425 \times 500$$
$$= 14.78 - 0.331 = 14.449 \, \text{W/m}^2 \, {}^\circ\text{C}$$

From Eq. 3.38, we have an expression for solar cell

$$T_c = \frac{\left[\tau_g \alpha_c - \eta_{m0}\right] \beta_c I(t) + U_L T_a - \eta_{m0} \beta_0 \beta_c I(t) \times 25}{[U_L - \eta_{m0} \beta_0 \beta_c I(t)]}$$
$$= \frac{652.64}{14.44} = 45.20\,{}^\circ\text{C}.$$

## (b) Without wind velocity

Now,

$$[\tau_g \alpha_c - \eta_{m0}]\beta_c I(t) + U_L T_a - \eta_{m0}\beta_c \beta_0 I(t) \times 25$$
$$= \{[0.95 \times 0.9 - 0.14] \times 1 \times 500 + 8.33 \times 25 - 0.14 \times 1 \times 0.00425 \times 500 \times 25\}$$
$$= [357.5 + 208.25 - 74.36] = 491.38 \text{ W/m}^2 \text{ and}$$
$$[U_L - \eta_{m0}\beta_c \beta_0 I(t)] = 8.33 - 0.14 \times 1 \times 0.00425 \times 500 = 8.33 - 0.2974$$
$$= 8.03 \text{ W/m}^2 \,^\circ\text{C}$$

From Eq. 3.38, we have

$$T_c = \frac{[\tau_g \alpha_c - \eta_{m0}]\beta_c I(t) + U_L T_a - \eta_{m0}\beta_0 \beta_c I(t) \times 25}{[U_L - \eta_{m0}\beta_0 \beta_c I(t)]}$$

$$= \frac{491.38}{8.03} = 61.19 \,^\circ\text{C}$$

Here, one can conclude that if packing factor is increased, then solar cell temperature is also increased. In this case too, the solar cell temperature of semi-transparent PV module is always less than the solar cell temperature of opaque PV module (Example 3.28).

**Example 3.43** Calculate the electrical efficiency of semi-transparent PV module for packing factor of one by using the solar cell temperature of Example 3.42.

### Solution

From Example 3.42, we have

**Solar cell temperature**: $T_c = 45.20 \,^\circ\text{C}$, with wind velocity and $T_c = 61.19 \,^\circ\text{C}$, without wind velocity.

From Eq. 3.20, we have an expression for electrical efficiency of PV module as

$$\eta_m = \eta_{m0}[1 - \beta_0(T_c - 25)], \text{ in fraction}$$

## (a) With wind velocity

$$\eta_m = 0.14\big[1 - 0.00425(45.20 \,^\circ\text{C}, -25)\big] = 0.1280, \text{ in fraction}$$

## (b) Without wind velocity

$$\eta_m = 0.14[1 - 0.00425(61.19 - 25)] = 0.1185, \text{ in fraction}$$

### 3.3.7  Energy Balance of Flexible PV Module

In this case, the back sheet of opaque PV module is copper (Cu)/aluminum (Al) sheet in the place of tedlar which is also opaque. The thickness of copper (Cu)/aluminum (Al) back sheet ($L_{bs}$) is about 0.2 mm. The thermal conductivity of pure copper (Cu)/aluminum (Al) material ($K_{bs}$) is 386 and 204 W/m K, respectively (Appendix Table E.3). Due to small thickness of copper (Cu)/aluminum (Al) back sheet, it is referred as flexible PV module unlike opaque PV module with packing factor of one ($\beta_c = 1$).

The energy balance equation for flexible PV module with packing factor ($\beta_c$) = 1 will be exactly same as Eq. 3.27 for opaque PV module with packing factor ($\beta_c$) = 1. So, the energy balance equation for flexible PV module can be written as follows:

$$\left[\tau_g \alpha_c I(t) A_m\right] = \left[U_{tc,a}(T_c - T_a) + U_{bc,a}(T_c - T_a)\right]A_m + \eta_m I(t) A_m \quad (3.41a)$$

or,

$$T_c = \frac{\tau_g \alpha_c I(t) + \left(U_{tc,a} + U_{bc,a}\right)T_a - \eta_m I(t)}{\left(U_{tc,a} + U_{bc,a}\right)} \quad (3.41b)$$

where, $U_{tc,a} = 9.12$ W/m$^2$ °C with wind velocity of 1 m/s and 5.56 W/m$^2$ °C without wind velocity, Example 3.22, an overall top loss coefficient from solar cell to ambient air through toughened glass will be exactly same (Example 3.22).

Following Eq. 2.27b and Example 3.23, we can write an expression for an overall bottom loss transfer coefficient from solar cell to ambient air through copper (Cu)/aluminum (Al) back sheet ($L_{bs}$) as

$$U_{b,ca} = \left[\frac{L_{bs}}{K_{bs}} + \frac{1}{h_i}\right]^{-1} \quad (3.42)$$

**Example 3.44**  Evaluate an overall bottom loss transfer coefficient $\left(U_{b,ca}\right)$ from solar cell to ambient air through copper (Cu)/aluminum (Al) back sheet ($L_{bs}$) for the following parameter with and without wind velocity of 1 m/s:

$L_{bs} = 0.2$ mm and $K_{bs} = 386$ W/m K for copper and 204 W/m K for aluminum.

**Solution**

By using Eq. 3.42, we have

(a)  **With wind velocity**

$$U_{b,ca} = \left[\frac{L_{bs}}{K_{bs}} + \frac{1}{h_i}\right]^{-1} = \left[\frac{0.0002}{386} + \frac{1}{2.8 + 3 \times 1}\right]^{-1}$$
$$= \left[5.18 \times 10^{-7} + 0.1724\right]^{-1} = [0.1724]^{-1}$$

$$= 5.8 \text{ W/m}^2 \, {}^\circ\text{C}$$

## (b) Without wind velocity

$$U_{b,\text{ca}} = \left[\frac{L_{bs}}{K_{bs}} + \frac{1}{h_i}\right]^{-1} = \left[5.18 \times 10^{-7} + \frac{1}{2.8}\right]^{-1}$$
$$= [0.01515 + 0.3571]^{-1} = [0.3571]^{-1}$$
$$= 2.8 \text{ W/m}^2 \, {}^\circ\text{C}.$$

In this case too, effect of copper back sheet has no effect on $U_{b,\text{ca}}$. Similar effect can be seen with aluminum (Al) back sheet on $U_{b,\text{ca}}$. Further, it can be noted that the bottom loss in the case of flexible PV module is more than opaque, Example 3.22, and semi-transparent PV module, Example 3.33 which indicates faster cooling of solar cell in flexible PV module and hence increased electrical efficiency.

**Example 3.45** Calculate the total overall heat transfer coefficient from solar cell to ambient air from top as well as bottom of flexible PV module, i.e., $U_L$ with and without wind velocity of 1 m/s.

## Solution

The total overall heat transfer coefficient from solar cell to ambient air for flexible PV module can be sum of top and bottom overall heat transfer coefficient, and it is given by

$$U_L = U_{t,\text{ca}} + U_{b,\text{ca}}$$

From Eqs. 3.41a, 3.41b and Example 3.22, we have
$U_{tc,a} = 9.12 \text{ W/m}^2 \, {}^\circ\text{C}$ with wind velocity of 1 m/s and 5.8 W/m$^2$ $^\circ$C with wind velocity.

From Example 3.44, we have

$U_{t,\text{ca}} = 9.12 \text{ W/m}^2 \, {}^\circ\text{C}$ with wind velocity of 1m/s and $U_{b,\text{ca}} = 2.8 \text{ W/m}^2 \, {}^\circ\text{C}$ without wind velocity.

Hence, the total overall heat transfer coefficient from solar cell to ambient air for flexible PV module can be obtained as

$$U_L = U_{t,\text{ca}} + U_{b,\text{ca}} = 9.12 + 5.8 = 14.92 \text{ W/m}^2 \, {}^\circ\text{C with wind velocity}$$

and,

$$U_L = U_{t,\text{ca}} + U_{b,\text{ca}} = 9.12 + 2.8 = 11.92 \text{ W/m}^2 \, {}^\circ\text{C without wind velocity.}$$

**Example 3.46** Calculate solar cell temperature of flexible solar cell with and without wind velocity for Examples 3.29 and 3.45.

**Solution**

From Example 3.29 [for multi-crystalline (mc-Si) solar cell]: packing factor $(\beta_c) = 1$, absorptivity of solar cell $= 0.9$ and solar intensity of 500 W/m$^2$, $\tau_g = 0.95$.

From Eq. 3.30, we have an expression for solar cell with packing factor of one as

(a)  **With wind velocity:** $U_L = 14.92$ W/m$^2$ °C, Example 3.45

$$
\begin{aligned}
T_c &= \frac{\tau_g \alpha_c I(t) + U_L T_a - [\eta_{m0} I(t)](1 + 25\beta_0)}{[U_L - [\eta_{m0} I(t)\beta_0]]} \\
&= \frac{0.95 \times 0.9 \times 500 + 14.92 \times 25 - 0.14 \times 500(1 + 25 \times 0.00425)}{14.92 - (0.14 \times 500 \times 0.00425)} \\
&= \frac{800.50 - 77.44}{14.62} = 49.46\,°C
\end{aligned}
$$

(b)  **Without wind velocity:** $U_L = 11.92$ W/m$^2$ °C, Example 3.45

$$
\begin{aligned}
T_c &= \frac{\tau_g \alpha_c I(t) + U_L T_a - [\eta_{m0} I(t)](1 + 25\beta_0)}{[U_L - [\eta_{m0} I(t)\beta_0]]} \\
&= \frac{0.95 \times 0.9 \times 500 + 11.92 \times 25 - 0.14 \times 500(1 + 25 \times 0.00425)}{11.92 - (0.14 \times 500 \times 0.00425)} \\
&= \frac{725.50 - 77.44}{11.62} = 55.77\,°C.
\end{aligned}
$$

**Example 3.47** Calculate the electrical efficiency and electrical power of flexible PV module for Example 3.46 with and without wind velocity.

**Solution**

From Eq. 3.20, we have an expression for electrical efficiency of PV module as
$\eta_m = \eta_{m0}[1 - \beta_0(T_c - 25)]$, in fraction.

(a)  **With wind velocity**

From Example 3.46, $T_c = 49.46\,°C$, then

$$\eta_m = 0.14[1 - 0.00425(49.46 - 25)] = 0.1254, \text{ in fraction}$$

Electrical power: $\left(\dot{E}_{pv}\right) = \eta_m \times I(t) \times \beta_c \times A_m = 0.1254 \times 500 \times 1 \times 1 = 62.7$ W.

(b)  **Without wind velocity**

From Example 3.46, $T_c = 55.77\,°C$, then
$$\eta_m = 0.14[1 - 0.00425(55.77 - 25)] = 0.1217, \text{ in fraction}$$

Electrical power $\left(\dot{E}_{pv}\right) = \eta_m \times I(t) \times \beta_c \times A_m = 0.1217 \times 500 \times 1 \times 1 = 60.85$ W.

The electrical efficiency of flexible PV module with wind velocity is more due to cooling effect as per expectation and in agreement with other results obtained earlier.

So finally, we can conclude that with wind velocity of 1 m/s and same design parameters

$$\eta_m(opaque) < \eta_m(flexible) < \eta_m(semi-transparent) \tag{3.43}$$

## 3.4 Energy Balance for PV-TEC Module [9]

As we have seen that the opaque and flexible PV module has similar behavior and design parameters in comparison with semi-transparent PV module. Both opaque and flexible PV modules do not allow solar radiation to pass on other side through back sheet. If the temperature of flexible PV module is further reduced, then its electrical efficiency can be improved. It can be achieved by cooling back sheet of flexible PV module. In order to do this, one has to attach, thermoelectric cooler (TEC) based on *the Peltier effect* at back of flexible PV module as shown in Fig. 3.8 with packing factor of $\beta_{tec}$ which can be defined as follows:

$$\beta_{tec} = \frac{\text{Number of TEC strip} \times \text{area of one strip}}{\text{Back sheet area of PV module}}$$
$$= \frac{\text{Total area of TEC}}{\text{Area of PV module}} = \frac{A_{tec}}{A_m} \leq 1 \tag{3.44}$$

The unit as shown in Fig. 3.8 will be referred as PV = TEC module. The packing factor of PV-TEC module for the present calculation has been taken as $\beta_{tec} = 0.4324$ and area of each TEC strip is 4 cm × 4 cm = 0.04 m × 0.04 m.

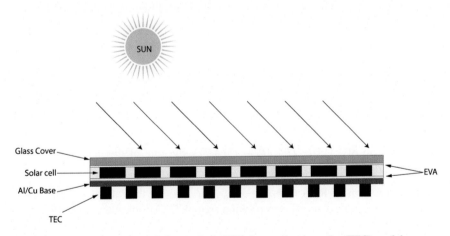

**Fig. 3.8** Cross-sectional view of photo-voltaic (PV)–thermoelectric cooler (TEC) module

There is another advantage of thermoelectric cooler (TEC). It also produces DC electric power similar to PV module, and it is proportional to temperature difference between top and bottom temperature of TEC strip. Mathematically it is expressed as follows:

$$\dot{E}_{tec} = \eta_{tec} U_{tec} \left( T_{tec,top} - T_{tec,bottom} \right) A_{tec} \tag{3.45}$$

where, $\eta_{tec}$ is electrical efficiency of thermoelectric cooler (TEC) having value between 0.08 and 0.15 and $T_{tec,top}$ and $T_{tec,bottom}$ are temperature of top and bottom of TEC. The top temperature is mostly equal to copper (Cu)/aluminum (Al) back sheet of flexible PV module.

Further, $U_{tec}$ is TEC conductive heat transfer coefficient from top of TEC to bottom of TEC through TEC material, and it is expressed as

$$U_{tec} = \frac{K_{tec}}{L_{tec}} = \frac{1.8}{0.004} = 450 \, \text{W/m}^2 \, {}^{\circ}\text{C} \tag{3.46}$$

Here, $L_{tec} = 0.004$ m and $K_{tec} = 1.8$ W/m °C are thickness and thermal conductivity of TEC. Area of one TEC strip

Referring to Fig. 3.8, the energy balance of each component of PV-TEC is as follows:

(a) **PV module**

$$\tau_g \alpha_{sc} I(t) A_m = U_{t,ca}(T_c - T_a) A_m + h_t \left( T_c - T_{tec,top} \right) \beta_{tec} A_m$$
$$+ U_{b,ca}(T_c - T_a)(1 - \beta_{tec}) A_m + \eta_m I(t) A_m \tag{3.47a}$$

Here,

$\tau_f = \mathbf{0.9}$: transmittivity of transparent flexible PVC (Polyvinyl chloride) sheet, $L_{pvc}$ = 1–5 mm. Thickness and 0.3 W/m °C thermal conductivity, $K_{pvc}$ of transparent flexible sheet.

$$\beta_{tec} = 0.4324, \text{packing factor of TEC}$$

The LHS $= \left[ \tau_g \alpha_{sc} I(t) A_m \right] =$ the rate of thermal energy absorbed by solar cell after transmission from flexible transparent sheet, $W$

The first term of RHS $= U_{t,ca}(T_c - T_a) A_m =$ the rate of thermal energy lost to ambient through top transparent sheet, $W$

The second term of RHS $= h_t \left( T_c - T_{tec,top} \right) \beta_{tec} A_m =$ the rate of thermal energy conducted through solar cell over packing area of TEC from solar cell to top of TEC in $W$.

The third term of RHS $= U_{b,ca}(T_c - T_a)(1 - \beta_{tec}) A_m =$ The rate of thermal energy lost to ambient air from back of solar cell through non-packing area of TEC, $W$

The fourth term of RHS $= \eta_m \tau_g I(t) A_m =$ the rate of DC power produced by solar cell.

For $\beta_{tec} = 1$, the base of flexible PV module is fully covered by TEC, then Eq. 3.47a reduces to

$$\tau_g \alpha_c I(t) A_m = U_{t,ca}(T_c - T_a) A_m + h_t \left(T_c - T_{tec,top}\right) A_m + \eta_m I(t) A_m. \quad (3.47b)$$

**Example 3.48** Evaluate top overall heat transfer coefficient, $U_{t,ca}$; conductive solar heat transfer coefficient, $h_t$; bottom overall heat transfer coefficient from solar cell to ambient, $U_{b,ca}$; of flexible PV-TEC module with and without wind velocity for the following parameters:

$L_{pvc} = 1$ mm; $K_{pvc} = 0.3$ W/m °C; $V = 1$ m/s; $L_{tec} = 0.004$ m and $K_{tec} = 1.8$ W/m$^2$ °C (Appendix Tables E.3 and E.5).

$L_{bs} = 0.2$ mm and $K_{bs} = 386$ W/m K for copper and 204 W/m K for aluminum (Example 3.44).

Thickness of solar cell, $L_c = 0.0003$ m; thermal conductivity of solar cell, $K_c = 0.036$ W/m °C (Appendix Table E.4 [8]).

**Solution**

**Top overall heat transfer coefficient, $U_{t,ca}$**

Following Eq. 2.27b and Example 3.1/3.22a and 3.22b, we can write an expression for an overall top heat transfer coefficient from solar cell ($T_c$) to ambient air through flexible PVC sheet as

(a) **With wind velocity**

$$U_{t,ca} = \left[\frac{L_{pvc}}{K_{pvc}} + \frac{1}{h_o}\right]^{-1} = \left[\frac{0.001}{0.3} + \frac{1}{5.7 + 3.8 \times 1}\right]^{-1}$$
$$= [0.0033 + 0.1042]^{-1} = [0.1129]^{-1}$$
$$= 9.30(9.12) \text{ W/m}^2 \text{ °C}$$

(b) **Without wind velocity**

$$U_{t,ca} = \left[\frac{L_{pvc}}{K_{pvc}} + \frac{1}{h_o}\right]^{-1} = \left[\frac{0.001}{0.3} + \frac{1}{5.7}\right]^{-1}$$
$$= [0.0033 + 0.1754]^{-1} = [0.1787]^{-1}$$
$$= 5.59(5.56) \text{ W/m}^2 \text{ °C.}$$

The above calculations show that an overall top heat transfer coefficient from solar cell ($T_c$) to ambient air $(U_{t,ca})$ through flexible PVC sheet is slightly higher than opaque, semi-transparent and flexible PV module due to different thickness and thermal conductivity; however it will not affect overall either thermal or electrical performance.

**Conductive solar cell heat transfer coefficient, $h_t$**

Following Eq. 2.2e, the conductive solar cell heat transfer coefficient, $h_t$ will be calculated as

$$h_t = \frac{K_c}{L_c} = \frac{0.036}{0.0003} = 120 \, \text{W/m}^2 \, ^\circ\text{C}$$

Following Eq. 2.27b and Example 3.1/3.22a and 3.22b, we can write an expression for an overall bottom loss transfer coefficient from solar cell ($T_c$) to ambient air through Cu/Al back sheet as

**Bottom overall heat transfer coefficient from solar cell to ambient, $U_{b,ca}$.**

(c) **With wind velocity**

$$U_{b,ca} = \left[ \frac{L_{bs}}{K_{bs}} + \frac{1}{h_i} \right]^{-1} = \left[ \frac{0.0002}{386} + \frac{1}{2.8 + 3 \times 1} \right]^{-1}$$
$$= \left[ 5.18 \times 10^{-7} + 0.1724 \right]^{-1} = [0.1724]^{-1}$$
$$= 5.80 \, \text{W/m}^2 \, ^\circ\text{C}$$

(d) **Without wind velocity**

$$U_{b,ca} = \left[ \frac{L_{bs}}{K_{bs}} + \frac{1}{h_i} \right]^{-1} = \left[ \frac{0.0002}{386} + \frac{1}{2.8} \right]^{-1}$$
$$= \left[ 5.18 \times 10^{-7} + 0.3571 \right]^{-1} = [0.3571]^{-1}$$
$$= 2.80 \, \text{W/m}^2 \, ^\circ\text{C}$$

In this case of flexible PV-TEC module, an overall bottom loss transfer coefficient from solar cell ($T_c$) to ambient air through Cu/Al back sheet is exactly same as Example 3.44 due to same material in back sheet.

**Example 3.49** Evaluate the overall heat transfer coefficient, $U_{bt}$, from bottom ambient air to top ambient air or vice versa through TEC, back copper/aluminum sheet, solar cell and flexible PVC top sheet of PV-TEC module for packing factor of TEC equal to one (Eq. 3.47b) for following parameters with and without wind velocity:

$L_{pvc} = 1$ mm; $K_{pvc} = 0.3$ W/m °C; $V = 1$ m/s; $L_{tec} = 0.004$ m and $K_{tec} = 1.8$ W/m °C (Appendix Tables E.3 and E.5).

$L_{bs} = 0.2$ mm and $K_{bs} = 386$ W/m K for copper and 204 W/m K for aluminum (Example 3.44).

Thickness of solar cell, $L_{sc} = 0.0003$ m; Thermal conductivity of solar cell, $K_{sc} = 0.036$ W/m °C (Appendix Table E.4 [8]).

**Solution**

Following Eq. 2.27b, one has an expression for $U_{bt}$ for $\beta_{tec} = 1$ as

(a) **With wind velocity**

$$U_{bt} = \left[\frac{1}{h_o} + \frac{L_{pvc}}{K_{pvc}} + \frac{L_{sc}}{K_{sc}} + \frac{L_{bs}}{K_{bs}} + \frac{L_{tec}}{K_{tec}} + \frac{1}{h_i}\right]^{-1}$$

$$= \left[\frac{1}{5.7 + 3.8 \times 1} + \frac{0.001}{0.3} + \frac{0.0003}{0.036} + \frac{0.0002}{386} + \frac{0.004}{1.8} + \frac{1}{2.8 + 3 \times 1}\right]^{-1}$$

$$= [0.1053 + 0.0033 + 0.00909 + 5.18 \times 10^{-7} + 0.0022 + 0.1754]^{-1}$$

$$= [0.3093]^{-1} = 3.38(3.233) \text{ W/m}^2 \,°\text{C}$$

(b) **Without wind velocity**

$$U_{bt} = \left[\frac{1}{h_o} + \frac{L_{pvc}}{K_{pvc}} + \frac{L_{sc}}{K_{sc}} + \frac{L_{bs}}{K_{bs}} + \frac{L_{tec}}{K_{tec}} + \frac{1}{h_i}\right]^{-1}$$

$$= \left[\frac{1}{5.7} + \frac{0.001}{0.3} + \frac{0.0003}{0.036} + \frac{0.0002}{386} + \frac{1}{2.8}\right]^{-1}$$

$$= [0.1754 + 0.0033 + 0.00909 + 5.18 \times 10^{-7} + 0.0022 + 0.3572]^{-1}$$

$$= [0.4472]^{-1} = 2.24(1.78) \text{ W/m}^2 \,°\text{C}$$

In the case of flexible PV-TEC, the overall heat transfer coefficient, $U_{bt}$, from bottom ambient air to top ambient air or vice versa is higher the value obtained in Example 3.23. The flexible PV-TEC will be useful for cooling the building. It can be used as façade.

(c) **For copper/aluminum:**

$$h_t\left(T_c - T_{tec,top}\right)\beta_{tec} A_m = U_{tec}\left(T_{tec,top} - T_{tec,bottom}\right)\beta_{tec} A_m \qquad (3.48a)$$

Here,

The term of LHS $= h_t\left(T_c - T_{tec,top}\right)\beta_{tec} A_m =$ the rate of thermal energy conducted through solar cell over packing area of TEC from solar cell to top of TEC in W.

The term of RHS $= U_{tec}\left(T_{tec,top} - T_{tec,bottom}\right)\beta_{tec} A_m =$ the rate of heat conducted from top of TEC to bottom of TEC through conductance.

$U_{tec} = 450 \text{ W/m}^2 \,°\text{C}$; conductive heat transfer coefficient from top of TEC to bottom of TEC, Eq. 3.46.

For $\beta_{tec} = 1$, then Eq. 3.48a reduces to

$$h_t\left(T_c - T_{tec,top}\right) A_m = U_{tec}\left(T_{tec,top} - T_{tec,bottom}\right) A_m \qquad (3.48b)$$

(d)  **For TEC module**:

$$U_{tec}\left(T_{tec,top} - T_{tec,bottom}\right)\beta_{tec}A_m = h_{teca}\left(T_{tec,bottom} - T_a\right)\beta_{tec}A_m$$
$$+ \eta_{tec}U_{tec}\left(T_{tec,top} - T_{tec,bottom}\right)\beta_{tec}A_m \tag{3.49a}$$

Here, $h_{teca} = 2.8 + 3V$; $V$ is the wind velocity in m/s.

The term of LHS $= U_{tec}\left(T_{tec,top} - T_{tec,bottom}\right)\beta_{tec}A_m = $ the rate of heat conducted from top of TEC to bottom of TEC through conductance, $W$.

The first term of RHS $= h_{teca}\left(T_{tec,bottom} - T_a\right)\beta_{tec}A_m = $ The rate of heat transfer by convection from the bottom of TEC to ambient.

The second term of RHS $= \eta_{tec}U_{tec}\left(T_{tec,top} - T_{tec,bottom}\right)\beta_{tec}A_m = $ The DC power produced by TEC, $W$.

For $\beta_{tec} = 1$, then Eq. 3.49a reduces to

$$U_{tec}\left(T_{tec,top} - T_{tec,bottom}\right)A_m = h_{ta}\left(T_{tec,bottom} - T_a\right)A_m$$
$$+ \eta_{tec}U_{tec}\left(T_{tec,top} - T_{tec,bottom}\right)A_m \tag{3.49b}$$

### 3.4.1   Analytical Expression for Solar Cell and TEC Temperatures

On solving Eqs. 3.47a, 3.48a and 3.49a by elimination process, one gets an analytical solution for solar cell $(T_c)$, top thermoelectric cooler $\left(T_{tec,top}\right)$ and bottom thermoelectric cooler $\left(T_{tec,bottom}\right)$ as

$$T_c = \frac{\tau_g\alpha_{sc}I(t) + \left[U_{t,ca} + U_{b,ca}(1 - \beta_{tec})\right]T_a + h_t\beta_{tec}T_{tec,top} - \eta_m I(t)}{\left[U_{t,ca} + h_t\beta_{tec} + U_{b,ca}(1 - \beta_{tec})\right]} \tag{3.50}$$

$$T_{tec,top} = \frac{h_{p1}\tau_g\alpha_{sc}I(t) + U_{tec,top-a}T_a + U_{tec}\beta_{tec}T_{tec,bottom}}{\left[U_{tec}\beta_{tec} + U_{tec,top-a}\right]} \tag{3.51}$$

and,

$$T_{tec,bottom} = \frac{(1 - \eta_{tec})\left[h_{p2}h_{p1}\tau_g\alpha_{sc}I(t) + U_{tec,bootom-a}T_a\right] + h_{teca}\beta_{tec}T_a}{\left[h_{teca} + (1 - \eta_{tec})U_{tec,bootom-a}\right]} \tag{3.52}$$

where,

$h_{p1} = \frac{h_t\beta_{tec}}{\left[U_{t,ca}+h_t\beta_{tec}+U_{b,ca}(1-\beta_{tec})\right]} < 1$; the penalty factor due to the presence of flexible Cu/Al back sheet between solar cell and top of TEC.

$U_{\text{tec,top}-a} = \dfrac{h_t \beta_{\text{tec}}\left[U_{t,\text{ca}}+U_{b,\text{ca}}(1-\beta_{\text{tec}})\right]}{\left[U_{t,\text{ca}}+h_t\beta_{\text{tec}}+U_{b,\text{ca}}(1-\beta_{\text{tec}})\right]}$; an overall heat transfer coefficient from TEC top to ambient through flexible Cu/Al back sheet, and flexible PVC top sheet.

$h_{p2} = \dfrac{(U_{\text{tec}}\beta_{\text{tec}})}{\left[U_{\text{tec}}\beta_{\text{tec}}+U_{\text{tec,top}-a}\right]} < 1$; the penalty factor due to presence of TEC between top and bottom of TEC.

and, $U_{\text{tec,bootom}-a} = \dfrac{(U_{\text{tec}}\beta_{\text{tec}})U_{\text{tec,top}-a}}{\left[U_{\text{tec}}\beta_{\text{tec}}+U_{\text{tec,top}-a}\right]}$; an overall heat transfer coefficient from bottom of TEC to ambient air through TEC, Cu/Al back sheet, solar cell and flexible top PVC.

**Example 3.50** Evaluate an overall heat transfer coefficient from top $\left(U_{\text{tec,top}-a}\right)$ and bottom $\left(U_{\text{tec,bootom}-a}\right)$ of TEC to ambient air through TEC, Cu/Al back sheet, solar cell and flexible top PVC of PV-TEC module for wind velocity of 1 m/s for the following parameters:

$\beta_{\text{tec}} = 0.4324$; $U_{t,\text{ca}} = 9.30\,\text{W/m}^2\,°\text{C}$; $h_t = 120\,\text{W/m}^2\,°\text{C}$; $U_{b,\text{ca}} = 5.80\,\text{W/m}^2\,°\text{C}$ (Example 3.48) and $U_{\text{tec}} = 450\,\text{W/m}^2\,°\text{C}$ (Eq. 3.46).

**Solution**

From Eqs. 3.51 and 3.52, we have the following expression for an overall heat transfer coefficient from top $\left(U_{\text{tec,top}-a}\right)$ and bottom $\left(U_{\text{tec,bootom}-a}\right)$ of TEC to ambient air.

$$
\begin{aligned}
U_{\text{tec, top}-a} &= \frac{h_t\beta_{\text{tec}}\left[U_{t,\text{ca}}+U_{b,\text{ca}}(1-\beta_{\text{tec}})\right]}{\left[U_{t,\text{ca}}+h_t\beta_{\text{tec}}+U_{b,\text{ca}}(1-\beta_{\text{tec}})\right]} \\
&= \frac{120\times 0.4324[9.30+5.80(1-0.4325)]}{[9.30+120\times 0.4324+5.80(1-0.4325)]} = \frac{653.50}{59.69} \\
&= 10.95\,\text{W/m}^2\,°\text{C}
\end{aligned}
$$

and,

$$
\begin{aligned}
U_{\text{tec,bootom}-a} &= \frac{(U_{\text{tec}}\beta_{\text{tec}})U_{\text{tec,top}-a}}{\left[U_{\text{tec}}\beta_{\text{tec}}+U_{\text{tec,top}-a}\right]} = \frac{450\times 0.4324\times 10.95}{450\times 0.4324+10.95} \\
&= \frac{2131.14}{205.575} = 10.37\ \text{W/m}^2\,°\text{C}
\end{aligned}
$$

**Example 3.51** Repeat Example 3.50 for $\beta_{\text{tec}} = 1$.

**Solution**

From Example 3.50 we have

$$
\begin{aligned}
U_{\text{tec,top}-a} &= \frac{h_t\beta_{\text{tec}}\left[U_{t,\text{ca}}+U_{b,\text{ca}}(1-\beta_{\text{tec}})\right]}{\left[U_{t,\text{ca}}+h_t\beta_{\text{tec}}+U_{b,\text{ca}}(1-\beta_{\text{tec}})\right]} = \frac{120\times 1\times[9.30]}{[9.30+120\times 1]} \\
&= \frac{1116}{129.3} = 8.31\,\text{W/m}^2\,°\text{C}
\end{aligned}
$$

and,

$$U_{\text{tec,bootom}-a} = \frac{(U_{\text{tec}}\beta_{\text{tec}})U_{\text{tec,top}-a}}{[U_{\text{tec}}\beta_{\text{tec}} + U_{\text{tec,top}-a}]} = \frac{450 \times 1 \times 8.31}{450 \times 1 + 8.31}$$

$$= \frac{3739.5}{458.31} = 8.18 \, \text{W/m}^2 \, ^\circ\text{C}$$

This shows that by increasing $\beta_{\text{tec}}$ from 0.4325 to 1, the top loss coefficient is reduced from 10.95 to 8.31 W/m² °C and bottom loss coefficient is increased from 10.37 to 8.31 W/m² °C which can be used for faster cooling of a room of a building.

**Example 3.52** Calculate the first $(h_{p1})$ and second $(h_{p2})$ penalty factor of flexible PV-TEC module for wind velocity of 1 m/s for the following parameters:

$\beta_{\text{tec}} = 0.4324$; $U_{t,\text{ca}} = 9.30 \, \text{W/m}^2 \, ^\circ\text{C}$; $h_t = 120 \, \text{W/m}^2 \, ^\circ\text{C}$; $U_{b,\text{ca}} = 5.80 \, \text{W/m}^2 \, ^\circ\text{C}$ (Example 3.48); $U_{\text{tec}} = 450 \, \text{W/m}^2 \, ^\circ\text{C}$ (Eq. 3.46) and $U_{\text{tec,top}-a} = 10.95 \, \text{W/m}^2 \, ^\circ\text{C}$ (Example 3.50).

**Solution**

The first $(h_{p1})$ and second $(h_{p2})$ penalty factor of flexible PV-TEC module from Eqs. 3.51 and 3.52, respectively, as

$$h_{p1} = \frac{h_t \beta_{\text{tec}}}{[U_{t,\text{ca}} + h_t \beta_{\text{tec}} + U_{b,\text{ca}}(1 - \beta_{\text{tec}})]}$$

$$= \frac{120 \times 0.4324}{9.30 + 120 \times 0.4324 + 5.80 \times (1 - 0.4324)} = \frac{51.89}{64.48} = 0.805$$

and,

$$h_{p2} = \frac{(U_{\text{tec}}\beta_{\text{tec}})}{[U_{\text{tec}}\beta_{\text{tec}} + U_{\text{tec,top}-a}]} = \frac{450 \times 0.4324}{450 \times 0.4324 + 10.95}$$

$$= \frac{194.62}{205.57} = 0.95$$

**Example 3.53** Repeat Example 3.52 for $\beta_{\text{tec}} = 1$.

**Solution**

From Example 3.51, we have the following

$$h_{p1} = \frac{h_t \beta_{\text{tec}}}{[U_{t,\text{ca}} + h_t \beta_{\text{tec}} + U_{b,\text{ca}}(1 - \beta_{\text{tec}})]} = \frac{120 \times 1}{9.30 + 120 \times 1 + 0}$$

$$= \frac{120}{129.3} = 0.928$$

and,

$$h_{p2} = \frac{(U_{tec}\beta_{tec})}{[U_{tec}\beta_{tec} + U_{tec,top-a}]} = \frac{450 \times 1}{450 \times 1 + 10.95}$$

$$= \frac{450}{460.95} = 0.98$$

In this case of $\beta_{tec} = 1$, both penalties factor is increased.

**Example 3.54** Evaluate solar cell, top and bottom temperature of TEC of flexible PV-TEC module for Examples 3.50 and 3.52 for wind velocity of 1 m/s for the following parameters:

$I(t) = 500\,\text{W/m}^2$; $\tau_g = 0.95$; $\alpha_{sc} = 0.9$; $T_a = 25\,°\text{C}$; $\beta_{tec} = 0.4324$; $U_{t,ca} = 9.30\,\text{W/m}^2\,°\text{C}$; $h_t = 120\,\text{W/m}^2\,°\text{C}$; $U_{b,ca} = 5.80\,\text{W/m}^2\,°\text{C}$ (Example 3.48); $U_{tec} = 450\,\text{W/m}^2$ (Eq. 3.46); $U_{tec,bootom-a} = 10.95\,\text{W/m}^2\,°\text{C}$ and $U_{tec,top-a} = 10.37\,\text{W/m}^2\,°\text{C}$ (Example 3.50).

$h_{p1} = 0.805$ and $h_{p2} = 0.97$ for $V = 1$ m/s (Example 3.52).

$h_{teca} = 2.8 + 3V = 5.8$ for $V = 1$ m/s; $h_{p1} = 0.805$ and $h_{p2} = 0.97$ (Example 3.52).

$\eta_{tec} = 0.08$ (Eq. 3.45).

**Solution**

From Eqs. 3.50, 3.51 and 3.52, we have to determine first bottom TEC temperature from Eq. 3.52 as

(a) **Bottom surface temperature of TEC**

$$T_{tec,bottom} = \frac{(1-\eta_{tec})[h_{p2}h_{p1}\tau_g\alpha_{sc}I(t) + U_{tec,bootom-a}T_a] + h_{teca}\beta_{tec}T_a}{[h_{teca} + (1-\eta_{tec})U_{tec,bootom-a}]}$$

$$= \frac{(1-0.08)[0.97 \times 0.805 \times 0.95 \times 0.9 \times 500 + 10.37 \times 25]}{5.8 + (1-0.08) \times 10.37}$$

$$= \frac{545.52 + 62.71}{15.34} = \frac{608.23}{15.34} = 39.64\,°\text{C}$$

After knowing bottom surface temperature of TEC, we have to evaluate top surface temperature of TEC from Eq. 3.51 as

(b) **Top surface temperature of TEC**

$$T_{tec,top} = \frac{h_{p1}\tau_g\alpha_{sc}I(t) + U_{tec,top-a}T_a + U_{tec}\beta_{tec}T_{tec,bottom}}{[U_{tec}\beta_{tec} + U_{tec,top-a}]}$$

$$= \frac{0.805 \times 0.95 \times 0.9 \times 500 + 10.95 \times 25 + 450 \times 0.4325 \times 39.64}{450 \times 0.4325 + 10.95}$$

$$= \frac{344.14 + 273.75 + 7714.94}{205.58} = \frac{8{,}332.83}{205.58} = 40.53\,^\circ\text{C}$$

After knowing top and bottom surface temperature of TEC, we have to evaluate the solar cell temperature from Eq. 3.50 as

(c)   **Solar cell temperature**

$$T_c = \frac{\tau_g \alpha_{sc} I(t) + \left[ U_{t,ca} + U_{b,ca}(1 - \beta_{tec}) \right] T_a + h_t \beta_{tec} T_{tec,top} - \eta_m I(t)}{\left[ U_{t,ca} + h_t \beta_{tec} + U_{b,ca}(1 - \beta_{tec}) \right]}$$

$$= \frac{\begin{array}{c} 0.95 \times 0.9 \times 500 + [9.30 + 5.80(1 - 0.4325)] \times 25 + 120 \times 0.4325 \\ \times 40.53 - 0.14 \times 500 \end{array}}{9.30 + 120 \times 0.4325 + 5.80 \times (1 - 0.4325)}$$

$$= \frac{427.5 + 314.79 + 2103.68 - 66.5}{64.49} = \frac{2779.47}{64.49} = 43.10\,^\circ\text{C}$$

From the above example one has the following

$$T_c > T_{tec,top} > T_{tec,bottom}$$

The results are as per our expectation.

**Example 3.55**   Repeat Example 3.54 for $\beta_{tec} = 1$.

**Solution**

**For $\beta_{tec} = 1$.**

$h_{p1} = 0.928$ and $h_{p2} = 0.98$ (Example 3.53).

$U_{tec,top-a} = 8.31$ W/m$^2\,^\circ$C and $U_{tec,bootom-a} = 8.18$ W/m$^2\,^\circ$C.

(a)   **Bottom surface temperature of TEC**

$$T_{tec,bottom} = \frac{(1 - \eta_{tec})\left[ h_{p2} h_{p1} \tau_g \alpha_{sc} I(t) + U_{tec,bootom-a} T_a \right] + h_{teca} \beta_{tec} T_a}{\left[ h_{teca} + (1 - \eta_{tec}) U_{tec,bootom-a} \right]}$$

$$= \frac{\begin{array}{c} (1 - 0.08)[0.98 \times 0.928 \times 0.95 \times 0.9 \times 500 + 8.18 \times 25] \\ + 5.8 \times 0.4325 \times 25 \end{array}}{5.8 + (1 - 0.08) \times 8.18}$$

$$= \frac{545.82 + 62.71}{13.33} = \frac{608.53}{13.33} = 45.65\,^\circ\text{C}$$

After knowing bottom surface temperature of TEC, we have to evaluate top surface temperature of TEC from Eq. 3.51 as

(b)  **Top surface temperature of TEC**

$$
\begin{aligned}
T_{\text{tec,top}} &= \frac{h_{p1}\tau_g\alpha_{\text{sc}}I(t) + U_{\text{tec,top}-a}T_a + U_{\text{tec}}\beta_{\text{tec}}T_{\text{tec,bottom}}}{\left[U_{\text{tec}}\beta_{\text{tec}} + U_{\text{tec,top}-a}\right]} \\
&= \frac{0.928 \times 0.95 \times 0.9 \times 500 + 8.31 \times 25 + 450 \times 0.4325 \times 45.65}{450 \times 0.4325 + 8.31} \\
&= \frac{396.72 + 207.75 + 8884.63}{202.93} = \frac{9489.11}{202.93} = 46.76\,^{\circ}\text{C}
\end{aligned}
$$

After knowing top and bottom surface temperature of TEC, we have to evaluate the solar cell temperature from Eq. 3.50 as

(c)  **Solar cell temperature**

$$
\begin{aligned}
T_c &= \frac{\tau_g\alpha_{\text{sc}}I(t) + \left[U_{t,\text{ca}} + U_{b,\text{ca}}(1 - \beta_{\text{tec}})\right]T_a + h_t\beta_{\text{tec}}T_{\text{tec,top}} - \eta_m I(t)}{\left[U_{t,\text{ca}} + h_t\beta_{\text{tec}} + U_{b,\text{ca}}(1 - \beta_{\text{tec}})\right]} \\
&= \frac{\begin{array}{c}0.95 \times 0.9 \times 500 + [9.30 + 5.80(1 - 0.4325)] \times 25 + 120 \\ \times 0.4325 \times 46.76 - 0.14 \times 500\end{array}}{9.30 + 120 \times 0.4325 + 5.80 \times (1 - 0.4325)} \\
&= \frac{427.5 + 314.79 + 2426.87 - 66.5}{64.49} = \frac{3102.36}{64.49} = 48.11\,^{\circ}\text{C}
\end{aligned}
$$

In this case too, one has the following conclusion

$$
T_c > T_{\text{tec,top}} > T_{\text{tec,bottom}}
$$

However, all temperatures are increased, and hence, PV-TEC with $\beta_{\text{tec}} = 1$ can be used for thermal heating of a building.

## 3.4.2  DC Electrical Power

(a)  **From PV module**

From Eq. 3.20, we have an expression for electrical efficiency of PV module as

$$
\eta_m = \eta_{m0}[1 - \beta_0(T_c - 298)]
$$

where, the values of $\eta_{m0}$ and $\beta_0$ for various PV modules are given in Table 3.2.
The electrical output from PV module is given by

$$
\dot{E}_{pv} = \eta_m I(t)\beta_c A_m \tag{3.53}
$$

**Example 3.56** Evaluate the electrical output of mc-Si (multi-crystalline silicon) PV module in flexible PV-TEC module for Example 3.54 with wind velocity of 1 m/s for the following parameters:

$A_m = 1 \, m^2$; $\beta_c = 1$ (packing factor of PV module); $\eta_{m0} = 0.14$, $\beta_0 = 0.00425$ for mc-Si PV module (Table 3.2); $T_c = 43.10 \, °C$ (Example 3.54).

**Solution**

As we know from Eq. 3.20, the electrical efficiency of PV module is

$$\eta_m = \eta_{m0}[1 - \beta_0(T_c - 298)] = 0.14[1 - 0.00425(43.10 - 25)] = 0.1292$$

where, the values of $\eta_{m0}$ and $\beta_0$ for various PV modules are given in Table 3.2.

The electrical output from PV module, Eq. 3.53, is given by

$$\dot{E}_{pv} = \eta_m I(t)\beta_c A_m = 0.1292 \times 500 \times 1 \times 1 = 64.61 \, W$$

**(b)  From TEC module**

From Eq. 3.45, we have an expression for electrical output from TEC as

$$\dot{E}_{tec} = \eta_{tec} U_{tec}\left(T_{tec,top} - T_{tec,bottom}\right) A_{tec}$$

**Example 3.57** Calculate DC power from TEC of PV-TEC module for $\beta_{tec} = 0.4325$; $A_m = 1 \, m^2$ for Example 3.54.

**Solution**

From Example 3.54, we have

$T_{tec,bottom} = 39.64 \, °C$; $T_{tec,top} = 40.53 \, °C$, and Area of TEC surface in PV-TEC module $(A_{tec}) = \beta_{tec} \times A_m = 0.4325 \times 1 = 0.4325 \, m^2$; $\eta_{tec} = 0.08$; $U_{tec} = 450 \, W/m^2 \, °C$.

From Eq. 3.45, we have an expression for DC power from TEC of PV-TEC module as

$$\dot{E}_{tec} = \eta_{tec} U_{tec}\left(T_{tec,top} - T_{tec,bottom}\right) A_{tec}$$
$$= 0.08 \times 450[40.53 - 39.64] \times 0.4325 = 13.85 \, W$$

**(c)  From PV-TEC module**

The total DC electrical output of PV-TEC module is obtained by summing of DC power from PV and TEC output of flexible PV-TEC module as

$$\dot{E}_{pv-tec} = \dot{E}_{pv} + \dot{E}_{tec} = 64.61 + 13.85 = 78.46 \, W \tag{3.54}$$

From Example 3.47, we have 62.7 W DC power from flexible PV module. This shows that there is enhancement of 25% in DC power due to integration of TEC in flexible PV module.

### 3.4.3 Special Case: Energy Balance of PV-TEC Module with $\beta_{tec} = 0$ (Without TEC)

**In this case,** $h_{p1}, h_{p2}, U_{tec,top-a}$ and $U_{tec,bootom-a}$ of Eqs. 3.50–3.52 becomes

$$h_{p1} = \frac{h_t \beta_{tec}}{\left[U_{t,ca} + h_t \beta_{tec} + U_{b,ca}(1 - \beta_{tec})\right]} = \frac{h_t \times 0}{\left[U_{t,ca} + h_t \beta_{tec} + U_{b,ca}(1 - \beta_{tec})\right]} 0$$

$$h_{p2} = \frac{(U_{tec}\beta_{tec})}{\left[U_{tec}\beta_{tec} + U_{tec,top-a}\right]} = \frac{(U_{tec} \times 0)}{\left[U_{tec}\beta_{tec} + U_{tec,top-a}\right]} 0$$

$$U_{tec,top-a} = \frac{h_t \beta_{tec}\left[U_{t,ca} + U_{b,ca}(1 - \beta_{tec})\right]}{\left[U_{t,ca} + h_t \beta_{tec} + U_{b,ca}(1 - \beta_{tec})\right]}$$

$$= \frac{h_t \times 0 \times \left[U_{t,ca} + U_{b,ca}(1 - \beta_{tec})\right]}{\left[U_{t,ca} + h_t \beta_{tec} + U_{b,ca}(1 - \beta_{tec})\right]} 0 \text{ and}$$

$$U_{tec,bootom-a} = \frac{(U_{tec} \times 0)U_{tec,top-a}}{\left[U_{tec}\beta_{tec} + U_{tec,top-a}\right]} = 0$$

Then Eqs. 3.50–3.52 reduce to

$$T_{sc} = \frac{\tau_g \alpha_{sc} I(t) + \left[U_{t,ca} + U_{b,ca}\right]T_a + -\eta_m I(t)}{\left[U_{t,ca} + U_{b,ca}\right]} \tag{3.55}$$

$$T_{tec,top} = \frac{h_{p1}\tau_g \alpha_{sc} I(t) + U_{tec,top-a}T_a + U_{tec}\beta_{tec}T_{tec,bottom}}{\left[U_{tec}\beta_{tec} + U_{tec,top-a}\right]}$$

$$= \frac{0 \times \tau_g \alpha_{sc} I(t) + 0 \times T_a + U_{tec}\beta_{tec} \times 0}{\left[U_{tec}\beta_{tec} + U_{tec,top-a}\right]} = 0 \tag{3.56}$$

and,

$$T_{tec,bottom} = \frac{(1 - \eta_{tec})\left[h_{p2}h_{p1}\tau_g \alpha_{sc} I(t) + U_{tec,bottom-a}T_a\right] + h_{teca}\beta_{tec}T_a}{\left[h_{teca} + (1 - \eta_{tec})U_{tec,bottom-a}\right]}$$

$$= \frac{(1 - \eta_{tec})\left[0 \times \tau_g \alpha_{sc} I(t) + 0 \times T_a\right] + h_{teca} \times 0 \times T_a}{\left[h_{teca} + (1 - \eta_{tec})U_{tec,bootom-a}\right]} = 0 \tag{3.57}$$

Equation 3.53 shows that PV-TEC becomes only flexible PV module (Sect. 3.3.7). Equation 3.50 is exactly same as Eq. 3.41b.

**Additional Examples**

**Example 3.58** Evaluate the top overall heat transfer coefficient, $U_t$, and sol–air temperature for blackened and double glazed solid surface, $T_{sa}$, for the following parameters:

$\alpha = 0.90$; $\tau = 0.9$; $I(t) = 500$ W/m$^2$; $T_a = 25\,°C$; $h_o = 5.7$ W/m$^2\,°C$ (Eq. 2.13a for $V = 0$); $L_g = 5$ mm and $K_g = 0.78$ W/m $°C$, Appendix Table E.5; $C = 4.75$ W/m$^2\,°C$ (for 5 cm air gap, Fig. 2.3), $h_o = 2.8$ W/m$^2\,°C$.

**Solution**

From Eq. 3.12b, one has an expression for the top overall heat transfer coefficient, $U_t$, for double glazed as

$$U_t = \left[\frac{1}{h_o} + \frac{L_g}{K_k} + \frac{1}{C} + \frac{1}{C}\right]^{-1} = \left[\frac{1}{5.7} + \frac{0.005}{0.78} + \frac{1}{4.75} + \frac{1}{4.75}\right]^{-1}$$
$$= [0.175 + 0.0064 + 0.21 + 0.21]^{-1} = [0.6019]^{-1} = 1.66\ \text{W/m}^2\,°C$$

The sol–air temperature for single blackened and glazed surface, $T_{sa}$, can be obtained from Eq. 3.9 as

$$T_{sa} = \frac{\alpha\tau^2 I(t)}{U_t} + T_a = \frac{0.9 \times 0.9 \times 0.9}{1.66} \times 500 + 25 = 244.57\,°C$$

Due to double glazed horizontal surface, an overall heat transfer coefficient as well as solar temperature is significantly increased in comparison with single glazed (Example 3.8). The double glazed wall is most useful for in winter condition.

**Example 3.59** Find out electrical efficiency of PV module for solar cell temperature is less than 298 K.

**Solution**

From Eq. 3.20, we have an expression for electrical efficiency of PV module as

$$\eta_m = \eta_{m0}[1 - \beta_0(T_c - 298)]$$

If $T_c < 298$ K, then the above equation reduces to

$$\eta_m = \eta_{m0}$$

Because, the electrical efficiency of PV module cannot be more than electrical efficiency obtained at standard test condition (STC).

# References

1. Tiwari, G. N. (2004). *Solar energy: Fundamental, design, modelling and applications.* Narosa Publishing House, New Delhi and CRC Press.
2. Tiwari, G. N. (2016). *Arvind Tiwari and Shyam, hand book of solar energy.* Springer.
3. Kern, J., & Harris, I. (1975). *Solar Energy, 17*(2), 97.
4. Evans, D. L. (1981). *Solar Energy, 27,* 555.
5. Durisch, W., Bitnar, B., Mayor, J. C., Kiess, H., Lam, K., & Close, J. (2007). *Solar Energy Materials & Solar Cells, 91,* 79–84. https://doi.org/10.1016/j.solmat.2006.05.011
6. Virtuani, A., Pavanello, D., Friesen, G. (2010). *25th European Photovoltaic Solar Energy Conference and Exhibition / 5th World Conference on Photovoltaic Energy Conversion* (pp. 6–10), Valencia, Spain. https://doi.org/10.4229/25thEUPVSEC2010-4AV.3.83
7. Tiwari, G. N., & Mishra, R. K. (2011). *Advanced renewable energy sources.* Royal Society of Chemistry.
8. Chaibi, Y., Malvoni, M., El Rhafiki, T., Kousksou, T., & Zeraouli, Y. (2021). *Cleaner Engineering and Technology, 4,* 100132. https://doi.org/10.1016/j.clet.2021.100132
9. Dimri, N., Tiwari, A., & Tiwari, G. N. (2018). Effect of thermoelectric cooler (TEC) integrated at the base of opaque photovoltaic (PV) module to enhance an overall electrical efficiency. *Solar Energy, 166,* 159–170.

# Chapter 4
# Liquid Flat Plate Collector

## 4.1 Introduction

The solar radiation available in terrestrial region incident on blackened surface is absorbed by blackened flat plate surface commonly known as absorber flat plate after reflection. The temperature of blackened flat plate surface/absorber flat plate $(T_p)$ increases due to absorption of thermal energy available in solar energy. If $\alpha_p$ is the absorptance of absorber flat plate and $I(t)$ is the incidence solar radiation on absorber flat plate, then following Eqs. 2.1a and 2.1b, the net rate of useful thermal energy available to absorber flat plate in $W$ can be written as

$$\dot{q}_u = \alpha_p I(t) - \dot{q}_L \tag{4.1}$$

where, $\dot{q}_L = h_o(T_p - T_a) + h_i(T_p - T_a) = U_L(T_p - T_a)$; the total rate of upward (top) and down ward (bottom) heat loss from absorber flat plate $(T_p)$ to ambient air $(T_a)$ and $U_L = h_o + h_i$ is the total top and bottom thermal/heat loss coefficient from absorber flat plate to ambient air. The $h_o$ and $h_i$ are the upward and down ward thermal/heat loss coefficient in W/m² °C from bare exposed surface of absorber flat plate to an ambient air, and their expressions are given by Eqs. 2.13a and 2.13b, respectively, as

$$h_o = 5.7 + 3.8 \ V \tag{4.2a}$$

and,

$$h_i = 2.8 + 3.0 \ V \tag{4.2b}$$

**Example 4.1** Evaluate the total top and bottom thermal/heat loss coefficient from absorber flat plate to ambient air $(U_L)$ with wind velocity of 1 m/s and Example 4.2, without wind velocity.

© Bag Energy Research Society 2024
G. N. Tiwari, *Fundamental of Mathematical Tools for Thermal Modeling of Solar Thermal and Photo-voltaic Systems-Part-I*,
https://doi.org/10.1007/978-981-99-7085-8_4

**Solution**

From Eqs. 4.1 and 4.2, we have the followings:

$U_L = h_o + h_i$ with $h_o = 5.7 + 3.8$ V and $h_i = 2.8 + 3.0$ V

(a)  **With wind velocity, Example** 4.1

$$U_L = h_o + h_i = 9.5 + 8.5 = 15.3 \text{ W/m}^2 \, °\text{C}$$

(b)  **Without wind velocity, Example** 4.2

$$U_L = h_o + h_i = 5.7 + 2.8 = 8.5 \text{ W/m}^2 \, °\text{C}$$

In this case, the heat/thermal loss coefficient with wind velocity is more as expected.

Example 4.1 indicates that the absorber flat plate will cool faster with wind velocity due to its direct exposure to an ambient air. In order to retain thermal energy in absorber flat plate, one should reduce both upward and down ward heat loss from the absorber flat plate. This can be achieved as follows:

(i)  **Glazed surface**: The absorber flat plate should be covered by transparent glass cover which is referred as glazed surface as shown in Fig. 4.1. In this case, short wavelength solar radiation as shown in Fig. 4.2 is transmitted through glass cover, and it is absorbed by absorber of flat plate and its temperature starts increasing. So, there is emission of radiation in long wavelength (according to due to Wien's displacement law, Eq. 1.4) from absorber flat plate due to its low temperature (Fig. 1.3) in comparison with sun's temperature of 6000 K. The emitted long wavelength radiation is blocked by transparent glass unlike bare surface, and hence, upward heat loss is reduced and its expression for glazed blackened flat plate surface can be written as

$$U_t = \left[ \frac{1}{h_0} + \frac{L_g}{K_g} + \frac{1}{h_{bc}} \right]^{-1}, \text{W/m}^2 \, °\text{C} \tag{4.3}$$

where, $h_0$ is convective and radiative heat transfer coefficient from outside glass cover to ambient air, $\frac{K_g}{L_g}$ is conductive heat transfer coefficient due to glass cover, Eq. 2.2e, and $h_{bc}$ is convective and radiative heat transfer coefficient from blackened surface to inner glass cover.

(ii)  **Bottom insulated**: The back surface of absorber flat plate should be covered with insulating layer of thickness, $L_i$ to reduce thermal losses from below the back surface of absorber. Due to its low thermal conductivity, $K_i$, the value of bottom heat loss coefficient is significantly reduced. In this case, an expression for bottom heat loss coefficient (Eq. 2.27b) becomes

$$U_b = \frac{1}{R} = \left[ \frac{L_i}{K_i} + \frac{1}{h_i} \right]^{-1}, \text{W/m}^2 \, °\text{C} \tag{4.4}$$

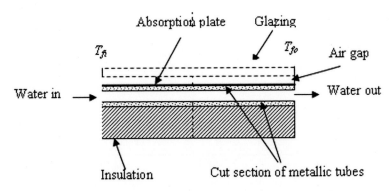

**Fig. 4.1** Cross-sectional view of blackened absorber flat plate with transparent glass covers on top and insulation at back surface

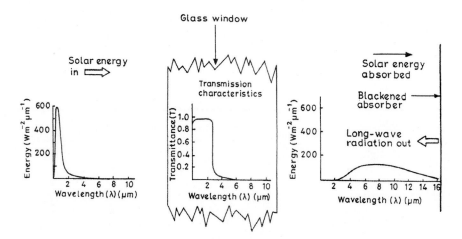

**Fig. 4.2** Propagation of short wave length solar radiation through transparent glass cover

where, $h_i = h_{bc} + h_{br}$ = sum of convective and radiative heat transfer coefficient. Since back of FPC faces ground surface, and hence, one can neglect the radiative heat transfer coefficient.

The thermal circuit diagram of FPC in terms of thermal resistance, Eq. 2.3c, has been shown in Fig. 4.3.

Now, the rate of useful thermal energy for blackened and single glazed flat plate available to absorber plate in W becomes

$$\dot{q}_u = \left[\alpha_p \tau_g I(t) - U_L(T_p - T_a)\right] = \left[\dot{q}_{ab} - U_L(T_p - T_a)\right] \text{ in W/m}^2$$

or,

$$\dot{Q}_u = A_c\left[\alpha_p \tau_g I(t) - U_L(T_p - T_a)\right] = A_c\left[\dot{q}_{ab} - U_L(T_p - T_a)\right] \text{ in W} \quad (4.5a)$$

**Fig. 4.3** Thermal resistance circuit of single glazed flat plate collector (The heat transfer coefficient is inversely proportional to thermal resistance and vice versa, Sect. 2.3)

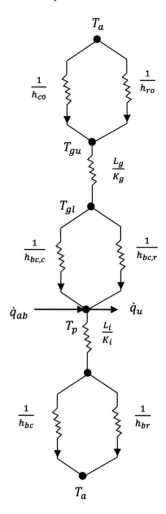

The blackened and single glazed flat plate surface is referred as flat plate collector (FPC) as shown in Fig. 4.1.

where, $U_L = U_t + U_b$, W/m$^2$ °C

Also, $\dot{q}_{ab} = \alpha_p \tau_g I(t)$, Fig. 4.3

## 4.2  Overall Total Loss Coefficient ($U_L$)

As mentioned above, the overall total loss coefficient ($U_L$) is sum of top ($U_t$) and bottom ($U_b$) loss coefficient from absorber plate to ambient air for single glazed collector.

## 4.2.1  Top Loss Coefficient ($U_t$) for Single Glazed Flat Plate Collector

From Eq. 4.3, we have an expression for overall top loss coefficient as

$$U_t = \frac{1}{r_t} = \frac{1}{r_o + r_k + r_{bc}} = [r_o + r_k + r_{bc}]^{-1} = \left[\frac{1}{h_0} + \frac{L_g}{K_g} + \frac{1}{h_{bc}}\right]^{-1} \quad (4.5b)$$

**(a) Convective and radiative heat transfer coefficient from outside glass cover to ambient air ($h_0$)**

As per Watmuff et al. [1], the outside convective heat transfer coefficient from top glass cover to ambient can be considered from Eq. 2.13b as

$$h_{co} = 2.8 + 3.0\,V \quad \text{with V as wind velocity over the glass cover} \quad (4.6a)$$

From Eq. 2.17f, the radiative heat transfer coefficient ($h_r$) from glass cover to sky is given by

$$h_{ro} = \left\{ \frac{\varepsilon\sigma\left[T_{x=0}^4 - T_{sky}^4\right]}{(T_{x=0} - T_a)} \right\} \quad (4.6b)$$

where, $T_{sky}$ (sky temperature in K) $= T_a - 12$ (Eq. 2.16c), emissivity of glass $\varepsilon = 0.88$ and Stefan–Boltzmann constant ($\sigma$) $= 5.67 \times 10^{-8}$ W/m$^2$ K and here $T_{x=0} = T_p$ in K, the absorber temperature in K.

So, the total convective and radiative heat transfer coefficient $h_0$ will be sum of Eqs. 4.6a and 4.6b, respectively, as

$$h_0 = h_{co} + h_{ro} \quad (4.6c)$$

**Example 4.3** Calculate the outside convective ($h_{co}$) with wind velocity of 1 m/s and without wind velocity, radiative ($h_{ro}$) and total heat transfer coefficient from single glass cover having temperature of 35 °C and ambient air temperature of 16 °C to sky of glazed flat plate surface.

**Solution**

From Eq. 4.6a, we have the outside convective heat transfer coefficient as

$$h_{co} = 2.8 + 3.0\,V = 5.8 \text{ W/m}^2\,°\text{C with 1 m/s wind velocity and}$$

$$h_{co} = 2.8 + 3.0 \times 0 = 2.8 \text{ W/m}^2\,°\text{C without wind velocity}$$

From Eq. 2.16 and 4.6b, the radiative heat transfer coefficient $(h_r)$ from glass cover to sky is given by

$$
h_{ro} = \left\{ \frac{\varepsilon\sigma\left[(T_p + 273)^4 - (T_{sky} + 273)^4\right]}{(T_p - T_{sky})} \right\}
$$
$$
= \frac{0.88 \times \times 5.67 \times 10^{-8}\left[(35 + 273)^4 - (4 + 273)^4\right]}{(35 - 16)}
$$
$$
= \frac{4.99 \times 10^{-8}\left[89.99 \times 10^8 - 58.87 \times 10^8\right]}{19} = \frac{155.29}{19} = 8.17 \text{ W/m}^2 \,^\circ\text{C}
$$

where, $T_{sky}$ (sky temperature in K) $= 16 - 12 = 4$ (Eq. 2.16c), and here $T_{x=0} = T_p$, the absorber temperature in K.

From Eq. 4.6c, we have

$$
h_0 = h_{co} + h_{ro} = 5.7 + 8.17 = 13.87 \text{ W/m}^2 \,^\circ\text{C, with wind velocity and}
$$

$$
h_0 = h_{co} + h_{ro} = 2.8 + 8.17 = 10.97 \text{ W/m}^2 \,^\circ\text{C; without wind velocity}
$$

### (b)  Conductive heat transfer coefficient

Following Eq. 2.2e, the conductive heat transfer is given by

$$
h_k = \frac{K_g}{L_g} = \frac{0.78}{0.003} = 260, \text{ W/m}^2 \,^\circ\text{C} \tag{4.7}
$$

### (c)  Convective and radiative heat transfer coefficient from blackened absorber plate to single glass cover outside $(h_{bc})$

The convective heat transfer coefficient $(h_{bc,c})$ between blackened and glazed absorber flat plate and inner glass cover can be evaluated by using the following dimensionless Nusselt number:

Following Holland et al. [2], the expression for Nusselt number (Nu) for blackened and glazed absorber flat plate with inclination of $\beta$ $(0 - 75^\circ)$ can be expressed as

$$
\text{Nu} = \frac{h_{bc,c}L}{K_a} = 1 + 1.44\left[1 - \frac{1708}{\text{Ra} \cos\beta}\right]^+ \left(1 - \frac{\sin(1.8\beta)^{1.6} \times 1708}{\text{Ra} \cos\beta}\right)
$$
$$
+ \left[\left\{\frac{\text{Ra} \cos\beta}{5830}\right\}^{1/3} - 1\right]^+ > 0 \tag{4.8a}
$$

The ' + ' exponent means that only the positive value of the term in bracket should be considered otherwise zero. The $h_{bc,c}$ is convective heat transfer coefficient between blackened absorber flat plate to inner glass cover in W/m$^2$ $^\circ$C and L is distance between absorber flat plate and the glass cover and $K_a$ is thermal conductivity of air at averge temperature of absorber plate and glass cover. The Rayleigh number (Ra) is given by Eq. 2.5e as

$$\mathrm{Ra} = \mathrm{Gr}.\mathrm{Pr} = \frac{g\beta' \Delta T L^3}{v\alpha}$$

If, $75° < \beta < 90°$, then

$$\mathrm{Nu} = \frac{h_{bc,c}L}{K_a} = \left[1, 0.288\left(\frac{\mathrm{Ra} \times \sin\beta}{A}\right)^{1/4}, 0.039(\mathrm{Ra} \times \sin\beta)^{1/3}\right]_{\max} \quad (4.8b)$$

where, the subscript 'max' indicates that the maximum of the three quantities separated by commas should be taken for a given value of Rayleigh number. $A$ is the ratio of length of inclined blackened and glazed flat plate ($L_0$) to spacing ($L$) between absorber flat plate cover and absorbing plate of FPC ($A = L_0/L$).

The following three region correlations for a convective heat transfer losses for an inclined flat plate collector have also proposed by Buchberg et al. [3]

$$\mathrm{Nu} = \frac{h_{bc,c}L}{K_a} = 1 + 1.446\left[1 - \frac{1708}{\mathrm{Ra}\cos\beta}\right] \text{for } 1708 < \mathrm{Ra}\cos\beta < 5900 \quad (4.9a)$$

$$\mathrm{Nu} = \frac{h_{bc,c}L}{K_a} = 0.229(\mathrm{Ra}\cos\beta)^{0.252} \text{ for } 5900 < \mathrm{Ra}\cos\beta < 9.23 \times 10^4 \quad (4.9b)$$

$$\mathrm{Nu} = \frac{h_{bc,c}L}{K_a} = 0.157(\mathrm{Ra}\cos\beta)^{0.285} \text{for } 9.23 \times 10^4 < \mathrm{Ra}\cos\beta < 10^6 \quad (4.9c)$$

**Example 4.4** Evaluate convective heat transfer coefficient $\left(h_{bc,c}\right)$ between blackened and single glazed absorber flat plate and inner glass cover which is separated by 0.02 m with an inclination of 45°. The blackened and glazed absorber flat plate and inner glass cover are at 50 °C and 30 °C, respectively.

**Solution**

The mean air temperature between two plate $= \frac{50+30}{2} = 40°\mathrm{C}$, So, air properties at 40 °C can be obtained from Appendix E(a) as follows:

$$K = 0.0272 \text{ W/m} °\mathrm{C}, T = 313 \text{ K so } \beta' = \frac{1}{T} = \frac{1}{313}$$
$$= 0.003195/\mathrm{K}; v = 1.70 \times 10^{-5} \mathrm{m}^2/\mathrm{s};$$

$\alpha = 2.40 \times 10^{-5} \mathrm{m}^2/\mathrm{s}$ (see Appendix E).
From Eqs. 2.5e and 4.8a, Rayleigh number has been calculated as follows:

$$Ra = \frac{9.81 \times 0.003195 \times 20 \times (0.020)^3}{1.70 \times 10^{-5} \times 2.40 \times 10^{-5}} = \frac{6.74 \times 10^{-6}}{4.08 \times 10^{-10}} = 1.23 \times 10^4$$

Substituting $Ra = 1.23 \times 10^4$ and $\cos 45 = 0.707$ in Eq. 4.8a, one has

$$Nu = \frac{h_{bc,c}L}{K_a} = 1 + 1.44\left[1 - \frac{1708}{1.23 \times 10^4 \times 0.707}\right]^+ \left(1 - \frac{\sin(1.8 \times 45)^{1.6} \times 1708}{1.23 \times 10^4 \times 0.707}\right)$$
$$+ \left[\left\{\frac{1.23 \times 10^4 \times 0.707}{5830}\right\}^{1/3} - 1\right]^+$$

$$= 1 + 1.44\left[1 - \frac{1708}{8696}\right]^+ \left(1 - \frac{0.78 \times 1708}{8696}\right) + \left[\left\{\frac{8696}{5830}\right\}^{1/3} - 1\right]^+$$

$$= 1 + 1.44 \times 0.8305 \times 0.8468 + 0.1424$$
$$= 1 + 1.0127 + 0.1424 = 2.1555$$

or,

$$h_{bc,c} = 2.1555 \times \frac{K_a}{L} = 2.1555 \times \frac{0.0272}{0.02} = 2.93 \text{ W/m}^2 \text{ °C}$$

**Example 4.5** Repeat Example 4.4 for inclination of zero ($\beta = 0$).

**Solution**

From Example 4.4, we have the following Rayleigh number as

$$Ra = 1.23 \times 10^4 = 12{,}300$$

Now, the Nusselt number for $\beta = 0$ ($\cos 0 = 1$) is

$$Nu = \frac{h_{bc,c}L}{K_a} = 1 + 1.44\left[1 - \frac{1708}{12{,}300 \times 1}\right]^+ \left(1 - \frac{\sin(1.8 \times 0)^{1.6} \times 1708}{12{,}300 \times 1}\right)$$
$$+ \left[\left\{\frac{12{,}300 \times 1}{5830}\right\}^{1/3} - 1\right]^+$$

$$= 1 + 1.44\left[1 - \frac{1708}{12{,}300}\right]^+ \left(1 - \frac{(0) \times 1708}{12{,}300}\right) + \left[\left\{\frac{12{,}300}{5830}\right\}^{1/3} - 1\right]^+$$
$$= 1 + 1.44 \times 0.9123 \times 1 + 0.2822 = 1 + 1.3137 + 0.2822 = 2.5959$$

or,

$$h_{bc,c} = 2.5959 \times \frac{K_a}{L} = 2.5959 \times \frac{0.0272}{0.02} = 3.53 \text{ W/m}^2 \text{ °C}$$

**Example 4.6** Repeat Example 4.4 for inclination of 75° ($\beta = 75°$).

**Solution**

From Example 4.4, we have the following Rayleigh number as

$$Ra = 1.23 \times 10^4 = 12,300$$

Substituting Ra = $1.65 \times 10^4$ and cos75 = 0.9659 in Eq. 4.8a, one has

$$Nu = \frac{h_{bc,c}L}{K_a} = 1 + 1.44\left[1 - \frac{1708}{12,300 \times 0.9659}\right]^+\left(1 - \frac{\sin(1.8 \times 75)^{1.6} \times 1708}{12,300 \times 0.9659}\right)$$

$$+ \left[\left\{\frac{12,300 \times 0.9659}{5830}\right\}^{1/3} - 1\right]^+$$

$$= 1 + 1.44\left[1 - \frac{1708}{11,880.57}\right]^+\left(1 - \frac{0.6656 \times 1708}{11,880.57}\right)$$

$$+ \left[\left\{\frac{11,880.5}{5830}\right\}^{1/3} - 1\right]^+ = 1 + 1.44 \times 0.8562 \times 0.9043 + 0.2675$$

$$= 1 + 1.1149 + 0.2675 = 2.3824$$

$$h_{bc,c} = 2.3824 \times \frac{K_a}{L} = 2.3824 \times \frac{0.0272}{0.02} = 3.24 \text{ W/m}^2 \text{ °C}$$

From Examples 4.4–4.6, one can observe that the convective heat transfer coefficient first deceases from 3.53 W/m² °C to 2.93 W/m² °C upto 45° and after that it starts increasing from 2.93 W/m² °C to 3.24 W/m² °C up to 75°. However, there is marginal effect of inclination. An average value is 3.23 W/m² °C.

**Example 4.7** Repeat Example 4.4 for inclination of 0°, 45° and 75° ($\beta = 0°$, 45° and 75°).

**Solution**

From Example 4.4, we have the following Rayleigh number as

$$Ra = 1.23 \times 10^4 = 12,300$$

For Ra = $1.23 \times 10^4 = 12,300$, Eq. 4.9b is applicable and hence

$$Nu = \frac{h_{bc,c}L}{K_a} = 0.229(Ra \cos \beta)^{0.252} \text{ for } 5900 < Ra \cos \beta < 9.23 \times 10^4$$

(a)  **For $\beta = 0^\circ$**

The Nusselt number is

$$\text{Nu} = \frac{h_{bc,c}L}{K_a} = 0.229(\text{Ra})^{0.252} = 0.229(12,300)^{0.252} = 2.457$$

Now, for Nu = 2.457, convective heat transfer coefficient becomes

$$h_{bc,c} = 2.457 \times \frac{K_a}{L} = 2.457 \times \frac{0.0272}{0.02} = 3.34 \text{ W/m}^2 \,^\circ\text{C}$$

(b)  **For $\beta = 45^\circ$**

The Nusselt number is

$$\text{Nu} = \frac{h_{bc,c}L}{K_a} = 0.229(\text{Ra} \times \cos 45)^{0.252} = 0.229(12,300 \times 0.707)^{0.252} = 3.0377$$

Now, for Nu = 3.0377 convective heat transfer coefficient becomes

$$h_{bc,c} = 3.0377 \times \frac{K_a}{L} = 3.0377 \times \frac{0.0272}{0.02} = 4.13 \text{ W/m}^2 \,^\circ\text{C}$$

(c)  **For $\beta = 75^\circ$**

The Nusselt number is

$$\text{Nu} = \frac{h_{bc,c}L}{K_a} = 0.229(\text{Ra})^{0.252} = 0.229(12,300 \times \cos 75)^{0.252}$$
$$= 0.229(12,300 \times 0.2588)^{0.252} = 1.748$$

Now, for Nu = 1.748 convective heat transfer coefficient becomes

$$h_{bc,c} = 1.748 \times \frac{K_a}{L} = 1.748 \times \frac{0.0272}{0.02} = 2.37 \text{ W/m}^2 \,^\circ\text{C}$$

Average value of $h_{bc,c} = 3.28$ W/m$^2$ °C

Now, the value of convective heat transfer for 45° inclination from Example 4.4 is $h_{bc,c} = 2.93$ W/m$^2$ °C and from Example 4.7 for the same parameters is $h_{bc,c} = 4.13$ W/m$^2$ °C. One can say it is nearly same. Hence for smaller spacing one can use Eq. 4.9b.

**Example 4.8** Evaluate convective heat transfer coefficient $\left(h_{bc,c}\right)$ between blackened and single glazed absorber flat plate with length ($L_0 = 2$m) of 2 m and inner glass cover which is separated by 0.02 m with an inclination of 75°, 80° and 90°. The blackened and glazed absorber flat plate and inner glass cover are at 50 °C and 30 °C, respectively.

**Solution**

From Example 4.4, we have the following Rayleigh number for same design parameters as

$$\text{Ra} = 1.23 \times 10^4 = 12{,}300$$

For inclination of 75°, 80° and 90°, we can use the expression given by Eq. 4.8b as

$$\text{Nu} = \frac{h_{bc,c}L}{K_a} = \left[1, 0.288 \left(\frac{\text{Ra} \times \sin \beta}{A}\right)^{1/4}, 0.039(\text{Ra} \times \sin \beta)^{1/3}\right]_{\max}$$

(a)  **For inclination of $\beta = 75°$, $A = \frac{L_0}{L} = \frac{2}{0.02} = 100$**

We have the following Nusselt number:

$$\text{Nu} = \frac{h_{bc,c}L}{K_a} = \left[1, 0.288 \left(\frac{12300 \times 0.9659}{100}\right)^{1/4}, 0.039(12300 \times 0.9659)^{1/3}\right]_{\max}$$

or,

$$\text{Nu} = \frac{h_{bc,c}L}{K_a} = [1, 0.9508, 0.8871]_{\max} = 0.9508 \text{ Now, for Nu} = 0.9555$$

convective heat transfer coefficient becomes

$$h_{bc,c} = 0.9508 \times \frac{K_a}{L} = 0.9508 \times \frac{0.0272}{0.02} = 1.29 \text{ W/m}^2 \text{ °C}$$

(b)  **For inclination of $\beta = 80°$, $A = \frac{L_0}{L} = \frac{2}{0.02} = 100$**

We have the following Nusselt number:

$$\text{Nu} = \frac{h_{bc,c}L}{K_a} = \left[1, 0.288 \left(\frac{12300 \times 0.9849}{100}\right)^{1/4}, 0.039(12300 \times 0.9849)^{1/3}\right]_{\max}$$

or,

$$\text{Nu} = \frac{h_{bc,c}L}{K_a} = [1, 0.9555, 0.8871]_{\max} = 0.9555$$

Now, for Nu = 0.9555 convective heat transfer coefficient becomes

$$h_{bc,c} = 0.9555 \times \frac{K_a}{L} = 0.9555 \times \frac{0.0272}{0.02} = 1.3 \text{ W/m}^2 \text{ °C}$$

**(c)  For inclination of $\beta = 90°$, $A = \frac{L_0}{L} = \frac{2}{0.02} = 100$**

We have the following Nusselt number:

$$\text{Nu} = \frac{h_{bc,c}L}{K_a} = \left[1, 0.288\left(\frac{12{,}300 \times 1}{100}\right)^{1/4}, 0.039(12{,}300 \times 1)^{1/3}\right]_{\max}$$

or,

$$\text{Nu} = \frac{h_{bc,c}L}{K_a} = [1, 0.9551, 0.8974]_{\max} = 0.9551$$

Now for $\text{Nu} = 0.9555$, convective heat transfer coefficient becomes

$$h_{bc,c} = 0.9551 \times \frac{K_a}{L} = 0.9551 \times \frac{0.0272}{0.02} = 1.299 \text{ W/m}^2 \text{ °C}$$

Average value of $h_{bc,c} = 1.29$ W/m$^2$ °C for $\beta \geq 75°$.

**Example 4.9** Repeat Example 4.8 with blackened and single glazed absorber flat plate with length of $L_0 = 1$ m for inclination of $\beta = 75°$.

**Solution**

**For inclination of $\beta = 75°$, $A = \frac{L_0}{L} = \frac{1}{0.02} = 50$**

We have the following Nusselt number:

$$\text{Nu} = \frac{h_{bc,c}L}{K_a} = \left[1, 0.288\left(\frac{12{,}300 \times 0.9659}{50}\right)^{1/4}, 0.039(12{,}300 \times 0.9659)^{1/3}\right]_{\max}$$

or,

$\text{Nu} = \frac{h_{bc,c}L}{K_a} = [1, 1.13, 0.8871]_{\max} = 1.13$. Now for $\text{Nu} = 0.9555$ convective heat transfer coefficient becomes

$$h_{bc,c} = 1.13 \times \frac{K_a}{L} = 1.13 \times \frac{0.0272}{0.02} = 1.53 \text{ W/m}^2 \text{ °C}$$

With decrease of collector length from 2 to 1 m, convective heat transfer coefficient increases.

The radiative heat transfer coefficient $(h_{bc,r})$ between blackened and glazed absorber flat plate and inner glass cover (Eq. 2.17f) can be evaluated by using the following expression:

$$h_{bc,r} = \left\{ \frac{\varepsilon_{\text{eff}}\sigma\left[T_p^4 - T_c^4\right]}{(T_p - T_c)} \right\}, \tag{4.10}$$

where, $T_c$(glass cover temperature + in K) $= T_a - 12$ (Eq. 2.16c), emissivity of glass $\varepsilon = 0.88$ and Stefan–Boltzmann constant ($\sigma$) $= 5.67 \times 10^{-8}$ W/m$^2$ K and absorber plate temperature ($T_p$) in K, and $\varepsilon_{\text{eff}}$ is effective emittance between absorber and glass cover.

**Example 4.10** Calculate effective emittance between parallel absorber flat plate and glass cover for glazed surface flat plate for the following parameters:

Absorber plate emittance ($\varepsilon_p$) $= 0.95$ and glass cover emittance ($\varepsilon_g$) $= 0.88$.

**Solution**

The effective emittance of a parallel surface ($\varepsilon_{\text{eff}}$) is determined as

$$\varepsilon_{\text{eff}} = \left[ \frac{1}{0.88} + \frac{1}{0.95} - 1 \right]^{-1} = 0.84$$

**Example 4.11** Calculate radiative heat transfer coefficient ($h_{bc,r}$) between the blackened and single glazed absorber flat plate and inner glass cover are at 50 °C and 30 °C, respectively, for Example 4.10.

**Solution**

From Example 4.10: $\varepsilon_{\text{eff}} = 0.84$
 Now, the radiative heat transfer coefficient ($h_{bc,r}$) is calculated from Eq. 4.10 as

$$h_{bc,r} = \left\{ \frac{\varepsilon_{eff}\sigma\left[T_p^4 - T_c^4\right]}{(T_p - T_c)} \right\} = \frac{0.84 \times \times 5.67 \times 10^{-8}\left[(50 + 273)^4 - (30 + 273)^4\right]}{(50 - 30)}$$

$$= \frac{4.99 \times 10^{-8}\left[108.8 \times 10^8 - 84.29 \times 10^8\right]}{20} = \frac{122.30}{20}$$

$$= 6.11 \text{ W/m}^2 \text{ °C}$$

**Example 4.12** Find our total convective and radiative heat transfer coefficient for Examples 4.4 and 4.11.

**Solution**

From Example 4.4: $h_{bc,c} = 2.93$ W/m$^2$ °C
 From Example 4.11: $h_{bc,r} = 6.11$ W/m$^2$ °C
 The total convective and radiative ($h_{bc}$) heat transfer coefficient will be sum of convective ($h_{bc,c}$) and radiative ($h_{bc,r}$) heat transfer coefficient, and it is given by

$$h_{bc} = h_{bc,c} + h_{bc,r} = 2.93 + 6.11 = 9.04 \text{ W/m}^2 \text{ °C}$$

**Example 4.13** Evaluate an overall top heat transfer coefficient $U_t$ from blackened and single glazed flat plate absorber to ambient air inclined at 45° for Examples 4.3 and 4.12 with and without wind velocity.

**Solution**

From Eq. 4.3, we have an expression for an overall heat transfer coefficient ($U_t$) as

$$U_t = \left[ \frac{1}{h_0} + \frac{L_g}{K_g} + \frac{1}{h_{bc}} \right]^{-1} = \left[ \frac{1}{h_0} + \frac{1}{h_k} + \frac{1}{h_{bc}} \right]^{-1}, W/m^2 \,^\circ C$$

**For Example** 4.3

The outside heat transfer coefficient from the glass cover to an ambient is given by $h_0 = 13.87 \, W/m^2 \,^\circ C$, with wind velocity and $h_0 = 10.97, W/m^2 \,^\circ C$; without wind velocity.

**From Eq.** 4.7:

The conductive heat transfer due to glass cover is given by

$$h_k = 260, W/m^2 \,^\circ C$$

**From Example** 4.12:

The total heat transfer coefficient from the blackened surface to glass cover is given by

$$h_{bc} = 9.04 \, W/m^2 \,^\circ C$$

From Sect. 4.2.1, an overall top heat transfer coefficient $U_t$ from blackened and glazed flat plate absorber to ambient air can be obtained as

(a) **With wind velocity**

$$U_t = \left[ \frac{1}{h_0} + \frac{L_g}{K_g} + \frac{1}{h_{bc}} \right]^{-1} = \left[ \frac{1}{h_0} + \frac{1}{h_k} + \frac{1}{h_{bc}} \right]^{-1} = \left[ \frac{1}{13.87} + \frac{1}{260} + \frac{1}{9.04} \right]^{-1}$$
$$= [0.0721 + 0.003846 + 0.1106]^{-1} = [0.1865]^{-1} = 5.36 \, W/m^2 \,^\circ C$$

(b) **Without wind velocity**

(c) $U_t = \left[ \frac{1}{h_0} + \frac{L_g}{K_g} + \frac{1}{h_{bc}} \right]^{-1} = \left[ \frac{1}{h_0} + \frac{1}{h_k} + \frac{1}{h_{bc}} \right]^{-1} = \left[ \frac{1}{10.97} + \frac{1}{260} + \frac{1}{9.04} \right]^{-1} =$
$[0.0912 + 0.003846 + 0.1106]^{-1} = [0.2056]^{-1} = 4.86 \, W/m^2 \,^\circ C$

**Example 4.14** Repeat Example 4.13 without considering conductive heat transfer coefficient.

**Solution**

Without conductive heat transfer coefficient, an overall top heat transfer coefficient $U_t$ from blackened and glazed flat plate absorber to ambient air inclined at 45° can be written as follows:

(a) **With wind velocity**

$$U_t = \left[\frac{1}{h_0} + \frac{1}{h_{bc}}\right]^{-1} = \left[\frac{1}{13.87} + \frac{1}{9.04}\right]^{-1} = [0.0721 + 0.1106]^{-1}$$
$$= [0.1827]^{-1} = 5.47 \text{ W/m}^2 \text{ °C}$$

(b) **With wind velocity**

$$U_t = \left[\frac{1}{h_0} + \frac{1}{h_{bc}}\right]^{-1} = \left[\frac{1}{10.97} + \frac{1}{9.04}\right]^{-1} = [0.0912 + 0.1106]^{-1}$$
$$= [0.2018]^{-1} = 4.95 \text{ W/m}^2 \text{ °C}$$

From Examples 4.13 and 4.14, one concludes that in determining an overall top heat transfer coefficient $U_t$ from blackened and glazed flat plate absorber to ambient air, an expression for $U_t$ can be finally considered as

$$U_t = \left[\frac{1}{h_0} + \frac{1}{h_{bc}}\right]^{-1} \tag{4.11}$$

It is due to very small thermal resistance $r_k = 0.003846 \text{ W/m}^2 \text{ °C}$ (Eq. 2.3c) created by glass cover.

**Example 4.15** Evaluate convective heat transfer coefficient $(h_{bc,c})$ between blackened and glazed absorber flat plate and inner glass cover which is separated by 0.10 m with an inclination of 45°. The blackened absorber flat plate and inner glass cover are at 50 °C and 30 °C respectively.

**Solution**

All parameters of Example 4.4 will be same except the spacing between blackened absorber and glazed surface as 0.10 m in place of 0.02m. In fact in the flat plate collector, the optimized spacing is 0.10 m. Hence in this case, Rayleigh number will be changed due to spacing.

From Eq. 2.5e and Eq. 4.8a, Rayleigh number has been calculated as follows:

$$\text{Ra} = \text{Gr.Pr} = \frac{9.81 \times 0.003195 \times 20 \times (0.10)^3}{1.70 \times 10^{-5} \times 2.40 \times 10^{-5}} = \frac{6.2685 \times 10^{-4}}{4.08 \times 10^{10}} = 1.54 \times 10^6$$

Substituting Ra $= 1.23 \times 10^4$ and cos 45 $= 0.707$ in Eq. 4.8a, one has

$$\text{Nu} = \frac{h_{bc,c}L}{K_a} = 1 + 1.44 \left[ 1 - \frac{1708}{1.54 \times 10^6 \times 0.707} \right]^+ \left( 1 - \frac{\sin(1.8 \times 45)^{1.6} \times 1708}{1.54 \times 10^6 \times 0.707} \right)$$

$$+ \left[ \left[ \left\{ \frac{1.54 \times 10^6 \times 0.707}{5830} \right\}^{1/3} - 1 \right] \right]$$

$$= 1 + 1.44 \left[ 1 - \frac{1708}{1.09 \times 10^6} \right]^+ \left( 1 - \frac{0.78 \times 1708}{1.09 \times 10^6} \right) + \left[ \left\{ \frac{1.09 \times 10^6}{5830} \right\}^{1/3} - 1 \right]^+$$

$$= 1 + 1.44 \times 0.9984 \times 0.9988 + 4.708 = 1 + 1.4359 + 4.708 = 8.1441$$

or,

$$h_{bc,c} = 8.1441 \times \frac{K_a}{L} = 8.1441 \times \frac{0.0272}{0.02} = 11.08 \text{ W/m}^2 \text{ °C}$$

**Example 4.16** Repeat Example 4.15 by using Eq. 4.9c.

**Solution**

From Example 4.15, we have Ra $= 1.54 \times 10^6$ and then Eq. 4.9c can be used to evaluate the Nusselt number as

$$Nu = \frac{h_{bc,c}L}{K_a} = 0.157(\text{Ra} \cos \beta)^{0.285} = 0.157\left(1.54 \times 10^6 \times 0.707\right)^{0.285}$$

$$= 0.157 \times \left(1.09 \times 10^6\right)^{0.285} = 0.157 \times 52.54 = 8.25$$

or,

$$h_{bc,c} = 8.25 \times \frac{K_a}{L} = 8.25 \times \frac{0.0272}{0.02} = 11.22 \text{ W/m}^2 \text{ °C}$$

So, one can see that for optimum spacing of 0.10 m between blackened surface and glass cover both Examples 4.15 and 4.16 give same results of convective heat transfer of 11.08 W/m$^2$ °C and 11.22 W/m$^2$ °C respectively. So it is better to use Eq. 4.9c instead of complex Eq. 4.8a.

**Example 4.17** Evaluate an overall top heat transfer coefficient $U_t$ from blackened and single glazed flat plate absorber to ambient air inclined at 45° for spacing of 0.10 m between absorber and glass cover by using the data of Examples 4.3, 4.7, 4.11 and 4.16 with and without wind velocity.

**Solution**

From Eq. 4.3, we have an expression for an overall heat transfer coefficient ($U_t$) as

$$U_t = \left[ \frac{1}{h_0} + \frac{L_g}{K_g} + \frac{1}{h_{bc}} \right]^{-1} = \left[ \frac{1}{h_0} + \frac{1}{h_k} + \frac{1}{h_{bc}} \right]^{-1}, \text{ W/m}^2 \text{ °C}$$

**From Example** 4.3:

The outside heat transfer coefficient from the glass cover to an ambient air is given by $h_0 = 13.87$ W/m$^2$ °C, with wind velocity and $h_0 = 10.97$, W/m$^2$ °C; without wind velocity.

**From Eq.** 4.7:

The conductive heat transfer due to glass cover is given by

$$h_k = 260, \text{W/m}^2 \text{ °C}$$

**From Examples 4.11 and 4.16:**

The total heat transfer coefficient from the blackened surface to glass cover is given by

$$h_{bc} = h_{bc,c} + h_{bc,r} = 11.22 + 6.11 = 17.33 \text{ W/m}^2 \text{ °C}$$

From Sect. 4.2.1, an overall top heat transfer coefficient $U_t$ from blackened and glazed flat plate absorber to ambient air can be obtained as

(a) **With wind velocity**

$$U_t = \left[ \frac{1}{h_0} + \frac{L_g}{K_g} + \frac{1}{h_{bc}} \right]^{-1} = \left[ \frac{1}{13.87} + \frac{1}{260} + \frac{1}{17.33} \right]^{-1}$$
$$= [0.0721 + 0.003846 + 0.0577]^{-1} = [0.1336]^{-1} = 7.49 \text{ W/m}^2 \text{ °C}$$

(b) **Without wind velocity**

$$U_t = \left[ \frac{1}{h_0} + \frac{L_g}{K_g} + \frac{1}{h_{bc}} \right]^{-1} = \left[ \frac{1}{10.97} + \frac{1}{260} + \frac{1}{17.33} \right]^{-1}$$
$$= [0.0912 + 0.003846 + 0.0577]^{-1} = [0.1527]^{-1} = 6.55 \text{ W/m}^2 \text{ °C}$$

Now, we can further evaluate an overall top heat transfer coefficient $U_t$ from blackened and glazed flat plate absorber to ambient air without glass cover then:

(a) **With wind velocity**

$$U_t = \left[ \frac{1}{h_0} + \frac{1}{h_{bc}} \right]^{-1} = \left[ \frac{1}{13.87} + \frac{1}{17.33} \right]^{-1} = [0.0721 + 0.0577]^{-1} = [0.1278]^{-1}$$
$$= 7.82 \text{ W/m}^2 \text{ °C}$$

(b) **Without wind velocity**

$$U_t = \left[\frac{1}{h_0} + \frac{1}{h_{bc}}\right]^{-1} = \left[\frac{1}{10.97} + \frac{1}{17.33}\right]^{-1} = [0.0912 + 0.0577]^{-1} = [0.1489]^{-1}$$
$$= 6.72 \text{ W/m}^2 \text{ }°\text{C}$$

In this case too, one can observe that there is marginal effect on $U_t$ due to presence of glass cover, and hence finely one can consider Eq. 4.11 to evaluate $U_t$.

### 4.2.2  Top Loss Coefficient ($U_t$) for Double Glazed Flat Plate Collector

In the case of blackened and double glazed flat plate collector (FPC) as shown in Fig. 4.4, the rate of useful thermal energy available to absorber plate in W becomes

$$\dot{q}_u = \left[\alpha_p \tau_g^2 I(t) - U_L(T_p - T_a)\right] \text{ in W/m}^2 \tag{4.12a}$$

where, $U_L = U_t + U_b$ and Eq. 4.3 for single glazed flat plate collector will be modified for double glazed flat plate collector for overall top loss coefficient ($U_t$) (Eq. 2.28b) as

$$U_t = \left[\frac{1}{h_0} + \frac{L_g}{K_g} + \frac{1}{C} + \frac{L_g}{K_g} + \frac{1}{h_{bc}}\right]^{-1}, \text{W/m}^2 \text{ }°\text{C} \quad \text{with two glass} \tag{4.12b}$$

$$U_t = \left[\frac{1}{h_0} + \frac{1}{C} + \frac{1}{h_{bc}}\right]^{-1}, \text{W/m}^2 \text{ }°\text{C} \quad \text{without glass cover} \tag{4.12c}$$

where, $C$ is the air conductance of space between two glazed cover having same temperature and its variation with spacing is given in Fig. 2.3. Generally, one can consider space between two glass cover as 2 cm $= 0.02$ m.

The numerical values of $h_0$ and $h_{bc}$ will remain the same as it is in single glazed flat plate collector (Sect. 4.2.1) for same operating and design parameters.

**Example 4.18** Compute an overall top loss coefficient ($U_t$) for double glazed flat plate collector for parameters given in Example 4.17 and Fig. 2.3.

**Solution**

From Eq. 4.12b, we have an expression for $U_t$ of double glazed flat plate collector as

**Fig. 4.4** Exposed vertical cross-sectional view of double glazed flat plate collector

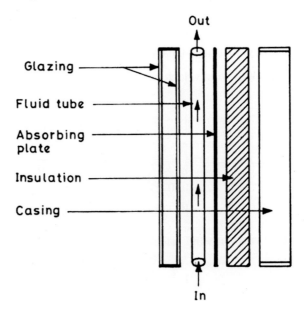

$$U_t = \left[ \frac{1}{h_0} + \frac{L_g}{K_g} + \frac{1}{C} + \frac{L_g}{K_g} + \frac{1}{h_{bc}} \right]^{-1}$$

$$= \left[ \frac{1}{h_0} + \frac{1}{h_k} + \frac{1}{C} + \frac{1}{h_k} + \frac{1}{h_{bc}} \right]^{-1}, W/m^2 \,°C$$

**Here,**

**From Example** 4.3:

The outside heat transfer coefficient from the glass cover to an ambient air is given by $h_0 = 13.87$ W/m$^2$ °C, with wind velocity and $h_0 = 10.97$, W/m$^2$ °C; without wind velocity.

**From Eq.** 4.7:

The conductive heat transfer due to both glass cover will be same, and it is given by

$$h_k = \frac{K_k}{L_g} = 260, W/m^2 \,°C$$

**From Examples 4.11 and 4.16:**

The total heat transfer coefficient from the blackened surface to glass cover is given by

$$h_{bc} = h_{bc,c} + h_{bc,r} = 11.22 + 6.11 = 17.33 \, W/m^2 \,°C$$

From Fig. 2.3, we have

$C = 5.7\,\text{W/m}^2\,°\text{C (about)}$

Now, we can do the following calculation:

(a) **With wind velocity**

$$U_t = \left[\frac{1}{h_0} + \frac{L_g}{K_g} + \frac{1}{C} + \frac{L_g}{K_g} + \frac{1}{h_{bc}}\right]^{-1}$$

$$= \left[\frac{1}{13.87} + \frac{1}{260} + \frac{1}{5.7} + \frac{1}{260} + \frac{1}{17.33}\right]^{-1}$$

$$= [0.0721 + 0.003846 + 0.1754 + 0.003846 + 0.0577]^{-1}$$

$$= [0.3129]^{-1} = 3.19\,\text{W/m}^2\,°\text{C}$$

(b) **Without wind velocity**

$$U_t = \left[\frac{1}{h_0} + \frac{L_g}{K_g} + \frac{1}{C} + \frac{L_g}{K_g} + \frac{1}{h_{bc}}\right]^{-1}$$

$$= \left[\frac{1}{10.97} + \frac{1}{260} + \frac{1}{5.7} + \frac{1}{260} + \frac{1}{17.33}\right]^{-1}$$

$$= [0.0912 + 0.003846 + 0.1754 + 0.003846 + 0.0577]^{-1}$$

$$= [0.332]^{-1} = 3.01\,\text{W/m}^2\,°\text{C}$$

Now, we can further evaluate an overall top heat transfer coefficient $U_t$ from blackened and double glazed flat plate absorber to ambient air without glass cover then

(a) **With wind velocity**

$$U_t = \left[\frac{1}{h_0} + \frac{1}{C} + \frac{1}{h_{bc}}\right]^{-1} = \left[\frac{1}{13.87} + \frac{1}{5.7} + \frac{1}{17.33}\right]^{-1}$$

$$= [0.0721 + 0.1754 + 0.0577]^{-1} = [0.3052]^{-1} = 3.28\,\text{W/m}^2\,°\text{C}$$

(b) **Without wind velocity**

$$U_t = \left[\frac{1}{h_0} + \frac{1}{C} + \frac{1}{h_{bc}}\right]^{-1} = \left[\frac{1}{10.97} + \frac{1}{5.7} + \frac{1}{17.33}\right]^{-1}$$

$$= [0.0912 + 0.1754 + 0.0577]^{-1} = [0.3243]^{-1} = 3.08\,\text{W/m}^2\,°\text{C}$$

In this case too, one can observe that there is marginal effect on $(U_t)$ due to presence of both glass cover, and hence finely one can consider Eq. 4.12c to evaluate $U_t$.

It is important to note that the effect of wind velocity on $U_t$ of double glazed flat plate collector, (Example 4.18), is marginal unlike single glazed flat plate collector (Example 4.17).

### 4.2.3 An Overall Bottom Loss Coefficient ($U_b$)

An expression for an overall bottom loss coefficient is given by Eq. 4.4.

**Example 4.19** Evaluate the bottom overall heat loss coefficient, $U_b$ if back of absorber plate is insulated by cork board Brick [$K_1 = 0.04$ W/mK and $L_1 = 0.05$ m] (Example 2.69) to reduce down ward loss with wind velocity of 1 m/s and without wind velocity.

**Solution**

(a) **With wind velocity**

From Eq. 4.4, we have bottom overall heat loss coefficient, $U_b$ as

$$U_b = \frac{1}{R} = \left[\frac{L_i}{K_i} + \frac{1}{h_i}\right]^{-1} = \left[\frac{0.05}{0.04} + \frac{1}{2.8 + 3.0 \times 1}\right]^{-1} = [1.25 + 0.1725]^{-1}$$
$$= [1.4224]^{-1} = 0.70 \text{ W/m}^2 \, {}^\circ\text{C}$$

(b) **Without wind velocity**

In this case, we have

$$U_b = \frac{1}{R} = \left[\frac{L_i}{K_i} + \frac{1}{h_i}\right]^{-1} = \left[\frac{0.05}{0.04} + \frac{1}{2.8}\right]^{-1} = [1.25 + 0.3571]^{-1}$$
$$= [1.607]^{-1} = 0.622 \text{ W/m}^2 \, {}^\circ\text{C}$$

One can infer that wind does not have significant effect on bottom overall heat loss coefficient, ($U_b$).

### 4.2.4 An Overall Total Loss Coefficient ($U_L$) of Single and Double Glazed Flat Plate Collector

An expression for an overall heat transfer coefficient is given by Eq. 4.5a.

**Example 4.20** Evaluate the total overall heat transfer coefficient for single and double glazed flat plate collector for wind velocity of 1 m/s for Examples 4.17, 4.18 and 4.19.

**Solution**

From Eq. 4.5b, we have

$$U_L = U_t + U_b, \text{W/m}^2 \, °\text{C}$$

(a)  **With single glazed FPC wind velocity**

From Examples 4.17 and 4.19: $U_t = 7.49$ W/m$^2$ °C and $U_b = 0.70$ W/m$^2$ °C.
So,

$$U_L = U_t + U_b = 7.49 + 0.70 = 8.19 \text{ W/m}^2 \, °\text{C}$$

(b)  **With double glazed FPC wind velocity**

From Examples 4.18 and 4.19: $U_t = 3.19$ W/m$^2$ °C and $U_b = 0.70$ W/m$^2$ °C.
So,

$$U_L = U_t + U_b = 3.19 + 0.70 = 3.89 \text{ W/m}^2 \, °\text{C}$$

The total overall heat transfer coefficient in case of double glazed flat plate collector is reduced by 63%.


## 4.3   Liquid Flat Plate Collector (FPC) Under Natural Mode of Operation

Refer to Eq. 4.5, there is no provision of fluid (water) used to extract the thermal energy associated with blackened absorber plate $(\dot{q}_u)$ under forced mode. If the fluid is allowed to pass below blackened absorber plate as shown in Fig. 4.1, then the thermal energy/heat available in absorber plate can be carried out from flat plate collector under natural mode (thermosyphon mode) to be used for any hot water applications namely domestic purposes. Such flat plate collector will be referred as **liquid flat plate collector**. The rate of heat transferred from absorber plate to fluid (water) will depend on the heat transfer coefficient $(\overline{h}_c)$ between inverted plates to the fluid below absorber plate (see Example 2.24). If flat plate collector is inclined then Example 2.25 should be used. For inclined flat plate, the hot fluid (water) moves upward direction due to its low density and hot water is collected at out let as shown in Fig. 4.5 absorber plate.

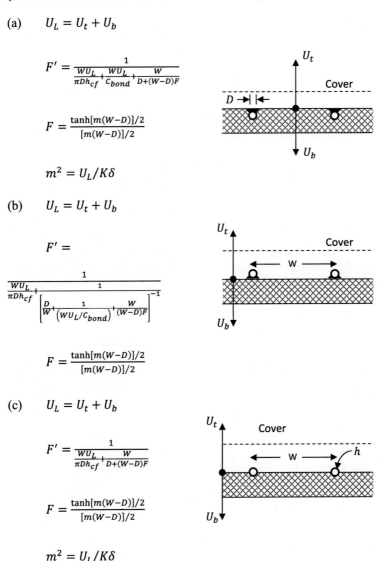

(a)   $U_L = U_t + U_b$

$$F' = \frac{1}{\frac{WU_L}{\pi Dh_{cf}} + \frac{WU_L}{C_{bond}} + \frac{W}{D+(W-D)F}}$$

$$F = \frac{\tanh[m(W-D)]/2}{[m(W-D)]/2}$$

$$m^2 = U_L/K\delta$$

(b)   $U_L = U_t + U_b$

$$F' =$$

$$\frac{1}{\frac{WU_L}{\pi Dh_{cf}} + \left[\frac{D}{W} + \frac{1}{(WU_L/C_{bond})} + \frac{W}{(W-D)F}\right]^{-1}}$$

$$F = \frac{\tanh[m(W-D)]/2}{[m(W-D)]/2}$$

(c)   $U_L = U_t + U_b$

$$F' = \frac{1}{\frac{WU_L}{\pi Dh_{cf}} + \frac{W}{D+(W-D)F}}$$

$$F = \frac{\tanh[m(W-D)]/2}{[m(W-D)]/2}$$

$$m^2 = U_L/K\delta$$

**Fig. 4.5** Various combinations of absorber plate and tube containing fluid (water) of single flat plate collectors

## 4.3.1   Convective Heat Transfer Between Absorber Plates to Working Fluid (Water) Through Riser

If the heat capacity (mass × specific heat)) of water is reduced by flowing of water through conductive tube as shown in Fig. 4.4, then the temperature rise will be increased significantly. In this case, the rate of heat will be transferred from absorber plate to fluid through cylindrical copper tube. The overall heat transfer coefficient $(U_k)$ for such configuration [4] can be determined by

$$\frac{1}{U_k} = \frac{1}{\frac{D_1}{D_2}(U_h)} + \frac{1}{U_c} + \frac{1}{\frac{2K_s}{D_2 \ln(D_2/D_1)}} \tag{4.13a}$$

Since, the thickness copper tube is very small and hence one can consider $D_1 \approx D_2 = D$, diameter of copper tube and $\ln(D_2/D_1) = \ln 1 = 0$, then Eq. 4.13 reduces to

$$\frac{1}{U_k} = \frac{1}{U_h} + \frac{1}{U_c}$$

with $U_h = U_c = \text{Nu}\frac{K_w}{D} = 3.66\frac{K_w}{D}$, Nu $= 3.66$ [case (i) of Table 2.5]

where, $D$ is the diameter of the copper tube used as a riser in FPC and $U_h$ and $U_c$ are an overall heat transfer coefficient from inner and outer wall of copper tube to hot and cold fluid surrounding copper tube. Since, there is no fluid outside copper tube hence effect of $U_c$ should not be considered, and in this case,

$$U_k = U_h = \text{Nu}\frac{K_w}{D} = 3.66\frac{K_w}{D} \text{ in W/m}^2 \,°\text{C} \tag{4.13b}$$

The $U_k$ and $U_h$ are convective heat transfer coefficient between inner wall of copper tube to working fluid, and it can also be referred as film heat transfer coefficient.

$$U_h = h_{cf} \tag{4.13c}$$

**Example 4.21** Find out convective heat transfer coefficient $(h_{cf})$ between inner wall of copper tube (riser of FPC) and fluid flowing at 60 °C through it per unit length of tube of a FPC with an effective area of 2 m$^2$ and riser length of 2 m. Diameter of copper tube is 16 mm (0.016 m).

**Solution**

Thermal conductivity of water at 60 °C $= 0.654$ W/m °C (Appendix E(b)).

From Eqs. 4.13b and 4.13c, one can evaluate the convective heat transfer coefficient $(h_{cf})$ between inner wall of copper tube (riser of FPC) and fluid flowing at 60 °C through it per unit length of tube of a FPC as

$$h_{cf} = U_k = U_h = \text{Nu}\frac{K_w}{D} = 3.66\frac{K_w}{D} = 3.66 \times \frac{0.654}{0.016} = 149.6 \text{ W/m}^2 \,°\text{C}$$

**Example 4.22** Repeat Example 4.21 with riser tube of diameter of 4 mm (0.004 m).

**Solution**

From Example 3.21, we have

$$h_{cf} = 3.66\frac{K_w}{D} = 3.66 \times \frac{0.654}{0.008} = 299.21 = 299.21 \text{ W/m}^2 \text{ }^\circ\text{C}$$

So, one can see an importance of diameter of riser. With increase of diameter, the convective heat transfer coefficient $(h_{cf})$ decreases.

## 4.3.2 Fin Efficiency Factor (F)

Figure 4.6 shows a configuration of half of one tube below plate (absorber) and tube in plate. There are two way to combine tube with plate, namely:

(i) The bonding conducting material has been used between plate and tube for fast transfer of thermal energy (first figure in LHS of Fig. 4.6).
(ii) The spot wielding is done in tube and plate configuration (second figure in RHS of Fig. 4.6).

The distance between the centers of two tubes is $W = 10$ cm (0.10 m) which is same as distance between centers of two absorber sheets. Hence the distance between centers of absorber sheet and tube is $W/2$ as shown in LHS of Fig. 4.6. If the outer and inner tubes (riser) diameters are $D$ and $d$, respectively, then solar radiation absorbed over $(W - D)/2$ length on both side of absorber sheet will be conducted toward the tube/riser. The region between the centerline separating the tubes and the tube bond base can be considered as a fin problem. The copper absorber sheet having thermal conductivity of 386 W/m °C has a thickness of $\delta$ [0.35 mm = 0.00035 m). The copper sheet material being a good conductor, the temperature gradient through the thickness of sheet is negligible.

In order to see the effectiveness of a fin to transfer a given quantity of heat, a new parameter is defined which is known as **fin efficiency factor**. It is expressed as

$$\text{Fin efficiency}(F) = \frac{\text{Actual the rate of heat transferred to near bond}}{\text{Heat would have been transferred if entire fin area were at } T_b} \tag{4.14}$$

The fin as shown in Fig. 4.6 is of length $(W - D)/2$. An expression for fin efficiency factor is given by

$$F = \frac{\tanh m(W - D)/2}{m(W - D)/2} \text{ with } m^2 = \frac{U_L}{K\delta} \tag{4.15}$$

The rate of heat conducted toward the tube by conduction due to fin per unit length can be written as

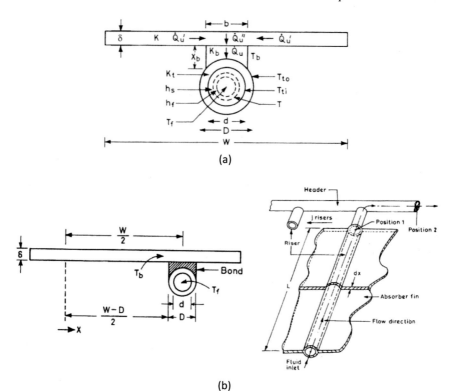

**Fig. 4.6** Configuration of tube below plate (sheet) and tube in absorber plate **a** first case of Fig. 4.5 and **b** half portion of first case of Fig. 4.5

$$\dot{q}_{\mathrm{fin}} = F(W - D)[(\alpha\tau)I(t) - U_L(T_b - T_a)] \qquad (4.16a)$$

The rate of heat conducted toward the tube by conduction due to fin for length $L_r$ in W can be written as

$$\dot{Q}_{\mathrm{fin}} = F(W - D) \times L_r \times [(\alpha\tau)I(t) - U_L(T_b - T_a)] \qquad (4.16b)$$

The derivation of Eqs. 4.15 and 4.16 is given by Tiwari [4] and Tiwari et al. [5]. If there are $n$ risers then Eq. 4.16b becomes as

$$\dot{Q}_{\mathrm{fin}} = n \times F(W - D) \times L_r \times [(\alpha\tau)I(t) - U_L(T_b - T_a)] \,\mathrm{in}\, W \qquad (4.16c)$$

**Example 4.23** Evaluate the fin efficiency factor $(F)$ of copper tube in copper plate absorber for single glazed FPC for the following parameters:

$W = 0.10$ m; $\delta = [0.35\,\text{mm} = 0.00035\,\text{m})$; $K = 386$ W/m °C; $U_L = 8.19$ W/m$^2$ °C with wind velocity of 1 m/s (Example 4.20); $D = 10$ mm $= 0.01$ m and $d = 8$ mm $= 0.008$ m.

**Solution**

From Eq. 4.15, one has expression for fin efficiency of flat plate collector as

$$F = \frac{\tanh m(W - D)/2}{m(W - D)/2} \text{ with } m^2 = \frac{U_L}{K\delta}$$

Here, $m = \sqrt{\frac{U_L}{K\delta}} = \sqrt{\frac{8.19}{386 \times 0.00035}} = \sqrt{\frac{8.19}{0.1351}} = \sqrt{\frac{8.19}{0.1351}} = \sqrt{60.62} = 7.79$

Now,

$$F = \frac{\tanh m(W - D)/2}{m(W - D)/2} = \frac{\tanh \frac{7.79(0.10 - 0.01)}{2}}{\frac{7.79(0.10 - 0.01)}{2}} = \frac{\tanh(0.35055)}{(0.35055)}$$

$$= \frac{0.33686}{(0.35055)} = 0.96.$$

**Example 4.24** Repeat Example 4.23 with aluminum (Al) absorber plate.

**Solution**

Thermal conductivity of aluminum (Al) absorber plate $= 204$ W/m °C

In this case,

$$m = \sqrt{\frac{U_L}{K\delta}} = \sqrt{\frac{8.19}{204 \times 0.00035}} = \sqrt{\frac{8.19}{0.0714}} = \sqrt{114.71} = 10.71$$

Now,

$$F = \frac{\tanh m(W - D)/2}{m(W - D)/2} = \frac{\tanh \frac{10.71(0.10 - 0.01)}{2}}{\frac{10.71(0.10 - 0.01)}{2}} = \frac{\tanh(0.482)}{(0.482)} = \frac{0.4478}{(0.482)} = 0.93$$

From Examples 4.23 and 4.24, one can see that the fin efficiency factor ($F$) of copper plate is higher (0.96) than aluminum (Al) one (0.93); hence copper plate is preferred in flat plate collector.

**Example 4.25** Evaluate the fin efficiency factor ($F$) of copper tube in copper plate absorber for double glazed FPC for the following parameters:
$W = 0.10$ m; $\delta = [0.35\,\text{mm} = 0.00035\,\text{m})$; $K = 386$ W/m °C; $U_L = 3.89$ W/m °C with wind velocity of 1 m/s (Example 4.20); $D = 10$ mm $= 0.01$ m and $d = 8$mm $= 0.008$ m.

**Solution**

From Eq. 4.15, one has expression for fin efficiency of flat plate collector as

$$F = \frac{\tanh m(W - D)/2}{m(W - D)/2} \text{ with } m^2 = \frac{U_L}{K\delta}$$

Here, $m = \sqrt{\frac{U_L}{K\delta}} = \sqrt{\frac{3.89}{386 \times 0.00035}} = \sqrt{\frac{3.89}{0.1351}} = \sqrt{\frac{3.89}{0.1351}} = \sqrt{28.79} = 5.37$

Now,

$$F = \frac{\tanh m(W - D)/2}{m(W - D)/2} = \frac{\tanh \frac{5.37(0.10-0.01)}{2}}{\frac{5.37(0.10-0.01)}{2}} = \frac{\tanh(0.2414)}{(0.2414)} = \frac{0.2368}{(0.2414)} = 0.98$$

Here, one concludes that there is enhancement in the value of fin efficiency factor ($F$) for lower value of an overall heat transfer coefficient $\left[U_L = 3.89 \text{ W/m}^2 \text{ °C}\right]$.

**Example 4.26** Evaluate the rate of heat conducted through copper fins to copper tube in W for effective area of 2 m × 1 m of single and double glazed FPC if there are 10 risers for $F = 0.99$ (Example 4.24); $I(t) = 500 \text{ W/m}^2$ and $T_b = 60 \text{ °C}$ (Example 4.21); $T_a = 25 \text{ °C}$ [Summer climatic condition].

**Solution**

From Eq. 4.16c, we have the rate of heat conducted through fins to tube in $W$.

(a) **For single glazed FPC**: $U_L = 8.19 \text{ W/m}^2 \text{ °C}$ (Example 4.20)

$$\dot{Q}_{\text{fin}} = n \times F(W - D) \times L_r \times [(\alpha\tau)I(t) - U_L(T_b - T_a)]$$
$$= 10 \times 0.96 \times (0.10 - 0.01) \times 2 \times [0.9 \times 0.9 \times 500 - 8.19 \times (60 - 25)]$$
$$= 1.728 \times [405 - 286.65] = 204.51 \text{ W}$$

(b) **For double glazed FPC**: $U_L = 3.89 \text{ W/m}^2 \text{ °C}$ (Example 4.20)

$$\dot{Q}_{\text{fin}} = n \times F(W - D) \times L_r \times \left[(\alpha\tau^2)I(t) - U_L(T_b - T_a)\right]$$
$$= 10 \times 0.98 \times (0.10 - 0.01) \times 2$$
$$\times [0.9 \times 0.9 \times 0.9 \times 500 - 3.89 \times (60 - 25)]$$
$$= 1.764 \times [364.5 - 136.15] = 402.81 \text{ W}$$

Here, one should observe that an overall heat loss coefficient plays ($U_L$) an important role in determining of the rate of heat conducted through copper fins to copper tube. For double glazed FPC, there is enhancement in $\dot{Q}_{\text{fin}}$ by nearly 100%.

### 4.3.3  Flat Plate Collector Efficiency ($F'$)

Equation 4.5a is applicable for flat surface and fluid below it. and there is direct convective heat transfer from flat plate to fluid as shown in Fig. 4.1. So, there is direct contact between flat plate and working fluid. If the fluid (water) is allowed to flow through metallic tube generally known as riser which is arranged in different configuration as shown in Fig. 4.4, then there is indirect contact between absorber plate to working fluid and the heat transfer from absorber plate to working fluid in tube/riser will be reduced unlike Eq. 4.5a. This reduction will depend upon configuration/arrangement of tube with absorber plate. A new parameters has been introduced as flat plate collector efficiency ($F'$) which will depend on design parameters of plate and tube configuration.

Now, the flat plate collector efficiency ($F'$) is defined as the ratio of actual rate of useful heat collection ($\dot{Q}_u$) under natural mode of operation of flat plate collector (FPC) to the rate of useful heat collection when the collector's absorbing plate ($T_p$) is at the local fluid temperature ($T_f$), Eq. 4.5a, and it is expressed as

$$F' = \frac{\dot{Q}_u}{\dot{Q}_u|_{T_p=T_f}} = \frac{\dot{Q}_u}{A_c\left[\dot{q}_{ab} - U_L\left(T_f - T_a\right)\right]} \tag{4.17a}$$

or,

$$\dot{Q}_u = F'A_c\left[\dot{q}_{ab} - U_L\left(T_f - T_a\right)\right] = F'A_c\left[\alpha_p\tau_g I(t) - U_L\left(T_f - T_a\right)\right] \tag{4.17b}$$

The expression for flat plate collector efficiency ($F'$) in terms of fin efficiency ($F$) for three configurations has been shown in Fig. 4.5.

### 4.3.4  Threshold Intensity

The solar intensity starts producing net positive thermal energy, i.e., $\dot{Q}_u \geq 0$.
   For Threshold Intensity for Single Glazed FPC
   From Eq. 4.17b, we have

$$\dot{Q}_u = F'A_c\left[\alpha_p\tau_g I(t) - U_L\left(T_f - T_a\right)\right] \geq 0 \cong I(t)_{\text{threesold}} \geq \frac{U_L\left(T_f - T_a\right)}{\left(\alpha_p\tau_g\right)} \tag{4.18a}$$

The threshold mainly depends on an overall heat transfer coefficient, $U_L$ and $\left(\alpha_p\tau_g\right)$.

**For threshold intensity for double glazed FPC**

In this case, $\tau_g$ will be replaced by $\tau_g^2$, and the numerical value of $U_L$ in Eq. 4.18a will be less than single glazed FPC:

$$I(t)_{threesold} \geq \frac{U_L(T_f - T_a)}{(\alpha_p \tau_g^2)} \tag{4.18b}$$

The value of $U_L$ will be different for single and double glazed FPC. Threshold solar intensity provides an information about net energy obtained by FPC at given fluid temperature, $T_f$ and ambient air temperature, $(T_a)$.

If **m** flat plate collectors (FPC's) are connected in parallel as shown in Fig. 4.7 (**referred as m-FPC**) then Eq. 4.17b becomes as

$$\dot{Q}_{um} = mF'A_c[\dot{q}_{ab} - U_L(T_f - T_a)] = F'A_{cm}[\dot{q}_{ab} - U_L(T_f - T_a)] \tag{4.19}$$

where, $A_{cm} = m \times A_c = m \times$ area of one flat plate collector. It will be referred as m-FPC.

**Equations** 4.17b and 4.17c **can be used for natural mode of operation of Flat plate collector.**

**Example 4.27** Evaluate flat plate collector (FPC) efficiency for case (a) [Tube below flat absorber plate] of Fig. 4.5 for single and double glazed collector using copper as absorber material with bond conductance [6] ($C_{bond}$) of 6.3 W/m² °C by using the data of Examples 4.23 and 4.25, respectively.

**Solution**

Here: $W = 0.10$ m; $D = 0.01$ m; $\pi = 3.14$; $h_{cf} = U_k/L_0 = 149.6$ W/m² °C for copper tube (Example 4.21); $C_{bond} = 6.3$ W/m² °C.

An expression for flat plate collector (FPC) efficiency for case (a) of Fig. 4.5 is given by $F' = \dfrac{1}{\frac{WU_L}{\pi Dh_{cf}} + \frac{WU_L}{C_{bond}} + \frac{W}{D+(W-D)F}}$.

From Examples 4.24 and 4.25, we have

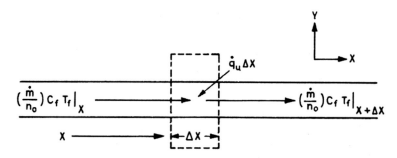

**Fig. 4.7** Elemental length ($\Delta x$) of one riser of length ($L_0$) for liquid flow in x-direction

(a) **For single glazed FPC**: $U_L = 8.19$ W/m$^2$ °C (Example 4.20) and $F = 0.96$ (Example 4.23).

Now,

$$\text{FPC efficiency for single glazed} = F' = \cfrac{1}{\cfrac{WU_L}{\pi Dh_{cf}} + \cfrac{WU_L}{C_{bond}} + \cfrac{W}{D+(W-D)F}}$$

$$= \cfrac{1}{\cfrac{0.10\times8.19}{3.14\times0.01\times149.6} + \cfrac{0.10\times8.19}{6.3} + \cfrac{0.10}{0.01+(0.10-0.01)\times0.96}}$$

or,

$$F' = \cfrac{1}{\cfrac{0.10\times8.19}{3.14\times0.01\times149.6} + \cfrac{0.10\times8.19}{6.3} + \cfrac{0.10}{0.01+(0.10-0.01)\times0.96}}$$

$$= \frac{1}{0.174 + 0.13 + 1.037} = 0.75$$

(b) **For double glazed**: $U_L = 3.89$ W/m$^2$ °C (Example 4.20) and $F = 0.98$ (Example 4.25).

$$\text{FPC efficiency for double glazed} = F'$$

$$= \cfrac{1}{\cfrac{0.10\times3.89}{3.14\times0.01\times149.6} + \cfrac{0.10\times3.89}{6.3} + \cfrac{0.10}{0.01+(0.10-0.01)\times0.98}}$$

$$= \frac{1}{0.083 + 0.062 + 1.018} = 0.86$$

**Hence, one can conclude that FPC efficiency for double glazed is increased in comparison with single glazed FPC.**

**Example 4.28** Evaluate flat plate collector (FPC) efficiency for case (b) [Tube above flat absorber plate] of Fig. 4.5 for single and double glazed collector using copper as absorber material with bond conductance [6] ($C_{bond}$) of 6.3 W/m$^2$ °C by using the data of Examples 4.23 and 4.25, respectively.

**Solution**

Here: $W = 0.10$ m; $D = 0.01$ m; $\pi = 3.14$; $h_{cf} = U_k/L_0 = 149.6$ W/m$^2$ °C for copper tube (Example 4.21); $C_{bond} = 6.3$ W/m$^2$ °C.

An expression for flat plate collector (FPC) efficiency for case (b) of Fig. 4.5 is given by

$$F' = \cfrac{1}{\cfrac{WU_L}{\pi Dh_{cf}} + \cfrac{1}{\left[\frac{D}{W} + \frac{1}{(WU_L/C_{bond})} + \frac{W}{(W-D)F}\right]^{-1}}}$$

(a) **For single glazed FPC:** $U_L = 8.19$ W/m² °C (Example 4.20) and $F = 0.96$ (Example 4.23).

Now,

FPC efficiency for single glazed =

$$F' = \cfrac{1}{\cfrac{WU_L}{\pi Dh_{cf}} + \cfrac{1}{\left[\frac{D}{W} + \frac{1}{(WU_L/C_{bond})} + \frac{W}{(W-D)F}\right]^{-1}}}$$

$$= \cfrac{1}{\cfrac{0.10 \times 8.19}{3.14 \times 0.01 \times 149.6} + \cfrac{1}{\left[\frac{0.01}{0.10} + \frac{1}{(0.10 \times 8.19/6.3)} + \frac{0.10}{(0.10-0.01) \times 0.96}\right]^{-1}}}$$

$$= \cfrac{1}{0.174 + \cfrac{1}{[0.001+0.13+1.1574]^{-1}}} = \cfrac{1}{0.174 + 1.288} = \cfrac{1}{1.4624} = 0.68$$

(b) **For double glazed:** $U_L = 3.89$ W/m² °C (Example 4.20) and $F = 0.98$ (Example 4.25). In double glazed FPC,

FPC efficiency for single glazed =

$$F' = \cfrac{1}{\cfrac{WU_L}{\pi Dh_{cf}} + \cfrac{1}{\left[\frac{D}{W} + \frac{1}{(WU_L/C_{bond})} + \frac{W}{(W-D)F}\right]^{-1}}}$$

$$= \cfrac{1}{\cfrac{0.10 \times 3.89}{3.14 \times 0.01 \times 149.6} + \cfrac{1}{\left[\frac{0.01}{0.10} + \frac{1}{\left(0.10 \times \frac{3.89}{6.3}\right)} + \frac{0.10}{(0.10-0.01) \times 0.98}\right]^{-1}}}$$

$$= \cfrac{1}{0.083 + \cfrac{1}{[0.001+0.062+1.1338]^{-1}}} = \cfrac{1}{0.174 + 1.1967} = \cfrac{1}{1.371} = 0.73$$

From Examples 4.27 and 4.28, it is inferred that the tube below flat absorber plate gives better flat plate collector (FPC) efficiency by 10% for single glazed and by 17.8% for double glazed FPC.

**Example 4.29** Evaluate flat plate collector (FPC) efficiency for case (c) [Tube fitted below in semi-curved absorber plate] of Fig. 4.5 for single and double glazed collector using copper as absorber material with bond conductance[6] ($C_{bond}$) of 6.3 W/m² C by using the data of Examples 4.23 and 4.25, respectively.

**Solution**

Here: $W = 0.10$ m; $D = 0.01$ m; $\pi = 3.14$; $h_{cf} = U_k/L_0 = 149.6$ W/m² °C for copper tube (Example 4.21); $C_{bond} = 6.3$ W/m² °C.

An expression for flat plate collector (FPC) efficiency for case (c) of Fig. 4.5 is given by

$$F' = \frac{1}{\frac{WU_L}{\pi Dh_{cf}} + \frac{W}{D+(W-D)F}}$$

(a)  **For single glazed FPC**: $U_L = 8.19$ W/m$^2$ °C (Example 4.20) and $F = 0.96$ (Example 4.23).

Now,

  FPC efficiency for single glazed $=$

$$F' = \frac{1}{\frac{WU_L}{\pi Dh_{cf}} + \frac{W}{D+(W-D)F}} = \frac{1}{\frac{0.10\times8.19}{3.14\times0.01\times149.6} + \frac{0.10}{0.01+(0.10-0.01)\times0.96}}$$

or,

$$F' = \frac{1}{\frac{0.10\times8.19}{3.14\times0.01\times149.6} + \frac{0.10}{0.01+(0.10-0.01)\times0.96}} = \frac{1}{0.174+1.0373} = \frac{1}{1.2113} = 0.82$$

(b)  **For double glazed**: $U_L = 3.89$ W/m$^2$ °C (Example 4.20) and $F = 0.98$ (Example 4.25).

  In double glazed FPC,
  FPC efficiency for single glazed $=$

$$F' = \frac{1}{\frac{0.10\times3.89}{3.14\times0.01\times149.6} + \frac{0.10}{0.01+(0.10-0.01)\times0.98}} = \frac{1}{0.083+1.0183} = \frac{1}{1.1013} = 0.91$$

**From Examples** 4.27–4.29 , **one concludes that case (c) of Fig.** 4.5, **i.e., tube fitted below in semi-curved absorber plate is the best configuration for absorber cum tube combination.**

**Example 4.30**  Repeat Example 4.29 for value of $h_{cf} = 299.21$ obtained in Example 4.11 $\left(h_{cf} = 299.21 \text{ W/m}^2 \text{ °C}\right)$ with riser tube of diameter of 4 mm (0.004 m).

**Solution**

In Example 4.29, one has to change only $h_{cf} = 299.21$ W/m$^2$ °C in place of 149.6 W/m$^2$ °C.

(a)  **For single glazed FPC**: $U_L = 8.19$ W/m$^2$ °C (Example 4.20) and $F = 0.96$ (Example 4.23)

Now,

  FPC efficiency for single glazed $=$

$$F' = \frac{1}{\frac{WU_L}{\pi Dh_{cf}} + \frac{W}{D+(W-D)F}} = \frac{1}{\frac{0.10\times8.19}{3.14\times0.01\times149.6} + \frac{0.10}{0.01+(0.10-0.01)\times0.96}}$$

or,

$$F' = \cfrac{1}{\cfrac{0.10\times8.19}{3.14\times0.01\times299.21} + \cfrac{0.10}{0.01+(0.10-0.01)\times0.96}} = \frac{1}{0.097 + 1.0373} = \frac{1}{1.1245} = 0.96$$

(b)  **For double glazed**: $U_L = 3.89\,\text{W/m}^2\,°\text{C}$ (Example 4.20) and $F = 0.98$
(Example 4.25)

In double glazed FPC,
FPC efficiency for single glazed =

$$F' = \cfrac{1}{\cfrac{0.10\times3.89}{3.14\times0.01\times299.21} + \cfrac{0.10}{0.01+(0.10-0.01)\times0.98}} = \frac{1}{0.0414 + 1.0183}$$

$$= \frac{1}{1.059} = 0.94$$

In single glazed FPC, there is significant effect of $h_{cf}$ on flat plate collector efficiency ($F'$) but marginal effect on double glazed FPC.

**Example 4.31** Find out threshold solar intensity for single and double glazed flat plate collector (FPC) for Example 4.29 and $\alpha_p = \tau_g = 0.9$ and $T_f = 60\,°\text{C}$ and $T_a = 25\,°\text{C}$.

**Solution**

**For single glazed FPC**: $U_L = 8.19\,\text{W/m}^2\,°\text{C}$ (Example 4.20).
From Eq. 4.18a, we have

$$I(t)_{\text{threesold}} \geq \frac{U_L(T_f - T_a)}{\alpha_p\tau_g} = \frac{8.19\times(60-25)}{0.9\times0.9} = \frac{262.08}{0.81} = 323.55\,\text{W/m}^2$$

**For double glazed FPC**: $U_L = 3.89\,\text{W/m}^2\,°\text{C}$ (Example 4.20) (Example 4.20).
From Eq. 4.18b, we have

$$I(t)_{\text{threesold}} \geq \frac{U_L(T_f - T_a)}{(\alpha_p\tau_g^2)} = \frac{3.89\times(60-25)}{0.9\times0.9\times0.91} = \frac{136.15}{0.729} = 186.76\,\text{W/m}^2$$

This example shows that the threshold solar intensity is significantly lower by 73% for double glazed FPC for the same design and operating parameters.

**Example 4.32** Repeat Example 4.31 for operating temperature of fluid $(T_f)$ as $40\,°\text{C}$.

**Solution**

**For single glazed FPC**: $U_L = 8.19\,\text{W/m}^2\,°\text{C}$ (Example 4.20).

$$I(t)_{threesold} \geq \frac{U_L(T_f - T_a)}{\alpha_p \tau_g} = \frac{8.19 \times (40 - 25)}{0.9 \times 0.9} = \frac{131.0s}{0.81} = 135 \text{ W/m}^2$$

**For double glazed FPC:** $U_L = 3.89$ W/m$^2$ °C (Example 4.20).
From Eq. 4.18b, we have

$$I(t)_{threesold} \geq \frac{U_L(T_f - T_a)}{(\alpha_p \tau_g^2)} = \frac{3.89 \times (40 - 25)}{0.9 \times 0.9 \times 0.91} = \frac{62.24}{0.729} = 85.37 \text{ W/m}^2$$

From Examples 4.30 and 4.31, one see that the threshold solar intensity is lowered at low operating fluid temperature due to low thermal losses $[U_L(T_f - T_a)]$.

**Example 4.33** Evaluate the rate of useful energy in natural mode of operation for single and double glazed 2 m$^2$ plate collector for solar intensity of 500 W/m$^2$ and ambient air temperature of 25 °C and operating at $T_f = 60$ °C for Example 4.29 [Summer condition].

**Solution**

(a) **For single glazed FPC:** $U_L = 8.19$ W/m$^2$ °C; $\alpha_p = \tau_g = 0.9$; $A_c = 2$ m$^2$; $F' = 0.82$

From Eq. 4.17b, we have an expression for the rate of useful energy for single glazed flat plate collector:

$$\dot{Q}_u = F'A_c[\dot{q}_{ab} - U_L(T_f - T_a)] = F'A_c[\alpha_p \tau_g I(t) - U_L(T_f - T_a)]$$
$$= 0.82 \times 2 \times [0.9 \times 0.9 \times 500 - 8.19(60 - 25)]$$
$$= 1.64[405 - 286.65] = 194.09 \text{ W}$$

(b) **For double glazed FPC:** $U_L = 3.19$ W/m$^2$ °C; $\alpha_p = \tau_g = 0.9$; $A_c = 2$ m$^2$; $F' = 0.91$

**Following Eq. 4.12a , we have**

$$\dot{Q}_u = A_c \dot{q}_u = F'A_c[\alpha_p \tau_g^2 I(t) - U_L(T_f - T_a)]$$
$$= 0.91 \times 2[0.9 \times 0.9 \times 0.9 \times 500 - 3.89(60 - 25)]$$
$$= 1.82[364.5 - 136.15] = 415.6 \text{ W}$$

## 4.3.5  Instantaneous Thermal Efficiency

It is the ration of the rate of useful thermal energy to the rate of incident solar radiation, and it can be expressed as follows:

$$\eta_i = \frac{\text{The rate of useful thermal energy}}{\text{The rate of incident solar radiation}} \qquad (4.18)$$

Following Eq. 4.18, one has an expression for an instantaneous thermal efficiency as

(a) **Single glazed flat plate collector**

$$\eta_i = \frac{\dot{Q}_u}{I(t) \times A_c} = \frac{F'A_c[\dot{q}_{ab} - U_L(T_f - T_a)]}{I(t) \times A_c} = F'\left[(\alpha_p\tau_g) - U_L\frac{(T_f - T_a)}{I(t)}\right]$$

$$(4.19)$$

with $\dot{q}_{ab} = (\alpha_p\tau_g)I(t)$ (Eq. 4.5a)

(b) **Double glazed flat plate collector**

$$\eta_i = \frac{\dot{Q}_u}{I(t) \times A_c} = \frac{F'A_c[\dot{q}_{ab} - U_L(T_f - T_a)]}{I(t) \times A_c} = F'\left[(\alpha_p\tau_g^2) - U_L\frac{(T_f - T_a)}{I(t)}\right]$$

$$(4.20)$$

with $\dot{q}_{ab} = (\alpha_p\tau_g^2)I(t)$ (Eq. 4.5a)

Equations 4.19 and 4.20 are known as Hottel–Whillier–Bliss (HWB equation for natural mode of operation.

**It is important to note that the numerical values of $U_L$ and $F'$ will be different for single and double glazed flat plate collectors (Examples 4.20 and 4.29).**

**Example 4.34** Evaluate instantaneous thermal efficiency ($\eta_i$) for Example 4.33 for summer condition.

**Solution**

From Eqs. 4.19 and 4.20 we have

(a) **For single glazed FPC:** $U_L = 8.19$ W/m$^2$ °C and $F' = 0.82$

**First method:**

$$\eta_i = F'\left[(\alpha_p\tau_g) - U_L\frac{(T_f - T_a)}{I(t)}\right] = 0.82\left[0.81 - 8.19 \times \frac{(60 - 25)}{500}\right]$$
$$= 0.194(19.4\%)$$

**Second Method:**

$$\eta_i = \frac{194.09}{2 \times 500} = 0.194(19.4\%)$$

(b)  **For double glazed FPC:** $U_L = 3.89$ W/m$^2$ °C and $F' = 0.91$

**First method:**

$$\eta_i = F'\left[(\alpha_p \tau_g^2) - U_L \frac{(T_f - T_a)}{I(t)}\right] = 0.91\left[0.729 - 3.89 \times \frac{(60 - 25)}{500}\right]$$

$$= 0.4156(41.56\%)$$

**Second Method:**

$$\eta_i = \frac{415.6}{2 \times 500} = 0.4156(41.56\%)$$

In double glazed FPC, an instantaneous thermal efficiency is more due to low thermal losses.

## 4.4  Liquid Flat Plate Collector (FPC) Under Forced Mode of Operation

In forced mode of operation, the fluid (water) is fed through inlet of flat plate collector as shown in Fig. 4.4 with mass flow rate of $\dot{m}_f$ and specific heat of fluid as $C_f$. There are $n$ riser then mass flow rate in one riser will be $\frac{\dot{m}_f}{n}$.

### 4.4.1  The Outlet Temperature from One Flat Plate Collector (FPC)

Referring to Fig. 4.7, an elemental length of $dx$ of riser has been considered to write energy balance equation. If temperature of fluid at $x$ and $(x + dx)$ are $T_f$ and $(T_f + dT_f)$, then temperature rise per unit length will be $\frac{dT_f}{dx}$.

The rate of thermal energy/heat carried by flowing fluid in elemental length of $dx$ of one riser $=$

$$\left[\frac{\dot{m}_f}{n}C_f\frac{dT_f}{dx}dx\right] \tag{4.21}$$

Further, following Eq. 4.19, the rate of solar radiation received by flowing fluid (water) in elemental area of $(Wdx)$ will be

$$\dot{Q}_{udx} = F'Wdx\left[\dot{q}_{ab} - U_L(T_f - T_a)\right] \tag{4.22}$$

where, $W \mathrm{d}x$ = breadth of one riser and fin combination $(W)$ × $\mathrm{d}x$ = an area of elemental length of one riser of flat plate collector

Equations 4.21 and 4.22 will be equal as per first of thermodynamics, i.e., energy conservation and hence

$$\left[\frac{\dot{m}_f}{n} C_f \frac{\mathrm{d}T_f}{\mathrm{d}x} \mathrm{d}x\right] = F'W \mathrm{d}x \left[\dot{q}_{ab} - U_L(T_f - T_a)\right] \text{ with } \dot{q}_{ab} = \alpha_p \tau_g I(t)$$

or,

$$\dot{m}_f C_f \frac{\mathrm{d}T_f}{\mathrm{d}x} - nWF'\left[\dot{q}_{ab} - U_L(T_f - T_a)\right] = 0 \tag{4.23}$$

By assuming $F'$ and $U_L$ to be constant during operating temperature, the solution of Eq. 4.23 with the boundary condition $T_f = T_{fi}$ at $x = 0$, is given by

$$\frac{T_f - T_a - (\dot{q}_{ab}/U_L)}{T_{fi} - T_a - (\dot{q}_{ab}/U_L)} = \exp\left[-\frac{U_L nWF'x}{\dot{m}_f C_f}\right] \tag{4.24}$$

The outlet fluid temperatures $T_{fo}$ at $x = L_0$, can be obtained as,

$$T_{fo} = T_f\big|_{x=L_0} = \left[\left(\frac{\dot{q}_{ab}}{U_L}\right) + T_a\right] + \left[T_{fi} - T_a - \frac{\dot{q}_{ab}}{U_L}\right] \exp\left[-A_c U_L F'/(\dot{m}_f C_f)\right]$$

or,

$$T_{fo} = \left[\left(\frac{\dot{q}_{ab}}{U_L}\right) + T_a\right]\left[1 - \exp\left[-A_c U_L F'/(\dot{m}_f C_f)\right]\right]$$
$$+ T_{fi} \exp\left[-A_c U_L F'/(\dot{m}_f C_f)\right] \tag{4.25}$$

## 4.4.2   The Rate of Thermal Energy $\left(\dot{Q}_u\right)$

Now, the rate of thermal energy carried out by flowing water in forced mode of operation is given by

$$\dot{Q}_u = \dot{m}_f C_f (T_{fo} - T_{fi})$$
$$= \frac{\dot{m}_f C_f}{A_c U_L}\left\{1 - \exp\left[-A_c U_L F'/(\dot{m}_f C_f)\right]\right\} A_c \left[\alpha_p \tau_g I(t) - U_L(T_{fi} - T_a)\right]$$

or,

$$\dot{Q}_u = F_R A_c \left[\alpha_p \tau_g I(t) - U_L(T_{fi} - T_a)\right] \tag{4.26}$$

where,

$$FR = \frac{\dot{m}_f C_f}{A_c U_L}\{1 - \exp[-A_c U_L F'/(\dot{m}_f C_f)]\} = \text{mass flow rate factor} \quad (4.27)$$

### 4.4.3 Flat Plate Heat Removal Factor/Mass Flow Rate Factor (F$_R$)

The **flat plate heat removal factor (F$_R$)** is defined as the ratio of the actual rate of useful energy gain to the rate of useful energy gain if the entire collector was at the fluid inlet temperature $(T_{fi})$ in a forced circulation mode. Mathematically it can be expressed as,

$$F_R$$
$$= \frac{\text{the actual rate of useful energy gain}}{\text{the rate of useful energy gain if the entire collector were at the fluid inlet temperature}(T_{fi})}$$

$$F_R = \frac{\dot{Q}_u}{\{A_c[\dot{q}_{ab} - U_L(T_p - T_a)]\}_{T_p=T_{fi}}} = \frac{\dot{Q}_u}{A_c[\dot{q}_{ab} - U_L(T_{fi} - T_a)]}$$
$$= \frac{\dot{m}_f C_f}{A_c U_L}\{1 - \exp[-A_c U_L F'/(\dot{m}_f C_f)]\} \quad (4.28)$$

**Example 4.35** Calculate mass flow rate for single and double glazed flat plate collector with effective area of 2 m$^2$ for mass flow rate $(\dot{m}_f)$ of 0.35 kg/s and specific heat of water is 4190 J/kg °C.

**Solution**

From Eq. 4.28, we have an expression for mass flow rate as

$$F_R = \frac{\dot{m}_f C_f}{A_c U_L}\{1 - \exp[-A_c U_L F'/(\dot{m}_f C_f)]\}$$

(a) **For single glazed FPC:** $U_L = 8.19$ W/m$^2$ °C (Example 4.20) and $F' = 0.82$ (Example 4.29).

Now,

$$F_R = \frac{0.35 \times 4190}{2 \times 8.19}\left\{1 - \exp\left[-\frac{0.82 \times 2 \times 8.19}{0.35 \times 4190}\right]\right\} = 89.53\{1 - \exp[-0.009158]\}$$
$$= 89.53\{1 - 0.9909\} = 89.53 \times 0.00916 = 0.82$$

(b)  **For double glazed FPC**: $U_L = 3.89$ W/m² °C (Example 4.20) and $F' = 0.91$ (Example 4.29).

Now,

$$F_R = \frac{0.35 \times 4190}{2 \times 3.89} \left\{ 1 - \exp\left[ -\frac{0.91 \times 2 \times 3.89}{0.35 \times 4190} \right] \right\} = 188.496\{1 - \exp[-0.004828]\}$$

$$= 188.496\{1 - 0.9951\} = 188.496 \times 0.0049 = 0.92$$

**For double glazed FPC, the flow rate factor is increased from 0.82 to 0.92. Further, it is to be noted that mass flow rate factor ($F_R$) remains equal to collector efficiency factor ($F'$) for mass flow rate of 0.35 kg/s. However, it will be different for lower mass flow rate, i.e., 0.035 kg/s shown in next example.**

**Example 4.36** Estimate mass flow rate factor ($F_R$) for Example 4.35 for mass flow rate of 0.035 kg/s.

**Solution**

(a)  **For single glazed FPC**: $U_L = 8.19$ W/m² °C (Example 4.20) and $F' = 0.82$ (Example 4.29).

Now,

$$F_R = \frac{0.035 \times 4190}{2 \times 8.19} \left\{ 1 - \exp\left[ -\frac{0.82 \times 2 \times 8.19}{0.035 \times 4190} \right] \right\} = 8.953\{1 - \exp[-0.09158]\}$$

$$= 8.953\{1 - 0.9124\} = 89.53 \times 0.00916 = 0.78$$

(b)  **For double glazed FPC**: $U_L = 3.89$ W/m² °C (Example 4.20) and $F' = 0.91$ (Example 4.29).

Now,

$$F_R = \frac{0.035 \times 4190}{2 \times 3.89} \left\{ 1 - \exp\left[ -\frac{0.91 \times 2 \times 3.89}{0.035 \times 4190} \right] \right\}$$

$$= 18.8496\{1 - \exp[-0.04828]\} = 18.8496\{1 - 0.9528\}$$

$$= 18.8496 \times 0.00472 = 0.89$$

**One can see, as mass flow rate is reduced, mass flow rate factor is also reduced as explained earlier.**

**Example 4.37** Evaluate the outlet temperature of fluid (water) for single and double glazed flat plate collector with effective area with inlet temperature ($T_{fi}$) of 60 °C [industrial waste] and ambient air temperature ($T_a$) of 25 °C [summer condition], solar intensity [$I(t)$] of 500 W/m² and $\alpha_p = \tau_p = 0.9$ by using the data of Example 4.35.

**Solution**

From Eq. 4.25, we have an expression for $T_{fo}$ as

$$T_{fo} = \left[\left(\frac{\dot{q}_{ab}}{U_L}\right) + T_a\right]\left[1 - \exp\left[-A_c U_L F'/(\dot{m}_f C_f)\right]\right]$$
$$+ T_{fi} \exp\left[-A_c U_L F'/(\dot{m}_f C_f)\right]$$

(a) **For single glazed FPC:** $U_L = 8.19$ W/m² °C (Example 4.20) and $F' = 0.82$ (Example 4.29).

$$\exp\left[-A_c U_L F'/(\dot{m}_f C_f)\right] = \exp\left[-\frac{0.82 \times 2 \times 8.19}{0.35 \times 4190}\right] = \exp[-0.009158] = 0.9908$$

Now,

$$T_{fo} = \left[\left(\frac{0.81 \times 500}{8.19}\right) + 25\right][1 - 0.9908] + 60 \times 0.9908$$
$$= 74.45 \times 0.0092 + 59.448 = 60.13\ °C$$

(b) **For double glazed FPC:** $U_L = 3.89$ W/m² °C (Example 4.20) and $F' = 0.92$ (Example 4.28).

$$\exp\left[-A_c U_L F'/(\dot{m}_f C_f)\right] = \exp\left[-\frac{0.91 \times 2 \times 3.89}{0.35 \times 4190}\right] = \exp[-0.004828] = 0.9952$$

$$T_{fo} = \left[\left(\frac{0.81 \times 500}{3.89}\right) + 25\right][1 - 0.9951] + 60 \times 0.9951 = 129.113 \times 0.004816$$
$$+ 59.71 = 60.33\ °C$$

So, there is marginal effect of double glazed cover on the outlet of 2 m² of FPC which has no practical use of FPC for such design parameters for industrial waste hot water.

**Example 4.38** Repeat Example 4.37 with mass flow rate of 0.035 kg/s.

**Solution**

(a) **For single glazed FPC:** $U_L = 8.19$ W/m² °C (Example 4.20) and $F' = 0.82$ (Example 4.29).

$$\exp\left[-A_c U_L F'/(\dot{m}_f C_f)\right] = \exp\left[-\frac{0.82 \times 2 \times 8.19}{0.035 \times 4190}\right] = \exp[-0.09158] = 0.9125$$

Now,

$$T_{fo} = \left[\left(\frac{0.81 \times 500}{8.19}\right) + 25\right][1 - 0.9125] + 60 \times 0.9125$$

$$= 74.45 \times 0.0875 + 54.75 = 61.26 \,°C$$

(b) **For double glazed FPC:** $U_L = 3.89$ W/m² °C (Example 4.20) and $F' = 0.92$
(Example 4.29)

$$\exp\left[-A_c U_L F'/(\dot{m}_f C_f)\right] = \exp\left[-\frac{0.92 \times 2 \times 3.89}{0.035 \times 4190}\right] = \exp[-0.0488] = 0.9524$$

$$T_{fo} = \left[\left(\frac{0.81 \times 500}{3.89}\right) + 25\right][1 - 0.9524] + 60 \times 0.9529$$
$$= 129.113 \times 0.0471 + 57.14 = 63.31 \,°C$$

In this case too, there is increased in outlet fluid (water) temperature by 1–3 °C
even **for use of industrial waste water for summer condition** due to reduced mass
flow rate because there is more heat transfer from tube to working fluid.

**Example 4.39** Repeat Example 4.37 for $T_{fi} = 15$ °C and $T_a = 16$ °C for winter
condition.

**Solution**

(a) **For single glazed FPC:** $U_L = 8.19$ W/m² °C (Example 4.20) and $F' = 0.82$
(Example 4.29).

$$\exp\left[-A_c U_L F'/(\dot{m}_f C_f)\right] = \exp\left[-\frac{0.82 \times 2 \times 8.19}{0.035 \times 4190}\right] = \exp[-0.09158] = 0.9125$$

Now,

$$T_{fo} = \left[\left(\frac{0.81 \times 500}{8.19}\right) + 16\right][1 - 0.9125] + 15 \times 0.9125$$
$$= 49.45 \times 0.0875 + 14.86 = 19.18 \,°C$$

(b) **For double glazed FPC:** $U_L = 3.89$ W/m² °C (Example 4.20) and $F' = 0.92$
(Example 4.29).

$$\exp\left[-A_c U_L F'/(\dot{m}_f C_f)\right] = \exp\left[-\frac{0.92 \times 2 \times 3.89}{0.035 \times 4190}\right] = \exp[-0.0488] = 0.9524$$

$$T_{fo} = \left[\left(\frac{0.81 \times 500}{3.89}\right) + 16\right][1 - 0.9524] + 15 \times 0.9524$$
$$= 120.11 \times 0.0471 + 14.29 = 20 \,°C$$

So, there is increase of up to 5 °C in outlet temperature in winter in comparison
with summer by reducing the mass flow rate for such design parameters.

**Example 4.40** Evaluate the rate of net useful energy available to the fluid (water) for single and double glazed flat plate collector with effective area with inlet temperature $(T_{fi})$ of 60 °C [industrial waste water] and ambient air temperature $(T_a)$ of 25 °C, solar intensity $[I(t)]$ of 500 W/m² [summer condition] and $\alpha_p = \alpha_p = 0.9$ by using the data of Example 4.35.

**Solution**

From Eq. 4.26, we have the following expression:

$$\dot{Q}_u = F_R A_c \left[ \alpha_p \tau_g I(t) - U_L (T_{fi} - T_a) \right]$$

(a) **For Single glazed FPC:** $U_L = 8.19$ W/m² °C (Example 4.20) and $F_R = 0.82$ (Example 4.35)

Now,

$$\dot{Q}_u = 0.82 \times 2[0.81 \times 500 - 8.19(60 - 25)] = 1.64[405 - 286.65] = 194.09 \text{ W}$$

(b) **For double glazed FPC:** $U_L = 3.89$ W/m² °C (Example 4.20) and $F_R = 0.92$ (Example 4.29)

Now,

$$\dot{Q}_u = 0.92 \times 2[0.81 \times 0.9 \times 500 - 3.89\,3.89(60 - 25)]$$
$$= 1.84[364.5 - 136.15] = 420.16 \text{ W}$$

**Example 4.41** Repeat Example 4.40 for mass flow rate of 0.035 kg/s.

**Solution**

(a) **For Single glazed FPC:** $U_L = 8.19$ W/m² °C (Example 4.20) and $F' = 0.82$ (Example 4.29)

Here, $\exp\left[ -A_c U_L F'/(\dot{m}_f C_f) \right] = \exp\left[ -\frac{0.82 \times 2 \times 8.19}{0.035 \times 4190} \right] = \exp[-0.09158] = 0.9125$ and,

$F_R = \frac{0.035 \times 4190}{2 \times 8.19}\left\{ 1 - \exp\left[ -\frac{0.91 \times 2 \times 8.19}{0.035 \times 4190} \right] \right\} = 8.95[1 - 0.9125] = 0.78$ (Example 4.36).

Now,

$$\dot{Q}_u = 0.78 \times 2[0.81 \times 500 - 8.19(60 - 25)] = 1.56[405 - 286.65] = 184.63 \text{ W}$$

(b) **For double glazed FPC:** $U_L = 3.89$ W/m² °C (Example 4.20) and $F' = 0.92$ (Example 4.29)

Here,

$$\exp[-A_cU_LF'/(\dot{m}_fC_f)] = \exp\left[-\frac{0.92 \times 2 \times 3.89}{0.035 \times 4190}\right] = \exp[-0.04828] = 0.9529$$

and,

$$F_R = \frac{0.035 \times 4190}{2 \times 3.89}\{1 - \exp[-\frac{0.91 \times 2 \times 8.19}{0.035 \times 4190}]\} = 18.85[1 - 0.9529] = 0.89$$

(Example 4.36)

Now,

$$\dot{Q}_u = 0.89 \times 2[0.81 \times 0.9 \times 500 - 3.89(60 - 25)] = 1.78[364.5 - 136.15]$$
$$= 406.46 \text{ W}$$

**From Examples** 4.40 and 4.41 , **one can say that the rate of useful thermal energy decreases with decrease of mass flow rate for industrial waste hot water due to loss of increase thermal energy and mass flow rate factor $(F_R)$ is reduced in comparison with collector efficiency factor $(F')$ for mass flow rate of 0.035 kg/ s.**

**Example 4.42** Find out condition for design parameters for maximum $(T_{fo} - T_{fi})$ for a given climatic parameters.

**Solution**

From Eq. 4.25, we have an expression for $T_{fo}$ as

$$T_{fo} = \left[\left(\frac{\dot{q}_{ab}}{U_L}\right) + T_a\right][1 - \exp[-A_cU_LF'/(\dot{m}_fC_f)]]$$
$$+ T_{fi} \exp[-A_cU_LF'/(\dot{m}_fC_f)]$$

Further,

$$(T_{fo} - T_{fi}) = [1 - \exp[-A_cU_LF'/(\dot{m}_fC_f)]]\left[\left(\frac{\dot{q}_{ab}}{U_L}\right) - (T_{fi} - T_a)\right]$$

To maximize $(T_{fo} - T_{fi})$ for a given climatic parameters namely $\left[\left(\frac{\dot{q}_{ab}}{U_L}\right) - (T_{fi} - T_a)\right]$,
Following steps are needed:

**Step 1**:

One should maximize $[1 - \exp[-A_cU_LF'/(\dot{m}_fC_f)]]$.

**Step 2:**

To maximize $[1 - \exp[-A_c U_L F'/(\dot{m}_f C_f)]]$, one should minimize $\exp[-A_c U_L F'/(\dot{m}_f C_f)]$.

**Step 3:**

To minimize $\exp[-A_c U_L F'/(\dot{m}_f C_f)]$, one has to maximize $(A_c U_L F')/(\dot{m}_f C_f)$.

**Step 4:**

To maximize $(A_c U_L F')/(\dot{m}_f C_f)$: The numerator namely area of FPC (Example 4.44), total heat loss coefficient and collector efficiency factor should be maximum and denominator namely mass flow rate should be minimum (Example 4.36).

**Step 5:**

For a given design parameters namely total heat loss coefficient $(U_L)$ and collector efficiency factor $(F')$, it easier to increase the number of FPC and reduce the mass flow rate $(\dot{m}_f)$.

### 4.4.4 An Instantaneous Thermal Efficiency $(\eta_i)$

An instantaneous thermal efficiency $(\eta_i)$ of FPC under forced mode of operation is given by

(a) **for Single Glazed FPC**

$$\eta_i = \frac{\dot{Q}_u \text{ in W}}{I(t) \times A_c \text{ in W}} = F_R\left[\alpha_p \tau_g - U_L \frac{T_{fi} - T_a}{I(t)}\right]$$

$$= \left[F_R(\alpha_p \tau_g) - (F_R U_L)\frac{T_{fi} - T_a}{I(t)}\right] \qquad (4.28a)$$

(b) **For double glazed**

$$\eta_i = \frac{\dot{Q}_u \text{ in W}}{I(t) \times A_c \text{ in W}} = F_R\left[\alpha_p \tau_g^2 - U_L \frac{T_{fi} - T_a}{I(t)}\right]$$

$$= \left[F_R(\alpha_p \tau_g) - (F_R U_L)\frac{T_{fi} - T_a}{I(t)}\right] \qquad (4.28b)$$

However, the values of $F_R$ and $U_L$ will be different for single and double glazed FPV (Example 4.38 and 4.39).

**Equation 4.28 is Known as Hottel, Whillier and Bliss (HWB) Equation.**

**Example 4.43** Calculate instantaneous thermal efficiency for single and double glazed collector under forced mode of operation for Example 4.41.

**Solution**

(a) **For single glazed FPC:** $F_R = 0.78$ (Example 4.41); $U_L = 8.19$ W/m² °C (Example 4.20).

**First method:**
From Eq. 4.28, we have

$$\eta_i = \left[ F_R(\alpha_p \tau_g) - (F_R U_L) \frac{T_{fi} - T_a}{I(t)} \right]$$
$$= \left[ 0.78 \times 0.81 - (0.78 \times 8.19) \frac{60 - 25}{500} \right] = 0.184(18.4\%)$$

**Second method**

$$\eta_i = \frac{\dot{Q}_u \text{ in } W}{I(t) \times A_c \text{ in } W} = \frac{184.63}{500 \times 2} = 0.1846(18.46\%)$$

(b) **For double glazed surface:**

$$F_R = 0.89 \text{ [Example 4.41]} U_L = 3.89 \text{ W/m}^2 \text{ °C} \left[ \text{Example4.20} \right]$$

**First method**

$$\eta_i = \left[ F_R(\alpha_p \tau_g^2) - (F_R U_L) \frac{T_{fi} - T_a}{I(t)} \right]$$
$$= \left[ 0.89 \times 0.81 \times 0.9 - (0.89 \times 3.89) \frac{60 - 25}{500} \right] = [0.6488 - 0.2423]$$
$$= 0.406.51(40.65\%)$$

**Second method**

$$\eta_i = \frac{\dot{Q}_u \text{ in } W}{I(t) \times A_c \text{ in } W} = \frac{406.46}{500 \times 2} = 0.4064(40.64\%)$$

## 4.5 The N-flat Plate Collectors (N-FPC) Connected in Series

Since most of requirement of flat plate collector is single glazed and hence onwards we will only consider and discuss single glazed flat plate collector and it will be referred as **FPC**.

### 4.5.1 The Outlet Fluid Temperature $(T_{foN})$ at $N^{th}$ FPC Connected in Series

If N similar flat plate collector (FPC) having effective area of $A_c$ and total overall heat transfer coefficient of $U_L$ have been considered and all are connected in series. So, in this case, the outlet of first FPC $(T_{fo1})$ will be inlet of second FPC $(T_{fi2})$, i.e., $T_{fo1} = T_{fi2}$ and outlet of second FPC $(T_{fo1})$ will be inlet of third one, i.e., $T_{fo2} = T_{fi3}$ and it continues till $N$th FPC.

From Eq. 4.25, one can write an expression for the outlet of first collector as

$$T_{fo1} = \left[\left(\frac{\dot{q}_{ab}}{U_L}\right) + T_a\right]\left[1 - \exp[-A_c U_L F'/(\dot{m}_f C_f)]\right]$$
$$+ T_{fi} \exp[-A_c U_L F'/(\dot{m}_f C_f)] \qquad (4.29)$$

Similarly, the outlet of second FPC can be written as

$$T_{fo2} = \left[\left(\frac{\dot{q}_{ab}}{U_L}\right) + T_a\right]\left[1 - \exp[-A_c U_L F'/(\dot{m}_f C_f)]\right]$$
$$+ T_{fi2} \exp[-A_c U_L F'/(\dot{m}_f C_f)] \qquad (4.30)$$

Since, $T_{fo1} = T_{fi2}$, hence substituting $T_{fo1}$ as $T_{fi2}$ in Eq. 4.30, one gets

$$T_{fo2} = \left[\left(\frac{\dot{q}_{ab}}{U_L}\right) + T_a\right]\left[1 - \exp[-A_c U_L F'/(\dot{m}_f C_f)]\right]$$
$$+ \left\{\left[\left(\frac{\dot{q}_{ab}}{U_L}\right) + T_a\right]\left[1 - \exp[-A_c U_L F'/(\dot{m}_f C_f)]\right]\right.$$
$$\left. + T_{fi} \exp[-A_c U_L F'/(\dot{m}_f C_f)]\right\} \exp[-A_c U_L F'/(\dot{m}_f C_f)]$$

On solving the above equation, one gets as

$$T_{fo2} = \left[\left(\frac{\dot{q}_{ab}}{U_L}\right) + T_a\right]\left[1 - \exp[-2A_c U_L F'/(\dot{m}_f C_f)]\right]$$
$$+ T_{fi} \exp[-2A_c U_L F'/(\dot{m}_f C_f)]$$

Similarly, the outlet at $N^{th}$ FPC can be obtained as follows:

$$T_{foN} = \left[\left(\frac{\dot{q}_{ab}}{U_L}\right) + T_a\right]\left[1 - \exp\left[-NA_cU_LF'/(\dot{m}_fC_f)\right]\right]$$
$$+ T_{fi}\exp\left[-NA_cU_LF'/(\dot{m}_fC_f)\right] \tag{4.31}$$

**Another important point to be noted**: The FPC connected in series is not appropriate for use of industrial hot water as inlet fluid. In this, the heat loss is maximum and hence there is insignificant gain in thermal energy. Industrial hot water can be used directly for applications such as domestic and space heating, etc.

### 4.5.2   The Outlet Fluid Temperature $(T_{fomN})$ at Connected $N^{th}$ (m – FPC, Fig. 4.8) Connected in Series

The m-FPCs are connected in parallel as shown in Fig. 4.8, known as one module. The inlet in m-FPC collector connected in parallel will be $\frac{\dot{m}_f}{m}$ in each FPC. So, in this case, the outlet fluid temperature at end of individual FPC, Eq. 4.29 will be expressed as

$$T_{fo1} = \left[\left(\frac{\dot{q}_{ab}}{U_L}\right) + T_a\right]\left[1 - \exp\left[-A_cU_LF'/\left(\frac{\dot{m}_f}{m}C_f\right)\right]\right]$$
$$+ T_{fi}\exp\left[-A_cU_LF'/\left(\frac{\dot{m}_f}{m}C_f\right)\right]$$

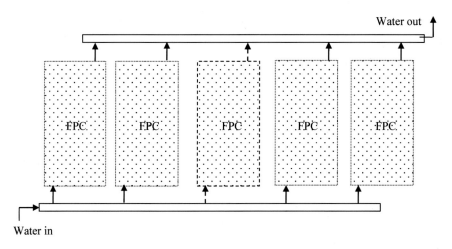

**Fig. 4.8** Flat plate collectors (FPC's) connected in parallel makes one module

The above expression will be same for each FPC connected in parallel and hence the above equation can be rewritten as

$$T_{fom} = \left[\left(\frac{\dot{q}_{ab}}{U_L}\right) + T_a\right]\left[1 - \exp\left[-mA_cU_LF'/(\dot{m}_fC_f)\right]\right]$$
$$+ T_{fi}\exp\left[-mA_cU_LF'/(\dot{m}_fC_f)\right]$$

or,

$$T_{fom} = \left[\left(\frac{\dot{q}_{ab}}{U_L}\right) + T_a\right]\left[1 - \exp\left[-A_{cm}U_LF'/(\dot{m}_fC_f)\right]\right]$$
$$+ T_{fi}\exp\left[-A_{cm}U_LF'/(\dot{m}_fC_f)\right] \tag{4.32a}$$

with $A_{cm} = m \times A_c$.

If m-parallel flat plate collector (m-FPC module, Eq. 4.19 and Fig. 4.8) having area of $A_{cm}$ as shown in Fig. 4.8 and Eq. 4.32a are connected in series, then it will be referred as one array. Figure 4.9 shows the one array of parallel and series connection. Then, Eq. 4.31 for one array will be as follows:

$$T_{fom \times N} = \left[\left(\frac{\dot{q}_{ab}}{U_L}\right) + T_a\right]\left[1 - \exp\left[-NA_{cm}U_LF'/(\dot{m}_fC_f)\right]\right]$$
$$+ T_{fi}\exp\left[-NA_{cm}U_LF'/(\dot{m}_fC_f)\right] \tag{4.32b}$$

**This Combination is Referred as Parallel and Series Combination of Flat Plate Collector (FPC) as Shown in** Fig. 4.9.

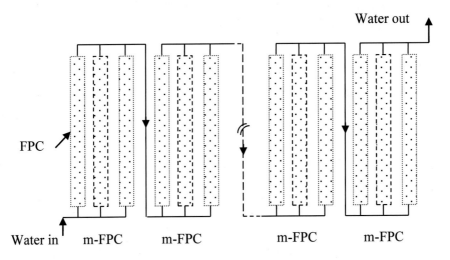

**Fig. 4.9** The m-FPC modules are connected in series (one array) [parallel and series connection]

### 4.5.3 Heat Removal Factor/mass Flow Rate Factor for N-FPC Collectors Connected in Series

The rate of thermal energy at $N$th FPC, $\dot{Q}_{uN}$ can be obtained as

$$\dot{Q}_{uN} = \dot{m}_f C_f \left(T_{foN} - T_{fi}\right)$$

$$= \dot{m}_f C_f \left[1 - \exp\left[-N A_c U_L F'/(\dot{m}_f C_f)\right]\right]\left\{\left(\frac{\dot{q}_{ab}}{U_L}\right) - \left(T_{fi} - T_a\right)\right\} \quad (4.33)$$

### 4.5.4 Instantaneous Thermal Efficiency $(\eta_{iN})$ for N-FPC Collectors Connected in Series

$$\eta_{iN} = \frac{\dot{Q}_{uN}}{N A_c \times I(t)}$$

$$= \frac{\dot{m}_f C_f}{N A_c U_L}\left[1 - \exp\left[-N A_c U_L F'/(\dot{m}_f C_f)\right]\right]\left\{\alpha_p \tau_g - U_L \frac{(T_{fi} - T_a)}{I(t)}\right\}$$

$$(4.33)$$

Equation 4.33 can also be referred as HWB equation for N-FPC connected in series.

### 4.5.5 Heat Removal Factor/mass Flow Rate Factor for N-FPC Connected in Series

Equation 4.33 can be also rewritten as

$$\dot{Q}_{uN} = \dot{m}_f C_f \left(T_{foN} - T_{fi}\right) = F_{RN} A_{cN}\left\{\dot{q}_{ab} - U_L\left(T_{fi} - T_a\right)\right\} \quad (4.34)$$

where,

$$F_{RN} = \frac{\dot{m}_f C_f}{(N A_c) U_L}\left[1 - \exp\left[-N A_c U_L F'/(\dot{m}_f C_f)\right]\right] \text{ with } A_{cN} = (N A_c) \quad (4.35)$$

For $N = 1$, Eq. 4.35 reduces to

$$F_R = \frac{\dot{m}_f C_f}{(A_c) U_L}\left[1 - \exp\left[-A_c U_L F'/(\dot{m}_f C_f)\right]\right]$$

This is exactly same as Eq. 4.28 for single glazed flat plate collector.

### 4.5.6 Heat Removal Factor/mass Flow Rate Factor for $N - (m - FPC)$ Connected in Series

From Eq. 4.32, we have

$$T_{fom \times N} = \left[ \left( \frac{\dot{q}_{ab}}{U_L} \right) + T_a \right] [1 - \exp[-NA_{cm}U_L F'/(\dot{m}_f C_f)]]$$
$$+ T_{fi} \exp[-NA_{cm}U_L F'/(\dot{m}_f C_f)]$$

Now,

$$\dot{Q}_{u-m \times N} = \dot{m}_f C_f (T_{fomN} - T_{fi}) = F_{RmN} N A_{cm} \{\dot{q}_{ab} - U_L (T_{fi} - T_a)\} \quad (4.36)$$

where,

$$F_{R-m \times N} = \frac{\dot{m}_f C_f}{(N A_{cm}) U_L} [1 - \exp[-NA_{cm}U_L F'/(\dot{m}_f C_f)]] \text{ with } A_{c-m \times N} = (mN A_c)$$
$$(4.37)$$

For $m = N = 1$,
  Equation 4.37 reduces to

$$F_R = \frac{\dot{m}_f C_f}{(A_c) U_L} [1 - \exp[-A_c U_L F'/(\dot{m}_f C_f)]]$$

**Example 4.44** Evaluate the outlet temperature at 5th and 10th of single glazed FPC with effective area of 2 m² connected in series for the data of Example 4.39 for winter condition with mass flow rate of 0.035 kg/s.

**Solution**

Required data from Example 4.39: $I(t) = 500$ W/m²; $\alpha_p = \tau_p = 0.9$; $T_{fi} = 15$ °C; $T_a = 16$ °C; $U_L = 8.19$ W/m² °C (Example 4.20) and $F' = 0.82$ (Example 4.29).

**Solution**

From Eq. 4.31, we have an expression for $T_{foN}$ as

$$T_{foN} = \left[ \left( \frac{\dot{q}_{ab}}{U_L} \right) + T_a \right] [1 - \exp[-NA_c U_L F'/(\dot{m}_f C_f)]]$$
$$+ T_{fi} \exp[-NA_c U_L F'/(\dot{m}_f C_f)]$$

(a) **For N = 1, $T_{fo1} = 19.18$ °C (Example 4.39)**
(b) **For N = 5 of single glazed FPC**

**Here,**

$$\exp\left[-NA_cU_LF'/(\dot{m}_fC_f)\right] = \exp\left[-\frac{5 \times 0.82 \times 2 \times 8.19}{0.035 \times 4190}\right]$$
$$= \exp[-0.4579] = 0.6326$$

Now,

$$T_{fo5} = \left[\left(\frac{0.81 \times 500}{8.19}\right) + 16\right][1 - 0.6326] + 15 \times 0.6326$$
$$= 65.46 \times 0.3874 + 9.49 = 34.85 \,^\circ C$$

### (c)  For N = 10 of single glazed FPC

Here,

$$\exp\left[-A_cU_LF'/(\dot{m}_fC_f)\right] = \exp\left[-\frac{10 \times 0.82 \times 2 \times 8.19}{0.035 \times 4190}\right]$$
$$= \exp[-0.9158] = 0.4002$$

Now,

$$T_{fo10} = \left[\left(\frac{0.81 \times 500}{8.19}\right) + 16\right][1 - 0.4002] + 15 \times 0.4002$$
$$= 65.46 \times 0.5998 + 6.003 = 45.26 \,^\circ C$$

From Examples 4.39 and 4.44, one can see that there is significant increase in outlet temperature from 19.18 °C to 45.26 °C [$\Delta T = 26.08$ °C] by varying $N$ from 1 to 10 for given winter climatic and design parameters.

**Example 4.45** Repeat Example 4.44 for $I(t) = 900$ W/m$^2$ for $N = 5$ and 10 for winter condition.

**Solution**

### (a)  For N = 5 of single glazed FPC

**Here,**

$$\exp\left[-A_cU_LF'/(\dot{m}_fC_f)\right] = \exp\left[-\frac{5 \times 0.82 \times 2 \times 8.19}{0.035 \times 4190}\right]$$
$$= \exp[-0.4579] = 0.6326$$

Now,

$$T_{fo5} = \left[\left(\frac{0.81 \times 900}{8.19}\right) + 16\right][1 - 0.6326] + 15 \times 0.6326 = 105 \times 0.3674$$
$$+ 9.489 = 38.577 + 9.489 = 48.07 \, °C$$

**(b) For N = 10 of single glazed FPC**

**Here,**

$$\exp\left[-A_c U_L F'/(\dot{m}_f C_f)\right] = \exp\left[-\frac{10 \times 0.82 \times 2 \times 8.19}{0.035 \times 4190}\right]$$
$$= \exp[-0.9158] = 0.4002$$

Now,

$$T_{fo10} = \left[\left(\frac{0.81 \times 900}{8.19}\right) + 16\right][1 - 0.4002] + 15 \times 0.4002$$
$$= 105 \times 0.5998 + 6.003 = 68.98 \, °C$$

**So, One Can Concludes that the Solar Intensity and Mass Flow Rate and Winter Climatic Play an Important Role in Determining the Outlet Fluid Temperature.**

**Example 4.46** The mass flow rate factor, the rate of useful thermal energy and an instantaneous thermal efficiency for $N = 5$ for Example 4.45.

**Solution**

We have the following formulas for mass flow rate factor $(F_{RN})$, the rate of useful thermal energy $(\dot{Q}_{uN})$ and an instantaneous thermal efficiency $(\eta_{iN})$ for N-FPC connected in series as follows:

**(a) Mass flow rate factor $(F_{RN})$.**

From Eq. 4.35:

$$F_{RN} = \frac{\dot{m}_f C_f}{(N A_c) U_L}\left[1 - \exp\left[-N A_c U_L F'/(\dot{m}_f C_f)\right]\right]$$
$$= \frac{0.035 \times 4190}{5 \times 2 \times 8.19}\left[1 - \exp\left[-\frac{5 \times 2 \times 8.19 \times 0.82}{0.035 \times 4190}\right]\right]$$
$$= 1.79 \times \left[1 - \exp[-0.4579]\right] = 1.79[1 - 0.6326] = 0.6578$$

As the number of FPC increases from $N = 1$–$5$, the mass flow rate factor $(F_{RN})$ decreases from 0.78 to 0.65 as discussed in Example 4.42 (step 4).

(b)  **The rate of useful thermal energy $(\dot{Q}_{uN})$.**

From Eq. 4.34:

$$\dot{Q}_{uN} = \dot{m}_f C_f (T_{foN} - T_{fi}) = F_{RN} A_{cN} \{\dot{q}_{ab} - U_L (T_{fi} - T_a)\}$$
$$= 0.6578 \times 10[0.81 \times 900 - 8.19(15 - 16)] = 6.578 \times [729 + 8.19]$$
$$= 6.576 \times 737.19 = 4791.17 \text{ W} = 4.849.24 \text{ kW}$$

**Other Method**

$$\dot{Q}_{uN} = \dot{m}_f C_f (T_{foN} - T_{fi}) = 0.035 \times 4190(48.07 - 15)$$
$$= 4849.71 \text{ W} \approx 4.850 \text{ kW}$$

(c)  **An instantaneous thermal efficiency $(\eta_{iN})$.**

From Eq. 4.33:

$$\eta_{iN} = F_{RN} \left\{ \alpha_p \tau_g - U_L \frac{(T_{fi} - T_a)}{I(t)} \right\} = 0.6578 \left[ 0.81 - 8.19 \times \frac{(15 - 16)}{900} \right]$$
$$= 0.6578 \times 0.8191 = 0.5388(53.88\%)$$

**Other Method**

$$\eta_{iN} = \frac{4849.71}{10 \times 900} = 0.5388(53.88\%)$$

**Example 4.47**  Estimate the outlet fluid temperature and the rate of thermal energy for winter condition of Example 4.45 with $I(t) = 900$ W/m² for following cases:

(a)  $m = 4$ [4-FPC connected in parallel; $A_{cm} = 4 \times 2 = 8$ m²] and $N = 3$ [Three m-FPC collectors connected in series] and
(b)  $4 \times 3 = 12 -$ FPC connected in series and $A_c = 2$ m²

**Solution**

**Case (a):** $A_{cm} = 8$ m²; $N = 3$.

From Eq. 4.32, we have an expression for N-(m-FOC) connected in series as

$$T_{fo-m \times N} = \left[ \left( \frac{\dot{q}_{ab}}{U_L} \right) + T_a \right] [1 - \exp[-N A_{cm} U_L F' / (\dot{m}_f C_f)]]$$
$$+ T_{fi} \exp[-N A_{cm} U_L F' / (\dot{m}_f C_f)]$$

Here,

$$\exp\left[-NA_{cm}U_LF'/(\dot{m}_fC_f)\right] = \exp\left\{-\frac{3 \times 8 \times 8.19 \times 0.82}{0.035 \times 4190}\right\} = \exp\left(-\frac{161.18}{146.65}\right)$$
$$= \exp(-1.099) = 0.3332$$

Now, the outlet fluid temperature is

$$T_{fo-m \times N} = T_{fo4 \times 3} = 105[1 - 0.3332] + 15 \times 0.333275 = 75\,°C$$

(see Example 4.45).
The rate of thermal energy has been calculated as

$$\dot{Q}_{u-m \times N} = \dot{Q}_{u4 \times 3} = \dot{m}_fC_f(T_{foN} - T_{fi}) = 0.035 \times 4190(75 - 15)$$
$$= 8.799 = 8.788\ \text{kW}$$

Further, the flow rate factor, $F_{R-m \times N}$, Eq. 4.37 is obtained as

$$F_{Rm \times N} = F_{R4 \times 3} = \frac{\dot{m}_fC_f}{(NA_{cm})U_L}\left[1 - \exp\left[-NA_{cm}U_LF'/(\dot{m}_fC_f)\right]\right]$$
$$= \frac{0.035 \times 4190}{3 \times 8 \times 8.19}(1 - 0.3332) = 0.746 \times 0.6668 = 0.49$$

**Case (b):** $A_c = 2\ \text{m}^2$; $N = 12$.
Here,

$$\exp\left[-NA_cU_LF'/(\dot{m}_fC_f)\right] = \exp\left\{-\frac{12 \times 2 \times 8.19 \times 0.82}{0.035 \times 4190}\right\} = \exp\left(-\frac{161.18}{146.65}\right)$$
$$= \exp(-1.099) = 0.3332$$

Now, the outlet fluid temperature is

$$T_{foN} = 105[1 - 0.3332] + 15 \times 0.3332 = 75\,°C$$

(see Example 4.45).
The rate of thermal energy has been calculated as

$$\dot{Q}_{u-m \times N} = \dot{m}_fC_f(T_{foN} - T_{fi}) = 0.035 \times 4190(75 - 15) = 8799 = 8.788\ \text{kW}$$

Further, the flow rate factor, $F_{R-m \times N}$, Eq. 4.37 is obtained as

$$F_{R-m \times N} = \frac{\dot{m}_fC_f}{(NA_c)U_L}\left[1 - \exp\left[-NA_{cm}U_LF'/(\dot{m}_fC_f)\right]\right]$$

$$= \frac{0.035 \times 4190}{12 \times 2 \times 8.19}(1 - 0.3332) = 0.746 \times 0.6668 = 0.49$$

The results in both cases are same.

**Example 4.48** Calculate the outlet temperature of 4-parallel FPC module connected in series for $N = 3$ for Example 4.47 for $I(t) = 900$ W/m$^2$.

**Solution**

In the case of 4-parallel FPC, the mass flow in each FPC will be distributed equally $(\dot{m}_f) = \frac{0.035}{4} = 0.00875$ kg/s.

So, here

$$\exp[-A_cU_LF'/(\dot{m}_fC_f)] = \exp\left\{-\frac{2 \times 8.19 \times 0.82}{0.00875 \times 4190}\right\} = \exp\left(-\frac{13.43}{36.66}\right)$$
$$= \exp(-0.3663) = 0.6933$$

Now, the outlet fluid temperature at first 4-parallel FPC has been calculated as

$$T_{fo1} = 105[1 - 0.6933] + 15 \times 0.6933 = 42\,°C$$

(see Example 4.45).

This becomes inlet fluid temperature for next 4-parallel FPC, so the outlet at second set of 4-parallel FPC will be determined as

$$T_{fo2} = 105[1 - 0.6933] + 42 \times 0.6933 = 61.32\,°C$$

Now, this temperature (61.32) becomes inlet fluid temperature for the 3rd 4-parallel FPC connected in parallel, then

$$T_{fo3} = 105[1 - 0.6933] + 61.32 \times 0.6933 = 74.71\,°C$$

This is approximately same as $T_{fo-m \times N} = 75\,°C$ of Example 4.47.

Other results for the rate of thermal energy and mass flow rate factor will be also same as reported in Example 4.47.

### 4.5.7   Series and Parallel Combination

In this case, the N-FPCs are connected in series known as one module and then m (N-FPC module) are connected in parallel as shown in Fig. 4.10. The whole combination will be referred as one array. The inlet and outlet mass flow rate of array will be same. The mass flow rate $(\dot{m}_f)$ entering at inlet of array is equally divided into $\frac{\dot{m}_f}{m}$ for each N-FPCs connected in aeries.

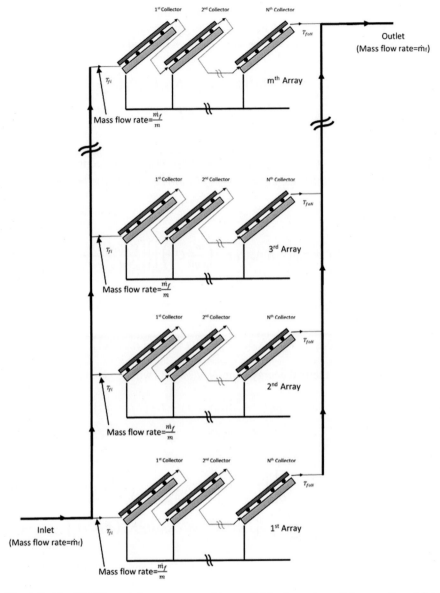

**Fig. 4.10** The N-FPCs modules are connected in parallel [series and parallel connection of flat plate collectors]

**Such combination is referred as series and parallel combination** (Fig. 4.10).

Referring to Fig. 4.10 and Eq. 4.31, the outlet temperature of one module of N-FPCs connected in series will be same for each row and can be expressed as

$$T_{foN} = \left[\left(\frac{\dot{q}_{ab}}{U_L}\right) + T_a\right]\left[1 - \exp\left[-NA_cU_LF'/\left(\frac{\dot{m}_f}{m}C_f\right)\right]\right]$$
$$+ T_{fi}\exp\left[-NA_cU_LF'/\left(\frac{\dot{m}_f}{m}C_f\right)\right] \tag{4.38}$$

where, the $\frac{\dot{m}_f}{m}$ is mass flow rate in each row of N-FPCs connected in series.

The rate of useful thermal energy $(\dot{Q}_{uN})$ from one module can be obtained as

$$\dot{Q}_{uN} = \frac{\dot{m}_f}{m}C_f(T_{foN} - T_{fi})$$
$$= \frac{\dot{m}_f}{m}C_f\left[1 - \exp\left[-NA_cU_LF'/\left(\frac{\dot{m}_f}{m}C_f\right)\right]\right]\left[\frac{\alpha_p\tau_gI(t)}{U_L} - (T_{fi} - T_a)\right] \tag{4.39}$$

Equation 4.39 can be rearranged as follows:

$$\dot{Q}_{uN} = F_{RN}(NA_c)\left[\alpha_p\tau_gI(t) - U_L(T_{fi} - T_a)\right] \tag{4.40a}$$

The total rate of thermal energy from $N \times m$ FPC will be given by

$$\dot{Q}_{u-N\times m} = m \times F_{RN}(NA_c)\left[\alpha_p\tau_gI(t) - U_L(T_{fi} - T_a)\right] \tag{4.40b}$$

where,

$$F_{RN} = \frac{\left(\frac{\dot{m}_f}{m}C_f\right)}{(NA_c)U_L}\left[1 - \exp\left[-NA_{cm}U_LF'/\left(\frac{\dot{m}_f}{m}C_f\right)\right]\right]$$
$$= \frac{(\dot{m}_fC_f)}{(mNA_c)U_L}\left[1 - \exp[-mNA_{cm}U_LF'/(\dot{m}_fC_f)]\right] \tag{4.41}$$

**Example 4.49** Calculate the outlet temperature $(T_{foN})$ and the rate of useful thermal energy $\dot{Q}_{uN}$ of 4-FPC connected in series with mass flow rate of $\frac{\dot{m}_f}{m} = \frac{0.035}{3} = 0.0117$ for $m = 3$ as shown in Fig. 4.9 for data of Example 4.48.

**Solution**

So here, mass flow rate of $\frac{\dot{m}_f}{m} = 0.0117$ and $N = 4$

$$\exp\left[-NA_cU_LF'/\left(\frac{\dot{m}_f}{m}C_f\right)\right] = \exp\left\{-\frac{4 \times 2 \times 8.19 \times 0.82}{0.01167 \times 4190}\right\} = \exp\left(-\frac{53.73}{48.88}\right)$$

$$= \exp(-1.099) = 0.3332$$

Now, the outlet fluid temperature at first 4-parallel FPC has been calculated as

$$T_{fo4} = 105[1 - 0.3332] + 15 \times 0.3332 = 75 \,°\text{C}$$

(see Example 4.45)
Now, the rate of useful thermal energy $\left( \dot{Q}_{uN} \right)$ from one module at $N = 4$ is

$$\dot{Q}_{u4} = \frac{\dot{m}_f}{m} C_f \left( T_{fo4} - T_{fi} \right) = 0.0117 \times 4190(75 - 15) = 2.933.838 \,\text{W}$$

The total thermal energy at end of one array, i.e., $m = 3$ will be calculated as

$$\dot{Q}_{u-N \times m} = \dot{Q}_{u-4 \times 3} = 3 \times \dot{Q}_{u4} = 3 \times 2933.838 = 8801.51 \,\text{W} = 8.801 \,\text{kW}$$

which is approximately same as Example 4.47 for parallel and series combination?

**Example 4.50** Calculate mass flow rate factor for Example 4.49 for $I(t) = 900 \,\text{W/}$
$\text{m}^2$

**Solution**

From Example 4.49, we have

$$\exp\left[ -N A_c U_L F' / \left( \frac{\dot{m}_f}{m} C_f \right) \right] = \exp\left\{ -\frac{4 \times 2 \times 8.19 \times 0.82}{0.01167 \times 4190} \right\}$$
$$= \exp\left( -\frac{53.73}{48.88} \right) = \exp(-1.099)$$

Now, an expression for mass flow rate is

$$F_{R-N \times m} = \frac{\left( \frac{\dot{m}_f}{m} C_f \right)}{N A_c U_L} \left\{ 1 - \exp\left[ -N A_c U_L F' / \left( \frac{\dot{m}_f}{m} C_f \right) \right] \right\}$$
$$= \frac{0.01167 \times 4190}{4 \times 2 \times 8.19} \left[ 1 - \exp(-1.099) \right]$$
$$= 0.7463 \times [1 - 0.3332] = 0.4976$$

which is also approximately same as reported in Example 4.47.

**Example 4.51** Calculate the rate of useful thermal energy for $3(m) \times 4(N)$ FPC
combination by using Eq. 4.40b for Examples 4.49 and 4.50.

**Solution**

From Eq. 4.40b, we have an expression for parallel and series combination as

$$\dot{Q}_{u-N\times m} = m \times F_{RN}(NA_c)\left[\alpha_p\tau_g I(t) - U_L\left(T_{fi} - T_a\right)\right]$$
$$= 3 \times 0.4976 \times 4 \times 2[729 - 8.19(15 - 16)]$$
$$= 11.94 \times 413.19 = 8802 \text{ W} = 8.802 \text{ kW}$$

The result is same as reported in Example 4.49.

## 4.6  Constant Collection Temperature ($T_{00}$)

In earlier section, we have considered liquid flat plate collector (FPC) under constant mass flow rate in kg/s. In this case, the outlet fluid (water) temperature varies with time of the day due to hourly variation in climatic condition namely solar radiation, ambient air temperature and wind velocity which influences outside through an overall top heat loss coefficient ($U_t$) heat transfer coefficient. Further, there is lot of application of hot water in industries like cotton industries where the temperature plays an important role during various coloring of cotton fiber. Under mentioned situation, the operating temperature should be constant, and hence in this section we will analyze liquid FPC under constant collection temperature.

### 4.6.1  Parallel and Series Combination

In parallel and series combination, an expression for the outlet air temperature, $T_{fom\times N}$ is given by Eq. 4.32b as follows:

$$T_{fom\times N} = \left[\left(\frac{\dot{q}_{ab}}{U_L}\right) + T_a\right]\left[1 - \exp\left[-NA_{cm}U_L F'/(\dot{m}_f C_f)\right]\right]$$
$$+ T_{fi}\exp\left[-NA_{cm}U_L F'/(\dot{m}_f C_f)\right] \tag{4.42}$$

The above equation is valid only for constant mass flow rate mode to determine the hourly variation of outlet water temperature $(T_{fom\times N})$. This means the mass flow rate $(\dot{m}_f)$ is constant even hourly variation of climatic parameters. If one desires to have constant outlet water temperature, then $T_{fom\times N} = T_{00}$ with variable mass flow rate $\left[\dot{m}_f(t)\right]$.

Further, for constant outlet fluid (water) temperature, Eq. 4.42 be rewritten as

$$T_{00} = \left[\left(\frac{\dot{q}_{ab}}{U_L}\right) + T_a\right]\left[1 - \exp\left[-NA_{cm}U_L F'/(\dot{m}_f C_f)\right]\right]$$
$$+ T_{fi}\exp\left[-NA_{cm}U_L F'/(\dot{m}_f C_f)\right] \tag{4.43}$$

Equation 4.43 can be solved for variable mass flow rate $\left[\dot{m}_f(t)\right]$ as

$$\dot{m}_f(t) = \frac{F'NA_{cm}U_L}{C_f}\left[\ln\left\{\frac{\left(\frac{\dot{q}_{ab}}{U_L}\right)+T_a-T_{fi}}{\left(\frac{\dot{q}_{ab}}{U_L}\right)+T_a-T_{00}}\right\}\right]^{-1} \tag{4.43}$$

For detail derivation see Sect. 5.10.1.

**Example 4.52** Evaluate mass flow rate for constant collection temperature of 75 °C for winter condition of Example 4.45 with $I(t) = 900$ W/m² for following cases:

(a)  $m = 4$ [4-FPC connected in parallel; $A_{cm} = 4 \times 2 = 8$ m²] and $N = 3$ [Three m-FPC collectors connected in series] and
(b)  $4(N) \times 3(m) = 12 - $ FPC connected in series and $A_c = 2$ m²

**Solution**

**Case (a):** $A_{cm} = 8m^2$; $N = 3$ and here $\dot{q}_{ab} = 0.9 \times 0.9 \times 900 = 729$ W/m².
From Eq. 4.43, we have an expression for $\dot{m}_f(t)$ as

$$\dot{m}_f(t) = \frac{F'NA_{cm}U_L}{C_f}\left[\ln\left\{\frac{\left(\frac{\dot{q}_{ab}}{U_L}\right)+T_a-T_{fi}}{\left(\frac{\dot{q}_{ab}}{U_L}\right)+T_a-T_{00}}\right\}\right]^{-1}$$

$$= \frac{0.82 \times 3 \times 8 \times 8.19}{4190}\left[\ln\left\{\frac{88.01}{29.01}\right\}\right]^{-1}$$

$$= 0.03847 \times \frac{1}{0.4818} = 0.0798 \text{ kg/s}$$

**Case (b):** $4(N) \times 3(m) = 12 - $ FPC connected in series and $A_c = 2$ m²

$$\dot{m}_f(t) = \frac{F'NA_{cm}U_L}{C_f}\left[\ln\left\{\frac{\left(\frac{\dot{q}_{ab}}{U_L}\right)+T_a-T_{fi}}{\left(\frac{\dot{q}_{ab}}{U_L}\right)+T_a-T_{00}}\right\}\right]^{-1}$$

$$= \frac{0.82 \times 4 \times 6 \times 8.19}{4190}\left[\ln\left\{\frac{88.01}{29.01}\right\}\right]^{-1}$$

$$= 0.03847 \times \frac{1}{0.4818} = 0.0798 \text{ kg/s}$$

Similar equation for variable mass flow rate and examples can be evaluated for each combination namely (a) for parallel, (b) for series only and (c) for series and parallel. The detail calculations for solar air collectors have been given in next Chap. 5.

**Additional Examples**

**Example 5.53** What can be the maximum value of mass flow rate of liquid FPC for a given area of FPC?

## Solution

From Eq. 4.35, one has an analytical expression for mass flow rate factor as

$$F_{RN} = \frac{\dot{m}_f C_f}{(NA_c)U_L}\left[1 - \exp\left[-NA_c U_L F'/(\dot{m}_f C_f)\right]\right] \text{ with } A_{cN} = (NA_c)$$

The condition for maximum $F_{RN}$, for a given $(NA_c)$, an exponential term should be minimum which can be achieved by

(i)   Decreasing the mass flow rate $\dot{m}_f$ and hence the outlet temperature will also increase.
(ii)  Increasing an overall heat loss coefficient $(U_L)$.
(iii) Increasing FPC efficiency factor $(F')$.

From Example 4.46, we have the mass flow rate factor for $N = 5$ and $\dot{m}_f = 0.035$

$$
\begin{aligned}
F_{R5} &= \frac{\dot{m}_f C_f}{(NA_c)U_L}\left[1 - \exp\left[-NA_c U_L F'/(\dot{m}_f C_f)\right]\right] \\
&= \frac{0.035 \times 4190}{5 \times 2 \times 8.19}\left[1 - \exp\left[-\frac{5 \times 2 \times 8.19 \times 0.82}{0.035 \times 4190}\right]\right] \\
&= 1.79 \times \left[1 - \exp[-0.4579]\right] = 1.79[1 - 0.6326] = 0.6578
\end{aligned}
$$

Now, for $\dot{m}_f = 0.0175$,

$$
\begin{aligned}
F_{R5} &= \frac{0.0175 \times 4190}{5 \times 2 \times 8.19}\left[1 - \exp\left[-\frac{5 \times 2 \times 8.19 \times 0.82}{0.0175 \times 4190}\right]\right] \\
&= 0.8953 \times \left[1 - \exp[-0.9159]\right] = 0.8953[1 - 0.3987] = 0.3569
\end{aligned}
$$

This supports the above condition.

**Example 5.54** Repeat Example 5.53 for outlet fluid temperature for different mass flow rate

## Solution

From Example 5.45, we have the outlet temperature for $N = 5$ and $\dot{m}_f = 0.035$

$$
\begin{aligned}
T_{fo5} &= \left[\left(\frac{0.81 \times 900}{8.19}\right) + 16\right][1 - 0.6326] + 15 \times 0.6326 \\
&= 105 \times 0.3674 + 9.489 = 38.577 + 9.489 = 48.07\,^\circ\text{C}
\end{aligned}
$$

Now, for $\dot{m}_f = 0.0175$, $T_{fo5}$ can be obtained as

$$T_{fo5} = \left[\left(\frac{0.81 \times 900}{8.19}\right) + 16\right][1 - 0.3987] + 15 \times 0.3987 = 105 \times 0.6013 + 5.9805$$
$$= 63.14 + 5.98 = 69.12\,°C$$

In this case too, the outlet temperature is increased with decrease of mass flow rate. Here, it is important to note that the first term which depends on solar radiation plays an important role in comparison with second term.

# References

1. Watmuff, J. H., Charters, W. W. S., & Proctor, D. (1977). *Complex, 2,* 56.
2. Holland, K. G. T., Unny, T. E., Raithby, G. D., & Konicek, L. (1976). *Journal of Heat Transfer, 98*(2), 189.
3. Buchberg, H., Catton, I., & Edwards, D. K. (1976). *Journal of Heat Transfer, 98*(2), 182.
4. Tiwari, G. N. (2002). *Solar energy.* Narosa Publishing House.
5. Tiwari, G. N., Tiwari, A., & Shyam. (2016). *Handbook of solar energy.* Springer (2016)
6. Badran, A. A., Mustafa, M., Dawood, W., & Ghazzawi, Z. (2008). On the measurement of bond conductance in solar collector absorber plate. *Energy Conversion and Management, 29*(11), 3305–3310. https://doi.org/10.1016/j.enconman.2008.01.041

# Chapter 5
# Flat Plate Air Collector

## 5.1 Introduction

The flat plate air collector is similar to flat plate liquid collector (Chap. 4) except liquid (water) which is replaced by air. The application of flat plate air collector is mostly space heating and crop/vegetables drying. It is also referred as **solar air collector**. It consists of single/double glazed transparent glass cover having transmittivity of $\tau_g$ and insulation at the bottom and on the sides as shown in Fig. 5.1 The flat plate air collector is inclined to receive the maximum solar radiation. Incident solar radiation is first reflected (8%) before transmission by glass cover and further reflected (10%) from absorber plate and loss due to absorption by 10%. Further, there is 4% loss through bottom insulation. So finally, there are 40% losses. Hence, solar air heater can provide 60% thermal energy for application of hot air. The net rate of thermal energy $(\dot{q}_u)$ in W/m² is withdrawn by flowing air either above or below absorber plate. It is also referred as **conventional solar air heater**. The whole assembly is encased in a sheet metal container and system is referred as **flat plate air collector**. Sometimes, we have also referred it as **solar air collector**.

The **flat plate air collector/solar air collector** is classified on the basis of its design as follows:

(a) **Non-porous solar air heaters**: In this case, the air is allowed to flow either above or below absorber plate as shown in Fig. 5.1. It is most economical with less maintenance and acceptable solar air heater.

(b) **Porous solar air heaters**: In this case, the air is allowed to pass through porous media as shown in Fig. 5.2 for maximum heat transfer from porous media to flowing air, but porous media create resistance to flowing air. The porous solar air heater is not so popular as non-porous solar air heater.

Therefore, we will only discuss non-porous solar air heaters with different designs.

© Bag Energy Research Society 2024      279
G. N. Tiwari, *Fundamental of Mathematical Tools for Thermal Modeling of Solar Thermal and Photo-voltaic Systems-Part-I*,
https://doi.org/10.1007/978-981-99-7085-8_5

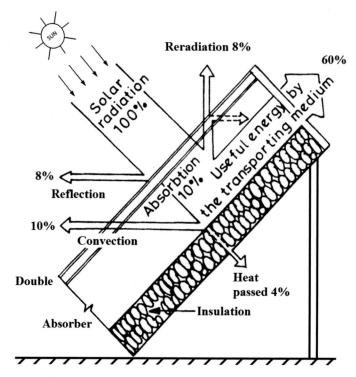

**Fig. 5.1** Cross-sectional view of flat plate air collector with various losses and gains in percentage

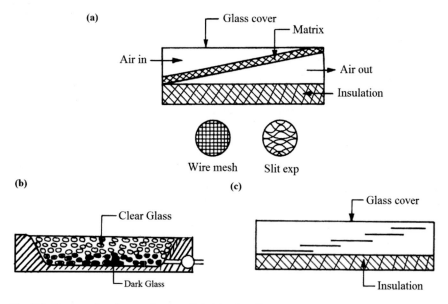

**Fig. 5.2** Cross-sectional view of porous flat plate air collector

## 5.2 Steady-State Analysis of Non-porous Flat Plate Collector with Air [1]

### 5.2.1 Stagnant Working Fluid Air Above the Absorber

In this case, there is no airflow over absorber plate similar to flat plate liquid collector (Eq. 4.5a). Following first law of thermodynamics, the rate of thermal energy at absorber plate in a steady-state condition will be as follows:

(a) **For single glazed solar air collector**

$$\dot{q}_u = \left[(\alpha_p \tau_g) I(t) - U_L (T_p - T_a)\right] \text{ in W/m}^2 \tag{5.1a}$$

$$\dot{Q}_u = A_c \left[(\alpha_p \tau_g) I(t) - U_L (T_p - T_a)\right] \text{ in W} \tag{5.1b}$$

where, $U_L$ is an overall total heat loss coefficient (Eq. 4.5a) from absorber to ambient through top glass and bottom insulation.

Equation 5.1b can be written as

$$\frac{\dot{Q}_u}{A_c U_L} = \left[\frac{(\alpha_p \tau_g)}{U_L} I(t) - (T_p - T_a)\right] \tag{5.2}$$

where, $I(t)$ is solar intensity, $A_c$ is area of solar air heater and $T_p$ and $T_a$ are plate/absorber and ambient air temperatures, respectively.

If the working fluid as air can be above the absorber plate as shown in Fig. 5.1 in a stagnant position, the rate of thermal energy transferred from absorber plate to air will be written as follows:

$$\dot{Q}_u = A_c \left[h_{pf} (T_p - T_f)\right] \tag{5.3}$$

where, $h_{pf} = h_2$ of Fig. 5.3 is convective heat transfer coefficient from absorber plate to the stagnant fluid (air). Its value will be half in the case of flowing fluid (air) which is below absorber plate.

Further, Eq. 5.3 can be rewritten as

$$\frac{\dot{Q}_u}{A_c h_{pf}} = (T_p - T_f) \tag{5.4}$$

From Eqs. 5.3 and 5.4, one get

$$\dot{Q}_u \left[\frac{1}{A_c U_L} + \frac{1}{A_c h_{pf}}\right] = \frac{(\alpha_p \tau_g)}{U_L} I(t) - (T_f - T_a)$$

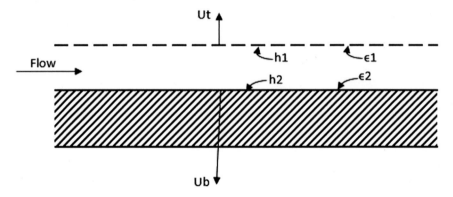

**Fig. 5.3** Flowing of air between single glass cover and absorber plate

or,

$$\dot{Q}_u = \frac{1/U_L}{\left[\frac{1}{U_L} + \frac{1}{h_{pf}}\right]} A_c\left[(\alpha_p \tau_g)I(t) - U_L(T_f - T_a)\right]$$
$$= F' A_c\left[(\alpha_p \tau_g)I(t) - U_L(T_f - T_a)\right] \tag{5.5}$$

where,

$$F' = \frac{1/U_L}{\left[\frac{1}{U_L} + \frac{1}{h_{pf}}\right]} = \frac{1/U_L}{\frac{1}{U_0}} = \frac{U_0}{U_L} = \frac{h_{pf}}{\left(h_{pf} + U_L\right)} \tag{5.6}$$

**(b)   For double glazed solar air collector**

Equation 5.6 becomes

$$\dot{Q}_u = F' A_c\left[(\alpha_p \tau_g^2)I(t) - U_L(T_f - T_a)\right] \quad \text{and}$$
$$F' = \frac{1/U_L}{\left[\frac{1}{U_L} + \frac{1}{h_{pf}}\right]} = \frac{1/U_L}{\frac{1}{U_0}} = \frac{U_0}{U_L} = \frac{h_{pf}}{\left(h_{pf} + U_L\right)} < 1 \tag{5.7}$$

where, numerical value of $U_t$ will be different for double glazed glass cover, so $U_L$ and $F'$ for double glazed air collector will be different than single glazed air collector. This gives different values of $F'$ for double glazed solar air heater.

The expression for $F'$, (Eq. 5.6) is also applicable for flat plate liquid collector (FPC) except the change in value of $h_{pf}$ from absorber plate to fluid (water).

**Example 5.1** Evaluate the solar air collector efficiency ($F'$) for single and double glazed solar air heaters for spacing between blackened and glazed absorber flat plate and inner glass cover of 0.10 m with an inclination of 45° for Fig. 5.3 for stagnant air. The blackened absorber flat plate and inner glass cover are maintained at 50 and 30 °C, respectively.

**Solution**

From Examples 4.15 and 4.20, we have the following data:

**Case (i): For single glazed:** $h_{bc,c} = h_{pf} = 11.08 \, \text{W/m}^2 \, °C$ (Example 4.15) and $U_L = 8.19 \, \text{W/m}^2 \, °C$ (Example 4.20).

From Eq. 5.6, we can get an expression for the solar air collector efficiency $(F')$ as

The solar air collector efficiency $(F') = \frac{h_{pf}}{(h_{pf}+U_L)} = \frac{11.08}{11.08+8.19} \approx 0.58.$

**Case (ii): For double glazed:** $h_{bc,c} = h_{pf} = 11.08 \, \text{W/m}^2 \, °C$ (Example 4.15) and $U_L = 3.89 \, \text{W/m}^2 \, °C$ (Example 4.20).

The solar air collector efficiency $(F') = \frac{h_{pf}}{(h_{pf}+U_L)} = \frac{11.08}{11.08+3.89} \approx 0.74$

**Example 5.2** Repeat the Example 5.1 for the absorber facing downward with stagnant air below absorber plate as shown in Fig. 5.4.

**Solution**

In this case, the value of $h_{pf}$ becomes half due to inverted hot plate absorber $= \frac{11.08}{2}$ $= 5.54 \, \text{W/m}^2 \, °C$ (case (iv) of Table 2.3).

**Case (i): For single glazed:** $h_{bc,c} = h_{pf} = 5.54 \, \text{W/m}^2 \, °C$ (Example 4.15) and $U_L = 8.19 \, \text{W/m}^2 \, °C$ (Example 4.20).

From Eq. 5.6, we can get an expression for the solar air collector efficiency $(F')$ as

The solar air collector efficiency $(F') = \frac{h_{pf}}{(h_{pf}+U_L)} = \frac{5.54}{5.54+8.19} = 0.40.$

**Case (ii): For double glazed:** $h_{bc,c} = h_{pf} = 5.54 \, \text{W/m}^2 \, °C$ (Example 4.15) and $U_L = 3.89 \, \text{W/m}^2 \, °C$ (Example 4.20).

The solar air collector efficiency $(F') = \frac{h_{pf}}{(h_{pf}+U_L)} = \frac{5.54}{5.54+3.89} \approx 0.59.$

**Example 5.3** Evaluate the rate of thermal energy in W for 2 m² single and double glazed solar air collectors with solar intensity of 500 W/m², ambient air temperature of 30 °C and air at 50 °C and $(\alpha_p \tau_g) = 0.81$ for Example 5.1.

**Solution**

**Case (i): For single glazed solar air collector:** $U_L = 8.19 \, \text{W/m}^2 \, °C$ (Example 4.20) and $F' = 0.58$ (Example 5.1).

From Eq. 5.5, we have an expression for the rate of thermal energy in W for single glazed air collector and it can be evaluated as

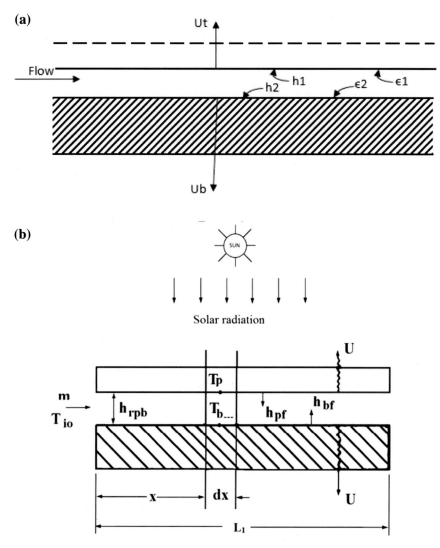

**Fig. 5.4  a** Flowing of air between absorber plate and insulated bottom, **b** flowing of air between absorber plate and insulated bottom through elemental distance 'd*x*'

$$\dot{Q}_u = F'A_c\left[(\alpha_p\tau_g)I(t) - U_L(T_f - T_a)\right]$$
$$= 0.58 \times 2[0.81 \times 500 - 8.19(50 - 30)] = 279.7\,\text{W}$$

**Case (ii): For double glazed solar air collector:** $U_L = 3.89\,\text{W/m}^2\,°\text{C}$ (Example 4.20) and

$F' = 0.74$ (Example 5.1).

From Eq. 5.7, we have an expression for the rate of thermal energy in W for single glazed air collector and it can be evaluated as

$$\dot{Q}_u = F' A_c \left[ (\alpha_p \tau_g^2) I(t) - U_L (T_f - T_a) \right]$$
$$= 0.74 \times 2[0.81 \times 0.9 \times 500 - 3.89(50 - 30)] = 424.32 \, \text{W}$$

The rate of useful thermal energy for double glazed solar air collector is increased due to increased solar collector efficiency.

**Example 5.4** Repeat Example 5.3 for Example 5.2 for air below absorber flat plate.

**Solution**

**Case (i): For single glazed solar air collector:** $U_L = 8.19 \, \text{W/m}^2 \, ^\circ\text{C}$ (Example 4.20) and $F' = 0.40$ (Example 5.2).

From Eq. 5.5, we have an expression for the rate of thermal energy in W for single glazed air collector and it can be evaluated as

$$\dot{Q}_u = F' A_c \left[ (\alpha_p \tau_g) I(t) - U_L (T_f - T_a) \right]$$
$$= 0.40 \times 2[0.81 \times 500 - 8.19(50 - 30)] = 192.96 \, \text{W}$$

**Case (ii): For double glazed solar air collector:** $U_L = 3.89 \, \text{W/m}^2 \, ^\circ\text{C}$ (Example 4.20) and $F' = 0.59$ (Example 5.2).

From Eq. 5.7, we have an expression for the rate of thermal energy in W for single glazed air collector and it can be evaluated as

$$\dot{Q}_u = F' A_c \left[ (\alpha_p \tau_g^2) I(t) - U_L (T_f - T_a) \right]$$
$$= 0.59 \times 2[0.81 \times 0.9 \times 500 - 3.89(50 - 30)]$$
$$= 338.31 \, \text{W}$$

The rate of useful thermal energy for double glazed solar air collector is increased due to increased solar collector efficiency.

### 5.2.1.1 Flat Plate Air Collector Efficiency

It is defined as ratio of actual rate of useful energy to the rate of useful energy when the plate is placed at fluid (air) temperature and it is expressed as

$$F' = \frac{\dot{Q}_u}{[\dot{Q}_u]_{T_p = T_f}} = \frac{\dot{Q}_u}{A_c \left[ (\alpha_p \tau_g) I(t) - U_L (T_f - T_a) \right]} < 1 \, \text{Examples 5.1 and 5.2}$$

$$(5.7)$$

The $F'$ is solar air heater collection efficiency factor for single glazed solar air heater which is always less than 1. Equation 5.6 has been derived under assumption that there is only heat transfer from absorber to fluid (air).

### 5.2.1.2  An Instantaneous Thermal Efficiency

It is defined as ration of the rate of useful thermal energy to the rate of incident solar energy/radiation, and it can be obtained as

(a)  **Single glazed solar air heater**

$$\eta_i = \frac{\dot{Q}_u}{A_c I(t)} = F' \left[ (\alpha_p \tau_g) - U_L \frac{(T_f - T_a)}{I(t)} \right] \tag{5.8}$$

(b)  **Double glazed solar air heater**

$$\eta_i = \frac{\dot{Q}_u}{A_c I(t)} = F' \left[ (\alpha_p \tau_g^2) - U_L \frac{(T_f - T_a)}{I(t)} \right] \tag{5.9}$$

Equations 5.8 and 5.9 are known as characteristic equation of solar air heater similar to HWB equation of FPC (Eq. 4.20).

**Example 5.5**  Calculate an instantaneous thermal efficiency ($\eta_i$) of single and double glazed solar air heaters for Example 5.3 and Eqs. 5.8 and 5.9, respectively.

**Solution**

From Eqs. 5.8 and 5.9, we have an expression for an instantaneous thermal efficiency ($\eta_i$) of single and double glazed solar air heaters as

(a)  **Single glazed solar air heater:** $U_L = 8.19\,\text{W/m}^2\,{}^\circ\text{C}$ (Example 4.20) and $F' = 0.58$ (Example 5.1).

**First method**

$$\eta_i = F' \left[ (\alpha_p \tau_g) - U_L \frac{(T_f - T_a)}{I(t)} \right] = 0.58 \left[ 0.81 - 8.19 \times \frac{(50 - 30)}{500} \right] = 0.2797$$

**Second method**

$$\eta_i = \frac{\dot{Q}_u}{A_c I(t)} = \frac{279.7}{2 \times 500} = 0.2797$$

(b)  **Double glazed solar air heater:** $U_L = 3.89\,\text{W/m}^2\,{}^\circ\text{C}$ (Example 4.20) and $F' = 0.74$ (Example 5.1).

**First method**

$$\eta_i = F'\left[(\alpha_p \tau_g^2) - U_L \frac{(T_f - T_a)}{I(t)}\right] = 0.74\left[0.81 \times 0.9 - 3.89 \times \frac{(50 - 30)}{500}\right]$$

$$= 0.4243$$

**Second method**

$$\eta_i = \frac{\dot{Q}_u}{A_c I(t)} = \frac{424.32}{2 \times 500} = 0.4243$$

**Example 5.6** Calculate an instantaneous thermal efficiency $(\eta_i)$ of single and double glazed solar air heaters for Example 5.4 for inverted absorber for Example 5.4 and Eqs. 5.8 and 5.9, respectively.

**Solution**

From Eqs. 5.8 and 5.9, we have an expression for an instantaneous thermal efficiency $(\eta_i)$ of single and double glazed solar air heaters as

**Case (i): For single glazed solar air collector:** $U_L = 8.19 \, \text{W/m}^2 \, ^\circ\text{C}$ (Example 4.20) and $F' = 0.40$ (Example 5.2).

**First method**

$$\eta_i = F'\left[(\alpha_p \tau_g) - U_L \frac{(T_f - T_a)}{I(t)}\right] = 0.40\left[0.81 - 8.19 \times \frac{(50 - 30)}{500}\right] = 0.193$$

**Second method**

$$\eta_i = \frac{\dot{Q}_u}{A_c I(t)} = \frac{192.96}{2 \times 500} = 0.193$$

**Case (ii): For double glazed solar air collector:** $U_L = 3.89 \, \text{W/m}^2 \, ^\circ\text{C}$ (Example 4.20) and

$F' = 0.59$ (Example 5.2).

**First method**

$$\eta_i = F'\left[(\alpha_p \tau_g^2) - U_L \frac{(T_f - T_a)}{I(t)}\right] = 0.59\left[0.81 \times 0.9 - 3.89 \times \frac{(50 - 30)}{500}\right]$$

$$= 0.338$$

**Second method**

$$\eta_i = \frac{\dot{Q}_u}{A_c I(t)} = \frac{4338.31}{2 \times 500} = 0.338$$

From Examples 5.5 and 5.6, one can observe that for an inverted absorber solar air flat plate collector, an instantaneous thermal efficiency is reduced. However, solar air collector efficiency is increased in double glazed solar air collector in both cases for stagnant air either above or below absorber plate. Further, the stagnant air solar collector has no practical application.

If one needs to use the net thermal energy available, Eq. 5.5/Fig. 5.6 for either space heating or solar drying, then air should be flown either above (Fig. 5.3) or (Fig. 5.4a).

If airflow above absorber plate then its temperature will be higher due to higher heat transfer in upward direction unlike downward direction after absorption of solar energy by blackened plate. But, there will be more thermal energy/heat losses too. Now, we have concluded (Chap. 4) through Examples 4.27–4.29 that

**The fluid flowing below absorber flat plate gives better performance due to higher value of collector efficiency $(F')$.**

In this case, an overall top loss coefficient $(U_t)$ is also less due to airflow below absorber plate due to low temperature, and hence in this chapter, we will discuss only air flowing below the blackened absorber flat plate as shown in Fig. 5.4a.

### 5.2.2  Flowing Working Fluid Air Below Absorber Plate [2]

Following Eqs. 4.22 and 5.5, the rate of solar radiation received by flowing fluid air in elemental area of $(bdx)$ can be written as

$$\dot{Q}_{udx} = F'bdx\left[(\alpha_p \tau_g)I(t) - U_L(T_f - T_a)\right] \tag{5.10}$$

where, an expression for $F'$ is given by Eq. 5.7 in terms of $U_L$. The numerical value of $U_L$ can be obtained by following Sect. 4.2.4.

If $\dot{m}_f$ is mass flow rate of working fluid air and $dT_f$ is temperature rise during elemental length $dx$, then

The rate of thermal energy carried out by flowing working fluid air in elemental area of $(bdx) =$

$$\dot{m}_f C_f \frac{dT_f}{dx}dx \tag{5.11}$$

where, $C_f$ is specific heat of working fluid air.

As per energy conservation law, Eqs. 5.10 and 5.11 will be equal, and hence

$$\dot{m}_f C_f \frac{dT_f}{dx} dx = F' b dx \left[ (\alpha_p \tau_g) I(t) - U_L (T_f - T_a) \right] \tag{5.12}$$

The solution of Eq. 5.12 with initial condition of $T_f(x = 0) = T_{fi}$ and $T_{fo}$ at $x = L_0$ (Sect. 4.4.1) for single glazed solar air collector will be expressed as

$$T_{fo} = \left[ \left( \frac{(\alpha_p \tau_g) I(t)}{U_L} \right) + T_a \right] \left[ 1 - \exp\left[ -A_c U_L F' / (\dot{m}_f C_f) \right] \right]$$
$$+ T_{fi} \exp\left[ -A_c U_L F' / (\dot{m}_f C_f) \right] \tag{5.13a}$$

with $A_c = b L_0$.

For double glazed solar air collector, Eq. 5.13a will be

$$T_{fo} = \left[ \left( \frac{(\alpha_p \tau_g^2) I(t)}{U_L} \right) + T_a \right] \left[ 1 - \exp\left[ -A_c U_L F' / (\dot{m}_f C_f) \right] \right]$$
$$+ T_{fi} \exp\left[ -A_c U_L F' / (\dot{m}_f C_f) \right]$$

The rate of thermal energy carried out by working fluid air $(\dot{Q}_u)$ efficiency $(\eta_i)$ for single and double glazed solar air collectors is given by

(a) **Single glazed solar air collector**

$$\dot{Q}_u = F_R A_c \left[ (\alpha_p \tau_g) I(t) - U_L (T_{fi} - T_a) \right] \tag{5.14a}$$

(b) **Double glazed solar air collector**

$$\dot{Q}_u = F_R A_c \left[ (\alpha_p \tau_g^2) I(t) - U_L (T_{fi} - T_a) \right] \tag{5.14b}$$

where, $F_R = \frac{\dot{m}_f C_f}{A_c U_L} \left\{ 1 - \exp\left[ -A_c U_L F' / (\dot{m}_f C_f) \right] \right\} = $ mass flow rate factor

The instantaneous thermal efficiency of single and double glazed solar air collectors is given by

(a) **Single glazed solar air collector**

$$\eta_i = \frac{\dot{Q}_u \text{in W}}{I(t) \times A_c \text{in W}} = F_R \left[ \alpha_p \tau_g - U_L \frac{T_{fi} - T_a}{I(t)} \right]$$
$$= \left[ F_R (\alpha_p \tau_g) - (F_R U_L) \frac{T_{fi} - T_a}{I(t)} \right] \tag{5.15a}$$

(b) **Double glazed solar air collector**

$$\eta_i = \frac{\dot{Q}_u \text{in W}}{I(t) \times A_c \text{in W}} = F_R \left[ \alpha_p \tau_g^2 - U_L \frac{T_{fi} - T_a}{I(t)} \right]$$

$$= \left[ F_R(\alpha_p \tau_g^2) - (F_R U_L)\frac{T_{fi} - T_a}{I(t)} \right] \tag{5.15b}$$

The numerical value of $F_R$ in terms of $F'$ and $U_L$ will be different for single and double glazed solar air collectors as shown in Fig. 5.3.

### 5.2.3  Other Expression for $U_L$ and $F'$ [Tiwari [2]] for Constant $h_{bc,c} = h_{pf} = h_1 = h_2 = 11.08\,\text{W/m}^2\,°C$

There is another expression for $U_L$ and $F'$ in addition to Eq. 5.7. The expression for $U_L$ is in terms of

(i)  An overall top ($U_t$) and bottom ($U_b$) loss coefficient (Example 4.20) which will not vary for flowing air with any velocity.
(ii)  Convective heat transfer coefficient between absorber plate and flowing fluid $h_{bc,c} = h_{pf} = h_2$ over it in terms of $U_L$ which has been discussed below.

Referring to Fig. 5.3, an expression for an overall heat loss coefficient ($U_L$) and solar air collector efficiency[1] ($F'$)[2] has been given as

$$U_L = \frac{(U_t + U_b)(h_1 h_2 + h_1 h_r + h_2 h_r) + U_t U_b(h_1 h_2)}{h_1 h_2 + h_1 h_r + h_2 h_r + h_2 U_t} \tag{5.16}$$

$$F' = \frac{h_1 h_2 + h_1 h_r + h_2 h_r + h_2 U_t}{(U_t + h_1 + h_r)(U_b + h_2 + h_r) - h_r^2} < 1 \tag{5.17}$$

$$h_r = \frac{\sigma(T_1^2 + T_2^2)(T_1 + T_2)}{\frac{1}{\varepsilon_1} + \frac{1}{\varepsilon_2} - 1}, \quad T \text{ in Kelvin (K)}, \tag{5.18}$$

or, $h_r = \varepsilon_{\text{eff}} \times 4\sigma \left(\overline{T}\right)^3$ with $\overline{T} = \frac{T_1 + T_2}{2}$ and

$$\varepsilon_{\text{eff}} = \left[ \frac{1}{\varepsilon_1} + \frac{1}{\varepsilon_2} - 1 \right]^{-1} \text{ (refer Eq. 2.17e)} \tag{5.19}$$

In Eqs. 5.16 and 5.17, there can be more possibility of relation between both convective heat transfer coefficients, i.e., $h_1$ and $h_2$.

(i)  $h_1 = h_2$.
(ii)  $h_1 = \frac{h_2}{2}$ due to glass cover facing down ward for same average temperature of flowing air.

In such case, Eqs. 5.16–5.17 are unchanged.

(iii)  There is insignificant heat transfer from inner glass cover to flowing air ($h_1 = 0$).

In such case, Eqs. 5.16–5.17 become

$$U_L = \frac{(U_t + U_b)h_2 h_r}{h_2 h_r + h_2 U_t} = \frac{h_r}{h_r + U_t}(U_t + U_b) \qquad (5.20)$$

$$F' = \frac{h_2 h_r + h_2 U_t}{(U_t + h_r)(U_b + h_2 + h_r) - h_r^2} < 1 \qquad (5.21)$$

(iv)  In case of air flowing between glass cover and absorber plate, the temperature difference between inner glass cover and absorber plate becomes negligible, i.e., $h_r = 0$. This can be explained in another way as there is no heat transfer in air by radiation in flowing air. Then, Eqs. 5.16–5.17 become as

$$U_L = \frac{(h_1 h_2)}{h_1 h_2 + h_2 U_t}[(U_t + U_b) + U_t U_b] \qquad (5.22)$$

and,

$$F' = \frac{h_1 h_2 + + h_2 U_t}{(U_t + h_1)(U_b + h_2)} < 1 \qquad (5.23)$$

**Example 5.7**  Estimate an overall heat transfer coefficient and solar air collector efficiency by using Eqs. 5.16–5.17 for absorber plate and glass cover temperature as 50 °C and 30 °C, respectively, for single and double glazed glass covers for $h_1 = h_2$.

**Solution**

From Eq. 5.19, we have

$$h_r = \varepsilon_{\text{eff}} \times 4\sigma \left(\overline{T}\right)^3 \text{ with } \overline{T} = \tfrac{T_1 + T_2}{2}, \text{Eq. 4.17e, and } \varepsilon_{\text{eff}} = \left[\tfrac{1}{\varepsilon_1} + \tfrac{1}{\varepsilon_2} - 1\right]^{-1} = 0.84$$

(Example 4.10).

Here, $\overline{T} = 40 \,°C$, then $h_r = \varepsilon_{\text{eff}}(4\sigma\overline{T})^3 = 0.84 \times 4 \times 5.67 \times 10^{-8}(273 + 40)^3 = 5.84 \,\text{W/m}^2 \,°C$.

**Case (i): For single glazed solar air collector:**  $U_t = 7.49 \,\text{W/m}^2 \,°C$ and $U_b = 0.70 \,\text{W/m}^2 \,°C$ (Example 4.20) and $h_{bc,c} = h_{pf} = h_1 = h_2 = 11.08 \,\text{W/m}^2 \,°C$ (Example 4.15).

From Eq. 5.16, we have the following expression for $U_L$ and $F'$

$$
\begin{aligned}
U_L &= \frac{(U_t + U_b)(h_1 h_2 + h_1 h_r + h_2 h_r) + U_t U_b (h_1 h_2)}{h_1 h_2 + h_1 h_r + h_2 h_r + h_2 U_t} \\
&= \frac{8.19 \times \{11.08 \times 11.08 + 11.08 \times 5.84 + 11.08 \times 5.84\} + 7.49 \times 0.70 \times 11.08 \times 11.08}{\{11.08 \times 11.08 + 11.08 \times 5.84 + 11.08 \times 5.84 + 11.08 \times 7.49\}} \\
&= \frac{8.19 \times 252.18 + 643.66}{\{252.18 + 82.98\}} = \frac{2709.14}{338.16} = 8.01 \,\text{W/m}^2 \,°C
\end{aligned}
$$

From Eq. 5.17, we have

$$
\begin{aligned}
F' &= \frac{h_1h_2 + h_1h_r + h_2h_r + h_2U_t}{(U_t + h_1 + h_r)(U_b + h_2 + h_r) - h_r^2} \\
&= \frac{\{11.08 \times 11.08 + 11.08 \times 5.84 + 11.08 \times 5.84\} + 11.08 \times 7.49}{(7.49 + 11.08 + 5.84)(0.70 + 11.08 + 5.84) - (5.84)^2} \\
&= \frac{252.18 + 82.98}{24.41 \times 17.62 - 34.11} = \frac{335.16}{395.99} = 0.84
\end{aligned}
$$

**Case (i): For double glazed solar air collector**

$U_t = 3.19\,\mathrm{W/m^2\,°C}$ and $U_b = 0.70\,\mathrm{W/m^2\,°C}$ (Example 4.20) and $h_{bc,c} = h_{pf} = 11.08\,\mathrm{W/m^2\,°C}$ (Example 4.15).

From Eqs. 5.16 and 5.17, we have the following expression for $U_L$ and $F'$:

$$
\begin{aligned}
U_L &= \frac{(U_t + U_b)(h_1h_2 + h_1h_r + h_2h_r) + U_tU_b(h_1h_2)}{h_1h_2 + h_1h_r + h_2h_r + h_2U_t} \\
&= \frac{3.89 \times \{11.08 \times 11.08 + 11.08 \times 5.84 + 11.08 \times 5.84\} + 3.19 \times 0.70 \times 11.08 \times 11.08}{\{11.08 \times 11.08 + 11.08 \times 5.84 + 11.08 \times 5.84 + 11.08 \times 3.19\}} \\
&= \frac{3.89 \times 252.18 + 274.14}{\{252.18 + 35.34\}} = \frac{1255.12}{287.52} = 4.37\,\mathrm{W/m^2\,°C}
\end{aligned}
$$

and,

$$
\begin{aligned}
F' &= \frac{h_1h_2 + h_1h_r + h_2h_r + h_2U_t}{(U_t + h_1 + h_r)(U_b + h_2 + h_r) - h_r^2} \\
&= \frac{\{11.08 \times 11.08 + 11.08 \times 5.84 + 11.08 \times 5.84\} + 11.08 \times 3.19}{(3.19 + 11.08 + 5.84)(0.70 + 11.08 + 5.84) - (5.84)^2} \\
&= \frac{252.18 + 35.34}{20.11 \times 17.62 - 34.11} = \frac{287.52}{320.22} = 0.89
\end{aligned}
$$

**Example 5.8** Estimate an overall heat transfer coefficient and solar air collector efficiency by using Eqs. 5.16–5.17 for absorber plate and glass cover temperature as 50 °C and 30 °C, respectively, for single and double glazed glass covers for $h_1 = \frac{h_2}{2}$.

**Solution**

From Eq. 5.18, we have

$h_r = \varepsilon_{\mathrm{eff}} \times 4\sigma\left(\overline{T}\right)^3$ with $\overline{T} = \frac{T_1 + T_2}{2}$, Eq. 4.17e, and $\varepsilon_{\mathrm{eff}} = \left[\frac{1}{\varepsilon_1} + \frac{1}{\varepsilon_2} - 1\right]^{-1} = 0.84$ (Example 4.10).

Here, $\overline{T} = 40\,°C$, then $h_r = \varepsilon_{\mathrm{eff}}\left(4\sigma\overline{T}\right)^3 = 0.84 \times 4 \times 5.67 \times 10^{-8}(273 + 40)^3 = 5.84\,\mathrm{W/m^2\,°C}$.

**Case (i): For single glazed solar air collector:** $U_t = 7.49 \, \text{W/m}^2 \, ^\circ\text{C}$ and $U_b = 0.70 \, \text{W/m}^2 \, ^\circ\text{C}$ (Example 4.20) and $h_{bc,c} = h_{pf} = h_2 = 11.08 \, \text{W/m}^2 \, ^\circ\text{C}$; $h_1 = \frac{11.08}{2} = 5.54 \, \text{W/m}^2 \, ^\circ\text{C}$ (Example 4.15).

From Eqs. 5.16 and 5.17, we have the following expression for $U_L$ and $F'$:

$$
\begin{aligned}
U_L &= \frac{(U_t + U_b)(h_1 h_2 + h_1 h_r + h_2 h_r) + U_t U_b (h_1 h_2)}{h_1 h_2 + h_1 h_r + h_2 h_r + h_2 U_t} \\
&= \frac{8.19 \times \{5.54 \times 11.08 + 5.54 \times 5.84 + 11.08 \times 5.84\} + 7.49 \times 0.70 \times 5.54 \times 11.08}{\{5.54 \times 11.08 + 5.54 \times 5.84 + 11.08 \times 5.84 + 11.08 \times 7.49\}} \\
&= \frac{8.19 \times 158.44 + 321.83}{\{158.44 + 82.98\}} = \frac{1615.85}{241.42} = 6.69 \, \text{W/m}^2 \, ^\circ\text{C}
\end{aligned}
$$

and,

$$
\begin{aligned}
F' &= \frac{h_1 h_2 + h_1 h_r + h_2 h_r + h_2 U_t}{(U_t + h_1 + h_r)(U_b + h_2 + h_r) - h_r^2} \\
&= \frac{\{5.54 \times 11.08 + 5.54 \times 5.84 + 11.08 \times 5.84\} + 11.08 \times 7.49}{(7.49 + 5.54 + 5.84)(0.70 + 11.08 + 5.84) - (5.84)^2} \\
&= \frac{158.44 + 82.98}{18.87 \times 17.62 - 34.11} = \frac{241.42}{298.38} = 0.81
\end{aligned}
$$

**Case (ii): For double glazed solar air collector**

$U_t = 3.19 \, \text{W/m}^2 \, ^\circ\text{C}$ and $U_b = 0.70 \, \text{W/m}^2 \, ^\circ\text{C}$ (Example 4.20) and $h_{bc,c} = h_{pf} = 5.54 \, \text{W/m}^2 \, ^\circ\text{C}$ (Example 4.15).

From Eqs. 5.16 and 5.17, we have the following expression for $U_L$ and $F'$:

$$
\begin{aligned}
U_L &= \frac{(U_t + U_b)(h_1 h_2 + h_1 h_r + h_2 h_r) + U_t U_b (h_1 h_2)}{h_1 h_2 + h_1 h_r + h_2 h_r + h_2 U_t} \\
&= \frac{3.89 \times \{5.54 \times 11.08 + 5.54 \times 5.84 + 11.08 \times 5.84\} + 3.19 \times 0.70 \times 5.54 \times 11.08}{\{5.54 \times 11.08 + 5.54 \times 5.84 + 11.08 \times 5.84 + 11.08 \times 3.19\}} \\
&= \frac{3.89 \times 158.44 + 137.07}{\{193.78 + 35.34\}} = \frac{753.40}{229.13} = 3.29 \, \text{W/m}^2 \, ^\circ\text{C}
\end{aligned}
$$

and,

$$
\begin{aligned}
F' &= \frac{h_1 h_2 + h_1 h_r + h_2 h_r + h_2 U_t}{(U_t + h_1 + h_r)(U_b + h_2 + h_r) - h_r^2} \\
&= \frac{\{5.54 \times 11.08 + 5.54 \times 5.84 + 11.08 \times 5.84\} + 11.08 \times 3.19}{(3.19 + 5.54 + 5.84)(0.70 + 11.08 + 5.84) - (5.84)^2} \\
&= \frac{158.44 + 35.34}{14.57 \times 17.62 - 34.11} = \frac{193.78}{222.61} = 0.87
\end{aligned}
$$

**Example 5.9** Estimate an overall heat transfer coefficient and solar air collector efficiency by using Eqs. 5.20–5.21 for absorber plate and glass cover temperature as 50 °C and 30 °C, respectively, for single and double glazed glass covers without considering $h_1$.

**Solution**

From Eq. 5.19, we have

$$h_r = \varepsilon_{\text{eff}} \times 4\sigma \left(\overline{T}\right)^3 \text{ with } \overline{T} = \tfrac{T_1 + T_2}{2}, \text{ Eq. 4.17e, and } \varepsilon_{\text{eff}} = \left[\tfrac{1}{\varepsilon_1} + \tfrac{1}{\varepsilon_2} - 1\right]^{-1} = 0.84$$

(Example 4.10).

Here, $\overline{T} = 40 \,°\text{C}$, then $h_r = \varepsilon_{\text{eff}} \left(4\sigma \overline{T}\right)^3 = 0.84 \times 4 \times 5.67 \times 10^{-8}(273 + 40)^3 = 5.84 \,\text{W/m}^2\,°\text{C}$.

**Case (i): For single glazed solar air collector:** $U_t = 7.49 \,\text{W/m}^2\,°\text{C}$ and $U_b = 0.70 \,\text{W/m}^2\,°\text{C}$ (Example 4.20) and $h_{bc,c} = h_{pf} = h_2 = 11.08 \,\text{W/m}^2\,°\text{C}$ (Example 4.15).

From Eqs. 5.20 and 5.21, we have the following expression for $U_L$ and $F'$:

$$U_L = (U_t + U_b) = 7.49 + 0.7 = 8.19 \,\text{W/m}^2\,°\text{C}$$

$$U_L = \frac{(U_t + U_b)h_2 h_r}{h_2 h_r + h_2 U_t} = \frac{h_r}{h_r + U_t}(U_t + U_b) = \frac{5.84}{5.84 + 7.49} \times 8.19 = 3.58$$

and,

$$\begin{aligned}
F' &= \frac{h_2 h_r + h_2 U_t}{(U_t + h_r)(U_b + h_2 + h_r) - h_r^2} \\
&= \frac{11.08 \times 5.84 + 11.08 \times 7.49}{(7.49 + 5.84) \times (0.7 + 11.08 + 5.84) - (5.84)^2} \\
&= \frac{147.69}{234.87 - 34.11} = \frac{147.69}{200.76} = 0.73
\end{aligned}$$

**Case (i): For double glazed solar air collector**

$U_t = 3.19 \,\text{W/m}^2\,°\text{C}$ and $U_b = 0.70 \,\text{W/m}^2\,°\text{C}$ (Example 4.20) and $h_{bc,c} = h_{pf} = h_2 = 11.08 \,\text{W/m}^2\,°\text{C}$; (Example 4.15).

From Eqs. 5.20 and 5.21, we have the following expression for $U_L$ and $F'$:

$$\begin{aligned}
U_L &= \frac{(h_2 h_r)}{h_2 h_r + h_2 U_t}(U_t + U_b) \\
&= \frac{11.08 \times 5.84}{11.08 \times 5.84 + 11.08 \times 3.19} \times 3.89 \\
&= \frac{64.71}{100.05} \times 3.89 = 2.52 \,\text{W/m}^2\,°\text{C}
\end{aligned}$$

and,

$$F' = \frac{h_2 h_r + h_2 U_t}{(U_t + h_r)(U_b + h_2 + h_r) - h_r^2}$$

$$= \frac{11.08 \times 5.84 + 11.08 \times 3.19}{(3.19 + 5.84) \times (0.7 + 11.08 + 5.84) - (5.84)^2}$$

$$= \frac{100.05}{159.11 - 34.11} = \frac{100.05}{124.99} = 0.80$$

**Example 5.10** Estimate an overall heat transfer coefficient and solar air collector efficiency $(F')$ by using Eqs. 5.22–5.23 for absorber plate and glass covers temperature as 50 °C and 30 °C, respectively, for single and double glazed glass covers without considering radiative heat transfer coefficient $(h_r)$.

**Solution**

We have from Eqs. 5.22 and 5.23.

$$U_L = \frac{(U_t + U_b) + U_t U_b}{h_1 h_2 + h_2 U_t} = \frac{(h_1 h_2)}{h_1 h_2 + h_2 U_t}[(U_t + U_b) + U_t U_b]$$

and,

$$F' = \frac{h_1 h_2 + + h_2 U_t}{(U_t + h_1)(U_b + h_2)} < 1$$

**Case (ii): For single glazed solar air collector**: $U_t = 7.49\,\text{W/m}^2\,°\text{C}$ and $U_b = 0.70\,\text{W/m}^2\,°\text{C}$ (Example 4.20) and $h_{bc,c} = h_{pf} = h_2 = 11.08\,\text{W/m}^2\,°\text{C}$; (Example 4.15).

From Eqs. 5.22 and 5.31, we have the following expression for $U_L$ and $F'$:

$$U_L = \frac{(U_t + U_b) + U_t U_b}{h_1 h_2 + h_2 U_t} = \frac{(h_1 h_2)}{h_1 h_2 + h_2 U_t}[(U_t + U_b) + U_t U_b]$$

$$= \frac{11.08 \times 11.08}{11.08 \times 11.08 + 11.08 \times 7.49}[8.19 + 7.49 \times 0.7]$$

$$= \frac{11.08 \times 11.08}{11.08 \times 11.08 + 11.08 \times 7.49}[8.19 + 7.49 \times 0.7]$$

$$= \frac{122.27}{205.76} \times 13.433 = 7.94\,\text{W/m}^2\,°\text{C}$$

and,

$$F' = \frac{h_1 h_2 + h_2 U_t}{(U_t + h_1)(U_b + h_2)} = \frac{205.76}{(7.49 + 11.08) \times (0.7 + 11.08)}$$

$$= \frac{205.76}{218.75} = 0.94 < 1$$

$$F' = \frac{h_1h_2 + + h_2U_t}{(U_t + h_1)(U_b + h_2)} = \frac{205.76}{(7.49 + 11.08) \times (0.7 + 11.08)}$$

$$= \frac{205.76}{218.75} = 0.94$$

**Case (i): for double glazed solar air collector**

$U_t = 3.19 \, \text{W/m}^2 \, ^\circ\text{C}$ and $U_b = 0.70 \, \text{W/m}^2 \, ^\circ\text{C}$ (Example 4.20) and $h_{bc,c} = h_{pf} = h_2 = 11.08 \, \text{W/m}^2 \, ^\circ\text{C}$; (Example 4.15).

From Eqs. 5.22 and 5.23, we have the following expression for $U_L$ and $F'$:

$$U_L = \frac{(h_1h_2)}{h_1h_2 + h_2U_t}[(U_t + U_b) + U_tU_b]$$

$$= \frac{11.08 \times 11.08}{11.08 \times 11.08 + 11.08 \times 3.19}[3.89 + 3.19 \times 0.7]$$

$$= \frac{122.27}{158.11} \times 6.123 = 4.73 \, \text{W/m}^2 \, ^\circ\text{C}$$

and,

$$F' = \frac{h_1h_2 + + h_2U_t}{(U_t + h_1)(U_b + h_2)} = \frac{158.11}{(3.19 + 11.08) \times (0.7 + 11.08)}$$

$$= \frac{158.11}{168.10} = 0.94 < 1$$

**Example 5.11** Estimate an average overall heat transfer coefficient ($U_L$) and solar air collector efficiency ($F'$) for Examples 5.7–5.10 except Example 5.9 for single and double glazed solar air collectors.

**Solution**

(a) **For single glazed solar air collector**

Average value of $U_L = \frac{8.01 + 6.69 + 7.94}{3} = 7.54 \, \text{W/m}^2 \, ^\circ\text{C}$ and
Average value of $F' = \frac{0.84 + 0.81 + 0.94}{3} = 0.86$

(b) **For double glazed solar air collector**

Average value of $U_L = \frac{4.37 + 3.29 + 4.73}{3} = 4.13 \, \text{W/m}^2 \, ^\circ\text{C}$ and
Average value of $F' = \frac{0.89 + 0.87 + 0.94}{3} = 0.90$.

This example shows that an overall heat transfer coefficient ($U_L$) is decreased from single glazed to double glazed by 45%. However, solar air collector efficiency ($F'$) is increased by 4.7%.

Further, one can observe from Examples 5.1–5.10 except Example 5.8 that the overall heat transfer coefficient ($U_L$) and solar air collector efficiency ($F'$) for single and double glazed air collectors are approximately closed to each other within accuracy of about 17%. Hence, the method given in Example 5.1 is most simple and accurate method to determine the overall heat transfer coefficient ($U_L$) and solar air collector efficiency ($F'$).

**Example 5.12** Evaluate the outlet air temperature flowing air over absorber plate with velocity of 0.25 m/s at inlet temperature of 17 °C with ambient air temperature of 16 °C for parameters of Example 5.11 for single and double glazed solar air collectors. Given solar intensity = 500 W/m² and $\alpha_p = \tau_g = 0.9$.

**Solution**

From Eq. 5.13a, we have an expression for outlet air temperature for single glazed and double glazed solar air collectors as

$$T_{fo} = \left[\left(\frac{(\alpha_p\tau_g)I(t)}{U_L}\right) + T_a\right]\left[1 - \exp[-A_cU_LF'/(\dot{m}_fC_f)]\right]$$
$$+ T_{fi}\exp[-A_cU_LF'/(\dot{m}_fC_f)]$$

and,

$$T_{fo} = \left[\left(\frac{(\alpha_p\tau_g^2)I(t)}{U_L}\right) + T_a\right]\left[1 - \exp[-A_cU_LF'/(\dot{m}_fC_f)]\right]$$
$$+ T_{fi}\exp[-A_cU_LF'/(\dot{m}_fC_f)]$$

(a) **For single glazed solar air collector:** $U_L = 7.54\,\text{W/m}^2\,°\text{C}$; $F' = 0.86$ (Example 5.11).

And mass flow rate $(\dot{m}_f)$ = cross sectional area × depth × density = 1 m × $0.10\,\text{m} \times 0.25\frac{\text{m}}{\text{s}} \times 1.17\,\frac{\text{kg}}{\text{m}^3} = 0.029\,\text{kg/s}$ and $C_f == 1005\,\text{j/kg}\,°\text{C}$ [Appendix E(a)] and $A_c = 2\,\text{m} \times 1\,\text{m} = 2\,\text{m}^2$.

Here,

$$\exp[-A_cU_LF'/(\dot{m}_fC_f)] = \exp\left[-\frac{2 \times 7.54 \times 0.86}{0.029 \times 1005}\right]$$
$$= \exp\left[-\frac{12.97}{29.145}\right] = \exp[-0.4450] = 0.6408$$

Now,

$$T_{fo} = \left[\left(\frac{0.81 \times 500}{7.54}\right) + 16\right][1 - 0.6408] + 17 \times 0.6408$$
$$= 69.71 \times 0.3592 + 10.89 = 35.93\,°\text{C}$$

(b) **For double glazed solar air collector:** $U_L = 4.13\,\text{W/m}^2\,°\text{C}$; $F' = 0.90$ (Example 5.11).

Here,

$$
\exp\left[-A_c U_L F'/\left(\dot{m}_f C_f\right)\right] = \exp\left[-\frac{2 \times 4.13 \times 0.90}{0.029 \times 1005}\right]
$$

$$
= \exp\left[-\frac{7.434}{29.145}\right] = \exp[-0.255] = 0.7749
$$

Now,

$$
T_{fo} = \left[\left(\frac{0.81 \times 0.9 \times 500}{4.13}\right) + 16\right][1 - 0.7749] + 17 \times 0.7749
$$

$$
= 104.26 \times 0.2251 + 13.17 = 36.42\,°C
$$

**Example 5.13** Evaluate the outlet air temperature flowing air over absorber plate with velocity of 0.5 m/s at inlet temperature of 17 °C with ambient air temperature of 16 °C for parameters of Examples 5.11 and 5.12 for single and double glazed solar air collectors. Given solar intensity = 500 W/m² and $\alpha_p = \tau_g = 0.9$.

**Solution**

(a) **For single glazed solar air collector:** $U_L = 7.54\,W/m^2\,°C$; $F' = 0.86$ (Example 5.11).

And mass flow rate $(\dot{m}_f)$ = cross sectional area × depth × density = 1 m × 0.10 m × 0.5 $\frac{m}{s}$ × 1.17 $\frac{kg}{m^3}$ = 0.058 kg/s and $C_f$ = 1005 j/kg °C [Appendix E(a)] and $A_c = 2\,m \times 1\,m = 2\,m^2$.

Here,

$$
\exp\left[-A_c U_L F'/\left(\dot{m}_f C_f\right)\right] = \exp\left[-\frac{2 \times 7.54 \times 0.86}{0.058 \times 1005}\right]
$$

$$
= \exp\left[-\frac{12.97}{58.28}\right] = \exp[-0.2225] = 0.8005
$$

Now,

$$
T_{fo} = \left[\left(\frac{0.81 \times 500}{7.54}\right) + 16\right][1 - 0.8005] + 17 \times 0.8005 = 27.52\,°C
$$

(b) **For double glazed solar air collector:** $U_L = 4.13\,W/m^2\,°C$; $F' = 0.90$ (Example 5.11).

Here,

$$\exp\left[-A_c U_L F'/(\dot{m}_f C_f)\right] = \exp\left[-\frac{2 \times 4.13 \times 0.90}{0.058 \times 1005}\right]$$

$$= \exp\left[-\frac{7.434}{58.29}\right] = \exp[-0.1275] = 0.8803$$

Now,

$$T_{fo} = \left[\left(\frac{0.81 \times 0.9 \times 500}{4.13}\right) + 16\right][1 - 0.8803] + 17 \times 0.8803 = 27.44\,^{\circ}C$$

**Example 5.14** Evaluate the outlet air temperature flowing air over absorber plate with velocity of 1 m/s at inlet temperature of 17 °C with ambient air temperature of 16 °C for parameters of Examples 5.11 and 5.12 for single and double glazed solar air collectors. Given solar intensity = 500 W/m² and $\alpha_p = \tau_g = 0.9$.

**Solution**

(a) **For single glazed solar air collector:** $U_L = 7.54\,\text{W/m}^2\,^{\circ}C$; $F' = 0.86$ (Example 5.11).

And mass flow rate $(\dot{m}_f)$ = cross-sectional area × depth × density = 1 m × 0.10 m × 0.5 $\frac{m}{s}$ × 1.17 $\frac{kg}{m^3}$ = 0.117 kg/s and $C_f = 1005\,\text{j/kg}^{\circ}C$ [Appendix E(a)] and $A_c = 2\,\text{m} \times 1\,\text{m} = 2\,\text{m}^2$.

Here,

$$\exp\left[-A_c U_L F'/(\dot{m}_f C_f)\right] = \exp\left[-\frac{2 \times 7.54 \times 0.86}{0.117 \times 1005}\right]$$

$$= \exp\left[-\frac{12.9}{117.6}\right] = \exp[-0.1097] = 0.8961$$

Now,

$$T_{fo} = \left[\left(\frac{0.81 \times 500}{7.54}\right) + 16\right][1 - 0.8961] + 17 \times 0.8961 = 22.48^{\circ}C$$

(b) **For double glazed solar air collector:** $U_L = 4.13\,\text{W/m}^2\,^{\circ}C$; $F' = 0.90$ (Example 5.11).

Here,

$$\exp\left[-A_c U_L F'/(\dot{m}_f C_f)\right] = \exp\left[-\frac{2 \times 4.13 \times 0.90}{0.117 \times 1005}\right]$$

$$= \exp\left[-\frac{7.434}{117.58}\right] = \exp[-0.06322] = 0.9387$$

Now,

$$T_{fo} = \left[\left(\frac{0.81 \times 0.9 \times 500}{4.13}\right) + 16\right][1 - 0.9387] + 17 \times 0.9387 = 22.35^\circ\text{C}$$

From Examples 5.12–5.14, one can see that the outlet temperature $(T_{fo})$ decreases from 35.93 to 22.48 °C as mass flow rate of air increases from 0.029 kg/s (0.25 m/s) to 0.117 kg/s(1 m/s) for single glazed solar air heater. However, there is not much difference between the outlet fluid temperatures for single and double glazed solar air collectors after mass flow rate of 0.058 kg/s (0.5 m/s air velocity). Further, it is also important to note that the convective heat transfer coefficient between absorber plates to flowing air over it $\{h_{bc,c} = h_{pf} = h_2 = 11.08\,\text{W/m}^2\,^\circ\text{C}$ Example 4.15$\}$ has been considered constant in evaluating $U_L$ and $F'$ (Examples 5.7–5.11).

**Example 5.15** Evaluate mass flow rate factor $(F_R)$ for Examples 5.12–5.14.

**Solution**

**Case (i): For mass flow rate = 0.029 kg/s (0.25 m/s air velocity)**

(a)  **Single glazed solar air collector**

From Eqs. 5.14a, 5.14b and Example 5.12, we have
Mass flow rate factor

$$(F_R) = \frac{\dot{m}_f C_f}{A_c U_L}\{1 - \exp[-A_c U_L F'/(\dot{m}_f C_f)]\}$$

$$= \frac{0.029 \times 1005}{2 \times 7.54}\left[1 - \exp\left[-\frac{2 \times 7.54 \times 0.86}{0.029 \times 1005}\right]\right]$$

$$= \frac{29.145}{15.08} \times [1 - 0.6408] = 0.69$$

(b)  **Double glazed solar air collector**

From Eqs. 5.14a, 5.14b and Example 5.12, we have
Mass flow rate factor

$$(F_R) = \frac{\dot{m}_f C_f}{A_c U_L}\{1 - \exp[-A_c U_L F'/(\dot{m}_f C_f)]\}$$

$$= \frac{0.029 \times 1005}{2 \times 4.13}\left[1 - \exp\left[-\frac{2 \times 4.13 \times 0.90}{0.029 \times 1005}\right]\right]$$

$$= \frac{29.145}{8.26} \times [1 - 0.7749] = 0.794$$

**Case (ii): For mass flow rate = 0.058 kg/s (0.5 m/s air velocity).**

**(a)  Single glazed solar air collector**

From Eqs. 5.14a, 5.14b and Example 5.13, we have
Mass flow rate factor

$$(F_R) = \frac{\dot{m}_f C_f}{A_c U_L}\left\{1 - \exp\left[-A_c U_L F'/(\dot{m}_f C_f)\right]\right\}$$

$$= \frac{0.058 \times 1005}{2 \times 7.54}\left[1 - \exp\left[-\frac{2 \times 7.54 \times 0.86}{0.058 \times 1005}\right]\right]$$

$$= \frac{58.29}{15.08} \times \left[1 - \exp\left[-\frac{12.97}{58.28}\right]\right] = \frac{58.29}{15.08} \times [1 - 0.8005] = 0.77$$

**(b)  Double glazed solar air collector**

From Eqs. 5.14a, 5.14b and Example 5.12, we have
Mass flow rate factor

$$(F_R) = \frac{\dot{m}_f C_f}{A_c U_L}\left\{1 - \exp\left[-A_c U_L F'/(\dot{m}_f C_f)\right]\right\}$$

$$= \frac{0.058 \times 1005}{2 \times 4.13}\left[1 - \exp\left[-\frac{2 \times 4.13 \times 0.90}{0.058 \times 1005}\right]\right]$$

$$= \frac{58.29}{8.26} \times \left[1 - \exp\left[-\frac{7.434}{58.29}\right]\right] = \frac{58.29}{8.26} \times [1 - 0.8803] = 0.84$$

**Case (iii): For mass flow rate = 0.117 kg/s (1 m/s air velocity)**

**(a)  Single glazed solar air collector**

From Eqs. 5.14a, 5.14b and Example 5.14, we have
Mass flow rate factor

$$(F_R) = \frac{\dot{m}_f C_f}{A_c U_L}\left\{1 - \exp\left[-A_c U_L F'/(\dot{m}_f C_f)\right]\right\}$$

$$= \frac{0.117 \times 1005}{2 \times 7.54}\left[1 - \exp\left[-\frac{2 \times 7.54 \times 0.86}{0.117 \times 1005}\right]\right]$$

$$= \frac{117.59}{15.08} \times \left[1 - \exp\left[-\frac{12.9}{117.6}\right]\right] = \frac{117.59}{15.08} \times [1 - \exp[-0.1097]]$$

$$= \frac{117.59}{15.08} \times [1 - 0.8961] = 0.81$$

### (b)  Double glazed solar air collector

From Eqs. 5.14a, 5.14b and Example 5.12, we have
Mass flow rate factor

$$
\begin{aligned}
(F_R) &= \frac{\dot{m}_f C_f}{A_c U_L}\left\{1 - \exp\left[-A_c U_L F'/(\dot{m}_f C_f)\right]\right\}\\
&= \frac{0.117 \times 1005}{2 \times 4.13}\left[1 - \exp\left[-\frac{2 \times 4.13 \times 0.90}{0.117 \times 1005}\right]\right]\\
&= \frac{117.6}{8.26} \times \left[1 - \exp\left[-\frac{7.434}{117.6}\right]\right]\\
&= \frac{117.6}{8.26} \times \left[1 - \exp[-0.0632]\right] = \frac{117.6}{8.26} \times [1 - 0.9387] = 0.87
\end{aligned}
$$

**Example 5.16** Evaluate the rate of heat energy available $(\dot{Q}_u)$ at outlet of flowing air over absorber plate with velocity of 0.25, 0.5 and 1 m/s at inlet temperature of 17 °C with ambient air temperature of 16 °C for parameters of Example 5.15 for single and double glazed solar air collectors. Given solar intensity $= 500 \text{ W/m}^2$ and $\alpha_p = \tau_g = 0.9$.

**Solution**

The rate of thermal energy carried out by working fluid air $(\dot{Q}_u)$ for single and double glazed solar air collectors from Eq. 5.14a, 5.14b is given by

### (a)  Single glazed solar air collector

$$\dot{Q}_u = F_R A_c\left[(\alpha_p \tau_g)I(t) - U_L(T_{fi} - T_a)\right]$$

### (b)  Double glazed solar air collector

$$\dot{Q}_u = F_R A_c\left[(\alpha_p \tau_g^2)I(t) - U_L(T_{fi} - T_a)\right]$$

**Case (i): For mass flow rate = 0.029 kg/s (0.25 m/s air velocity).**

(a)  **Single glazed solar air collector:** $F_R = 0.69$ and $U_L = 7.54 \text{ W/m}^2 \,°\text{C}$ (Example 5.15).

**First method**

$$\dot{Q}_u = 0.69 \times 2[0.81 \times 500 - 7.54(17 - 16)] = 548.49 \text{ W}$$

**Second method**

$$\dot{Q}_u = \dot{m}_f C_f(T_{fo} - T_{fi}) = 0.029 \times 1005 \times (35.93 - 17) = 551.71 \text{ W}$$

(b) **Double glazed solar air collector:** $F_R = 0.794$ and $U_L = 4.13\,\text{W/m}^2\,^\circ\text{C}$ (Example 5.15).

**First method**

$$\dot{Q}_u = 0.794 \times 2[0.81 \times 0.9 \times 500 - 4.13(17 - 16)] = 572.26\,\text{W}$$

**Second method**

$$Q_u = \dot{m}_f C_f (T_{fo} - T_{fi}) = 0.029 \times 1005 \times (36.64 - 17) = 572.45\,\text{W}$$

**Case (ii): For mass flow rate = 0.058 kg/s (0.5 m/s air velocity).**

(a) **Single glazed solar air collector:** $F_R = 0.77$ and $U_L = 7.54\,\text{W/m}^2\,^\circ\text{C}$ (Example 5.15).

**First method**

$$\dot{Q}_u = 0.77 \times 2[0.81 \times 500 - 7.54(17 - 16)] = 612.08\,\text{W}$$

**Second method**

$$Q_u = \dot{m}_f C_f (T_{fo} - T_{fi}) = 0.058 \times 1005 \times (27.52 - 17) = 613.21\,\text{W}$$

(b) **Double glazed solar air collector:** $F_R = 0.84$ and $U_L = 4.13\,\text{W/m}^2\,^\circ\text{C}$ (Example 5.15).

**First method**

$$\dot{Q}_u = 0.84 \times 2[0.81 \times 0.9 \times 500 - 4.13(17 - 16)] = 605.42\,\text{W}$$

**Second method**

$$Q_u = \dot{m}_f C_f (T_{fo} - T_{fi}) = 0.058 \times 1005 \times (27.44 - 17) = 608.55\,\text{W}$$

**Case (iii): For mass flow rate = 0.117 kg/s (1 m/s air velocity)**

(a) **Single glazed solar air collector:** $F_R = 0.81$ and $U_L = 7.54\,\text{W/m}^2\,^\circ\text{C}$ (Example 5.15).

**First method**

$$\dot{Q}_u = 0.81 \times 2[0.81 \times 500 - 7.54(17 - 16)] = 643.88\,\text{W}$$

(b)  **Second method:**

$$Q_u = \dot{m}_f C_f \left( T_{fo} - T_{fi} \right) = 0.117 \times 1005 \times (22.48 - 17) = 644.36\,\text{W}$$

(c)  **Double glazed solar air collector:** $F_R = 0.87$ and $U_L = 4.13\,\text{W/m}^2\,^\circ\text{C}$ (Example 5.15).

**First method**

$$\dot{Q}_u = 0.87 \times 2[0.81 \times 0.9 \times 500 - 4.13(17 - 16)] = 627.04\,\text{W}$$

**Second method**

$$Q_u = \dot{m}_f C_f \left( T_{fo} - T_{fi} \right) = 0.117 \times 1005 \times (22.35 - 17) = 629.08\,\text{W}$$

This example shows that the rate of useful heat/thermal energy by both methods is approximately same. The small difference is only due to consideration of approximate value in calculation.

**Example 5.17**  Evaluate the instantaneous thermal efficiency ($\eta_i$) for Example 5.16 for single as well as double glazed solar air collector.

**Solution**

From Eqs. 5.15a, 5.15b, one has an expression for instantaneous thermal efficiency of single and double glazed solar air collectors as

(a)  **Single glazed solar air collector**

$$\eta_i = \frac{\dot{Q}_u}{I(t) \times A_c} = F_R \left[ \alpha_p \tau_g - U_L \frac{T_{fi} - T_a}{I(t)} \right]$$

(b)  **Double glazed solar air collector**

$$\eta_i = \frac{\dot{Q}_u}{I(t) \times A_c} = F_R \left[ \alpha_p \tau_g^2 - U_L \frac{T_{fi} - T_a}{I(t)} \right]$$

**Case (i): For mass flow rate = 0.029 kg/s (0.25 m/s air velocity).**

(c)  **Single glazed solar air collector:** $F_R = 0.69$ and $U_L = 7.54\,\text{W/m}^2\,^\circ\text{C}$ (Example 5.15).

**First method**

$$\eta_i = 0.69 \left[ 0.81 - 7.54 \times \frac{(17 - 16)}{500} \right] = 0.54$$

**Second method**

$$\eta_i = \frac{\dot{Q}_u}{I(t) \times A_c} = \frac{551.71}{500 \times 2} = 0.55 \text{ (Example 5.16)}$$

(d)  **Double glazed solar air collector:** $F_R = 0.794$ and $U_L = 4.13 \text{ W/m}^2 \,°\text{C}$ (Example 5.15).

**First method**

$$\eta_i = 0.794 \left[ 0.81 \times 0.9 - 4.13 \times \frac{(17 - 16)}{500} \right] = 0.57$$

**Second method**

$$\eta_i = \frac{\dot{Q}_u}{I(t) \times A_c} = \frac{572.26}{500 \times 2} = 0.57, \text{ (Example 5.16)}$$

**Case (ii): For mass flow rate = 0.058 kg/s (0.5 m/s air velocity)**

(a)  **Single glazed solar air collector:** $F_R = 0.77$ and $U_L = 7.54 \text{ W/m}^2 \,°\text{C}$ (Example 5.15).

**First method**

$$\eta_i = \frac{\dot{Q}_u}{I(t) \times A_c} = 0.77 \left[ 0.81 - 7.54 \frac{(17 - 16)}{500} \right] = 0.61$$

**Second method**

$$\eta_i = \frac{\dot{Q}_u}{I(t) \times A_c} = \frac{613.21}{500 \times 2} = 0.61. \text{ (Example 5.16)}$$

(b)  **Double glazed solar air collector:** $F_R = 0.84$ and $U_L = 4.13 \text{ W/m}^2 \,°\text{C}$ (Example 5.15).

**First method**

$$\eta_i = \frac{\dot{Q}_u}{I(t) \times A_c} = 0.84_R \left[ 0.81 \times 0.9 - 4.13 \frac{(17 - 16)}{500} \right] = 0.605$$

**Second method**

$$\eta_i = \frac{\dot{Q}_u}{I(t) \times A_c} = \frac{608.55}{500 \times 2} = 0.609, \text{ (Example 5.16)}$$

**Case (iii): For mass flow rate = 0.117 kg/s (1 m/s air velocity)**

(a)  **Single glazed solar air collector**: $F_R = 0.81$ and $U_L = 7.54 \, \text{W/m}^2 \, ^\circ\text{C}$ (Example 5.15).

**First method**

$$\eta_i = \frac{\dot{Q}_u}{I(t) \times A_c} = 0.81\left[0.81 - 7.54\frac{(17 - 16)}{500}\right] = 0.543$$

**Second method**

$$\eta_i = \frac{\dot{Q}_u}{I(t) \times A_c} = \frac{644.36}{500 \times 2} = 0.644, \text{ (Example 5.16)}$$

(b)  **Double glazed solar air collector**: $F_R = 0.87$ and $U_L = 4.13 \, \text{W/m}^2 \, ^\circ\text{C}$ (Example 5.15).

**First method**

$$\eta_i = \frac{\dot{Q}_u}{I(t) \times A_c} = 0.87\left[0.81 \times 0.9 - 4.13\frac{(17 - 16)}{500}\right] = 0.627$$

**Second method**

$$\eta_i = \frac{\dot{Q}_u}{I(t) \times A_c} = \frac{629.08}{500 \times 2} = 0.629, \text{ (Example 5.16)}$$

All results are as per expectation. Here, it is important to mention that the users have to select as per their requirement either temperature or efficiency. Then, one has to optimize the design parameters.

## 5.2.4 Other Expression for $U_L$ and $F'$ [Tiwari [2]] for Varying $h_{bc,c} = h_{pf} = h_1 = h_2$ with Velocity of Flowing Air $(u_0)$

**Example 5.18** Evaluate an overall heat transfer coefficient $(U_L)$ and solar air collector efficiency $(F')$ by using Eqs. 5.22 and 5.23 by varying air velocity for $u_0 = 0.25$ m/s; 0.50 and 1.0 m/s, respectively.

**Solution**

In this case, we have considered varying convective heat transfer coefficients with air velocity unlike previous one.

We have an expression for $U_L$ and $F'$ from Eqs. 5.22 and 5.23 and Example 5.10 as

$$U_L = \frac{(h_1 h_2)}{h_1 h_2 + h_2 U_t}[(U_t + U_b) + U_t U_b]$$

and,

$$F' = \frac{h_1 h_2 + + h_2 U_t}{(U_t + h_1)(U_b + h_2)} < 1$$

**Case (a): For air velocity $(u_0)$ of $= 0.25$ m/s**

The average temperature of hot plate and air, $\overline{T}_f = 40\,°C \ (= 313\,K)$. The values of physical properties of air from Appendix [V(a)] are as follows:

$$v = 1.6 \times 10^{-5}\,m^2/s, Pr = 0.708, K = 0.026\,W/mk$$

The value of Reynolds number for above physical properties of air at $u_0 = 0.25$ m/s is given by:

$$Re_L = \frac{u_0 L}{v} == \frac{0.25 \times 2}{1.6 \times 10^{-5}} = 3.125 \times 10^5$$

The flow is laminar, therefore, with help of Eq. 2.11b and Example. 2.33; one can have an expression for an average convective heat transfer coefficient as

$$h_{bc,c} = h_{pf} = h_2 = \frac{0.026}{2} \times 0.664 \times (3.125 \times 10^5)^{1/2} \times (0.708)^{1/3}$$
$$= 0.008632 \times 0.559 \times 10^3 \times 0.8914 = 4.30\,W/m^2\,°C$$

**Case (i): For single glazed solar air collector:** $U_t = 7.49\,\mathrm{W/m^2\,°C}$ and $U_b = 0.70\,\mathrm{W/m^2\,°C}$ (Example 4.20) and $h_{bc,c} = h_{pf} = h_2 = 4.30\,\mathrm{W/m^2\,°C}$; (Example 4.15).

From Eqs. 5.22 and 5.31, we have the following expression for $U_L$ and $F'$:

$$U_L = \frac{(U_t + U_b) + U_t U_b}{h_1 h_2 + h_2 U_t} = \frac{(h_1 h_2)}{h_1 h_2 + h_2 U_t}[(U_t + U_b) + U_t U_b]$$

$$= \frac{4.30 \times 4.30}{4.30 \times 4.30 + 4.30 \times 7.49}[8.19 + 7.49 \times 0.7]$$

$$= \frac{18.49}{50.69} \times 13.433 = 4.90\,\mathrm{W/m^2\,°C}$$

and,

$$F' = \frac{h_1 h_2 + + h_2 U_t}{(U_t + h_1)(U_b + h_2)} = \frac{50.69}{(7.49 + 4.30) \times (0.7 + 4.30)} = \frac{50.69}{58.96} = 0.86 < 1$$

**Case (ii): For double glazed solar air collector**

$U_t = 3.19\,\mathrm{W/m^2\,°C}$ and $U_b = 0.70\,\mathrm{W/m^2\,°C}$ (Example 4.20) and $h_{bc,c} = h_{pf} = h_2 = 4.30\,\mathrm{W/m^2\,°C}$ (same as single glazed collector).

From Eqs. 5.22 and 5.23, we have the following expression for $U_L$ and $F'$:

$$U_L = \frac{(U_t + U_b) + U_t U_b}{h_1 h_2 + h_2 U_t} = \frac{(h_1 h_2)}{h_1 h_2 + h_2 U_t}[(U_t + U_b) + U_t U_b]$$

$$= \frac{4.30 \times 4.30}{4.30 \times 4.30 + 4.30 \times 3.19}[3.89 + 3.19 \times 0.7]$$

$$= \frac{18.49}{32.21} \times 6.123 = 3.51\,\mathrm{W/m^2\,°C}$$

and,

$$F' = \frac{h_1 h_2 + + h_2 U_t}{(U_t + h_1)(U_b + h_2)} = \frac{32.21}{(3.19 + 4.30) \times (0.7 + 4.30)}$$

$$= \frac{32.21}{37.45} = 0.86 < 1$$

**Case (b): For air velocity ($u_0$) of = 0.5 m/s.**

The Reynolds number is given by

$$\mathrm{Re}_L = \frac{u_0 L}{\nu} = \frac{0.5 \times 2}{1.6 \times 10^{-5}} = 6.25 \times 10^5$$

Again, the flow is laminar, therefore, with help of Eq. 2.11b and Example 2.33; one can have a numerical value of an average convective heat transfer coefficient $\left[h_{bc,c} = h_{pf} = h_2\right]$ between absorber plate and flowing air above it as

$$h_{bc,c} = h_{pf} = h_2 = \frac{0.026}{2} \times 0.664 \times (6.25 \times 10^5)^{1/2} \times (0.708)^{1/3}$$
$$= 0.008632 \times 0.79 \times 10^3 \times 0.8914 = 6.07 \, \text{W/m}^2 \, ^\circ\text{C}$$

**Case (i): For single glazed solar air collector**: $U_t = 7.49 \, \text{W/m}^2 \, ^\circ\text{C}$ and $U_b = 0.70 \, \text{W/m}^2 \, ^\circ\text{C}$ (Example 4.20) and $h_{bc,c} = h_{pf} = h_2 = 6.07 \, \text{W/m}^2 \, ^\circ\text{C}$; (Example 4.15)

From Eqs. 5.22 and 5.31, we have the following expression for $U_L$ and $F'$:

$$U_L = \frac{(h_1 h_2)}{h_1 h_2 + h_2 U_t}[(U_t + U_b) + U_t U_b]$$
$$= \frac{6.07 \times 6.07}{6.07 \times 6.07 + 6.07 \times 7.49}[8.19 + 7.49 \times 0.7]$$
$$= \frac{36.85}{82.31} \times 13.433 = 6.01 \, \text{W/m}^2 \, ^\circ\text{C}$$

and,

$$F' = \frac{h_1 h_2 + +h_2 U_t}{(U_t + h_1)(U_b + h_2)} = \frac{82.31}{(7.49 + 6.07) \times (0.7 + 6.07)} = \frac{82.31}{91.80} = 0.897 < 1$$

**Case (ii): For double glazed solar air collector**

$U_t = 3.19 \, \text{W/m}^2 \, ^\circ\text{C}$ and $U_b = 0.70 \, \text{W/m}^2 \, ^\circ\text{C}$ (Example 4.20) and $h_{bc,c} = h_{pf} = h_2 = 6.07 \, \text{W/m}^2 \, ^\circ\text{C}$ (Example 4.15).

From Eqs. 5.22 and 5.23, we have the following expression for $U_L$ and $F'$:

$$U_L = \frac{(U_t + U_b) + U_t U_b}{h_1 h_2 + h_2 U_t} = \frac{(h_1 h_2)}{h_1 h_2 + h_2 U_t}[(U_t + U_b) + U_t U_b]$$
$$= \frac{6.07 \times 6.07}{6.07 \times 6.07 + 6.07 \times 3.19}[3.89 + 3.19 \times 0.7]$$
$$= \frac{36.85}{56.21} \times 6.123 = 4.01 \, \text{W/m}^2 \, ^\circ\text{C}$$

and,

$$F' = \frac{h_1 h_2 + +h_2 U_t}{(U_t + h_1)(U_b + h_2)} = \frac{56.21}{(3.19 + 6.07) \times (0.7 + 6.07)} = \frac{56.21}{62.69} = 0.897 < 1$$

**Case (c): For air velocity ($u_0$) of 1 m/s.**

The value of Reynolds number for $u_0 = 1\,\frac{m}{s}$ is given by:

$$Re_L = \frac{u_0 L}{\nu} == \frac{1 \times 2}{1.6 \times 10^{-5}} = 12.5 \times 10^5$$

This is also laminar flow, therefore, with the help of Eq. 2.11b and Example 2.33; one can have an average convective heat transfer coefficient $\left[h_{bc,c} = h_{pf} = h_2\right]$ between absorber plate and flowing air above it as

$$h_{bc,c} = h_{pf} = h_2 = h_1 = \frac{0.026}{2} \times 0.664 \times \left(12.5 \times 10^5\right)^{1/2} \times (0.708)^{1/3}$$
$$= 0.008632 \times 1.11 \times 10^3 \times 0.8913 = 8.54\,W/m^2\,°C$$

**Case (i): For single glazed solar air collector:** $U_t = 7.49\,W/m^2\,°C$ and $U_b = 0.70\,W/m^2\,°C$ (Example 4.20) and $h_{bc,c} = h_{pf} = h_2 = 8.54\,W/m^2\,°C$ (Example 4.15).

From Eqs. 5.22 and 5.31, we have the following expression for $U_L$ and $F'$:

$$U_L = \frac{(h_1 h_2)}{h_1 h_2 + h_2 U_t}[(U_t + U_b) + U_t U_b]$$
$$= \frac{8.54 \times 8.54}{8.54 \times 8.54 + 8.54 \times 7.49}[8.19 + 7.49 \times 0.7]$$
$$= \frac{72.93}{136.896} \times 13.433 = 7.156\,W/m^2\,°C$$

and,

$$F' = \frac{h_1 h_2 + + h_2 U_t}{(U_t + h_1)(U_b + h_2)} = \frac{136.896}{(7.49 + 8.54) \times (0.7 + 8.54)}$$
$$= \frac{136.896}{148.12} = 0.924 < 1$$

**Case (ii): For double glazed solar air collector**

$U_t = 3.19\,W/m^2\,°C$ and $U_b = 0.70\,W/m^2\,°C$ (Example 4.20) and $h_{bc,c} = h_{pf} = h_2 = 8.54\,W/m^2\,°C$ (Example 4.15).

From Eqs. 5.22 and 5.23, we have the following expression for $U_L$ and $F'$:

$$U_L = \frac{(U_t + U_b) + U_t U_b}{h_1 h_2 + h_2 U_t} = \frac{(h_1 h_2)}{h_1 h_2 + h_2 U_t}[(U_t + U_b) + U_t U_b]$$
$$= \frac{8.54 \times 8.54}{8.54 \times 8.54 + 8.54 \times 3.19}[3.89 + 3.19 \times 0.7]$$

$$= \frac{72.93}{100.17} \times 6.123 = 4.45 \, \text{W/m}^2 \, ^\circ\text{C}$$

and,

$$F' = \frac{h_1 h_2 + + h_2 U_t}{(U_t + h_1)(U_b + h_2)} = \frac{100.17}{(3.19 + 8.54) \times (0.7 + 8.54)}$$

$$= \frac{100.17}{108.385} = 0.924 < 1$$

**From this example, one can infer that there is significant variation in $U_L$ and $F'$ with varying convective heat transfer coefficients $\left[h_{bc,c} = h_{pf} = h_2\right]$ with air flowing velocity. However, solar collector efficiency $(F')$ is mostly unchanged for single and double glazed solar air collectors.**

**Example 5.19** Evaluate the outlet air temperature $(T_{fo})$, mass flow rate factor $(F_R)$, the rate of useful thermal energy $(\dot{Q}_u)$ and instantaneous thermal efficiency $(\eta_i)$ for single glazed air collector with air flowing with 0.25 m/s over the absorber for Example 5.18 and design parameters of Example 5.16.

**Solution**

**From Examples 5.16 and 5.18, we have**

$U_L = 4.90 \, \text{W/m}^2 \, ^\circ\text{C}$, $F' = 0.86$, case (i) of case (a) (Example 5.18), $(\dot{m}_f C_f) = 0.029 \, \frac{\text{kg}}{\text{s}} \times 1005 \, \text{j/kg} \, ^\circ\text{C} = 29.145 \, \text{W/}^\circ\text{C}$.

(a)  **Outlet air temperature $(T_{fo})$**

$$T_{fo} = \left[\left(\frac{(\alpha_p \tau_g) I(t)}{U_L}\right) + T_a\right]\left[1 - \exp\left(-\frac{F' A_c U_L}{\dot{m}_f C_f}\right)\right] + T_{fi} \exp\left(-\frac{F' A_c U_L}{\dot{m}_f C_f}\right)$$

$$= \left[\frac{0.81 \times 500}{4.90} + 16\right]\left[1 - \exp(-0.2892)\right] + 17 \times \exp(-0.2892)$$

$$= 24.77 + 17 \times 0.7489 = 37.50 \, ^\circ\text{C}$$

(b)  **Mass flow rate factor $(F_R)$**

$$F_R = \frac{\dot{m}_f C_f}{A_c U_L}\left\{1 - \exp\left[-A_c U_L F' / (\dot{m}_f C_f)\right]\right\}$$

$$= \frac{29.145}{2 \times 4.9}\left[1 - \exp(-0.2892)\right]$$

$$= 2.9740 \times (1 - 0.7489) = 0.7468$$

(c)  **The rate of useful thermal energy $(\dot{Q}_u)$**

$$\dot{Q}_u = F_R A_c\left[(\alpha_p \tau_g)I(t) - U_L(T_{fi} - T_a)\right]$$
$$= 0.7468 \times 2[0.81 \times 500 - 4.9(17 - 16)] = 597\,\text{W}$$

(d)  **Instantaneous thermal efficiency $(\eta_i)$**

$$\eta_i = \frac{\dot{Q}_u}{I(t) \times A_c} = F_R\left[\alpha_p \tau_g - U_L\frac{T_{fi} - T_a}{I(t)}\right] = 0.597$$

**Example 5.20**  Evaluate the outlet air temperature $(T_{fo})$, mass flow rate factor $(F_R)$, **the** rate of useful thermal energy $(\dot{Q}_u)$ and instantaneous thermal efficiency $(\eta_i)$ for double glazed air collector with air flowing with 0.25 m/s over the absorber for Example 5.18 and design parameters of Example 5.16.

**Solution**

**From Examples 5.16 and 5.18, we have**

$U_L = 3.51\,\text{W/m}^2\,°\text{C}$  $F' = 0.86$, case (i) of case (a) (Example 5.18), $(\dot{m}_f C_f) = 0.029\,\frac{\text{kg}}{\text{s}} \times 1005\text{j/kg}\,°\text{C} = 29.145\,\text{W/}°\text{C}$.

(a)  **Outlet air temperature $(T_{fo})$**

$$T_{fo} = \left[\left(\frac{(\alpha_p \tau_g)I(t)}{U_L}\right) + T_a\right]\left[1 - \exp\left(-\frac{F'A_c U_L}{\dot{m}_f C_f}\right)\right] + T_{fi}\exp\left(-\frac{F'A_c U_L}{\dot{m}_f C_f}\right)$$

$$= \left[\frac{0.81 \times 500}{3.51} + 16\right]\left[1 - \exp(-0.2071)\right] + 17 \times \exp(-0.2071)$$

$$= 24.77 + 17 \times 0.8129 = 38.40\,°\text{C}$$

(b)  **Mass flow rate factor $(F_R)$**

$$F_R = \frac{\dot{m}_f C_f}{A_c U_L}\{1 - \exp[-A_c U_L F'/(\dot{m}_f C_f)]\}$$

$$= \frac{29.145}{2 \times 3.51}[1 - \exp(-0.2892)] = 4.15 \times (1 - 0.8129) = 0.777$$

(c)  **The rate of useful thermal energy $(\dot{Q}_u)$**

$$\dot{Q}_u = F_R A_c\left[(\alpha_p \tau_g)I(t) - U_L(T_{fi} - T_a)\right]$$
$$= 0.777 \times 2[0.81 \times 0.9 \times 500 - 3.51(17 - 16)] = 560.98\,\text{W}$$

(d)  **Instantaneous thermal efficiency** $(\eta_i)$

$$\eta_i = \frac{\dot{Q}_u}{I(t) \times A_c} = F_R\left[\alpha_p\tau_g^2 - U_L\frac{T_{fi} - T_a}{I(t)}\right] = 0.561$$

In the double glazed solar air collector, the outlet temperature, $T_{fo}$ is increased due to less overall heat loss coefficient, $U_L$; however, the rate of useful thermal energy, $\dot{Q}_u$, is reduced solar intensity by transmission by two glass covers.

**This example shows that the value of convective heat transfer coefficient from flat plate absorber to working fluid (air) should be always taken as a function of air velocity {Example 5.18} which improves the thermal performance of solar air heater.**

## 5.3  Quasi-Steady-State Analysis of Non-porous Flat Plate Collector with Flowing Air [1] Below Absorber Plate

There are many thermal models given by researchers for flowing air below the absorber plate, Fig. 5.4a, in terms of an overall heat transfer coefficient ($U_L$) and solar air collector efficiency ($F'$). Some of them will be discussed in following sections.

**Case (A): With radiation between absorber plate and insulated bottom surface**

The thermal performance of conventional non-porous solar air heater, Fig. 5.4a, was also studied analytically by Whillier[3]. In this case, air is flowing between absorber and insulated bottom surface. The solar radiation after transmission from the glass cover is absorbed by blackened absorber and then transferred to flowing air below it and gets heated. Thermal energy is carried away at outlet/exit.

For an infinitesimal small area $bdx$ along the flow direction at $x$ (Fig. 5.4b), the thermal energy balance equations for various components are given as following:

(i)  **The absorber flat plate**

$$(\alpha_p\tau_g)I(t)bdx = U_t(T_p - T_a)bdx$$
$$+ h_{pf}(T_p - T_f)bdx + h_{rpb}(T_p - T_b)bdx \qquad (5.24)$$

(ii)  **Insulated bottom plate**

In this case, it can be written as:

$$h_{rpb}(T_p - T_b)bdx = h_{bf}(T_b - T_f)bdx + U_b(T_b - T_a)\,bdx \qquad (5.25)$$

Here, $b$ is width of the absorber plate.

(iii) **Flowing fluid air between absorber plate and bottom insulated surface, Fig. 5.4**

The rate of heat given by bottom and plate to fluid is carried away, and it can be written as

$$\dot{m}_f C_f \frac{dT_f}{dx} dx = \left[h_{pf}(T_p - T_f) + h_{bf}(T_b - T_f)\right] b dx \qquad (5.26)$$

These equations can be solved for the outlet fluid (air) temperature by elimination process and using the technique for one order differential equation in a way similarly done in the first method.

### 5.3.1    The Outlet Air Temperature

Following Eqs. (5.24–5.26), the outlet fluid air temperatures $T_{fo}$ at $x = L$ can be written as,

$$T_{fo} = T_f|_{x=L} = \left[\left(\frac{\dot{q}_{ab}}{U_L}\right) + T_a\right] + \left[T_{fi} - T_a - \left(\frac{\dot{q}_{ab}}{U_L}\right)\right] \exp\left(-\frac{F' A_c U_L}{\dot{m}_f C_f}\right)$$

or,

$$T_{fo} = \left[\left(\frac{\dot{q}_{ab}}{U_L}\right) + T_a\right]\left[1 - \exp\left(-\frac{F' A_c U_L}{\dot{m}_f C_f}\right)\right] + T_{fi} \exp\left(-\frac{F' A_c U_L}{\dot{m}_f C_f}\right) \qquad (5.27)$$

where, $\dot{q}_{ab} = (\alpha_p \tau_g) I(t)$

### 5.3.2    The Rate of Useful Thermal Energy $\left(\dot{Q}_u\right)$ and Mass Flow Rate $(F_R)$

The rate of useful heat gain of the conventional non-porous single glazed solar air collector, Tiwari et al. [2] can also be written as:

$$\dot{Q}_u = \dot{m}_f C_f (T_{fo} - T_{fi}) = F_R A_c \left[(\alpha_p \tau_g) I(t) - U_L (T_{fi} - T_a)\right] \qquad (5.28a)$$

Similarly, for double glazed solar air collector

$$\dot{Q}_u = \dot{m}_f C_f (T_{fo} - T_{fi}) = F_R A_c \left[\left(\alpha_p (\tau_g)^2\right) I(t) - U_L (T_{fi} - T_a)\right] \qquad (5.28b)$$

where, $F_R$ is the collector heat removal factor $(F_R)$ and is given by

$$F_R = \frac{\dot{m}_f C_f}{A_c U_L}\left[1 - \exp\left(-\frac{F' A_c U_L}{\dot{m}_f C_f}\right)\right] \tag{5.29}$$

where,

$$U_L = U' + (1/F')[U_b h_{bf}/(h_{rpb} + h_{bf} + U_b)]$$

$$U' = U_t + [h_{rpb} U_b/(h_{rpb} + h_{bf} + U_b)]$$

and, the collector efficiency factor $(F')$ is given by

$$F' = \left[1 + \frac{U'}{h_e}\right]^{-1} \text{with } h_e = h_{pf} + \left[\frac{h_{bf} h_{rpb}}{(h_{rpb} + h_{bf} + U_b)}\right] \tag{5.30}$$

### 5.3.3   An Instantaneous Thermal Efficiency $(\eta_i)$

It is defined in similar way as done in Sect. 5.2.2. An instantaneous thermal efficiency $(\eta_i)$ for single and double glazed solar collectors can be obtained as follows:

(a)   **Single glazed solar air heater**

$$\eta_i = \frac{\dot{Q}_u}{A_c I(t)} = F_R\left[(\alpha_p \tau_g) - U_L \frac{(T_{fi} - T_a)}{I(t)}\right] \tag{5.31}$$

(b)   **Double glazed solar air heater**

$$\eta_i = \frac{\dot{Q}_u}{A_c I(t)} = F_R\left[(\alpha_p \tau_g^2) - U_L \frac{(T_{fi} - T_a)}{I(t)}\right] \tag{5.32}$$

The equation derived for the outlet air temperature $(T_{fo})$, the rate of thermal energy $(\dot{Q}_u)$, mass flow rate factor $(F_R)$, collector efficiency $(F')$ and an instantaneous thermal efficiency $(\eta_i)$ is exactly same as done for liquid flat plate collector (Chap. 4) except the change in numerical value of (i) an overall heat transfer $(U_L)$, (ii) solar collector efficiency $(F')$ and (iii) convective heat transfer coefficient due to change in physical properties of air instead of water and design of air collector.

**Example 5.21** Estimate an overall heat transfer coefficient $(U_L)$ and solar air collector efficiency $(F')$ for air flowing with velocity of 0.25 m/s below the absorber plate (Fig. 5.4a) for single and double glazed solar air collectors (Fig. 5.4) for design and climatic parameters of Example 5.12.

Given other parameters:

Single glazed air collector: $U_t = 7.49\,\text{W/m}^2\,°\text{C}$ and $U_b = 0.70\,\text{W/m}^2\,°\text{C}$ (Example 4.20).

Double glazed air collector: $U_t = 3.19\,\text{W/m}^2\,°\text{C}$ and $U_b = 0.70\,\text{W/m}^2\,°\text{C}$ (Example 4.20).

**Solution**

**For air velocity ($u_0$) of $= 0.25\,\text{m/s}$**

The average temperature of hot plate and air, $\overline{T}_f = 40°\text{C} = (313\,\text{K})$. The values of physical properties of air from Appendix [E(a)] are as follows:

$$v = 1.6 \times 10^{-5}\,\text{m}^2/\text{s}, \ \text{Pr} = 0.708, \ K = 0.026\,\text{W/mK}$$

The value of Reynolds number for above physical properties of air at $u_0 = 0.25\,\text{m/s}$ is given by:

$$\text{Re}_L = \frac{u_0 L}{v} == \frac{0.25 \times 2}{1.6 \times 10^{-5}} = 3.125 \times 10^5$$

The flow is laminar, therefore, with the help of Eq. 2.11b and Example. 2.33; one can have an expression for an average convective heat transfer coefficient as
$h_{bc,c} = h_{pf} = h_{bf} = \frac{0.026}{2} \times 0.332 \times (3.125 \times 10^5)^{1/2} \times (0.708)^{1/3} = 0.004316 \times$
$0.559 \times 10^3 \times 0.8913 = 2.15\,\text{W/m}^2\,°\text{C}$; Eq. 2.11c and Example. 2.33.
Here $\overline{T} = 40°\text{C}$, then $h_r = h_{rpb} = \varepsilon_{\text{eff}}(4\sigma \overline{T})^3 = 0.84 \times 4 \times 5.67 \times 10^{-8}(273 + 40)^3 = 5.84\,\text{W/m}^2\,°\text{C}$ (Example 5.7).
Here: $\dot{m}_f = $ cross sectional area $\times$ depth $\times$ density $= 1\,\text{m} \times 0.10\,\text{m} \times 0.25\,\frac{\text{m}}{\text{s}} \times 1.17\,\frac{\text{kg}}{\text{m}^3} = 0.029\,\text{kg/s}$ and $C_f = 1005\,\text{j/kg}°\text{C}$ [Appendix V(a)] and $A_c = 2\,\text{m} \times 1\,\text{m} = 2\,\text{m}^2$.
From Eqs. 5.29 and 5.30, we have the following expression for an overall heat transfer coefficient ($U_L$) solar air heater collector efficiency ($F'$) as

$$U_L = U' + (1/F')\left[U_b h_{bf}/(h_{rpb} + h_{bf} + U_b)\right]$$

and,

$$F' = \left[1 + \frac{U'}{h_e}\right]^{-1} \text{with } h_e = h_{pf} + \left[\frac{h_{bf}h_{rab}}{(h_{rbp} + h_{bf} + U_b)}\right]$$

with

$$U' = U_t + \left[h_{rpb}U_b/(h_{rpb} + h_{bf} + U_b)\right].$$

Now,

**Case (a): For single glazed solar air collector**

$$U' = U_t + \frac{h_{rpb}U_b}{(h_{rpb} + h_{bf} + U_b)} = 7.49 + \frac{5.84 \times 0.7}{5.84 + 2.15 + 0.7}$$
$$= 7.49 + 0.4715 = 7.96 \,\text{W/m}^2\,°\text{C};$$

$$h_e = h_{pf} + \left[\frac{h_{bf}h_{rab}}{(h_{rbp} + h_{bf} + U_b)}\right] = 2.15 + \frac{2.15 \times 5.84}{5.84 + 2.15 + 0.7}$$
$$= 2.15 + 1.445 = 3.595 \,\text{W/m}^2\,°\text{C};$$

$$F' = \left[1 + \frac{U'}{h_e}\right]^{-1} = \left[1 + \frac{7.96}{3.595}\right]^{-1} = [3.214]^{-1} = 0.31$$

$$U_L = U' + \frac{1}{F'} \times \frac{U_b h_{bf}}{h_{rpb} + h_{bf} + U_b} = 7.96 + \frac{1}{0.31} \times \frac{0.7 \times 2.15}{5.84 + 2.15 + 0.7}$$
$$= 7.78 + 3.226 \times 0.1732 = 8.34 \,\text{W/m}^2\,°\text{C};$$

**Case (b): For double glazed solar air collector**

**In this case, numerical value of $U_L$ is reduced.**

$$U' = U_t + \frac{h_{rpb}U_b}{(h_{rpb} + h_{bf} + U_b)} = 3.19 + \frac{5.84 \times 0.7}{5.84 + 2.15 + 0.7}$$
$$= 3.19 + 0.4715 = 3.66 \,\text{W/m}^2\,°\text{C};$$

$$h_e = h_{pf} + \left[\frac{h_{bf}h_{rab}}{(h_{rbp} + h_{bf} + U_b)}\right] = 2.13 + \frac{2.15 \times 5.84}{5.84 + 2.13 + 0.7}$$
$$= 2.13 + 1.445 = 3.575 \,\text{W/m}^2\,°\text{C};$$

$$F' = \left[1 + \frac{U'}{h_e}\right]^{-1} = \left[1 + \frac{3.66}{3.575}\right]^{-1} = [2.02]^{-1} = 0.495$$

$$U_L = U' + \frac{1}{F'} \times \frac{U_b h_{bf}}{h_{rpb} + h_{bf} + U_b} = 3.66 + \frac{1}{0.495} \times \frac{0.7 \times 2.15}{5.84 + 2.15 + 0.7}$$
$$= 3.66 + 2.02 \times 0.1732 = 4.01 \,\text{W/m}^2\,°\text{C};$$

**Case (c): without radiation between absorber plate and insulated bottom surface**

In this case, the energy balance for absorber plate and flowing air above absorber plate for an elemental area of '$bdx$' can be written as follows:

$$(\alpha_p \tau_g) I(t) b dx = U_t (T_p - T_a) b dx + h_{pf} (T_p - T_f) b dx \qquad (5.33)$$

and,

$$\dot{m}_f C_f \frac{dT_f}{dx} dx = \left[ h_{pf} (T_p - T_f) - U_{bfa} (T_f - T_a) \right] b dx \qquad (5.34)$$

From Eq. 5.33, one can have

$$h_{pf} (T_p - T_f) = F' \dot{q}_{ab} - \frac{h_{bf} U'}{[U' + h_e]} (T_f - T_a) \qquad (5.35)$$

With help of Eqs. 5.35, 5.34 can be rearranged as

$$\dot{m}_f C_f \frac{dT_f}{dx} dx = \left[ F' \dot{q}_{ab} - \frac{h_{bf} U'}{[U' + h_e]} (T_f - T_a) - U_{bfa} (T_f - T_a) \right] b dx$$

$$= F' [ \dot{q}_{ab} - U_L (T_f - T_a) ] b dx \qquad (5.36)$$

where,

$$F' = \frac{h_{pf}}{(U_t + h_{pf})} \text{ and } U_L = \frac{1}{F'} \left[ U_{bfa} + \frac{h_{pf} \times U_t}{(U_t + h_{pf})} \right] \qquad (5.37)$$

The numerical values of $U_L$ and $F'$ will be different for single and double glazed solar air collectors. All other expressions remain the same as in Sects. 5.3.1–5.3.3.

**Example 5.22** Estimate an overall heat transfer coefficient ($U_L$) and solar air collector efficiency ($F'$) for air flowing with velocity of 0.25 m/s below the absorber plate (Fig. 5.4a) for single and double glazed solar air collectors (Fig. 5.4) for design and climatic parameters of Example 5.21 by using Eq. 5.37.

**Solution**

(a) **Single glazed solar air collector:** $U_t = 7.49 \, \text{W/m}^2 \, {}^\circ\text{C}$ and $U_b = 0.70 \, \text{W/m}^2 \, {}^\circ\text{C}$ (Example 4.20).

From Eq. 5.37, we have

$$F' = \frac{h_{pf}}{(U_t + h_{pf})} = \frac{2.15}{7.49 + 2.15} = 0.22$$

and,

$$U_L = \frac{1}{F'}\left[U_{bfa} + \frac{h_{pf} \times U_t}{(U_t + h_{pf})}\right] = \frac{1}{0.22}\left[0.53 + \frac{2.15 \times 7.49}{7.49 + 2.15}\right]$$

$$= 10.00\,\text{W/m}^2\,^\circ\text{C}.$$

(b) **Double glazed solar air collector:** $U_t = 3.19\,\text{W/m}^2\,^\circ\text{C}$ and $U_b = 0.70\,\text{W/m}^2\,^\circ\text{C}$ (Example 4.20).

From Eq. 5.37, we have

$$F' = \frac{h_{pf}}{(U_t + h_{pf})} = \frac{2.15}{3.19 + 2.15} = 0.60$$

and,

$$U_L = \frac{1}{F'}\left[U_{bfa} + \frac{h_{pf} \times U_t}{(U_t + h_{pf})}\right] = \frac{1}{0.60}\left[0.53 + \frac{2.15 \times 3.19}{3.19 + 2.15}\right] = 3.06\,\text{W/m}^2\,^\circ\text{C}.$$

It is to be noted from Examples 5.21 and 5.22 that the value of product of $U_L$ and $F'$ $(F'U_L)$ for both cases is approximately same. This product plays an important role in calculating the outlet air temperature, $T_{fo}$; mass flow rate factor, $F_R$; the rate of useful thermal energy, $\dot{Q}_u$, and instantaneous thermal efficiency, $\eta_i$.

(c) **There is another expression for an overall heat transfer coefficient and collector efficiency for flat plate solar air collector with air flowing below absorber plate which is given by**

$$U_L = U_t + U_b \quad \text{(same as in Eq. 5.1)} \tag{5.38a}$$

$$F' = \frac{1}{1 + \dfrac{U_L}{h_1 + \left(\frac{1}{h_2} + \frac{1}{h_r}\right)^{-1}}} < 1 \tag{5.38b}$$

$$h_r = \frac{\sigma\left(T_1^2 + T_2^2\right)(T_1 + T_2)}{\frac{1}{\varepsilon_1} + \frac{1}{\varepsilon_2} - 1}; \quad T\,\text{in Kelvin (K)},$$

or,

$$h_r = \varepsilon_{\text{eff}}\left(4\sigma \overline{T}\right)^3 \text{ with } \overline{T} = \frac{T_1 + T_2}{2} \text{ and}$$

$$\varepsilon_{\text{eff}} = \left[\frac{1}{\varepsilon_1} + \frac{1}{\varepsilon_2} - 1\right]^{-1} \quad \text{[refer Eq. 2.17e]} \tag{5.38c}$$

If there is air flowing below absorber plate, then there will be heat transfer from absorber plate to flowing air by only convection and the radiative heat loss can be neglected. Hence an expression for collector efficiency becomes

$$F' = \frac{1}{1 + \frac{U_L}{h_1}} < 1 \tag{5.39}$$

**Example 5.23** Estimate an overall heat transfer coefficient $(U_L)$ and solar air collector efficiency $(F')$ for air flowing with velocity of 0.25 m/s below the absorber plate (Fig. 5.4a) for single and double glazed solar air collectors (Fig. 5.4) for design and climatic parameters of Example 5.21 by using Eq. 5.38a–5.38c.

**Solution**

(a) **Single glazed solar air collector:** $U_t = 7.49 \, \text{W/m}^2 \, {}^\circ\text{C}$, $U_b = 0.70 \, \text{W/m}^2 \, {}^\circ\text{C}$ :
$U_L = U_t + U_b = 7.49 + 0.7 = 8.19 \, \text{W/m}^2 \, {}^\circ\text{C}$ (Example 4.20); $h_r = 5.84 \, \text{W/m}^2 \, {}^\circ\text{C}$ (Example 5.21).

From Eq. 5.38b, we have

$$F' = \frac{1}{1 + \frac{U_L}{h_1 + \left(\frac{1}{h_2} + \frac{1}{h_r}\right)^{-1}}} = \frac{1}{1 + \frac{8.19}{2.15 + \left(\frac{1}{2.15} + \frac{1}{5.84}\right)^{-1}}}$$

$$= \frac{1}{1 + \frac{8.19}{2.15 + 1.57}} = \frac{1}{1 + \frac{8.19}{2.15 + 1.57}} = \frac{1}{1 + 2.20} = \frac{1}{3.2} = 0.31$$

(b) **Double glazed solar air collector:** $U_t = 3.19 \, \text{W/m}^2 \, {}^\circ\text{C}$, $U_b = 0.70 \, \text{W/m}^2 \, {}^\circ\text{C}$; $U_L = U_t + U_b = 3.19 + 0.7 = 3.89 \, \text{W/m}^2 \, {}^\circ\text{C}$ (Example 4.20); $h_r = 5.84 \, \text{W/m}^2 \, {}^\circ\text{C}$ (Example 5.21).

From Eq. 5.38b, we have

$$F' = \frac{1}{1 + \frac{U_L}{h_1 + \left(\frac{1}{h_2} + \frac{1}{h_r}\right)^{-1}}} = \frac{1}{1 + \frac{3.89}{2.15 + \left(\frac{1}{2.15} + \frac{1}{5.84}\right)^{-1}}}$$

$$= \frac{1}{1 + \frac{3.89}{2.15 + 1.57}} = \frac{1}{1 + \frac{3.89}{2.15 + 1.57}} = \frac{1}{1 + 1.04} = \frac{1}{3.2} = 0.49$$

**Example 5.24** Repeat Example 5.23 without radiative loss between absorber plate and bottom insulated surface.

**Solution**

From Eq. 5.39, we have the following expression:

(a) Single glazed solar air collector

$$F' = \frac{1}{1 + \frac{U_L}{h_1}} = \frac{1}{1 + \frac{8.19}{2.15}} = \frac{1}{1 + 3.8} = 0.20$$

(b) Double glazed solar air collector

$$F' = \frac{1}{1 + \frac{U_L}{h_1}} = \frac{1}{1 + \frac{3.89}{2.15}} = \frac{1}{1 + 1.81} = 0.35$$

Selcuk [4] has reported the performance of non-porous type of air heaters as shown in Fig. 5.4a. It is inferred from the figure that transparent covers are not always necessary for low-temperature applications. However, in single glass cover, heat loss reduction due to single glass compensates the transmission losses and becomes beneficial beyond a certain temperature. However, choice of number of glass cover has always economical constraints.

**Example 5.25** Evaluate the outlet temperature, $T_{fo}$, and mass flow rate, $F_{R;}$, of single glazed solar air collector for air velocity of 0.25 m/a (0.029 kg s) and specific heat of 1005 j/kg °C flowing below absorber plate for data of Example 5.20.

**Solution**

From Eq. 5.27, we have an expression for an outlet air temperature as

$$T_{fo} = \left[ \left( \frac{\dot{q}_{ab}}{U_L} \right) + T_a \right]\left[ 1 - \exp\left( -\frac{F'A_cU_L}{\dot{m}_fC_f} \right) \right] + T_{fi}\exp\left( -\frac{F'A_cU_L}{\dot{m}_fC_f} \right)$$

$$= \left[ \left( \frac{0.81 \times 500}{10.00} \right) + 16 \right]\left[ 1 - \exp\left( -\frac{0.22 \times 2 \times 10.00}{0.029 \times 1005} \right) \right]$$

$$+ 17 \times \exp\left( -\frac{0.22 \times 2 \times 10.00}{0.029 \times 1005} \right)$$

$$= 56.6 \times \left[ 1 - \exp(-0.15097) \right] + 17 \times \exp(-0.15097)$$

$$= 56.6 \times \left[ 1 - 0.85987 \right] + 17 \times 0.85987 = 22.55\,°C$$

From Eq. 5.29, we have an expression for mass flow rate factor as

$$F_R = \frac{\dot{m}_fC_f}{A_cU_L}\left[ 1 - \exp\left( -\frac{F'A_cU_L}{\dot{m}_fC_f} \right) \right]$$

$$= \frac{0.029 \times 1005}{2 \times 10}\left[ 1 - \exp\left( -\frac{0.22 \times 2 \times 10.00}{0.029 \times 1005} \right) \right]$$

$$= 1.4573 \times \left[ 1 - 0.85987 \right] = 0.204$$

**Example 5.26** Evaluate the rate of thermal energy and instantaneous thermal efficiency of single glazed solar air collector for air velocity of 0.25 m/a (0.029 kg s) and specific heat of 1005 j/kg °C flowing below absorber plate for data of Examples 5.20 and 5.23.

**Solution**

From Eq. 5.28a, we have the following:

$$\dot{Q}_u = F_R A_c \big[ (\alpha_p \tau_g) I(t) - U_L (T_{fi} - T_a) \big]$$
$$= 0.204 \times 2[405 - 10(17 - 16)] = 161\ \text{W}$$

$$\dot{Q}_u = \dot{m}_f C_f (T_{fo} - T_{fi}) = 0.029 \times 1005(22.55 - 17) = 161.72$$

and, we have from Eq. 5.31

$$\eta_i = F_R \left[ (\alpha_p \tau_g) - U_L \frac{(T_{fi} - T_a)}{I(t)} \right]$$
$$= 0.204 \left[ 0.81 - 10 \times \frac{(17 - 16)}{500} \right] = 0.161$$

## 5.4  Other Configuration of Solar Air Collector

Since, air is a poor heat media transfer in comparison with liquid (water) as discussed in Chap. 2. Due to this, many attempts have been made to increase heat transfer by redesigning the absorber plate. Some of popular designs are discussed below.

### 5.4.1  Fins' Arrangement Below Absorber

In order to have maximum heat transfer from absorber plate to flowing air below absorber, many fins with width of $W_2$ at equidistance of W are attached below absorber plate to increase the surface area of heat transfer as shown in Fig. 5.5. It is required due to less heat transfer coefficient from flat absorber plate as shown in Fig. 5.4a to fluid. With fin attached to flat absorber as shown in Fig. 5.5, air solar collector efficiency ($F'$) is given [2] as

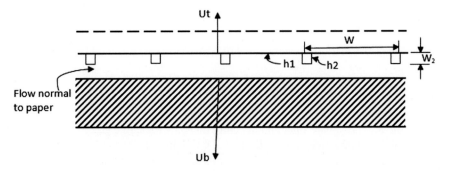

**Fig. 5.5** Flowing of air between absorber plate with fin and insulated bottom

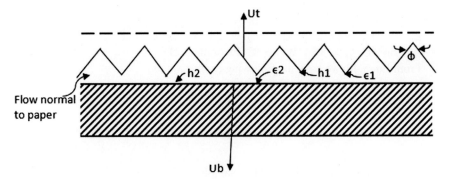

**Fig. 5.6** Flowing of air between corrugated absorber plate and insulated bottom

$$F' = F_0' \left[ 1 + \frac{\left(1 - F_0'\right)}{\dfrac{F_0'}{F_p} + \dfrac{Wh_1}{2W_2h_2F_f}} \right] \tag{5.40}$$

where, $F_0' = \dfrac{1}{1 + \dfrac{U_L}{h_1 + \left(\frac{1}{h_2} + \frac{1}{h_r}\right)^{-1}}} < 1$;

$F_p$ = fin efficiency of absorber flat plate = $\dfrac{h_{pf}}{(h_{pf} + U_L)}$, Eq. 5.6, same as flat plate
$<1$; $F_f$ = fin efficiency of fin = $F = \dfrac{\tanh m(W - D)/2}{m(W - D)/2}$ with $m^2 = \dfrac{U_L}{K\delta}$ (Eq. 4.15) and $h_{bc,c} = h_{pf} = h_{bf} = h_1 = h_2$(assumed), and it will depend on airflow velocity below absorber flat plate (Example 5.18).

If $W_2$ = thickness of fin = 0, this means flat absorber. In this case, the collector efficiency ($F'$) reduces to

$$F' = F_0' = \frac{1}{1 + \dfrac{U_L}{h_1 + \left(\frac{1}{h_2} + \frac{1}{h_r}\right)^{-1}}} < 1 \tag{5.41}$$

Equation 5.41 is exactly same as Eq. 5.38b for flat plate solar air collector with inverted absorber.

The expression for mass flow rate will remain same as given by Eq. 5.29 as

$$F_R = \frac{\dot{m}_f C_f}{A_c U_L}\left[1 - \exp\left(-\frac{F' A_c U_L}{\dot{m}_f C_f}\right)\right] \tag{5.42}$$

**Example 5.27** Evaluate solar air collector efficiency for single glazed flat plate with copper fin with effective area of (0.01 m × 0.008 m) attached at $W = 0.10$ m from center to center for air velocity of 0.25 m/s with effective area of 2 m² by using Eq. 5.37 with design parameters of Example 5.18.

**Solution**

From Example 5.18, we have the following data:

$h_{bc,c} = h_{pf} = h_{bf} = h_1 = h_2 = 2.15\,\text{W/m}^2\,°\text{C}$; and $U_L = 7.49 + 0.7 = 8.19\,\text{W/m}^2\,°\text{C}$; $h_r = 5.84\,\text{W/m}^2\,°\text{C}$ (Example 5.21).

From Eq. 5.40, we have the following expression as:

$$F' = F_0'\left[1 + \frac{(1 - F_0')}{\frac{F_0'}{F_p} + \frac{Wh_1}{2W_2 h_2 F_f}}\right], \text{ where } F_0' = \frac{1}{1 + \frac{U_L}{h_1 + \left(\frac{1}{h_2} + \frac{1}{h_r}\right)^{-1}}}$$

Here, $F_0' = \frac{1}{1 + \frac{U_L}{h_1 + \left(\frac{1}{h_2} + \frac{1}{h_r}\right)^{-1}}} = 0.31$ (Example 5.23); $F_p = \frac{h_{pf}}{(h_{pf} + U_L)} = \frac{2.15}{2.15 + 8.9} = 0.21$ and $F_f = \frac{\tanh m(W-D)/2}{m(W-D)/2} = 0.96$ (Eq. 4.15 and Example 4.23) assume in $W = 0.10$ m and $D = 10\,\text{mm} = 0.01$ m and $W_2 = 0.008\,\text{mm}$ (square fin); $D$ is breadth of fin in $m$.

Now,

$$F' = F_0'\left[1 + \frac{(1 - F_0')}{\frac{F_0'}{F_p} + \frac{Wh_1}{2W_2 h_2 F_f}}\right] = 0.31\left[1 + \frac{(1 - 0.31)}{\frac{0.31}{0.21} + \frac{0.10 \times 2.15}{2 \times 0.008 \times 2.15 \times 0.96}}\right]$$

$$= 0.31\left[1 + \frac{0.61}{1.43 + 6.51}\right] = 0.31[1 + 0.077] = 0.334$$

The value of solar collector efficiency $(F')$ is increased from 0.31 (Example 5.23) to 0.334 for single glazed solar air collector due to attachment of square (0.008 mm × 0.008 m) fin below absorber. The fin increases the surface area of heat transfer.

**Example 5.28** Repeat the Example 5.27 for copper fin breadth $(D)$ of 0.020 m.

## Solution

Given: $W = 0.10$ m; thickness of copper absorber $(\delta) = (0.35$ mm $= 0.00035$ m); $K = 386$ W/m °C; $U_L = 8.19$ W/m$^2$ °C with wind velocity of 1 m/s (Example 4.20); $D = 20$ mm $= 0.02$ m; and $d = 8$ mm $= 0.008$ m.

## Solution

From Eq. 4.15, one has expression for fin efficiency of flat plate collector as

$$F_f = \frac{\tanh m(W-D)/2}{m(W-D)/2} \text{ with } m^2 = \frac{U_L}{K\delta}.$$

Here $m = \sqrt{\frac{U_L}{K\delta}} = \sqrt{\frac{8.19}{386 \times 0.00035}} = \sqrt{\frac{8.19}{0.1351}} = \sqrt{\frac{8.19}{0.1351}} = \sqrt{60.62} = 7.79.$

Now,

$$F_f = \frac{\tanh m(W-D)/2}{m(W-D)/2} = \frac{\tanh \frac{7.79(0.10-0.02)}{2}}{\frac{7.79(0.10-0.02)}{2}}$$

$$= \frac{\tanh(0.3116)}{(0.3116)} = \frac{0.30189}{0.3116} = 0.95$$

Now,

$$F' = F'_0\left[1 + \frac{(1-F'_0)}{\frac{F'_0}{F_p} + \frac{Wh_1}{2W_2h_2F_f}}\right] \text{ with } F'_0 = 0.31 \text{ and } F_p = 0.21 \text{ (Example 5.27).}$$

So,

$$F' = 0.31\left[1 + \frac{(1-0.31)}{\frac{0.31}{0.21} + \frac{0.10 \times 2.15}{2 \times 0.008 \times 2.15 \times 0.95}}\right]$$

$$= 0.31\left[1 + \frac{0.61}{1.43 + 6.58}\right] = 0.31[1 + 0.076] = 0.334$$

There is not much change in $F'$ by increasing the breadth of fin from 0.01 to 0.02 m.

**Example 5.29** Evaluate solar air collector efficiency for single glazed flat plate with copper fin length equal to absorber plate length $(W = D)$ for air velocity of 0.25 m/s with effective area of 2 m$^2$ by using Eq. 5.37 with design parameters of Example 5.18.

## Solution

For, $W = D$, $F_f = \frac{\tanh m(W-D)/2}{m(W-D)/2} = 0$, then $F' = F'_0 = 0.31$.

The value of solar collector efficiency $(F')$ is exactly same as in the case of flat plate solar air collector. In this case, the fin surface area is equal to flat plate surface are as expected.

### 5.4.2   Corrugated Absorber

In this design, the surface area of absorber plate is increased by having absorber in corrugated shape as shown in Fig. 5.6. The surface area will depend on angle $\varphi$ between two inclined surfaces of absorber.

The expression for solar collector efficiency $(F')$ for corrugated shape absorber [2] is given by

$$F' = \frac{1}{1 + \frac{U_L}{\frac{h_1}{\sin\frac{\phi}{2}} + \left(\frac{1}{h_2} + \frac{1}{h_r}\right)^{-1}}} < 1 \tag{5.43}$$

where, $U_L = U_t + U_b$; $h_r = \varepsilon_{\text{eff}}\left(4\sigma\overline{T}\right)^3$ with $\overline{T} = \frac{T_1+T_2}{2}$ and $\varepsilon_{\text{eff}} = \left[\frac{1}{\varepsilon_1} + \frac{1}{\varepsilon_2} - 1\right]^{-1}$
(refer Eq. 2.17e).

For $\phi = 180°$, $\sin\frac{\phi}{2} = \sin90 = 1$, then
The collector efficiency $(F')$ reduces to

$$F' = \frac{1}{1 + \frac{U_L}{h_1 + \left(\frac{1}{h_2} + \frac{1}{h_r}\right)^{-1}}} \tag{5.44}$$

Equation 5.44 is exactly same as Eqs. 5.38b and 5.41 for flat plate solar air collector with inverted absorber.

**Example 5.30**  Evaluate solar air collector efficiency for single glazed flat plate with corrugated shape absorber plate length with $\phi = 90°$, for air velocity of 0.25 m/s with effective area of 2 m$^2$ by using Eq. 5.37 with design parameters of Example 5.18.

**Solution**

From Eq. 5.42, we have

$$F' = \frac{1}{1 + \frac{U_L}{\frac{h_1}{\sin\frac{\phi}{2}} + \left(\frac{1}{h_2} + \frac{1}{h_r}\right)^{-1}}} = \frac{1}{1 + \frac{8.19}{\frac{2.15}{\sin45} + \left(\frac{1}{2.15} + \frac{1}{5.84}\right)^{-1}}}$$

$$= \frac{1}{1 + \frac{8.19}{2.527+1.57}}$$

$$= \frac{1}{1 + \frac{8.19}{2.527+1.57}} = \frac{1}{1 + 1.999} = \frac{1}{2.99} = 0.334$$

The other parameters are same as in Example 5.27.

**Example 5.31**  Evaluate solar air collector efficiency for single glazed flat plate with corrugated shape absorber plate length with $\phi = 45°$, for air velocity of 0.25 m/s with effective area of 2 m$^2$ by using Eq. 5.37 with design parameters of Example 5.18.

**Solution**

From Eq. 5.42, we have

$$F' = \cfrac{1}{1 + \cfrac{U_L}{\frac{h_1}{\sin\frac{\phi}{2}} + \left(\frac{1}{h_2} + \frac{1}{h_r}\right)^{-1}}} = \cfrac{1}{1 + \cfrac{8.19}{\frac{2.15}{\sin22.5} + \left(\frac{1}{2.15} + \frac{1}{5.84}\right)^{-1}}}$$

$$= \cfrac{1}{1 + \cfrac{8.19}{4.4129 + 1.57}} = \frac{1}{1 + 1.3689} = \frac{1}{2.37} = 0.42$$

This indicates that solar air collector efficiency $(F')$ is increased due to reduction of angle, $(\phi)$, due to increased surface area of absorber.

Now, the outlet air temperature $(T_{fo})$, the rate of useful thermal energy $(\dot{Q}_u)$, mass flow rate factor $(F_R)$ and an instantaneous thermal efficiency $(\eta_i)$ for each case of Examples 5.27–5.31 can be evaluated for design and climatic parameter of Examples 5.25 and 5.26 for a given $(F'U_L)$.

## 5.5 The m-Solar Air Collector (m-SAC) Connected in Parallel

In this case, the inlet and outlet air of individual solar air collector connected in parallel will be same with same mass flow rate $(\dot{m}_f)$ of air as shown in Fig. 4.7 [m-flat plate liquid collector: m-FPC]. The m-solar air collector [m-SAC] connected in parallel, generally referred as one module, can be discussed in two ways, namely.

**Case (a):** If inlet mass flow rate $(\dot{m}_f)$ in individual solar air collector (SAC) is same, then the outlet air temperature will be exactly same for all SACs. But the mass flow rate at inlet of m-SAC should be $(m\dot{m}_f)$. The expression for evaluating outlet air temperature of individual solar air collector will be exactly same as Eq. 5.27 **as**

$$T_{fo} = \left[\left(\frac{\dot{q}_{ab}}{U_L}\right) + T_a\right]\left[1 - \exp\left(-\frac{F'A_cU_L}{\dot{m}_fC_f}\right)\right] + T_{fi}\exp\left(-\frac{F'A_cU_L}{\dot{m}_fC_f}\right) \quad (5.45)$$

The rate of useful energy of m-SAC for single glazed will be

$$\dot{Q}_{um} = m \times \dot{m}_fC_f(T_{fo} - T_{fi}) = F_RA_c[(\alpha_p\tau_g)I(t) - U_L(T_{fi} - T_a)] \quad (5.46)$$

**Case (b):** If inlet mass flow rate of m-SAC is $\dot{m}_f$ instead of $(m\dot{m}_f)$ as in case (a), then mass flow rate in individual SAC will be $\frac{\dot{m}_f}{m}$. The expression for outlet air temperature becomes as

$$T_{fom} = \left[\left(\frac{\dot{q}_{ab}}{U_L}\right) + T_a\right]\left[1 - \exp\left(-\frac{F'A_cU_L}{\frac{\dot{m}_f}{m} \times C_f}\right)\right] + T_{fi}\exp\left(-\frac{F'F'A_cU_L}{\frac{\dot{m}_f}{m} \times C_f}\right)$$

$$= \left[\left(\frac{\dot{q}_{ab}}{U_L}\right) + T_a\right]\left[1 - \exp\left(-\frac{F'(mA_c)U_L}{\dot{m}_fC_f}\right)\right] + T_{fi}\exp\left(-\frac{F'(mA_c)U_L}{\dot{m}_fC_f}\right)$$

$$(5.47a)$$

In this case, the outlet air temperature will be higher than case (a) due to reduction in mass flow rate.

$$T_{fom} = \left[\left(\frac{\dot{q}_{ab}}{U_L}\right) + T_a\right]\left[1 - \exp\left(-\frac{F'(mA_c)U_L}{\dot{m}_fC_f}\right)\right] + T_{fi}\exp\left(-\frac{F'(mA_c)U_L}{\dot{m}_fC_f}\right)$$

$$(5.47b)$$

The rate of useful energy of m-SAC for single glazed will be

$$\dot{Q}_{um} = m \times \frac{\dot{m}_f}{m} \times C_f(T_{fom} - T_{fi})$$

$$= F_{Rm}(mA_c)\left[(\alpha_p\tau_g)I(t) - U_L(T_{fi} - T_a)\right] \qquad (5.48)$$

where,

$$F_{Rm} = \frac{\frac{\dot{m}_f}{m} \times C_f}{A_cU_L}\left[1 - \exp\left(-\frac{F'A_cU_L}{\frac{\dot{m}_f}{m} \times C_f}\right)\right]$$

$$= \frac{\dot{m}_f \times C_f}{(mA_c)U_L}\left[1 - \exp\left(-\frac{F'(mA_c)U_L}{\dot{m}_f \times C_f}\right)\right] \qquad (5.49)$$

**Example 5.32** Evaluate the outlet temperature, $T_{fom}$, and mass flow rate, $F_{Rm}$, of single glazed solar air collector for air velocity of 0.25 m/s (0.029 kg s) and specific heat of 1005 j/kg °C flowing below absorber plate for data of Example 5.25 for case (b) of Sect. 5.5 for 5 ($m = 5$) solar air collector connected in parallel.

**Solution**

In this case, all five ($m = 5$) SACs connected in parallel.

From Eqs. 5.47a, 5.47b, we have an expression for an outlet air temperature at $m^{\text{th}}$ SAC as

$$T_{fom} = \left[\left(\frac{\dot{q}_{ab}}{U_L}\right) + T_a\right]\left[1 - \exp\left(-\frac{F'(mA_c)U_L}{\dot{m}_fC_f}\right)\right] + T_{fi}\exp\left(-\frac{F'(mA_c)U_L}{\dot{m}_fC_f}\right)$$

$$= \left[\left(\frac{0.81 \times 500}{10.00}\right) + 16\right]\left[1 - \exp\left(-\frac{0.22 \times 5 \times 2 \times 10.00}{0.029 \times 1005}\right)\right]$$

$$+ 17 \times \exp\left(-\frac{0.22 \times 5 \times 2 \times 10.00}{0.029 \times 1005}\right)$$

$$= 56.6 \times \left[1 - \exp(-0.7545)\right] + 17 \times \exp(-0.7545)$$
$$= 56.6 \times [1 - 0.4702] + 17 \times 0.4702 = 37.98\,°C$$

From Eq. 5.49, we have an expression for mass flow rate factor as

$$
F_{Rm} = \frac{\dot{m}_f C_f}{(m A_c) U_L}\left[1 - \exp\left(-\frac{F'(m A_c)U_L}{\dot{m}_f C_f}\right)\right]
$$
$$
= \frac{0.029 \times 1005}{5 \times 2 \times 10}\left[1 - \exp\left(-\frac{0.22 \times 5 \times 2 \times 10.00}{0.029 \times 1005}\right)\right]
$$
$$
= 0.29146 \times [1 - 0.4702] = 0.1544
$$

This shows that the outlet air temperature of SAC increases in comparison with case (a) [case (a) results is same as Example 5.25] as expected.

**Example 5.33** Calculate the rate of useful thermal energy $\left(\dot{Q}_{um}\right)$ and instantaneous thermal efficiency $(\eta_{im})$ of m-SAC connected in parallel for Example 5.32 for both case (a) and case (b), respectively.

**Solution**

**Case (a)**

By using Eq. 5.46, the rate of thermal energy is

**First method**

$$
\dot{Q}_{um} = m \times F_R A_c\left[(\alpha_p \tau_g)I(t) - U_L(T_{fi} - T_a)\right] = 5 \times 161 = 805\,W
$$

**Second method**

$$
\dot{Q}_{um} = m \times \dot{m}_f C_f\left(T_{fo} - T_{fi}\right) = 5 \times \dot{Q}_u = m \times \dot{m}_f C_f\left(T_{fo} - T_{fi}\right)
$$
$$
= 5 \times 161.72 = 808.6\,W \text{ (Example 5.26)}
$$

And we have from Eq. 5.31, we have

**First method**

$$
\eta_{im} = F_R\left[(\alpha_p \tau_g) - U_L\frac{(T_{fi} - T_a)}{I(t)}\right]
$$
$$
= 0.204\left[0.81 - 10 \times \frac{(17 - 16)}{500}\right] = 0.161
$$

**Second method**

$$
\eta_{im} = \frac{\dot{Q}_{um}}{m \times A_c \times I(t)} = \frac{805}{5 \times 2 \times 500} = 0.16
$$

**Case (b)**

From Eq. 5.48, we have the following.

**First method**

$$\dot{Q}_{um} = m \times \frac{\dot{m}_f}{m} \times C_f\left(T_{fom} - T_{fi}\right)$$
$$= F_{Rm}(m\,A_c)\left[\left(\alpha_p\tau_g\right)I(t) - U_L\left(T_{fi} - T_a\right)\right]$$
$$= 0.1544 \times 2 \times 5[405 - 10(17 - 16)] = 609.88 \text{ W}$$

**Second method**

$$\dot{Q}_{um} = m \times \frac{\dot{m}_f}{m} \times C_f\left(T_{fom} - T_{fi}\right) = 0.029 \times 1005(37.98 - 17) = 611.14$$

From Eq. 5.31, we have

**First method**

$$\eta_{im} = F_R\left[\left(\alpha_p\tau_g\right) - U_L\frac{\left(T_{fi} - T_a\right)}{I(t)}\right]$$
$$= 0.1544\left[0.81 - 10 \times \frac{(17 - 16)}{500}\right] = 0.122$$

**Second method**

$$\eta_{im} = \frac{\dot{Q}_{um}}{m \times A_c \times I(t)} = \frac{611}{5 \times 2 \times 500} = 0.122$$

It is to be noted that in case (b), the rate of useful thermal energy and instantaneous thermal efficiency is less due to high operating temperature of air.

## 5.6   The N-Solar Air Collector (N-SAC) Connected in Series

In this case, the outlet of first solar air collector (SAC), $T_{fo1}$, is connected to the inlet of second solar air collector, $T_{fi2}$, and the outlet of second SAC, $T_{fo1}$, is connected to inlet of third SAC, $T_{fi3}$, and it continues up to $N^{\text{th}}$ SAC as discussed in Sect 4.5.1 of Chap. 4. By using the same methodology of Sect 4.5.1, one can get the exact same equation for the outlet air temperature at $N^{\text{th}}$ solar air collector $\left(T_{foN}\right)$ as

$$T_{foN} = \left[\left(\frac{\dot{q}_{ab}}{U_L}\right) + T_a\right]\left[1 - \exp\left[-N\,A_c U_L F'/\left(\dot{m}_f C_f\right)\right]\right]$$

$$+ T_{fi} \exp\left[-N A_c U_L F'/(\dot{m}_f C_f)\right] \tag{5.50}$$

where the numerical values of $U_L$, $F'$, $\dot{m}_f$, and $C_f$ will be different due to air as a working fluid.

The rate of thermal energy at $N^{\text{th}}$ solar air collector, $(T_{foN})$, will be given by

$$\dot{Q}_{uN} = \dot{m}_f C_f (T_{foN} - T_{fi}) = F_{RN} \times N A_c \left[\dot{q}_{ab} - U_L (T_{fi} - T_a)\right] \tag{5.51}$$

where,

$$F_{RN} = \frac{\dot{m}_f C_f}{N A_c U_L} \left[1 - \exp\left[-N A_c U_L F'/(\dot{m}_f C_f)\right]\right] \tag{5.52}$$

and, an instantaneous thermal efficiency, $\eta_i$, is given by

$$\eta_i = \frac{\dot{Q}_{uN}}{N \times A_c \times I(t)} = F_{RN} \left[(\alpha_p \tau_g) - U_L \frac{(T_{fi} - T_a)}{I(t)}\right] \tag{5.53}$$

**Example 5.34** Evaluate the outlet temperature, $T_{foN}$, and mass flow rate, $F_{RN;}$, of glazed solar air collector for air velocity of 0.25 m/s (0.029 kg.s) and specific heat of 1005 j/kg °C flowing below absorber plate for data of Example 5.25 for five solar air collectors connected in series.

**Solution**

From Eq. 5.50, we have an expression for an outlet air temperature as

$$
\begin{aligned}
T_{fo5} &= \left[\left(\frac{\dot{q}_{ab}}{U_L}\right) + T_a\right]\left[1 - \exp\left(-\frac{F' N A_c U_L}{\dot{m}_f C_f}\right)\right] + T_{fi} \exp\left(-\frac{F' N A_c U_L}{\dot{m}_f C_f}\right) \\
&= \left[\left(\frac{0.81 \times 500}{10.00}\right) + 16\right]\left[1 - \exp\left(-\frac{0.22 \times 2 \times 5 \times 10.00}{0.029 \times 1005}\right)\right] \\
&\quad + 17 \times \exp\left(-\frac{0.22 \times 2 \times 5 \times 10.00}{0.029 \times 1005}\right) \\
&= 56.6 \times \left[1 - \exp(-0.7545)\right] + 17 \times \exp(-0.7545) \\
&= 56.6 \times [1 - 0.4702] + 17 \times 0.4702 = 37.98\,°C
\end{aligned}
$$

From Eq. 5.29, we have an expression for mass flow rate factor as

$$
\begin{aligned}
F_{RN} &= \frac{\dot{m}_f C_f}{N A_c U_L}\left[1 - \exp\left(-\frac{F' N A_c U_L}{\dot{m}_f C_f}\right)\right] \\
&= \frac{0.029 \times 1005}{2 \times 5 \times 10}\left[1 - \exp\left(-\frac{0.22 \times 2 \times 5 \times 10.00}{0.029 \times 1005}\right)\right] \\
&== 0.29146 \times [1 - 0.4702] = 0.1544
\end{aligned}
$$

**Example 5.35** Calculate the rate of useful thermal energy $(\dot{Q}_{uN})$ and instantaneous thermal efficiency $(\eta_{iN})$ of N-SAC connected in series for Example 5.32 for both case (a) and case (b), respectively.

From Eq. 5.51, we have the following:

**First method**

$$\dot{Q}_{uN} = \dot{m}_f \times C_f (T_{foN} - T_{fi}) = F_{RN}(NA_c)[(\alpha_p \tau_g)I(t) - U_L(T_{fi} - T_a)]$$
$$= 0.1544 \times 2 \times 5[405 - 10(17 - 16)] = 609.88\,\text{W}$$

**Second method**

$$\dot{Q}_{uN} = \dot{m}_f \times C_f (T_{foN} - T_{fi}) = 0.029 \times 1005(37.98 - 17) = 611.14$$

From Eq. 5.53, we have

**First method**

$$\eta_{iN} = F_{RN}\left[(\alpha_p \tau_g) - U_L \frac{(T_{fi} - T_a)}{I(t)}\right]$$
$$= 0.1544\left[0.81 - 10 \times \frac{(17 - 16)}{500}\right] = 0.122$$

**Second method**

$$\eta_{iN} = \frac{\dot{Q}_{uN}}{N \times A_c \times I(t)} = \frac{611}{5 \times 2 \times 500} = 0.122$$

It is very important to mention here that case (b) of 5-SAC connected in parallel is exactly same as 5-SAC connected in series.

## 5.7   The N-[m-SAC)] Connected in Series, Fig. 4.8

Here, m-SAC connected in parallel with inlet mass flow rate $(\dot{m}_f)$ [case (b) of Sect. 5.5] of single glazed solar air collector will be considered as one module and if $N$ modules (m-SAC) are connected in series will be referred as one array. Following Sect 4.5.2 for flat plate liquid collector, the outlet air temperature at $N^{\text{th}}$ m-SAC [end of one array of solar air collectors] can be written as

$$T_{fom \times N} = \left[\left(\frac{\dot{q}_{ab}}{U_L}\right) + T_a\right][1 - \exp[-NA_{cm}U_LF'/(\dot{m}_fC_f)]]$$

$$+ T_{fi} \exp\left[-NA_{cm}U_L F'/(\dot{m}_f C_f)\right] \tag{5.54}$$

The rate of thermal energy at end of array will be determined as follows:

$$\dot{Q}_{u-m\times N} = \dot{m}_f C_f \left(T_{fom\times N} - T_{fi}\right) = F_{R-m\times N}\left[\dot{q}_{ab} - U_L\left(T_{fi} - T_a\right)\right] \tag{5.55}$$

where,

$$F_{R-m\times N} = \frac{\dot{m}_f C_f}{NA_{cm}U_L}\left[1 - \exp\left[-NA_{cm}U_L F'/(\dot{m}_f C_f)\right]\right] \tag{5.56}$$

An instantaneous thermal efficiency of array will be determined by

$$\eta_{i-m\times N} = \frac{\dot{Q}_{u-m\times N}}{NA_{cm}I(t)} = F_{R-m\times N}\left[(\alpha_p\tau_g) - U_L\frac{(T_{fi} - T_a)}{I(t)}\right] \tag{5.57}$$

**This array will be referred as parallel and series combination**

**Example 5.36** Evaluate the outlet temperature, $T_{fo-m\times N}$, and mass flow rate, $F_{RN;}$, of single glazed solar air collector of one array for air velocity of 0.25 m/a (0.029 kg s) and specific heat of 1005 j/kg °C flowing below absorber plate for data of Example 5.34 for 5 × 5 parallel and series-connected solar air collector.

**Solution**

From Eq. 5.54, we have an expression for an outlet air temperature as

$$
\begin{aligned}
T_{fo-m\times N} &= \left[\left(\frac{\dot{q}_{ab}}{U_L}\right) + T_a\right]\left[1 - \exp\left(-\frac{F'N(mA_c)U_L}{\dot{m}_f C_f}\right)\right] \\
&\quad + T_{fi}\exp\left(-\frac{F'N(mA_c)U_L}{\dot{m}_f C_f}\right) \\
&= \left[\left(\frac{0.81 \times 500}{10.00}\right) + 16\right]\left[1 - \exp\left(-\frac{0.22 \times 2 \times 5 \times 5 \times 10.00}{0.029 \times 1005}\right)\right] \\
&\quad + 17 \times \exp\left(-\frac{0.22 \times 2 \times 5 \times 5 \times 10.000}{0.029 \times 1005}\right) \\
&= 56.6 \times \left[1 - \exp(-3.7725)\right] + 17 \times \exp(-3.7725) \\
&= 56.6 \times [1 - 0.023] + 17 \times 0.023 = 55.10\,°C
\end{aligned}
$$

From Eq. 5.56, we have an expression for mass flow rate factor as

$$
\begin{aligned}
F_{R-m\times N} &= \frac{\dot{m}_f C_f}{N(mA_c)U_L}\left[1 - \exp\left(-\frac{F'N(mA_c)U_L}{\dot{m}_f C_f}\right)\right] \\
&= \frac{0.029 \times 1005}{2 \times 5 \times 5 \times 10}\left[1 - \exp\left(-\frac{0.22 \times 2 \times 5 \times 5 \times 10}{0.029 \times 1005}\right)\right]
\end{aligned}
$$

$$= 0.0583 \times [1 - 0.023] = 0.057$$

**Example 5.37** Calculate the rate of useful thermal energy $\left(\dot{Q}_{u-m\times N}\right)$ and instantaneous thermal efficiency $\left(\eta_{i-m\times N}\right)$ of N-(m-SAC) connected in series for Example 5.37 for case (b) only.

**Solution**

From Eq. 5.51, we have the following:

**First method**

$$\dot{Q}_{u-m\times N} = \dot{m}_f \times C_f\left(T_{fo-m\times N} - T_{fi}\right)$$
$$= F_{R-m\times N}(mN)A_c\left[(\alpha_p\tau_g)I(t) - U_L\left(T_{fi} - T_a\right)\right]$$
$$= 0.057 \times 2 \times 5 \times 5[405 - 10(17 - 16)] = 1125.75\,\text{W}$$

**Second method**

$$\dot{Q}_{u-m\times N} = \dot{m}_f \times C_f\left(T_{foN} - T_{fi}\right) = 0.029 \times 1005(55.10 - 17) = 1110.32$$

From Eq. 5.53, we have

**First method**

$$\eta_{i-m\times N} = F_{RN}\left[(\alpha_p\tau_g) - U_L\frac{\left(T_{fi} - T_a\right)}{I(t)}\right]$$

$$= 0.057\left[0.81 - 10 \times \frac{(17 - 16)}{500}\right] = 0.045$$

**Second method**

$$\eta_{i-m\times N} = \frac{\dot{Q}_{u-m\times N}}{N \times A_c \times I(t)} = \frac{1107.80}{5 \times 2 \times 5 \times 500} = 0.0443$$

## 5.8   The m-(N-solar Air Collectors: SAC) Connected in Parallel

Referring to Fig. 5.7, the N-solar air collectors are connected in series and it will be considered one N-SAC module. Further, m (N-SAC) modules are connected in parallel which will be known as one array.

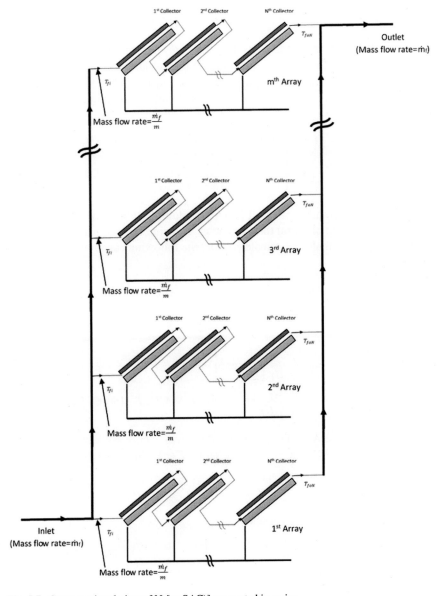

**Fig. 5.7** Cross-sectional view of N-[m-SAC)] connected in series

From Eq. 5.50, one can write an expression for the outlet air temperature $(T_{foN})$ of one module (N-SAC connected in series) as

$$T_{foN} = \left[\left(\frac{\dot{q}_{ab}}{U_L}\right) + T_a\right]\left[1 - \exp\left[-NA_cU_LF'/(\dot{m}_fC_f)\right]\right]$$

$$+ T_{fi} \exp\left[-NA_cU_LF'/\left(\dot{m}_fC_f\right)\right]$$

In the above equation, the mass flow rate $(\dot{m}_f)$ is same throughout N-SAC. However, the mass flow rate $(\dot{m}_f)$ entering at inlet of array is equally divided into $\frac{\dot{m}_f}{m}$ for each N-FPC connected in series as shown in Fig. 5.7. Then, the outlet air temperature at $N$th solar air collector (Sect 4.5.7) will be as follows:

$$T_{foN \times m} = \left[\left(\frac{\dot{q}_{ab}}{U_L}\right) + T_a\right]\left[1 - \exp\left[-NA_cU_LF'/\left(\frac{\dot{m}_f}{m}C_f\right)\right]\right]$$
$$+ T_{fi} \exp\left[-NA_cU_LF'/\left(\frac{\dot{m}_f}{m}C_f\right)\right] \tag{5.58}$$

The outlet air temperature of all N-SAC modules will be same and determined by Eq. 5.58. The rate of useful thermal energy available for each row/one module of (N-SAC) $(\dot{Q}_{uN})$ will also be calculated from the following equation:

$$\dot{Q}_{uN \times m} = \frac{\dot{m}_f}{m}C_f\left(T_{foN} - T_{fi}\right) = F_{RN} \times NA_c\left[\dot{q}_{ab} - U_L\left(T_{fi} - T_a\right)\right] \tag{5.59}$$

where,

$$F_{RN \times m} = \frac{\frac{\dot{m}_f}{m}C_f}{NA_cU_L}\left[1 - \exp\left[-NA_cU_LF'/\left(\frac{\dot{m}_f}{m}C_f\right)\right]\right]$$
$$= \frac{\dot{m}_fC_f}{mNA_cU_L}\left[1 - \exp\left[-mNA_cU_LF'/\left(\dot{m}_fC_f\right)\right]\right] \tag{5.60}$$

For m parallel of N-SAC, the rate of useful thermal energy available for each row/one module of (N-SAC) $(\dot{Q}_{u-N \times m})$ with same $(F_{RN})$ will be as follows:

$$\dot{Q}_{u-N \times m} = F_{RN} \times mNA_c\left[\dot{q}_{ab} - U_L\left(T_{fi} - T_a\right)\right] \tag{5.61}$$

The instantaneous thermal efficiency

$$\eta_{i-N \times m} = \frac{\dot{Q}_{u-N \times m}}{N \times m \times A_c \times I(t)} = F_{RN}\left[(\alpha_p\tau_g) - -U_L\frac{\left(T_{fi} - T_a\right)}{I(t)}\right] \tag{5.62}$$

**Example 5.38** Evaluate the outlet temperature, $T_{fo-N \times m}$, and mass flow rate factor, $F_{R-N \times m;}$, of single glazed solar air collector of one array for air velocity of 0.25 m/a (0,029 kg s) and specific heat of 1005 j/kg °C flowing below absorber plate for data of Example 5.34 for 5 × 5 series and parallel [$N \times m$] connected solar air collector.

**Solution**

From Eq. 5.58, we have an expression for an outlet air temperature at individual N-SAC connected in series as

$$T_{foN \times m} = \left[ \left( \frac{\dot{q}_{ab}}{U_L} \right) + T_a \right] \left[ 1 - \exp\left( -\frac{F'(NA_c)U_L}{\dot{m}_f C_f} \right) \right] + T_{fi} \exp\left( -\frac{F'(NA_c)U_L}{\dot{m}_f C_f} \right)$$

$$= \left[ \left( \frac{0.81 \times 500}{10.00} \right) + 16 \right] \left[ 1 - \exp\left( -\frac{0.22 \times 5 \times 2 \times 10.00}{0.029 \times 1005} \right) \right]$$

$$+ 17 \times \exp\left( -\frac{0.22 \times 5 \times 2 \times 10.00}{0.029 \times 1005} \right)$$

$$= 56.6 \times \left[ 1 - \exp(-0.7545) \right] + 17 \times \exp(-0.7545)$$

$$= 56.6 \times [1 - 0.4702] + 17 \times 0.4702 = 37.98 \,^{\circ}\text{C}$$

From Eq. 5.49, we have an expression for mass flow rate factor for N-SAC connected in series as

$$F_{RN \times m} = \frac{\dot{m}_f C_f}{(NA_c)U_L} \left[ 1 - \exp\left( -\frac{F'(NA_c)U_L}{\dot{m}_f C_f} \right) \right]$$

$$= \frac{0.029 \times 1005}{5 \times 2 \times 10} \left[ 1 - \exp\left( -\frac{0.22 \times 5 \times 2 \times 10.00}{0.029 \times 1005} \right) \right]$$

$$= 0.29146 \times [1 - 0.4702] = 0.1544$$

The outlet air temperature $(T_{foN})$ at $N$th SAC connected in series and its mass flow rate factor $(F_{RN})$ will be same for each N-SAC connected in series for each row.

**Example 5.39** Calculate the rate of useful thermal energy $(\dot{Q}_{u-N \times m})$ and instantaneous thermal efficiency $(\eta_{i-N \times m})$ of m-(N-SAC) connected in parallel for Example 5.38.

**Solution**

From Eq. 5.61, we have the following:

**First method**

$$\dot{Q}_{u-N \times m} = F_{RN}(mA_c)\left[ (\alpha_p \tau_g) I(t) - U_L \left( T_{fi} - T_a \right) \right]$$

$$= 0.1544 \times 2 \times 5[405 - 10(17 - 16)] = 609.88 \,\text{W}$$

**Second method**

$$\dot{Q}_{um} = m \times \frac{\dot{m}_f}{m} \times C_f \left( T_{fom} - T_{fi} \right) = 0.029 \times 1005(37.98 - 17) = 611.14$$

From Eq. 5.62, we have

**First method**

$$\eta_{i-N\times m} = F_{RN}\left[\left(\alpha_p\tau_g\right) - U_L\frac{\left(T_{fi} - T_a\right)}{I(t)}\right]$$

$$= 0.1544\left[0.81 - 10 \times \frac{(17 - 16)}{500}\right] = 0.122$$

**Second method**

$$\eta_{i-N\times m} = \frac{\dot{Q}_{u-N\times m}}{N \times A_c \times I(t)} = \frac{611}{5 \times 2 \times 500} = 0.122$$

Here '$m$' stands for number of module/collectors connected in parallel and '$N$' stands for number of module/collectors connected in series. Further, it is very important to note that

(a) The results for the outlet air temperature $\left(T_{fo}\right)$, the rate of useful thermal energy $\left(\dot{Q}_u\right)$, mass flow rate factor $(F_R)$ and an instantaneous thermal efficiency $(\eta_i)$ for parallel and series connection $(m \times N)$ (Examples 5.36 and 5.37) are better than series and parallel $(N \times m)$ connection (Examples 5.38 and 5.39) for given mass flow rate.

(b) The results for the outlet air temperature $\left(T_{fo}\right)$, the rate of useful thermal energy $\left(\dot{Q}_u\right)$, mass flow rate factor $(F_R)$ and an instantaneous thermal efficiency $(\eta_i)$ for N-solar air collectors (N-SAC) (Examples 5.34 and 5.35) are exactly same as the results for parallel and series connections [m-(N-SAC)] (Examples 5.36 and 5.37) with higher results in comparison with either parallel or series connection only.

## 5.9   Threshold Intensity of Solar Air Collector

It is a minimum value of solar energy/radiation after that value of the system (solar air collector) starts functioning in terms of net gain of thermal energy.

(a) **Threshold intensity for single glazed solar air collector**

**From Eq. 5.14a, we have**

$$\dot{Q}_u = F_R A_c\left[\left(\alpha_p\tau_g\right)I(t) - U_L\left(T_{fi} - T_a\right)\right] > 0.$$

$$\text{This gives, } I(t)_{\text{threesold}} \geq \frac{U_L\left(T_{fi} - T_a\right)}{\left(\alpha_p\tau_g\right)} \tag{5.63}$$

(b) **Threshold intensity for double glazed solar air collector**

$$\dot{Q}_u = F_R A_c \left[ (\alpha_p \tau_g^2) I(t) - U_L (T_{fi} - T_a) \right] > 0;$$

$$\text{This gives, } I(t)_{\text{threesold}} \geq \frac{U_L (T_{fi} - T_a)}{(\alpha_p \tau_g^2)} \tag{5.64}$$

The threshold mainly depends on an overall heat transfer coefficient, $U_L$ and $(\alpha_p \tau_g)$. The value of $U_L$ will be different for single and double glazed solar air collectors. Threshold solar intensity provides information about net energy obtained by solar air collector at given fluid inlet air temperature, $T_{fi}$ and ambient air temperature $(T_a)$.

Equations 5.63 and 5.64 are exactly same as threshold intensity for flat plate liquid collector, Eq. 4.18.

**Example 5.40** Find out threshold solar intensity for single and double glazed flat plate collectors (FPCs) for Example 4.29. and $\alpha_p = \tau_g = 0.9$ and $T_{fi} = 17\,°C$ and $T_a = 15\,°C$ (winter condition).

**Solution**

**For single glazed FPC:** $U_L = 8.19\,W/m^2\,°C$ (Example 4.20).

From Eq. 5.63, we have

$$I(t)_{\text{threesold}} \geq \frac{U_L (T_{fi} - T_a)}{\alpha_p \tau_g} = \frac{8.19 \times (17 - 16)}{0.9 \times 0.9} = \frac{262.08}{0.81} = 10\,W/m^2$$

**For double glazed FPC:** $U_L = 3.89\,W/m^2\,°C$ (Example 4.20).

From Eq. 4.18b, we have

$$I(t)_{\text{threesold}} \geq \frac{U_L (T_{fi} - T_a)}{(\alpha_p \tau_g^2)} = \frac{3.89 \times (17 - 16)}{0.9 \times 0.9 \times 0.9} = \frac{136.15}{0.729} = 5.33\,W/m^2$$

This example shows that the threshold solar intensity is significantly lowered by 46.7% for double glazed FPC for the same design and operating parameters.

**Example 5.41 Repeat Example 5.40 for** $T_{fi} = 25\,°C$ and $T_a = 24\,°C$ (winter condition).

**Solution**

**For single glazed FPC:** $U_L = 8.19\,W/m^2\,°C$ (Example 4.20).

From Eq. 5.63, we have

$$I(t)_{\text{threesold}} \geq \frac{U_L (T_{fi} - T_a)}{\alpha_p \tau_g} = \frac{8.19 \times (25 - 24)}{0.9 \times 0.9} = \frac{262.08}{0.81} = 10\,W/m^2$$

**For double glazed FPC:** $U_L = 3.89\,\text{W/m}^2\,°\text{C}$ (Example 4.20).

From Eq. 4.18b, we have

$$I(t)_{\text{threesold}} \geq \frac{U_L(T_{fi} - T_a)}{(\alpha_p \tau_g^2)} = \frac{3.89 \times (25 - 24)}{0.9 \times 0.9 \times 0.9} = \frac{136.15}{0.729} = 5.33\,\text{W/m}^2$$

This example shows that the threshold solar intensity only depends on design parameter and $(T_{fi} - T_a)$.

## 5.10  Effect of Mass Flow Rate on Performance of Parallel (m-Number) and Series (N-Number) Connection of Solar Air Collector

As discussed earlier that the outlet air temperature of parallel and series ($m \times N$) connection [m-(N-SAC)] is higher (Examples 5.36 and 5.37) than in comparison with other combination, and hence, we will discuss the effect of mass flow rate on the performance of parallel and series connections [m-(N-SAC)]. Further, there are mainly two applications of solar air heater (SAC), namely solar crop drying and space heating. The air temperature in solar crop drying needs higher temperature than direct space heating. So in this section, we will discuss effect of mass flow rate on performance of parallel and series connection [m-(N-SAC)] combination through following examples.

**Example 5.42** Evaluate the outlet temperature, $T_{fo-m \times N}$, and mass flow rate, $F_{RN;}$, of single glazed solar air collector of one array for air velocity of 0.50 m/a (0,029 kg.s) and specific heat of 1005 j/kg °C flowing below absorber plate for data of Example 5.34 for $5 \times 5$ ($m \times N$) parallel and series-connected solar air collector.

**Solution**

Here, $\dot{m}_f = 1\,\text{m} \times 0.10\,\text{m} \times 0.5\,\frac{\text{m}}{\text{s}} \times 1.17\,\frac{\text{kg}}{\text{m}^3} = 0.058\,\text{kg/s}$

From Eq. 5.54, we have an expression for an outlet air temperature as

$$T_{fo-m \times N} = \left[\left(\frac{\dot{q}_{ab}}{U_L}\right) + T_a\right]\left[1 - \exp\left(-\frac{F'N(mA_c)U_L}{\dot{m}_f C_f}\right)\right]$$

$$+ T_{fi} \exp\left(-\frac{F'N(mA_c)U_L}{\dot{m}_f C_f}\right)$$

$$= \left[\left(\frac{0.81 \times 500}{10.00}\right) + 16\right]\left[1 - \exp\left(-\frac{0.22 \times 2 \times 5 \times 5 \times 10.00}{0.058 \times 1005}\right)\right] + 17$$

$$\times \exp\left(-\frac{0.22 \times 2 \times 5 \times 5 \times 10.000}{0.058 \times 1005}\right)$$

$$= 56.6 \times \left[1 - \exp(-1.99625)\right] + 17 \times \exp(-1.99625)$$

$$= 56.6 \times [1 - 0.1358] + 17 \times 0.1358 = 51.22\,^\circ C$$

From Eq. 5.56, we have an expression for mass flow rate factor as

$$
\begin{aligned}
F_{R-m \times N} &= \frac{\dot{m}_f C_f}{N(m A_c) U_L}\left[1 - \exp\left(-\frac{F'N(m A_c)U_L}{\dot{m}_f C_f}\right)\right] \\
&= \frac{0.058 \times 1005}{2 \times 5 \times 5 \times 10}\left[1 - \exp\left(-\frac{0.22 \times 2 \times 5 \times 5 \times 10}{0.058 \times 1005}\right)\right] \\
&= 0.1166 \times [1 - 0.1358] = 0.1007
\end{aligned}
$$

**Example 5.43** Calculate the rate of useful thermal energy $\left(\dot{Q}_{u-m \times N}\right)$ and instantaneous thermal efficiency $(\eta_{i-m \times N})$ of N-(m-SAC) connected in series for Example 5.37 for case (b) only.

**Solution**

From Eq. 5.51, we have the following

**First method**

$$
\begin{aligned}
\dot{Q}_{u-m \times N} &= \dot{m}_f \times C_f\left(T_{fo-m \times N} - T_{fi}\right) \\
&= F_{R-m \times N}(m N) A_c\left[(\alpha_p \tau_g)I(t) - U_L\left(T_{fi} - T_a\right)\right] \\
&= 0.1007 \times 2 \times 5 \times 5[405 - 10(17 - 16)] = 1989\,W
\end{aligned}
$$

**Second method**

$$\dot{Q}_{u-m \times N} = \dot{m}_f \times C_f\left(T_{foN} - T_{fi}\right) = 0.058 \times 1005(51.22 - 17) = 1994.68\,W$$

From Eq. 5.53, we have

**First method**

$$
\begin{aligned}
\eta_{im \times N} &= F_{RN}\left[(\alpha_p \tau_g) - U_L\frac{\left(T_{fi} - T_a\right)}{I(t)}\right] \\
&= 0.1007\left[0.81 - 10 \times \frac{(17 - 16)}{500}\right] = 0.079
\end{aligned}
$$

**Second method:**

$$\eta_{i-m \times N} = \frac{\dot{Q}_{u-m \times N}}{N \times A_c \times I(t)} = \frac{1994.68}{5 \times 2 \times 5 \times 500} = 0.079$$

Now, one can observe that if mass flow rate is increased from 0.029 to 0.058 kg/s, there is drop of 3.88 °C and the rate of useful thermal energy and instantaneous thermal efficiency is nearly doubled.

**Example 5.44** Evaluate the outlet air temperature, $T_{fo}$, and mass flow rate factor, $F_R$, with constant mass airflow rate $\left(\dot{m}_f = 0.029\,\text{kg/s}\right)$ for single glazed solar air collector (SAC). The design and climatic parameters of Example 5.32 for a given number of SAC can be considered $[F' = 0.22, U_L = 10\,\text{W/m}^2\,°\text{C}, I(t) = 500\,\text{W/m}^2, T_a = 16\,°\text{C}$ and $T_{fi} = 17\,°\text{C}$ and $\alpha_p = \tau_g = 0.9]$. The following conditions are given:

(a) Six SACs are connected in parallel.
(b) Six SACs are connected in series.
(c) Two SACs are in parallel and three m-SAC connected in series.
(d) Three SACs are connected in series and two N-SAC are connected in parallel.

**Solution**

**Case (a):** Six SACs are connected in parallel
From Eqs. 5.47a, 5.47b, we have an expression for an outlet air temperature at $m^{\text{th}}$ SAC as

$$
\begin{aligned}
T_{fom} &= \left[\left(\frac{\dot{q}_{ab}}{U_L}\right) + T_a\right]\left[1 - \exp\left(-\frac{F'(mA_c)U_L}{\dot{m}_f C_f}\right)\right] + T_{fi}\exp\left(-\frac{F'(mA_c)U_L}{\dot{m}_f C_f}\right) \\
&= \left[\left(\frac{0.81 \times 500}{10.00}\right) + 16\right]\left[1 - \exp\left(-\frac{0.22 \times 6 \times 2 \times 10.00}{0.029 \times 1005}\right)\right] \\
&\quad + 17 \times \exp\left(-\frac{0.22 \times 6 \times 2 \times 10.00}{0.029 \times 1005}\right) \\
&= 56.6 \times \left[1 - \exp(-0.9058)\right] + 17 \times \exp(-0.9058) \\
&= 56.6 \times [1 - 0.4042] + 17 \times 0.4042 = 40.59\,°\text{C}
\end{aligned}
$$

From Eq. 5.49, we have an expression for mass flow rate factor as

$$
\begin{aligned}
F_{Rm} &= \frac{\dot{m}_f C_f}{(mA_c)U_L}\left[1 - \exp\left(-\frac{F'(mA_c)U_L}{\dot{m}_f C_f}\right)\right] \\
&= \frac{0.029 \times 1005}{6 \times 2 \times 10}\left[1 - \exp\left(-\frac{0.22 \times 6 \times 2 \times 10.00}{0.029 \times 1005}\right)\right] \\
&= 0.2429 \times [1 - 0.4042] = 0.1447
\end{aligned}
$$

**Case (b): Six ($n = 6$) SACs are connected in series**

According to Example 5.34, the values of $T_{foN}$ and $F_{RN}$ will be exactly same as done in case (a).

**Case (c): two ($m = 2$) sacs are in parallel and three ($n = 3$) m-SACs are connected in series**

From Eq. 5.54, we have an expression for an outlet air temperature as

$$
\begin{aligned}
T_{fo-m \times N} &= \left[ \left( \frac{\dot{q}_{ab}}{U_L} \right) + T_a \right] \left[ 1 - \exp\left( -\frac{F'N(mA_c)U_L}{\dot{m}_f C_f} \right) \right] \\
&\quad + T_{fi} \exp\left( -\frac{F'N(mA_c)U_L}{\dot{m}_f C_f} \right) \\
&= \left[ \left( \frac{0.81 \times 500}{10.00} \right) + 16 \right] \left[ 1 - \exp\left( -\frac{0.22 \times 2 \times 2 \times 3 \times 10.00}{0.029 \times 1005} \right) \right] \\
&\quad + 17 \times \exp\left( -\frac{0.22 \times 2 \times 2 \times 3 \times 10.000}{0.029 \times 1005} \right) \\
&= 56.6 \times \left[ 1 - \exp(-0.9058) \right] + 17 \times \exp(-0.9058) \\
&= 56.6 \times [1 - 0.4042] + 17 \times 0.4042 = 40.59\,^{\circ}\mathrm{C}
\end{aligned}
$$

From Eq. 5.56, we have an expression for mass flow rate factor as

$$
\begin{aligned}
F_{R-m \times N} &= \frac{\dot{m}_f C_f}{N(mA_c)U_L} \left[ 1 - \exp\left( -\frac{F'N(mA_c)U_L}{\dot{m}_f C_f} \right) \right] \\
&= \frac{0.029 \times 1005}{2 \times 5 \times 5 \times 10} \left[ 1 - \exp\left( -\frac{0.22 \times 2 \times 5 \times 5 \times 10}{0.029 \times 1005} \right) \right] \\
&= 0.0583 \times [1 - 0.023] = 0.057
\end{aligned}
$$

**Case (d): Three ($n = 3$) SACs are connected in series and two ($m = 2$) N-SACs are connected in parallel**

In this case too, the results will remain same.

**Finally, one can conclude that there is no effect of parallel, series, parallel and series and series and parallel combinations on the outlet air temperature and mass flow rate factor for a given number of solar air collectors.**

## 5.11 Constant Collection Temperature (CCT)

It has been seen earlier that solar air collectors are working in a constant mass flow rate condition generally known as constant mass flow rate mode. In this case, the outlet air temperature varies with time of the day due to hourly variation in solar radiation and ambient air temperature. However, in constant collection temperature mode, there will be hourly variation of mass flow rate unlike constant mass flow rate. It is further to be noted that hot air at constant collection mode has same application as constants mass flow rate condition. Because, there are some industries like crop/vegetables' drying industries which are very sensitive to temperature in dying process. In such case, solar air heaters should be analyzed under constant collection temperature ($T_{00}$).

### 5.11.1   Solar Air Collectors Connected in Parallel

Following Eq. 5.47b, the outlet air temperature of m-solar air collectors connected in parallel under constant mass flow rate condition can be written as follows:

$$T_{fom} = \left[\left(\frac{\dot{q}_{ab}}{U_L}\right) + T_a\right]\left[1 - \exp\left(-\frac{F'(mA_c)U_L}{\dot{m}_f C_f}\right)\right] + T_{fi}\exp\left(-\frac{F'(mA_c)U_L}{\dot{m}_f C_f}\right)$$

For constant collector temperature (CCT) mode, the above equation can be written as

$$T_{00} = T_{fom} = \left[\left(\frac{\dot{q}_{ab}}{U_L}\right) + T_a\right]\left[1 - \exp\left(-\frac{F'(mA_c)U_L}{\dot{m}_f(t) \times C_f}\right)\right]$$
$$+ T_{fi}\exp\left(-\frac{F'(mA_c)U_L}{\dot{m}_f(t) \times C_f}\right) \tag{5.65}$$

In this case, mass flow rate $(\dot{m}_f)$ will vary with time of the day, and hence, an analytical expression should be obtained in terms of constant collection temperature ($T_{00}$), solar intensity [$I(t)$] and ambient air temperature ($T_a$).

From Eq. 5.65, one can have the following after algebraic manipulation:

$$\frac{T_{00} - T_{fi}}{\left[\left(\frac{\dot{q}_{ab}}{U_L}\right) + T_a - T_{fi}\right]} = \left[1 - \exp\left(-\frac{F'(mA_c)U_L}{\dot{m}_f(t) \times C_f}\right)\right]$$

or,

$$\exp\left(-\frac{F'(mA_c)U_L}{\dot{m}_f(t) \times C_f}\right) = 1 - \frac{T_{00} - T_{fi}}{\left[\left(\frac{\dot{q}_{ab}}{U_L}\right) + T_a - T_{fi}\right]} = \frac{\left(\frac{\dot{q}_{ab}}{U_L}\right) + T_a - T_{00}}{\left[\left(\frac{\dot{q}_{ab}}{U_L}\right) + T_a - T_{fi}\right]}$$

Taking log of both sides, then rearrange and one gets

$$-\frac{F'(mA_c)U_L}{\dot{m}_f(t) \times C_f} = \ln\frac{\left(\frac{\dot{q}_{ab}}{U_L}\right) + T_a - T_{00}}{\left[\left(\frac{\dot{q}_{ab}}{U_L}\right) + T_a - T_{fi}\right]}$$

From above equation, an expression for mass flow rate for constant collection temperature is obtained as follows:

$$\dot{m}_f(t) = \frac{F'(mA_c)U_L}{C_f}\left[\ln\left\{\frac{\left(\frac{\dot{q}_{ab}}{U_L}\right) + T_a - T_{fi}}{\left(\frac{\dot{q}_{ab}}{U_L}\right) + T_a - T_{00}}\right\}\right]^{-1} \quad (5.66a)$$

or,

$$\dot{m}_f(t) = -\frac{F'(mA_c)U_L}{C_f}\left[\ln\left\{\frac{\left(\frac{\dot{q}_{ab}}{U_L}\right) + T_a - T_{00}}{\left(\frac{\dot{q}_{ab}}{U_L}\right) + T_a - T_{fi}}\right\}\right]^{-1} \quad (5.66b)$$

**Example 5.45** Evaluate mass flow rate $(\dot{m}_f)$ for constant collector temperature of 40 °C for five $(m = 5)$ single glazed solar air collectors (SACs) for design and climatic parameters of Example 5.32 $[F' = 0.22, U_L = 10\,\text{W/m}^2\,°\text{C}, I(t) = 500\,\text{W/m}^2, T_a = 16\,°\text{C}$ and $T_{fi} = 17\,°\text{C}$ and $\alpha_p = \tau_g = 0.9].$

**Solution**

From Eq. 5.66a, we get an expression for mass flow rate as

$$\dot{m}_f(t) = \frac{F'(mA_c)U_L}{C_f}\left[\ln\left\{\frac{\left(\frac{\dot{q}_{ab}}{U_L}\right) + T_a - T_{fi}}{\left(\frac{\dot{q}_{ab}}{U_L}\right) + T_a - T_{00}}\right\}\right]^{-1}$$

$$= 0.0219 \times \left[\ln\left\{\frac{39.5}{16.5}\right\}\right]^{-1}$$

$$= 0.0219 \times \frac{1}{0.3791} = 0.058\,\text{kg/s}$$

and, also from Eq. 5.66b, we get an expression for mass flow rate as

$$\dot{m}_f(t) \times C_f = -\frac{F'(mA_c)U_L}{C_f}\left[\ln\left\{\frac{\left(\frac{\dot{q}_{ab}}{U_L}\right) + T_a - T_{00}}{\left(\frac{\dot{q}_{ab}}{U_L}\right) + T_a - T_{fi}}\right\}\right]^{-1}$$

$$= \frac{0.22 \times 5 \times 2 \times 10}{1005}\left[\ln\left\{\frac{40.5 + 16 - 40}{40.5 + 16 - 17}\right\}\right]^{-1}$$

$$= -0.0219\left[\ln\frac{16.5}{39.5}\right]^{-1} = -0.0219[\ln 0.4177]^{-1}$$

$$= -0.0215 \times \frac{1}{-0.3791} = 0.058\,\text{kg/s}$$

The results are same. Further, it is to be noted that for same operating temperature ($\approx 40\,°\text{C}$) (Example 5.32), the mass flow rate is doubled in the case of constant collection temperature (Example 5.44).

**Example 5.46** Repeat the Example 5.44 by using Eqs. 5.47a, 5.47b for the same operating condition of 40 °C and design and climatic parameters.

**Solution**

From Eqs. 5.47a, 5.47b, we have an expression for an outlet air temperature at $m$th SAC as

$$T_{fom} = \left[\left(\frac{\dot{q}_{ab}}{U_L}\right) + T_a\right]\left[1 - \exp\left(-\frac{F'(mA_c)U_L}{\dot{m}_f(t) \times C_f}\right)\right] + T_{fi}\exp\left(-\frac{F'(mA_c)U_L}{\dot{m}_f(t) \times C_f}\right)$$

Substitute $T_{fom} = 40\,°C$, then

$$40 = [56.5]\left[1 - \exp\left(-\frac{0.0219}{\dot{m}_f(t)}\right)\right] + 17 \times \exp\left(-\frac{0.0219}{\dot{m}_f(t)}\right)$$
$$= 56.6 \times \left[1 - \exp(-0.7545)\right] + 17 \times \exp(-0.7545)$$
$$= 56.6 \times [1 - 0.4702] + 17 \times 0.4702$$

Subtract $T_{fi} = 17\,°C$ from both sides, one gets

$$23 = [56.5 - 17]\left[1 - \exp\left(-\frac{0.0219}{\dot{m}_f(t)}\right)\right] = 39.5\left[1 - \exp\left(-\frac{0.0219}{\dot{m}_f(t)}\right)\right]$$

or,

$$\left[1 - \exp\left(-\frac{0.0219}{\dot{m}_f(t)}\right)\right] = \frac{23}{39.5}\text{or} - \frac{0.0219}{\dot{m}_f(t)} = -0.3791\text{or}\,\dot{m}_f(t) = 0.058\text{kg/s}$$

**Example 5.47** Evaluate mass flow rate $(\dot{m}_f(t))$ for constant collector temperature of 40 °C for six $(m = 6)$ single glazed solar air collectors (SACs) connected in parallel for design and climatic parameters of Example 5.32 [$m = 5, F' = 0.22, U_L = 10\,\text{W/m}^2\,°C, I(t) = 500\,\text{W/m}^2, T_a = 16\,°C$ and $T_{fi} = 17\,°C$ and $\alpha_p = \tau_g = 0.9$].

**Solution**

Here, $m = 6$ (given). In this case, the mass flow rate in each SAC will be $\frac{\dot{m}_f}{6}$.

From Eq. 5.66a, we have an expression for mass flow rate for SAC connected in parallel as

$$\dot{m}_f(t) = \frac{F'(mA_c)U_L}{C_f}\left[\ln\left\{\frac{\left(\frac{\dot{q}_{ab}}{U_L}\right) + T_a - T_{fi}}{\left(\frac{\dot{q}_{ab}}{U_L}\right) + T_a - T_{00}}\right\}\right]^{-1}$$

$$= \frac{0.22 \times 6 \times 2 \times 10}{1005}\left[\ln\left\{\frac{\left(\frac{405}{10}\right) + 16 - 17}{\left(\frac{405}{10}\right) + 16 - 40}\right\}\right]^{-1}$$

$$= 0.02627 \times \left[ \ln \left\{ \frac{39.5}{16.5} \right\} \right]^{-1} = 0.02627 \times \frac{1}{0.3791} = 0.069 \, \text{kg/s}$$

## 5.11.2 Solar Air Collectors Connected in Series

From Eq. 5.50, we have expression for SAC connected in series for constant mass flow rate as

$$T_{foN} = \left[ \left( \frac{\dot{q}_{ab}}{U_L} \right) + T_a \right] [1 - \exp[-N A_c U_L F'/(\dot{m}_f C_f)]]$$
$$+ T_{fi} \exp[-N A_c U_L F'/(\dot{m}_f C_f)]$$

Following Sect. 5.10.1, one can derive an expression for mass flow rate of SAC connected in series for constant collection temperature, $T_{00}$, as

$$\dot{m}_f(t) = \frac{F' N A_c U_L}{C_f} \left[ \ln \left\{ \frac{\left( \frac{\dot{q}_{ab}}{U_L} \right) + T_a - T_{fi}}{\left( \frac{\dot{q}_{ab}}{U_L} \right) + T_a - T_{00}} \right\} \right]^{-1} \tag{5.67}$$

**Example 5.48** Evaluate mass flow rate $\left( \dot{m}_f(t) \right)$ for constant collector temperature of 40 °C for six ($N = 6$) single glazed solar air collectors (SACs) connected in series for design and climatic parameters of Example 5.32 [$m = 5$, $F' = 0.22$, $U_L = 10 \, \text{W/m}^2 \, °\text{C}$, $I(t) = 500 \, \text{W/m}^2$, $T_a = 16 \, °\text{C}$ and $T_{fi} = 17 \, °\text{C}$ and $\alpha_p = \tau_g = 0.9$].

**Solution**

Here, $m = 6$ (given). In this case, the mass flow rate in each SAC will be $\frac{\dot{m}_f}{6}$.

From Eq. 5.67, we have an expression for mass flow rate for SAC connected in parallel as

$$\dot{m}_f(t) = \frac{F' N A_c U_L}{C_f} \left[ \ln \left\{ \frac{\left( \frac{\dot{q}_{ab}}{U_L} \right) + T_a - T_{fi}}{\left( \frac{\dot{q}_{ab}}{U_L} \right) + T_a - T_{00}} \right\} \right]^{-1}$$

$$= \frac{0.22 \times 6 \times 2 \times 10}{1005} \left[ \ln \left\{ \frac{\left( \frac{405}{10} \right) + 16 - 17}{\left( \frac{405}{10} \right) + 16 - 40} \right\} \right]^{-1}$$

$$= 0.02627 \times \left[ \ln \left\{ \frac{39.5}{16.5} \right\} \right]^{-1} = 0.02627 \times \frac{1}{0.3791} = 0.069 \, \text{kg/s}$$

Examples 5.46 and 5.47 show that the mass flow rate $\left[ \dot{m}_f(t) \right]$ will be same irrespective of their connection either in parallel or series.

### 5.11.3  Solar Air Collectors Connected in Parallel and Series

In the case of parallel and series connections, the first m-solar air collectors are connected in parallel known as one module of SAC, with inlet mass flow rate of $\frac{\dot{m}_f}{m}$, and then, N such modules are connected in series. For this configuration, an expression for outlet air temperature under constant flow rate, Eq. 5.54, is given by

$$T_{fom \times N} = \left[ \left( \frac{\dot{q}_{ab}}{U_L} \right) + T_a \right] \left[ 1 - \exp\left[ -N A_{cm} U_L F' / \left( \dot{m}_f C_f \right) \right] \right]$$
$$+ T_{fi} \exp\left[ -N A_{cm} U_L F' / \left( \dot{m}_f C_f \right) \right]$$

In the case of constant collection temperature, $T_{00}$, $T_{fom \times N} = T_{00}$ and mass flow rate becomes $\dot{m}_f(t)$, then the above equation becomes

$$T_{00} = \left[ \left( \frac{\dot{q}_{ab}}{U_L} \right) + T_a \right] \left[ 1 - \exp\left[ -N A_{cm} U_L F' / \left( \dot{m}_f(t) \times C_f \right) \right] \right]^{\cdot}$$
$$+ T_{fi} \exp\left[ -N A_{cm} U_L F' / \left( \dot{m}_f(t) \times C_f \right) \right]$$

Further, following Sect. 5.10.1, an expression for mass flow rate for constant collection temperature for parallel and series connections can be obtained as follows:

$$\dot{m}_f(t) = \frac{F' N A_{cm} U_L}{C_f} \left[ \ln \left\{ \frac{\left( \frac{\dot{q}_{ab}}{U_L} \right) + T_a - T_{fi}}{\left( \frac{\dot{q}_{ab}}{U_L} \right) + T_a - T_{00}} \right\} \right]^{-1} \text{ with } A_{cm} = m \times A_c \quad (5.68)$$

**Example 5.49** Evaluate mass flow rate $\left( \dot{m}_f(t) \right)$ for constant collector temperature of 40 °C for parallel ($m = 2$) and series ($N = 3$) for six single glazed solar air collectors (SACs) for design and climatic parameters of Example 5.32 [$m = 5$, $F' = 0.22$, $U_L = 10 \text{ W/m}^2 \,°C$, $I(t) = 500 \text{ W/m}^2$, $T_a = 16 \,°C$ and $T_{fi} = 17 \,°C$ and $\alpha_p = \tau_g = 0.9$].

**Solution**

Here, $m = 2$ (given). In this case, the mass flow rate in each parallel SAC will be $\frac{\dot{m}_f}{2}$ and $N = 3$.

From Eq. 5.68, we have an expression for mass flow rate for SAC connected in parallel $A_{cm} = m \times A_c$ as

$$\dot{m}_f(t) = \frac{F' \times N \times m \times A_c \times U_L}{C_f} \left[ \ln \left\{ \frac{\left( \frac{\dot{q}_{ab}}{U_L} \right) + T_a - T_{fi}}{\left( \frac{\dot{q}_{ab}}{U_L} \right) + T_a - T_{00}} \right\} \right]^{-1}$$

$$= \frac{0.22 \times 3 \times 2 \times 2 \times 10}{1005} \left[ \ln \left\{ \frac{\left( \frac{405}{10} \right) + 16 - 17}{\left( \frac{405}{10} \right) + 16 - 40} \right\} \right]^{-1}$$

$$= 0.02627 \times \left[ \ln \left\{ \frac{39.5}{16.5} \right\} \right]^{-1} = 0.02627 \times \frac{1}{0.3791} = 0.069 \, \text{kg/s}$$

Examples 5.46–5.48 show that the mass flow rate $\left[ \dot{m}_f(t) \right]$ will be same irrespective of their connection either in parallel or series or parallel and series.

### 5.11.4  Solar Air Collectors Connected in Series and Parallel

In the case of series and parallel combination, an expression for the outlet air temperature is given by Eq. 5.58 as

$$T_{foN \times m} = \left[ \left( \frac{\dot{q}_{ab}}{U_L} \right) + T_a \right] \left[ 1 - \exp\left[ -NA_cU_LF' \Big/ \left( \frac{\dot{m}_f}{m} C_f \right) \right] \right]$$
$$+ T_{fi} \exp\left[ -NA_cU_LF' \Big/ \left( \frac{\dot{m}_f}{m} C_f \right) \right]$$

By adopting same procedure, the mass flow rate expression will be

$$\dot{m}_f(t) = \frac{F' \times N \times m \times A_c \times U_L}{C_f} \left[ \ln\left\{ \frac{\left( \frac{\dot{q}_{ab}}{U_L} \right) + T_a - T_{fi}}{\left( \frac{\dot{q}_{ab}}{U_L} \right) + T_a - T_{00}} \right\} \right]^{-1} \qquad (5.69)$$

In this case too, the results for series and parallel combination will be same as reported in Examples 5.46–5.48 for fixed number of SAC.

## References

1. Tiwari, G. N., & Tiwari, A., & Shyam. (2016). *Handbook of solar energy*. Springer.
2. Tiwari, G. N. (2002). *Solar energy: Fundamentals, design, modelling and applications*. CRC Press (USA) and Alpha Science International (UK) also published by Narosa Publishing House.
3. Whiller, A. (1964). *Solar energy, 8*(1), 31.
4. Selcuk, M. K. (1977). In A.A.M. Sayigh (Ed.), *Solar air heaters and their applications Chapter 8 in Solar energy engineering*. Academic Press, New York

# Chapter 6
# Photo-Voltaic Thermal Liquid Collector

## 6.1 Introduction

As we have seen in Chaps. 4 and 5 and also known too that liquid and air flat plate collector gives better performance in forced mode of operation in comparison with natural circulation of fluid (water) and air as a working fluid. It is due to low operating temperature condition in forced mode of operation with less heat loss. In forced mode of operation, one needs electric power to operate electric motor. In remote area of developing and under developing countries, there is shortage of electric power during day/sunshine hours. In order to solve the shortage of electric power during sunshine hour, an integration of PV module with liquid and air flat plate collector was suggested in the past by many scientists [1, 2]. Such photo-voltaic integrated collector is known as photo-voltaic thermal (PVT) collector due to simultaneous production of thermal (low grade power) and electrical (high grade power) power. In this case, PVT liquid and air collector will be analyzed in terms of an overall thermal (thermal plus equivalent thermal from electrical) and an overall exergy (electrical power plus thermal exergy, Chap. 2).

In this chapter, PVT liquid and liquid PVT collector is also same. Further, the fluid as a water has been used.

## 6.2 Liquid PVT Collector

Liquid flat photo-voltaic thermal (PVT) collector is similar to conventional liquid flat plate collector (FPC) as shown in Fig. 4.1 except the glass cover/glazing is replaced by either tedlar-based/aluminum based (opaque) PV module or semi-transparent PV module as shown in Fig. 3.6. Further, if the glass cover/glazing is replaced by PV module as shown in Fig. 6.1, then it is referred as PVT collector as mentioned in introduction.

© Bag Energy Research Society 2024

G. N. Tiwari, *Fundamental of Mathematical Tools for Thermal Modeling of Solar Thermal and Photo-voltaic Systems-Part-I*,
https://doi.org/10.1007/978-981-99-7085-8_6

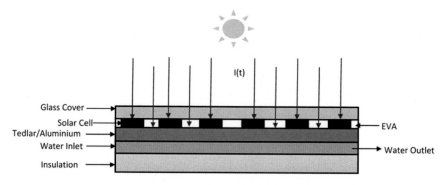

**Fig. 6.1**  Cross-sectional view of opaque liquid PVT collector with water flowing below PV module

There are some assumptions to be made before writing energy balance equations for liquid PVT collector and these are as follows:

(i)   The PVT collector is in quasi-steady-state condition.
(ii)  Each component of PVT collector and working fluid has negligible heat capacity.
(iii) There is no electrical losses between two consecutive solar cells.

With above assumptions, these are following energy balances.

### 6.2.1  Opaque Unglazed Opaque PV (OPV) Module

Following Eq. 3.26, one has energy balance of opaque PV module above the absorber plate as

$$\tau_g[\alpha_c\beta_c I(t) + (1 - \beta_c)\alpha_T I(t)]A_m = \left[U_{tc,a}(T_c - T_a) + U_{bc,p}(T_c - T_p)\right]A_m$$
$$+ \eta_m\beta_c I(t)A_m \tag{6.1a}$$

In Eq. 6.1a, the rate of bottom loss from solar cell to absorber plate is $\left[U_{bc,p}(T_c - T_p)\right]$ which is going to be transferred to working fluid flowing through riser of absorber plate as shown in Fig. 4.6. The numerical value of $U_{tc,a} = 9.12\,\text{W/m}^2\,°\text{C}$ with wind velocity of 1 m/s and $5.56\,\text{W/m}^2\,°\text{C}$ without wind velocity (Example 3.22).

***Example 6.1***  Evaluate the top loss with wind velocity of 1 m/s and bottom loss coefficient, $U_{bc,p}$ from solar cell to absorber plate for (a) tedlar-based ($K_t = 0.033\,\text{W/m}\,°\text{C}$ and $L_t = 0.0005\,\text{m}$) and (b) copper-based back sheet ($K_{bs} = 386\,\text{W/m}\,°\text{C}$ [Appendix V(c)] and $L_{bs} = 0.0005\,\text{m}$) opaque PV module.

**Solution**

In this case, there is no movement of wind between PV module and absorber plate; hence $V = 0$.

(a) Referring Example 3.22, we have the top loss $(U_{t,ca})$ and bottom loss $(U_{b,cp})$ coefficient as

$$U_{t,ca} = \left[\frac{L_g}{K_g} + \frac{1}{h_o}\right]^{-1} = \left[\frac{0.003}{0.688} + \frac{1}{5.7 + 3.8 \times 1}\right]^{-1}$$
$$= [0.00436 + 0.10416]^{-1} = [0.1096]^{-1}$$
$$= 9.12 \text{ W/m}^2 \,^\circ\text{C} \tag{2.13a}$$

and,

$$U_{b,cp} = \left[\frac{L_t}{K_t} + \frac{1}{h_i}\right]^{-1} = \left[\frac{0.0005}{0.033} + \frac{1}{2.8}\right]^{-1}$$
$$= [0.01515 + 0.3571]^{-1} = [0.3722]^{-1}$$
$$= 2.686 \text{ W/m}^2 \,^\circ\text{C} \tag{2.13b}$$

(b) Further, for copper-based PV module

$$U_{b,cp} = \left[\frac{L_{bs}}{K_{bs}} + \frac{1}{h_i}\right]^{-1} = \left[\frac{0.0005}{386} + \frac{1}{2.8}\right]^{-1}$$
$$= \left[1.295 \times 10^{-6} + 0.3571\right]^{-1} = [0.3571]^{-1}$$
$$= 2.8 \text{ W/m}^2 \,^\circ\text{C}$$

The copper-based PV module, the overall heat transfer between solar cell and absorber plate is higher by 4.2%.

### 6.2.2 Unglazed Semi-transparent PV (SPV) Module

Referring Eq. 3.34 and Fig. 6.2, the energy balance for semi-transparent PV module above the absorber plate can be written as

$$\tau_g[\alpha_c \beta_c I(t)]A_m = \left[U_{tc,a}(T_c - T_a) + U_{bc,p}(T_c - T_p)\right]A_m + \eta_m \beta_c I(t) A_m \tag{6.2}$$

The solar radiation incident on non-packing area of PV module is further transmitted toward to absorber plate $\left[\tau_g^2(1 - \beta_c)I(t)A_m\right]$ of liquid flat plate collector.

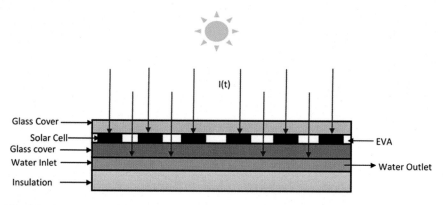

**Fig. 6.2** Cross-sectional view of semi-transparent liquid PVT collector with water flowing below PV module

**Example 6.2** Evaluate the bottom loss coefficient, $U_{bc,p}$ from solar cell to absorber plate for semi-transparent PV module ($K_g = 0.688\,\text{W/m}\,^\circ\text{C}$ and $L_g = 0.003\,\text{m}$) above the absorber plate.

**Solution**

In this case, the tedlar and aluminum are replaced by transparent toughen glass. Referring to Example 6.1,

$$U_{b,cp} = \left[\frac{L_g}{K_g} + \frac{1}{h_i}\right]^{-1} = \left[\frac{0.003}{0.688} + \frac{1}{2.8}\right]^{-1} = [0.00436 + 0.3571]^{-1} = [0.3615]^{-1}$$
$$= 2.7663\,\text{W/m}^2\,^\circ\text{C}$$

So, the value of $U_{b,cp}$ for semi-transparent PV module is higher by about 3% in comparison with opaque tedlar due to higher thermal conductivity of glass.

### 6.2.3  Absorber Plate with Opaque PV (OPV) Module

Here, two cases can be considered, namely

**Case (a): For constant PV module electrical efficiency $(\eta_m)$**

From Eq. 6.1a, one gets an expression for the rate of heat transfer from the solar cell to absorber plate $\left[U_{bc,p}(T_c - T_p)\right]$ as

**Step 1**

The expression for solar cell temperature ($T_c$) is given by

$$T_c = \frac{\{\tau_g[\alpha_c\beta_c + (1-\beta_c)\alpha_T] - \eta_m\beta_c\}I(t) + U_{tc,a}T_a + U_{bc,p}T_p}{(U_{tc,a} + U_{bc,p})} \quad (6.3)$$

**Step 2**

The rate of heat transfer $[U_{bc,p}(T_c - T_p)]$ can be expressed as follows:

$$U_{bc,p}(T_c - T_p) = F_0[(\alpha_c\tau_g)_{\text{meff}}I(t) - U_{tc,a}(T_p - T_a)] \quad (6.4a)$$

where,

$(\alpha_c\tau_g)_{\text{meff}} = \{\tau_g[\alpha_c\beta_c + (1-\beta_c)\alpha_T] - \eta_m\beta_c\}$. It depends on packing factor $(\beta_c)$, and $F_0 = \frac{U_{bc,p}}{(U_{tc,a}+U_{bc,p})}$ is independent of packing factor $(\beta_c)$.

The net rate of heat available per m$^2$ of opaque PVT collector at absorber $(\acute{q}_u)$ will be obtained as

$$\begin{aligned}
\acute{q}_u &= [U_{bc,p}(T_c - T_p)] - U_b(T_p - T_a) \\
&= F_0[(\alpha_c\tau_g)_{\text{meff}}I(t) - U_{tc,a}(T_p - T_a)] - U_b(T_p - T_a)
\end{aligned}$$

or,

$$\begin{aligned}
\acute{q}_u &= F_0\left\{\left[(\alpha_c\tau_g)_{\text{meff}}I(t) - \left(U_{tc,a} + \frac{U_b}{F_0}\right)(T_p - T_a)\right]\right\} \\
&= F_0\{[(\alpha_c\tau_g)_{\text{meff}}I(t) - U_{Lm}(T_p - T_a)]\} \text{ in W/m}^2 \quad (6.4b)
\end{aligned}$$

Here, $U_{Lm} = \left(U_{tc,a} + \frac{U_b}{F_0}\right)$ is an overall heat transfer coefficient from top and bottom of PVT collector similar to $U_L$ of conventional FPC (Eq. 4.5b and Example 4.20), and $U_b$ is an overall bottom heat loss coefficient of PVT collector (Eq. 4.4 and Example 4.19.)

**Case (b): For temperature dependent PV module electrical efficiency $(\eta_m)$, Eq. 3.20**

In this case, Eq. 6.1 with help of Eq. 3.20 can be solved for solar cell temperature as

**Step 1**

$$T_c = \frac{\tau_g[\tau_g\{\alpha_c\beta_c + (1-\beta_c)\alpha_T\} - [\eta_{m0}\beta_c](1 + 25\beta_0)]I(t) + U_{tc,a}T_a + U_{bc,p}T_p}{[U_{tc,a} + U_{bc,p} - [\eta_{m0}\beta_c I(t)\beta_0]]}$$

If $(\alpha_c\tau_g)_{\text{meff}} = [\tau_g\{\alpha_c\beta_c + (1-\beta_c)\alpha_T\} - [\eta_{m0}\beta_c](1 + 25\beta_0)]$, then

$$T_c = \frac{(\alpha_c\tau_g)_{\text{meff}}I(t) + U_{tc,a}T_a + U_{bc,p}T_p}{[U_{tc,a} + U_{bc,p} - [\eta_{m0}\beta_c I(t)\beta_0]]} \quad (6.5)$$

Further,

$$U_{bc,p}\left(T_c - T_p\right) = F_0\left(\alpha_c \tau_g\right)_{meff} I(t) - F_0\left\{\left[U_{tc,a} - [\eta_{m0}\beta_c I(t)\beta_0]\right]T_p - U_{tc,a}T_a\right\}$$

$$(6.6a)$$

where, $F_0 = \frac{U_{bc,p}}{[U_{tc,a}+U_{bc,p}-[\eta_{m0}\beta_c I(t)\beta_0]]}$ depends on packing factor $(\beta_c)$.

**Step 2**

Now, the net rate of heat available per $m^2$ of semi-transparent PVT collector at absorber $(\acute{q}_u)$ will be written as

$$\acute{q}_u = U_{bc,p}\left(T_c - T_p\right) - U_b\left(T_p - T_a\right)$$
$$= F_0\left(\alpha_c \tau_g\right)_{meff} I(t) - F_0\left\{\left[U_{tc,a} - [\eta_{m0}\beta_c I(t)\beta_0]\right]T_p - U_{tc,a}T_a\right\}$$
$$- U_b\left(T_p - T_a\right)$$

or,

$$\acute{q}_u = F_0\left(\alpha_c \tau_g\right)_{meff} I(t) - \left(F_0 U_{tc,a} + U_b\right)\left(T_p - T_a\right)$$
$$+ F_0\eta_{m0}\beta_c I(t)\beta_0 T_p \text{ in W/m}^2$$

or,

$$\acute{q}_u = F_0\left[\left(\alpha_c \tau_g\right)_{meff} I(t) - U_{Lm}\left(T_p - T_a\right) + \left(\eta_{m0}\beta_c I(t)\beta_0\right)T_p\right] \text{ in W/m}^2 \quad (6.6b)$$

where, $U_{Lm} = \left(U_{tc,a} + \frac{U_b}{F_0}\right)$ is an overall heat transfer coefficient from top and bottom of PVT collector similar to $U_L$ of conventional FPC (Eq. 4.5b). There is another terms as $(\eta_{m0}\beta_c I(t)\beta_0)$ due to consideration of temperature dependent PV module efficiency as mentioned earlier.

For $\beta_0 = 0$, in Eq. 3.20, i.e., constant PV module efficiency, then Eqs. 6.5, 6.6a and 6.6b reduce to Eqs. 6.3, 6.4a and 6.4b.

***Example 6.3*** Evaluate penalty factor $(F_0)$ and effective absorptance and transmittance product $\left[\left(\alpha_c \tau_g\right)_{meff}\right]$ for opaque multi-crystalline silicon PV (OPV) module **for constant** $(\eta_m)$ given in Eq. 6.4a for the following given design parameters:

$\tau_g = 0.95$, $\alpha_T = 0.20$, $\alpha_c = 0.90$, packing factor $(\beta_c) = 0.89$, $\eta_m = 0.12$, $U_{t,ca} = 9.12 \text{ W/m}^2 \,°C$, Example 6.1, and $U_{b,ca} = 2.7663 \text{ W/m}^2 \,°C$, Example 6.2.

**Solution**

From Eq. 6.4a, we have the following expressions:

$$F_0 = \frac{U_{bc,p}}{(U_{tc,a} + U_{bc,p})} \quad \text{and} \quad (\alpha_c \tau_g)_{meff}$$
$$= \{\tau_g[\alpha_c\beta_c + (1 - \beta_c)\alpha_T] - \eta_m\beta_c\}$$

Now,

Penalty factor $(F_0) = \frac{U_{bc,p}}{(U_{tc,a}+U_{bc,p})} = \frac{2.7663}{9.12+2.7663} = 0.233$, independent of $\eta_m$.

and,

Effective absorptance and transmittance product $[(\alpha_c\tau_g)_{meff}]$ of opaque PV module.

(a) **For $\beta_c = 0.89$**

$$[(\alpha_c\tau_g)_{meff}] = \{0.95[0.90 \times 0.89 + (1 - 0.89) \times 0.20] - 0.12 \times 0.89\}$$
$$= 0.95 \times (0.801 + 0.022) - 0.1068 = 0.675$$

(b) For $\beta_c = 0.50$

$$[(\alpha_c\tau_g)_{meff}] = \{0.95[0.90 \times 0.50 + (1 - 0.50) \times 0.20] - 0.12 \times 0.50\}$$
$$= 0.95 \times (0.454 + 0.10) - 0.06 = 0.4663$$

(c) For $\beta_c = 0.25$

$$[(\alpha_c\tau_g)_{meff}] = \{0.95[0.90 \times 0.25 + (1 - 0.25) \times 0.20] - 0.12 \times 0.25\}$$
$$= 0.95 \times (0.225 + 0.15) - 0.03 = 0.3263$$

***Example 6.4*** Evaluate the penalty factor $(F_0)$ and effective absorptance and transmittance product $[(\alpha_c\tau_g)_{meff}]$ for opaque multi-crystalline silicon PV (OPV) module **for variable $(\eta_m)$** given in Eqs. 6.5 and 6.6a for the following given design parameters: $\tau_g = 0.95$, $\alpha_T = 0.20$, $\alpha_c = 0.90$, packing factor $(\beta_c) = 0.89$, $\eta_{m0} = 0.12$, $U_{t,ca} = 9.12 \text{ W/m}^2 \,°C$, Example 6.1 and $U_{b,ca} = 2.7663 \text{ W/m}^2 \,°C$, Example 6.2, $I(t) = 500 \text{ W/m}^2$ and $\beta_0 = 0.00425 \,°C^{-1}$ (Table 3.2).

**Solution**

From Eqs. 6.5 and 6.6a, we have the following expressions:

$$F_0 = \frac{U_{bc,p}}{[U_{tc,a} + U_{bc,p} - [\eta_{m0}\beta_c I(t)\beta_0]]} \quad \text{and} \quad (\alpha_c\tau_g)_{meff}$$
$$= \tau_g[\{\alpha_c\beta_c + (1 - \beta_c)\alpha_T\} - [\eta_{m0}\beta_c](1 + 25\beta_0)]$$

Both $F_0$ and $(\alpha_c \tau_g)_{meff}$ depend on packing factor $(\beta_c)$.
Now,

(a) **For $\beta_c = 0.89$**

$$\text{Penalty factor } (F_0) = \frac{2.7663}{9.12 + 2.7663 - [0.12 \times 0.89 \times 500 \times 0.00425]}$$

$$= \frac{2.7663}{11.8863 - 0.22695} = 0.2373$$

(b) **For $\beta_c = 0.50$**

$$\text{Penalty factor}(F_0) = \frac{2.7663}{9.12 + 2.7663 - [0.12 \times 0.50 \times 500 \times 0.00425]}$$

$$= \frac{2.7663}{11.8863 - 0.1275} = 0.2353$$

(c) **For $\beta_c = 0.25$**

$$\text{Penalty factor}(F_0) = \frac{2.7663}{9.12 + 2.7663 - [0.12 \times 0.25 \times 500 \times 0.00425]}$$

$$= \frac{2.7663}{11.8863 - 0.06375} = 0.2340$$

and,
from Eq. 6.5, we have

$$(\alpha_c \tau_g)_{meff} = \left[ \tau_g \{ \alpha_c \beta_c + (1 - \beta_c) \alpha_T \} - [\eta_{m0} \beta_c](1 + 25\beta_0) \right]$$

(a) **For $\beta_c = 0.89$**

$(\alpha_c \tau_g)_{meff} = [0.95[0.90 \times 0.89 + (1 - 0.89) \times 0.20] - [0.12 \times 0.89](1 + 25 \times 0.0045)]$

$\qquad = 0.95 \times (0.801 + 0.022) - 0.1068 \times (1 + 25 \times 0.00425)$

$\qquad = 0.78185 - 0.1181 = 0.66375$

(b) **For $\beta_c = 0.50$**

$(\alpha_c \tau_g)_{meff} = [0.95[0.90 \times 0.50 + (1 - 0.50) \times 0.20] - [0.12 \times 0.50](1 + 25 \times 0.0045)]$

$\qquad = 0.95 \times (0.475 + 0.10) - 0.06 \times (1 + 25 \times 0.00425)$

$\qquad = 0.54625 - 0.0663751 = 0.4796$

(c) **For $\beta_c = 0.25$**

$$(\alpha_c \tau_g)_{meff} = [0.95[0.90 \times 0.25 + (1 - 0.25) \times 0.20] - [0.12 \times 0.25](1 + 25 \times 0.0045)]$$
$$= 0.95 \times (0.2375 + 0.046875) - 0.03 \times (1 + 25 \times 0.00425)$$
$$= 0.23453 - 0.0331875 = 0.2013$$

Examples 6.3 and 6.4 show that the penalty factor $(F_0)$ is increased by 1.9%, and an effective absorptance and transmittance product $\left[(\alpha_c \tau_g)_{meff}\right]$ is reduced by 1.7% due to consideration of temperature dependent of opaque PV (OPV) module efficiency for $\beta_c = 0.89$. So, one can say that there is not significant effect of temperature dependent electrical efficiency of opaque PV module on thermal modeling of opaque liquid PVT collector. But, effect of packing factor on $(\alpha_c \tau_g)_{meff}$ has significant effect. Further for $\beta_c = 0$, the results in both cases are same.

### 6.2.4 The Opaque Liquid PVT (OPVT) Collector Efficiency Factor $\left(F'\right)$ Under Natural Mode of Operation

Following Sect. 4.3.3, the opaque liquid PVT collector efficiency $(F')$ is defined as the ratio of actual rate of useful heat collection $\left(\dot{Q}_u\right)$ in W under natural mode of operation of PVT collector to the rate of useful heat collection when the PVT collector's absorbing plate $\left(T_p\right)$ is at the local fluid temperature $\left(T_f\right)$, Eqs. 6.4b and 6.6b, and it is expressed as

$$F' = \frac{\dot{Q}_u}{\dot{Q}_u\big|_{T_p=T_f}} \quad \text{or} \quad \dot{Q}_u = F' \dot{Q}_u\big|_{T_p=T_f} = F' A_m \dot{q}_u\big|_{T_p=T_f} \tag{6.7}$$

where, from Example 4.29, we have

$$F' = \cfrac{1}{\cfrac{WU_L}{\pi D h_{cf}} + \cfrac{W}{D+(W-D)F}}$$
$$= \cfrac{1}{\cfrac{0.10 \times U_L}{3.14 \times 0.01 \times 149.6} + \cfrac{0.10}{0.01+(0.10-0.01) \times 0.96}}$$

In this case, $U_{t,ca} = 9.12\,\text{W/m}^2\,°\text{C}$ and $U_{b,ca} = 2.8\,\text{W/m}^2\,°\text{C}$ (without wind), then

Top loss coefficient from absorber to ambient $U_t = \left[\frac{1}{U_{t,ca}} + \frac{1}{U_{b,ca}}\right]^{-1} =$ $\left[\frac{1}{9.12} + \frac{1}{2.8}\right]^{-1} = 2.14\,\text{W/m}^2\,^{\circ}\text{C}.$

Hence, in the case of semi-transparent PV module integrated FPC, the total heat loss coefficient becomes

$$U_L = U_t + U_b = 2.14 + 0.7 = 2.84\,\text{W/m}^2\,^{\circ}\text{C}$$

Substitute the above value in expression for $F'$, then one gets

$$F' = \cfrac{1}{\cfrac{0.10 \times 2.84}{3.14 \times 0.01 \times 149.6} + \cfrac{0.10}{0.01 + (0.10 - 0.01) \times 0.96}}$$

$$= \frac{1}{0.06 + 1.0373} = 0.91$$

Equation 6.7 becomes for two cases, namely.

**Case (a): For constant PV module electrical efficiency**

$$\dot{Q}_u = F'_{\text{eff}} A_m \left\{ \left[ (\alpha_c \tau_g)_{m\text{eff}} I(t) - U_{Lm}(T_f - T_a) \right] \right\} \tag{6.8a}$$

**Case (b): For temperature dependent PV module electrical efficiency**

$$\dot{Q}_u = F'_{\text{eff}} A_m \left[ (\alpha_c \tau_g)_{m\text{eff}} I(t) - U_{Lm}(T_f - T_a) + (\eta_{m0}\beta_c I(t)\beta_0) T_f \right] \tag{6.8b}$$

The $F'_{\text{eff}} = F'F_0$ is effective PVT collector efficiency for PVT collector and $U_{Lm} = \left( U_{tc,a} + \frac{U_b}{F_0} \right)$. There are two loss terms in Eq. 6.8b due to variable $\eta_m$.

Expression for $F' = 0.82$ and $0.91$ for single and double glazed FPC respectively is exactly same as reported in Chap. 4 for liquid flat plate collector for tube in plate configuration, Fig. 4.5c (Example 4.29).

***Example 6.5*** Evaluate the effective PV module thermal efficiency factor $\left( F'_{\text{eff}} \right)$ and an overall top and bottom heat transfer coefficient for both cases, namely constant and varying $\eta_m$ of opaque multi-crystalline silicon PV (OPV) module for the following parameters.

$F' = 0.82$ (Example 4.29); and $0.2373$, Example 6.4; $U_{t,ca} = 9.12\,\text{W/m}^2\,^{\circ}\text{C}$, Example 6.1 and $U_b = 0.70\,\text{W/m}^2\,^{\circ}\text{C}$, Example 4.19; $\beta_0 = 0.00425\,^{\circ}\text{C}^{-1}$; $\eta_{m0} = 0.12$; packing factor $(\beta_c) = 0.89$ and $l(t) = 500\,\text{W/m}^2$ (Example 6.4.)

**Solution**

From Eqs. 6.8a and 6.8b we have

**Case (a):** $F'_{\text{eff}} = F'F_0$ and $U_{Lm} = \left( U_{tc,a} + \frac{U_b}{F_0} \right)$, $F_0 = 0.233$, Example 6.3 independent of $\eta_m$

The effective PV module thermal efficiency factor $\left(F'_{\text{eff}}\right) = F'F_0 = 0.82 \times 0.233 = 0.191$.

An overall top and bottom heat transfer coefficient $(U_{Lm}) = \left(9.12 + \frac{0.70}{0.233}\right) = 12.12 \, \text{W/m}^2 \, °\text{C}$.

**Case (b):** $F'_{\text{eff}} = F'F_0$ and first loss coefficient $(U_{Lm}) = \left(U_{tc,a} + \frac{U_b}{F_0}\right)$; and second loss coefficient $= (\eta_{m0}\beta_c I(t)\beta_0)$, $F_0 = 0.2373$, Example 6.4 dependent of $\eta_m$.

The effective PV module thermal efficiency factor $\left(F'_{\text{eff}}\right) = F'F_0 = 0.82 \times 0.2373 = 0.195$.

In addition to first loss coefficient in Eq. 6.8b $U_{Lm} = 12.12 \, \text{W/m}^2 \, °\text{C}$ same as case (a), there is another second loss coefficient $= \eta_{m0}\beta_c I(t)\beta_0$ in $\text{W/m}^2 \, °\text{C} = 0.12 \times 0.89 \times 500 \times 0.00425 = 0.227 \, \text{W/m}^2 \, °\text{C}$ which is negligible in comparison with $U_{Lm} = 12.12 \, \text{W/m}^2 \, °\text{C}$.

***Example 6.6*** Evaluate the second term of Eq. 6.8b for different packing factor $(\beta_c)$, namely 1, 0.5 and 0.

**Solution**

The second loss coefficient of Eq. 6.8b $= \eta_{m0}\beta_c I(t)\beta_0 \, \text{W/m}^2 \, °\text{C}$.

**For $\beta_c = 1$**

The second loss coefficient of Eq. 6.8b $= \eta_{m0}\beta_c I(t)\beta_0 = 0.12 \times 1 \times 500 \times 0.00425 = 0.225 \, \text{W/m}^2 \, °\text{C}$.

**For $\beta_c = 0.5$**

The second loss coefficient of Eq. 6.8b $= \eta_{m0}\beta_c I(t)\beta_0 = 0.12 \times 0.5 \times 500 \times 0.00425 = 0.1125 \, \text{W/m}^2 \, °\text{C}$.

**For $\beta_c = 0.0$**

The second loss coefficient of Eq. 6.8b $= \eta_{m0}\beta_c I(t)\beta_0 = 0.12 \times 0,0 \times 500 \times 0.00425 = 0.0 \, \text{W/m}^2 \, °\text{C}$.

It is simply a conventional flat plate collector. If this second term is neglected, then there is minor effect on the performance of opaque liquid flat plate collector.

### 6.2.5  The Outlet Fluid Temperature $\left(T_{fo}\right)$ of Opaque PVT (OPVT) Liquid Collector

For elemental area of '$bdx$' of opaque liquid PVT collector, the rate of useful thermal energy available to absorber can be written by using Eq. 6.8b with $A_c = bdx$ as

$$\dot{Q}_u = F'_{\text{eff}}\left[(\alpha_c\tau_g)_{\text{meff}}I(t) - U_{Lm}\left(T_f - T_a\right) + (\eta_{m0}\beta_c I(t)\beta_0)T_f\right]bdx \qquad (6.9)$$

If $\acute{m}_f$ is mass flow rate of fluid flowing through raisers of opaque liquid PVT collector similar to Fig. 4.6a and $C_f$ is specific heat of fluid and $dT_f$ is temperature rise of fluid with $dx$ length, then the rate of thermal energy carried from elemental length can be expressed as

$$\acute{m}_f C_f \frac{dT_f}{dx} dx \tag{6.10}$$

Based on Eqs. 4.23, 6.9 and 6.10 are equal, and hence

$$\acute{m}_f C_f \frac{dT_f}{dx} dx = F'_{\text{eff}} \big[ (\alpha_c \tau_g)_{\text{meff}} I(t) - U_{Lm} (T_f - T_a) + (\eta_{m0} \beta_c I(t) \beta_0) T_f \big] b dx \tag{6.11}$$

Equation 6.11 can be rearranged as

$$\frac{dT_f}{dx} + aT_f = f(t) \tag{6.12}$$

where, $a = a_0 b$, $a_0 = \frac{F'_{\text{eff}}[U_{Lm} - (\eta_{m0}\beta_c I(t)\beta_0)]b}{\acute{m}_f C_f}$ and $f(t) = \frac{F'_{\text{eff}}\big[(\alpha_c\tau_g)_{\text{meff}} I(t) + U_{Lm} T_a\big]b}{\acute{m}_f C_f}$.

The solution of Eq. 6.12 with initial condition of $T_f = T_{fi}$ at $x = 0$ will be as follows:

$$T_f = \frac{\big[(\alpha_c\tau_g)_{\text{meff}} I(t) + U_{Lm} T_a\big]}{[U_{Lm} - (\eta_{m0}\beta_c I(t)\beta_0)]} \big[1 - \exp(-a_0 bx)\big] + T_{fi} \exp(-a_0 bx) \tag{6.13a}$$

The average fluid temperature over length of liquid PVT collector, $L_0$, can be obtained as

$$\overline{T_f} = \frac{1}{L_0} \int_0^{L_0} T_f dx = \frac{\big[(\alpha_c\tau_g)_{\text{meff}} I(t) + U_{Lm} T_a\big]}{[U_{Lm} - (\eta_{m0}\beta_c I(t)\beta_0)]} \left[ 1 - \frac{1 - \exp(-a_0 bL_0)}{a_0 bL_0} \right]$$
$$+ T_{fi} \frac{1 - \exp(-a_0 bL_0)}{a_0 bL_0}$$

or,

$$\overline{T_f} = \frac{\big[(\alpha_c\tau_g)_{\text{meff}} I(t) + U_{Lm} T_a\big]}{[U_{Lm} - (\eta_{m0}\beta_c I(t)\beta_0)]} \left[ 1 - \frac{1 - \exp(-a_0 A_m)}{a_0 A_m} \right] + T_{fi} \frac{1 - \exp(-a_0 A_m)}{a_0 A_m} \tag{6.14}$$

According to Eq. 6.7 and definition of PVT collector efficiency, the average fluid temperature will be equal to average plate temperature, and hence

$$\overline{T_f} \approx \overline{T_p} \tag{6.15}$$

By using above condition, the average solar cell temperature, $\overline{T}_c$ can be obtained from Eq. 6.5.

$$\overline{T}_c = \frac{(\alpha_c \tau_g)_{\text{meff}} I(t) + U_{tc,a} T_a + U_{bc,p} \overline{T}_f}{\left[ U_{tc,a} + U_{bc,p} - \left[ \eta_{m0} \beta_c I(t) \beta_0 \right] \right]} \tag{6.16}$$

After knowing $\overline{T}_c$, the electrical efficiency and electrical power can be determined as follows:

$$\overline{\eta}_m = \eta_{m0} \left[ 1 - \beta_0 (\overline{T}_c - 25) \right] \tag{6.17a}$$

and,

$$\dot{E}_{el} = \overline{\eta}_m \times I(t) \times \beta_c A_m = \eta_{m0} \left[ 1 - \beta_0 (\overline{T}_c - 25) \right] \times I(t) \times \beta_c A_m \tag{6.17b}$$

The outlet fluid temperatures $T_{fo}$ at $x = L_0$ can be obtained as

$$T_{fo} = \frac{\left[ (\alpha_c \tau_g)_{\text{meff}} I(t) + U_{Lm} T_a \right]}{\left[ U_{Lm} - (\eta_{m0} \beta_c I(t) \beta_0) \right]} \left[ 1 - \exp\left( -\frac{F'_{\text{eff}} [U_{Lm} - (\eta_{m0} \beta_c I(t) \beta_0)] A_m}{\acute{m}_f C_f} \right) \right]$$
$$+ T_{fi} \exp\left( -\frac{F'_{eff} [U_{Lm} - (\eta_{m0} \beta_c I(t) \beta_0)] A_m}{\acute{m}_f C_f} \right) \tag{6.18}$$

where, $A_m = b L_0$.

For constant electrical efficiency of PV module $(\eta_m)$, $\beta_0 = 0$, then $a_0 = \frac{F'_{\text{eff}} U_{Lm}}{\acute{m}_f C_f}$; the temperature coefficient $(\beta_0)$ will be zero, then Eq. 6.14 reduces to

$$T_{fo} = \left( \frac{(\alpha_c \tau_g)_{\text{meff}}}{U_{Lm}} I(t) + T_a \right) \left[ 1 - \exp\left( -\frac{F'_{\text{eff}} U_{Lm} A_m}{\acute{m}_f C_f} \right) \right] + T_{fi} \exp\left( -\frac{F'_{\text{eff}} U_{Lm} A_m}{\acute{m}_f C_f} \right) \tag{6.19}$$

Equation 6.15 for liquid PVT collector is similar to Eq. 4.26 for FPC. Here, an expression for $F'_{\text{eff}}$, $(\alpha_c \tau_g)_{\text{meff}}$ and $U_{Lm}$ are given by Eqs. 6.4a, 6.4b for constant electrical efficiency of PV module.

***Example 6.7*** Evaluate the outlet temperature of liquid PVT (OPVT) collector with effective area of 2 m² for the data of Example 4.39 for winter condition with mass flow rate of 0.035 kg/s, and ambient air $(T_a)$ and fluid inlet $(T_{fi})$ temperatures are $T_a = 16\,°C$ and $T_{fi} = 15\,°C$ and $I(t) = 500$ W/m². Other data is same as obtained in Example 6.5 for case (b).

**Solution**

From Example 6.5: $F'_{\text{eff}} = 0.195$; first loss coefficient $(U_{Lm}) = 12.12\,\text{W/m}^2\,°\text{C}$ and second loss coefficient $(\eta_{m0}\beta_c I(t)\beta_0) = 0.227\,\text{W/m}^2\,°\text{C}$, $(\alpha_c\tau_g)_{\text{meff}} = 0.66375$ (Example 6.4) and $C_f = 4190$ [Appendix E(c)].

**Case (a): For constant $\eta_m$:** $F'_{\text{eff}} = 0.191$ (Example 6.5), $(\alpha_c\tau_g)_{\text{meff}} = 0.675$ (Example 6.3).

From Eq. 6.18 with $\beta_0 = 0$, we have an expression for $(T_{fo})$ as

$$T_{fo} = \left(\frac{(\alpha_c\tau_g)_{\text{meff}}}{U_{Lm}}I(t) + T_a\right)\left[1 - \exp\left(-\frac{F'_{\text{eff}}U_{Lm}A_m}{\acute{m}_f C_f}\right)\right] + T_{fi}\exp\left(-\frac{F'_{\text{eff}}U_{Lm}A_m}{\acute{m}_f C_f}\right)$$

Substitute the appropriate value in the above equation, one gets

$$T_{fo} = \left(\frac{0.675}{12.12} \times 500 + 16\right)\left[1 - \exp\left(-\frac{0.191 \times 12.12 \times 2}{0.035 \times 4190}\right)\right]$$
$$+ 15 \times \exp\left(-\frac{0.191 \times 12.12 \times 2}{0.035 \times 4190}\right)$$

or,

$$T_{fo} = 43.84 \times [1 - \exp(-0.03157)] + 15 \times \exp(-0.03157)$$
$$= 43.84 \times [1 - 0.9689] + 15 \times 0.9689 = 15.87\,°\text{C}$$

**Case (b): For varying $\eta_m$ with temperature:** $F'_{\text{eff}} = 0.195$ (Example 6.5), $(\alpha_c\tau_g)_{\text{meff}} = 0.66375$ (Example 6.4).

From Eq. 6.18, we have

$$T_{fo} = \frac{[(\alpha_c\tau_g)_{\text{meff}}I(t) + U_{Lm}T_a]}{[U_{Lm} - (\eta_{m0}\beta_c I(t)\beta_0)]}\left[1 - \exp\left(-\frac{F'_{\text{eff}}[U_{Lm} - (\eta_{m0}\beta_c I(t)\beta_0)]A_m}{\acute{m}_f C_f}\right)\right]$$
$$+ T_{fi}\exp\left(-\frac{F'_{eff}[U_{Lm} - (\eta_{m0}\beta_c I(t)\beta_0)]A_m}{\acute{m}_f C_f}\right)$$

Now, substitute the appropriate given value in the above equation, one gets

$$T_{fo} = \frac{[0.66375 \times 500 + 0.195 \times 16]}{[12.12 - 0.227]}\left[1 - \exp\left(-\frac{0.195 \times [12.12 - 0.227] \times 2}{0.035 \times 4190}\right)\right]$$
$$+ T_{fi}\exp\left(-\frac{0.195 \times [12.12 - 0.227] \times 2}{0.035 \times 4190}\right)$$

or,

$$T_{fo} = \frac{[0.66375 \times 500 + 0.195 \times 16]}{[12.12 - 0.227]}\left[1 - \exp\left(-\frac{0.195 \times [12.12 - 0.227] \times 2}{0.035 \times 4190}\right)\right]$$
$$+ T_{fi}\exp\left(-\frac{0.195 \times [12.12 - 0.227] \times 2}{0.035 \times 4190}\right)$$

or,

$$T_{fo} = \frac{[334.995]}{[11.893]}\left[1 - \exp\left(-\frac{4.6383}{146.65}\right)\right] + 15 \times \exp\left(-\frac{4.6383}{146.65}\right)$$
$$= 28.17\left[1 - \exp(-0.0316)\right] + 15 \times \exp(-0.0316)$$
$$= 28.17[1 - 0.9688] + 15 \times 0.9699 = 15.41\,°C$$

***Example 6.8*** Evaluate an average fluid and solar cell temperature for case (b) of Example 6.7 for varying $\eta_m$.

**Solution**

From Example 6.7, we have

**Step 1:** $\frac{[(\alpha_c\tau_g)_{meff}I(t)+U_{Lm}T_a]}{[U_{Lm}-(\eta_{m0}\beta_c I(t)\beta_0)]} = \frac{[0.66375\times500+0.195\times16]}{[12.12-0.227]} = \frac{[334.995]}{[11.893]} = 28.17$

**Step 2:** $a = a_0 b$, $a_0 = \frac{F'_{eff}[U_{Lm}-(\eta_{m0}\beta_c I(t)\beta_0)]}{\dot{m}_f C_f} = \frac{0.195\times[12.12-0.227]}{0.035\times4190} = 0.0158$

**Step 3:** $\frac{1-\exp(-a_0 A_m)}{aL_0} = \frac{1-\exp(-0.0158\times2)}{a_0 b L_0} = \frac{[1-0.9688]}{0.0158\times1\times2} = 0.9844$.
From Eq. 6.14, we have

$$\overline{T_f} = \frac{[(\alpha_c\tau_g)_{meff}I(t) + U_{Lm}T_a]}{[U_{Lm} + (\eta_{m0}\beta_c I(t)\beta_0)]}\left[1 - \frac{1 - \exp(-a_0 A_m)}{aL_0}\right] + T_{fi}\frac{1 - \exp(-a_0 A_m)}{aL_0}$$
$$= 28.17[1 - 0.9844] + 15 \times 0.9844 = 15.21\,°C$$

From Eq. 6.16, we have expression for an average solar cell temperature as

$$\overline{T_c} = \frac{(\alpha_c\tau_g)_{meff}I(t) + U_{tc,a}T_a + U_{bc,p}\overline{T}_f}{\left[U_{tc,a} + U_{bc,p} - [\eta_{m0}\beta_c I(t)\beta_0]\right]}$$
$$= \frac{0.66375 \times 500 + 9.12 \times 15 + 2.686 \times 15.21}{9.12 + 2.686 - 0.227} = \frac{509.45}{11.58} = 44\,°C$$

From Example 3.26, the solar cell temperature is 48.19 °C which is higher by 4.2 °C in comparison with 43.99 °C of liquid PVT collector due to cooling effect by flowing water through risers.

**Example 6.9** Evaluate an electrical efficiency and electrical energy power for Example 6.8 for varying for varying $\eta_m$.

**Solution**

From Example 6.8, we have $\overline{T_c} = 44\,°C$.
  Further,

$$\eta_{m0} = 0.12[1 - 0.00425(44 - 25)] = 0.11031(11.031)$$

From Eq. 6.17b, one has an expression of electrical energy power as

$$\dot{E}_{el} = \eta_{m0}\left[1 - \beta_0\left(\overline{T_c} - 25\right)\right] \times I(t) \times \beta_c A_m = 0.12[1 - 0.00425(44 - 25)]$$
$$\times 500 \times 0.89 \times 2 = 98.18\,W$$

**Example 6.10** Evaluate an average fluid, solar cell temperature and electrical power for case (a) of Example 6.7 for constant $\eta_m$.

**Solution**

From Example 6.7, we have

**Step 1:** $\frac{[(\alpha_c\tau_g)_{meff}I(t)+U_{Lm}T_a]}{U_{Lm}} = \frac{[0.675\times500+0.195\times16]}{12.12} = \frac{[340.62]}{12.12} = 28.10.$

**Step 2:** $a = a_0 b, a_0 = \frac{F'_{eff}U_{Lm}}{\dot{m}_f C_f} = \frac{0.195\times12.12}{0.035\times4190} = 0.0161.$

**Step 3:** $\frac{1-\exp(-a_0 A_c)}{aL_0} = \frac{1-\exp(-0.0161\times2)}{a_0 b L_0} = \frac{[1-0.9683]}{0.0161\times1\times2} = 0.9845.$
  From Eq. 6.14, we have

$$\overline{T_f} = \frac{\left[(\alpha_c\tau_g)_{meff}I(t) + U_{Lm}T_a\right]}{U_{Lm}}\left[1 - \frac{1 - \exp(-a_0 A_m)}{aL_0}\right] + T_{fi}\frac{1 - \exp(-a_0 A_m)}{aL_0}$$
$$= 28.10[1 - 0.9845] + 15 \times 0.9845 = 15.20\,°C$$

From Eq. 6.16, we have an expression for an average solar cell temperature  after using the data of Example 6.1 as

$$\overline{T_c} = \frac{(\alpha_c\tau_g)_{meff}I(t) + U_{tc,a}T_a + U_{bc,p}\overline{T}_f}{\left[U_{tc,a} + U_{bc,p}\right]}$$
$$= \frac{0.675 \times 500 + 9.12 \times 15 + 2.686 \times 15.20}{9.12 + 2.686} = \frac{515.12}{11.81} = 43.61\,°C$$

From Eq. 6.17b, one has an expression of electrical energy power as

$$\dot{E}_{el} = \eta_{m0}\left[1 - \beta_0\left(\overline{T_c} - 25\right)\right] \times I(t) \times \beta_c A_m$$
$$= 0.12[1 - 0.00425(43.61 - 25)] \times 500 \times 0.89 \times 2 = 110.5\,W$$

Here, one can conclude from Examples 6.9 and 6.10 that there is drop of 11.12% in electrical power due to consideration of temperature dependent electrical efficiency of opaque PV module.

## 6.2.6 The Rate of Useful Thermal Energy of Opaque Liquid PVT (OPVT) Collector Under Forced Mode of Operation

**Case (a): For constant $\eta_m$**

With help of Eq. 6.19, it can be derived as follows:

$$\acute{Q}_u = \acute{m}_f C_f \left(T_{fo} - T_{fi}\right) = F_{Rm} A_m \left[\left(\alpha_c \tau_g\right)_{meff} I(t) - U_{Lm}\left(T_{fi} - T_a\right)\right] \quad (6.20)$$

where, the mass flow rate factor $(F_{Rm}) = \frac{\acute{m}_f C_f}{U_{Lm} A_m}\left[1 - \exp\left(-\frac{F'_{eff} U_{Lm} A_m}{\acute{m}_f C_f}\right)\right]$.

**Case (b): For varying $\eta_m$ with solar cell temperature, Eq. 6.17b**

From Eq. 6.18, we have an expression for the outlet fluid temperature as

$$
T_{fo} = \frac{\left[\left(\alpha_c \tau_g\right)_{meff} I(t) + U_{Lm_{eff}} T_a\right]}{U_{Lm_{eff}}}\left[1 - \exp\left(-\frac{F'_{eff} U_{Lm_{eff}} A_m}{\acute{m}_f C_f}\right)\right]
$$
$$
+ T_{fi}\exp\left(-\frac{F'_{eff} U_{Lm_{eff}} A_m}{\acute{m}_f C_f}\right)
$$

Here, $U_{Lm_{eff}} = [U_{Lm} + (\eta_{m0}\beta_c I(t)\beta_0)]$.

Now, the rate of useful thermal energy with help of above equation can be obtained as

$$\acute{Q}_u = \acute{m}_f C_f \left(T_{fo} - T_{fi}\right) = F_{Rm} A_m \left[\left(\alpha_c \tau_g\right)_{meff} I(t) + U_{Lm} T_a - U_{Lm_{eff}} T_{fi}\right] \quad (6.21)$$

where, the mass flow rate factor $(F_{Rm}) = \frac{\acute{m}_f C_f}{U_{Lm_{eff}} A_m}\left[\left[1 - \exp\left(-\frac{F'_{eff} U_{Lm_{eff}} A_m}{\acute{m}_f C_f}\right)\right]\right]$.

If $\beta_0 = 0$, Eq. 6.20 reduces to Eq. 6.19.

**Example 6.11** Evaluate mass flow rate factor and the rate of thermal energy of opaque liquid PVT (OPVT) collector for case (a), i.e., constant $\eta_m$ and case (b) for varying $\eta_m$ of Example 6.7.

**Solution**

**Case (a): For constant $\eta_m$**

$T_{fo} = 15.87\,°C$ and $T_{fi} = 15\,°C$ (Example 6.8), $T_a = 16\,°C$ $F'_{\text{eff}} = 0.191$ (Example 6.5) $U_{Lm} = 12.12\,\text{W/m}^2\,°C$ (Example 6.5), $\acute{m}_f = 0.035\,\text{kg/s}$ $C_f = 4190\,\text{j/kg}\,°C$ and $(\alpha_c \tau_g)_{\text{meff}} = 0.675$ (Example 6.3).

From Eq. 6.20, one has the following:

The mass flow rate factor

$$
\begin{aligned}
(F_{Rm}) &= \frac{\acute{m}_f C_f}{U_{Lm} A_m}\left[1 - \exp\left(-\frac{F'_{\text{eff}} U_{Lm} A_m}{\acute{m}_f C_f}\right)\right] \\
&= \frac{0.035 \times 4190}{12.12 \times 2}\left[1 - \exp\left(-\frac{0.191 \times 12.12 \times 2}{0.035 \times 4190}\right)\right] \\
&= 6.05 \times \left[1 - \exp\left(-\frac{4.63}{146.65}\right)\right] = 6.05 \times [1 - 0.9689] = 0.1882
\end{aligned}
$$

The rate of thermal energy has been evaluated as

$$
\begin{aligned}
\acute{Q}_u &= \acute{m}_f C_f \left(T_{fo} - T_{fi}\right) = F_{Rm} A_m\left[(\alpha_c \tau_g)_{\text{meff}} I(t) - U_{Lm}\left(T_{fi} - T_a\right)\right] \\
&= 0.1882 \times 2[0.675 \times 500 - 12.12(15 - 16)] = 131.59\,\text{W}
\end{aligned}
$$

**Case (b): For varying $\eta_m$.**

**From Eq. 6.20, we have the following**

**Given:** $T_{fo} = 15.41\,°C$ and $T_{fi} = 15\,°C$ (Example 6.8), $T_a = 16\,°C$ $F'_{\text{eff}} = 0.195$ (Example 6.5), $U_{Lm} = 12.12\,\text{W/m}^2\,°C$ (Example 6.5), $\acute{m}_f = 0.035\,\text{kg/s}$ $C_f = 4190\,\text{j/kg}\,°C$ and $(\alpha_c \tau_g)_{\text{meff}} = 0.66375$ (Example 6.4).

In this case $U_{Lm_{\text{eff}}} = [U_{Lm} + (\eta_{m0} \beta_c I(t) \beta_0)] = 12.12 + 0.2269 = 12.35.$

The mass flow rate factor $(F_{Rm}) = \frac{\acute{m}_f C_f}{U_{Lm_{\text{eff}}} A_m}\left[\left[1 - \exp\left(-\frac{F'_{\text{eff}} U_{Lm_{\text{eff}}} A_m}{\acute{m}_f C_f}\right)\right]\right] = $
$\frac{0.035 \times 4190}{12.35 \times 2}\left[1 - \exp\left(-\frac{0.195 \times 12.35 \times 2}{0.035 \times 4190}\right)\right] = 5.93 \times \left[1 - \exp\left(-\frac{4.82}{146.65}\right)\right] = 5.93 \times [1 - 0.9676] = 0.1919.$

The rate of thermal energy has been evaluated from Eq. 6.21 as

$$
\begin{aligned}
\acute{Q}_f C_f\left(T_{fo} - T_{fi}\right) &= F_{Rm} A_m\left[(\alpha_c \tau_g)_{\text{meff}} I(t) + U_{Lm} T_a - U_{Lm\text{eff}} T_{fi}\right] \\
&= 0.1919 \times 2[0.66375 \times 500 - 12.35(15 - 16)] = 132.11\,\text{W}
\end{aligned}
$$

One can observe from this example that there is also marginal difference in the rate of thermal energy by considering either constant $\eta_m$ or varying $\eta_m$. It is due to fact that the value of smallest value of temperature thermal expansion value, i.e. $(\beta_0 = 0.00425\,°C^{-1})$.

**So, finally it is concluded that the performance of opaque liquid PVT (OPVT) collector is same in both cases constant and varying electrical efficiency.**

### 6.2.7 An Instantaneous Thermal Efficiency

It is defined as

$$\eta_i = \frac{\dot{Q}_u}{I(t) \times A_c} = \frac{132.11}{500 \times 2} = 0.13(13\%) \text{ for varying } \eta_m.$$

### 6.2.8 Exergy of Opaque Liquid PVT (OPVT) Collector

Following Eq. 2.35a, we have an expression for evaluating hourly exergy.

$$\dot{Ex}_u = \dot{m}_f C_f \left[ (T_{fo} - T_{fi}) - (T_a + 273)\ln \frac{T_{fo} + 273}{T_{fi} + 273} \right] \tag{6.22}$$

**Example 6.12** Evaluate hourly exergy for water mass flow rate of 0.035 kg/s through liquid PVT (OPVT) collector for case (a) of Example 6.7.

**Solution**

**From Example 6.7:** $T_{fo} = 15.85\,°C$ and $T_{fi} = 15\,°C$, $\dot{m}_f = 0.035$ kg s and specific heat of water, $C_f$, is 4190. Then from Eq. 2.35a, one gets

$$\dot{Ex}_{th} = \dot{m}_f C_f \left[ (T_{fo} - T_{fi}) - (T_a + 273)\ln \frac{T_{fo} + 273}{T_{fi} + 273} \right]$$

$$= 0.035 \times 4190 \left[ (15.85 - 15) - (287)\ln \frac{288.5}{288} \right]$$

$$= 146.65[(15.85 - 15) - 0.9169] = -\text{ve W}.$$

This indicates that there is no constrictive thermal energy due to low operating temperature, and in this case, the exergy of PVT collector will only electrical power obtained in Examples 6.9 and 6.10, respectively.

## 6.3 Analysis of Unglazed Semi-transparent Liquid PVT (SPVT) Collector for Variable $\eta_m$

Referring to Fig. 6.2 and Eq. 6.2, we can get, the energy balance for semi-transparent PV module with temperature dependent PV module electrical efficiency, i.e., $\eta_m = \eta_{m0}[1 - \beta_0(\overline{T}_c - 25)]$ above the absorber plate as

$$\tau_g[\alpha_c \beta_c I(t)]A_m = [U_{tc,a}(T_c - T_a) + U_{bc,p}(T_c - T_p)]A_m$$
$$+ \eta_{m0}[1 - \beta_0(T_c - 25)]\beta_c I(t)A_m \tag{6.23}$$

The solar radiation incident on non-packing area of PV module is further transmitted toward to absorber plate $\left[\tau_g^2(1 - \beta_c)I(t)A_m\right]$ of liquid flat plate collector.

The transmitted solar radiation through non-packing factor $\left[\tau_g^2(1 - \beta_c)I(t)A_m\right]$ is absorbed by absorber as direct gain as shown in Fig. 6.2. Further, there is indirect gain from back of solar cell to absorber plate. Hence, the rate of thermal energy in W available to absorber plate of unglazed semi-transparent PVT collector becomes as

$$\dot{Q}_u = \alpha_p\left[\tau_g^2(1 - \beta_c)I(t)A_m\right] + U_{b,cp}\left(T_c - T_p\right)A_m \tag{6.24}$$

where, $\alpha_p = 0.9$ $U_{b,cp} = 2.7663\,\text{W/m}^2\,°\text{C}$ (Example 6.2). In the present case, $\alpha_p = 1.0$ has been considered.

Following Sect. 6.2 for opaque liquid PV collector with temperature dependent electrical efficiency ($\eta_m$), one can derive the various formulas for semi-transparent liquid flat plate collector in following section.

### 6.3.1  Solar Cell Temperature

From Eq. 6.23, one can get solar cell temperature **for variable $\eta_m$** as

$$T_c = \frac{(\alpha_c \tau_g)_{\text{meff}}I(t) + U_{tc,a}T_a + U_{bc,p}T_p}{U_{tc,a} + U_{bc,p} - \eta_{mo}\beta_0\beta_c I(t)} \tag{6.25}$$

where, $\left[(\alpha_c \tau_g)_{\text{meff}}\right]_{\text{semi}} = \left[\tau_g\alpha_c - \eta_{mo}(1 + 25\beta_0)\right]\beta_c$.

***Example 6.13*** Calculate effective absorptance of solar cell and transmittance of semi-transparent PV module $\left[(\alpha_c \tau_g)_{\text{meff}}\right]$ for data of Example 6.4 for (a) constant $\eta_m$ and (b) variable $\eta_m$ for packing factor ($\beta_c$) of 0.89, 0.5 and 0.25.

**Solution**

**Case (a): For constant $\eta_m$ ($\beta_0 = 0$)**

Substitute $\beta_0 = 0$ in Eq. 6.25, we have

$\left[(\alpha_c \tau_g)_{\text{meff}}\right] = \left[\tau_g\alpha_c - \eta_{mo}\right]\beta_c = [0.95 \times 0.9 - 0.12] \times 0.89 = 0.654$ for $\beta_c = 0.89$

$\left[(\alpha_c \tau_g)_{\text{meff}}\right] = \left[\tau_g\alpha_c - \eta_{mo}\right]\beta_c = [0.95 \times 0.9 - 0.12] \times 0.50 = 0.3675$ for $\beta_c = 0.50$

$\left[(\alpha_c \tau_g)_{\text{meff}}\right] = \left[\tau_g\alpha_c - \eta_{mo}\right]\beta_c = [0.95 \times 0.9 - 0.12] \times 0.25 = 0.1838$ for $\beta_c = 0.25$.

which is reduced from 0.675 (Example 6.3) to 0.1838 due reduction of packing factor of semi-transparent PV module.

**Case (b): For variable $\eta_m$**

From Eq. 6.25, we have

$\left[(\alpha_c \tau_g)_{\text{meff}}\right] = \left[\tau_g \alpha_c - \eta_{mo}(1 + 25\beta_0)\right]\beta_c = [0.95 \times 0.9 - 0.12(1 + 25 \times 0.00425)] \times 0.89 = 0.6428$ for $\beta_c = 0.89$.

$\left[(\alpha_c \tau_g)_{\text{meff}}\right] = \left[\tau_g \alpha_c - \eta_{mo}(1 + 25\beta_0)\right]\beta_c = [0.95 \times 0.9 - 0.12(1 + 25 \times 0.00425)] \times 0.50 = 0.3611$ for $\beta_c = 0.50$.

$\left[(\alpha_c \tau_g)_{\text{meff}}\right] = \left[\tau_g \alpha_c - \eta_{mo}(1 + 25\beta_0)\right]\beta_c = [0.95 \times 0.9 - 0.12(1 + 25 \times 0.00425)] \times 0.25 = 0.1806$ for $\beta_c = 0.25$.

In all cases, it has lower value than 0.66375 (Example 6.4) due to nature of semi-transparent PV module.

## 6.3.2 The Indirect Rate of Heat Transfer from Solar Cell to Absorber Plate ($\dot{q}'_u$)

It is expressed as follows:

$$\dot{q}'_u = U_{bc,p}(T_c - T_p) = F_0\left[(\alpha_c \tau_g)_{\text{meff}} I(t) - U_{tc,a}(T_p - T_a) + [\eta_{mo}\beta_0\beta_c I(t)]T_p\right]$$
(6.26)

where, $F_0 = \frac{U_{bc,p}}{U_{tc,a}+U_{bc,p}-\eta_{mo}\beta_0\beta_c I(t)} = 0.2373$ (Example 6.4) $\left[(\alpha_c \tau_g)_{\text{meff}}\right]$

and, $F_0 = \frac{U_{bc,p}}{U_{tc,a}+U_{bc,p}} = 0.233$ (Example 6.3) **for constant $\eta_m$**.

***Example 6.14*** Compute the penalty factor ($F_0$) for semi-transparent PV module for (a) constant $\eta_m$ and (b) variable $\eta_m$ for packing factor ($\beta_c$) of 0.89, 0.5 and 0.25.

**Solution**

**Case (a): For constant $\eta_m$ ($\beta_0 = 0$).**

$F_0 = \frac{U_{bc,p}}{U_{tc,a}+U_{bc,p}} = 0.233$ (Example 6.3).

For constant $\eta_m$, it is independent of packing factor ($\beta_c$).

**Case (b): For variable $\eta_m$ with solar cell temperature**

From Eq. 6.26, we have the following packing factor ($F_0$) formula (Example 6.4):

$$F_0 = \frac{2.7663}{9.12 + 2.7663 - [0.12 \times 0.89 \times 500 \times 0.00425]}$$

$$= \frac{2.7663}{11.8863 - 0.22695} = 0.2373 \quad \textbf{for } \beta_c = \textbf{0.89}$$

$$F_0 = \frac{U_{bc,p}}{U_{tc,a} + U_{bc,p} - \eta_{mo}\beta_0\beta_c I(t)}$$

$$= \frac{2.7663}{11.8863 - 0.1275} = 0.2353 \quad \text{for } \boldsymbol{\beta_c = 0.50}$$

$$F_0 = \frac{U_{bc,p}}{U_{tc,a} + U_{bc,p} - \eta_{mo}\beta_0\beta_c I(t)}$$

$$= \frac{2.7663}{11.8863 - 0.06375} = 0.2339 \quad \text{for } \boldsymbol{\beta_c = 0.25}$$

**Therefore, one can see that there is marginal effect of $\beta_c$ on the penalty factor.**

### 6.3.3   The Rate of Thermal Energy Available to Absorber Plate ($\dot{q}_u$) in W/m² for Variable $\eta_m$

After substitution of Eq. 6.26 in to Eq. 6.24, one gets

$$\dot{q}_u = \left\{\left[(\alpha_c\tau_g)_{meff}\right]_{semi} I(t) - F_0 U_{tc,a}(T_p - T_a) + F_0[\eta_{mo}\beta_0\beta_c I(t)]T_p\right\} \quad (6.27a)$$

or,

$$\dot{Q}_u = \left\{\left[(\alpha_c\tau_g)_{meff}\right]_{semi} I(t) - F_0 U_{tc,a}(T_p - T_a) + F_0[\eta_{mo}\beta_0\beta_c I(t)]T_p\right\} A_m \tag{6.27b}$$

where, $\left[(\alpha_c\tau_g)_{meff}\right]_{semi} = \left[\tau_g^2(1 - \beta_c) + F_0(\alpha_c\tau_g)_{meff}\right]$.

**Example 6.15** Evaluate product of absorptance and transmittance, $\left[(\alpha_c\tau_g)_{meff}\right]_{semi}$ for absorber of semi-transparent PVT (SPVT) collector (a) constant $\eta_m$ and (b) variable $\eta_m$ for packing factor ($\beta_c$) of 0.89, 0.5 and 0.25.

**Solution**

**Case (a): For constant $\eta_m (\beta_0 = 0)$,**

(i) **For $\beta_c = 0.89$** : $(\alpha_c\tau_g)_{meff} = 0.654$, **Example 6.13 and $F_0 = 0.233$ (Example 6.3)**

From Eq. 6.27b, we have for constant $\eta_m$

$$\left[(\alpha_c\tau_g)_{meff}\right]_{semi} = \left[\tau_g^2(1 - \beta_c) + F_0(\alpha_c\tau_g)_{meff}\right]$$
$$= \left[(0.95)^2(1 - 0.89) + 0.233 \times 0.654\right] = 0.2516$$

(ii) **For $\beta_c = 0.50$:** $(\alpha_c\tau_g)_{meff} = 0.3675$, **Example 6.13 and $F_0 = 0.233$** (Example 6.3).

From Eq. 6.27b, we have for constant $\eta_m$

$$\left[(\alpha_c \tau_g)_{meff}\right]_{semi} = \left[\tau_g^2(1-\beta_c) + F_0(\alpha_c \tau_g)_{meff}\right]$$
$$= \left[(0.95)^2(1-0.50) + 0.233 \times 0.3675\right] = 0.5386$$

(iii) **For $\beta_c = 0.25$: $(\alpha_c \tau_g)_{meff} = 0.1838$, Example 6.13** and $F_0 = 0.233$ (Example 6.3).

$$\left[(\alpha_c \tau_g)_{meff}\right]_{semi} = \left[\tau_g^2(1-\beta_c) + F_0(\alpha_c \tau_g)_{meff}\right]$$
$$= \left[(0.95)^2(1-0.25) + 0.233 \times 0.1838\right] = 0.7197$$

**Case (b): For variable $\eta_m$,**

From Example 6.13, we have

$(\alpha_c \tau_g)_{meff} = 0.6428$ for $\beta_c = 0.89$

$(\alpha_c \tau_g)_{meff} = 0.3611$ for $\beta_c = 0.50$

$(\alpha_c \tau_g)_{meff} = 0.1806$ for $\beta_c = 0.25$

From Example 6.14, we have

$F_0 = 0.2373$ for $\beta_c = 0.89$

$F_0 = 0.2353$ for $\beta_c = 0.50$

$F_0 = 0.2339$ for $\beta_c = 0.50$

From Eq. 6.27b, we have for variable $\eta_m$ with solar cell temperature.

$$\left[(\alpha_c \tau_g)_{meff}\right]_{semi} = \left[\tau_g^2(1-\beta_c) + F_0(\alpha_c \tau_g)_{meff}\right]$$
$$= \left[(0.95)^2(1-0.89) + 0.2373 \times 0.6428\right] = 0.2518 \text{ for } \beta_c = 0.89$$

$$\left[(\alpha_c \tau_g)_{meff}\right]_{semi} = \left[(0.95)^2(1-0.50) + 0.2353 \times 0.3611\right] = 0.5362 \text{ for } \beta_c = 0.50$$

$$\left[(\alpha_c \tau_g)_{meff}\right]_{semi} = \left[(0.95)^2(1-0.25) + 0.2339 \times 0.1806\right] = 0.7191 \text{ for } \beta_c = 0.25$$

So, one can concludes that packing factor plays a significant role in determining the analysis of semi-transparent liquid PVT collector.

### 6.3.4  The Outlet Fluid Temperature $(T_{f0})$

Following Sect. 6.2.5, one gets an expression for the outlet temperature for semi-transparent liquid PVT (SPVT) collector for **variable electrical efficiency** $(\eta_m)$ as

$$T_{f0} = \frac{\left\{F'\left[(\alpha_c\tau_g)_{meff}\right]_{semi} I(t) + F'\left[F_0 U_{tc,a} + U_b\right]T_a\right\}}{F'\left[\{F_0 U_{tc,a} + U_b\} - F_0\eta_{mo}\beta_0\beta_c I(t)\right]}\left[1 - \exp(-a_0 A_m)\right]$$

$$+ T_{fi}\exp(-a_0 A_m) \tag{6.28a}$$

where, $a_0 = \frac{F'\left[\{F_0 U_{tc,a} + U_b\} - F_0\eta_{mo}\beta_0\beta_c I(t)\right]}{\acute{m}_f C_f}$.

The rate of thermal energy of PVT collector is given by

$$\dot{Q}_u = \acute{m}_f C_f\left(T_{f0} - T_{fi}\right) \tag{6.28b}$$

**Example 6.16** Evaluate the outlet fluid (water) temperature from liquid semi-transparent PVT collector for data of Example 6.5 with **variable electrical efficiency** $(\eta_m)$ of PV module for packing factor $(\beta_c) = 0.89$.

**Given parameters for variable electrical efficiency of PV module**

$F' = 0.82$ (Example 4.29); $F_0 = 0.2373$ (Example 6.14); $U_{t,ca} = 9.12\,\text{W/m}^2\,^\circ\text{C}$, Example 6.1 and $U_b = 0.70\,\text{W/m}^2\,^\circ\text{C}$, Example 4.19; $\beta_0 = 0.00425\,^\circ\text{C}^{-1}$; $\eta_{m0} = 0.12$; packing factor $(\beta_c) = 0.89$ and $I(t) = 500\,\text{W/m}^2$ (Example 6.4),$\left[(\alpha_c\tau_g)_{meff}\right]_{semi} = 0.2518$ (Example 6.15), $T_a = 16\,^\circ\text{C}$ and $T_{fi} = 15\,^\circ\text{C}$.

**Solution**

From Eqs. 6.28a, 6.28b, we have

$$a_0 = \frac{F'\left[\{F_0 U_{tc,a} + U_b\} - F_0\eta_{mo}\beta_0\beta_c I(t)\right]}{\acute{m}_f C_f}$$

$$= \frac{0.82[\{0.2373 \times 9.12 + 0.70\} - 0.2373 \times 0.12 \times 0.00425 \times 0.89 \times 500]}{0.035 \times 4190}$$

$$= \frac{0.82[2.8642 - 0.05386]}{146.65} = \frac{2.3045}{146.65} = 0.0158$$

and,

$$\frac{\left\{F'\left[(\alpha_c\tau_g)_{meff}\right]_{semi} I(t) + F'\left[F_0 U_{tc,a} + U_b\right]T_a\right\}}{F'\left[\{F_0 U_{tc,a} + U_b\} - F_0\eta_{mo}\beta_0\beta_c I(t)\right]}$$

$$= \frac{0.82 \times 0.2518 \times 500 + 0.82[0.2373 \times 9.12 + 0.7] \times 16}{0.82[\{0.2373 \times 9.12 + 0.70\} - 0.2373 \times 0.12 \times 0.00425 \times 0.89 \times 500]}$$

$$= \frac{103.238 + 37.578}{2.3045} = \frac{140.816}{2.3045} = 61.105$$

From Eqs. 6.28a, 6.28b, we have

$$T_{f0} = 61.105[1 - \exp(-0.0158 \times 2)] + 15 \exp(-0.0158 \times 2)$$
$$= 61.105 \times (1 - 0.9689) + 15 \times 0.9689 = 16.43°C$$

which is higher by 1.02 °C in comparison with opaque liquid PVT collector with variable $\eta_m$ (Example 6.7).

**Example 6.17** Repeat Example with **constant electrical efficiency** ($\beta_0 = 0$) of PV module ($\eta_m$) in semi-transparent liquid PVT collector.

**Solution**

Known: $F_0 = \frac{U_{bc,p}}{U_{tc,a}+U_{bc,p}} = 0.233$ (Example 6.3) and $(\alpha_c \tau_g)_{meff} = 0.6428$ for $\beta_c = 0.89$.

From Eq. 6.28a, 6.28b, we have for constant electrical efficiency of PV module ($\eta_m$), we have

$$a_0 = \frac{F'[\{F_0 U_{tc,a} + U_b\}]}{\acute{m}_f C_f}$$
$$= \frac{0.82[\{0.233 \times 9.12 + 0.70\}]}{0.035 \times 4190}$$
$$= \frac{0.82[2.82496]}{146.65} = \frac{2.3165}{146.65} = 0.01578$$

and,

$$\frac{\{F'(\alpha_c \tau_g)_{meff} I(t) + F'[F_0 U_{tc,a} + U_b]T_a\}}{F'[\{F_0 U_{tc,a} + U_b\}]}$$
$$= \frac{0.82 \times 0.6428 \times 500 + 0.82[0.233 \times 9.12 + 0.7] \times 16}{0.82[\{0.233 \times 9.12 + 0.70\}]}$$
$$= \frac{103.156 + 37.06}{2.3165} = \frac{140.216}{2.3156} = 60.55$$

From Eq. 6.28a, 6.28b, we have

$$T_{f0} = 60.55[1 - \exp(-0.01578 \times 2)] + 15 \times \exp(-0.01578 \times 2)$$
$$= 60.55 \times (1 - 0.9689) + 15 \times 0.9689 = 16.420°C$$

Both outlet temperatures are approximately same from Examples 6.16 and 6.17.

## 6.3.5   The Average Fluid $\left(\overline{T}_f\right)$ and Solar Cell $\left(\overline{T}_c\right)$ Temperatures

Following Eqs. 6.14–6.16, one can derive an average fluid and solar cell temperatures as

$$
\overline{T}_f = \frac{\left\{F'\left[(\alpha_c\tau_g)_{\text{meff}}\right]_{\text{semi}}I(t) + F'\left[F_0U_{tc,a} + U_b\right]T_a\right\}}{F'\left[\{F_0U_{tc,a} + U_b\} - F_0\eta_{mo}\beta_0\beta_c I(t)\right]}\left[1 - \frac{1 - \exp(-a_0A_m)}{a_0A_m}\right]
$$
$$
+ T_{fi}\frac{1 - \exp(-a_0A_m)}{a_0A_m} \tag{6.29}
$$

with $A_m = bL_0$

Since, $T_p = T_f$ as per definition of $F'$, then $\overline{T}_p = \overline{T}_f$

$$
\overline{T}_c = \frac{(\alpha_c\tau_g)_{\text{meff}}I(t) + U_{tc,a}T_a + U_{bc,p}\overline{T}_f}{U_{tc,a} + U_{bc,p} - \eta_{mo}\beta_0\beta_c I(t)} \tag{6.30}
$$

**Example 6.18** Evaluate an average fluid and solar cell temperature for Example 6.16 for **variable electrical efficiency of PVT (SPVT) collector.**

**Solution**

After substituting appropriate values in Eqs. 6.29 and 6.30 from Example 6.16, we have a numerical value for an average fluid $\left(\overline{T}_f\right)$ and solar cell temperatures $\left(\overline{T}_c\right)$ as

$$
\overline{T}_f = 61.105\left[1 - \frac{1 - \exp(-0.0158 \times 2)}{0.0158 \times 2}\right] + 15 \times \frac{1 - \exp(-0.0158 \times 2)}{0.0158 \times 2}
$$
$$
= 61.105\left[1 - \frac{1 - 0.9689}{0.0316}\right] + 15 \times \frac{1 - 0.9689}{0.0316}
$$
$$
= 61.105[1 - 0.9842] + 15 \times 0.9842 = 15.73
$$

and, $(\alpha_c\tau_g)_{\text{meff}} = 0.6428$ for $\beta_c = 0.89$ (Example 6.13); $U_{b,cp} = 2.7663\,\text{W/m}^2\,^\circ\text{C}$ (Example 6.2)

$$
\overline{T}_c = \frac{0.6428 \times 500 + 9.12 \times 15 + 2.7663 \times 15.73}{9.12 + 2.7663 - 0.12 \times 0.00425 \times 0.89 \times 500} = \frac{501.71}{11.66} = 43.02\,^\circ\text{C}
$$

which is less by 1 °C in comparison with opaque liquid PVT collector due to low operating temperature of semi-transparent PVT collector.

## 6.3.6 Electrical Efficiency $(\eta_m)$ and Electrical Power $\left(\dot{E}_{el}\right)$

The electrical efficiency is given by

$$\eta_m = \eta_{mo}\left[1 - \beta_0\left(\overline{T}_c - 25\right)\right] \qquad (6.31a)$$

So, an electrical power can be obtained as

$$\dot{E}_{el} = \eta_{mo}\left[1 - \beta_0\left(\overline{T}_c - 25\right)\right] \times \beta_c I(t) A_m \qquad (6.31b)$$

Further, if $\gamma$ is conversion factor of coal based thermal power plant then equivalent rate of thermal energy of electrical power will be determined as follows:

$$\dot{Q}_{u-el} = \frac{\dot{E}_{el}}{\gamma} \qquad (6.32a)$$

where, the numerical value of $\gamma$ varies between 0.20 and 0.35. It depends on quality of coal. For Indian coal, it has value of 0.22, and Australian coal, it has value of 0.35.

The rate of total thermal energy is sum of Eqs. 6.28b and 6.32a as

$$\dot{Q}_{uth-total} = \dot{Q}_u + \dot{Q}_{u-el} \qquad (6.32b)$$

**Example 6.19** Compute an electrical efficiency, an electrical power and an equivalent rate of thermal energy of electrical power of semi-transparent PVT collector for Example 6.18.

**Solution**

Substitute an appropriate value in Eq. 6.31a, 6.31b from Example 6.18 to get an electrical efficiency and an electrical power of semi-transparent PVT collector.

$$\eta_m = \eta_{mo}\left[1 - \beta_0\left(\overline{T}_c - 25\right)\right] = 0.12[1 - 0.00425(43.02 - 25)] = 0.1108(11.08\%)$$

which is up by 0.049% in comparison with opaque PV module (Example 6.9).

$$\dot{E}_{el} = \eta_{mo}\left[1 - \beta_0\left(\overline{T}_c - 25\right)\right] \times \beta_c I(t) A_m$$
$$= 0.1108 \times 0.89 \times 500 \times 2 = 98.612 \text{ W}$$

The electrical power is higher by 0.432 W in comparison with opaque PV module (Example 6.9).

An equivalent rate of thermal energy of electrical power $\left(\dot{Q}_{u-el}\right) = \frac{\dot{E}_{el}}{\gamma} = \frac{98.612}{0.35} = 281.77 \text{ W}$.

### 6.3.7  The Rate of Useful Thermal Energy, Mass Flow Rate Factor and Instantaneous Efficiency

From Eqs. 6.28a, 6.28b, we have an expression of an outlet fluid temperature as

$$
T_{f0} = \frac{\left\{F'\left[(\alpha_c\tau_g)_{meff}\right]_{semi}I(t) + F'\left[F_0U_{tc,a} + U_b\right]T_a\right\}}{F'\left[\left\{F_0U_{tc,a} + U_b\right\} - F_0\eta_{mo}\beta_0\beta_c I(t)\right]}\left[1 - \exp(-a_0A_m)\right]
$$
$$
+ T_{fi}\exp(-a_0A_m) \tag{6.33a}
$$

or,

$$
T_{f0} = \frac{\left\{\left[(\alpha_c\tau_g)_{meff}\right]_{semi}I(t) + \left[F_0U_{tc,a} + U_b\right]T_a\right\}}{U_{Lm-\text{eff}}}\left[1 - \exp(-a_0A_m)\right]
$$
$$
+ T_{fi}\exp(-a_0A_m) \tag{6.33b}
$$

or,

$$
T_{f0} = \frac{\left\{\left[(\alpha_c\tau_g)_{meff}\right]_{semi}I(t) + \left[F_0U_{tc,a} + U_b\right]T_a\right\}}{U_{Lm-\text{eff}}}\left[1 - \exp\left(-\frac{F'A_mU_{Lm-\text{eff}}}{\acute{m}_fC_f}\right)\right]
$$
$$
+ T_{fi}\exp\left(-\frac{F'A_mU_{Lm-\text{eff}}}{\acute{m}_fC_f}\right) \tag{6.33c}
$$

where,

$$
a_0 = \frac{F'\left[\left\{F_0U_{tc,a} + U_b\right\} - F_0\eta_{mo}\beta_0\beta_c I(t)\right]}{\acute{m}_fC_f} = \frac{F'U_{Lm-\text{eff}}}{\acute{m}_fC_f}\text{ with } U_{Lm-\text{eff}}
$$
$$
= \left[\left\{F_0U_{tc,a} + U_b\right\} - F_0\eta_{mo}\beta_0\beta_c I(t)\right].
$$

Equation 6.33c is similar to the outlet expression derived for liquid flat plate collector (Eq. 4.25).

The rate of useful thermal energy, Eq. 6.28b **for variable $\eta_m$** is given by

$$
\dot{Q}_u = \acute{m}_fC_f\left(T_{f0} - T_{fi}\right) = F_{Rm}A_c
$$
$$
\left[\left\{\left[(\alpha_c\tau_g)_{meff}\right]_{semi}I(t) + \left[F_0U_{tc,a} + U_b\right]T_a\right\} - \left[\left\{F_0U_{tc,a} + U_b\right\} - F_0\eta_{mo}\beta_0\beta_c I(t)\right]T_{fi}\right] \tag{6.34a}
$$

with mass flow rate factor $= F_{Rm} = \frac{\acute{m}_fC_f}{A_mU_{Lm-\text{eff}}}\left[1 - \exp\left(-\frac{F'A_mU_{Lm-\text{eff}}}{\acute{m}_fC_f}\right)\right]$.

**For constant $\eta_m$,** thermal expansion coefficient is zero, i.e., $\beta_0 = 0$, then Eq. 6.33a–6.33c reduces to

$$\dot{Q}_u = F_{Rm} A_c \left[ \left[ (\alpha_c \tau_g)_{\text{meff}} \right] I(t) - U_{Lm} \left( T_{fi} - T_a \right) \right] \tag{6.34b}$$

with mass flow rate factor $= F_{Rm} = \dfrac{\dot{m}_f C_f}{A_m U_{Lm-\text{eff}}} \left[ 1 - \exp \left( -\dfrac{F' A_m U_{Lm}}{\dot{m}_f C_f} \right) \right]$ and $U_{Lm} = \left[ \left\{ F_0 U_{tc,a} + U_b \right\} \right]$.

For **variable** $\eta_m$ means temperature dependent electrical efficiency of PV module, instantaneous thermal efficiency is given by

$$
\begin{aligned}
\eta_i &= \frac{\dot{Q}_u}{I(t) \times A_m} \\
&= F_{Rm} \left[ \left[ (\alpha_c \tau_g)_{\text{meff}} \right]_{\text{semi}} \right. \\
&\quad \left. + \frac{\left[ F_0 U_{tc,a} + U_b \right] T_a - \left[ \left\{ F_0 U_{tc,a} + U_b \right\} - F_0 \eta_{mo} \beta_0 \beta_c I(t) \right] T_{fi} + \left[ F_0 U_{tc,a} + U_b \right] T_a}{I(t)} \right]
\end{aligned}
\tag{6.35a}
$$

The instantaneous thermal efficiency, **for constant** $\eta_m$ means $\beta_0 = 0$ in Eq. 6.34a, can be derived as follows:

$$\eta_i = \frac{\dot{Q}_u}{I(t) \times A_m} = F_{Rm} \left[ \left[ (\alpha_c \tau_g)_{\text{meff}} \right] - U_{Lm-\text{eff}} \frac{\left( T_{fi} - T_a \right)}{I(t)} \right] \text{ with}$$

$$U_{Lm-\text{eff}} = \left[ \left\{ F_0 U_{tc,a} + U_b \right\} \right]$$

$$\text{and } F_{Rm} = \frac{\dot{m}_f C_f}{A_m U_{Lm-\text{eff}}} \left[ 1 - \exp \left( -\frac{F' A_m U_{Lm}}{\dot{m}_f C_f} \right) \right] \tag{6.35b}$$

So, overall instantaneous thermal energy efficiency of PVT collector can be obtained as

$$\eta_i = \frac{\dot{Q}_{\text{uth-total}}}{[I(t)]_{\text{ex}} \times A_m} \tag{6.35c}$$

An expression for $\dot{Ex}_{\text{sun}} = [I(t)]_{\text{ex}} = U_{ee}[I(t) \times A_m]$ is given by Eq. 2.34, $\dot{Ex}_{\text{sun}} = 0.9513[I(t) \times A_m]$ (Example 2.75) and $\dot{Q}_{\text{uth-total}}$ is given by Eq. 6.32b.

***Example 6.20*** Calculate the rate of thermal energy, thermal exergy and total rate of exergy for semi-transparent liquid PVT (SPVT) collector for packing factor of 0.89 for Example 6.17.

**Solution**

From Example 6.17, we have $T_{f0} = 16.40\,°\text{C}$ and $T_{f0} = 15\,°\text{C}$.

The rate of thermal energy available in W can be calculated as

$$\dot{Q}_u = \dot{m}_f C_f \left( T_{f0} - T_{fi} \right) = 0.035 \times 4190(16.40 - 15) = 205\,\text{W}$$

The rate of thermal energy is higher by 67% in comparison with case (b) of opaque PVT collector (122.63 W, Example 6.11).

The rate of thermal exergy available in W can be calculated.

$$\dot{E}x_{th} = \dot{m}_f C_f \left[ (T_{fo} - T_{fi}) - (T_a + 273)\ln \frac{T_{fo} + 273}{T_{fi} + 273} \right]$$

$$= 0.035 \times 4190 \left[ (16.40 - 15) - (287)\ln \frac{289.4}{288} \right]$$

$$= 146.65[(16.4 - 15) - 1.40] = 0 \, \text{W}$$

It is important to note that the rate of thermal exergy is zero for semi-transparent PVT collector (SPVT) and negative for opaque PVT collector (OPVT) (Example 6.12) as per expectation.

The total rate of overall exergy is sum of the rate of thermal exergy, Example 6.20 and electrical power, Example 6.19 from PVT collector given by

$$\dot{E}x_{total} = 0 + 98.612 \, \text{W} = 98.612$$

The overall exergy efficiency will be

$$\eta_{i-exergy} = \frac{\dot{E}x_{total}}{[I(t)]_{ex} \times A_m} = \frac{98.612}{0.9513 \times 500 \times 2} = 0.1037(10.37\%)$$

The overall total rate of thermal energy of semi-transparent PVT (SPVT) collector is sum of the rate of thermal energy and equivalent rate of thermal energy of electrical power, Eq. 6.32a and Example 6.19 which can be obtained as follows:

$$\dot{Q}_{uth-total} = \dot{Q}_u + \dot{Q}_{u-el} = 205 + 281.77 = 486.77 \, \text{W}$$

So, overall instantaneous thermal energy efficiency of PVT collector, Eq. 6.35c, can be obtained as

$$\eta_{ith-exergy} = \frac{\dot{Q}_{uth-total}}{[I(t)]_{ex} \times A_m} = \frac{486.77}{0.9513 \times 500 \times 2} = 0.5117[51.17\%]$$

An expression for $\dot{E}x_{sun} = [I(t)]_{ex} = U_{ee}[I(t) \times A_m]$ is given by Eq. 2.34 and $\dot{E}x_{sun} = 0.9513[I(t) \times A_c]$ (Example 2.75).

**Based on Examples 6.19 and 6.20, it has been concluded that semi-transparent PVT (SPVT) liquid collector is better in term of both thermal and electrical point of view, and hence we will only consider semi-transparent liquid PVT (SPVT) collector**

## 6.4 The Outlet Fluid Temperature $(T_{foN})$ at $N^{th}$ Semi-transparent Liquid PVT (SPVT) Connected in Series

Following Sect. 4.5.1, an expression for outlet fluid temperature at $n$th semi-transparent liquid PVT collector (SPVT) by using Eq. 6.32c **for variable electrical efficiency** $(\eta_m)$ can be derived as follows:

$$
T_{f0N} = \frac{\left\{\left[(\alpha_c\tau_g)_{meff}\right]_{semi}I(t) + \left[F_0U_{tc,a} + U_b\right]T_a\right\}}{U_{Lm-eff}}\left[1 - \exp\left(-\frac{NF'A_mU_{Lm-eff}}{\dot{m}_fC_f}\right)\right]
$$

$$
+ T_{fi}\exp\left(-\frac{NF'A_mU_{Lm-eff}}{\dot{m}_fC_f}\right)
\tag{6.35a}
$$

Here, $U_{Lm-eff} = \left[\{F_0U_{tc,a} + U_b\} - F_0\eta_{mo}\beta_0\beta_cI(t)\right]$ and $a_0 = $
$\frac{F'[\{F_0U_{tc,a}+U_b\}-F_0\eta_{mo}\beta_0\beta_cI(t)]}{\dot{m}_fC_f} = \frac{F'U_{Lm-eff}}{\dot{m}_fC_f}$.

Adopting the procedure given in Eqs. 6.14–6.16, one can derive an average fluid and solar cell temperatures of N-PVT collectors connected in series as

$$
\overline{T}_{fN} = \frac{\left\{\left[(\alpha_c\tau_g)_{meff}\right]_{semi}I(t) + \left[F_0U_{tc,a} + U_b\right]T_a\right\}}{U_{Lm-eff}}\left[1 - \frac{1 - \exp\left(-\frac{NF'A_mU_{Lm-eff}}{\dot{m}_fC_f}\right)}{\frac{NF'A_mU_{Lm-eff}}{\dot{m}_fC_f}}\right]
$$

$$
+ T_{fi}\frac{1 - \exp\left(-\frac{NF'A_mU_{Lm-eff}}{\dot{m}_fC_f}\right)}{\frac{NF'A_mU_{Lm-eff}}{\dot{m}_fC_f}}
\tag{6.35b}
$$

and,

$$
\overline{T}_{cN} = \frac{(\alpha_c\tau_g)_{meff}I(t) + U_{tc,a}T_a + U_{bc,p}\overline{T}_{fN}}{U_{tc,a} + U_{bc,p} - \eta_{mo}\beta_0\beta_cI(t)}
\tag{6.35c}
$$

**Example 6.21** Evaluate the outlet fluid (water) temperature at end of each PVT collector if five semi-transparent PVT (SPVT) collectors are connected in series for data of Example 6.16 for **variable electrical efficiency** $(\eta_m)$ of PVT collector.

**Given parameters for variable electrical efficiency of PV module**

$F' = 0.82$ (Example 4.29); $F_0 = 0.2373$ (Example 6.14); $U_{t,ca} = 9.12\,\text{W/m}^2\,°\text{C}$, Example 6.1 and $U_b = 0.70\,\text{W/m}^2\,°\text{C}$, Example 4.19; $\beta_0 = 0.00425\,°\text{C}^{-1}$; $\eta_{m0} = 0.12$; packing factor $(\beta_c) = 0.89$ and $I(t) = 500\,\text{W/m}^2$ (Example 6.4), $\left[(\alpha_c\tau_g)_{meff}\right]_{semi} = 0.2518$ (Example 6.15), $T_a = 16\,°\text{C}$ and $T_{fi} = 15\,°\text{C}$.

**Solution**

From Eqs. 6.28a, 6.28b and Example 6.16, we have

$$a_0 = \frac{F'\left[\left\{F_0 U_{tc,a} + U_b\right\} - F_0 \eta_{mo} \beta_0 \beta_c I(t)\right]}{\dot{m}_f C_f}$$

$$= \frac{0.82[\{0.2373 \times 9.12 + 0.70\} - 0.2373 \times 0.12 \times 0.00425 \times 0.89 \times 500]}{0.035 \times 4190}$$

$$= \frac{0.82[2.8642 - 0.05386]}{146.65} = \frac{2.3045}{146.65} = 0.0158$$

and,

$$\frac{\left\{F'\left[(\alpha_c \tau_g)_{meff}\right]_{semi} I(t) + F'\left[F_0 U_{tc,a} + U_b\right]T_a\right\}}{F'\left[\left\{F_0 U_{tc,a} + U_b\right\} - F_0 \eta_{mo} \beta_0 \beta_c I(t)\right]}$$

$$= \frac{\left\{F'\left[(\alpha_c \tau_g)_{meff}\right]_{semi} I(t) + F'\left[F_0 U_{tc,a} + U_b\right]T_a\right\}}{F' U_{Lm-eff}}$$

$$= \frac{0.82 \times 0.2518 \times 500 + 0.82[0.2373 \times 9.12 + 0.7] \times 16}{0.82[\{0.2373 \times 9.12 + 0.70\} - 0.2373 \times 0.12 \times 0.00425 \times 0.89 \times 500]}$$

$$= \frac{103.238 + 37.578}{2.3045} = \frac{140.816}{2.3045} = 61.105$$

(a)  **For $N = 1$**

From Eq. 6.35a, we have

$$T_{f01} = 61.105\left[1 - \exp(-0.0158 \times 2)\right] + 15 \exp(-0.0158 \times 2)$$
$$= 61.105 \times (1 - 0.9689) + 15 \times 0.9689 = 16.43 \,^{\circ}\mathrm{C}$$

(b)  **For $N = 2$**

$$T_{f02} = 61.105\left[1 - \exp(-2 \times 0.0158 \times 2)\right] + 15 \times \exp(-2 \times 0.0158 \times 2)$$
$$= 61.105 \times (1 - 0.9418) + 15 \times 0.9418 = 17.68 \,^{\circ}\mathrm{C}$$

(c)  **For $N = 3$**

$$T_{f03} = 61.105\left[1 - \exp(-3 \times 0.0158 \times 2)\right] + 15 \times \exp(-3 \times 0.0158 \times 2)$$
$$= 61.105 \times (1 - 0.9096) + 15 \times 0.9096 = 19.17 \,^{\circ}\mathrm{C}$$

(d)  **For $N = 4$**

$$T_{f04} = 61.105\left[1 - \exp(-4 \times 0.0158 \times 2)\right] + 15 \times \exp(-4 \times 0.0158 \times 2)$$
$$= 61.105 \times (1 - 0.8813) + 15 \times 0.8813 = 20.47 \,^{\circ}\mathrm{C}$$

(e) **For $N = 5$**

$$T_{f05} = 61.105\left[1 - \exp(-5 \times 0.0158 \times 2)\right] + 15 \times \exp(-5 \times 0.0158 \times 2)$$
$$= 61.105 \times (1 - 0.8539) + 15 \times 0.8539 = 21.72\,°C$$

**Example 6.22** Repeat Example 6.21 **for constant electrical efficiency** $(\eta_m)$ of SPVT collector for packing factor $(\beta_c)$ of 0.89 to determine the outlet fluid temperature at end of 5th PVT collector.

Given: $\left[(\alpha_c\tau_g)_{meff}\right]_{semi} = 0.2516$ (Example 6.15), $U_{Lm-eff} = \{F_0U_{tc,a} + U_b\}$.

**Solution**

From Eqs. 6.28a, 6.28b and Example 6.16, we have the following for constant electrical efficiency $(\eta_m)$:

$$a_0 = \frac{F'\left[\{F_0U_{tc,a} + U_b\}\right]}{\dot{m}_fC_f} = \frac{0.82[\{0.2373 \times 9.12 + 0.70\}]}{0.035 \times 4190}$$
$$= \frac{0.82[2.8642]}{146.65} = \frac{2.3486}{146.65} = 0.016$$

and,

$$\frac{\left\{F'\left[(\alpha_c\tau_g)_{meff}\right]_{semi}I(t) + F'\left[F_0U_{tc,a} + U_b\right]T_a\right\}}{F'\left[\{F_0U_{tc,a} + U_b\}\right]}$$

$$= \frac{\left\{F'\left[(\alpha_c\tau_g)_{meff}\right]_{semi}I(t) + F'\left[F_0U_{tc,a} + U_b\right]T_a\right\}}{F'U_{Lm}}$$

$$= \frac{0.82 \times 0.2518 \times 500 + 0.82[0.2373 \times 9.12 + 0.7] \times 16}{0.82[\{0.2373 \times 9.12 + 0.70\}]}$$

$$= \frac{103.238 + 37.578}{2.3486} = \frac{140.816}{2.3486} = 59.96$$

From Eq. 6.35a, we have **an expression for constant electrical efficiency** $(\eta_m)$ **as**

$$T_{foN} = \frac{\left\{\left[(\alpha_c\tau_g)_{meff}\right]_{semi}I(t) + U_{Lm}T_a\right\}}{U_{Lm-eff}}\left[1 - \exp\left(-\frac{NF'A_mU_{Lm-eff}}{\dot{m}_fC_f}\right)\right]$$
$$+ T_{fi}\exp\left(-\frac{NF'A_mU_{Lm-eff}}{\dot{m}_fC_f}\right)$$

Here, $U_{Lm-eff} = \left[\{F_0U_{tc,a} + U_b\}\right]$ and $a_0 = \frac{F'\left[\{F_0U_{tc,a}+U_b\}\right]}{\dot{m}_fC_f} = \frac{F'U_{Lm-eff}}{\dot{m}_fC_f}$.

After substituting the appropriate numerical value, one gets

$$
\begin{aligned}
T_{f05} &= 59.96\left[1 - \exp(-5 \times 0.016 \times 2)\right] + 15 \times \exp(-5 \times 0.016 \times 2) \\
&= 59.96 \times (1 - 0.8521) + 15 \times 0.8521 = 21.726
\end{aligned}
$$

There is not much difference in outlet fluid temperature evaluated in Examples 6.21 and 6.22.

**Example 6.23** Evaluate an average temperature of each PVT collector for $N = 1$–$5$ by using the data of Example 6.21.

**Solution**

The average fluid temperature of each PVT collector is as follows:

(a) **For $N = 1$:** $T_{f01} = 16.43\,°C$ and $T_{fi} = 15\,°C$ (Example 6.21), then

$$
\overline{T}_{f1} = \frac{T_{f01} + T_{fi}}{2} = \frac{16.43 + 15}{2} = 15.715\,°C
$$

(b) **For $N = 2$:** $T_{f02} = 17.68\,°C$ and $T_{f01} = 16.43\,°C$ (Example 6.21), then

$$
\overline{T}_{f2} = \frac{T_{f02} + T_{f01}}{2} = \frac{17.68 + 16.43}{2} = 17.005\,°C
$$

(c) **For $N = 3$:** $T_{f03} = 19.17\,°C$ and $T_{f02} = 17.68\,°C$ (Example 6.21), then

$$
\overline{T}_{f3} = \frac{T_{f03} + T_{f02}}{2} = \frac{19.17 + 17.68}{2} = 18.425\,°C
$$

(d) **For $N = 4$:** $T_{f04} = 20.47\,°C$ and $T_{f03} = 19.17\,°C$ (Example 6.21), then

$$
\overline{T}_{f4} = \frac{T_{f04} + T_{f03}}{2} = \frac{20.47 + 19.17}{2} = 19.82\,°C
$$

(e) **For $N = 5$:** $T_{f05} = 21.72\,°C$ and $T_{f04} = 20.47\,°C$ (Example 6.21), then

$$
\overline{T}_{f5} = \frac{T_{f05} + T_{f04}}{2} = \frac{21.72 + 20.47}{2} = 21.095\,°C
$$

**Example 6.24** Compute the average solar cell temperature for each PVT collector by using the average data of fluid temperature from Example 6.22.

**Solution**

From Eq. 6.35c, we have an expression for an average solar cell temperature as

$$
\overline{T}_{cN} = \frac{(\alpha_c \tau_g)_{meff} I(t) + U_{tc,a} T_a + U_{bc,p} \overline{T}_{fN}}{U_{tc,a} + U_{bc,p} - \eta_{mo} \beta_0 \beta_c I(t)}
$$

where, $(\alpha_c \tau_g)_{meff} = 0.6428$ for $\beta_c = 0.89$ (Example 6.13).

$U_{t,ca} = 9.12\,\text{W/m}^2\,^\circ\text{C}$ (Example 6.1), $U_{b,cp} = 2.7663\,\text{W/m}^2\,^\circ\text{C}$ (Example 6.2).

$\beta_0 = 0.00425\,^\circ\text{C}^{-1}$; $\eta_{m0} = 0.12$; packing factor $(\beta_c) = 0.89$ and $I(t) = 500\,\text{W/}$ $\text{m}^2$ (Example 6.4), $T_a = 16\,^\circ\text{C}$.

The average solar temperature of each PVT collector is as follows:

(a) **For $N = 1$:** $\overline{T}_{f1} = 15.715\,^\circ\text{C}$ (Example 6.22), then

Substitute an appropriate known value in the expression for average solar cell temperature, one gets

$$
\begin{aligned}
\overline{T}_{c1} &= \frac{0.6428 \times 500 + 9.12 \times 16 + 2.7663 \times 15.715}{9.12 + 2.7663 - 0.12 \times 0.00425 \times 0.89 \times 500} \\
&= \frac{467.32 + 2.7663 \times 15.715}{11.8863 - 0.227} = \frac{467.32 + 43.47}{11{,}6593} = 43.81\,^\circ\text{C}
\end{aligned}
$$

(b) **For $N = 2$:** $\overline{T}_{f2} = 17.005\,^\circ\text{C}$ (Example 6.22), then similarly

$$
\overline{T}_{c2} = \frac{467.32 + 2.7663 \times 17.005}{11.8863 - 0.227} = \frac{467.32 + 47.04}{11{,}6593} = 44.12\,^\circ\text{C}
$$

(c) **For $N = 3$:** $\overline{T}_{f3} = 18.425\,^\circ\text{C}$ (Example 6.22), then similarly

$$
\overline{T}_{c3} = \frac{467.32 + 2.7663 \times 18.425}{11.8863 - 0.227} = \frac{467.32 + 50.97}{11{,}6593} = 44.45\,^\circ\text{C}
$$

(d) **For $N = 4$:** $\overline{T}_{f4} = 19.82\,^\circ\text{C}$ (Example 6.22), then similarly

$$
\overline{T}_{c4} = \frac{467.32 + 2.7663 \times 19.82}{11.8863 - 0.227} = \frac{467.32 + 54.83}{11{,}6593} = 44.78\,^\circ\text{C}
$$

(e) **For $N = 5$:** $\overline{T}_{f5} = 21.095$ (Example 6.22), then similarly

$$
\overline{T}_{c5} = \frac{467.32 + 2.7663 \times 21.095}{11.8863 - 0.227} = \frac{467.32 + 58.35}{11{,}6593} = 45.08\,^\circ\text{C}
$$

The average solar cell temperature can be evaluated as follows:

$$
\begin{aligned}
\overline{T}_{cN} &= \frac{\overline{T}_{c1} + \overline{T}_{c2} + \overline{T}_{c3} + \overline{T}_{c4} + \overline{T}_{c1}}{5} \\
&= \frac{43.81 + 44.12 + 44.45 + 44.78 + 45.08}{5} \\
&= 44.45
\end{aligned}
$$

### 6.4.1 Electrical Power from N-SPVT Collectors Connected in Series

From Eqs. 6.31a, 6.31b, the electrical efficiency is given by

$$\eta_{mN} = \eta_{mo}\left[1 - \beta_0\left(\overline{T}_{cN} - 25\right)\right]$$

So, an electrical power can be obtained as

$$\dot{E}_{el} = \eta_{mo}\left[1 - \beta_0\left(\overline{T}_{cN} - 25\right)\right] \times \beta_c I(t) A_m$$

**Example 6.25** Compute an electrical efficiency and electrical power for each SPVT collector by using the average data of average solar cell temperature from Example 6.23.

**Solution**

(a) **For $N = 1$:** $\overline{T}_{c1} = 43.81\,°C$ (Example 6.23), then

From Eqs. 6.31a, 6.31b, the electrical efficiency is given by

$$\eta_{m1} = \eta_{mo}\left[1 - \beta_0\left(\overline{T}_{c1} - 25\right)\right] = 0.12[1 - 0.00425(43.81 - 25)]$$
$$= 0.1104(11.04\%)$$

So, an electrical power can be obtained as

$$\dot{E}_{el1} = \eta_m \times \beta_c I(t) A_m = 0.1104 \times 0.89 \times 500 \times 2 = 98.26\,W$$

(b) **For $N = 2$:** $\overline{T}_{c2} = 44.12\,°C$ (Example 6.23), then similarly

$$\eta_{m2} = 0.12[1 - 0.00425(44.12 - 25)] = 0.110254(11.025\%)$$

So, an electrical power can be obtained as

$$\dot{E}_{el2} = \eta_m \times \beta_c I(t) A_m = 0.110254 \times 0.89 \times 500 \times 2 = 98.12\,W$$

(c) **For $N = 3$:** $\overline{T}_{c3} = 44.45\,°C$ (Example 6.23), then similarly

$$\eta_{m3} = 0.12[1 - 0.00425(44.45 - 25)] = 0.11008(11.008\%)$$

So, an electrical power can be obtained as

$$\dot{E}_{el3} = \eta_m \times \beta_c I(t) A_m = 0.11008 \times 0.89 \times 500 \times 2 = 97.97\,W$$

(d) **For $N = 4$:** $\overline{T}_{c4} = 44.78\,^\circ\text{C}$ (Example 6.23), then similarly

$$\eta_{m4} = 0.12[1 - 0.00425(44.78 - 25)] = 0.1099(10.99\%)$$

So, an electrical power can be obtained as

$$\dot{E}_{el4} = \eta_m \times \beta_c I(t) A_m = 0.1099 \times 0.89 \times 500 \times 2 = 97.81\text{ W}$$

(e) **For $N = 5$:** $\overline{T}_{c5} = 45.08\,^\circ\text{C}$ (Example 6.23), then similarly

$$\eta_{m5} = 0.12[1 - 0.00425(45.08 - 25)] = 0.1098(10.99\%)$$

So, an electrical power can be obtained as

$$\dot{E}_{el5} = \eta_m \times \beta_c I(t) A_m = 0.1098 \times 0.89 \times 500 \times 2 = 97.68\text{ W}$$

The total electric power from five PVT collector is sum of five electrical power obtained above. Hence,

The total electrical power $(\dot{E}_{elT}) = \dot{E}_{el1} + \dot{E}_{el2} + \dot{E}_{el3} + \dot{E}_{el4} + \dot{E}_{el5} = 98.26 + 98.12 + 97.97 + 97.81 + 97.68 = 489.84$ W.

Electrical efficiency of 5-SPVT collector $(\overline{\eta}_{m5}) = \frac{489.84}{500 \times 2 \times 5} = 0.0979$.

***Example 6.26*** Calculate an average electrical efficiency and total electrical power by using Eqs. 6.35a, 6.35b and 6.35c and data of Example 6.21 for **variable electrical efficiency** $(\eta_m)$ of SPVT collector.

**Solution**

From Eq. 6.35a–6.35c we have an expression for an average fluid and solar cell temperature as follows:

$$\overline{T}_{fN} = \frac{\left\{\left[(\alpha_c \tau_g)_{\text{meff}}\right]_{\text{semi}} I(t) + [F_0 U_{tc,a} + U_b] T_a\right\}}{U_{Lm-\text{eff}}} \left[1 - \frac{1 - \exp\left(-\frac{N F' A_m U_{Lm-\text{eff}}}{\dot{m}_f C_f}\right)}{\frac{N F' A_m U_{Lm-\text{eff}}}{\dot{m}_f C_f}}\right]$$

$$+ T_{fi} \frac{1 - \exp\left(-\frac{N F' A_m U_{Lm-\text{eff}}}{\dot{m}_f C_f}\right)}{\frac{N F' A_m U_{Lm-\text{eff}}}{\dot{m}_f C_f}}$$

Substitute an appropriate value from Example 6.21, we have

$$\overline{T}_{fN} = 61.105\left[1 - \frac{1 - \exp(-5 \times 0.0158 \times 2)}{5 \times 0.0158 \times 2}\right] + 15 \times \frac{1 - \exp(-5 \times 0.0158 \times 2)}{5 \times 0.0158 \times 2}$$

$$= 61.105\left[1 - \frac{1 - 0.8539}{0.158}\right] + 15 \times \frac{1 - 0.8539}{0.158}$$

$$= 61.105 \times (1 - 0.9247) + 15 \times 0.9247 = 18.47\,^\circ\text{C}$$

and,

$$\overline{T}_{cN} = \frac{(\alpha_c \tau_g)_{meff} I(t) + U_{tc,a} T_a + U_{bc,p} \overline{T}_{fN}}{U_{tc,a} + U_{bc,p} - \eta_{mo} \beta_0 \beta_c I(t)}$$

$$= \frac{467.32 + 2.7663 \times 18.47}{11.8863 - 0.227} = 44.46\,°C$$

Now,

$$\eta_{m5} = 0.12[1 - 0.00425(44.46 - 25)] = 0.11(11\%)$$

So, an electrical power from five PVT collectors can be obtained as

$$\dot{E}_{el5} = \eta_m \times \beta_c I(t) A_m \times N = 0.11 \times 0.89 \times 500 \times 2 \times 5 = 489.83\,W$$

which is exactly same as obtained in Example 6.24.

### 6.4.2 The Rate of Thermal Energy and Mass Flow Rate Factor from Semi-transparent N-PVT (N-SPVT) Collector Connected in Series

The rate of thermal energy form N-PVT collector connected in series for **variable electrical efficiency** $(\eta_m)$ of PVT collector is given by

$$\dot{Q}_{uN} = \acute{m}_f C_f (T_{f0} - T_{fi}) = F_{Rm-N} N A_m \Big[ \{ [(\alpha_c \tau_g)_{meff}]_{semi} I(t) + [F_0 U_{tc,a} + U_b] T_a \}$$
$$- [\{F_0 U_{tc,a} + U_b\} - F_0 \eta_{mo} \beta_0 \beta_c I(t)] T_{fi} \Big] \tag{6.36a}$$

where, the  mass flow rate factor, $F_{Rm-N} = \frac{\acute{m}_f C_f}{N A_m U_{Lm-eff}} \Big[ 1 - \exp\Big( -\frac{N F' A_m U_{Lm-eff}}{\acute{m}_f C_f} \Big) \Big]$

**For constant** $\eta_m$, thermal expansion coefficient is zero, i.e., $\beta_0 = 0$, then Eqs. 6.33a–6.33c reduces to

$$\dot{Q}_{uN} = F_{Rm-N} A_c N \Big[ [(\alpha_c \tau_g)_{meff}] I(t) - U_{Lm}(T_{fi} - T_a) \Big] \tag{6.36b}$$

with mass flow rate factor $= F_{Rm-N} = \frac{\acute{m}_f C_f}{N A_m U_{Lm}} \Big[ 1 - \exp\Big( -\frac{N F' A_m U_{Lm}}{\acute{m}_f C_f} \Big) \Big]$ and $U_{Lm} = [\{F_0 U_{tc,a} + U_b\}]$.

***Example 6.27*** Evaluate the rate of thermal energy and mass flow rate factor for Example 6.21.

## Solution

For **variable electrical efficiency** $\left(\eta_m\right)$ of PVT collector.
   From Eq. 6.36a, we have an expression for mass flow rate as

$$F_{Rm-N} = \frac{\acute{m}_f C_f}{N A_m U_{Lm-\mathrm{eff}}}\left[1 - \exp\left(-\frac{N F' A_m U_{Lm-\mathrm{eff}}}{\acute{m}_f C_f}\right)\right]$$

Here,

$$
\begin{aligned}
U_{Lm-\mathrm{eff}} &= \left[\{F_0 U_{tc,a} + U_b\} - F_0 \eta_{mo}\beta_0\beta_c I(t)\right] \\
&= [\{0.2373 \times 9.12 + 0.70\} - 0.2373 \times 0.12 \times 0.00425 \times 0.89 \times 500] \\
&= 2.81 \,\mathrm{W/m^2\,°C}
\end{aligned}
$$

Then,

$$
\begin{aligned}
F_{Rm-N} &= \frac{0.035 \times 4190}{5 \times 2 \times 2.81}\left[1 - \exp\left(-\frac{5 \times 0.82 \times 2 \times 2.81}{0.035 \times 4190}\right)\right] \\
&= 5.219 \times (1 - 0.8545) = 0.76
\end{aligned}
$$

For constant $\eta_m$, mass flow rate will be
   $U_{Lm-\mathrm{eff}} = \left[\{F_0 U_{tc,a} + U_b\}\right] = \{0.2373 \times 9.12 + 0.70\} = 2.8642$ for constant
$\eta_m$

$$
\begin{aligned}
F_{Rm-N} &= \frac{\acute{m}_f C_f}{N A_m U_{Lm}}\left[1 - \exp\left(-\frac{N F' A_m U_{Lm}}{\acute{m}_f C_f}\right)\right] \\
&= \frac{0.035 \times 4190}{5 \times 2 \times 2.8642}\left[1 - \exp\left(-\frac{5 \times 0.82 \times 2 \times 2.8642}{0.035 \times 4190}\right)\right] \\
&= 5.12 \times (1 - 0.85197) = 0.7579
\end{aligned}
$$

(a)   **First method:** $T_{f05} = 21.72\,°\mathrm{C}$ and $T_{f04} = 20.47\,°\mathrm{C}$ (Example 6.21).

   The rate of useful thermal energy, $\dot{Q}_{uN} = \acute{m}_f C_f \left(T_{f05} - T_{fi}\right) = 0.035 \times 4190(21.72 - 15) = 985.49\,\mathrm{W}.$

(b)   Second Method: Here,

$$
\begin{aligned}
U_{Lm-\mathrm{eff}} &= \left[\{F_0 U_{tc,a} + U_b\} - F_0 \eta_{mo}\beta_0\beta_c I(t)\right] \\
&= [\{0.2373 \times 9.12 + 0.70\} - 0.2373 \times 0.12 \times 0.00425 \times 0.89 \times 500] \\
&= 2.81 \,\mathrm{W/m^2\,°C}
\end{aligned}
$$

$$\left[(\alpha_c \tau_g)_{m\mathrm{eff}}\right]_{\mathrm{semi}} = 0.2518 \,\mathrm{W/m^2\,°C} \text{ (Example 6.15)}$$

$$\left[\{F_0 U_{tc,a} + U_b\}\right] = \{0.2373 \times 9.12 + 0.70\} = 2.8642$$

$$\dot{Q}_{uN} = F_{Rm-N} N A_c$$

$$\left[ \left\{ \left[ (\alpha_c \tau_g)_{\text{meff}} \right]_{\text{semi}} I(t) + \left[ F_0 U_{tc,a} + U_b \right] T_a \right\} - \left[ \left\{ F_0 U_{tc,a} + U_b \right\} - F_0 \eta_{mo} \beta_0 \beta_c I(t) \right] T_{fi} \right]$$

$$= 0.7579 \times 5 \times 2[\{0.2518 \times 500 + 2.8642 \times 16\} - 2.81 \times 15]$$

$$= 984.5 \, \text{W}$$

By both methods, the rate of thermal energy from five PVT collectors connected in series is coming approximately equal.

Further, **for constant** $\eta_m$: $\left[ (\alpha_c \tau_g)_{\text{meff}} \right]_{\text{semi}} = 0.2516$ (Example 6.15) for $\beta_c = 0.89$ (Example 6.13), $U_{Lm-\text{eff}} = \left[ \{ F_0 U_{tc,a} + U_b \} \right] = 2.8642 \, \text{W/m}^2 \, °C$.

The rate of useful thermal energy will be

$$\dot{Q}_{uN} = F_{Rm-N} N A_m \left[ \left\{ \left[ (\alpha_c \tau_g)_{\text{meff}} \right]_{\text{semi}} I(t) + \left[ F_0 U_{tc,a} + U_b \right] T_a \right\} - \left[ \{ F_0 U_{tc,a} + U_b \} - \right] T_{fi} \right]$$

$$= 0.7579 \times 5 \times 2[\{0.2516 \times 500 - 2.8642 \times (15 - 16)\}]$$

$$= 0.7579 \times 5 \times 2[324.2642] = 975.15 \, \text{W}$$

There is difference of 9.3 W between $\dot{Q}_{uN}$ obtained by considering constant and variable $\eta_m$ for five PVT collector connected in series.

***Example 6.28*** Calculate an overall rate of exergy for five SPVT collectors connected in series by using the data of Examples 5.25 and 5.26.

**Solution**

Known: $T_{f05} = 21.72 \, °C$, $T_a = 16 \, °C$ and $T_{fi} = 15 \, °C$ (Example 6.21).

The rate of thermal exergy has been evaluated as follows:

$$\dot{Ex}_{th5} = \acute{m}_f C_f \left[ (T_{f05} - T_{fi}) - (273 + T_a) \ln \frac{273 + T_{f05}}{273 + T_{fi}} \right]$$

$$= 0.035 \times 4190 \left[ (21.72 - 15) - (273 + 16) \ln \frac{273 + 21.72}{273 + 15} \right]$$

$$= 146.65[6.72 - 289 \times 0.023] = 10.71 \, \text{W}$$

With higher number of PVT collectors connected in series thermal exergy is more than zero. It may be due to higher outlet temperature at end of PVT collectors unlike single PVT collector.

The electrical power $= \dot{E}_{el5} = 489.83 \, \text{W}$ (Example 6.25).

So, the overall rate of exergy will be sum of the rate of thermal exergy and electrical power which can be written as

$$\dot{Ex}_{\text{overall/total}} \dot{Ex}_{th5} + \dot{E}_{el5} = 10.71 + 489.83 = 500.53 \, \text{W}$$

Overall exergy efficiency of PVT collector can be obtained as follows:

$$\eta_{i5} = \frac{\dot{E}x_{\text{overall/total}}}{[I(t)]_{ex} \times A_m} = \frac{500.53}{0.9513 \times 500 \times 5 \times 2} = 0.10522[10.522\%]$$

The overall exergy of five PVT collector is higher by 1.44% in comparison with single SPVT collector (0.10522, Example 6.20) due to increase in rate of thermal exergy.

***Example 6.29*** Repeat the Example 6.16 for mass flow rate $\left(\dot{m}_f\right)$ of 0.0175 kg/s for **variable electrical efficiency** $\left(\eta_m\right)$ **for** $N = 1$.

**Solution:** From Eqs. 6.28a, 6.28b, we have

$$
\begin{aligned}
a_0 &= \frac{F'[\{F_0 U_{tc,a} + U_b\} - F_0 \eta_{mo} \beta_0 \beta_c I(t)]}{\dot{m}_f C_f} \\
&= \frac{0.82[\{0.2373 \times 9.12 + 0.70\} - 0.2373 \times 0.12 \times 0.00425 \times 0.89 \times 500]}{0.0175 \times 4190} \\
&= \frac{0.82[2.8642 - 0.05386]}{73.325} = \frac{2.3045}{73.325} = 0.03143 \text{ (Example 6.16)}
\end{aligned}
$$

and,

$$\frac{\{F'[(\alpha_c \tau_g)_{\text{meff}}]_{\text{semi}} I(t) + F'[F_0 U_{tc,a} + U_b] T_a\}}{F'[\{F_0 U_{tc,a} + U_b\} - F_0 \eta_{mo} \beta_0 \beta_c I(t)]} = 61.105 \text{ (Example 6.16)}$$

From Eqs. 6.28a, 6.28b, we have

$$
\begin{aligned}
T_{f0} &= 61.105[1 - \exp(-0.03143 \times 2)] + 15 \exp(-0.03143 \times 2) \\
&= 61.105 \times (1 - 0.9390) + 15 \times 0.9390 = 17.81\,°C
\end{aligned}
$$

The above outlet fluid temperature becomes higher due to decrease of mass flow rate. In this case, the heat transfer becomes more due to slow movement of water. So, the rate of thermal exergy will be

$$
\begin{aligned}
\dot{E}x_{th} &= \dot{m}_f C_f\left[(T_{f05} - T_{fi}) - (273 + T_a)\ln\frac{273 + T_{f05}}{273 + T_{fi}}\right] \\
&= 0.0175 \times 4190\left[(17.81 - 15) - (273 + 16)\ln\frac{273 + 17.81}{273 + 15}\right] \\
&= 73.325[2.81 - 289 \times 0.097] = 0.29\,W
\end{aligned}
$$

Average fluid temperature can be obtained as

$$\overline{T}_f = 61.105\left[1 - \frac{1 - \exp(-0.03143 \times 2)}{0.03143 \times 2}\right] + 15 \times \frac{1 - \exp(-0.03143 \times 2)}{0.03143 \times 2}$$

$$= 61.105 \times \left(1 - \frac{1 - 0.9390}{0.06286}\right) + 15 \times \frac{1 - 0.9390}{0.06286} = 16.34\,°C$$

Following Example 6.24, an average solar cell temperature for $N = 1$ is obtained as

$$\overline{T}_c = \frac{467.32 + 2.7663 \times 16.34}{11.8863 - 0.227} = \frac{467.32 + 45.20}{11{,}6593} = 43.95\,°C$$

The average fluid temperature becomes higher in comparison with Example 6.18 due to low mass flow rate.

Substitute an appropriate value in Eqs. 6.31a, 6.31b from Example 6.18 to get an electrical efficiency and an electrical power of semi-transparent PVT collector.

$$\eta_m = 0.12[1 - 0.00425(43.95 - 25)] = 0.1103(11.03\%)$$

which is up by 0.049% in comparison with opaque PV module (Example 6.9).

$$\dot{E}_{el} = 0.1103 \times 0.89 \times 500 \times 2 = 98.167\,W$$

Hence, the mass flow plays an important role in determining the outlet fluid temperature along with rate of thermal exergy.

**Example 6.30** Repeat Example 6.29 for packing factor of 0.5 **for variable electrical efficiency of PVT collector** for $N = 1$.

**Solution**

$(\alpha_c \tau_g)_{meff} = 0.3611$, $\left[(\alpha_c \tau_g)_{meff}\right]_{semi} = 0.5362$, $F_0 = 0.2353$ (Example 6.15).

From Example 6.29 for $\beta_c = 0.50$, we have

$$a_0 = \frac{F'[\{F_0 U_{tc,a} + U_b\} - F_0 \eta_{mo} \beta_0 \beta_c I(t)]}{\dot{m}_f C_f}$$

$$= \frac{0.82[\{0.2353 \times 9.12 + 0.70\} - 0.2353 \times 0.12 \times 0.00425 \times 0.50 \times 500]}{0.0175 \times 4190}$$

$$= \frac{0.82[2.8459 - 0.03]}{73.325} = \frac{2.309}{73.325} = 0.03149\,(\text{Example } 6.16)$$

and,

$$\frac{\left\{F'[(\alpha_c \tau_g)_{meff}]_{semi} I(t) + F'[F_0 U_{tc,a} + U_b] T_a\right\}}{F'[\{F_0 U_{tc,a} + U_b\} - F_0 \eta_{mo} \beta_0 \beta_c I(t)]}$$

$$= \frac{0.82 \times 0.5362 \times 500 + 0.82[0.2353 \times 9.12 + 0.7] \times 16}{0.82[\{0.2353 \times 9.12 + 0.70\} - 0.2353 \times 0.12 \times 0.00425 \times 0.50 \times 500]}$$

$$= \frac{219.84 + 37.339}{2.3036} = \frac{140.816}{2.3045} = 111.64$$

From Eq. 6.28a, 6.28b, we have

$$T_{f0} = 111.64[1 - \exp(-0.03149 \times 2)] + 15 \times \exp(-0.03149 \times 2)$$
$$= 111.64 \times (1 - 0.939) + 15 \times 0.939 = 20.89\,°C$$

The above outlet fluid temperature becomes higher due to decrease of packing factor. In this case, there is more direct gain to absorber due more transparent area in PVT collector.

So, the rate of thermal exergy will be

$$\dot{Ex}_{th} = \dot{m}_f C_f \left[ (T_{f05} - T_{fi}) - (273 + T_a) \ln \frac{273 + T_{f05}}{273 + T_{fi}} \right]$$
$$= 0.0175 \times 4190 \left[ (20.89 - 15) - (273 + 16) \ln \frac{273 + 20.89}{273 + 15} \right]$$
$$= 73.325[5.89 - 289 \times 0.0202] = 3.83\,W$$

Average fluid temperature can be obtained as

$$\overline{T}_f = 111.64 \left[ 1 - \frac{1 - \exp(-0.03149 \times 2)}{0.03143 \times 2} \right] + 15 \times \frac{1 - \exp(-0.03149 \times 2)}{0.03143 \times 2}$$
$$= 111.64 \times \left( 1 - \frac{1 - 0.9390}{0.06286} \right) + 15 \times \frac{1 - 0.9390}{0.06286} = 17.80\,°C$$

Following Example 6.24, an average solar cell temperature for $N = 1$ is obtained as

$$\overline{T}_c = \frac{467.32 + 2.7663 \times 17.80}{11.8863 - 0.227} = \frac{467.32 + 49.24}{11,6593} = 44.31\,°C$$

The average fluid temperature becomes higher in comparison with Example 6.18 due to low mass flow rate.

Substitute an appropriate value in Eqs. 6.31a, 6.31b from Example 6.18 to get an electrical efficiency and an electrical power of semi-transparent PVT collector.

$$\eta_m = 0.12[1 - 0.00425(44.31 - 25)] = 0.11015(11.015\%)$$

Now, an electrical power will be

$$\dot{E}_{el} = 0.11015 \times 0.50 \times 500 \times 2 = 55.08\,W$$

This shows that as packing factor decreases, the rate of thermal energy and its exergy increases: however, electrical power decreases.

### 6.4.3  Instantaneous Thermal Efficiency

**For variable** $\eta_m$ means temperature dependent electrical efficiency of PV module, instantaneous thermal efficiency is given by

$$\eta_{iN} = \frac{\dot{Q}_u}{I(t) \times A_m \times N} = F_{Rm-N}\left[\left[(\alpha_c\tau_g)_{meff}\right]_{semi}\right.$$
$$\left. + \frac{\left[F_0 U_{tc,a} + U_b\right]T_a - \left[\{F_0 U_{tc,a} + U_b\} - F_0\eta_{mo}\beta_0\beta_c I(t)\right]T_{fi} + \left[F_0 U_{tc,a} + U_b\right]T_a}{I(t)}\right]$$

$$(6.37a)$$

The instantaneous thermal efficiency, **for constant** $\eta_m$ means $\beta_0 = 0$ in Eq. 6.37a, can be derived as follows:

$$\eta_i = \frac{\dot{Q}_u}{N \times I(t) \times A_m} = F_{Rm-N}\left[\left[(\alpha_c\tau_g)_{meff}\right] - U_{Lm-\text{eff}}\frac{(T_{fi} - T_a)}{I(t)}\right]$$

$$\text{with } U_{Lm-\text{eff}} = \left[\{F_0 U_{tc,a} + U_b\}\right]$$

$$\text{and } F_{Rm-N} = \frac{\dot{m}_f C_f}{A_c U_{Lm-\text{eff}}}\left[1 - \exp\left(-\frac{N F' A_c U_{Lm-\text{eff}}}{\dot{m}_f C_f}\right)\right] \qquad (6.37b)$$

**Example 6.31** Compute the instantaneous thermal efficiency for Example 6.26.

**Solution**

From Eq. 6.27a, we have

$$\eta_{iN} = \frac{\dot{Q}_u}{I(t) \times A_m \times N} = \frac{984.5}{500 \times 2 \times 5} = 0.1969[19.69\%]$$

## 6.5  The Outlet Fluid Temperature $(T_{foN})$ at $m^{th}$ SPVT Collectors Connected in Parallel

Following Eq. 4.32a for parallel connection of m semi-transparent liquid PVT (SPVT) collector, the outlet of each SPVT collector will remain the same. The expression for outlet of each SPVT collector for variable $(\eta_m)$ will be written as

$$T_{fom} = \frac{\left\{\left[(\alpha_c\tau_g)_{meff}\right]_{semi}I(t) + \left[F_0 U_{tc,a} + U_b\right]T_a\right\}}{U_{Lm-\text{eff}}}\left[1 - \exp\left(-\frac{F'm A_m U_{Lm-\text{eff}}}{\dot{m}_f C_f}\right)\right]$$
$$+ T_{fi}\exp\left(-\frac{F'm A_m U_{Lm-\text{eff}}}{\dot{m}_f C_f}\right)$$

$$\text{with } A_{cm} = m \times A_m \qquad (6.38)$$

Here, two cases will be discussed, namely

**Case (a):** by considering mass flow rate same at inlet of m-liquid SPVT collectors connected in parallel:

In this case, the mass flow rate $(\dot{m}_f)$ is equally divided in each collector $\frac{\dot{m}_f}{m}$ if $m =$ PVT collectors are connected in parallel. In this case each PVT collector has mass flow rate of $\frac{\dot{m}_f}{m}$ and the results are similar to those PVT collectors connected in series. The total rate of thermal energy and electrical power is also same for given number of PVT collector (N).

and,

**Case (b):** by considering variable mass flow rate at inlet:

In this case, the mass flow rate becomes m times the given mass flow rate $\dot{m}_f$ and in this case mass flow rate in each collector is $\dot{m}_f$ unlike case (a). However, the rate of thermal energy and exergy becomes $m$ times of case (a).

***Example 6.31*** Calculate the outlet temperature and the rate of thermal energy if five liquid flat plate collectors are connected in series for case (a) of Eq. 6.38.

**Solution**

From Eq. 6.38, we have

$$
T_{fom} = \frac{\left\{ \left[ (\alpha_c \tau_g)_{meff} \right]_{semi} I(t) + \left[ F_0 U_{tc,a} + U_b \right] T_a \right\}}{U_{Lm-eff}} \left[ 1 - \exp\left( -\frac{F' m A_m U_{Lm-eff}}{\dot{m}_f C_f} \right) \right]
$$
$$
+ T_{fi} \exp\left( -\frac{F' m A_m U_{Lm-eff}}{\dot{m}_f C_f} \right)
$$

with $A_{cm} = m \times A_c$ where,

$$
\begin{aligned}
U_{Lm-eff} &= \left[ \left\{ F_0 U_{tc,a} + U_b \right\} - F_0 \eta_{mo} \beta_0 \beta_c I(t) \right] \\
&= \left[ \{0.2373 \times 9.12 + 0.70\} - 0.2373 \times 0.12 \times 0.00425 \times 0.89 \times 500 \right] \\
&= 2.810334
\end{aligned}
$$

and,

$$
a_0 = \frac{F' \left[ \left\{ F_0 U_{tc,a} + U_b \right\} - F_0 \eta_{mo} \beta_0 \beta_c I(t) \right]}{\dot{m}_f C_f} = \frac{F' U_{Lm-eff}}{\dot{m}_f C_f} = 0.0158 \text{ (Example 6.16)}
$$

$$
\frac{\left\{ \left[ (\alpha_c \tau_g)_{meff} \right]_{semi} I(t) + \left[ F_0 U_{tc,a} + U_b \right] T_a \right\}}{U_{Lm-eff}}
$$

$$= \frac{0.82 \times 0.2518 \times 500 + 0.82[0.2373 \times 9.12 + 0.7] \times 16}{0.82[\{0.2373 \times 9.12 + 0.70\} - 0.2373 \times 0.12 \times 0.00425 \times 0.89 \times 500]}$$

$$= \frac{103.238 + 37.578}{2.3045} = \frac{140.816}{2.3045} = 61.105 \text{ (Example 6.16)}$$

Substitute above appropriate value for $m = 5$ and $A_m = 2$ m$^2$, one gets

$$T_{\text{fom}} = 61.105\big[1 - \exp(-0.0158 \times 5 \times 2)\big] + T_{fi}\exp(-0.0158 \times 5 \times 2)$$
$$= 61.105[1 - 0.8538] + 15 \times 0.8538] = 21.74\,°C$$

which is approximately same as Example 6.22 for five PVT collectors connected in series.

The rate of thermal energy in W will be

$$\dot{Q}_{um} = \dot{m}_f C_f \big(T_{f05} - T_{fi}\big) = 0.035 \times 4190(21.74 - 15) = 985.50 \text{ W (Example 6.27)}$$

**Example 6.32** Calculate the outlet temperature and the rate of thermal energy if five liquid SPVT collectors are connected in series for case (b) of Eq. 6.38.

**Solution**
In the case (b), Eq. 6.38 reduces to

$$T_{fom} = \frac{\left\{\big[(\alpha_c\tau_g)_{\text{meff}}\big]_{\text{semi}} I(t) + \big[F_0 U_{tc,a} + U_b\big]T_a\right\}}{U_{Lm-\text{eff}}} \left[1 - \exp\left(-\frac{F' A_m U_{Lm-\text{eff}}}{\dot{m}_f C_f}\right)\right]$$
$$+ T_{fi}\exp\left(-\frac{F' A_m U_{Lm-\text{eff}}}{\dot{m}_f C_f}\right)$$

Here, each liquid PVT collector of m-PVT collectors connected in parallel will have mass flow rate of 0.03535 kg/s $(\dot{m}_f)$.

$$T_{fom} = 61.105\big[1 - \exp(-0.0158 \times 2)\big] + T_{fi}\exp(-0.0158 \times 2)$$
$$= 61.105[1 - 0.9689] + 15 \times 0.9689] = 16.43\,°C$$

The rate of thermal energy in W will be

$$\dot{Q}_{um} = \dot{m}_f C_f \big(T_{f05} - T_{fi}\big) = 5 \times 0.035 \times 4190(16.43 - 15) = 1048.55 \text{ W}$$

This is clear that for case (a), one gets higher outlet temperature with lower value of the rate of thermal energy in comparison with case (b).

## 6.6  Liquid OPVT-TEC Collector

Figure 6.3 shows the cross-sectional view of one Al-base (opaque) photo-voltaic thermal (OPVT)-thermoelectric cooler (TEC). In place of Al-base, one can also consider opaque tedlar too depending upon requirement. In this case, thermoelectric cooler (TEC) with proper thickness and length is attached to Al-base of OPV module for fast heat transfer to cool solar cell as shown in Fig. 6.3. The packing factor of Al-base PV module has been considered as one while packing factor of TEC is less than one (Eq. 3.44). There is a provision of tube in plate collector below PVT-TEC collector through which liquid/water is allowed to pass to cool bottom of TEC. So, the temperature difference between top and bottom of TEC is created to produce electric current (Eq. 3.45). The top of TEC is attached to Al plate. Since Al is conducting material, hence the temperature of Al plate and top of TEC is assumed to be same. In this case, the absorber plate of tube in plate collector gets thermal energy from non-packing area of OPVT-TEC collector as well as from bottom of TEC through conduction through TEC material. The rest is same as in the case of conventional flat plate collector. In this case, one gets (a) electrical power from solar cell, (b) electrical power from TEC and thermal power to make the OPVT-TEC system more economical.

### 6.6.1  Energy Balance of OPVT–TEC Liquid Collector [5–7]

Following Dimri et al. [5–7], referring to Fig. 6.3 and Sect. 3.4 with some basic assumptions namely (a) quasi-steady-state condition, (b) neglecting heat capacity of each component (c) bottom of TEC is in direct contact with absorber plate $\{T_{\text{tec,bottom}} = T_p = T_f\}$ and perfectly insulated bottom of collector and (d) flexible PVT module is fully covered with solar cell, etc., these are basic energy balance for elemental area '$bdx$' of PVT-TEC liquid collector:

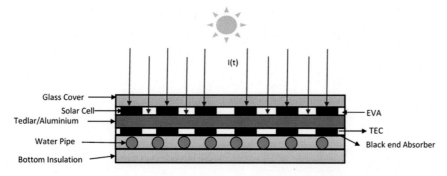

**Fig. 6.3**  Cross-sectional view of opaque tedlar/aluminum PV-TEC liquid collector

### (a)  For Solar Cell of Flexible Fully Covered with Solar Cell of OPV Module

Following Eq. 3.47a, the energy balance of PV module in W of PVT-TEC collector can be written as

$$\tau_g \alpha_{sc} I(t) b dx = U_{t,ca}(T_c - T_a)bdx + h_t\left(T_c - T_{\text{tec,top}}\right)\beta_{\text{tec}}bdx$$
$$+ U_{b,cp}\left(T_c - T_p\right)(1 - \beta_{tec})bdx + \eta_m I(t)bdx \qquad (6.39\text{a})$$

The third term of RHS in Eq. 6.39a $= U_{b,cp}\left(T_c - T_p\right)(1 - \beta_{tec})bdx =$ The rate of thermal energy lost to absorber plate from back of solar cell through non-packing area $((1 - \beta_{\text{tec}})A_m)$ of TEC, W.

For $\beta_{\text{tec}} = 0$ means $h_t = 0$, then Eq. 6.39a reduced to

$$\tau_g \alpha_{sc} I(t)bdx = U_{t,ca}(T_c - T_a)bdx + U_{b,cp}\left(T_c - T_p\right)bdx + \eta_m I(t)bdx \quad (6.39\text{b})$$

The numerical value of each design parameters of Eqs. 6.39a–6.39c is mostly similar to given in Sect. 3.4.

where, $U_{t,ca} = 9.30\,\text{W/m}^2\,^\circ\text{C}$ with wind velocity of 1 m/s; $h_t = 120\,\text{W/m}^2\,^\circ\text{C}$ and $U_{b,cp} = 2.80\,\text{W/m}^2\,^\circ\text{C}$ without wind velocity because the air between PVT-TEC and absorber plate is stagnate (Example 3.48).

From Eq. 6.39a, we have

$$T_c = \frac{\left(\tau_g\alpha_{sc} - \eta_m\right)I(t) + U_{t,ca}T_a + h_t T_{\text{tec,top}}\beta_{\text{tec}} + U_{b,cp}(1 - \beta_{\text{tec}})T_p}{U_{t,ca} + h_t\beta_{\text{tec}} + U_{b,cp}(1 - \beta_{\text{tec}})} \qquad (6.39\text{c})$$

### (b)  For Top of TEC

$$h_t\left(T_c - T_{\text{tec,top}}\right)\beta_{tec}bdx = U_{\text{tec}}\left(T_{\text{tec,top}} - T_{\text{tec,bottom}}\right)\beta_{\text{tec}}bdx \qquad (6.40\text{a})$$

From Eq. 6.39c, one can get

$$h_t\left(T_c - T_{\text{tec,top}}\right)\beta_{tec} = F_0(\alpha\tau)_{\text{eff}}I(t) - (U)_{\text{tec,top}-a}\left(T_{\text{tec,top}} - T_a\right)$$
$$- (U)_{\text{tec,top}-p}\left(T_{\text{tec,top}} - T_p\right) \qquad (6.40\text{b})$$

where,

The penalty factor, $F_0 = \frac{(h_t\beta_{\text{tec}})}{U_{t,ca}+h_t\beta_{\text{tec}}+U_{b,cp}(1-\beta_{\text{tec}})}$,

The product of absorptivity and transmittivity, $(\alpha\tau)_{\text{eff}} = \left(\tau_g\alpha_{sc} - \eta_m\right)$;

An overall heat transfer coefficient from top of TEC to ambient air, $(U)_{\text{tec,top}-a} = \frac{(h_t\beta_{\text{tec}})U_{t,ca}}{U_{t,ca}+h_t\beta_{\text{tec}}+U_{b,cp}(1-\beta_{\text{tec}})}$;

An overall heat transfer coefficient from top of TEC to absorber plate $(U)_{\text{tec,top}-p} = \frac{(h_t\beta_{\text{tec}})\{U_{b,cp}(1-\beta_{\text{tec}})\}}{U_{t,ca}+h_t\beta_{\text{tec}}+U_{b,cp}(1-\beta_{\text{tec}})}$.

Substitute $\left[h_t\left(T_c - T_{\text{tec,top}}\right)\beta_{\text{tec}}\right]$ from Eq. 6.40b into Eq. 6.40a with $T_{\text{tec,bottom}} = T_p$ as per assumption, we can have the following expression for top temperature of

TEC as

$$T_{\text{tec,top}} = \frac{F_0(\alpha\tau)_{\text{eff}}I(t) + \left\{(U)_{\text{tec,top}-p} + U_{\text{tec}}\beta_{\text{tec}}\right\}T_p + (U)_{\text{tec,top}-a}T_a}{(U)_{\text{tec,top}-a} + (U)_{\text{tec,top}-p} + U_{\text{tec}}\beta_{\text{tec}}} \quad (6.40\text{c})$$

From Eq. 6.40c, one gets

$$U_{\text{tec}}\beta_{\text{tec}}\left(T_{\text{tec,top}} - T_p\right) = F_{01}F_0(\alpha\tau)_{\text{eff}}I(t) - (U)_{\text{tec,top}-pa}\left(T_p - T_a\right) \quad (6.41)$$

where,

The penalty factor, $F_{01} = \frac{U_{\text{tec}}\beta_{\text{tec}}}{\left[(U)_{\text{tec,top}-a}+(U)_{\text{tec,top}-p}+U_{\text{tec}}\beta_{\text{tec}}\right]}$;

An overall heat transfer coefficient from plate to ambient through top of TEC,

$(U)_{\text{tec,top}-pa} = \frac{U_{\text{tec}}\beta_{\text{tec}}(U)_{\text{tec,top}-a}}{\left[(U)_{\text{tec,top}-a}+(U)_{\text{tec,top}-p}+U_{\text{tec}}\beta_{\text{tec}}\right]}$.

From Eq. 6.40b, we can also have

$$(h_t\beta_{\text{tec}})T_{\text{tec,top}} = F_0'\left[F_0(\alpha\tau)_{\text{eff}}I(t) + \left\{(U)_{\text{tec,top}-p} + U_{\text{tec}}\beta_{\text{tec}}\right\}T_p + (U)_{\text{tec,top}-a}T_a\right] \quad (6.42\text{a})$$

$$h_t\left(T_c - T_{\text{tec,top}}\right)\beta_{\text{tec}} = F_0(\alpha\tau)_{\text{eff}}I(t) - (U)_{\text{tec,top}-a}\left(T_{\text{tec,top}} - T_a\right) \\ - (U)_{\text{tec,top}-p}\left(T_{\text{tec,top}} - T_p\right) \quad (6.42\text{b})$$

where, the penalty factor, $F_0' = \frac{(h_t\beta_{\text{tec}})}{\left[(U)_{\text{tec,top}-a}+(U)_{\text{tec,top}-p}+U_{\text{tec}}\beta_{\text{tec}}\right]}$.

Substitute the expression for $(h_t\beta_{tec})T_{\text{tec,top}}$ from Eq. 6.42 into Eq. 6.39b as

$$T_c = \frac{\left(\tau_g\alpha_{sc} - \eta_m\right)I(t) + U_{t,ca}T_a + h_t T_{\text{tec,top}}\beta_{\text{tec}} + U_{b,cp}(1 - \beta_{\text{tec}})T_p}{U_{t,ca} + h_t\beta_{\text{tec}} + U_{b,cp}(1 - \beta_{\text{tec}})}$$

$$T_c = \frac{\left(\tau_g\alpha_{sc} - \eta_m\right)I(t) + U_{t,ca}T_a + F_0'\left[F_0(\alpha\tau)_{\text{eff}}I(t) + \left\{(U)_{\text{tec,top}-p} + U_{tec}\beta_{\text{tec}}\right\}T_p + (U)_{\text{tec,top}-a}T_a\right] + U_{b,cp}(1 - \beta_{\text{tec}})T_p}{U_{t,ca} + h_t\beta_{\text{tec}} + U_{b,cp}(1 - \beta_{\text{tec}})}$$

and,

$$U_{b,cp}\left(T_c - T_p\right) = F_0'' \\ \left[\left(\tau_g\alpha_{sc} - \eta_m\right)I(t) + U_{t,ca}T_a + F_0'\left[F_0(\alpha\tau)_{\text{eff}}I(t) + U_pT_p + (U)_{\text{tec,top}-a}T_a\right]\right] \quad (6.43)$$

where, the penalty factor, $F_0'' = \frac{U_{b,cp}}{U_{t,ca}+h_t\beta_{\text{tec}}+U_{b,cp}(1-\beta_{\text{tec}})}$

An overall net heat transfer coefficient, $U_p = \left\{(U)_{\text{tec,top}-p} + U_{\text{tec}}\beta_{\text{tec}} - \left(U_{t,ca} + h_t\beta_{\text{tec}}\right)\right\}$.

**Example 6.33** Evaluate the constants (a) penalty factor, $F_0$, (b) product of absorptivity and transmittivity, $(\alpha\tau)_{\text{eff}}$, an overall heat transfer coefficient from top of TEC to ambient air $(U)_{\text{tec,top}-a}$ and an overall heat transfer coefficient from top of TEC to absorber plate $(U)_{\text{tec,top}-p}$ in Eq. 6.40b.

**Solution**

From Example 3.48, we have the following parameters:

$U_{t,ca} = 9.30\,\text{W/m}^2\,{}^\circ\text{C}$ with wind velocity of 1 m/s; $h_t = 120\,\text{W/m}^2\,{}^\circ\text{C}$ and $U_{b,cp} = 2.80\,\text{W/m}^2\,{}^\circ\text{C}$ without wind velocity, $\beta_{tec} = 0.4324$, packing factor of TEC, $\tau_g = 0.95$, $\alpha_{sc} = 0.9$ (Eq. 3.47a, Example 3.54).

From Eq. 6.40b, we have the following expression for various constants:

$$F_0 = \frac{(h_t \beta_{tec})}{U_{t,ca} + h_t \beta_{tec} + U_{b,cp}(1 - \beta_{tec})}$$

$$= \frac{120 \times 0.4324}{9.30 + 120 \times 0.4324 + 2.80 \times (1 - 0.4324)} = \frac{51.888}{91.765} = 0.565$$

$$(\alpha\tau)_{\text{eff}} = (\tau_g \alpha_{sc} - \eta_m) = 0.95 \times 0.9 - 0.12 = 0.735$$

$$(U)_{\text{tec,top}-a} = \frac{(h_t \beta_{tec}) U_{t,ca}}{U_{t,ca} + h_t \beta_{tec} + U_{b,cp}(1 - \beta_{tec})}$$

$$= \frac{(120 \times 0.4324) \times 9.30}{9.30 + 120 \times 0.4324 + 2.80 \times (1 - 0.4324)} = 5.2545\,\text{W/m}^2\,{}^\circ\text{C}$$

$$(U)_{\text{tec,top}-p} = \frac{(h_t \beta_{tec})\{U_{b,cp}(1 - \beta_{tec})\}}{U_{t,ca} + h_t \beta_{tec} + U_{b,cp}(1 - \beta_{tec})}$$

$$= \frac{(120 \times 0.4324) \times 2.80 \times (1 - 0.4324)}{9.30 + 120 \times 0.4324 + 2.80 \times (1 - 0.4324)}$$

$$= \frac{82.465}{91.765} = 0.8989\,\text{W/m}^2\,{}^\circ\text{C}$$

***Example 6.34*** Evaluate the penalty factor, $F_{01}$, and an overall heat transfer coefficient from plate to ambient through top of TEC, $(U)_{\text{tec,top}-pa}$, for Eq. 6.41.

**Solution**

We have the following known parameters:

$(U)_{\text{tec,top}-a} = 5.2545\,\text{W/m}^2\,{}^\circ\text{C}$ and $(U)_{\text{tec,top}-p} = 0.8989\,\text{W/m}^2\,{}^\circ\text{C}$ and $\beta_{tec} = 0.4324$ (Example 6.33); $U_{tec} = 450\,\text{W/m}^2\,{}^\circ\text{C}$ (Eq. 3.48a).

From Eq. 6.41, we have
The penalty factor,

$$F_{01} = \frac{U_{tec}\beta_{tec}}{[(UA)_{\text{tec,top}-a} + (UA)_{\text{tec,top}-p} + U_{tec}\beta_{tec}]}$$

$$= \frac{450 \times 0.4324}{5.2545 + 0.8989 + 450 \times 0.4324} = \frac{194.58}{200.7334} = 0.9693.$$

An overall heat transfer coefficient from plate to ambient through top of TEC,

$$(U)_{tec,top-pa} = \frac{U_{tec}\beta_{tec}(U)_{tec,top-a}}{\left[(U)_{tec,top-a} + (U)_{tec,top-p} + U_{tec}\beta_{tec}\right]}$$

$$= \frac{(450 \times 0.4324) \times 5.2545}{5.2545 + 0.8989 + 450 \times 0.4324} = 5.093 \, W/m^2 \, °C.$$

**Example 6.35** Evaluate the penalty factors, $F_0'$, $F_0''$ and $U_p$ for Eq. 6.43 by using the parameters of Examples 6.33 and 6.34.

**Solution**

The penalty factor,

$$F_0' = \frac{(h_t\beta_{tec})}{\left\{(U)_{tec,top-a} + (U)_{tec,top-p} + U_{tec}\beta_{tec}\right\}}$$

$$= \frac{120 \times 0.4324}{5.2545 + 0.8989 + 450 \times 0.4324} = \frac{51.888}{200.7334} = 0.2585$$

The penalty factor,

$$F_0'' = \frac{U_{b,cp}}{U_{t,ca} + h_t\beta_{tec} + U_{b,cp}(1 - \beta_{tec})}$$

$$= \frac{2.80}{9.30 + 120 \times 0.4324 + 2.80 \times (1 - 0.4324)} = \frac{2.80}{91.765} = 0.0305$$

An overall net heat transfer coefficient,

$$U_p = \left\{(U)_{tec,top-p} + U_{tec}\beta_{tec} - \left(U_{t,ca} + h_t\beta_{tec}\right)\right\}$$

$$= 5.2545 + 450 \times 0.4324 - (9.30 + 120 \times 0.4324)$$

$$= 5.2545 + 194.58 - 61.188 = 138.6465 \, W/m^2 \, °C.$$

#### (c) For Absorber Plate

The rate of net thermal energy in W available to the absorber plate can be written as follows:

$$\dot{Q}_u = \left[U_{tec}\left(T_{tec,top} - T_{tec,bottom}\right)\beta_{tec}bdx + U_{b,cp}\left(T_c - T_p\right)(1 - \beta_{tec})bdx\right]$$
$$- \left[U_b\left(T_p - T_a\right)bdx + \eta_{tec}U_{tec}\left(T_{tec,top} - T_{tec,bottom}\right)\beta_{tec}bdx\right] \quad (6.44)$$

Here, the first two terms are gain to absorber plate from the top of TEC by conduction through packing factor of TEC and from solar cell through non-packing area of TEC. The last two terms are lost term from absorber plate to the ambient through bottom insulation and electrical power generation by TEC.

Since the bottom of TEC is in direct contact with absorber plate of collector, and hence

$$T_{tec,bottom} = T_p$$

Then, Eq. 6.44 becomes as

$$\dot{Q}_u = \left[ U_{tec}\left(T_{tec,top} - T_p\right)\beta_{tec}bdx + U_{b,cp}\left(T_c - T_p\right)(1 - \beta_{tec})bdx \right]$$
$$- \left[ U_b\left(T_p - T_a\right)\beta_{tec}bdx + \eta_{tec}U_{tec}\left(T_{tec,top} - T_p\right)\beta_{tec}bdx \right]$$

or,

$$\dot{Q}_u = \left[ (1 - \eta_{tec})(U_{tec}\beta_{tec})\left(T_{tec,top} - T_p\right) + U_{b,cp}\left(T_c - T_p\right)(1 - \beta_{tec}) - U_b\left(T_p - T_a\right)\right]bdx$$
$$\tag{6.45}$$

Substitute $(U_{tec}\beta_{tec})\left(T_{tec,top} - T_p\right)$ and $U_{b,cp}\left(T_c - T_p\right)$ from Eqs. 6.41 and 6.43 into Eq. 6.45, one gets

$$\dot{Q}_u = \left[ (1 - \eta_{tec})\{ F_{01} F_0(\alpha\tau)_{eff} I(t) - (U)_{tec,top-pa}\left(T_p - T_a\right)\} \right.$$
$$+ \{ F_0''[(\tau_g\alpha_{sc} - \eta_m)I(t) + U_{t,ca}T_a + $$
$$+ F_0'[F_0(\alpha\tau)_{eff}I(t) + U_pT_p + (U)_{tec,top-a}T_a]]\}(1 - \beta_{tec})$$
$$\left. - U_b\left(T_p - T_a\right)\right]bdx$$

After rearranging above equation, one has the following:

$$\dot{Q}_u = \left[ (\alpha\tau)_{meff} I(t) - U_{Lm}\left(T_p - T_a\right) + F_0'' F_0' U_p T_p + U_a T_a\right]bdx \tag{6.46}$$

where,   $(\alpha\tau)_{meff}$ $=$ $[(1 - \eta_{tec})F_{01} F_0(\alpha\tau)_{eff} + F_0''(1 - \beta_{tec})\{(\tau_g\alpha_{sc} - \eta_m) + F_0'F_0(\alpha\tau)_{eff}\}]$; $U_{Lm}$ $=$ $\{(1 - \eta_{tec})(U)_{tec,top-pa} + U_b\}$, $U_p$ $=$ $\{(U)_{tec,top-p} + U_{tec}\beta_{tec} - (U_{t,ca} + h_t\beta_{tec})\}$   (Eq.   6.43)   and   $U_a$ $=$ $\left[ F_0''\left(U_{t,ca} + F_0'(U)_{tec,top-a}\right)\right]$.

Following Eq. 6.7, one can define flat plate collector efficiency $(F')$ for elemental area of '$bdx$' as

$$F' = \frac{\dot{Q}_u}{\dot{Q}_u\big|_{T_p=T_f}} \quad [T_p = T_f]$$

or,

$$\dot{Q}_u = F' \dot{Q}_u\big|_{T_p=T_f} = F'\left[(\alpha\tau)_{meff}I(t) - U_{Lm}\left(T_f - T_a\right) + F_0'' F_0' U_p T_f + U_a T_a\right]bdx, \text{ W}$$
$$\tag{6.47}$$

Here, $T_p = T_f = T_{tec,bottom}$.

Equation 6.46 is the rate of net thermal energy in W available to the absorber plate which can be carried away if fluid flow below it.

***Example 6.36*** Evaluate the product of absorptivity and transmittivity, $(\alpha\tau)_{meff}$, and an overall heat transfer coefficient $U_{Lm}$ and $U_a$ of OPVT-TEC liquid collector for Eq. 6.44 by using the data of Example 6.33–6.35.

### Solution

**Know parameters:** $F_0 = 0.565$, $F_{01} = 0.9693$, $F_0' = 0.2585$, $F_0'' = 0.0305$, $(\alpha\tau)_{eff} = (\tau_g\alpha_{sc} - \eta_m) = (0.95 \times 0.9 - 0.12) = 0.735$, $\beta_{tec} = 0.4324$, $U_b = 0.7\,\text{W/m}^2\,°\text{C}$ and $\eta_{tec} = 0.15$ (Eq. 3.45).

$(U)_{tec,top-pa} = 5.093\,\text{W/m}^2\,°\text{C}$, $(U)_{tec,top-a} = 5.2545\,\text{W/m}^2\,°\text{C}$ (Example 6.34), $U_{t,ca} = 9.30\,\text{W/m}^2\,°\text{C}$.

From Eq. 6.44, we have the following:

$$(\alpha\tau)_{meff} = \left[(1 - \eta_{tec})F_{01}F_0(\alpha\tau)_{eff} + F_0''(1 - \beta_{tec})\{(\tau_g\alpha_{sc} - \eta_m) + F_0'F_0(\alpha\tau)_{eff}\}\right]$$

$$= (1 - 0.15) \times 0.9693 \times 0.565 \times 0.735 + 0.0305$$

$$\times (1 - 0.4324) \times \{0.735 + 0.2585 \times 0.565 \times 0.735\}$$

$$= 0.3422 + 0.0173 \times 1.146 = 0.362$$

$$U_{Lm} = \{(1 - \eta_{tec})(U)_{tec,top-pa} + U_b\} = (1 - 0.15) \times 5.093 + 0.7 = 5.029\,\text{W/m}^2\,°\text{C}$$

$$U_a = \left[F_0''(U_{t,ca} + F_0'(U)_{tec,top-a})\right] = 0.0305 \times (9.30 + 0.2585 \times 5.2545) = 0.3251\,\text{W/m}^2\,°\text{C}$$

### (d)  For Flowing Fluid (Water)

If fluid (water) is flowing with mass flow rate of $\dot{m}_f$ below absorber plate, then The rate of thermal energy carried out by fluid,

$$\dot{Q}_u = \dot{m}_f c_f \frac{dT_f}{dx}dx \tag{6.48}$$

According to first law of thermodynamics namely energy conservation, Eqs. 6.47 and 6.48 will be equal, and it can be expressed as follows:

$$\dot{m}_f c_f \frac{dT_f}{dx}dx = F'\left[(\alpha\tau)_{meff}I(t) - U_{Lm}(T_f - T_a) + F_0''F_0'U_pT_f + U_aT_a\right]bdx \tag{6.49}$$

**where,** $F' = 0.75$ is tube in plate for single glazed flat plate collector efficiency (Sect. 4.3.3 and Example 4.27).

### 6.6.2 The Outlet Fluid Temperature of OPVT-TEC Liquid Collector

Equation 6.49 can be rearranged as follows:

$$
\frac{dT_f}{dx} + \frac{F'(U_{Lm} - F_0''F_0'U_p)b}{\dot{m}_f c_f} T_f = \frac{F'[(\alpha\tau)_{meff} I(t) + (U_{Lm} + U_a)T_a]b}{\dot{m}_f c_f}
$$

(6.50)

If $U_{Lm-\text{eff}} = (U_{Lm} - F_0''F_0'U_p)$, $a = \frac{F'U_{Lm-\text{eff}}b}{\dot{m}_f c_f}$ and $f(t) = \frac{F'[(\alpha\tau)_{meff}I(t)+(U_{Lm}+U_a)T_a]b}{\dot{m}_f c_f}$.

The solution of one order differential Eq. 6.50 with initial condition of $T_f = T_{fi}$ at $x = 0$ will be as follows:

$$
T_f = \frac{[(\alpha\tau)_{meff} I(t) + (U_{Lm} + U_a)T_a]}{U_{Lm-\text{eff}}} \left[ 1 - \exp\left\{ -\frac{F'U_{Lm-\text{eff}}b}{\dot{m}_f c_f} x \right\} \right]
$$
$$
+ T_{fi}\exp\left\{ -\frac{F'U_{Lm-\text{eff}}b}{\dot{m}_f c_f} x \right\}
$$

(6.51)

The outlet fluid temperature $T_{fo} = T_f$ at $x = L$, the length of PVT-TEC liquid collector, and hence

$$
T_{fo} = \frac{[(\alpha\tau)_{meff} I(t) + (U_{Lm} + U_a)T_a]}{U_{Lm-\text{eff}}} \left[ 1 - \exp\left\{ -\frac{F'U_{Lm-\text{eff}}b}{\dot{m}_f c_f} L_0 \right\} \right]
$$
$$
+ T_{fi}\exp\left\{ -\frac{F'U_{Lm-\text{eff}}b}{\dot{m}_f c_f} L_0 \right\}
$$

or,

$$
T_{fo} = \frac{[(\alpha\tau)_{meff} I(t) + (U_{Lm} + U_a)T_a]}{U_{Lm-\text{eff}}} \left[ 1 - \exp\left\{ -\frac{F'U_{Lm-\text{eff}}A_m}{\dot{m}_f c_f} \right\} \right]
$$
$$
+ T_{fi}\exp\left\{ -\frac{F'U_{Lm-\text{eff}}A_m}{\dot{m}_f c_f} \right\} \text{ with } A_m = bL_0
$$

(6.52)

**Example 6.37** Calculate the outlet fluid (water) temperature from PVT-TEC liquid collector for the following climatic and design parameters:

Climatic parameters: $I(t) = 500 \text{ W/m}^2$, $T_a = 16\,°C$ and $T_{fi} = 15\,°C$.

Design parameters: $F' = 0.75$, $(\alpha\tau)_{meff} = 0.326$, $U_{Lm} = 5.029 \text{ W/m}^2\,°C$, $U_a = 0.3251 \text{ W/m}^2\,°C$, $A_m = 2 \times 1 \text{ m}^2$, the penalty factor, $F_0' = 0.2585$, the penalty factor, $F_0'' = 0.0305$, an overall net heat transfer coefficient, $U_p = 138.6465 \text{ W/m}^2\,°C$ (Examples 3.35–3.36) and $\dot{m}_f c_f = 0.035 \times 419 = 146.65\text{J/s}\,°C = 146.65 \text{ W/°C}$.

**Solution**

Now from Eq. 6.50, we have

$$U_{Lm-\text{eff}} = \left(U_{Lm} - F_0'' F_0' U_p\right) = 5.029 - 0.0305 \times 0.2585 \times 138.6465$$
$$= 3.9348 \text{ W/m}^2 \,^\circ\text{C}$$

Further,

$$\exp\left\{-\frac{F'U_{Lm-\text{eff}}A_{cm}}{\dot{m}_f c_f}\right\} = \exp\left\{-\frac{0.75 \times 3.9348 \times 2}{0.035 \times 4190}\right\}$$
$$= \exp\left\{-\frac{5.9022}{146.65}\right\} = \exp\{-0.04025\} = 0.9605$$

From Eq. 6.52, we have an expression for the outlet temperature of fluid as

$$T_{fo} = \frac{[0.326 \times 500 + (5.029 + 0.3251) \times 16]}{3.9348}[1 - 0.9605] + 15 \times 0.9605$$
$$= \frac{248.666}{3.9348} \times 0.0395 + 14.41 = 2.496 + 14.41 = 16.91 \,^\circ\text{C}$$

## 6.6.3   The Rate of Thermal Energy $\left(\dot{Q}_u\right)$ and Mass Flow Rate Factor $(F_R)$

With the help of Eq. 6.52, we can have the rate of thermal energy as follows:

$$\dot{Q}_u = \dot{m}_f c_f \left(T_{fo} - T_{fi}\right) = F_R A_m\left[\left[(\alpha\tau)_{\text{meff}} I(t) + (U_{Lm} + U_a)T_a\right] - U_{Lm-\text{eff}}T_{fi}\right] \tag{6.53}$$

$$F_R = \frac{\dot{m}_f c_f}{A_m U_{Lm-\text{eff}}}\left[1 - \exp\left\{-\frac{F'U_{Lm-\text{eff}}A_m}{\dot{m}_f c_f}\right\}\right]$$

**Example 6.38** Evaluate the rate of thermal energy and mass flow rate factor for data of Example 6.37 for PVT-TEC liquid collector. Also calculate an instantaneous thermal efficiency.

**Solution**

**From Example 6.37:** $T_{fo} = 16.91 \,^\circ\text{C}$, $T_{fi} = 15 \,^\circ\text{C}$ and $\dot{m}_f c_f = 146.65 \text{J/s} \,^\circ\text{C} = 146.65 \text{ W/}^\circ\text{C}$.
$\exp\left\{-\frac{F'U_{Lm-\text{eff}}A_{cm}}{\dot{m}_f c_f}\right\} = 0.9605$ and $U_{Lm-\text{eff}} = 3.9348 \text{ W/m}^2 \,^\circ\text{C}$.

The rate of thermal energy in W $\dot{Q}_u = \dot{m}_f c_f (T_{fo} - T_{fi}) = 146.65(16.91 - 15) = 280.10\,W$.

The mass flow rate factor,

$$FR = \frac{\dot{m}_f c_f}{A_m U_{Lm-\text{eff}}} \left[1 - \exp\left\{-\frac{F' U_{Lm-\text{eff}} A_m}{\dot{m}_f c_f}\right\}\right] = \frac{146.65}{2 \times 3.9348}[1 - 0.9605] = 0.736.$$

Another method to evaluate the rate of thermal energy is given by Eq. 6.53 and after substitution of appropriate value from Example 6.37, one gets

$$\dot{Q}_u = F_R A_m \left[[(\alpha\tau)_{\text{meff}} I(t) + (U_{Lm} + U_a)T_a] - U_{Lm-\text{eff}} T_{fi}\right]$$
$$= 0.736 \times 2[[\mathbf{0.326} \times \mathbf{500} + (5.029 + 0.3251) \times 16] - 3.934 \times 15]$$
$$= 1.472 \times [[\mathbf{163} + \mathbf{85.666}] - 59.01] = 279.17\,W$$

The above value is nearly same as evaluated earlier.

An instantaneous thermal efficiency $(\eta_i) = \frac{279.17}{1000} = 0.28$ (26%).

### 6.6.4 An Average Fluid, Solar Cell and TEC Top Surface Temperatures

Following Eq. 6.14, the average fluid temperature over length of liquid PVT collector, $L_0$, can be obtained as

$$\overline{T_f} = \frac{1}{L_0} \int_0^{L_0} T_f dx = \frac{[(\alpha\tau)_{\text{meff}} I(t) + (U_{Lm} + U_a)T_a]}{U_{Lm-\text{eff}}} \left[1 - \frac{1 - \exp\left\{-\frac{F' U_{Lm-\text{eff}} A_m}{\dot{m}_f c_f}\right\}}{\frac{F' U_{Lm-\text{eff}} A_m}{\dot{m}_f c_f}}\right]$$
$$+ T_{fi} \frac{1 - \exp\left\{-\frac{F' U_{Lm-\text{eff}} A_m}{\dot{m}_f c_f}\right\}}{\frac{F' U_{Lm-\text{eff}} A_m}{\dot{m}_f c_f}} \tag{6.54}$$

After knowing average fluid temperature, $(\overline{T_f})$ as per assumption $\overline{T}_f = \overline{T}_p = \overline{T}_{\text{tec,bottom}}$.

Then, an average TEC top surface temperature can be obtained from Eq. 6.40c as

$$\overline{T}_{\text{tec,top}} = \frac{F_0(\alpha\tau)_{\text{eff}} I(t) + \{(U)_{\text{tec,top}-p} + U_{\text{tec}}\beta_{\text{tec}}\}\overline{T}_f + (U)_{\text{tec,top}-a}T_a}{(UA)_{\text{tec,top}-a} + (UA)_{\text{tec,top}-p} + U_{\text{tec}}\beta_{\text{tec}}} \tag{6.55}$$

After an average TEC top surface temperature, an average solar cell temperature can be obtained from Eq. 6.39b as

$$\overline{T}_c = \frac{(\tau_g\alpha_{sc} - \eta_m)I(t) + U_{t,ca}T_a + h_t\overline{T}_{tec,top}\beta_{tec} + U_{b,cp}(1 - \beta_{tec})\overline{T}_f}{U_{t,ca} + h_t\beta_{tec} + U_{b,cp}(1 - \beta_{tec})}$$

[since $\overline{T}_f = \overline{T}_p = \overline{T}_{tec,bottom}$] (6.66)

***Example 6.39*** Compute an average fluid, TEC top surface and solar cell temperatures for data of Example 6.38.

**Solution**

**From Examples 6.37 and 6.38:** $\exp\left\{-\dfrac{F'U_{Lm-eff}A_m}{\dot{m}_fc_f}\right\} = 0.9605;$ $\dfrac{F'U_{Lm-eff}A_m}{\dot{m}_fc_f} = \dfrac{0.75\times3.9348\times2}{0.035\times4190} = 0.04025$ (Example 6.37).

$$\frac{[(\alpha\tau)_{meff}I(t) + (U_{Lm} + U_a)T_a]}{U_{Lm-eff}} = \frac{[0.326 \times 500 + (5.029 + 0.3251) \times 16]}{3.9348}$$

$$= \frac{248.666}{3.9348} = 63.197$$

$(U)_{tec,top-p} = 0.8989\,\text{W/m}^2\,°\text{C}$ (Example 6.34), $U_{tec} = 450\,\text{W/m}^2\,°\text{C}$ (Eq. 3.48a), $U_{t,ca} = 9.12\,\text{W/m}^2\,°\text{C}$ (Example 6.1), $U_{b,cp} = 2.7663\,\text{W/m}^2\,°\text{C}$ (Example 6.2), $h_t = 120\,\text{W/m}^2\,°\text{C}$ (Example 6.33), $F_0 = 0.565$ (Example 6.33).

Other's constants have been taken from Examples 6.34–6.36.

Substitute appropriate values in Eq. 6.54, we have

$$\overline{T}_f = 63.197\left[1 - \frac{1 - 0.9605}{0.04025}\right] + 15 \times \frac{1 - 0.9605}{0.04025}$$

$$- 63.107 \times (1 - 0.9814) + 15 \times 0.9814 = 15.89\,°\text{C}$$

From Eq. 6.55, we have an expression for TEC top surface temperature and after substitution of appropriate values from Examples 6.36 and 6.37, one gets

$$\overline{T}_{tec,top} = \frac{F_0(\alpha\tau)_{eff}I(t) + \{(U)_{tec,top-p} + U_{tec}\beta_{tec}\}\overline{T}_f + (U)_{tec,top-a}T_a}{(U)_{tec,top-a} + (U)_{tec,top-p} + U_{tec}\beta_{tec}}$$

$$= \frac{0.565 \times 0.326 \times 500 + (0.8989 + 450 \times 0.4324)15.89 + 5.2545 \times 15}{5.2545 + 0.8989 + 450 \times 0.4324}$$

$$= \frac{92.095 + 195.479 \times 15.89 + 78.82}{199.734} = 16.407\,°\text{C}$$

From Eq. 5.66, we have

$$\overline{T}_c = \frac{(\tau_g\alpha_{sc} - \eta_m)I(t) + U_{t,ca}T_a + h_t\overline{T}_{tec,top}\beta_{tec} + U_{b,cp}(1 - \beta_{tec})\overline{T}_f}{U_{t,ca} + h_t\beta_{tec} + U_{b,cp}(1 - \beta_{tec})}$$

$$= \frac{0.735 \times 500 + 9.12 \times 16 + 120 \times 16.407 \times 0.4324 + 2.7663 \times (1 - 0.4324) \times 15.89}{9.12 + 120 \times 0.4324 + 2.7663 \times (1 - 0.4324)}$$

$$= \frac{367.5 + 145.92 + 851{,}33 + 24.95}{9.12 + 51.89 + 1.57} = \frac{1389.7}{62.578} = 22.21\,°\text{C}$$

### 6.6.5   Total Electrical Power from OPVT-TEC Liquid Collector

**Example 6.40** Evaluate electrical power from OPV, TEC and total electrical power from OPVT-TEC liquid collector by using the data of Example 6.30.

**Solution**

**From Example 8.39:** $\overline{T_f} = \overline{T}_{\text{tec,bottom}} = 15.89\,°\text{C}$, $\overline{T}_c = 22.21\,°\text{C}$, $\overline{T}_{\text{tec,top}} = 16.407\,°\text{C}$.

Since, $\overline{T}_c < 25\,°\text{C}$, then following Eq. 3.21b, an electrical power can be obtained as

$$\dot{E}_{el} = \eta_{mo} \times \beta_c I(t) A_m = 0.12 \times 0.89 \times 500 \times 2 = 106\,\text{W}$$

Electrical power from REC can be evaluated as

$$\dot{E}_{\text{TEC}} = \eta_{\text{tec}} U_{\text{tec}} \left( T_{\text{tec,top}} - T_{\text{tec,bottom}} \right) \beta_{\text{tec}} A_m$$

**Here** $\beta_{\text{tec}} = 0.4324$ and $\eta_{\text{tec}} = 0.15$ (Example 6.33), $U_{\text{tec}} = 450\,\text{W/m}^2\,°\text{C}$ (Eq. 3.48a).

Then,

$$\dot{E}_{\text{TEC}} = 0.15 \times 450(16.407 - 15.89)0.4324 \times 2 = 30.18\,\text{W}$$

The total electrical power is the sum of electrical power from PV and TEC, and hence

$$\dot{E}_{\text{total}} = \dot{E}_{el} + \dot{E}_{\text{TEC}} = 89.74 + 30.18 = 119.92\,\text{W}$$

The above value is 21.74 W (22%) more than PVT liquid collector (98.18 W, Example 6.9). Further, it is important to note that solar cell temperature in PVT-TEC is more; hence electrical power due to PV module is reduced in PVT-TEC liquid collector, but it helps to give additional electrical power of 30.18 W due to TEC.

**Example 6.41** Evaluate the rate of thermal exergy and an overall exergy of OPVT-TEC liquid collector.

**Solution**

**Known parameters** $T_{fo} = 16.91\,°\text{C}$ (Example 6.37), $T_{fi} = 15\,°\text{C}$ and $T_a = 16\,°\text{C}$.

Following Example 6.12, the rate of thermal exergy can be obtained as

$$\dot{E}x_{th} = \dot{m}_f C_f \left[ (T_{fo} - T_{fi}) - (T_a + 273) \ln \frac{T_{fo} + 273}{T_{fi} + 273} \right]$$

$$= 0.035 \times 4190 \left[ (16.91 - 15) - (289) \ln \frac{289.91}{288} \right]$$

$$= 146.65[(16.91 - 15) - 1.91] = 0 \, \text{W}$$

In the case of PVT-TEC liquid collector, the rate of thermal exergy is zero unlike PVT collector with negative the rate of exergy.

So, the total exergy of PVT-TEC collector is exactly same as the rate of total electrical energy (electrical power), i.e., $\dot{E}_{\text{total}} = 119.92 \, \text{W}$ (Example 6.40).

## 6.7   The Analysis of N-PVT-TEC (N-OPVT-TEC) Liquid Collectors Connected in Series

In this case, the outlet of 1st PVT-TEC collector is connected to inlet of 2nd PVT-TEC, and the outlet of 2nd is connected to inlet of 3rd, and it continues till $n^{\text{th}}$ PVT-TEC collector as shown in Fig. 6.4.

### 6.7.1   The Outlet Fluid Temperature

Following Sect. 4.5.1 and Eq. 4.31 and with help of Eq. 6.52, one can derive an expression for the outlet fluid temperature at $n^{\text{th}}$ PVT-TEC collector as follows:

$$T_{foN} = \frac{[(\alpha\tau)_{meff} I(t) + (U_{Lm} + U_a)T_a]}{U_{Lm-\text{eff}}} \left[ 1 - \exp \left\{ -\frac{NF'U_{Lm-\text{eff}}A_m}{\dot{m}_f c_f} \right\} \right]$$

$$+ T_{fi} \exp \left\{ -\frac{NF'U_{Lm-\text{eff}}A_m}{\dot{m}_f c_f} \right\} \tag{6.67a}$$

By assuming same design including heat transfer coefficients and climatic parameters with same $A_{cm} = bL_0$.

If $A_m = bx$, then $T_{foN} = T_{fN}(x)$, then Eq. 6.67a becomes as follows:

$$T_{fN}(x) = \frac{[(\alpha\tau)_{meff} I(t) + (U_{Lm} + U_a)T_a]}{U_{Lm-\text{eff}}} \left[ 1 - \exp \left\{ -\frac{NF'U_{Lm-\text{eff}}bx}{\dot{m}_f c_f} \right\} \right]$$

$$+ T_{fi} \exp \left\{ -\frac{NF'U_{Lm-\text{eff}}bx}{\dot{m}_f c_f} \right\} \tag{6.67b}$$

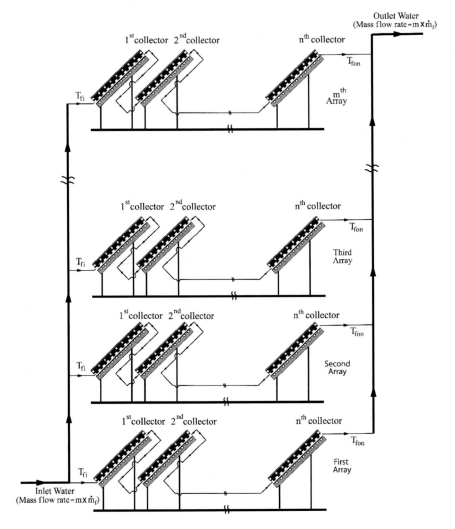

**Fig. 6.4** Cross-sectional view of PVT-TEC liquid collectors connected in series and parallel combination

***Example 6.42*** Calculate the outlet fluid temperature at end of 1st and 2nd flexible OPVT-TEC collectors connected in series by using the data of Example 6.37.

**Solution**

From Example 6.37, we have

$$U_{Lm-\text{eff}} = \left(U_{Lm} - F_0'' F_0' U_p\right)$$
$$= 5.029 - 0.0305 \times 0.2585 \times 138.6465 = 3.9348\,\text{W/m}^2\,°\text{C}$$

Further,

$$\exp\left\{-\frac{F'U_{Lm-\text{eff}}A_{cm}}{\dot{m}_f c_f}\right\} = \exp\left\{-\frac{0.75 \times 3.9348 \times 2}{0.035 \times 4190}\right\}$$

$$= \exp\left\{-\frac{5.9022}{146.65}\right\} = \exp\{-0.04025\} = 0.9605$$

From Eq. 6.52, we have an expression for the outlet temperature of fluid at end of 1st PVT-TEC collector as

$$T_{fo1} = \frac{[\mathbf{0.326} \times \mathbf{500} + (5.029 + 0.3251) \times 16]}{3.9348}[1 - 0.9605] + 15 \times 0.9605$$

$$= \frac{248.666}{3.9348} \times 0.0395 + 14.41 = 2.496 + 14.41 = 16.91\,°\text{C}$$

Now, $T_{fo1} = 16.91\,°\text{C}$, the outlet of first becomes inlet of second PVT-TEC collector, then

$$T_{fo2} = \frac{[\mathbf{0.326} \times \mathbf{500} + (5.029 + 0.3251) \times 16]}{3.9348}[1 - 0.9605] + 16.91 \times 0.9605$$

$$= \frac{248.666}{3.9348} \times 0.0395 + 14.41 = 2.496 + 16.24 = 18.73\,°\text{C}$$

Now, an average of two PVT-TEC collectors will be

$$\overline{T}_{f2} = \frac{T_{fo1} + T_{fo2}}{2} = \frac{15 + 18.73}{2} = 16.865\,°\text{C}$$

An average of 1st PVT-TEC collector $= \overline{T}_{f1} = \frac{T_{fo1}+T_{fi}}{2} = \frac{16.91+15}{2} = 15.995\,°\text{C}$.
An average of 2nd PVT-TEC collector $= \overline{T}_{f2} = \frac{T_{fo2}+T_{fo1}}{2} = \frac{18.73+16.91}{2} = 17.82\,°\text{C}$.

**Example 6.43** Evaluate the average TEC top surface temperature for 1st and 2nd OPVT-TEC collector by using the average temperature of fluid for 1st and 2nd OPVT-TEC collector from Example 6.42.

**Solution**

**Given:** $\overline{T}_{f1} = 15.995\,°\text{C}$ and $\overline{T}_{f2} = 17.82\,°\text{C}$ (Example 6.42).

From Eq. 6.55, we have an expression for TEC top surface temperature and after substitution of appropriate values from Example 6.39, one gets

$$\overline{T}_{\text{tec,top1}} = \frac{F_0(\alpha\tau)_{\text{eff}}I(t) + \{(U)_{\text{tec,top}-p} + U_{\text{tec}}\beta_{\text{tec}}\}\overline{T}_f + (U)_{\text{tec,top}-a}T_a}{(U)_{\text{tec,top}-a} + (U)_{\text{tec,top}-p} + U_{\text{tec}}\beta_{\text{tec}}}$$

$$= \frac{0.565 \times \mathbf{0.326} \times \mathbf{500} + (0.8989 + 450 \times 0.4324) \times 15.995 + 5.2545 \times 15}{5.2545 + 0.8989 + 450 \times 0.4324}$$

$$= \frac{92.095 + 195.479 \times 15.995 + 78.82}{199.734} = 16.51\,^{\circ}\mathrm{C}$$

and,

$$
\begin{aligned}
\overline{T}_{\text{tec,top2}} &= \frac{F_0(\alpha\tau)_{\text{eff}} I(t) + \{(U)_{\text{tec,top}-p} + U_{\text{tec}}\beta_{\text{tec}}\}\overline{T}_f + (U)_{\text{tec,top}-a} T_a}{(U)_{\text{tec,top}-a} + (U)_{\text{tec,top}-p} + U_{\text{tec}}\beta_{\text{tec}}} \\
&= \frac{0.565 \times \mathbf{0.326} \times \mathbf{500} + (0.8989 + 450 \times 0.4324) \times 17.82 + 5.2545 \times 15}{5.2545 + 0.8989 + 450 \times 0.4324} \\
&= \frac{92.095 + 195.479 \times 17.82 + 78.82}{199.734} = 18.296\,^{\circ}\mathrm{C}
\end{aligned}
$$

***Example 6.44*** Evaluate the average solar cell temperature for 1st and 2nd PVT-TEC collector by using the average TEC top surface temperature for 1st and 2nd OPVT-TEC collector from Example 6.43.

**Solution**

Given: $\overline{T}_{\text{tec,top1}} = 16.51\,^{\circ}\mathrm{C}$, $\overline{T}_{\text{tec,top2}} = 18.296\,^{\circ}\mathrm{C}$ (Example 6.43) and $\overline{T}_{f1} = 15.995\,^{\circ}\mathrm{C}$ and $\overline{T}_{f2} = 17.82\,^{\circ}\mathrm{C}$ (Example 6.42).

From Eq. 5.66, we have

$$
\begin{aligned}
\overline{T}_{c1} &= \frac{(\tau_g\alpha_{sc} - \eta_m)I(t) + U_{t,ca}T_a + h_t\overline{T}_{\text{tec,top1}}\beta_{\text{tec}} + U_{b,cp}(1 - \beta_{\text{tec}})\overline{T}_{f1}}{U_{t,ca} + h_t\beta_{\text{tec}} + U_{b,cp}(1 - \beta_{\text{tec}})} \\
&= \frac{0.735 \times 500 + 9.12 \times 16 + 120 \times 16.51 \times 0.4324 + 2.7663 \times (1 - 0.4324) \times 15.995}{9.12 + 120 \times 0.4324 + 2.7663 \times (1 - 0.4324)} \\
&= \frac{367.5 + 145.92 + 856.67 + 25.11}{9.12 + 51.89 + 1.57} = \frac{1395.2}{22.21} = 62.82\,^{\circ}\mathrm{C}
\end{aligned}
$$

and,

$$
\begin{aligned}
\overline{T}_{c2} &= \frac{(\tau_g\alpha_{sc} - \eta_m)I(t) + U_{t,ca}T_a + h_t\overline{T}_{\text{tec,top2}}\beta_{\text{tec}} + U_{b,cp}(1 - \beta_{\text{tec}})\overline{T}_{f2}}{U_{t,ca} + h_t\beta_{\text{tec}} + U_{b,cp}(1 - \beta_{\text{tec}})} \\
&= \frac{0.735 \times 500 + 9.12 \times 16 + 120 \times 18.296 \times 0.4324 + 2.7663 \times (1 - 0.4324) \times 17.82}{9.12 + 120 \times 0.4324 + 2.7663 \times (1 - 0.4324)} \\
&= \frac{367.5 + 145.92 + 949.34 + 27.98}{9.12 + 51.89 + 1.57} = \frac{1395.2}{22.21} = 67.12\,^{\circ}\mathrm{C}
\end{aligned}
$$

***Example 6.45*** Evaluate electrical power from OPV, TEC and total from OPVT-TEC collector by using the data of Examples 6.43 and 6.44.

**Solution**

Given: $\overline{T}_{\text{tec,top1}} = 16.51\,^{\circ}\mathrm{C}$, $\overline{T}_{\text{tec,top2}} = 18.296\,^{\circ}\mathrm{C}$ (Example 6.43); $\overline{T}_{f1} = T_{\text{tec,bottom1}} = 15.995\,^{\circ}\mathrm{C}$ and $\overline{T}_{f2} = T_{\text{tec,bottom2}} = 17.82\,^{\circ}\mathrm{C}$ (Example 6.42) and $\overline{T}_{c1} = 62.82\,^{\circ}\mathrm{C}$, $\overline{T}_{c2} = 67.12\,^{\circ}\mathrm{C}$ (Example 6.44).

**For 1st PVT-TEC Collector**

Following Eq. 6.31a, an electrical power can be obtained as

$$
\begin{aligned}
\dot{E}_{el1} &= \eta_{mo}\big[1 - \beta_0\big(\overline{T}_{c1} - 25\big)\big] \times \beta_c I(t) A_m \\
&= 0.12[1 - 0{,}00425(62.82 - 25)] \times 0.89 \times 500 \times 2 \\
&= 0.1010 \times 0.89 \times 500 \times 2 = 89.62\,\text{W}
\end{aligned}
$$

Electrical power from REC can be evaluated as

$$
\dot{E}_{TEC1} = \eta_{tec} U_{tec}\big(T_{tec,top1} - T_{tec,bottom1}\big)\beta_{tec} A_m
$$

**Here,** $\beta_{tec} = 0.4324$ and $\eta_{tec} = 0.15$ (Example 6.33), $U_{tec} = 450\,\text{W/m}^2\,{}^\circ\text{C}$ (Eq. 3.48a).
  Then,

$$
\dot{E}_{TEC1} = 0.15 \times 450 \times (16.51 - 15.995) \times 0.4324 \times 2 = 30.06\,\text{W}
$$

The total electrical power is the sum of electrical power from PV and TEC, and hence

$$
\dot{E}_{total1} = \dot{E}_{el1} + E_{TEC1} = 89.62 + 30.06 = 119.62\,\text{W}
$$

**For 2nd OPVT-TEC collector**

Following Eq. 6.31a, an electrical power can be obtained as

$$
\begin{aligned}
\dot{E}_{el2} &= \eta_{mo}\big[1 - \beta_0\big(\overline{T}_{c2} - 25\big)\big] \times \beta_c I(t) A_m \\
&= 0.12[1 - 0{,}00425(67.12 - 25)] \times 0.89 \times 500 \times 2 \\
&= 0.0985 \times 0.89 \times 500 \times 2 = 87.68\,\text{W}
\end{aligned}
$$

Electrical power from second TEC can be evaluated as

$$
\dot{E}_{TEC2} = \eta_{tec} U_{tec}\big(T_{tec,top2} - T_{tec,bottom2}\big)\beta_{tec} A_m
$$

**Here** $\beta_{tec} = 0.4324$ and $\eta_{tec} = 0.15$ (Example 6.33), $U_{tec} = 450\,\text{W/m}^2\,{}^\circ\text{C}$ (Eq. 3.48a).
  Then,

$$
\dot{E}_{TEC2} = 0.15 \times 450 \times (18.296 - 17.82) \times 0.4324 \times 2 = 27.79\,\text{W}
$$

The total electrical power is the sum of electrical power from PV and TEC, and hence

$$
\dot{E}_{total2} = \dot{E}_{el} + \dot{E}_{TEC} = 87.68 + 27.79 = 115.47\,\text{W}
$$

Therefore, the total electrical power from two PVT-TEC collector is obtained as follows:

$$\dot{E}_{\text{total}} = \dot{E}_{\text{total1}} + \dot{E}_{\text{total2}} = 119.62 + 115.47 = 235.09 \, \text{W}$$

An average electrical efficiency of OPV of two flexible OPVT-TEC $=$ $\frac{0.1010 + 0.0985}{2} = 0.09995.$

## 6.7.2  The Rate of Useful Thermal Energy $\left( \dot{Q}_{uN} \right)$ and Mass Flow Rate Factor $(F_{RN})$

The rate of useful thermal energy at $n$th PVT-TEC collector can be obtained as

$$\dot{Q}_{uN} = \dot{m}_f c_f \left( T_{foN} - T_{fi} \right)$$
$$= F_{RN} A_m \left[ \left[ (\alpha \tau)_{\text{meff}} I(t) + (U_{Lm} + U_a) T_a \right] - U_{Lm-\text{eff}} T_{fi} \right] \quad (6.68a)$$

where, $F_{RN} = \frac{\dot{m}_f c_f}{A_m U_{Lm-\text{eff}}} \left[ 1 - \exp \left\{ -\frac{N F' U_{Lm-\text{eff}} A_m}{\dot{m}_f c_f} \right\} \right]$ is mass flow rate factor of N-PVT-TEC collectors connected in series.

The other method to get the rate of useful thermal energy at $n$th PVT-TEC collector can be obtained by summing the rate of useful thermal energy available at end of each PVT-TEC collector as follows:

$$\dot{Q}_{uN} = \dot{Q}_{u1} + \underset{u2}{\dot{Q}} + \underset{u3}{\dot{Q}} \pm - - - \dot{-} - - - \underset{uN-2}{\dot{Q}} + \underset{uN-1}{\dot{Q}} + \underset{uN}{\dot{Q}}$$
$$= \dot{m}_f c_f \left( T_{fo1} - T_{fi} \right) + \dot{m}_f c_f \left( T_{fo2} - T_{fo1} \right) + \dot{m}_f c_f \left( T_{fo3} - T_{fo2} \right)$$
$$+ - - - - \dot{-} - - - m \, c_f \left( T_{foN-1} - T_{fN-2} \right) + \dot{m}_f c_f \left( T_{foN} - T_{N-1i} \right)$$
$$+ \dot{m}_f c_f \left( T_{foN} - T_{fi} \right) = \dot{m}_f c_f \left( T_{foN} - T_{fi} \right) \quad (6.68b)$$

Therefore, the rate of useful thermal energy at $n$th PVT-TEC collector $\left( \dot{Q}_{uN} \right)$ will be exactly same by both methods.

## 6.7.3  The Average Fluid Temperature of N-OPVT-TEC Collectors Connected in Series

Following Sect. 4.5.1 and Eq. 4.31 and with the help of Eq. 4.67b, one can derive an expression for an average fluid temperature of N-PVT-TEC collectors connected in series as

$$\overline{T}_{fN} = \frac{1}{L_0} \int_0^{L_0} T_{fN}(x)\mathrm{d}x$$

$$= \frac{[(\alpha\tau)_{meff} I(t) + (U_{Lm} + U_a)T_a]}{U_{Lm-eff}} \left[ 1 - \frac{1 - \exp\left\{-\frac{NF'U_{Lm-eff}A_m}{\dot{m}_f c_f}\right\}}{\frac{NF'U_{Lm-eff}A_m}{\dot{m}_f c_f}} \right]$$

$$+ T_{fi} \frac{1 - \exp\left\{-\frac{NF'U_{Lm-eff}A_m}{\dot{m}_f c_f}\right\}}{\frac{NF'U_{Lm-eff}A_m}{\dot{m}_f c_f}} \tag{6.69}$$

**Example 6.46** Repeat the Example 6.42 by using Eq. 6.69 for $N = 2$.

**Solution**

Here $N = 2$ and from Example 6.37, we have

$$U_{Lm-eff} = \left(U_{Lm} - F_0'' F_0' U_p\right) = 5.029 - 0.0305 \times 0.2585 \times 138.6465$$
$$= 3.9348 \ \mathrm{W/m^2 \, °C}$$

Further,

For $N = 1$: $\exp\left\{-\frac{F'U_{Lm-eff}A_{cm}}{\dot{m}_f c_f}\right\} = \exp\left\{-\frac{0.75 \times 3.9348 \times 2}{0.035 \times 4190}\right\} = \exp\left\{-\frac{5.9022}{146.65}\right\} = \exp\{-0.04025\} = 0.9605.$

For $N = 2$: $\exp\left\{-\frac{2 \times F'U_{Lm-eff}A_{cm}}{\dot{m}_f c_f}\right\} = \exp\left\{-\frac{2 \times 0.75 \times 3.9348 \times 2}{0.035 \times 4190}\right\} = \exp\left\{-\frac{2 \times 5.9022}{146.65}\right\} = \exp\{-0.0805\} = 0.92265.$

$$\frac{[(\alpha\tau)_{meff} I(t) + (U_{Lm} + U_a)T_a]}{U_{Lm-eff}} = \frac{[0.326 \times 500 + (5.029 + 0.3251) \times 16]}{3.9348}$$

$$= \frac{248.666}{3.9348} = 63.197$$

Substitute the above value into Eq. 6.69, one gets

$$\overline{T}_{f2} = \mathbf{63.197}\left[1 - \frac{1 - 0.92265}{0.0805}\right] + T_{fi}\frac{1 - 0.92265}{0.0805}$$

$$= \mathbf{63.197} \times [1 - 0.9609] + 15 \times 0.9609 = 16.885 \, °\mathrm{C}$$

The above value is nearly same as 16.865 as reported in Example 6.42.

**Example 6.47** Compute an average TEC top surface $\left(\overline{T}_{tec,top2}\right)$ and solar cell $\left(\overline{T}_{c2}\right)$ temperatures for data of an average fluid temperature $\left(\overline{T}_{f2}\right)$ of Example 6.46 for 2-OPVT-TEC collector.

**Solution**

**Given:** $\overline{T}_{f2} = 16.885 \, °\mathrm{C}$ (Example 6.46).

From Eq. 6.55, we have an expression for TEC top surface temperature and after substitution of appropriate values from Examples 6.36 and 6.37, one gets

$$
\begin{aligned}
\overline{T}_{\text{tec,top2}} &= \frac{F_0(\alpha\tau)_{eff}I(t) + \{(U)_{\text{tec,top}-p} + U_{\text{tec}}\beta_{\text{tec}}\}\overline{T}_{f2} + (U)_{\text{tec,top}-a}T_a}{(U)_{\text{tec,top}-a} + (U)_{\text{tec,top}-p} + U_{\text{tec}}\beta_{\text{tec}}} \\
&= \frac{0.565 \times \mathbf{0.326} \times 500 + (0.8989 + 450 \times 0.4324)16.885 + 5.2545 \times 15}{5.2545 + 0.8989 + 450 \times 0.4324} \\
&= \frac{92.095 + 195.479 \times 16.885 + 78.82}{199.734} = 17.38\,^{\circ}\text{C}
\end{aligned}
$$

From Eq. 5.66, we have

$$
\begin{aligned}
\overline{T}_{c2} &= \frac{(\tau_g\alpha_{sc} - \eta_m)I(t) + U_{t,ca}T_a + h_t\overline{T}_{\text{tec,top2}}\beta_{\text{tec}} + U_{b,cp}(1 - \beta_{\text{tec}})\overline{T}_{f2}}{U_{t,ca} + h_t\beta_{\text{tec}} + U_{b,cp}(1 - \beta_{\text{tec}})} \\
&= \frac{0.735 \times 500 + 9.12 \times 16 + 120 \times 17.38 \times 0.4324 + 2.7663 \times (1 - 0.4324) \times 16.885}{9.12 + 120 \times 0.4324 + 2.7663 \times (1 - 0.4324)} \\
&= \frac{367.5 + 145.92 + 901.81 + 26.51}{9.12 + 51.89 + 1.57} = \frac{1{,}441.74}{22.21} = 64.91\,^{\circ}\text{C}
\end{aligned}
$$

### 6.7.4  Total Electrical Power from OPVT-TEC Liquid Collector

**Example 6.48** Evaluate electrical power from PV, TEC and total electrical power from 2-PVT-TEC liquid collector by using the data of Example 6.47.

**Solution**

**From Example 8.39** $\overline{T}_{f2} = \overline{T}_{\text{tec,bottom2}} = 16.885\,^{\circ}\text{C}$ (Example 6.46), $\overline{T}_{c2} = 64.91\,^{\circ}\text{C}$, $\overline{T}_{\text{tec,top2}} = 17.38\,^{\circ}\text{C}$ (Example 6.47).

Following Eq. 6.31a, an electrical power from PV of 2-PVT-TEC collector can be obtained as

$$
\begin{aligned}
\dot{E}_{el2} &= \eta_{mo}\left[1 - \beta_0(\overline{T}_c - 25)\right] \times \beta_c I(t)A_m \times 2 \\
&= 0.12[1 - 0{,}00425(64.91 - 25)] \times 0.89 \times 500 \times 2 \times 2 \\
&= 177.36\,\text{W}
\end{aligned}
$$

Electrical power from TEC of 2-PVT-TEC collector can be evaluated as

$$
\dot{E}_{\text{TEC2}} = \eta_{\text{tec}}U_{\text{tec}}(T_{\text{tec,top}} - T_{\text{tec,bottom}})\beta_{\text{tec}}A_m \times 2
$$

**Here,** $\beta_{\text{tec}} = 0.4324$ and $\eta_{\text{tec}} = 0.15$ (Example 6.33), $U_{\text{tec}} = 450\,\text{W/m}^2\,^{\circ}\text{C}$ (Eq. 3.48a).

Then,

$$\dot{E}_{TEC2} = 0.15 \times 450(17.38 - 16.885)0.4324 \times 2 \times 2 = 57.78 \text{ W}$$

The total electrical power is the sum of electrical power from PV and TEC, and hence

$$\dot{E}_{total} = \dot{E}_{el} + \dot{E}_{TEC} = 177.36 + 57.78 = 235.14 \text{ W}$$

The above value is approximately same as 235.09 W obtained in Example 6.45.

### 6.7.5  The M-OPVT-TEC Collectors Connected in Parallel

Referring to Sect. 6.5, the outlet fluid temperature at end of $m$th PVT-TEC collector connected in parallel for given mass flow rate $\left(\dot{m}_f\right)$ will be expressed as follows:

$$T_{fom} = \frac{\left[(\alpha\tau)_{meff}I(t) + (U_{Lm} + U_a)T_a\right]}{U_{Lm-eff}}\left[1 - \exp\left(-\frac{F'A_m U_{Lm-eff}}{\left(\frac{\dot{m}_f}{m}\right)C_f}\right)\right]$$

$$+ T_{fi}\exp\left(-\frac{F'A_m U_{Lm-eff}}{\left(\frac{\dot{m}_f}{m}\right)C_f}\right) \tag{6.70a}$$

where, $U_{Lm-eff} = \left(U_{Lm} - F_0'' F_0' U_p\right)$.

In Eq. 6.70a, the mass flow rate is equally divided in each PVT-TEC collector. Equation 6.70a can also be written as

$$T_{fom} = \frac{\left[(\alpha\tau)_{meff}I(t) + (U_{Lm} + U_a)T_a\right]}{U_{Lm-eff}}\left[1 - \exp\left(-\frac{F'm A_m U_{Lm-eff}}{\dot{m}_f C_f}\right)\right]$$

$$+ T_{fi}\exp\left(-\frac{F'm A_m U_{Lm-eff}}{\dot{m}_f C_f}\right) \tag{6.70b}$$

with $A_{cm} = m \times A_m$.

In Eq. 6.70b, each PVT-TEC collector has same mass flow rate, and these are connected like series connection.

**Example 6.49** Calculate the outlet and the rate of useful thermal energy for six PVT-TEC collectors for the following combinations:

**Case (a):** All six PVT-TEC collectors are connected in parallel ($m = 6$).

**Case (b):** All six PVT-TEC collectors are connected in series ($N = 6$).

**Case (c):** Three PVT-TEC collectors connected in series ($N = 3$) as one module and each module is connected in parallel ($m = 2$), Fig. 6.4.

**Case (d):** Two PVT-TEC collectors are connected in parallel ($m = 2$) as one module, and then, each module is connected in series ($N = 3$).

Given mass flow rate at inlet of each case ($\dot{m}_f$) $= 0.035$ kg/s and other parameters are same as given in Example 6.37. These are as follows:

Climatic parameters: $I(t) = 500$ W/m$^2$, $T_a = 16\,°C$ and $T_{fi} = 15\,°C$.

Design parameters: $F' = 0.75$, $(\alpha\tau)_{meff} = \mathbf{0.326}$, $U_{Lm} = 5.029$ W/m$^2$ $°C$, $U_a = 0.3251$ W/m$^2$ $°C$, $A_m = 2\times1$ m$^2$, the penalty factor, $F_0' = 0.2585$, the penalty factor, $F_0'' = 0.0305$, an overall net heat transfer coefficient, $U_p = 138.6465$ W/m$^2$ $°C$ (Examples 3.35–3.36) and $\dot{m}_f c_f = 0.035 \times 419 = 146.65$ J/s $°C = 146.65$ W/ $°C$.

**Solution**

Now from Eq. 6.50, we have

$$U_{Lm-\text{eff}} = \left(U_{Lm} - F_0'' F_0' U_p\right) = 5.029 - 0.0305 \times 0.2585 \times 138.6465$$
$$= 3.9348 \text{ W/m}^2\,°C$$

**Case (a):** The six PVT-TEC collectors are connected in parallel, and hence $N = 1$ and $m = 6$. In $N = 1$ means single PVT-TEC collector, the mass flow rate will be $\frac{\dot{m}_f}{6} = \frac{0.035}{6} = 0.0058$ kg/s. For each PVT-TEC collector, the outlet fluid temperature will be same.

In this case, for one PVT-TEC collector, we have

$$\exp\left\{-\frac{F'U_{Lm-\text{eff}}A_m}{\left(\frac{\dot{m}_f}{6}\right)c_f}\right\} = \exp\left\{-\frac{0.75 \times 3.9348 \times 2}{0.0058 \times 4190}\right\}$$
$$= \exp\left\{-\frac{5.9022}{24.302}\right\} = \exp\{-0.2429\} = 0.7843$$

and,

$$\frac{\left[(\alpha\tau)_{meff}I(t) + (U_{Lm} + U_a)T_a\right]}{U_{Lm-\text{eff}}} = \frac{[\mathbf{0.326} \times 500 + (5.029 + 0.3251) \times 16]}{3.9348}$$
$$= \frac{248.666}{3.9348} = 63.197$$

From Eq. 6.52, we have an expression for the outlet temperature of fluid for each PVT-TEC collector as

$$T_{fo} = \frac{[\mathbf{0.326} \times 500 + (5.029 + 0.3251) \times 16]}{3.9348}[1 - 0.7843] + 15 \times 0.7843$$

$$= 63.197 \times 0.2157 + 11.765 = 13.63 + 11.765 = 25.396\,^\circ\text{C}$$

The rate of thermal energy from one PVT-TEC collector $= \dot{m}_f C_f \left(T_{fo} - T_{fi}\right) = 0.0058 \times 4190(25.396 - 15) = 252.64\,\text{W}$.

Therefore, from six PVT-TEC collectors connected in parallel will be six time of 252.64 W; hence

$$\dot{Q}_{u-6\times1} = 6 \times 252.64 = 1.515.86\,\text{W} = 1.516\,\text{kW}$$

**Case (b):** In this case all six PVT-TEC collectors are connected in series, and hence, the outlet fluid temperature at 6th PVT-TEC collector can be obtained from Eq. 6.67a as

$$T_{foN} = \frac{\left[(\alpha\tau)_{meff} I(t) + (U_{Lm} + U_a)T_a\right]}{U_{Lm-\text{eff}}}\left[1 - \exp\left\{-\frac{N F' U_{Lm-\text{eff}} A_m}{\dot{m}_f c_f}\right\}\right]$$
$$+ T_{fi}\exp\left\{-\frac{N F' U_{Lm-\text{eff}} A_m}{\dot{m}_f c_f}\right\}$$

In above equation,

$$\frac{\left[(\alpha\tau)_{meff} I(t) + (U_{Lm} + U_a)T_a\right]}{U_{Lm-\text{eff}}} = \frac{[0.326 \times 500 + (5.029 + 0.3251) \times 16]}{3.9348}$$
$$= \frac{248.666}{3.9348} = 63.197$$

and,

For $N = 6$: $\exp\left\{-\frac{N F' U_{Lm-\text{eff}} A_{cm}}{\dot{m}_f c_f}\right\} = \exp\left\{-\frac{6\times0.75\times3.9348\times2}{0.035\times4190}\right\} = \exp\left\{-\frac{35.4132}{146.65}\right\} = \exp\{-0.2415\} = 0.7854$.

Substitute the above value in expression for $T_{foN}$ as

$$T_{fo6} = 63.197[1 - 0.7854] + 15 \times 0.7854 = 63.197 \times 0.2146 + 11.781 = 25.34\,^\circ\text{C}$$

The rate of thermal energy from six PVT-TEC collector connected in series will be as follows:

$$\dot{Q}_{u6} = \dot{m}_f C_f \left(T_{fo6} - T_{fi}\right) = 0.035 \times 4190(25.34 - 15) = 1516.81\,\text{W} = 1.517\,\text{kW}$$

**Case (c):** In this case, the three OPVT-TEC collectors are connected in series $(N = 3)$ as one module and mass flow rate will be $\frac{\dot{m}_f}{2} = \frac{0.035}{2} = 0.0175\,\text{kg/s}$; hence

$$T_{fo3} = \frac{[(\alpha\tau)_{meff} I(t) + (U_{Lm} + U_a)T_a]}{U_{Lm-\text{eff}}}\left[1 - \exp\left\{-\frac{3F'U_{Lm-\text{eff}}A_m}{\dot{m}_f c_f}\right\}\right]$$
$$+ T_{fi}\exp\left\{-\frac{3F'U_{Lm-\text{eff}}A_m}{\dot{m}_f c_f}\right\}$$

**Here,**

$$\exp\left\{-\frac{3F'U_{Lm-\text{eff}}A_m}{\dot{m}_f c_f}\right\} = \exp\left\{-\frac{3 \times 0.75 \times 3.9348 \times 2}{0.0175 \times 4190}\right\}$$
$$= \exp\left\{-\frac{17.7066}{73.325}\right\}$$
$$= \exp\{-0.2415\} = 0.7854.$$

Substitute this value in $T_{fo3}$, one gets

$$T_{fo-2\times3} = 63.197 \times [1 - 0.7854] + 15 \times 0.7854 = 25.34\,^\circ\text{C}$$

So, the rate of useful thermal energy for m = 2 will be as

$$\dot{Q}_{u-2\times3} = 2 \times \left(\frac{\dot{m}_f}{2}\right)c_f\left(T_{fo-2\times3} - T_{fi}\right) = 0.035 \times 4190 \times (25.34 - 15)$$
$$= 1516.36\,\text{W} = 1.516\text{kW}$$

**Case (d): In this case, the first two collectors are connected in parallel as one module having an area of 2 × 2 = 4 m², then mass flow rate will be** $\frac{\dot{m}_f}{2} = \frac{0.035}{2} = 0.0175$ **for each OPVT-TEC collector, and hence**

$$T_{fo2} = \frac{[(\alpha\tau)_{meff} I(t) + (U_{Lm} + U_a)T_a]}{U_{Lm-\text{eff}}}\left[1 - \exp\left\{-\frac{F'U_{Lm-\text{eff}}A_m}{\left(\frac{\dot{m}_f}{2}\right)c_f}\right\}\right]$$
$$+ T_{fi}\exp\left\{-\frac{F'U_{Lm-\text{eff}}A_m}{\left(\frac{\dot{m}_f}{2}\right)c_f}\right\}$$

**Here,**

$$\exp\left\{-\frac{F'U_{Lm-\text{eff}}A_m}{\left(\frac{\dot{m}_f}{2}\right)c_f}\right\} = \exp\left\{-\frac{0.75 \times 3.9348 \times 2}{0.0175 \times 4190}\right\} = \exp\left\{-\frac{5.9022}{73.325}\right\}$$
$$= \exp\{-0.0805\} = 0.9226$$

**Substitute the above value in** $T_{fo2}$**, one gets**

$$T_{fo2} = 63.197[1 - 0.9226] + 15 \times 0.9226 = 48.89 + 13.839 = 18.729\,°C$$

This will be inlet temperature for 2nd PVT-TEC module having area of 4 m$^2$, and the temperature will be

$$T_{fo-2\times2} = 63.197[1 - 0.9226] + 18.729 \times 0.9226 = 4.89 + 17.28 = 22.17\,°C$$

Further, this temperature will be inlet of 3rd PVT-TEC collector having area of 4 m$^2$, and hence

$$T_{fo-3\times2} = 63.197[1 - 0.9226] + 22.17 \times 0.9226 = 4.89 + 20.45 = 25.34\,°C$$

So, the rate of useful thermal energy will be as

$$\dot{Q}_{u-2\times3} = \acute{m}_f C_f(T_{fo-3\times2} - T_{fi}) = 0.035 \times 4190 \times (25.34 - 15)$$
$$= 1516.36\,W = 1.516\,kW$$

From this example, one can conclude that

(a) The outlet fluid temperature is same irrespective of any combination.
(b) The rate of thermal energy also remains the same in all cases.
(c) The exergy will again remain the same because it depends on operating temperature.

## 6.7.6 The Mass Flow Rate Factor of $N^{th}$ OPVT-TEC Collector

Expression for the rate of thermal energy $(\dot{Q}_{u,n^{th}})$ obtained at n$^{th}$ PVT-TEC collector can written as

$$\dot{Q}_{u,n^{th}} = \dot{m}_f C_f(T_{fo,n} - T_{fo,n-1}) \tag{6.71a}$$

With the help of Eqs. 6.67a and 6.71a and taking the term $\exp\left\{-\frac{(n-1)F'U_{Lm-\text{eff}}A_m}{\dot{m}_f c_f}\right\}$ outside bracket, one can rewritten $\dot{Q}_{u,n^{th}}$ as

$$\dot{Q}_{u,n^{th}} = m_f C_f(T_{fon} - T_{fo,n-1})$$
$$= \dot{m}_f C_f \exp\left\{-\frac{(n-1)F'U_{Lm-\text{eff}}A_m}{\dot{m}_f c_f}\right\}$$
$$\left\{\frac{[(a\tau)_{m\text{eff}}I(t) + (U_{Lm} + U_\alpha)T_a]}{U_{Lm-\text{eff}}}\left[\exp\left\{\frac{(n-1)F'U_{Lm-\text{eff}}A_m}{\dot{m}_f c_f}\right\}\right.\right.$$
$$\left.\left. - \exp\left\{-\frac{F'U_{Lm-\text{eff}}A_m}{\dot{m}_f c_f}\right\}\right] + T_{fi}\exp\left\{-\frac{F'U_{Lm-\text{eff}}A_m}{\dot{m}_f c_f}\right\}\right\}$$

$$-\left\{\frac{[(\alpha\tau)_{\mathbf{meff}}I(t)+(U_{Lm}+U_{\alpha})T_a]}{U_{Lm-\text{eff}}}\left[\exp\left\{\frac{(n-1)F'U_{Lm-\text{eff}}A_m}{\dot{m}_fc_f}\right\}-1\right]+T_{fi}\right\}\right\}$$

**In the above equation, the term** $\exp\left\{\frac{(n-1)F'U_{Lm-\text{eff}}A_m}{\dot{m}_fc_f}\right\}$ **will be cancelled out as** follows:

$$\dot{Q}_{u,n^{th}}=m_fC_f\left(T_{fon}-T_{fo,n-1}\right)$$

$$=\dot{m}_fC_f\exp\left\{-\frac{(n-1)F'U_{Lm-\text{eff}}A_m}{\dot{m}_fc_f}\right\}$$

$$\left\{\frac{[(\alpha\tau)_{meff}I(t)+(U_{Lm}+U_a)T_a]}{U_{Lm-\text{eff}}}\left[\exp\left\{\frac{(n-1)F'U_{Lm-\text{eff}}A_m}{\dot{m}_fc_f}\right\}\right.\right.$$

$$\left.-\exp\left\{-\frac{F'U_{Lm-\text{eff}}A_m}{\dot{m}_fc_f}\right\}-\left[\exp\left\{\frac{(n-1)F'U_{Lm-\text{eff}}A_m}{\dot{m}_fc_f}\right\}\right]-1\right]$$

$$\left.+T_{fi}\exp\left\{-\frac{F'U_{Lm-\text{eff}}A_m}{\dot{m}_fc_f}\right\}-T_{fi}\right\}$$

Further, the above equation can be written as

$$\dot{Q}_{u,n^{th}}=m_fC_f\left(T_{fon}-T_{fo,n-1}\right)\quad=F_{R,n^{th}}A_m\{[(\alpha\tau)_{meff}I(t)+(U_{Lm}+U_a)T_a]-U_{Lm-\text{eff}}T_{fi}\}$$

$$(6.71\text{b})$$

Here, $F_{R,n^{th}}=\frac{\dot{m}_fc_f}{U_{Lm-\text{eff}}A_m}\exp\left\{-\frac{(n-1)F'U_{Lm-\text{eff}}A_m}{\dot{m}_fc_f}\right\}\left[1-\exp\left\{-\frac{F'U_{Lm-\text{eff}}A_m}{\dot{m}_fc_f}\right\}\right]$; **mass flow rate factor.**

For $N=1$, Eqs. 6.71a, 6.71b will be reduced to

$$\dot{Q}_u=\dot{m}_fc_f\left(T_{fo}-T_{fi}\right)$$

$$=F_RA_m\left[[(\alpha\tau)_{\mathbf{meff}}I(t)+(U_{Lm}+U_a)T_a]-U_{Lm-\text{eff}}T_{fi}\right]$$

with, $F_R=\frac{\dot{m}_fc_f}{A_mU_{Lm-\text{eff}}}\left[1-\exp\left\{-\frac{F'U_{Lm-\text{eff}}A_m}{\dot{m}_fc_f}\right\}\right]$

which is exactly same as Eq. 6.53.

The expression for various constants including $U_{Lm-\text{eff}}$ for different case is given in Table 6.1 as

(a) $U_{Lm-\text{eff}}$ $=$ $\left(U_{Lm}-F_0''F_0'U_p\right)$ with $U_p$ $=$ $\{(U)_{\text{tec,top}-p}+U_{\text{tec}}\beta_{\text{tec}}-(U_{t,ca}+h_t\beta_{\text{tec}})\}$, for $\beta_{\text{tec}}\neq0$;

(b) $U_{Lm-\text{eff}}=\left(U_{Lm}-F_0''F_0'U_p\right)=U_{Lm}$ with $U_p=\{(U)_{\text{tec,top}-p}-U_{t,ca}\}$, $F_0'=$ $0$ for $\beta_{\text{tec}}=0$

(c) $U_{Lm-\text{eff}}$ $=$ $\left(U_{Lm}-F_0''F_0'U_p\right)$ with $U_p$ $=$ $\{(U)_{\text{tec,top}-p}+U_{\text{tec}}-(U_{t,ca}+h_t)\}$, $F_0'=\frac{h_t}{\{(U)_{\text{tec,top}-a}+(U)_{\text{tec,top}-p}+U_{\text{tec}}\}}$ and $F_0''=$ $\frac{U_{b,cp}}{U_{t,ca}+h_t\beta_{\text{tec}}}$ for $\beta_{\text{tec}}=1$

**Table 6.1** Formulas for various constant

### (A) For PVT collector

| | |
|---|---|
| (i) | $F_0 = \frac{U_{bc,p}}{(U_{tc,a}+U_{bc,p})}$ for constant $\eta_m$ |
| (ii) | $F_0 = \frac{U_{bc,p}}{[U_{tc,a}+U_{bc,p}-[\eta_{m0}\beta_c I(t)\beta_0]]}$ for variable $\eta_m$ |
| (iii) | $(\alpha_c \tau_g)_{meff} = \{\tau_g[\alpha_c \beta_c + (1-\beta_c)\alpha_T] - \eta_m \beta_c\}$ for constant $\eta_m$ |
| (iv) | $[(\alpha_c \tau_g)_{meff}]_{semi} = [\tau_g^2(1-\beta_c) + F_0(\alpha_c \tau_g)_{meff}]$ for constant $\eta_m$ and semi-transparent PVT collector |
| (v) | $U_{Lm} = \left(U_{tc,a} + \frac{U_b}{F_0}\right)$ |
| (vi) | $(\alpha_c \tau_g)_{meff} = [\tau_g\{\alpha_c \beta_c + (1-\beta_c)\alpha_T\} - [\eta_{m0}\beta_c](1+25\beta_0)]$ for variable $\eta_m$ |
| (vii) | $F'_{eff} = F' F_0$ |
| (viii) | $U_{Lm-eff} = [\{F_0 U_{tc,a} + U_b\} - F_0\eta_{mo}\beta_0\beta_c I(t)]$ for variable $\eta_m$ |
| (ix) | $U_{Lm-eff} = \{F_0 U_{tc,a} + U_b\}$ for constant $\eta_m$ |
| (x) | $F_{Rm} = \frac{\dot{m}_f C_f}{A_c U_{Lm-eff}}\left[1 - \exp\left(-\frac{F' A_c U_{Lm-eff}}{\dot{m}_f C_f}\right)\right]$ for variable $\eta_m$ |
| (xi) | $F_{Rm} = \frac{\dot{m}_f C_f}{A_c U_{Lm-eff}}\left[1 - \exp\left(-\frac{F' A_c U_{Lm}}{\dot{m}_f C_f}\right)\right]$ and $U_{Lm-eff} = [\{F_0 U_{tc,a} + U_b\}]$ for constant $\eta_m$ |

### For PVT-TEC collector

| | |
|---|---|
| (i) | $F_0 = \frac{(h_t \beta_{tec})}{U_{t,ca}+h_t\beta_{tec}+U_{b,cp}(1-\beta_{tec})}$ for $\beta_{tec} \neq 0$; |
| | $F_0 = 0$ for $\beta_{tec} = 0$; |
| | $F_0 = \frac{h_t}{U_{t,ca}+h_t}$ for $\beta_{tec} = 1$ |
| (ii) | $(U)_{tec,top-a} = \frac{(h_t \beta_{tec})U_{t,ca}}{U_{t,ca}+h_t\beta_{tec}+U_{b,cp}(1-\beta_{tec})}$ for $\beta_{tec} \neq 0$; |
| | $(U)_{tec,top-a} = 0$ for $\beta_{tec} = 0$ |
| | $(U)_{tec,top-a} = \frac{h_t U_{t,ca}}{U_{t,ca}+h_t}$ for $\beta_{tec} = 1$ |
| (iii) | $(U)_{tec,top-p} = \frac{(h_t \beta_{tec})\{U_{b,cp}(1-\beta_{tec})\}}{U_{t,ca}+h_t\beta_{tec}+U_{b,cp}(1-\beta_{tec})}$ for $\beta_{tec} \neq 0$; |
| | $(U)_{tec,top-p} = \frac{(h_t \beta_{tec})\{U_{b,cp}(1-\beta_{tec})\}}{U_{t,ca}+h_t\beta_{tec}+U_{b,cp}(1-\beta_{tec})} = 0$ for $\beta_{tec} = 0$; |
| | $(U)_{tec,top-p} = \frac{(h_t \beta_{tec})\{U_{b,cp}(1-\beta_{tec})\}}{U_{t,ca}+h_t\beta_{tec}+U_{b,cp}(1-\beta_{tec})} = 0$ for $\beta_{tec} = 1$; |
| (iv) | $F_{01} = \frac{U_{tec}\beta_{tec}}{[(U)_{tec,top-a}+(U)_{tec,top-p}+U_{tec}\beta_{tec}]}$ for $\beta_{tec} \neq 0$; |
| | $F_{01} = 0$ for $\beta_{tec} = 0$ |
| | $F_{01} = \frac{U_{tec}}{[(U)_{tec,top-a}+(U)_{tec,top-p}+U_{tec}]}$ for $\beta_{tec} = 1$ |
| (v) | $(U)_{tec,top-pa} = \frac{U_{tec}\beta_{tec}(U)_{tec,top-a}}{[(U)_{tec,top-a}+(U)_{tec,top-p}+U_{tec}\beta_{tec}]}$ for $\beta_{tec} \neq 0$; |
| | $(U)_{tec,top-pa} = 0$ for $\beta_{tec} = 0$ |
| | $(U)_{tec,top-pa} = \frac{U_{tec}(U)_{tec,top-a}}{[(U)_{tec,top-a}+(U)_{tec,top-p}+U_{tec}]}$ for $\beta_{tec} = 1$ |

(continued)

**Table 6.1** (continued)

**(A) For PVT collector**

| | |
|---|---|
| (vi) | $F_0' = \dfrac{(h_t \beta_{tec})}{\{(U)_{tec,top-a}+(U)_{tec,top-p}+U_{tec}\beta_{tec}\}}$ for $\beta_{tec} \neq 0$; |
| | $F_0' = 0$ for $\beta_{tec} = 0$ |
| | $F_0' = \dfrac{h_t}{\{(U)_{tec,top-a}+(U)_{tec,top-p}+U_{tec}\}}$ for $\beta_{tec} = 1$ |
| (vii) | $F_0'' = \dfrac{U_{b,cp}}{U_{t,ca}+h_t\beta_{tec}+U_{b,cp}(1-\beta_{tec})}$ for $\beta_{tec} \neq 0$; |
| | $F_0'' = \dfrac{U_{b,cp}}{U_{t,ca}+U_{b,cp}}$ for $\beta_{tec} = 0$ |
| | $F_0'' = \dfrac{U_{b,cp}}{U_{t,ca}+h_t}$ for $\beta_{tec} = 1$ |
| (viiii) | $U_p = \{(U)_{tec,top-p} + U_{tec}\beta_{tec} - (U_{t,ca} + h_t\beta_{tec})\}$ (Eq. 6.43) for $\beta_{tec} \neq 0$ |
| | $U_p = \{(U)_{tec,top-p} - U_{t,ca}\}$ for $\beta_{tec} = 0$ |
| | $U_p = \{(U)_{tec,top-p} + U_{tec} - (U_{t,ca} + h_t)\}$ for $\beta_{tec} = 1$ |
| (ix) | $(\alpha\tau)_{meff} =$<br>$[(1-\eta_{tec})F_{01}F_0(\alpha\tau)_{eff} + F_0''(1-\beta_{tec})\{(\tau_g\alpha_{sc}-\eta_m)+F_0'F_0(\alpha\tau)_{eff}\}]$, $\beta_{tec} \neq 0$ |
| | $(\alpha\tau)_{meff} = \left[\dfrac{U_{b,cp}}{U_{t,ca}+U_{b,cp}}\{(\tau_g\alpha_{sc}-\eta_m)\}\right]$ for $\beta_{tec} = 0, \eta_{tec} = F_{01} = F_0 = F_0 = 0$ |
| | $(\alpha\tau)_{meff} = [(1-\eta_{tec})F_{01}F_0(\alpha\tau)_{eff} + +F_0'F_0(\alpha\tau)_{eff}]$ for $\beta_{tec} = 1$ |
| (x) | (a) $U_{Lm} = \{(1-\eta_{tec})(U)_{tec,top-pa}+U_b\}$; |
| | $(U)_{tec,top-pa} = \dfrac{U_{tec}(U)_{tec,top-a}}{[(U)_{tec,top-a}+(U)_{tec,top-p}+U_{tec}]}$; $(U)_{tec,top-a} = \dfrac{h_t U_{t,ca}}{U_{t,ca}+h_t}$ for $\beta_{tec} = 1$ |
| | (b) $U_{Lm} = \{(1-\eta_{tec})(U)_{tec,top-pa}+U_b\} = U_b$ for $\beta_{tec} = \eta_{tec} = 0$ |
| (xi) | $U_a = [F_0''(U_{t,ca}+F_0'(U)_{tec,top-a})]$ |
| (xii) | $(U)_{tec,top-a} = \dfrac{h_t U_{t,ca}}{U_{t,ca}+h_t}$ |
| (xiii) | $F_0'' = \dfrac{U_{b,cp}}{U_{t,ca}+h_t\beta_{tec}}$ |
| (xiv) | (a) $U_{Lm-eff} = (U_{Lm} - F_0''F_0'U_p)$ with $U_p =$<br>$\{(U)_{tec,top-p} + U_{tec}\beta_{tec} - (U_{t,ca}+h_t\beta_{tec})\}$, for $\beta_{tec} \neq 0$; |
| | (b) $U_{Lm-eff} = (U_{Lm} - F_0''F_0'U_p) = U_{Lm}$ with $U_p = \{(U)_{tec,top-p} - U_{t,ca}\}$ and $F_0' = 0$ for $\beta_{tec} = 0$ |
| | (c) $U_{Lm-eff} = (U_{Lm} - F_0''F_0'U_p)$ with<br>$U_p = \{(U)_{tec,top-p} + U_{tec} - (U_{t,ca}+h_t)\}$ for $\beta_{tec} = 1$ |
| (xv) | $a = \dfrac{F'U_{Lm-eff}b}{\dot{m}_f c_f}$; |

Further, the basic constants involved in various examples are also given in Table 6.2.

***Example 6.50*** Compute the rate of thermal energy and mass flow rate factor for data of Example 6.38 for 6th OPVT-TEC liquid flexible collector along with an instantaneous thermal efficiency.

**Table 6.2** Some basic physical constants

| S. No | Constants | Numerical value | Units |
|---|---|---|---|
| 1 | $A_c$ | 2 | $m^2$ |
| 2 | $A_m$ | 2 | $m^2$ |
| 3 | b | 1 | m |
| 4 | $C_f$ | 4190 | j/kg °C |
| | $F'$ | 0.82–0.91 | Equation 6.7, Example 4.29 |
| 5 | I(t) | 500 | $W/m^2$ |
| 6 | $K_{bs}$ | 386 | W/m °C |
| 7 | $K_g$ | 0.688 | W/m °C |
| 8 | $K_t$ | 0.033 | W/m °C |
| 9 | $L_0$ | 2 | m |
| 10 | $L_{bs}$ | 0.0005 | m |
| 11 | $L_g$ | 0.003 | m |
| 12 | $L_t$ | 0.0005 | m |
| 13 | $\dot{m}_f$ | 0.035 | Kg/s |
| 14 | $T_a$ | 16 | °C |
| 15 | $T_{fi}$ | 15 | °C |
| 16 | $U_b$ | 0.70 | $W/m^2$ °C |
| | **Greek letters** | | |
| 17 | $\alpha_c$ | 0.9 | |
| 18 | $\alpha_T$ | 0.2 | |
| 19 | $\beta_0$ | 0.00425 | $°C^{-1}$ |
| 20 | $\beta_c$ | 0.89 $\eta_m$ | |
| 21 | $\beta_{tec}$ | 0.4324 | |
| 22 | $\gamma$ | 0.20–0.35 | |
| 23 | $\tau_g$ | 0.95 | |
| 24 | $\eta_{m0}$ | 0.12 | |
| 25 | $\eta_m$ | 0.12 | |

**Solution**
**Here $n = 6$. from Eq. 5.71b, we have**
Mass flow rate factor

$$
F_{R,6} = \frac{\dot{m}_f c_f}{U_{Lm-\text{eff}} A_m} \exp\left\{ -\frac{(n-1) F' U_{Lm-\text{eff}} A_m}{\dot{m}_f c_f} \right\} \left[ 1 - \exp\left\{ -\frac{F' U_{Lm-\text{eff}} A_m}{\dot{m}_f c_f} \right\} \right]
$$

$$
= \frac{146.65}{2 \times 3.9348} [1 - 0.9605] \exp\left\{ -\frac{5 \times 0.75 \times 3.9348 \times 2}{146.65} \right\}
$$

$$
= 18.635 \times 0.0395 \times \exp(-0.1932) = 18.635 \times 0.0395 \times 0.8243 = 0.61
$$

which is less than the value obtained in the first flexible PVT-TEC collector as expected (Example 6.38).

From Eq. 6.71b, we have

$$\dot{Q}_{u,6^{th}} = F_{R,6^{th}} A_m \left[ \left[ (\alpha\tau)_{meff} I(t) + (U_{Lm} + U_a) T_a \right] - U_{Lm-\text{eff}} T_{fi} \right]$$
$$= 0.61 \times 2[[\mathbf{0.326} \times \mathbf{500} + (5.029 + 0.3251) \times 16] - 3.934 \times 15]$$
$$= 1.22 \times [[\mathbf{163} + \mathbf{85.666}] - 59.01] = 231.38\,\text{W}$$

An instantaneous thermal efficiency is obtained as

$$\eta_i^{6^{th}} = \frac{\dot{Q}_{u,6^{th}}}{I(t) \times 2} = \frac{231.38}{500 \times 2} = 0.2314(23.14\%)$$

# References

1. Agrawal, B., & Tiwari, G. N. (2008). *Developments in environmental durability for photovoltaics*. Pira International Ltd.
2. Tiwari, G. N., & Dubey, S. (2010). *Fundamentals of photovoltaic modules and their applications*. Royal Society of Chemistry (RSC).
3. Tiwari, G. N., & Mishra, R. K. (2011). *Advanced renewable energy sources*. Royal Society of Chemistry (RSC).
4. Agrawal, B., & Tiwari, G. N. (2010). *Building integrated photovoltaic thermal systems*. Royal Society of Chemistry (RSC)
5. Tiwari, A., Alashqar, O. A., & Dimri, N. (2020). Modeling and validation of photovoltaic thermoelectric flat plate collector (PV-TE-FPC). *Energy Conversion and Management, 205*, 112378. https://doi.org/10.1016/j.enconman.2019.112378
6. Dimri, N., & Tiwari, A., & Tiwari, G. N. (2018). Effect of thermoelectric cooler (TEC) integrated at the base of opaque photovoltaic (PV) module to enhance an overall electrical efficiency. *Solar Energy, 166*, 159–170. https://doi.org/10.1016/j.solener.2018.03.030
7. Dimri, N., Tiwari, A., & Tiwari, G. N. (2017). Thermal modelling of semitransparent photovoltaic thermal (PVT) with thermoelectric cooler (TEC) collector. *Energy Conversion and Management, 146*, 68–77.https://doi.org/10.1016/j.enconman.2017.05.017

# Chapter 7
# PVT-CPC Liquid and Air Collectors

## 7.1 Introduction

As discussed in Chap. 4 about photo-voltaic thermal (PVT) and photo-voltaic thermal-thermoelectrical cooler (PVT-TEC) collectors to produce thermal energy and electrical power to make the collector unit self-sustained. However, the design based on optimization is an important issue. The parameters should be optimized as per requirement of user's demands on priority basis, namely either thermal energy or electrical energy. For example, (i) the packing factor of PV module should be minimum if demand of thermal energy is more and (ii) the packing factor of PV module and TEC should be more for higher demand of electrical power.

In order to meet both requirement like thermal energy and electrical power, one can integrate the suitable concentrator unit with PVT and PVT-TEC collector to obtain higher thermal energy as well higher electrical power. The suitable concentrator enhances the incident solar radiation on collector's surface, and hence both thermal energy and electrical power are increased due to increase in input energy.

In this chapter, a suitable concentrator cum PVT and PVT-TEC liquid and air will be discussed with examples.

## 7.2 The Semi-transparent PVT-CPC (SPVT-CPC) Liquid Collector [1]

In this case, the compound parabolic concentrator (CPC) is most suitable to concrete solar radiation incident larger aperture area ($A_{am}$) in comparison with PV module surface area ($A_m = A_{rm}$) as shown in Fig. 7.1. Here PV module surface area will be referred as receiver area of SPVT-CPC liquid collector. The system which consists of SPVT and CPC will be referred as SPVT-CPC liquid collector solar radiation incident on aperture area ($A_{am}$) is reflected back to receiver/glass of PV module is

© Bag Energy Research Society 2024
G. N. Tiwari, *Fundamental of Mathematical Tools for Thermal Modeling of Solar Thermal and Photo-voltaic Systems-Part-I*,
https://doi.org/10.1007/978-981-99-7085-8_7

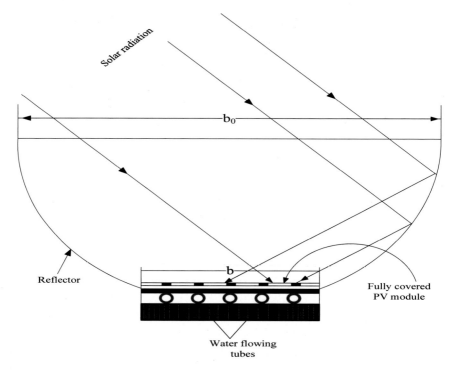

**Fig. 7.1** Cross-sectional view of single semi-transparent PVT-CPC liquid collector

transmitted to solar cell which is partially absorbed and raises its temperature. The part of thermal energy is lost to ambient air, and the reamaining is transferred to absorber plate of SPVT-CPC liquid collector. The design parameters of PVT-CPC are given in Table 7.1.

## 7.2.1 Energy Balance Equation for Solar Cell

The energy balance of solar cell for SPVT-CPC in W can be written as follows:

$$\rho\alpha_c\tau_g\beta_c I_u A_{\text{am}} = U_{t,\text{ca}}(T_c - T_a)A_{\text{rm}} + U_{b,\text{cp}}(T_c - T_p)A_{\text{rm}} + \rho I_u \eta_m \beta_c A_{\text{am}}, \quad (7.1)$$

where,

The rate of solar radiation received by solar cells of SPV module after reflection from CPC $= \rho\alpha_c\tau_g\beta_c I_u A_{\text{am}}$.

The solar radiation received by solar cell after reflection from CPC $(I_u) = I_b + \frac{I_d}{C}$, Prapas et al. [2] with $C = \frac{A_{\text{am}}}{A_{\text{rm}}}$, concentration ratio.

**Table 7.1**  Some basic physical constants for PVT-CPC liquid collector

| S. No. | Constants | Numerical value | Units |
|--------|-----------|-----------------|-------|
| 1 | $A_{am}$ | 4 | $m^2$ |
| 2 | $A_{rm}$ | 2 | $m^2$ |
| 3 | b | 1 | m |
| 4 | $C_f$ | 4190 | j/kg °C |
| 5 | $F'$ | 0.82 | Example 4.29 |
| 6 | $F_0$ | 0.233 | Example 7.3 |
| 7 | $F_{01}$ | 0.97 | Example 7.6 |
| 8 | $h_{pf}$ | 100 | $W/m^2$ °C |
| 9 | I(t) | 500 | $W/m^2$ |
| 10 | $K_{bs}$ | 386 | $W/m^2$ °C |
| 11 | $K_g$ | 0.688 | $W/m^2$ °C |
| 12 | $K_t$ | 0.033 | $W/m^2$ °C |
| 13 | $L_0$ | 2 | m |
| 14 | $L_{bs}$ | 0.0005 | m |
| 15 | $L_g$ | 0.003 | m |
| 16 | $L_t$ | 0.0005 | m |
| 17 | $m'_f$ | 0.035 | kg/s |
| 18 | $T_a$ | 16 | °C |
| 19 | $T_{fi}$ | 15 | °C |
| 20 | $U_b$ | 0.70 | $W/m^2$ °C |
| 21 | $U_{b,cp}$ | 2.7663 | $W/m^2$ °C |
| 22 | $U_{fa}$ | 2.76 | $W/m^2$ °C |
| 23 | $U_{pa}$ | 2.1225 | $W/m^2$ °C |
| 24 | $U_{t,ca}$ | 9.12 (Example 6.1) | $W/m^2$ °C |
| | **Greek letters** | | |
| 25 | $\alpha_c$ | 0.9 | |
| 26 | $\alpha_p$ | 0.9 | |
| 27 | $(\alpha\tau)_{m,eff}$ | 0.4575 | PVT-CPC ($C = 2$) |
| 28 | $(\alpha\tau)_{m,eff}$ | 0.2110 | PVT-CPC ($C = 1$) |
| 29 | $(\alpha\tau)_{eff}$ | 0.5887 | PVT ($C = 1$) |
| 30 | $(\alpha\tau)_{eff}$ | 1.2736 | PVT-CPC ($C = 2$) |
| 31 | $\beta_0$ | 0.00425 | °C$^{-1}$ |
| 32 | $\beta_c$ | 0.89 | |
| 33 | $\gamma$ | 0.20–0.35 | |
| 34 | $\rho$ | 0.9 | |
| 35 | $\tau'_g$ | 0.95 | |
| 36 | $\eta_{m0}$ | 0.12 | |
| 37 | $\eta_m$ | 0.12 | |

The rate of heat loss from solar cells to ambient through top glass surface through receiver area $(A_{rm}) = U_{t,ca}(T_c - T_a)A_{rm}$.

The rate of heat transferred from solar cells to absorber pate through the glass through receiver area $(A_{rm}) = U_{b,cp}(T_c - T_p)A_{rm}$.

The rate of electrical power generated by PV module $= \rho I_u \eta_m \beta_c A_{am}$.

If $A_{am} = A_{rm} = A_m$, $I_u = I(t)$ and $\rho = 1$, then Eq. 7.1 reduces to exactly same as Eq. 6.2 for unglazed semi-transparent PV (SPV) module which is given by

$$\alpha_c \tau_g \beta_c I(t) A_m = U_{t,ca}(T_c - T_a)A_m + U_{b,cp}(T_c - T_p)A_m + \eta_m I(t)\beta_c A_{am}.$$

**Example 7.1**  Compute the solar radiation on receiver of SPVT-CPC liquid collector for beam radiation $(I_b)$ of 450 W/m$^2$ and diffuse radiation $(I_d)$ of 50 W/m$^2$ and concentration ratio for $A_{am} = 4$m$^2$ and $A_{rm} = 2$ m$^2$.

**Solution**

Concentration ratio $C = \frac{A_{am}}{A_{rm}} = \frac{4}{2} = 2$ (Table 7.1).

The solar radiation on receiver of SPVT-CPC liquid collector $(I_u) = I_b + \frac{I_d}{C} = 450 + \frac{50}{2} = 475$ W/m$^2$ (Eq. 7.1).

From Eq. 7.1, one gets an expression for solar cell temperature $(T_c)$ as follows:

$$T_c = \frac{(\alpha\tau)_{eff}I_u + U_{t,ca}T_a + U_{t,cp}T_p}{U_{t,ca} + U_{t,cp}} \tag{7.2}$$

where,

$(\alpha\tau)_{eff} = \left(\alpha_c \tau_g \frac{A_{am}}{A_{rm}} - \eta_m\right)\rho\beta_c$. For SPVT-CPC, an effective product of absorptivity and transmittivity of semi-transparent SPV module of PVT-CPC collector.

$(\alpha\tau)_{eff} = \left(\alpha_c \tau_g - \eta_m\right)\rho\beta_c$. For PVT, $A_{am} = A_{rm}$, an effective product of absorptivity and transmittivity of semi-transparent PV module of SPVT collector.

**Example 7.2**  Evaluate an effective product of absorptivity and transmittivity of semi-transparent PV module of (a) SPVT-CPC and (b) SPVT liquid collectors.

**Solution**

From Table 7.1 $\tau_g = 0.95$, $\alpha_c = 0.9$, $\beta_c = 0.89$, $\rho = 0.9$, $A_{am} = 4$ m$^2$ and $A_{rm} = 2$ m$^2$.

From Eq. 7.2, we have

(a) $(\alpha\tau)_{eff} = \left(\alpha_c \tau_g \frac{A_{am}}{A_{rm}} - \eta_m\right)\rho\beta_c = \left(0.9 \times 0.95 \times \frac{4}{2} - 0.12\right)0.9 \times 0.89 = 1.2736(1.1775)$ for SPVT-CPC

(b) $(\alpha\tau)_{eff} = \left(\alpha_c \tau_g - \eta_m\right)\rho\beta_c = (0.9 \times 0.95 - 0.12)0.9 \times 0.89 = 0.5887$ for SPVT.

So, one can see that $(\alpha\tau)_{eff}$ is doubled in PVT-CPC which enhances incident solar radiation.

## 7.2.2 Energy Balance Equation for Absorber Plate of SPVT

The energy balance of the absorber in $W$ can be written as

$$
\rho \alpha_p \tau_g^2 (1 - \beta_c) I_u A_{am} + U_{b,cp} (T_c - T_p) A_{rm}
$$
$$
= F' h_{pf} (T_p - T_f) A_{rm} + U_b (T_p - T_a) A_{rm}. \tag{7.3}
$$

Here,

First thermal gain to absorber $= \rho \alpha_p \tau_g^2 (1 - \beta_c) I_u A_{am} =$ The rate of solar energy available on absorber plate through non-packing area of SPV module.

Second thermal gain to absorber $U_{b,cp}(T_c - T_p)A_{rm} =$ The rate of heat transferred from solar cell to absorber plate.

The rate of thermal energy gain to working fluid $= F' h_{pf}(T_p - T_f)A_{rm}$.

The tube in plate collector efficiency $(F') = 0.82$(Example 4.29).

The rate of thermal energy lost from absorber pate to ambient through bottom insulation $= U_b(T_p - T_a)A_{rm}$.

The conductive heat transfer coefficient for copper absorber plate and tube to fluid $h_{pf} = 100 \text{ W/m}^2 {}^\circ\text{C}$, Tiwari et al. [1]

**Example 7.3** Derive an expression for $U_{b,cp}(T_c - T_p)$ by using Eq. 7.2.

**Solution**

From Eq. 7.2, we have

$$
T_c = \frac{(\alpha\tau)_{eff} I_u + U_{t,ca} T_a + U_{b,cp} T_p}{U_{t,ca} + U_{b,cp}}.
$$

Now,

$$
U_{b,cp}(T_c - T_p) = U_{b,cp} \left[ \frac{(\alpha\tau)_{eff} I_u + U_{t,ca} T_a + U_{b,cp} T_p}{U_{t,ca} + U_{b,cp}} - T_p \right]
$$

or,

$$
U_{b,cp}(T_c - T_p) = \frac{U_{b,cp}}{U_{t,ca} + U_{b,cp}} (\alpha\tau)_{eff}, I_u - \frac{U_{t,ca} U_{b,cp}}{U_{t,ca} + U_{b,cp}} (T_p - T_a)
$$
$$
= F_0 (\alpha\tau)_{eff} I_u - U_{pa}(T_p - T_a)
$$

where, $F_0 = \frac{U_{bc,a}}{U_{tc,p} + U_{bc,a}} = \frac{2.7663}{9.12 + 2.7663} = 0.233$; an efficiency/penalty factor for semi-transparent PV module (Example 6.3) and $(\alpha\tau)_{eff}$ is given in Eq. 7.2 and Example 7.2

and,

$U_{pa} = \frac{U_{t,ca} U_{b,cp}}{U_{t,ca} + U_{b,cp}}$, an overall heat transfer coefficient from absorber plate to ambient air through semi-transparent PV module.

***Example 7.4*** Evaluate an overall heat transfer coefficient from absorber plate to ambient air $(U_{pa})$ through semi-transparent PV (SPV) module derived in Example 7.3.

**Solution**

The expression for $U_{pa}$ (Example 7.3) is given by

$$U_{pa} = \frac{U_{t,ca}U_{b,cp}}{U_{t,ca} + U_{b,cp}} = \frac{9.12 \times 2.7663}{9.12 + 2.7663} = 2.1225 \text{ W/m}^2\,{}^{\circ}\text{C}$$

***Example 7.5*** Derive an expression for the rate of thermal energy transferred from absorber plate to working fluid, i.e., $F'h_{pf}(T_p - T_f)$ by using the results of Example 7.3.

**Solution**

Substitute $U_{b,cp}(T_c - T_p)$ from Example 7.3 into Eq. 7.3, one gets

$$\rho\alpha_p\tau_g^2(1 - \beta_c)I_u A_{am} + F_0(\alpha\tau)_{1,\text{eff}}I_u A_{rm} - U_{pa}(T_p - T_a)A_{rm}$$
$$= F'h_{pf}(T_p - T_f)A_{rm} + U_b(T_p - T_a)A_{rm}$$

From above equation, one may get an expression for $T_p$ as

$$T_p = \frac{\left[\rho\alpha_p\tau_g^2(1 - \beta_c)C + F_0(\alpha\tau)_{,\text{eff}}\right]I_u + (U_{pa} + U_b)T_a + F'h_{pf}T_f}{(U_{pa} + F'h_{pf} + U_b)} \qquad (7.4)$$

Now, the rate of thermal energy transferred from absorber plate to working fluid can be written as

$$F'h_{pf}(T_p - T_f) = F'\left[(\alpha\tau)_{m,\text{eff}}I_u - U_{Lm}(T_f - T_a)\right] \qquad (7.5)$$

where,

$(\alpha\tau)_{m,\text{eff}} = F_{01}\left[\rho\alpha_p\tau_g^2(1 - \beta_c)C + F_0(\alpha\tau)_{\text{eff}}\right]$, an effective product of absorptivity and transmittivity of SPVT-CPC collector

$F_{01} = \frac{h_{pf}}{(U_{pa}+F'h_{pf}+U_b)}$, an efficiency/penalty factor due to fluid

and,

$U_{Lm} = \frac{h_{pf}(U_{pa}+U_b)}{(U_{pa}+F'h_{pf}+U_b)}$; an overall heat transfer coefficient from working fluid to ambient air through top surface and bottom insulation.

***Example 7.6*** Evaluate efficiency/penalty factor $(F_{01})$, an effective product of absorptivity and transmittivity of SPVT-CPC collector $\left[(\alpha\tau)_{m,\text{eff}}\right]$ and an overall heat transfer coefficient from working fluid to ambient air through bottom insulation $(U_{Lm})$ for Example 7.5.

**Solution**

From Table 7.1 $\tau_g = 0.95$, $\alpha_c = 0.9$, $\beta_c = 0.89$, $\rho = 0.9$, $A_{am} = 4\ m^2$, $\alpha_p = 0.9$, $C = 2$ and $A_{rm} = 2\ m^2$.

From Example 7.5, we have the following expression as per requirement and after substitution of appropriate vale from Table 7.1, one gets

An efficiency/penalty factor due to fluid, $F_{01} = \dfrac{h_{pf}}{(U_{pa}+F'h_{pf}+U_b)} = \dfrac{100}{2.1225+0.82\times100+0.7} = \dfrac{100}{102.8225} = 0.97.$

An effective product of absorptivity and transmittivity of SPVT-CPC collector,

$$(\alpha\tau)_{m,\text{eff}} = F_{01}\left[\rho\alpha_p\tau_g^2(1-\beta_c)C + F_0(\alpha\tau)_{\text{eff}}\right]$$

$$= 0.97\left[0.9\times0.9\times(0.95)^2(1-0.89)\times2 + 0.233\times1.2736\right]$$

$$= 0.97\times[0.1608+0.2967] = 0.4575.$$

An overall heat transfer coefficient from working fluid to ambient air through bottom insulation,

$$U_{Lm} = \frac{h_{pf}(U_{pa}+U_b)}{(U_{pa}+F'h_{pf}+U_b)} = \frac{100\times(2.1225+0.7)}{2.1225+0.82\times100+0.7}$$

$$= \frac{282.26}{102.8225} = 2.76\ \text{W/m}^2\,°\text{C}$$

### 7.2.3 Energy Balance for Flowing Fluid (Water) in Tube Below the Absorber Plate of SPVT-CPC Collector

Referring to Fig. 7.2 and Sect. 4.4.1, the rate of thermal energy carried away through elemental area of 'bdx' will be equal to the rate of thermal energy available in the copper blackened absorber plate derived in Example 7.5 through Eq. 7.5

$$\dot{m}_f C_f \frac{dT_f}{dx}dx = F'h_{pf}(T_p - T_f)bdx \tag{7.6}$$

Substitute the expression for $F'h_{pf}(T_p - T_f)$ from Eq. 7.5 into Eq. 7.6, one gets

$$\dot{m}_f C_f \frac{dT_f}{dx}dx = F'\left[(\alpha\tau)_{m,\text{eff}}I_u - U_{Lm}(T_f - T_a)\right]bdx \tag{7.7a}$$

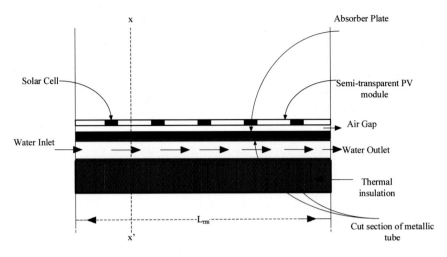

**Fig. 7.2** Cut-section along the length of single semi-transparent PVT-CPC liquid collector

### 7.2.4 The Outlet Fluid Temperature

The above equation can also be written as

$$\frac{dT_f}{dx} + \frac{F'U_{\mathrm{Lm}}b}{\dot{m}_f C_f} T_f = \frac{F'\big[(\alpha\tau)_{m,\mathrm{eff}} I_u + U_{\mathrm{Lm}} T_a\big]b}{\dot{m}_f C_f} \qquad (7.7\mathrm{b})$$

Equation 7.7b can be solved by using initial condition; $T_f|_{x=0} = T_{\mathrm{fi}}$ and the solution can be obtained as

$$T_f = \left[\frac{\big[(\alpha\tau)_{m,\mathrm{eff}} I_u + U_{\mathrm{Lm}} T_a\big]}{U_{\mathrm{Lm}}} + T_a\right]\left[1 - \exp\left\{\frac{-F'U_{\mathrm{Lm}}bx}{\dot{m}_f C_f}\right\}\right]$$

$$+ T_{\mathrm{fi}} \exp\left\{\frac{-F'U_{\mathrm{Lm}}bx}{\dot{m}_f C_f}\right\} \qquad (7.8)$$

Further, the outlet temperature of fluid at end of the SPVT-CPC combination can be obtained as:

$$T_{\mathrm{fo}} = T_f|_{x=L_0} = \left[\frac{\big[(\alpha\tau)_{m,\mathrm{eff}} I_u + U_{\mathrm{Lm}} T_a\big]}{U_{\mathrm{Lm}}} + T_a\right]\left[1 - \exp\left\{\frac{-F'U_{\mathrm{Lm}}bL_0}{\dot{m}_f C_f}\right\}\right]$$

$$+ T_{\mathrm{fi}} \exp\left\{\frac{-F'U_{\mathrm{Lm}}bL_0}{\dot{m}_f C_f}\right\}$$

or,

$$T_{\text{fo}} = \left[ \frac{\left[ (\alpha\tau)_{m,\text{eff}} I_u + U_{\text{Lm}} T_a \right]}{U_{\text{Lm}}} + T_a \right] \left[ 1 - \exp\left\{ \frac{-F' U_{\text{Lm}} A_m}{\dot{m}_f C_f} \right\} \right]$$
$$+ T_{\text{fi}} \exp\left\{ \frac{-F' U_{\text{Lm}} A_m}{\dot{m}_f C_f} \right\} \tag{7.9}$$

where, $A_m = bL_0$.

***Example 7.7*** Evaluate the outlet fluid temperature of single semi-transparent SPVT-CPC collector for the following climatic data's of $I(t) = 500$ W/m$^2$, $I_u = 475$ W/m$^2$ (Example 7.1) $T_a$ 16 °C and $T_a$ 15 °C and design parameters of $(\alpha\tau)_{m,\text{eff}} = 0.4575$ [C = 2], $U_{\text{Lm}} = 2.76$ W/m °C, $\dot{m}_f = 0.035 \frac{\text{kg}}{\text{s}}$ and $C_f = 4190 f/\text{kg}$ °C and $F' = 0.82$ (Table 7.1 and Example 7.6).

**Solution**

From Eq. 7.9, we have an expression for the outlet fluid temperature as

$$T_{\text{fo}} = \left[ \frac{\left[ (\alpha\tau)_{m,\text{eff}} I_u + U_{\text{Lm}} T_a \right]}{U_{\text{Lm}}} + T_a \right] \left[ 1 - \exp\left\{ \frac{-F' U_{\text{Lm}} A_m}{\dot{m}_f C_f} \right\} \right]$$
$$+ T_{\text{fi}} \exp\left\{ \frac{-F' U_{\text{Lm}} A_m}{\dot{m}_f C_f} \right\}.$$

Here,

$$\left[ \frac{\left[ (\alpha\tau)_{m,\text{eff}} I_u + U_{\text{Lm}} T_a \right]}{U_{\text{Lm}}} + T_a \right] = \frac{0.4575 \times 475 + 2.76 \times 16}{2.76} = \frac{244.65}{2.76} = 94.75$$

and,

$$\exp\left\{ \frac{-F' U_{\text{Lm}} A_m}{\dot{m}_f C_f} \right\} = \exp\left\{ \frac{-0.82 \times 2.76 \times 2}{0.035 \times 4190} \right\}$$
$$= \exp\left\{ \frac{-4.5264}{146.65} \right\} = \exp\{-0.0309\} = 0.9696.$$

Substitute the above value in the expression for $T_{\text{fo}}$, one gets

$$T_{\text{fo}} = 94.75 \times [1 - 0.9696] + 15 \times 0.9696 = 17.42 \text{ °C}$$

**Example 7.8** Repeat the Example 7.7 for SPVT liquid collector, i.e., ($A_{am} = A_{rm} = A_m$ and $C = 1$).

**Solution**

For $C = 1$, $I_u = I(t) = 500 \text{ W/m}^2$, we have from Table 7.1 and Examples 7.1–7.6.
$(\alpha\tau)_{\text{eff}} = 0.5887$ for PVT (Example 7.2), $F_0 = 0.233$, (Example 7.3), $U_{pa} = 2.1225 \text{ W/m}^{2\,°C}$, (Example 7.4), $U_{Lm} = 2.76 \text{ W/m}^2\,°C$, $F_{01} = 0.97$ (Example 7.6).

$$(\alpha\tau)_{m,\text{eff}} = F_{01}\left[\rho\alpha_p\tau_g^2(1 - \beta_c)C + F_0(\alpha\tau)_{\text{eff}}\right]$$
$$= 0.97\left[0.9 \times 0.9 \times (0.95)^2(1 - 0.89) \times 1 + 0.233 \times 0.5887\right]$$
$$= 0.97 \times [0.0804 + 0.2743] = 0.2110(C = 1).$$
$$\exp\left\{\frac{-F'U_{Lm}A_m}{\dot{m}_f C_f}\right\} = 0.9696.$$

Now,

$$\left[\frac{[(\alpha\tau)_{m,\text{eff}}I(t) + U_{Lm}T_a]}{U_{Lm}} + T_a\right] = \frac{0.2110 \times 500 + 2.76 \times 16}{2.76} = \frac{149.68}{2.76} = 54.23.$$

Substitute the above value in the expression for $T_{fo}$, one gets

$$T_{fo} = 54.23 \times [1 - 0.9696] + 15 \times 0.9696 = 16.19\,°C$$

This shows that there is an increase of 1.04 $°C$ due to the integration of CPC with SPVT collector. Further, it is to be noted that the outlet fluid temperature of SPVT-CPC collector is only influence by the term $(\alpha\tau)_{m,\text{eff}}$ due to concentration ratio (C).

**Example 7.9** Repeat Example 7.7 with mass flow rate $(\dot{m}_f)$ of 0.0175.

**Solution**

Only exponential term will be changed as follows:

$$\exp\left\{\frac{-F'U_{Lm}A_m}{\dot{m}_f C_f}\right\} = \exp\left\{\frac{-0.82 \times 2.76 \times 2}{0.0175 \times 4190}\right\}$$
$$= \exp\left\{\frac{-4.5264}{74.582}\right\} = \exp\{-0.04618\} = 0.9549.$$

The outlet temperature will be obtained by using data's of Example 7.7 as

$$T_{fo} = 94.75 \times [1 - 0.9549] + 15 \times 0.9549 = 18.60\,°C.$$

The outlet fluid temperature is further increased due to decrease in mass flow rate.

## 7.2.5 An Average Temperature of Fluid $\left(\overline{T}_f\right)$, Plate $\left(\overline{T}_p\right)$ and Solar Cell $\left(\overline{T}_c\right)$ Temperatures

From Eq. 7.8, we can derive an average fluid temperature of SPVT collector as follows:

$$
\overline{T}_f = \frac{1}{L_0} \int\limits_0^{L_0} T_f dx = \left[ \frac{[(\alpha\tau)_{m,\text{eff}} I_u + U_{\text{Lm}} T_a]}{U_{\text{Lm}}} + T_a \right] \left[ 1 - \frac{1 - \exp\left\{\frac{-F' U_{\text{Lm}} b L_0}{\dot{m}_f C_f}\right\}}{\frac{F' U_{\text{Lm}} b L_0}{\dot{m}_f C_f}} \right]
$$

$$
+ T_{\text{fi}} \frac{1 - \exp\left\{\frac{-F' U_{\text{Lm}} b L_0}{\dot{m}_f C_f}\right\}}{\frac{F' U_{\text{Lm}} b L_0}{\dot{m}_f C_f}} \tag{7.10}
$$

After evaluating an average value of fluid temperature $\left(\overline{T}_f\right)$ from Eq. 7.10, one can get an average value of plate temperature $\left(\overline{T}_p\right)$ from Eq. 7.4 as

$$
\overline{T}_p = \frac{\left[\rho\alpha_p \tau_g^2 (1 - \beta_c) C + F_0 (\alpha\tau)_{1,\text{eff}}\right] I_u + (U_{\text{pa}} + U_b) T_a + F' h_{\text{pf}} \overline{T}_f}{(U_{\text{pa}} + F' h_{\text{pf}} + U_b)} \tag{7.11}
$$

Once, we get an average value of plate temperature $\left(\overline{T}_p\right)$ **from Eq. 7.11**, one can determine the average temperature of solar cell of SPVT collector from Eq. 7.2 as

$$
\overline{T}_c = \frac{(\alpha\tau)_{\text{eff}} I_u + U_{t,\text{ca}} T_a + U_{b,\text{cp}} \overline{T}_p}{U_{t,\text{ca}} + U_{b,\text{cp}}} \tag{7.12}
$$

**Example 7.10** Evaluate an average temperature of fluid $\left(\overline{T}_f\right)$, absorber plate $\left(\overline{T}_p\right)$ and solar cell $\left(\overline{T}_c\right)$ by using the data's of Table 7.1 and Example 7.7 for single semi-transparent PVT-CPC (SPVT-CPC) liquid collector.

**Solution**

(a)  Average fluid temperature of SPVT-CPC collector.

From Example 7.7: $\left[\frac{[(\alpha\tau)_{m,\text{eff}} I_u + U_{\text{Lm}} T_a]}{U_{\text{Lm}}} + T_a\right] = 94.75$, $\frac{F' U_{\text{Lm}} A_m}{\dot{m}_f C_f} = 0.0309$ and $\exp\left\{\frac{-F' U_{\text{Lm}} A_m}{\dot{m}_f C_f}\right\} = 0.9696$ and $I_u = 475$ W/m$^2$ (Example 7.1).

Substitute the above values in Eq. 7.10 to get an average value of fluid temperature of SPVT-CPC collector

$$
\overline{T}_f = 94.75\left[1 - \frac{1 - 0.9696}{0.0309}\right] + 15 \times \frac{1 - 0.9696}{0.0309}
$$

$$
= 94.75[1 - 0.9838] + 15 \times 0.9838 = 16.28\,^{\circ}C.
$$

(b)   Average absorber plate temperature of SPVT-CPC collector.

Now, $\left[\rho\alpha_p\tau_g^2(1-\beta_c)C + F_0(\alpha\tau)_{1,\text{eff}}\right] = \left[0.9 \times 0.9 \times (0.95)^2(1-0.89) \times 2 + 0.233 \times 1.2736\right] = [0.1608 + 0.2967] = 0.4575[C = 2]$, $F' = 0.82$, $h_{\text{pf}} = 100 \text{ W/m}^2 \,^\circ\text{C}$, $\overline{T}_{varvecf} = 16.19\,^\circ\text{C}$.

$U_{\text{pa}} = 2.1225 \text{ W/m}^2 \,^\circ\text{C}$, $U_b = 0.7 \text{ W/m}^2 \,^\circ\text{C}$ (Table 7.1).

Substitute the above values in Eq. 7.11 to get an average value of plate temperature of SPVT-CPC collector

$$\overline{T}_p = \frac{0.4575 \times 475 + (2.1225 + 0.7) \times 16 + 0.82 \times 100 \times 16.28}{(2.1225 + 0.82 \times 100 + 0.7)}$$

$$= \frac{217.34 + 45.16 + 1334.96}{84.8225} = \frac{1579.41}{84.8225} = 18.83\,^\circ\text{C}$$

(c)   Average solar cell temperature of SPVT-CPC collector

In this case, $(\alpha\tau)_{\text{eff}} = 1.1775$, $U_{t.\text{ca}} = 9.12 \text{ W/m}^2 \,^\circ\text{C}$, $U_{b,\text{cp}} = 2.7663 \text{ W/m}^2 \,^\circ\text{C}$.

Substitute the above value in Eq. 7.12, one gets

$$\tau_c = \frac{1.2736 \times 475 + 9.12 \times 16 + 2.7663 \times 18.83}{9.12 + 2.7663} = \frac{803.61}{11.8863} = 67.61\,^\circ\text{C}$$

***Example 7.11*** Evaluate an electrical efficiency of PV module, electrical power, thermal exergy and total exergy of single SPVT-CPC liquid collector for Example 7.10 ($C = 2$).

**Solution**

**From Example 7.10:** $\overline{T} = 67.61\,^\circ\text{C}$; $T_{\text{fo}} = 17.23\,^\circ\text{C}$ and $T_{fi} = 15\,^\circ\text{C}$ (Example 7.7), $I_u = 475 \text{ W/m}^2$ (Example 7.1), $\beta_c = 0.89$, $\beta_0 = 0.00425(^\circ\text{C})^{-1}$, $A_m = 2 \text{ m}^2$(Table 7.1).

Following Eq. 6.17 with known $\overline{T}_c$, the electrical efficiency and electrical power from of SPVT-CPC collector can be determined as follows:

$$\overline{\eta}_m = \eta_{m0}\left[1 - \beta_0\left(\overline{T}_c - 25\right)\right] = 0.12[1 - 0.00425(67.61 - 25)] = 0.0983$$

and,

$$\dot{E}_{\text{el}} = \overline{\eta}_m \times I_u \times \beta_c A_m = 0.0983 \times 475 \times 0.89 \times 2 = 83.09 \text{ W}.$$

Following Eqs. 2.35a and 6.22, an expression for evaluating hourly exergy of SPVT-CPC collector is given by

$$\dot{E}x_u = \dot{m}_f C_f\left[(T_{\text{fo}} - T_{\text{fi}}) - (T_a + 273)\ln\frac{T_{\text{fo}} + 273}{T_{\text{fi}} + 273}\right] = 0.035$$

$$\times 4190\left[(18.60 - 15) - (16 + 273)\ln\frac{18.60 + 273}{15 + 273}\right]$$

$$= 145.65[3.60 - 289 \ln 1.0125] = 145 \times 0.01 = 1.45 \text{ W}$$

Total exergy of PVT-CPC liquid collector $(\dot{E}x_T) = $ thermal exergy $+$ electrical power $= 1.45 + 83.09 = 84.54 \text{ W}$

**Example 7.12** Repeat the Example 7.9 for $C = 1$ and $I_u = I(t) = 500 \text{ W/m}^2$ for SPVT liquid collector for data's of Example 7.8.

**Solution**

**From Example 7.8, We have**
For $C = 1, I_u = I(t) = 500 \text{ W/m}^2$, we have from Table 7.1 and Examples 7.1–7.6.
$(\alpha\tau)_{\text{eff}} = 0.5887$ for PVT (Example 7.2), $F_0 = 0.233$, (Example 7.3), $U_{\text{pa}} = 2.1225 \text{ W/m}^2 \,^\circ\text{C}$, (Example 7.4), $U_{\text{Lm}} = 2.76 \text{ W/m} \,^\circ\text{C}$, $F_{01} = 0.97$ (Example 7.6).

$$(\alpha\tau)_{m,\text{eff}} = F_{01}\left[\rho\alpha_p\tau_g^2(1 - \beta_c)C + F_0(\alpha\tau)_{1,\text{eff}}\right]$$
$$= 0.97\left[0.9 \times 0.9 \times (0.95)^2(1 - 0.89) \times 1 + 0.233 \times 0.5887\right]$$
$$= 0.97 \times [0.0804 + 0.1372] = 0.2176$$

$$\exp\left\{\frac{-F'U_{\text{Lm}}A_m}{\dot{m}_fC_f}\right\} = 0.9696, \quad \frac{F'U_{\text{Lm}}A_m}{\dot{m}_fC_f} = 0.0309, \text{ and}$$

$$\left[\frac{[(\alpha\tau)_{m,\text{eff}}I(t) + U_{\text{Lm}}T_a]}{U_{\text{Lm}}} + T_a\right] = \left[\frac{[0.2176 \times 500 + 2.76 \times 15]}{2.76} + 15\right] = 69.42$$

In Eqs. 7.10–7.12, $I_u = I(t) = 500 \text{ W/m}^2$,
(a) Average fluid temperature of SPVT collector as follows:

$$\overline{T}_f = \left[\left[\frac{[(\alpha\tau)_{\text{meff}}I(t) + U_{\text{Lm}}T_a]}{U_{\text{Lm}}} + T_a\right]\left[1 - \frac{1 - \exp\left\{\frac{-F'U_{\text{Lm}}bL_0}{\dot{m}_fC_f}\right\}}{\frac{F'U_{\text{Lm}}bL_0}{\dot{m}_fC_f}}\right]\right.$$

$$\left. + T_{\text{fi}}\exp\left\{\frac{-F'U_{\text{Lm}}bL_0}{\dot{m}_fC_f}\right\}\right] = 69.42 \times [1 - 0.9838] + 15 \times 0.9838 = 15.88 \,^\circ\text{C}$$

(b) Average value of plate temperature $(\overline{T}_p)$ of SPVT collector.
**In this case, for $C = 1$**

$$\left[\rho\alpha_p\tau_g^2(1 - \beta_c)C + F_0(\alpha\tau)_{1,\text{eff}}\right]$$
$$= \left[[0.9 \times 0.9 \times (0.95)^2(1 - 0.89) \times 1 + 0.233 \times 0.5887]\right]$$
$$= [0.0804 + 0.1372] = 0.2176, \; F' = 0.82, \; h_{\text{pf}}$$
$$= 100 \text{ W/m}^2 \,^\circ\text{C}, \overline{varvecT}_f = 15.88 \,^\circ\text{C}.$$

$U_{\text{pa}} = 2.1225 \text{ W/m}^2 \,^\circ\text{C}, U_b = 0.7 \text{ W/m}^2 \,^\circ\text{C}$ (Table 7.1).
Then,

$$\overline{T}_p = \frac{\left[\rho \alpha_p \tau_g^2 (1 - \beta_c)C + F_0(\alpha\tau)_{1,\text{eff}}\right]I(t) + (U_{\text{pa}} + U_b)T_a + F'h_{\text{pf}}\overline{T}_f}{(U_{\text{pa}} + F'h_{\text{pf}} + U_b)}$$

$$= \frac{0.2176 \times 500 + (2.1225 + 0.7) \times 16 + 0.82 \times 100 \times 15.88}{(2.1225 + 0.82 \times 100 + 0.7)}$$

$$= \frac{108.8 + 45.16 + 1302.16}{84.8225} = \frac{1456.12}{84.8225} = 17.17\,°C$$

(c) Average temperature of solar cell of SPVT collector from Eq. 7.12 as
**Here,** $(\alpha\tau)_{\text{eff}} = 0.5887$, $U_{b,\text{cp}} = 2.7663$ W/m °C, (Table 7.1)

$$\overline{T}_c = \frac{(\alpha\tau)_{\text{eff}}I(t) + U_{t,\text{ca}}T_a + U_{b,\text{cp}}\overline{T}_p}{U_{t,\text{ca}} + U_{b,\text{cp}}}$$

$$= \frac{0.5887 \times 500 + 9.12 \times 16 + 2.7663 \times 17.17}{9.12 + 2.7663}$$

$$= \frac{535.26}{11.8863} = 45.03\,°C$$

**Example 7.13** Repeat the Example 7.11 for Example 7.12 for $C = 1$ (semi-transparent SPVT collector).

**Solution**

**From Example 7.12:** $\overline{T}_c = 45.03\,°C$; $T_{\text{fo}} = 16.19\,°C$ and $T_{\text{fi}} = 15\,°C$ (Example 7.8), $I(t) = 500$ W/m$^2$ (Example 7.1), $\beta_c = 0.89$, $\beta_0 = 0.00425(°C)^{-1}$, $A_m = 2$ m$^2$(Table 7.1).

Following Eq. 6.17 and Example 7.11 with known $\overline{T}_c$, the electrical efficiency and electrical power can be determined as follows:

$$\overline{\eta}_m = \eta_{m0}\left[1 - \beta_0\left(\overline{T}_c - 25\right)\right] = 0.12[1 - 0.00425(45.03 - 25)] = 0.1098$$

and,

$$\dot{E}_{\text{el}} = \overline{\eta}_m \times I(t) \times \beta_c A_m = 0.1098 \times 500 \times 0.89 \times 2 = 97.77\text{ W}.$$

Following Eqs. 2.35a and 6.22, an expression for evaluating hourly exergy is given by

$$\dot{E}x_u = \dot{m}_f C_f\left[(T_{\text{fo}} - T_{\text{fi}}) - (T_a + 273)\ln\frac{T_{\text{fo}} + 273}{T_{\text{fi}} + 273}\right]$$

$$= 0.035 \times 4190\left[(16.19 - 15) - (16 + 273)\ln\frac{16.19 + 273}{15 + 273}\right]$$

$$= 146.65[1.19 - 289\ln 1.00413] = 146 \times 1.0389 = -0.163\text{ W}$$

Total exergy of SPVT liquid collector is determined by

$$\left(\dot{Ex}_T\right) = \text{thermal exergy} + \text{electrical power} = 0 + 97.77 = 97.77 \text{ W}$$

The negative value of thermal exergy obtained in Example 7.13 for semi-transparent PVT collector shows that all thermal energy obtained in Example 7.13 is destroyed unlike PVT-CPC liquid collector ($\dot{Ex}_u = 1.94$ W, Example 7.11).

### 7.2.6   The Outlet Temperature at $N^{th}$ PVT-CPC Liquid Collector

Following Sect. 4.5.1 and Fig. 7.3, one can write an expression for the outlet temperature at $N^{th}$ PVT-CPC liquid collector as

$$T_{\text{foN}} = \left[\frac{\left[(\alpha\tau)_{m,\text{eff}}I_u + U_{\text{Lm}}T_a\right]}{U_{\text{Lm}}} + T_a\right]\left[1 - \exp\left\{\frac{-NF'U_{\text{Lm}}A_m}{\dot{m}_f C_f}\right\}\right]$$
$$+ T_{\text{fi}}\exp\left\{\frac{-NF'U_{\text{Lm}}A_m}{\dot{m}_f C_f}\right\} \tag{7.13a}$$

For any x length of PVT-CPC, $A_m = bx$ and then we can rewrite Eq. 7.13a as

$$T_{\text{foN}}(x) = \left[\frac{\left[(\alpha\tau)_{m,\text{eff}}I_u + U_{\text{Lm}}T_a\right]}{U_{\text{Lm}}} + T_a\right]\left[1 - \exp\left\{\frac{-NF'U_{\text{Lm}}bx}{\dot{m}_f C_f}\right\}\right]$$
$$+ T_{\text{fi}}\exp\left\{\frac{-NF'U_{\text{Lm}}bx}{\dot{m}_f C_f}\right\} \tag{7.13b}$$

Following Sect. 4.5 and using Eqs. 7.10 and 7.13b, one can derive an average fluid temperature of N-PVT-CPC liquid collector as follows:

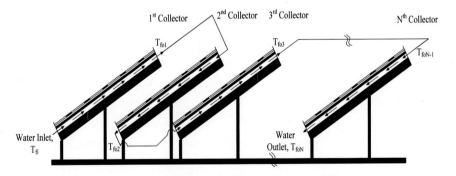

**Fig. 7.3** N-semi-transparent PVT-CPC liquid collectors connected in series

$$\overline{T}_{fN} = \frac{1}{L_0} \int_0^{L_0} T_{foN}(x)dx = \left[ \frac{\left[(\alpha\tau)_{m,\text{eff}} I_u + U_{\text{Lm}} T_a\right]}{U_{\text{Lm}}} + T_a \right] \left[ 1 - \frac{1 - \exp\left\{\frac{-NF'U_{\text{Lm}}bL_0}{\dot{m}_f C_f}\right\}}{\frac{NF'U_{\text{Lm}}bL_0}{\dot{m}_f C_f}} \right]$$

$$+ T_{\text{fi}} \frac{1 - \exp\left\{\frac{-NF'U_{\text{Lm}}bL_0}{\dot{m}_f C_f}\right\}}{\frac{NF'U_{\text{Lm}}bL_0}{\dot{m}_f C_f}} \tag{7.14}$$

After evaluating an average value of N-PVT-CPC fluid temperature $\left(\overline{T}_{fN}\right)$ from Eq. 7.14, one can get an average value of plate temperature $\left(\overline{T}_{pN}\right)$ from Eq. 7.4 as

$$\overline{T}_{pN} = \frac{\left[\rho\alpha_p\tau_g^2(1-\beta_c)C + F_0(\alpha\tau)_{1,\text{eff}}\right]I_u + \left(U_{\text{pa}}+U_b\right)T_a + F'h_{\text{pf}}\overline{T}_{fN}}{\left(U_{\text{pa}}+F'h_{\text{pf}}+U_b\right)} \tag{7.15}$$

By evaluating an average value of plate temperature $\left(\overline{T}_{pN}\right)$ from Eq. 7.15, the average temperature of solar cell of N + PVT-CPC collector from Eq. 7.2 as

$$\overline{T}_{cN} = \frac{(\alpha\tau)_{\text{eff}} I_u + U_{t,\text{ca}} T_a + U_{b,\text{cp}} \overline{T}_{pN}}{U_{t,\text{ca}} + U_{b,\text{cp}}} \tag{7.16}$$

**Example 7.14** Determine the average fluid $\left(\overline{T}_{fN}\right)$, the plate $\left(\overline{T}_{pN}\right)$ and the solar cell $\left(\overline{T}_{cN}\right)$ temperature of N-PVT-CPC collectors connected in series by using the data's of Example 7.10 for $N = 6$ and $C = 2$.

**Solution**

(a) **Average fluid temperature**

From Example 7.7: $\left[\frac{\left[(\alpha\tau)_{m,\text{eff}} I_u + U_{\text{Lm}} T_a\right]}{U_{\text{Lm}}} + T_a\right] = 94.75$, $\frac{NF'U_{\text{Lm}}A_m}{\dot{m}_f C_f} = 6 \times 0.0309 = 0.1854$ and $\exp\left\{\frac{-NF'U_{\text{Lm}}A_m}{\dot{m}_f C_f}\right\} = 0.8308$ and $I_u = 475$ W/m$^2$ (Example 7.1).

Substitute the above values in Eq. 7.14 to get an average value of fluid temperature.

$$\overline{T}_{f6} = 94.75\left[1 - \frac{1 - 0.8308}{0.1854}\right] + 15 \times \frac{1 - 0.8308}{0.1854}$$

$$= 94.75[1 - 0.9126] + 15 \times 0.9126 = 21.97\,°C$$

(b) **Average absorber plate temperature**

Now, $\left[\rho\alpha_p\tau_g^2(1-\beta_c)C + F_0(\alpha\tau)_{\text{eff}}\right] = [0.9 \times 0.9 \times (0.95)^2(1-0.89) \times 2 + 0.233 \times 1.2736] = [0.1608 + 0.2967] = 0.4575$, $F' = 0.82$, $h_{\text{pf}} = 100$ W/m$^2$ °C, $\overline{T}_f = 16.19\,°C$.

$U_{\text{pa}} = 2.1225$ W/m$^2$ °C, $U_b = 0.7$ W/m °C (Table 7.1).

Substitute the above values in Eq. 7.11 to get an average value of plate temperature

$$\overline{T}_{p6} = \frac{0.4575 \times 475 + (2.1225 + 0.7) \times 16 + 0.82 \times 100 \times 21.97}{(2.1225 + 0.82 \times 100 + 0.7)}$$

$$= \frac{217.31 + 45.16 + 1801.54}{84.8225} = \frac{2064.01}{84.8225} = 24.33\,°C$$

### (c) Average solar cell temperature

In this case, $(\alpha\tau)_{\text{eff}} = 1.1775$, $U_{t,ca} = 9.12$ W/m$^2$ °C, $U_{b,cp} = 2.7663$ W/m$^2$ °C. Substitute the above value in Eq. 7.12, one gets

$$\overline{T}_{c6} = \frac{1.2736 \times 475 + 9.12 \times 16 + 2.7663 \times 24.33}{9.12 + 2.7663} = \frac{818.18}{11.8863} = 68.83\,°C$$

**Example 7.15** Find out the outlet fluid temperature and mass flow rate of six PVT-CPC collectors connected in series for Example 7.14 for $C = 2$.

**Solution**

From Eq. 7.13a, We have

$$T_{\text{foN}} = \left[\frac{[(\alpha\tau)_{m,\text{eff}}I_u + U_{\text{Lm}}T_a]}{U_{\text{Lm}}} + T_a\right]\left[1 - \exp\left\{\frac{-NF'U_{\text{Lm}}A_m}{\dot{m}_f C_f}\right\}\right]$$
$$+ T_{\text{fi}}\exp\left\{\frac{-NF'U_{\text{Lm}}A_m}{\dot{m}_f C_f}\right\}$$

From Eq. 7.14, we know $\left[\frac{[(\alpha\tau)_{m,\text{eff}}I_u + U_{\text{Lm}}T_a]}{U_{\text{Lm}}} + T_a\right] = 94.75$, $T_{\text{fi}} = 15\,°C$ and $\exp\left\{\frac{-6U_{\text{Lm}}A_m}{\dot{m}_f C_f}\right\} = 0.8308$

Substitute the above values in the expression for $T_{\text{foN}}$, one gets

$$T_{\text{foN}} = 94.75[1 - 0.8308] + 15 \times 0.8308 = 16.03 + 12.462 = 28.49\,°C.$$

The rate of thermal energy is determined as follows:

$$\dot{Q}_{u6} = \dot{m}_f C_f(T_{\text{fo6}} - T_{\text{fi}}) = NA_m F_{\text{RN}}\left[[(\alpha\tau)_{m,\text{eff}}I_u + U_{\text{Lm}}T_a] - U_{\text{Lm}}(T_{\text{fi}} - T_a)\right]$$

where,

Mass flow rate factor, $F_{R6} = \frac{\dot{m}_f C_f}{NA_m U_{\text{Lm}}}\left[1 - \exp\left\{\frac{-NF'U_{\text{Lm}}A_m}{\dot{m}_f C_f}\right\}\right] = \frac{0.035 \times 4190}{6 \times 2 \times 2.76}[1 - 0.8308] = 0.749$.

**Example 7.16** Find out an electrical efficiency of PV module, electrical power, thermal exergy and total exergy of six PVT-CPC liquid collectors connected in series for Example 7.14 for $C = 2$.

**Solution**

**From Example 7.10:** $\overline{T}_{c6} = 68.83\,°C$; $T_{fo6} = 28.49\,°C$ and $T_{fi} = 15\,°C$ (Example 7.7), $I_u = 475\ \text{W/m}^2$ (Example 7.1), $\beta_c = 0.89$, $\beta_0 = 0.00425(°C)^{-1}$, $A_m = 2\ \text{m}^2$(Table 7.1).

Following Eq. 6.17 with known $\overline{T}_c$, the electrical efficiency and electrical power can be determined as follows:

$$\overline{\eta}_{m6} = \eta_{m0}\left[1 - \beta_0\left(\overline{T}_{c6} - 25\right)\right] = 0.12[1 - 0.00425(68.83 - 25)] = 0.0976$$

and,

$$\dot{E}_{el6} = \overline{\eta}_{m6} \times I_u \times \beta_c A_m \times N = 0.0976 \times 475 \times 0.89 \times 2 \times 6 = 495.12\ \text{W}.$$

Following Eqs. 2.35a and 6.22, an expression for evaluating hourly exergy is given by

$$\dot{Ex}_{u,\text{th}6} = \dot{m}_f C_f\left[(T_{fo6} - T_{fi}) - (T_a + 273)\ln\frac{T_{fo} + 273}{T_{fi} + 273}\right]$$

$$= 0.035 \times 4190\left[(28.49 - 15) - (16 + 273)\ln\frac{28.49 + 273}{15 + 273}\right]$$

$$= 146.65[13.90 - 289\ \text{In}\ 1.0458] = 146.65 \times 0.6817 - 99.97\ \text{W}$$

Total exergy of six PVT-CPC liquid collector $(\dot{Ex}_T)$ = thermal exergy + electrical power $= 99.97 + 495.12 = 595.10\ \text{W}$

### 7.2.7  The PVT Liquid Collector

The PVT liquid collector is a device which provides thermal (heat) energy along with electrical power. It is sustainable device and has been discussed in detail in Chap. 7.

***Example 7.17*** Determine the average fluid $(\overline{T}_{fN})$, the plate $(\overline{T}_{pN})$ and the solar cell $(\overline{T}_{cN})$ temperature of N-SPVT-collectors connected in series by using the data's of Example 7.10 for $N = 6$ and $C = 1$ (PVT liquid collector).

**Solution**

For $C = 1$, $I_u = I(t) = 500\ \text{W/m}^2$, we have from Table 7.1 and Examples 7.1–7.6.
$(\alpha\tau)_{\text{eff}} = 0.5887$ for PVT(Example 7.2), $F_0 = 0.233$, (Example 7.3), $U_{pa} = 2.1225\ \text{W/m}^2\,°C$, (Example 7.4), $U_{Lm} = 2.76\ \text{W/m}^2\,°C$, $F_{01} = 0.97$ (Example 7.6)

$$(\alpha\tau)_{m,\text{eff}} = F_{01}\left[\rho\alpha_p\tau_g^2(1 - \beta_c)C + F_0(\alpha\tau)_{1,\text{eff}}\right]$$

$$= 0.97\left[0.9 \times 0.9 \times (0.95)^2(1 - 0.89) \times 1 + 0.233 \times 0.5887\right]$$

$$= 0.97 \times [0.0804 + 0.1372] = 0.2176$$

$\exp\left\{\dfrac{-NF'U_{\text{Lm}}A_m}{\dot{m}_f C_f}\right\} = 0.8308,\ \dfrac{NF'U_{\text{Lm}}A_m}{\dot{m}_f C_f} = 0.1854$ and $\left[\dfrac{[(\alpha\tau)_{m,\text{eff}}I(t)+U_{\text{Lm}}T_a]}{U_{\text{Lm}}}+T_a\right] =$ 69.42 (Example 7.12).

### (a) Average fluid temperature for C = 1 (SPVT collector)

Substitute the above values in Eq. 7.14 to get an average value of fluid temperature

$$\overline{T}_{fN} = 69.42\left[1 - \frac{1-0.8308}{0.1854}\right] + 15 \times \frac{1-0.8308}{0.1854} = 69.42[1-0.9126]$$
$$+ 15 \times 0.9126 = 19.76\,°C$$

### (b) Average absorber plate temperature

Now, $\left[\rho\alpha_p\tau_g^2(1-\beta_c)C + F_0(\alpha\tau)_{,eff}\right] = [0.9 \times 0.9 \times (0.95)^2(1-0.89) \times 1 + 0.233 \times 0.5887] = [0.0804 + 0.1372] = 0.2176$, $F' = 0.82$, $h_{pf} = 100$ W/m$^2\,°$C, $\overline{T}_{fN} = 19.76\,°$C.

$U_{pa} = 2.1225$ W/m$^2\,°$C, $U_b = 0.7$ W/m$^2\ °$C. (Table 7.1).

Substitute the above values in Eq. 7.11 to get an average value of plate temperature

$$\overline{T}_{pN} = \frac{0.2176 \times 500 + (2.1225 + 0.7) \times 16 + 0.82 \times 100 \times 21.44}{(2.1225 + 0.82 \times 100 + 0.7)}$$
$$= \frac{108.8 + 45.16 + 1758.08}{84.8225} = \frac{200.91}{84.8225} = 22.54\,°C$$

### (c) Average solar cell temperature

In this case, $(\alpha\tau)_{\text{eff}} = 1.1775$, $U_{t,ca} = 9.12$ W/m$^2\,°$C, $U_{b,cp} = 2.7663$ W/m$^2\,°$C.

Substitute the above value in Eq. 7.12, one gets

$$\overline{T}_{cN} = \frac{0.5887 \times 500 + 9.12 \times 16 + 2.7663 \times 22.54}{9.12 + 2.7663} = \frac{487.62}{11.8863} = 41.02\,°C.$$

**Example 7.18** Find out the outlet fluid temperature and mass flow rate of six SPVT collectors connected in series for Example 7.15 for $C = 1$.

### Solution

From Eq. 7.13a, we have

$$T_{\text{foN}} = \left[\frac{[(\alpha\tau)_{m,\text{eff}}I_u + U_{\text{Lm}}T_a]}{U_{\text{Lm}}} + T_a\right]\left[1 - \exp\left\{\frac{-NF'U_{\text{Lm}}A_m}{\dot{m}_f C_f}\right\}\right]$$
$$+ T_{\text{fi}}\exp\left\{\frac{-NF'U_{\text{Lm}}A_m}{\dot{m}_f C_f}\right\}$$

From Eq. 7.14, we know $\left[\frac{[(\alpha\tau)_{m,\text{eff}}I(t)+U_{\text{Lm}}T_a]}{U_{\text{Lm}}} + T_a\right] = 69.42$ $T_{\text{fi}} = 15\,°\text{C}$ and $\exp\left\{\frac{-NF'U_{\text{Lm}}A_m}{\dot{m}_f C_f}\right\} = 0.8308$, $\frac{NF'U_{\text{Lm}}A_m}{\dot{m}_f C_f} = 0.1854$ (Example 7.17).

Substitute the above values in the expression for $T_{\text{foN}}$, one gets

$$T_{\text{fo6}} = 69.42[1 - 0.8308] + 15 \times 0.8308 = 14.99 + 12.462 = 24.20\,°\text{C}$$

The rate of thermal energy is determined as follows:

$$\dot{Q}_{u6} = \dot{m}_f C_f(T_{\text{foN}} - T_{\text{fi}}) = NA_m F_{\text{RN}}\big[[(\alpha\tau)_{m,\text{eff}}I_u + U_{\text{Lm}}T_a] - U_{\text{Lm}}(T_{\text{fi}} - T_a)\big]$$

where,

Mass flow rate factor, $F_{\text{RN}} = \frac{\dot{m}_f C_f}{NA_m U_{\text{Lm}}}\left[1 - \exp\left\{\frac{-NF'U_{\text{Lm}}A_m}{\dot{m}_f C_f}\right\}\right] = \frac{0.035 \times 4190}{6 \times 2 \times 2.76}$

$[1 - 0.8308] = 0.749$.

The mass flow rate is same for both $C = 1$ and 2, respectively.

**Example 7.19** Find out an electrical efficiency of SPV module, electrical power, thermal exergy and total exergy of six SPVT-CPC liquid collectors connected in series for Example 7.14 for $C = 1$ (SPVT liquid collector).

**Solution**

**From Example 7.10:** $\overline{T}_{\text{cN}} = 41.02\,°\text{C}$; $T_{\text{fo}} = 24.20\,°\text{C}$ and $T_{\text{fi}} = 15\,°\text{C}$ (Example 7.7), $I(t) = 500\ \text{W/m}^2$ (Example 7.1), $\beta_c = 0.89$, $\beta_0 = 0.00425\ (°\text{C})^{-1}$, $A_m = 2\text{m}^2$ (Table 7.1).

Following Eq. 6.17 with known $\overline{T}_c$, the electrical efficiency and electrical power can be determined as follows:

$$\overline{\eta}_m = \eta_{m0}\big[1 - \beta_0(\overline{T}_{\text{cN}} - 25)\big] = 0.12[1 - 0.00425(41.02 - 25)] = 0.1118$$

and,

$$\dot{E}_{\text{el}} = \overline{\eta}_m \times I(t) \times \beta_c A_m \times N = 0.1118 \times 500 \times 0.89 \times 2 \times 6 = 597.17\ \text{W}.$$

Following Eqs. 2.35a and 6.22, an expression for evaluating hourly exergy is given by

$$\dot{E}x_{u\text{th}} = \dot{m}_f C_f\left[(T_{\text{fo}} - T_{\text{fi}}) - (T_a + 273)\ln\frac{T_{\text{fo}} + 273}{T_{\text{fi}} + 273}\right]$$

$$= 0.035 \times 4190\left[(24.20 - 15) - (16 + 273)\ln\frac{24.20 + 273}{15 + 273}\right]$$

$$= 146.65[12.452 - 289\ \text{In}\ 1.0319] - 146.65 \times [9.2 - 9.0875] = 16.498\ \text{W}$$

Total exergy of six PVT liquid collector $(\dot{E}x_T)$ = the rate of thermal exergy + electrical power $= 16.498 + 597.17 = 613.67\ \text{W}$

This shows that there is 3.12% $\left[\frac{613.67-595.10}{595.10}=0.0312\right]$, (Example 7.16) more exergy in PVT liquid collector than PVT-CPC liquid collector.

## 7.2.8   The FPC-CPC Liquid Collector

The FPC-CPC liquid is a device which consists of conventional flat plate collector and compound parabolic concentrator and both has been integrated in one unit. Hence it is referred as FPC-CPC liquid collector. It only provides thermal (heat) energy at higher operating temperature than conventional flat plate collector. It is also not sustainable.

In this case, the FPC refers to conventional flat plate collector (Chap. 4). The packing factor $(\beta_c)$, absorptivity $(\alpha_c)$ and electrical efficiency $(\eta_m)$ of semi-transparent PV module will be zero, i.e., $\beta_c = \alpha_c = \eta_m = (\alpha\tau)_{\text{eff}}0$. However, there will not be any electrical power available from FPC-CPC collector.

Equation 7.13a will be reduced to

$$T_{\text{foN}} = \left[\frac{\left[F_{01}\rho\alpha_p\tau_g^2 C \times I_u + U_{\text{Lm}}T_a\right]}{U_{\text{Lm}}} + T_a\right]\left[1 - \exp\left\{\frac{-NF'U_{\text{Lm}}A_m}{\dot{m}_f C_f}\right\}\right]$$

$$+ T_{\text{fi}}\exp\left\{\frac{-NF'U_{\text{Lm}}A_m}{\dot{m}_f C_f}\right\} \tag{7.17}$$

where,   $(\alpha\tau)_{m,\text{eff}} = F_{01}\left[\rho\alpha_p\tau_g^2 C\right]$, $F_{01} = \frac{h_{\text{pf}}}{(U_{\text{pa}}+F'h_{\text{pf}}+U_b)}$, $U_{\text{Lm}} = 0.1065$

$\frac{h_{\text{pf}}(U_{\text{pa}}+U_b)}{(U_{\text{pa}}+F'h_{\text{pf}}+U_b)}$

**Example 7.20** Evaluate the outlet fluid temperature for six $(N = 6)$-FPC-CPC air collectors connected in series for $C = 2$ for the following parameters:

$\rho = 0.9, \alpha_p = 0.9, \tau_g = 0.95, F_{01} = 0.97, U_{\text{Lm}} = 2.76$ W/m$^2$ °C, $U_{\text{pa}} = 2.1225$ (Example 7.4), $T_a = 16$ °C, $T_{\text{fi}} = 15$ °C, $N = 6, I_u = 475$ W/m$^2$, $\dot{m}_f C_f = 146.65$ W/°C.

**Solution**

Here, $\left\{\frac{NF'U_{\text{Lm}}A_m}{\dot{m}_f C_f}\right\} = \frac{6\times0.82\times2.76\times2}{146.65} = 0.1852$; $\exp(-0.1852) = 0.8309$.

Substitute the appropriate values in Eq. 7.17, one gets

$$T_{\text{fo6}} = \left[\frac{\left[0.97 \times 0.9 \times 0.9 \times (0.95)^2 \times 2 \times 475 + 2.76 \times 16\right]}{2.76} + 16\right][1 - 0.8309]$$

$$+ 15 \times 0.8309 = \left[\frac{673.64 + 44.16}{2.76} + 16\right] \times 0.1691 + 12.46$$

$$= 46.68 + 12.46 = 59.14\,°C.$$

An instantaneous thermal efficiency is obtained as

$$\eta_i = \frac{\dot{m}_f C_f (T_{fo6} - T_{fi})}{I(t) \times N \times A_{am}} = \frac{146.65 \times 44.14}{500 \times 6 \times 4} = \frac{6473.13}{12{,}000} = 0.5394.$$

Now, the thermal exergy of six FPC-CPC collectors can be obtained as

$$
\begin{aligned}
\dot{Ex}_{uth6} &= 146.65 \left[ (T_{fo} - T_{fi}) - (T_a + 273) \ln \frac{T_{fo} + 273}{T_{fi} + 273} \right] \\
&= 146.65 \left[ (59.14 - 15) - (16 + 273) \ln \frac{59.14 + 273}{15 + 273} \right] \\
&= 146.65[44.14 - 41.21] = 429.68 \text{ W}
\end{aligned}
$$

### 7.2.9  The Liquid Flat Plat Collector (FPC)

The liquid flat plate collector (FPC) is generally known as conventional flat plate collector which has been discussed in detail in Chap. 4. This collector only provides thermal (heat) energy, and it is not sustainable.

**Example 7.21**  Repeat Example 7.20 for $C = 1$ (only FPC).

**Solution**

For $C = 1$,

$$
\begin{aligned}
T_{foN} &= \left[ \frac{[0.97 \times 0.9 \times 0.9 \times (0.95)^2 \times 1 \times 475 + 2.76 \times 16]}{2.76} + 16 \right] [1 - 0.8309] \\
&+ 15 \times 0.8309 = \left[ \frac{336.82 + 44.16}{2.76} + 16 \right] \times 0.1691 \\
&+ 12.46 = 26.05 + 12.46 = 38.51 \,^\circ\text{C}
\end{aligned}
$$

Now, the thermal exergy of FPC-CPC collector can be obtained as

$$
\begin{aligned}
\dot{Ex}_{uth} &= \dot{m}_f C_f \left[ (T_{fo} - T_{fi}) - (T_a + 273) \ln \frac{T_{fo} + 273}{T_{fi} + 273} \right] \\
&= 146.65 \left[ (38.51 - 15) - (16 + 273) \ln \frac{38.51 + 273}{17 + 273} \right] \\
&= 146.65 \times [23.51 - 22.68] - 122.0 \text{ W}
\end{aligned}
$$

Now one can have the following conclusions:

$$\dot{Ex}_T = 613.67 \ W(PVT)[\text{Example 7.19}] >$$
$$\dot{Ex}_T = 595.10 \ W(PVT - CPC0)[\text{Example 7.16}] >$$
$$Ex_{uth} = 429.68W(FPC - CPC)[\text{Example 7.20}] >$$
$$\dot{Ex}_{uth} = 122.0 \ W(FPC)[\text{Example 7.21}] \tag{7.18}$$

Here, one can observe that the FPC and FPC-CPC are not sustainable and PVT and PVT-CPC are a sustainable device.

## 7.3 The PVT-CPC Air Collector

Figure 7.4 shows a cross-sectional view of a semi-transparent photo-voltaic thermal (PVT) air collector. The semi-transparent PVT air collector consists of semi-transparent PV module as cover with a conventional flat plate air collector as shown in Fig. 5.4a (Chap. 5). The glass of a conventional flat plate air collector as shown in Fig. 5.4a is replaced by semi-transparent photo-voltaic (PV) module, and hence it is referred as PVT air collector. Further, if PVT air collector is integrated with compound parabolic concentrator as shown in Fig. 7.1 for PVT-CPC liquid collector, then it is known as PVT-CPC air collector. It is also self-sustained device like PVT-CPC liquid collector (Sect. 7.2).

The thermal modeling of PVT-CPC air collector is same as thermal modeling of PVT-CPC liquid collector with following modifications:

(a) The collector efficiency $(F')$ factor for solar air collector has been given in Chap. 5.
(b) The absorber is a flat blackened surface for indirect transfer of absorbed solar energy to flowing air below plate.
(c) The specific heat capacity of working fluid (air) is very small with no storage capacity and hence mass flow rate has smaller value (Chap. 5).

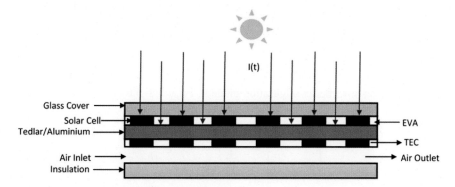

**Fig. 7.4** Cross-sectional view of PVT air collector

(d) The heat transfer coefficient from absorber to flowing air ($h_{pf}$ = 11.08 W/m$^2$ °C, Examples 4.15 and 5.1) is also very small in comparison with PVT liquid collector.

### 7.3.1 Energy Balances

(a) **Solar cell of PVT-CPC air collector**

Following Eq. 7.1, an energy balance of semi-transparent PV module of PVT-CPC air collector will be same

$$\rho \alpha_c \tau_g \beta_c I_u A_{am} = U_{t,ca}(T_c - T_a)A_{rm} + U_{b,cp}(T_c - T_p)A_{rm} + \rho I_u \eta_m \beta_c A_{am}$$
(7.18a)

From Eq. 7.18a, we have an expression for $T_c$ in terms of $T_p$ as

$$T_c = \frac{[\alpha_c \tau_g \left(\frac{A_{am}}{A_{rm}}\right) - \eta_m]\beta_c \rho I_u + U_{t,ca}T_a + U_{b,cp}T_p}{(U_{t,ca} + U_{b,cp})}$$
(7.18b)

(b) **Absorber flat late of PVT-CPC air collector**

Following Eq. 7.3 with collector efficiency $(F')$ factor, the energy balance of the absorber plate in W can be written as

$$\rho \alpha_p \tau_g^2 (1 - \beta_c) I_u A_{am} + U_{b,cp}(T_c - T_p)A_{rm}$$
$$= F' h_{pf}(T_p - T_f)A_{rm} + U_b(T_p - T_a)A_{rm}$$
(7.19a)

or,

$$\rho \alpha_p \tau_g^2 (1 - \beta_c) I_u \frac{A_{am}}{A_{rm}} + U_{b,cp}(T_c - T_p) = F' h_{pf}(T_p - T_f) + U_b(T_p - T_a)$$
(7.19b)

where, the numerical value of $h_{pf}$ will be half ($h_{pf}$ = 5.54 W/m$^2$ °C) as pointed out in Eq. 5.3 due to stagnant air below absorber plate. The numerical values of $F'$ will depend on absorber configuration given in Figs. 5.4–5.6 along with $U_L$, and it is given by

$$F' = \frac{1/U_L}{\left[\frac{1}{U_L} + \frac{1}{h_{pf}}\right]} = \frac{1/U_L}{\frac{1}{U_0}} = \frac{U_0}{U_L} = \frac{h_{pf}}{(h_{pf} + U_L)}.$$

**Example 7.22** Compute PVT-CPC air collector efficiency $(F')$.

**Solution**

For semi-transparent PV module, $U_L$ is given by Eq. 6.7

$$U_L = U_t + U_b = 2.14 + 0.7 = 2.84 \text{ W/m}^2 \text{ °C}$$

Substitute the above value in expression for $F'$ and one gets

$$F' = \frac{h_{pf}}{(h_{pf} + U_L)} = \frac{5.54}{5.54 + 2.84} = 0.66.$$

From Examples 7.3–7.6, we have

$$U_{b,cp}(T_c - T_p) = F_0(\alpha\tau)_{,eff} I_u - U_{pa}(T_p - T_a) \tag{7.20}$$

where, $(\alpha\tau)_{eff} = \left(\alpha_c \tau_g \frac{A_{am}}{A_{rm}} - \eta_m\right)\rho\beta_c$

**Example 7.23** Compute the product of absorptivity–transmittivity of semi-transparent PV module for PVT-CPC $(C = 2)$ and PVT $(C = 1)$ air collector.

**Solution**

**For $C = \frac{A_{am}}{A_{rm}} = 2$ (PVT-CPC air collector)**

$$(\alpha\tau)_{eff} = \left(\alpha_c \tau_g \frac{A_{am}}{A_{rm}} - \eta_m\right)\rho\beta_c = \left(0.9 \times 0.95 \times \frac{4}{2} - 0.12\right)0.9 \times 0.89 = 1.2736$$

**For $C = \frac{A_{am}}{A_{am}} = 1$ (PVT air collector)**

$$(\alpha\tau)_{eff} = \left(\alpha_c \tau_g \frac{A_{am}}{A_{rm}} - \eta_m\right)\rho\beta_c$$

$$= (0.9 \times 0.95 \times 1 - 0.12)0.9 \times 0.89 = 0.5887 \text{ for PVT for } C = \frac{A_{am}}{A_{rm}} = 1$$

This shows that $(\alpha\tau)_{eff}$ is same for both liquid and air collectors (Tables 7.1 and 7.2).

$F_0 = 0.233$ (Example 7.3) and $U_{pa} = 2.1225 \text{ W/m}^2\text{C}$ (Example 7.4) Table 7.2
Substitute $U_{b,cp}(T_c - T_p)$ from Eq. 7.20 into Eq. 7.19b, one gets.

$$\left[\rho\alpha_p\tau_g^2(1 - \beta_c)\frac{A_{am}}{A_{rm}} + F_0(\alpha\tau)_{,eff}\right]I_u - U_{pa}(T_p - T_a)$$

$$= F'h_{pf}(T_p - T_f) + U_b(T_p - T_a) \tag{7.21a}$$

**Table 7.2**  Some additional constants for PVT-CPC air collector

| S. No. | Constants | Numerical value | Units |
|---|---|---|---|
| 1 | $A_{am}$ | 4 | $m^2$ |
| 2 | $A_{rm}$ | 2 | $m^2$ |
| 3 | b | 1 | m |
| 4 | $C_a$ | 1009 | j/kg °C |
| 5 | $F'$ | 0.66 | Example 7.22 |
| 6 | $F_0$ | 0.233 | Example 7.3 |
| 7 | $F_{01}$ | 0.56 | Section 7.3.1 |
| 8 | $h_{pf}$ | 5.54 | $W/m^2$ °C Section 7.3.1 |
| 9 | $I_u$ | 475 | $W/m^2$ |
| 10 | $K_{bs}$ | 386 | $W/m^2$ °C |
| 11 | $K_g$ | 0.688 | $W/m^2$ °C |
| 12 | $K_t$ | 0.033 | $W/m^2$ °C |
| 13 | $L_0$ | 2 | m |
| 14 | $L_{bs}$ | 0.0005 | m |
| 15 | $L_g$ | 0.003 | m |
| 16 | $L_t$ | 0.0005 | m |
| 17 | $m'_f$ | 0.035 | kg/s |
| 18 | $T_a$ | 16 | °C |
| 19 | $T_{fi}$ | 15 | °C |
| 20 | $U_b$ | 0.70 | $W/m^2$ °C |
| 21 | $U_{b,cp}$ | 2.7663 | $W/m^2$ °C |
| 22 | $U_{fa}$ | 2.76 | $W/m^2$ °C |
| 23 | $U_L$ | 2.84 | $W/m^2$ °C |
| 24 | $U_{Lm}$ | 2.05 | Section 7.3.1 |
| 25 | $U_{pa}$ | 2.1125 | $W/m^2$ °C |
| 26 | $U_{t,ca}$ | 9.12 (Exan. 6.1) | $W/m^2$ °C |
| | **Greek letters** | | |
| 27 | $\alpha_c$ | 0.9 | |
| 28 | $\alpha_p$ | 0.9 | |
| | $(\alpha\tau)_{eff}$ | 1.2736 | PVT-CPC ($C = 2$) |
| | $(\alpha\tau)_{eff}$ | 0.5887 | PVT ($C = 1$) |
| 29 | $(\alpha\tau)_{m,eff}$ | 0.2561 | PVT-CPC ($C = 2$) |
| 30 | $(\alpha\tau)_{m,eff}$ | 0.2178 | PVT ($C = 1$) |
| 31 | $\beta_0$ | 0.00425 | °C$^{-1}$ |
| 32 | $\beta_c$ | 0.89 | |
| 33 | $\gamma$ | 0.20–0.35 | |
| 34 | $\rho$ | 0.9 | |

(continued)

**Table 7.2** (continued)

| S. No. | Constants | Numerical value | Units |
|--------|-----------|-----------------|-------|
| 35 | $\tau_g$ | 0.95 | |
| 36 | $\eta_{m0}$ | 0.12 | |
| 37 | $\eta_m$ | 0.12 | |

From Eq. 7.21a, we get an expression for $T_p$ as

$$T_p = \frac{\left[\rho\alpha_p\tau_g^2(1-\beta_c)\frac{A_{am}}{A_{rm}} + F_0(\alpha\tau)_{,eff}\right]I_u + U_{pa}T_a + F'h_{pf}T_f}{U_{pa} + F'h_{pf} + U_b} \tag{7.21b}$$

The rate of thermal energy transferred from absorber plate to working fluid in W can be written as

$$F'h_{pf}(T_p - T_f) = F'\left[(\alpha\tau)_{m,eff}I_u - U_{Lm}(T_f - T_a)\right] \tag{7.21}$$

where, $(\alpha\tau)_{m,eff} = F_{01}\left[\rho\alpha_p\tau_g^2(1-\beta_c)C + F_0(\alpha\tau)_{eff}\right]$ and $F_{01} = \frac{F'h_{pf}}{(U_{pa}+F'h_{pf}+U_b)}$

**Example 7.24** Compute $(\alpha\tau)_{m,eff}$ for both PVT-CPC ($C = 2$) and PVT air collector ($C = 1$).

**Solution**

From Eq. 7.21, we have the followings.

$$F_{01} = \frac{F'h_{pf}}{(U_{pa} + F'h_{pf} + U_b)} = \frac{0.66 \times 5.54}{2.1225 + 0.66 \times 5.54 + 0.7} = \frac{3.6564}{6.4789} = 0.56 \text{ with } h_{pf}$$
$$= 5.54 \text{ W/m}°\text{C}$$

**For C = 2 (PVT-CPC air collector)**

$$(\alpha\tau)_{m,eff} = F_{01}\left[\rho\alpha_p\tau_g^2(1-\beta_c)C + F_0(\alpha\tau)_{eff}\right]$$
$$= 0.56\left[0.9 \times 0.9 \times (0.95)^2(1-0.89) \times 2 + 0.233 \times 1.2736\right]$$
$$= 0.56 \times [0.1608 + 0.2965] = 0.2561$$

**For C = 1 (PVT Air Collector)**

$$(\alpha\tau)_{m,eff} = 0.56\left[0.9 \times 0.9 \times (0.95)^2(1-0.89) \times 1 + 0.233 \times 0.5887\right]$$
$$= 0.56 \times [0.0804 + 0.1372] = 0.2178$$

The $(\alpha\tau)_{m,eff}$ for liquid and air collectors are different.
Further,

$$U_{\text{Lm}} = \frac{F' h_{\text{pf}}(U_{\text{pa}} + U_b)}{(U_{\text{pa}} + F' h_{\text{pf}} + U_b)} = \frac{0.56 \times 5.54 \times (2.1225 + 0.7)}{2.1225 + 0.66 \times 5.54 + 0.7}.$$

$$= \frac{13.2674}{6.4789} = 2.05 \text{ W/m}^2\,^\circ\text{C}$$

These constants of Examples 7.22–7.24 have also been summarized in Table 7.2 along with other parameters.

### (c)  Flowing air below absorber plate

The rate of thermal energy transferred from absorber plate to working fluid given by Eq. 7.21 in W in an elemental area of 'bdx' will be carried out by mass flow rate which can be expressed as

$$\dot{m}_f C_f \frac{\mathrm{d}T_f}{\mathrm{d}x} \mathrm{d}x = F'\big[(\alpha\tau)_{m,\text{eff}} I_u - U_{\text{Lm}}(T_f - T_a)\big] b \mathrm{d}x \tag{7.22}$$

The above equation can be also written as

$$\frac{\mathrm{d}T_f}{\mathrm{d}x} + aT_f = \mathrm{f}(t) \tag{7.23}$$

where, $a = \frac{F' U_{\text{Lm}} b}{\dot{m}_f C_f}$ and $f(t) = \frac{F'[(\alpha\tau)_{m,\text{eff}} I_u + U_{\text{Lm}} T_a] b}{\dot{m}_f C_f}$.

Following Sect. 7.2.4, the solution of Eq. 7.23 with initial condition $T_f = T_{\text{fi}}$ at $x = 0$ can be written as

$$T_f = \left[\frac{(\alpha\tau)_{m,\text{eff}} I_u}{U_{\text{Lm}}} + T_a\right]\left[1 - \exp\left(-\frac{F' U_{\text{Lm}} b}{\dot{m}_f C_f} x\right)\right] + T_{\text{fi}} \exp\left(-\frac{F' U_{\text{Lm}} b}{\dot{m}_f C_f} x\right) \tag{7.24}$$

### 7.3.2  The Outlet Temperature at $N^{th}$ PVT-CPC Air Collector

Following Sect. 7.2.5, the outlet air temperature at $N$th PVT-CPC air collector as a function of $x$ by using Eq. 7.24 can be obtained as

$$T_{fN}(x) = \left[\frac{(\alpha\tau)_{m,\text{eff}} I_u}{U_{\text{Lm}}} + T_a\right]\left[1 - \exp\left(-\frac{N F' U_{\text{Lm}} b}{\dot{m}_f C_f} x\right)\right] + T_{\text{fi}} \exp\left(-\frac{N F' U_{\text{Lm}} b}{\dot{m}_f C_f} x\right) \tag{7.25}$$

The outlet of air at $N$th PVT-CPC air collector will be obtained as

$$T_{\text{foN}} = \left[\frac{(\alpha\tau)_{m,\text{eff}} I_u}{U_{\text{Lm}}} + T_a\right]\left[1 - \exp\left(-\frac{NF'U_{\text{Lm}}bL_0}{\dot{m}_f C_f}\right)\right] + T_{\text{fi}}\exp\left(-\frac{NF'U_{\text{Lm}}bL_0}{\dot{m}_f C_f}\right)$$

(7.26a)

Since, $bL_0 = A_m$, then Eq. 7.26a becomes as

$$T_{\text{foN}} = \left[\frac{(\alpha\tau)_{m,\text{eff}} I_u}{U_{\text{Lm}}} + T_a\right]\left[1 - \exp\left\{\frac{-NF'U_{\text{Lm}}A_m}{\dot{m}_f C_f}\right\}\right] + T_{\text{fi}}\,\exp\left\{\frac{-NF'U_{\text{Lm}}A_m}{\dot{m}_f C_f}\right\}$$

(7.26b)

***Example 7.25*** Evaluate the outlet air temperature for $N = 1$ and $N = 6$ of N-PVT-CPC air collectors connected in series for mass flow rate $(\dot{m}_f)$ of 0.029 kg/s (0.25 m/s) and specific heat $(C_f)$ of 1005 j/kg °C by using the data's of Table 7.2 for $C = 2$.

**Solution**

**Given:** $T_a = 15\,°\text{C}$ and $I_u = 475\ \text{W/m}^2$.

For $N = 1$

(a) $\left[\frac{\left[(\alpha\tau)_{m,\text{eff}} I_u + U_{\text{Lm}} T_a\right]}{U_{\text{Lm}}}\right] = \frac{0.2561\times475+2.05\times16}{2.05} = \frac{154.44}{2.05} = 75.37$

(b) $\exp\left\{\frac{-NF'U_{\text{Lm}}A_m}{\dot{m}_f C_f}\right\} = \exp\left\{\frac{-1\times0.66\times2.05\times2}{0.029\times1005}\right\}$

$= \exp\left\{-\frac{2.706}{29.145}\right\} = \exp(-0.0928) = 0.9114.$

Substitute the above value in Eq. 7.26b, one gets

$$T_{\text{fo1}} = 75.37[1 - 0.9114] + 15\times0.9114 = 20.35\,°\text{C}$$

**For $N = 6$**

(a) $\left[\frac{\left[(\alpha\tau)_{m,\text{eff}} I_u + U_{\text{Lm}} T_a\right]}{U_{\text{Lm}}}\right] = \frac{0.2561\times475+2.05\times16}{2.05} = \frac{154.44}{2.05} = 75.34$

(b) $\exp\left\{\frac{-NF'U_{\text{Lm}}A_m}{\dot{m}_f C_f}\right\} = \exp\left\{\frac{-6\times0.66\times2.05\times2}{0.029\times1005}\right\}$

$= \exp\left\{-\frac{16.236}{29.145}\right\} = \exp(-0.5570) = 0.5729$

Substitute the above value in Eq. 7.26b, one gets

$$T_{\text{fo6}} = 75.34[1 - 0.5729] + 15\times0.5729 = 40.78\,°\text{C}.$$

This shows that there is an increase of about 20.43 °C temperature of air from $N = 1$ to $N = 6$.

### 7.3.3 An Average Fluid $\left(\overline{T}_{fN}\right)$, Absorber Plate $\left(\overline{T}_{pN}\right)$ and Solar Cell $\left(\overline{T}_c\right)$ Temperatures of N-PVT-CPC Air Collectors Connected in Series

The average air temperature for N-PVT-CPC air collector can be obtained as

$$
\overline{T}_{fN} = \frac{1}{L_0} \int_0^{L_0} T_{fN}(x)dx = \left[\frac{(\alpha\tau)_{m,\mathrm{eff}}I_u}{U_{\mathrm{Lm}}} + T_a\right]\left[1 - \frac{1 - \exp\left\{\frac{-NF'U_{\mathrm{Lm}}A_m}{\dot{m}_f C_f}\right\}}{\frac{NF'U_{\mathrm{Lm}}A_m}{\dot{m}_f C_f}}\right]
$$

$$
+ T_{\mathrm{fi}}\left[\frac{1 - \exp\left\{\frac{-NF'U_{\mathrm{Lm}}A_m}{\dot{m}_f C_f}\right\}}{\frac{NF'U_{\mathrm{Lm}}A_m}{\dot{m}_f C_f}}\right] \tag{7.27}
$$

After knowing $\overline{T}_{fN}$ from Eq. 7.27, the average plate temperature $\left(\overline{T}_p\right)$ can be evaluated from Eq. 7.21a as

$$
\overline{T}_{pN} = \frac{\left[\rho\alpha_p\tau_g^2(1 - \beta_c)\frac{A_{\mathrm{am}}}{A_{\mathrm{rm}}} + F_0(\alpha\tau)_{,eff}\right]I_u + U_{\mathrm{pa}}T_a + F'h_{\mathrm{pf}}\overline{T}_{fN}}{U_{\mathrm{pa}} + F'h_{\mathrm{pf}} + U_b} \tag{7.28}
$$

Further Eq. 7.28 can be used to evaluate the average solar cell temperature from Eq. 7.18b as

$$
\overline{T}_{cN} = \frac{[\alpha_c\tau_g\left(\frac{A_{\mathrm{am}}}{A_{\mathrm{rm}}}\right) - \eta_m]\beta_c\rho I_u + U_{t,\mathrm{ca}}T_a + U_{b,\mathrm{cp}}\overline{T}_{pN}}{\left(U_{t,\mathrm{ca}} + U_{b,\mathrm{cp}}\right)} \tag{7.29}
$$

**Example 7.26** Find out an average fluid $\left(\overline{T}_{fN}\right)$ of N-PVT-CPC air collectors connected in series for Example 7.25 for $N = 1$ and $N = 6$, respectively.

**Solution**

**For N = 1**

(a) $\left[\frac{[(\alpha\tau)_{m,\mathrm{eff}}I_u + U_{\mathrm{Lm}}T_a]}{U_{\mathrm{Lm}}}\right] = \frac{0.2561\times475 + 2.05\times16}{2.05} = \frac{154.44}{2.05} = 75.34$

(b) $\exp\left\{\frac{-NF'U_{\mathrm{Lm}}A_m}{\dot{m}_f C_f}\right\} = \exp\left\{\frac{-1\times0.66\times2.05\times2}{0.029\times1005}\right\}$

$= \exp\left\{-\frac{2.706}{29.145}\right\} = \exp(-0.0928) = 0.9114$

c. $\frac{NF'U_{\mathrm{Lm}}A_m}{\dot{m}_f C_f} = 0.0928$

From Eq. 7.27, one gets

$$\overline{T}_{f1} = 75.34\left[1 - \frac{1 - 0.9114}{0.0928}\right] + T_{fi}\left[\frac{1 - 0.9114}{0.0928}\right]$$

$$= 75.34[1 - 0.9547] + 15 \times 0.9547 = 17.73\,°C$$

**For N = 6**

(a) $\left[\frac{(\alpha\tau)_{m,\text{eff}}I_u + U_{Lm}T_a}{U_{Lm}}\right] = \frac{0.2561 \times 475 + 2.05 \times 16}{2.05} = \frac{154.44}{2.05} = 75.34$

(b) $\exp\left\{\frac{-NF'U_{Lm}A_m}{\dot{m}_f C_f}\right\} = \exp\left\{\frac{-6 \times 0.66 \times 2.05 \times 2}{0.029 \times 1005}\right\}$

$$= \exp\left\{-\frac{16.236}{29.145}\right\} = \exp(-0.5568) = 0.5730$$

(c) $\frac{NF'U_{Lm}A_m}{\dot{m}_f C_f} = 0.5570.$

From Eq. 7.27, we have

$$\overline{T}_{f6} = 75.34\left[1 - \frac{1 - 0.5730}{0.5568}\right] + 15 \times \left[\frac{1 - 0.5730}{0.5568}\right]$$

$$= 75.34[1 - 0.76688] + 15 \times 0.76688 = 29.07\,°C$$

**Example 7.27** Find out an average absorber plate $\left(\overline{T}_{pN}\right)$ temperature of N-PVT-CPC air collectors connected in series for Example 7.26 for $N = 1$ and $N = 6$ by using the data's of Table 7.2.

**Solution**

**For N = 1:** $\overline{T}_{f1} = 17.73\,°C$, $\left[\rho\alpha_p\tau_g^2(1 - \beta_c)\frac{A_{am}}{A_{rm}} + F_0(\alpha\tau)_{,\text{eff}}\right] = [0.9 \times 0.9 \times (0.95)^2(1 - 0.89) \times 2 + 0.233 \times 1.2736] = [0.1608 + 0.2967] = 0.4575.$

From Eq. 7.28, we have an expression for $\overline{T}_{p1}$

$$\overline{T}_{p1} = \frac{0.4575 \times 475 + 2.1225 \times 16 + 0.66 \times 5.54 \times 17.73}{2.1225 + 0.66 \times 5.54 + 0.7}$$

$$= \frac{217.32 + 33.96 + 64.82}{6.4789} = \frac{318.59}{6.4789} = 48.79\,°C$$

**For N = 6:** $\overline{T}_{f6} = 29.07\ °C$

$$\overline{T}_{p6} = \frac{0.4575 \times 475 + 2.1225 \times 16 + 0.66 \times 5.54 \times 29.07}{2.1225 + 0.66 \times 5.54 + 0.7}$$

$$= \frac{217.32 + 33.96 + 106.29}{6.4789} = \frac{357.57}{6.4789} = 55.19\,°C$$

**Example 7.28** Find out an average solar cell $\left(\overline{T}_c\right)$ temperature of N-PVT-CPC air collectors connected in series for Example 7.25 for $N = 1$ and $N = 6$, respectively, for Example 7.17 by using the data's of Table 7.2.

**Solution**

**For  N  =  1:**  $\overline{T}_{p1}$  =  48.79 °C  $(\alpha\tau)_{\text{eff}}$  =  $\left(\alpha_c\tau_g\frac{A_{\text{am}}}{A_{\text{rm}}} - \eta_m\right)\rho\beta_c$  =

$\left(0.9 \times 0.95 \times \frac{4}{2} - 0.12\right)0.9 \times 0.89 = 1.2736$ (Example 7.23, Table 7.2).

From Eq. 7.29, we have an expression for $\overline{T}_c$ as

$$\overline{T}_{c1} = \frac{\left[\alpha_c\tau_g\left(\frac{A_{\text{can}}}{A_{nn}}\right) - \eta_m\right]\beta_c\rho I_u + U_{t,\text{ca}}T_a + U_{b,\text{cp}}\overline{T}_{p1}}{\left(U_{t,\text{ca}} + U_{b,\text{cp}}\right)}$$

$$= \frac{1.2736 \times 475 + 9.12 \times 16 + 2.7663 \times 48.79}{9.12 + 2.7663}$$

$$= \frac{604.96 + 145.92 + 134.96}{11.8863} = 74.53\,°C$$

**For  N  =  6:**  $\overline{T}_{p6}$  =  55.19 °C,  $(\alpha\tau)_{\text{eff}}$  =  $\left(\alpha_c\tau_g\frac{A_{\text{am}}}{A_{\text{rm}}} - \eta_m\right)\rho\beta_c$  =

$\left(0.9 \times 0.95 \times \frac{4}{2} - 0.12\right)0.9 \times 0.89 = 1.2736$ (Example 7.23, Table 7.2)

$$\overline{T}_{c6} = \frac{\left[\alpha_c\tau_g\left(\frac{A_{\text{am}}}{A_{\text{rm}}}\right) - \eta_m\right]\beta_c\rho I_u + U_{t,\text{ca}}T_a + U_{b,\text{cp}}\overline{T}_{p6}}{\left(U_{t,\text{ca}} + U_{b,\text{cp}}\right)}$$

$$= \frac{1.2736 \times 475 + 9.12 \times 16 + 2.7663 \times 55.19}{9.12 + 2.7663}$$

$$= \frac{604.96 + 145.92 + 152.67}{11.8863} = 76.02\,°C$$

***Example 7.29***  Evaluate an electrical efficiency, electrical power for $N = 1$ and $N = 6$ for PVT-CPC collectors connected in series for Example 7.28.

**Solution**

**For N = 1:** $\overline{T}_{c1} = 74.53\,°C$.

Following Eq. 6.17 with known $\overline{T}_c$, the electrical efficiency and electrical power can be determined as follows:

$$\overline{\eta}_{m1} = \eta_{m0}\left[1 - \beta_0\left(\overline{T}_{cN} - 25\right)\right] = 0.12[1 - 0.00425(74.53 - 25)] = 0.0947$$

and,

$\dot{E}_{\text{ell}} = \overline{\eta}_m \times I(t) \times \beta_c A_m \times N = 0.0947 \times 475 \times 0.89 \times 2 \times 1 = 80.07\,\text{W}.$

**For N = 6:** $\overline{T}_{c6} = 76.02\,°C$.

Following Eq. 6.17 with known $\overline{T}_c$, the electrical efficiency and electrical power can be determined as follows:

$$\overline{\eta}_{m6} = \eta_{m0}\big[1 - \beta_0\big(\overline{T}_{cN} - 25\big)\big] = 0.12[1 - 0.00425(76.02 - 25)] = 0.09373$$

and,

$$\dot{E}_{el6} = \overline{\eta}_m \times I(t) \times \beta_c A_m \times N = 0.09373 \times 475 \times 0.89 \times 2 \times 6 = 475.52\mathrm{W}$$

### 7.3.4 The Rate of Useful Thermal Energy ($\dot{Q}_{uN}$), Mass Flow Rate Factor ($F_{RN}$) and Thermal Exergy ($Ex_u$)

The rate of useful thermal energy from N-PVT-CPC air collector with the help of Eq. 7.26b is obtained as.

$$\dot{Q}_{uN} = \dot{m}_f C_f (T_{foN} - T_{fi}) = \dot{m}_f C_f \left\{ \left[ \frac{[(\alpha\tau)_{m,\mathrm{eff}} I_u + U_{\mathrm{Lm}} T_a]}{U_{\mathrm{Lm}}} + T_a \right] \right.$$
$$\left[ 1 - \exp\left\{ \frac{-NF'U_{\mathrm{Lm}} A_m}{\dot{m}_f C_f} \right\} \right] + T_{fi} \exp\left\{ \frac{-NF'U_{\mathrm{Lm}} A_m}{\dot{m}_f C_f} \right\} - T_{fi} \right\}$$

Further, the above equation can be rewritten as

$$\dot{Q}_{uN} = \dot{m}_f C_f \left[ 1 - \exp\left\{ \frac{-NF'U_{\mathrm{Lm}} A_m}{\dot{m}_f C_f} \right\} \right] \left\{ \left[ \frac{[(\alpha\tau)_{m,\mathrm{eff}} I_u + U_{\mathrm{Lm}} T_a]}{U_{\mathrm{Lm}}} + T_a \right] - T_{fi} \right\}$$
$$= F_{RN} N A_m \big[ (\alpha\tau)_{m,\mathrm{eff}} I_u - U_{\mathrm{Lm}}(T_{fi} - T_a) \big]. \tag{7.30a}$$

Mass flow rate factor of N-PVT-CPC air collector $=$

$$F_{RN} = \frac{\dot{m}_f C_f}{N U_{\mathrm{Lm}} A_m} \left[ 1 - exp\left\{ \frac{-NF'U_{\mathrm{Lm}} A_m}{\dot{m}_f C_f} \right\} \right] \tag{7.33b}$$

The rate of thermal exergy can be evaluated as follows:

$$\dot{Ex}_{uN,\,\mathrm{th}} = \dot{m}_f C_f \left[ (T_{foN} - T_{fi}) - (T_a + 273) \ln \frac{T_{foN} + 273}{T_{fi} + 273} \right] \tag{7.31}$$

## 7.3.5  Thermal Exergy $\left(\dot{Ex}_{u1, th}\right)$ and an Overall Total Exergy $\left(\dot{Ex}_T\right)$

**Example 7.30** Calculate the rate of thermal exergy and rate of total exergy of N-PVT-CPC air collectors connected in series for $N = 1$ and $N = 6$ for Example.

**Solution**

**For N = 1:** $T_{\text{fo}1} = 17.73\,°C$ and $\dot{m}_f C_f = 29.145\,W/°C$.
The rate of thermal exergy is obtained from Eq. 7.31 as

$$\dot{Ex}_{u1,th} = 29.145 \left[ (17.73 - 15) - (16 + 273)\ln \frac{17.73 + 273}{15 + 273} \right]$$

$$= 29.145[2.73 - 289\ln 1.02347] = -ve$$

Total exergy of one PVT-CPC air collector $\left(\dot{Ex}_T\right)$ = the rate of thermal exergy + electrical power $= 0 + 80.07 = 80.07\,W$
**For N = 6:** $T_{fo6} = 40.78\,°C$ and $\dot{m}_f C_f = 29.145\,W/°C$.
The rate of thermal exergy is obtained from Eq. 7.31 as

$$\dot{Ex}_{u1,th} = 29.145 \left[ (40.78 - 15) - (16 + 273)\ln \frac{47.61 + 273}{15 + 273} \right]$$

$$= 29.145[25.78 - 289\ln 1.1133] = -ve\ W$$

Total exergy of six PVT-CPC air collector $\left(\dot{Ex}_T\right)$ = the rate of thermal exergy + electrical power $= 0 + 475.52 = 475.52\,W$

## 7.4  The OPVT-TEC-CPC Liquid Collector

Refer to Sect. 6.6, the concept of integration of thermoelectric cooler (TEC) with back of flexible aluminum (Al)-based photo-voltaic unit has been discussed to increase the electrical power of PV-TEC system as shown in Fig. 6.3. This has been shown by having many examples in Chap. 6. The additional electrical power from TEC, i.e., $U_{\text{tec}}\left(T_{\text{tec,top}} - T_{\text{tec,bottom}}\right)\beta_{\text{tec}}$, (Eq. 3.44) depends on temperature difference between its top $\left(T_{\text{tec,top}}\right)$ and bottom $T_{\text{tec,bottom}}$ surfaces (Eq. 6.40a). In order to further increase this difference, one has to integrate compound parabolic concentrator (CPC) on the top of flexible PVT-TEC system as shown in Fig. 7.1. In this case, the incident solar radiation on aperture of CPC is reflected back to absorber/receiver of flexible PVT-TEC liquid collector which enhances the operating temperature.

In this section, we will discuss the performance of PVT-TEC-CPC liquid collector in terms of thermal exergy and an overall total exergy. Further, a comparison will be also carried out with ($C = 2$) and without ($C = 1$) CPC.

## 7.4.1 Energy Balance of OPVT-TEC-CPC Liquid Collector

In order to write energy balance of PVT-TEC-CPC liquid collector, the same assumptions as mentioned in Sect. 6.6.1 have been adopted. It is important to mention about an assumption that the bottom of TEC is equal to absorber plate $\{T_{\text{tec,bottom}} = T_p\}$ due to direct contact and having highest thermal conductivity of both materials. The basic energy balance of OPVT-TEC-CPC liquid collector having as aperture and receiver as $A_{am}$ and $A_{rm}$ can be written in the following section.

(a) **for Solar cell of flexible fully covered with solar cell of OPV module**

Following Eqs. 6.39a and 7.1, the energy balance of PV module in W of PVT-TEC-CPC liquid collector is as follows:

$$\rho \tau_g \alpha_{sc} I_u A_{am} = U_{t,ca}(T_c - T_a)A_{rm} + h_t(T_c - T_{\text{tec,top}})\beta_{\text{tec}}A_{rm}$$
$$+ U_{b,cp}(T_c - T_p)(1 - \beta_{\text{tec}})A_{rm} + \eta_m I_u A_{rm} \qquad (7.32)$$

In the above equation, we have assumed $\eta_m$ is constant for time interval of $\Delta t = t - 0$ and equal to the value obtained at standard test condition, i.e., 0.12.

Equation 7.32 can also be considered for two cases as follows:

**Case (a): OPVT-CPC liquid collector.** In this case, $\beta_{\text{tec}} = 0$ means $h_t = 0$, then Eq. 7.32 reduced to

$$\rho \tau_g \alpha_{sc} I_u A_{am} = U_{t,ca}(T_c - T_a)A_{rm} + U_{b,cp}(T_c - T_p)(1 - \beta_{\text{tec}})A_{rm} + \eta_m I_u A_{rm} \qquad (7.33)$$

**Case (b): OPVT-TEC liquid collector:** $C = \frac{A_{am}}{A_{rm}} = 1; A_{am} = A_{rm} = A_m$ and $I_u = I(t)$, Eq. 7.32 reduces to

$$\rho \tau_g \alpha_{sc} I(t), A_m = U_{t,ca}(T_c - T_a)A_m + h_t(T_c - T_{\text{tec,top}})\beta_{\text{tec}}A_m$$
$$+ U_{b,cp}(T_c - T_p)(1 - \beta_{\text{tec}})A_m + \eta_m I(t), A_m \qquad (7.34)$$

Equation 7.34 is exactly same as Eq. 6.39a in W.

The numerical value of each design parameters of Eq. 6.39 is mostly similar to given in Sect. 3.4.

From Eq. 7.32, we have

$$T_c = \frac{\left(\rho\tau_g\alpha_{sc}\frac{A_{am}}{A_{rm}} - \eta_m\right)I_u + U_{t,ca}T_a + h_tT_{tec,top}\beta_{tec} + U_{b,cp}(1 - \beta_{tec})T_p}{U_{t,ca} + h_t\beta_{tec} + U_{b,cp}(1 - \beta_{tec})} \tag{7.35}$$

**(b)  for Top of TEC**

$$h_t\left(T_c - T_{tec,top}\right)\beta_{tec}bdx = U_{tec}\left(T_{tec,top} - T_{tec,bottom}\right)\beta_{tec}bdx \tag{7.36a}$$

With the help of Eq. 7.35, one can get

$$h_t\left(T_c - T_{tec,top}\right)\beta_{tec} = F_0(\alpha\tau)_{eff}I_u$$
$$- (U)_{tec,top-a}\left(T_{tec,top} - T_a\right) - (U)_{tec,top-p}\left(T_{tec,top} - T_p\right) \tag{7.36b}$$

where,

The penalty factor, $F_0 = \frac{(h_t\beta_{tec})}{U_{t,ca}+h_t\beta_{tec}+U_{b,cp}(1-\beta_{tec})}$.

The product of absorptivity and transmittivity, $(\alpha\tau)_{eff} = \left(\rho\tau_g\alpha_{sc}\frac{A_{am}}{A_{rm}} - \eta_m\right)$;

An overall heat transfer coefficient from top of TEC to ambient air, $(U)_{tec,top-a} = \frac{(h_t\beta_{tec})U_{t,ca}}{U_{t,ca}+h_t\beta_{tec}+U_{b,cp}(1-\beta_{tec})}$;

The expressions for above constants are exactly same as given in Eq. 6.40b.

An overall heat transfer coefficient from top of TEC to absorber plate $(U)_{tec,top-p} = \frac{(h_t\beta_{tec})\{U_{b,cp}(1-\beta_{tec})\}}{U_{t,ca}+h_t\beta_{tec}+U_{b,cp}(1-\beta_{tec})}$.

Equate Eqs. 7.36a and 7.36b with $T_{tec,bottom} = T_p$ as per assumption; we can have the following expression for top temperature of TEC as

$$T_{tec,top} = \frac{F_0(\alpha\tau)_{eff}I_u + \{(U)_{tec,top-p} + U_{tec}\beta_{tec}\}T_p + (U)_{tec,top-a}T_a}{(U)_{tec,top-a} + (U)_{tec,top-p} + U_{tec}\beta_{tec}} \tag{7.37}$$

Further, Eq. 7.37 can be rewritten as follows:

$$U_{tec}\beta_{tec}\left(T_{tec,top} - T_p\right) = F_{01}F_0(\alpha\tau)_{eff}I_u - (U)_{tec,top-pa}\left(T_p - T_a\right) \tag{7.38}$$

where,

The penalty factor, $F_{01} = \frac{U_{tec}\beta_{tec}}{[(U)_{tec,top-a}+(U)_{tec,top-p}+U_{tec}\beta_{tec}]}$.

An overall heat transfer coefficient from plate to ambient through top of TEC, $(U)_{tec,top-pa} = \frac{U_{tec}\beta_{tec}(U)_{tec,top-a}}{[(U)_{tec,top-a}+(U)_{tec,top-p}+U_{tec}\beta_{tec}]}$.

Again, Eq. 7.37, we can also have

$$(h_t\beta_{tec})T_{tec,top} = F_0'\left[F_0(\alpha\tau)_{eff}I_u + \{(U)_{tec,top-p} + U_{tec}\beta_{tec}\}T_p + (U)_{tec,top-a}T_a\right] \tag{7.39}$$

where, the penalty factor, $F_0' = \frac{(h_t\beta_{tec})}{\{(U)_{tec,top-a}+(U)_{tec,top-p}+U_{tec}\beta_{tec}\}}$

Substitute the expression for $(h_t\beta_{tec})T_{tec,top}$ from Eq. 6.42 into Eq. 6.39b as

$$T_c = \frac{\left(\rho\tau_g\alpha_{sc}\frac{A_{am}}{A_{rm}} - \eta_m\right)I_u + U_{t,ca}T_a + F_0'\left[F_0(\alpha\tau)_{eff}I(t) + \{(U)_{tec,top-p} + U_{tec}\beta_{tec}\}T_p + (U)_{tec,top-a}T_a\right] + U_{b,cp}(1 - \beta_{tec})T_p}{U_{t,ca} + h_t\beta_{tec} + U_{b,cp}(1 - \beta_{tec})}$$

and,

$$U_{b,cp}(T_c - T_p) = F_0''$$

$$\left[\left(\rho\tau_g\alpha_{sc}\frac{A_{am}}{A_{rm}} - \eta_m\right)I_u + U_{t,ca}T_a + F_0'\left[F_0(\alpha\tau)_{eff}I_u + U_pT_p + (U)_{tec,top-a}T_a\right]\right]$$

$$(7.40)$$

where, the penalty factor, $F_0'' = \frac{U_{b,cp}}{U_{t,ca} + h_t\beta_{tec} + U_{b,cp}(1 - \beta_{tec})}$

An overall net heat transfer coefficient, $U_p = \{(U)_{tec,top-p} + U_{tec}\beta_{tec} - (U_{t,ca} + h_t\beta_{tec})\}$.

***Example 7.31*** Find out constants (a) penalty factor, $F_0$, (b) product of absorptivity and transmittivity, $(\alpha\tau)_{eff}$, an overall heat transfer coefficient from top of TEC to ambient air $(U)_{tec,top-a}$ and an overall heat transfer coefficient from top of TEC to absorber plate $(U)_{tec,top-p}$ in Eq. 7.36b for concentration ratio of $\frac{A_{am}}{A_{rm}} = 2$ and $\rho = 1$ assumed.

**Solution**

From Example 3.48, we have the following parameters.

$U_{t,ca} = 9.30$ W/m$^2$ °C with wind velocity of 1 m/s; $h_t = 120$ W/m$^2$ °C and $U_{b,cp} = 2.80$ W/m$^2$ °C without wind velocity, $\beta_{tec} = 0.4324$, packing factor of TEC, $\tau_g = 0.95$, $\alpha_{sc} = 0.9$ (Eq. 3.47a, Example 3.54).

From Eq. 7.36b, we have the following expression for various constants

$$F_0 = \frac{(h_t\beta_{tec})}{U_{t,ca} + h_t\beta_{tec} + U_{b,cp}(1 - \beta_{tec})} = \frac{120 \times 0.4324}{9.30 + 120 \times 0.4324 + 2.80 \times (1 - 0.4324)}$$

$$= \frac{51.888}{91.765} = 0.565$$

$$(\alpha\tau)_{eff} = \left[\rho\tau_g\alpha_{sc}\left(\frac{A_{am}}{A_{rm}}\right) - \eta_m\right] = 1 \times 0.95 \times 0.9 \times 2 - 0.12 = 1.59$$

$$(U)_{tec,top-a} = \frac{(h_t\beta_{tec})U_{t,ca}}{U_{t,ca} + h_t\beta_{tec} + U_{b,cp}(1 - \beta_{tec})}$$

$$= \frac{(120 \times 0.4324) \times 9.30}{9.30 + 120 \times 0.4324 + 2.80 \times (1 - 0.4324)} = 5.2545 \text{ W/m}^2 \text{ °C}$$

$$(U)_{tec,top-p} = \frac{(h_t\beta_{tec})\{U_{b,tep}(1 - \beta_{ac})\}}{U_{t,ca} + h_t\beta_{tcc} + U_{b,cp}(1 - \beta_{ac})}$$

$$= \frac{(120 \times 0.4324) \times 2.80 \times (1 - 0.4324)}{9.30 + 120 \times 0.4324 + 2.80 \times (1 - 0.4324)} = \frac{82.465}{91.765} = 0.8989 \text{ W/m}^2 \text{ °C}$$

One can observe that except $(\alpha\tau)_{\mathrm{eff}} = 1.59$, all parameters, namely $F_0$, $(U)_{\mathrm{tec,top}-a}$ and $(U)_{\mathrm{tec,top}-p}$, are exactly same as obtained Example 6.33.

The other parameters, namely $F_{01}$, $(U)_{\mathrm{tec,top}-\mathrm{pa}}$, $F_0'$, and $F_0''$, will be exactly same as obtained in Examples 6.34 and 6.35. However, for convenience of readers, we are giving here as Examples 7.32 and 7.33

**Example 7.32** Evaluate the penalty factor, $F_{01}$ and an overall heat transfer coefficient from plate to ambient through top of TEC, $(U)_{\mathrm{tec,top}-\mathrm{pa}}$ for Eq. 7.38.

### Solution

We have the following known parameters:

$(U)_{\mathrm{tec.top}-a} = 5.2545$ W/m$^2$ °C and $(U)_{\mathrm{tec,\,top}-p} = 0.8989$ W/m$^2$ °C and $\beta_{\mathrm{tec}} = 0.4324$ (Example 6.33); $U_{\mathrm{tcc}} = 450$ W/m$^2$ °C (Eq. 3.48a).

From Eq. 7.38, we have

The penalty factor, $F_{01} = \dfrac{U_{\mathrm{tec}}\beta_{\mathrm{tec}}}{\left[(UA)_{\mathrm{tec,top}-a}+(UA)_{\mathrm{tec,top}-p}+U_{\mathrm{tec}}\beta_{\mathrm{tec}}\right]} = $

$\dfrac{450\times0.4324}{5.2545+0.8989+450\times0.4324} = \dfrac{194.58}{200.7334} = 0.9693.$

An overall heat transfer coefficient from plate to ambient through top of TEC,

$$(U)_{\mathrm{tec,\,top}-\mathrm{pa}} = \frac{U_{\mathrm{tec}}\,\beta_{\mathrm{tec}}\,(U)_{\mathrm{tec,top}-a}}{\left[(U)_{\mathrm{tec,top}-a} + (U)_{\mathrm{tec,top}-p} + U_{\mathrm{tec}}\beta_{\mathrm{tec}}\right]}$$

$$= \frac{(450\times0.4324)\times5.2545}{5.2545+0.8989+450\times0.4324} = 5.093 \text{ W/m °C}$$

**Example 7.33** Evaluate the penalty factors, $F_0'$, $F_0''$ and $U_p$ for Eqs. 7.39 and 7.40 by using the parameters of Examples 6.33 and 6.34.

### Solution

The penalty factor, $F_0' = \dfrac{(h_t\beta_{\mathrm{tec}})}{\left\{(U)_{\mathrm{tec,top}-a}+(U)_{\mathrm{tec,top}-p}+U_{\mathrm{tec}}\beta_{\mathrm{tec}}\right\}} = \dfrac{120\times0.4324}{5.2545+0.8989+450\times0.4324} =$

$\dfrac{51.888}{200.7334} = 0.2585.$

The penalty factor, $F_0'' = \dfrac{U_{b,\mathrm{cp}}}{U_{t,\mathrm{ca}}+h_t\beta_{\mathrm{tec}}+U_{b,\mathrm{cp}}(1-\beta_{\mathrm{tec}})} = \dfrac{2.80}{9.30+120\times0.4324+2.80\times(1-0.4324)} =$

$\dfrac{2.80}{91.765} = 0.0305.$

An overall net heat transfer coefficient,

$$U_p = \left\{(U)_{\mathrm{tec,top}-p} + U_{\mathrm{tec}}\beta_{\mathrm{tec}} - \left(U_{t,\mathrm{ca}} + h_t\beta_{\mathrm{tec}}\right)\right\} = 5.2545 + 450\times0.4324$$

$$- (9.30 + 120\times0.4324) = 5.2545 + 194.58 - 61.188 = 138.6465 \text{ W/m °C}$$

### (c)  for absorber plate

Following Eq. 6.46, the rate of net thermal energy in W available to the absorber plate of PVT-TEC-CPC liquid collector will be exactly same as given below

$$\dot{Q}_u = \left[(\alpha\tau)_{\mathrm{meff}}I_u - U_{\mathrm{Lm}}\left(T_p - T_a\right) + F_0''F_0'U_pT_p + U_aT_a\right]bdx \qquad (7.41)$$

where,  $(\alpha\tau)_{\text{meff}} = [(1 - \eta_{\text{tec}})F_{01}F_0(\alpha\tau)_{eff} + F_0''(1 - \beta_{\text{tec}})\{(\rho\tau_g\alpha_{\text{sc}} - \eta_m)$
$+F_0'F_0(\alpha\tau)_{\text{eff}}\}];$   $U_{\text{Lm}} = \{(1 - \eta_{\text{tec}})(U)_{\text{tec,top}-pa} + U_b\},$   $U_p =$
$\{(U)_{\text{tec,top}-p} + U_{\text{tec}}\beta_{\text{tec}} -(U_{t,ca} + h_t\beta_{\text{tec}})\}$   (Eq.   6.43)   and   $U_a =$
$[F_0''(U_{t,ca} + F_0'(U)_{\text{tec,top}-a})]$ with $(\alpha\tau)_{\text{eff}} = [\rho\tau_g\alpha_{\text{sc}}(\frac{A_{\text{am}}}{A_{\text{rm}}}) - \eta_m].$

Following Eq. 6.7, one can define flat plate collector efficiency $(F')$ for elemental area of 'bdx' as

$$F' = \frac{\dot{Q}_u}{\dot{Q}_u|_{T_p=T_f}}$$

or,

$$\dot{Q}_u = F'\dot{Q}_u|_{T_p=T_f} = F'[(\alpha\tau)_{\text{meff}}I_u - U_{\text{Lm}}(T_f - T_a) + F_0''F_0'U_pT_f + U_aT_a]bdx, \text{ W}$$
(7.42)

Here, $T_p = T_f = T_{\text{tec,bottom}}$

Equation 7.42 is the rate of net thermal energy in W available to the absorber plate which can be carried away if fluid flow below it.

***Example 7.34*** Evaluate the product of absorptivity and transmittivity, $(\alpha\tau)_{\text{meff}}$ and an overall heat transfer coefficient $U_{\text{Lm}}$ and $U_a$ of OPVT-TEC-CPC liquid collector for Eq. 7.41 by using the data's of Examples 6.33–6.35. The $\rho = 1$ is assumed.

**Solution**

**Know parameters**: $F_0 = 0.565,$ $F_{01} = 0.9693,$ $F_0' = 0.2585,$ $F_0'' = 0.0305,$
$(\alpha\tau)_{\text{eff}} = [\rho\tau_g\alpha_{\text{sc}}(\frac{A_{\text{am}}}{A_{\text{rm}}}) - \eta_m] = 1.59$, (Example 7.31), $\beta_{\text{tec}} = 0.4324,$ $U_b =$
$0.7 \text{ W/m}^2 \,{}^\circ\text{C}$ and $\eta_{\text{tec}} = 0.15$ (Eq. 3.45).

$(U)_{\text{tec.top - pa}} = 5.093 \text{ W/m}^2 \,{}^\circ\text{C}, ,$ $(U)_{\text{tec.pp}} - a = 5.2545 \text{ W/m}^2 \,{}^\circ\text{C}$ (Example 6.34), $U_{t,ca} = 9.30 \text{ W/m}^2 \,{}^\circ\text{C}.$

From Eq. 6.44, we have the following:

$(\alpha\tau)_{\text{meff}} = \left[(1 - \eta_{\text{tec}})F_{01}F_0(\alpha\tau)_{\text{eff}} + F_0''(1 - \beta_{\text{tec}})\left\{\left(\rho\tau_g\alpha_{\text{sc}}\frac{A_{\text{am}}}{A_{\text{rm}}} - \eta_m\right) + F_0'F_0(\alpha\tau)_{\text{eff}}\right\}\right]$

$= (1 - 0.15) \times 0.9693 \times 0.565 \times 1.59 + 0.0305 \times (1 - 0.4324) \times \{1.59 + 0.2585 \times 0.565 \times 1.59\}$

$= 0.740 + 0.0173 \times 3.326 = 0.7975$

$U_{\text{Lm}} = \{(1 - \eta_{\text{tec}})(U)_{\text{tec,top}-pa} + U_b\} = (1 - 0.15) \times 5.093 + 0.7 = 5.029 \text{ W/m}^2 \,{}^\circ\text{C}$

$$U_a = [F_0''(U_{t,ca} + F_0(U)_{\text{tec,top}-a})]$$
$$= 0.0305 \times (9.30 + 0.2585 \times 5.2545) = 0.3251 \text{ W/m}^2 \,{}^\circ\text{C}$$

Here, also except $(\alpha\tau)_{\text{meff}} = 0.7975$, the numerical values of $U_{\text{Lm}}$ and $U_a$ are exactly same as obtained in Example 3.36.

### (d)  For flowing fluid (water)

According to first law of thermodynamics, namely energy conservation, the rate of thermal energy for 'bdx' available to absorber plate, Eq. 7.42, will be carried away by flowing water below the absorber plate and it can be expressed as follows:

$$\dot{m}_f c_f \frac{dT_f}{dx} dx = F'\left[(\alpha\tau)_{\text{meff}} I_u - U_{\text{Lm}}(T_f - T_a) + F_0'' F_0' U_p T_f + U_a T_a\right] b dx$$

$$(7.43)$$

where, $F' = 0.91$ (Eq. 6.7) is tube in plate for single glazed flat plate collector efficiency (Sect. 4.3.3 and Example 4.27). It is important to note that $F'$ depends on an overall total heat transfer coefficient as shown in Eq. 6.7. It has difference value of single- and double-glazed collector as well OPVT collector. However, for comparison with various results, we will consider the value of $F' = 0.75$ (Example 6.42)

### 7.4.2  The Outlet Fluid Temperature at End of N-OPVT-TEC-CPC Liquid Collectors Connected in Series

Following Sects. 4.1 and 4.5 and by using Eq. 7.43, one can derive an expression for an outlet temperature at $N^{\text{th}}$ OPVT-TEC-CPC liquid collector as a function of x $(T_{fN}(x))$. The expression for the same is given by

$$T_{fN}(x) = \frac{\left[(\alpha\tau)_{\text{meff}} I_u + (U_{\text{Lm}} + U_a)T_a\right]}{U_{\text{Lm-eff}}}\left[1 - \exp\left\{-\frac{N F' U_{\text{Lm-eff}} b}{\dot{m}_f c_f} x\right\}\right]$$

$$+ T_{fi} \exp\left\{-\frac{N F' U_{\text{Lm-eff}} b}{\dot{m}_f c_f} x\right\}$$

$$(7.44a)$$

where, $U_{\text{Lm-eff}} = \left(U_{\text{Lm}} - F_0'' F_0' U_p\right)$.

The outlet temperature at $x = L_0$ can be obtained from Eq. 7.44 as

$$T_{foN} = \frac{\left[(\alpha\tau)_{\text{meff}} I_u + (U_{\text{Lm}} + U_a)T_a\right]}{U_{\text{Lm-eff}}}\left[1 - \exp\left\{-\frac{N F' U_{\text{Lm-eff}} b}{\dot{m}_f c_f} L_0\right\}\right]$$

$$+ T_{fi} \exp\left\{-\frac{N F' U_{\text{Lm-eff}} b}{\dot{m}_f c_f} L_0\right\}$$

$$(7.44b)$$

**Example 7.35** Calculate the outlet fluid (water) temperature at $N$th OPVT-TEC-CPC liquid collectors connected in series for $N = 1$ for the following climatic and design parameters:

Climatic parameters: $I_u = 475$ W/m$^2$, $T_a = 16\,°C$ and $T_{fi} = 15\,°C$ for $C = 2$.

Design parameters: $N = 1$, $F' = 0.75$, $(\alpha\tau)_{meff} = 0.7975$ (Example 7.34), $U_{Lm} = 5.029$ W/m$\,°C$, $U_a = 0.3251$ W/m$^2\,°C$, $A_{rm} = b \times L_0 = 1 \times 2$ m$^2$. The penalty factor, $F'_0 = 0.2585$. The penalty factor, $F''_0 = 0.0305$ (Example 7.33). An overall net heat transfer coefficient, $U_p = 138.6465$ W/m$^2\,°C$ (Examples 3.35–3.36) and $\dot{m}_f c_f = 0.035 \times 419 = 146.65$J/s$\,°C = 146.65$W/$°C$

## Solution

Now, from Eq. 6.50, we have

$$U_{\text{Lm-eff}} = \left(U_{\text{Lm}} - F''_0 F'_0 U_p\right) = 5.029 - 0.0305 \times 0.2585$$
$$\times\, 138.6465 = 3.9348 \text{ W/m}^2 \ °C$$

Further, for $N = 1$, we have

$$\exp\left\{-\frac{F'U_{\text{Lm-eff}}A_{cm}}{\dot{m}_f c_f}\right\} = \exp\left\{-\frac{0.75 \times 3.9348 \times 2}{0.035 \times 4190}\right\}$$
$$= \exp\left\{-\frac{5.9022}{146.65}\right\} = \exp\{-0.04025\} = 0.9605.$$

From Eq. 7.44b, we have an expression for the outlet temperature of fluid $\left(T_{fo1}\right)$ after substitution of appropriate value given above as

$$T_{fo1} = T_{fo} = \frac{[0.7975 \times \mathbf{475} + (5.029 + 0.3251) \times 16]}{3.9348}[1 - 0.9605] + 15$$
$$\times\, 0.9605 = \frac{464.4781}{3.9348} \times 0.0385 + 14.44 = 4.54 + 14.44 = 18.98\,°C.$$

For $N = 1$, the outlet temperature of OPVT-TEC-CPC $[T_{fo1}]$ is 2.08 $°C$ higher than single OPVT-TEC liquid collector $[T_{fo}]$, (16.91 $°C$, Example 6.37) due to the integration of CPC with concentration ratio of 2 ($C = 2$).

**Example 7.36** Calculate the outlet fluid (water) temperature at $N$th OPVT-TEC-TEC liquid collectors connected in series for $C = \frac{A_{am}}{A_{rm}} = 2$ for the following climatic and design parameters:

Climatic parameters: $I_u = 475$ W/m$^2$, $T_a = 16\,°C$ and $T_{fi} = 15\,°C$.

Design parameters: $N = 6$, $F' = 0.75$, $(\alpha\tau)_{meff} = 0.7975$ (Example 7.34) , $U_{Lm} = 5.029$ W/m$^2\,°C$, $U_a = 0.3251$ W/m$^2\,°C$, $A_{rm} = b \times L_0 = 1 \times 2$m$^2$. The penalty factor, $F'_0 = 0.2585$. The penalty factor, $F''_0 = 0.0305$ (Example 7.33). An overall net heat transfer coefficient, $U_p = 138.6465$ W/m$^2\,°$ (Examples 3.35–3.36) and $\dot{m}_f c_f = 0.035 \times 419 = 146.65$ J/s$\,°C = 146.65$ W/C

## Solution

Now from Eq. 6.50, we have

$$U_{\text{Lm-eff}} = \left(U_{\text{Lm}} - F_0'' F_0' U_p\right) = 5.029 - 0.0305 \times 0.2585$$
$$\times 138.6465 = 3.9348 \text{ W/m}^2 \,{}^\circ\text{C}$$

Further,

$$\exp\left\{-\frac{N F' U_{\text{Lm-eff}} A_{\text{rm}}}{\dot{m}_f c_f}\right\} = \exp\left\{-\frac{6 \times 0.75 \times 3.9348 \times 2}{0.035 \times 4190}\right\}$$
$$= \exp\left\{-\frac{35.4132}{146.65}\right\} = \exp\{-0.2415\} = 0.7854.$$

From Eq. 7.44b, we have an expression for the outlet temperature of fluid $(T_{\text{foN}})$ of OPVT-TEC-CPC liquid collector after substitution of appropriate value given above as

$$T_{\text{fo6}} = \frac{[0.7975 \times 475 + (5.029 + 0.3251) \times 16]}{3.9348}[1 - 0.7854] + 15 \times 0.7854$$
$$= \frac{464.4781}{3.9348} \times 0.2146 + 11.781 = 25.33 + 11.781 = 37.11 \,{}^\circ\text{C}.$$

The above temperature is increased significantly by 18.13 °C in comparison with one OPVT-TEC-CPC liquid collector for N is increased from 1 to 6 due to series connection.

### 7.4.3   The Average Temperature of Fluid for N-OPVT-TEC-CPC Liquid Collectors Connected in Series

Further, an average fluid temperature $(\overline{T}_{\text{fN}})$ of N-PVT-TEC-CPC liquid collector connected in series can also be determined as

$$\overline{T}_{\text{fN}} = \frac{1}{L_0} \int_0^{L_0} T_{\text{fN}}(x)\mathrm{d}x = \frac{\left[(\alpha\tau)_{\text{meff}} I(t) + (U_{\text{Lm}} + U_a)T_a\right]}{U_{\text{Lm-eff}}}$$

$$\left[1 - \frac{1 - \exp\left\{-\frac{N F' U_{\text{Lm-eff}} b}{\dot{m}_f c_f} L_0\right\}}{\frac{N F' U_{\text{Lm-eff}} b}{\dot{m}_f c_f} L_0}\right] + T_{\text{fi}} \frac{1 - \exp\left\{-\frac{N F' U_{\text{Lm-eff}} b}{\dot{m}_f c_f} L_0\right\}}{\frac{N F' U_{\text{Lm-eff}} b}{\dot{m}_f c_f} L_0} \tag{7.45}$$

**Example 7.37** Evaluate an average fluid temperature for $N = 1$ of OPVT-TEC-CPC liquid collector for Example 7.35 by using Eq. 7.45 for $C = 2$.

## Solution

**Given and known:** $U_{\text{Lm-eff}}$ = 3.9348 W/m² °C,
$\exp\left\{-\dfrac{F'U_{\text{Lm-eff}}A_{\text{rm}}}{\dot{m}_f c_f}\right\}$ = 0.9605, $\dfrac{F'U_{\text{Lm-eff}}A_{\text{rm}}}{\dot{m}_f c_f}$ = 0.04025 and

$\dfrac{[(\alpha\tau)_{\text{meff}} I(t) + (U_{\text{Lm}} + U_a)T_a]}{U_{\text{Lm-eff}}}$ (Example 7.35).

$= \dfrac{[0.7975 \times 475 + (5.029 + 0.3251) \times 16]}{3.9348} = \dfrac{464.4781}{3.9348}$

Substitute the above appropriate value in Eq. 7.45, one gets

$$\overline{T}_{f1} = \frac{464.4781}{3.9348}\left[1 - \frac{1 - 0.9605}{0.04025}\right] + 15 \times \frac{1 - 0.9605}{0.04025}$$
$$= 118.0436 \times 0.0186 + 14.72 = 16.92\,°\text{C}.$$

The 16.92 °C can also be obtained by considering the average value of 18.98 °C (Example 7.35) and 15 °C.

**Example 7.38** Calculate an average fluid temperature for N-OPVT-TEC-CPC liquid collectors connected in series for $C = 2$ for Example 7.37 by using Eq. 7.45.

## Solution

**Given and known:** $N = 6$, $U_{\text{Lm-eff}} = 3.9348$ W/m² °C, $\exp\left\{-\dfrac{NF'U_{\text{Lm-eff}}A_{\text{rm}}}{\dot{m}_f c_f}\right\}$

$= \exp\left\{-\dfrac{6 \times 0.75 \times 3.9348 \times 2}{0.035 \times 4190}\right\} = \exp\left\{-\dfrac{35.4132}{146.65}\right\} = \exp\{-0.2415\} = 0.7854$ and

$\dfrac{6F'U_{\text{Lm-eff}}A_{\text{rm}}}{\dot{m}_f c_f} = 0.2415$ and $\dfrac{[(\alpha\tau)_{\text{meff}} I(t)+(U_{\text{Lm}}+U_a)T_a]}{U_{\text{Lm-eff}}} = \dfrac{[0.7975 \times 475 + (5.029 + 0.3251) \times 16]}{3.9348} =$
$\dfrac{464.4781}{3.9348}$ (Example 7.35).

Substitute the above appropriate value in Eq. 7.45, one gets

$$\overline{T}_{f6} = \frac{464.4781}{3.9348}\left[1 - \frac{1 - 0.7854}{0.2415}\right] + 15 \times \frac{1 - 0.7854}{0.2415}$$
$$= 118.0436 \times (1 - 0.8886) + 13.33 = 26.48\,°\text{C}$$

### 7.4.4 The Average Temperature of Top Surface of TEC for N-OPVT-TEC-CPC Liquid Collectors Connected in Series

Since, the fluid, absorber plate and bottom of TEC is in direct contact, hence $\overline{T}_{fN} = \overline{T}_p = \overline{T}_{tec,bottom}$ due to high thermal conductivity. After knowing, an average fluid temperature $(\overline{T}_{fN} = \overline{T}_{pN})$ of N-OPVT-TEC-CPC liquid collector connected in series, one can find out an average top surface of TEC from Eq. 7.37 as

$$\overline{T}_{\text{tec,top}N} = \frac{F_0(\alpha\tau)_{\text{eff}}I_u + \{(U)_{\text{tec,top}-p} + U_{\text{tec}}\beta_{\text{tec}}\}\overline{T}_p + (U)_{\text{tec,top}-a}\overline{T}_a}{(U)_{\text{tec,top}-a} + (U)_{\text{tec,top}-p} + U_{\text{tec}}\beta_{\text{tec}}} \qquad (7.46)$$

**Example 7.39** Calculate an average top surface of TEC for single OPVT-TEC-CPC liquid collectors $C = 1$ for Example 7.37 by using Eq. 7.46 for $N = 1$.

**Solution**

**Known constants:** $\overline{T}_{f1} = \overline{T}_{p1} = 16.92\,°C$ (Example 7.38), $F_0 = 0.565$, $(\alpha\tau)_{\text{eff}} = 1.59$, $\beta_{\text{tec}} = 0.4324$, $(U)_{\text{tec,top}-p} = 0.8989\,\text{W/m}^2\,°C$, $(U)_{\text{tec,top}-a} = 5.2545\,\text{W/m}^2\,°C$ (Example 7.31), $U_{\text{tcc}} = 450\,\text{W/m}^2\,°C$ (Eq. 3.48a).

Substitute the above values in Eq. 7.46 to get $\overline{T}_{\text{tec,top1}}$

$$\overline{T}_{\text{tec,top 1}} = \frac{0.565 \times 1.59 \times 475 + \{0.8989 + 450 \times 0.4324\} \times 16.92 + 5.2545 \times 16}{5.2545 + 0.8989 + 450 \times 0.4324}$$

$$= \frac{402.56 + 3,307.50 + 84.07}{200.73} = 18.9\,°C$$

**Example 7.40** Calculate an average top surface of TEC for N-OPVT-TEC-CPC liquid collectors connected in series for $C = 2$ for Example 7.38 by using Eq. 7.46 for $N = 6$.

**Solution**

**Known constants:** $\overline{T}_{f6} = \overline{T}_{p6} = 26.48\,°C$ (Example 7.38), $F_0 = 0.565$, $(\alpha\tau)_{\text{eff}} = 1.59$, $\beta_{\text{tec}} = 0.4324$, $(U)_{\text{tec,top}-p} = 0.8989\,\text{W/m}^2\,°C$, $(U)_{\text{tec,top}-a} = 5.2545\,\text{W/m}^2\,°C$ (Example 7.31), $U_{\text{tcc}} = 450\,\text{W/m}^2\,°C$ (Eq. 3.48a).

Substitute the above values in Eq. 7.46 to get $\overline{T}_{\text{tec,top6}}$, one gets

$$\overline{T}_{\text{tec,top6}} = \frac{0.565 \times 1.59 \times 475 + \{0.8989 + 450 \times 0.4324\} \times 26.48 + 5.2545 \times 16}{5.2545 + 0.8989 + 450 \times 0.4324}$$

$$= \frac{402.56 + 5176.28 + 84.07}{200.73} = 28.21\,°C.$$

### 7.4.5  The Average Temperature of Solar Cell for N-OPVT-TEC-CPC Liquid Collectors Connected in Series

Further, for known $\overline{T}_{\text{tec,top}}$, an average value of solar cell temperature of N-OPVT-TEC-CPC liquid collector connected in series can be determined from Eq. 7.35 as

$$\overline{T}_{cN} = \frac{\left(\tau_g \alpha_{sc} \frac{A_{am}}{A_{rm}} - \eta_m\right) I_u + U_{t,ca} T_a + h_t T_{tec,topN} \beta_{tec} + U_{b,cp}(1 - \beta_{tec})\overline{T}_{pN}}{U_{t,ca} + h_t \beta_{tec} + U_{b,cp}(1 - \beta_{tec})}$$

(7.47)

***Example 7.41*** Evaluate an average solar cell temperature for $N = 1$ for OPVT-TEC-CPC collector for design parameters of Example 7.39 by using Eq. 7.47.

**Solution**

**Given and Known parameters**: $N = 1$, $\overline{T}_{toc,top1} = 18.9\,°C$, (Example 7.39), $\overline{T}_{f1} = \overline{T}_{p1} = 16.92\,°C$, (Example 7.37), $= \left[\tau_g \alpha_{sc}\left(\frac{A_{am}}{A_{rm}}\right) - \eta_m\right] = 1.59$, $U_{t,ca} = 9.30\,W/m^2\,°C$ with wind velocity of 1 m/s; $h = 120\,W/m^2\,°C$ and $U_{b_b,cp} = 2.80\,W/m^2\,°C$ without wind velocity, $\beta_{tec} = 0.4324$, packing factor of TEC, $\tau_g = 0.95$, $\alpha_{sc} = 0.9$, $I_u = 475\,W/m^2$ (Eq. 3.47a, Example 3.54) (Example 7.31),

From Eq. 7.47, we get after substitution of above value for $\overline{T}_{c1}$ as

$$\overline{T}_{cN} = \frac{1.59 \times 475 + 9.12 \times 16 + 120 \times 18.9 \times 0.4324 + 2.80 \times (1 - 0.4324) \times 16.92}{9.12 + 120 \times \times 0.4324 + 2.80(1 - 0.4324)}$$

$$= \frac{755.25 + 148.8 + 980.68 + 26.89}{62.7773} = \frac{1911.62}{62.7773} = 30.45\,°C$$

***Example 7.42*** Evaluate an average solar cell temperature for $N = 6$ for 6-OPVT-TEC-CPC collectors connected in series for design parameters of Example 7.41 by using Eq. 7.47.

**Solution**

**Given and Known parameters**: $N = 6$, $\overline{T}_{tec,top\,6} = 28.21\,°C$ (Example 7.40), $\overline{T}_{f6} = \overline{T}_{p6} = 26.48\,°C$, (Example 7.38), $= \left[\tau_g \alpha_{sc}\left(\frac{A_{am}}{A_{rm}}\right) - \eta_m\right] = 1.59$, $U_{t,ca} = 9.30\,W/m^2\,°C$ with wind velocity of 1 m/s; $h_t = 120\,W/m^2\,°C$ and $U_{b,cp} = 2.80\,W/m^2\,°C$ without wind velocity, $\beta_{tec} = 0.4324$, packing factor of TEC, $\tau_g = 0.95$, $\alpha_{sc} = 0.9$, $I_u = 475\,W/m^2$ (Eq. 3.47a, Example 3.54) (Example 7.31).

From Eq. 7.47, we get after substitution of above value for $\overline{T}_{c1}$ as

$$T_{c6} = \frac{1.59 \times 475 + 9.30 \times 16 + 120 \times 28.21 \times 0.4324 + 2.80 \times (1 - 0.4324) \times 26.48}{9.30 + 120 \times \times 0.4324 + 2.80(1 - 0.4324)}$$

$$= \frac{755.25 + 148.8 + 1563.76 + 42.084}{62.7773} = \frac{2509.89}{62.7773} = 39.98\,°C$$

### 7.4.6  An Electrical Efficiency and Electrical Power for N-OPVT-TEC-CPC Liquid Collectors Connected in Series

Now, one can determine the average electrical efficiency and electrical power (Eq. 6.17) from N-PVT-TEC-CPC liquid collector connected in series as

$$\overline{\eta}_{mN} = \eta_{m0}\left[1 - \beta_0\left(\overline{T}_{cN} - 25\right)\right] \tag{7.48a}$$

and,

$$\dot{E}_{elN} = \overline{\eta}_m \times I_u \times \beta_c \times A_m \times N \tag{7.48b}$$

where, an expression for $\overline{T}_{cN}$ is given by Eq. 7.47.

**Example 7.43**  Evaluate an electrical efficiency and electrical power for single ($N = 1$) OPVT-TEC-CPC liquid collector for Example 7.41 for $C = 2$.

**Solution**

**Given parameters:** N $= 1$, $\overline{T}_{c1} = 30.45\,°C$, $I_u = 475\ W/m^2$ (Example 7.41), $\eta_{m0} = 0.12$.

From Eq. 7.48a, one gets the values of $\overline{\eta}_{m1}$ for N $= 1$ for given above parameters as

$\overline{\eta}_{m1} = \eta_{m0}\left[1 - \beta_0\left(\overline{T}_{c1} - 25\right)\right] = 0.12[1 - 0.00425(30.45 - 25)] = 0.1172\,(11.72\%)$ and,

$\dot{E}_{el1} = 0.1172 \times 475 \times 0.4324 \times 2 \times 1 = 48.15\ W$.

There is a drop of $\frac{106 - 48.15}{106} = 0.5457(54.57\%)$ in PV electrical power due to increase in operating temperature from 22.21 to 30.45 °C.

**Example 7.44**  Repeat Example 7.43 for N-6 OPVT-TEC-CPC collectors connected in series for $C = 2$.

**Solution**

**Given parameters:** N $= 6$, $\overline{T}_{c6} = 39.98\,°C$, $I_u = 475\ W/m^2$ (Example 7.41), $\eta_{m0} = 0.12$.

From Eq. 7.48a, one gets the values of $\overline{\eta}_{m1}$ for $N = 1$ for given above parameters as

$\overline{\eta}_{m6} = \eta_{m0}\left[1 - \beta_0\left(\overline{T}_{c6} - 25\right)\right] = 0.12[1 - 0.00425(39.98 - 25)] = 0.1124\,(11.24\%)$ and,

$\dot{E}_{el6} = 0.1124 \times 475 \times 0.4324 \times 2 \times 6 = 277.03\ W$.

### 7.4.7 An Electrical Efficiency and Electrical Power of N-OPVT-TEC-CPC Liquid Collectors Connected in Series

**Example 7.45** Evaluate the rate of thermal exergy of single OPVT-TEC-CPC liquid (water) collector for Example 7.35 for $N = 1$.

**Solution**

Given: $N = 1$, $T_{fol} = 18.98\,°C$ and $T_{fi} = 15\,°C$ (Example 7.35) and $\dot{m}_f C_f = 0.035 \times 4190 = 146.65\,W/°C$.

Following Eq. 7.31, the rate of thermal exergy for single OPVT-TEC-CPC collector has been evaluated as

$$\dot{Ex}_{u1,th} = \dot{m}_f C_f \left[ (T_{fol} - T_{fi}) - (T_a + 273) \ln \frac{T_{fol} + 273}{T_{fi} + 273} \right]$$

$$= 146.65 \left[ (18.98 - 15) - (16 + 273) \ln \frac{18.98 + 273}{15 + 273} \right]$$

$$= 146.65 \times [3.98 - 3.966] = 1.98\,W$$

In this case, the rate of thermal exergy becomes + ve (1.98 W) due to increase outlet fluid temperature from 16.91 °C (Example 6.37) to 18.98 °C (Example 7.35).

**Example 7.46** Evaluate the rate of thermal exergy of 6-OPVT-TEC-CPC collectors connected in series for Example 7.36.

**Solution**

Given: $N = 6$, $T_{fo6} = 37.11\,°C$ and $T_{fi} = 15\,°C$ (Example 7.36).

Following Eq. 7.31, the rate of thermal exergy for 6-PVT-TEC-CPC collectors connected in series has been evaluated as

$$\dot{Ex}_{u6,th} = \dot{m}_f C_f \left[ (T_{fo6} - T_{fi}) - (T_a + 273) \ln \frac{T_{fo6} + 273}{T_{fi} + 273} \right]$$

$$= 146.65 \left[ (37.11 - 15) - (16 + 273) \ln \frac{37.11 + 273}{15 + 273} \right]$$

$$= 146.65 \times [22.11 - 21.376] = 107.61\,W$$

### 7.4.8 The Electrical Power from Thermoelectric Cooler (TEC) of N-OPVT-TEC-CPC Liquid Collectors Connected in Series

From Example 6.40, an electrical power from TEC of N-OPVT-TEC-CPC liquid collector can be evaluated as

$$\dot{E}_{\text{TECN}} = \eta_{\text{tec}} U_{\text{tec}} \left( \overline{T}_{\text{tec,top}} - \overline{T}_{\text{tec,bottom}} \right) \beta_{\text{tec}} A_m \times N.$$

In the above equation $\overline{T}_{\text{tec,bottom}} = \overline{T}_{\text{pN}} = \overline{T}_{\text{fN}}$, so above equation becomes as

$$\dot{E}_{\text{TECN}} = \eta_{\text{tec}} U_{\text{tec}} \left( \overline{T}_{\text{tec,topN}} - \overline{T}_{\text{fN}} \right) \beta_{\text{tec}} A_m \times N \tag{7.49}$$

where, $\beta_{\text{tec}} = 0.4324$ and $\eta_{\text{tec}} = 0.15$ (Example 6.33), $U_{\text{tex}} = 450 \text{ W/m}^2 \text{ °C}$ (Eq. 3.48a)

**Example 7.47** Evaluate electrical power from TEC of N-OPVT-TEC-CPC liquid collectors connected in series for $N = 1$ and 6, respectively, for $C = 2$.

**Solution**

**For N = 1, Known:** $\overline{T}_{\text{tec, top1}} = 18.9 \text{ °C}$ (Example 7.39), $\overline{T}_{f1} = 16.92 \text{ °C}$ (Example 7.37).

From Eq. 7.49, we have

$$\dot{E}_{\text{TEC1}} = \eta_{\text{tec}} U_{\text{tec}} \left( \overline{T}_{\text{tec,top1}} - \overline{T}_{f1} \right) \beta_{\text{tec}} A_m \times N$$
$$= 0.15 \times 450 \times (18.9 - 16.92) \times 2 \times 0.4324 \times 1 = 115.58 \text{ W}$$

There is enhancement of about $\frac{115.58-30.18}{30.18} = 2.93$ (293%) in electric power due to CPC in comparison without CPC (30.18 W, Example 6.40).

**For N = 6, Known:** $\overline{T}_{\text{tec, top6}} = 28.21 \text{ °C}$ (Example 7.40), $\overline{T}_{f6} = 26.48 \text{ °C}$ (Example 7.38).

From Eq. 7.39, we have

$$\dot{E}_{TEC6} = \eta_{\text{tec}} U_{\text{tec}} \left( \overline{T}_{\text{tec,top6}} - \overline{T}_{f6} \right) \beta_{\text{tec}} A_m \times N$$
$$= 0.15 \times 450 \times (28.21 - 26.48) \times 2 \times 0.4324 \times 6 = 605 \text{ W}$$

### 7.4.9 An Overall Exergy (Total) of N-OPVT-TEC-CPC Liquid Collector Connected in Series

An overall exergy (total) of N-PVT-TEC-CPC liquid collectors connected in series is sum of the rate of thermal exergy, an electrical power from PV and TEC of the system.

**Example 7.48** Evaluate an overall exergy (total) for single and six flexible PVT-TEC-CPC collectors connected in series for $N = 1$ and 6 for $C = 2$.

## Solution

**For N = 1, Known**: $\dot{E}x_{ul, th} = 1.98$ W(Example 7.45), $\dot{E}_{el1} = 48.15$ W (Example 7.43) and $\dot{E}_{TEC1} = 115.58$ (Example 7.46).

Total exergy of single flexible PVT-TEC-CPC water collector $(Ex_T)$ = The rate of thermal exergy + Electrical power from PV + Electrical power from TEC = $1.98 + 48.15 + 115.58 = 165.63$ W.

**For N = 6, Known**: $\dot{E}x_{ul, th} = 107.61$ W (Example 7.46), $\dot{E}_{el6} = 277.03$ W (Example 7.44) and $\dot{E}_{TEC6} = 605$ W (Example 7.46).

Total exergy of six flexible PVT-TEC-CPC water collector $(\dot{E}x_T)$ = The rate of thermal exergy + Electrical power from PV + Electrical power from TEC = $107.61 + 277.03$ W $+ 605$ W $= 989.64$ W.

### 7.4.9.1 Analytical Optimization of Mass Flow Rate and Number of OPVT-TEC-CPC Collectors Connected in Series

Based on the above examples, one can observe that the mass flow rate $(\dot{m}_f)$ and number (N) of PVT-TEC-CPC collectors play an important role and it becomes necessary to find out analytical method to optimize the outlet fluid temperature. We will discuss this issue in the following examples.

Following Sect. 4.6.1 and with the help of Eq. 7.44b, we have an analytical expression for the outlet temperature at $N^{th}$ OPVT-TEC-CPC collector. If $T_{foN} = T_{f00}$, then we can rewrite Eq. 7.44b as

$$T_{foN} = T_{f00} = \frac{\left[(\alpha\tau)_{meff}I_u + (U_{Lm} + U_a)T_a\right]}{U_{Lm-eff}}\left[1 - \exp\left\{-\frac{NF'U_{Lm-eff}b}{\dot{m}_fc_f}L_0\right\}\right]$$
$$+ T_{fi}\exp\left\{-\frac{NF'U_{Lm-eff}b}{\dot{m}_fc_f}L_0\right\}$$

or,

$$\left\{T_{f00} - \frac{\left[(\alpha\tau)_{meff}I_u + (U_{Lm} + U_a)T_a\right]}{U_{Lm-eff}}\right\}$$
$$= \left\{T_{fi} - \frac{\left[(\alpha\tau)_{meff}I_u + (U_{Lm} + U_a)T_a\right]}{U_{Lm-eff}}\right\}\exp\left\{-\frac{NF'U_{Lm-eff}b}{\dot{m}_fc_f}L_0\right\}$$

or,

$$\exp\left\{-\frac{NF'U_{Lm-eff}b}{\dot{m}_fc_f}L_0\right\} = \left[\frac{\left\{T_{f00} - \frac{(\alpha\tau)_{meff}I_u+(U_{Lm}+U_a)T_a}{U_{Lm-eff}}\right\}}{\left\{T_{fi} - \frac{(\alpha\tau)_{meff}I_u+(U_{Lm}+U_a)T_a}{U_{Lm-eff}}\right\}}\right]$$

After considering log of both side, one can have an expression of $\dot{m}_f$ as

$$\dot{m}_f(t) = \frac{F'N(bL_0)U_{\text{Lm-eff}}}{C_f}\left[ln\left\{\frac{\frac{[(\alpha\tau)_{\text{meff}}I_u+(U_{\text{Lm}}+U_a)T_a]}{U_{\text{Lm-eff}}} - T_{\text{fi}}}{\frac{[(\alpha\tau)_{\text{meff}}I_u+(U_{\text{Lm}}+U_a)T_a]}{U_{\text{Lm-eff}}} - T_{f00}}\right\}\right]^{-1} \quad (7.50)$$

For a given mass flow rate $(\dot{m}_f)$ and constant working fluid temperature $(T_{f00})$, the number of OPVT-TEC-CPC collectors (N) connected in series can also be found from Eq. 7.50 as

$$N = \frac{\dot{m}_f c_f}{F'(bL_0)U_{\text{Lm-eff}}}ln\left\{\frac{\frac{[(\alpha\tau)_{\text{meff}}I_u+(U_{\text{Lm}}+U_a)T_a]}{U_{\text{Lm-eff}}} - T_{\text{fi}}}{\frac{[(\alpha\tau)_{\text{meff}}I_u+(U_{\text{Lm}}+U_a)T_a]}{U_{\text{Lm-eff}}} - T_{f00}}\right\} \quad (7.51)$$

***Example 7.49*** Find out the mass flow rate by using Eq. 7.50 for a given constant outlet temperature, i.e., $T_{f6N} = T_{f00} = 25\,°C$ for $N = 1$ for the data's of Example 7.35

**Solution**

**From Example** 7.35, we have
$$\frac{[(\alpha\tau)_{\text{meff}}I_u+(U_{\text{Lm}}+U_a)T_a]}{U_{\text{Lm-eff}}} = \frac{[0.7975\times475+(5.029+0.3251)\times16]}{3.9348} = \frac{464.4781}{3.9348} = 118.04,$$
$U_{\text{Lm-eff}} = 3.9348\frac{W}{m^2}\,°C$, $T_{\text{fi}} = 25\,°C$ and $\frac{F'U_{\text{Lm-eff}}A_{\text{cm}}}{c_f} = \frac{5.9022}{4190} = 0.00141.$
Substitute the above value in Eq. 7.50 with $N = 1$ and $A_m = bL_0 = 2\,m^2$, we get

$$\dot{m}_f(t) = 0.00141\left[ln\left\{\frac{118.04 - 15}{118.04 - 25}\right\}\right]^{-1} = 0.00141 \times [0.1021]^{-1} = 0.0137\,kg/s$$

***Example 7.50*** Repeat Example 7.49 for $N = 6$ by using Eq. 7.50 for the data's of Examples 7.36 and 7.49.

**Solution**

From Example 7.36, we have
$$\frac{[(\alpha\tau)_{\text{meff}}I_u+(U_{\text{Lm}}+U_a)T_a]}{U_{\text{Lm-eff}}} = \frac{[0.7975\times475+(5.029+0.3251)\times16]}{3.9348} = \frac{464.4781}{3.9348} = 118.04,$$
$U_{\text{Lm-eff}} = 3.9348\,W/m^2\,°C$, $T_{\text{fi}} = 25\,°C$ and $\frac{NF'U_{\text{Lm-eff}}A_{\text{cm}}}{c_f} = \frac{35.4132}{4190} = 0.008452.$
Substitute the above value in Eq. 7.50 with $N = 1$ and $A_m = bL_0 = 2\,m^2$, we get

$$\dot{m}_f(t) = 0.008452\left[ln\left\{\frac{118.04 - 15}{118.04 - 25}\right\}\right]^{-1} = 0.008452 \times [0.1021]^{-1} = 0.0826\,kg/s$$

***Example 7.51*** Estimate the number of OPVT-TEC-CPC liquid collator (N) for design parameters of Example 7.49 and mass flow rate $(\dot{m}_f)$ of 0.035 kg/s.

## Solution

Given: $\dfrac{\left[(\alpha\tau)_{\mathbf{meff}}I_u+(U_{\mathbf{Lm}}+U_a)T_a\right]}{U_{\mathbf{Lm-eff}}} = 118.04$, $\dfrac{\dot{m}_f c_f}{F'(bL_0)U_{\mathbf{Lm-eff}}} = \dfrac{0.035\times4190}{5.9022} = \dfrac{146.65}{5.9022} = 24.85$

$$N = 24.85 \times \ln\left\{\frac{118.04 - 15}{118.04 - 25}\right\} = 24.85 \times 0.1021 = 2.54 \approx 3 \qquad (7.52)$$

# References

1. Tiwari, G. N., Meraj, Md., Khan, M. E., Mishra, R. K., & Garg, V. (2018). Improved Hottel-Whillier-Bliss equation for N-photovoltaic thermal-compound parabolic concentrator (N-PVT-CPC) collector. *Solar Energy 166*, 203–212.
2. Prapas, D. E., Norton, B., & Probert, S. D. (1987). Thermal design of compound parabolic concentrating solar-energy collectors. *ASME Journal of Solar Energy Engineering, 109*, 161–168.

# Chapter 8
# Fresnel Lens (FL) Concentrated PVT (FL-PVT) Liquid Collector

## 8.1 Introduction

It has been observed that the self-sustained collectors are as follows:

(a) Photo-voltaic thermal (PVT) liquid/air collectors (Chap. 6).
(b) Photo-voltaic thermal and thermoelectric cooler (PVT-TEC) collector (Chap. 7).

Their operating thermal temperature range is lower than conventional liquid (Chap. 4) and air (Chap. 5) collectors which can be used for various low operating solar devices, namely water/air heating in swimming pool, biogas plant and space heating, active solar drying, etc.

Further, it is to be noted that the regular cleaning of compound parabolic concentrator (CPC) cum photo-voltaic thermal (PVT) liquid/air collectors is not practical, and it demands more maintenance cost which affects the feasibility to install for low capacity for thermal energy and electrical power production.

In view of this, a Fresnel lens (FL)-integrated PVT liquid collector can be an option which may require only cleaning the top surface of Fresnel lens (FL). The whole system will be referred as FL-PVT liquid collector. In this case, one can use other liquids which have boiling temperature more than water of 100 °C. We can also use a mixture of water and nanofluid as a working fluid for best heat transfer (Chap. 2, Sect. 2.4.4).

For comparison between solar thermal (ST), photo-voltaic thermal (PVT), PVT-CPC and FL-PVT, the most important parameter is thermal efficiency, and hence, thermal efficiency will be discussed in coming sections.

© Bag Energy Research Society 2024

479

G. N. Tiwari, *Fundamental of Mathematical Tools for Thermal Modeling of Solar Thermal and Photo-voltaic Systems-Part-I*,
https://doi.org/10.1007/978-981-99-7085-8_8

## 8.2  Optical and Thermal Efficiencies

Basically, there are two types of thermal efficiency, namely instantaneous efficiency and overall efficiency. These can be defined as follows:

### 8.2.1  Instantaneous Efficiency

It is a ratio of the rate of useful energy to the rate of incident solar radiation. It is further classified as optical and thermal efficiencies which are as follows:

(i)  **Optical instantaneous efficiency** $\left(\eta_0\right)$: It is mathematically defined as follows:

$$\eta_0 = \frac{\dot{Q}_{u,\text{th}}}{\gamma\left(\alpha\tau\right)I_b A_{\text{am}}} = \frac{\dot{q}_{u,\text{th}}}{\gamma\left(\alpha\tau\right)I_b}, \text{ in fraction,} \qquad (8.1a)$$

where, $\gamma$ is intercept factor, generally considered as one, the $(\alpha\tau)$ is a product of absorptivity of absorber plate and transmitivity of transparent glass, $A_{\text{am}}$ is aperture area of concentrator, the $\dot{Q}_{u,\text{th}}$ is the rate of useful thermal energy in W and $\dot{q}_{u,\text{th}} = \frac{\dot{Q}_{u,\text{th}}}{A_{\text{am}}}$ is the rate of useful thermal energy in W/m$^2$.

If intercept factor $(\gamma)$ is considered as one, then

$$\eta_0 = \frac{\dot{Q}_{u,\text{th}}}{(\alpha\tau)I_b A_{\text{am}}} = \frac{\dot{q}_{u,\text{th}}}{(\alpha\tau)I_b}, \text{ in fraction,} \qquad (8.1b)$$

In the case of without concentrator, the beam radiation is replaced by $I(t)$, the total incident radiation including diffuse radiation.

(ii)  **Thermal instantaneous efficiency** $\left(\eta_{\text{th}}\right)$: It is mathematically defined as follows:

$$\eta_{\text{th}} = \frac{\dot{Q}_{u,\text{th}}}{\gamma\, I_b A_{\text{am}}} = \frac{\dot{q}_{u,\text{th}}}{\gamma\, I_b}, \text{ in fraction,} \qquad (8.2a)$$

If intercept factor $(\gamma)$ is considered as one, then

$$\eta_{\text{th}} = \frac{\dot{Q}_{u,\text{th}}}{I_b A_{\text{am}}} = \frac{\dot{q}_{u,\text{th}}}{I_b}, \text{ in fraction,} \qquad (8.2b)$$

The optical instantaneous efficiency $(\eta_0)$ is always higher than thermal instantaneous efficiency $(\eta_{\text{th}})$ due to lower value of denominator.

## 8.2.2 An Overall Efficiency

It is a ratio of daily, monthly and yearly useful thermal energies to daily, monthly and yearly incident solar radiations. It is based on thermal energy in kW.

Mathematically, it defined as follows:

$$\eta_{th,ov} = \frac{[Q_{u,th}]_{\text{daily, monthly and yearly}}}{[I_b A_{am}]_{\text{daily, monthly and yearly}}}, \text{ in fraction,} \qquad (8.3)$$

Equations 8.1–8.3 are valid for all solar thermal (ST) (Chaps. 4 and 5), photo-voltaic thermal (PVT) (Chap. 7) with following limitations:

(a) For solar thermal (ST) (Chaps. 4 and 5), photo-voltaic thermal (PVT) (Chap. 7), the beam radiation $(I_b)$ is replaced by total beam, diffuse and reflected solar radiations given by Eq. 1.23 for inclined collector.
(b) For PVT-CPC (Chap. 7), the expression/value of $I_b$ will be replaced by $(I_u)$ (Example 7.1).

## 8.2.3 An Overall Thermal Efficiency of PVT Collector

In this case, one gets thermal based on first law of thermodynamics and electrical power based on second law of thermodynamics. In order to get an overall thermal efficiency of PVT system, one has to convert electrical power into equivalent thermal power by dividing it by conversion factor $(\gamma_c)$ of fossil fuel-based thermal power plant. Its values vary between 0.22 and 0.35. It depends upon quality of coal. The coal is most polluting fossil fuel.

(i) **An instantaneous overall thermal efficiency**: It is mathematically defined as follows:

$$\eta_{i,ov-th} = \frac{\left[\dot{Q}_{u,th} + \frac{E_{el}}{\gamma_c}\right]}{\text{Solar radiation in } W \times A_m} \qquad (8.4)$$

(ii) **An overall thermal efficiency**: In this case, it is defined as follows:

$$\eta_{ov-th,} = \frac{[Q_{u,th}]_{\text{daily, monthly and yearly}} + \frac{[E_{el,}]_{\text{daily, monthy and yearly}}}{\gamma_c}}{\left[\text{daily, monthly and yearly } A_{am}\right]_{\text{daily, monthly and yearly}}}, \text{ in fraction,} \qquad (8.5)$$

where, the solar radiation $[I(t)]$ in Eqs. 8.4 and 8.5 will be different for solar thermal (ST), photo-voltaic thermal (PVT), PVT-CPC and FL-PVT as discussed above.

### 8.2.4  An Overall Thermal Efficiency of PVT-TEC Collector

(i)  **An instantaneous overall thermal efficiency**: It is mathematically defined as
follows:

$$\eta_{i,\text{ov}-\text{th}} = \frac{\left[\dot{Q}_{u,\text{th}} + \frac{\dot{E}_{\text{el}} + \dot{E}_{\text{TEC}}}{\gamma_c}\right]}{\text{Solar radiation in } W \times A_m} \tag{8.6}$$

where, $\dot{E}_{\text{TEC}}$ is electrical power from TEC of PVT-TEC liquid collector.

(ii)  **An overall thermal efficiency**: In this case, it is defined as follows:

$$\eta_{\text{ov}-\text{th},} = \frac{\left[Q_{u,\text{th}}\right]_{\text{daily, monthly and yearly}} + \frac{\left[\dot{E}_{\text{el,}}\right]_{\text{daily, monthly and yearly}} + \left[\dot{E}_{\text{TEC}}\right]_{\text{daily, monthly and yearly}}}{\gamma_c}}{\left[\text{daily, monthly and yearly } A_{\text{am}}\right]_{\text{daily, montly and yearly}}} \tag{8.7}$$

## 8.3  An Instantaneous Thermal and Overall Thermal Efficiency

It has been a practice to optimize the solar system parameters on the basis of its
thermal efficiency. In this section, an example of thermal efficiency of solar systems,
which has been discussed in Chaps. 4–7, will be covered in brief and the results will
be compared.

### 8.3.1  Liquid Flat Plate Collector (FPC)

**Example 8.1** Evaluate the optical and an instantaneous thermal efficiencies of liquid
FPC connected in series for $N = 5$ for Example 4.44.

**Given parameters**: $\alpha_p = \tau_p = 0.9, \dot{m}_f C_f = 0.035 \times 4190 = 146.65 \frac{W}{°C}, A_c = 2 \text{ m}^2$
and $I(t) = 500 \text{ W/m}^2$.

**Solution**

**For N = 5 FPC**: $T_{\text{fo5}} = 34.85\,°\text{C}, T_{\text{fi}} = 15\,°\text{C}$.
    From Eq. 8.1a, we have an expression for optical efficiency as

$$\eta_0 = \frac{\dot{Q}_{u,\text{th}}}{\gamma \left(\alpha_p \tau_p\right) I(t) \text{NA}_c} = \frac{\dot{m}_f C_f (T_{\text{fo5}} - T_{\text{fi}})}{1 \times 0.81 \times 500 \times 2 \times 5}$$

$$= \frac{146.65 \times 19.85}{4050} = \frac{2764.35}{4050} = 0.6825$$

From Eq. 8.2a, we have an expression for thermal efficiency as

$$\eta_{th} = \frac{\dot{Q}_{u,th}}{\gamma I(t) N A_c} = \frac{146.65 \times 19.85}{1 \times 500 \times 2 \times 5} = \frac{2764.35}{5000} = 0.5529$$

This shows that the optical efficiency is more than thermal efficiency as expected earlier.

***Example 8.2*** Evaluate the optical and an instantaneous thermal efficiency of liquid FPC connected in series for $N = 10$ for Example 4.44.

**Given parameters:** $\alpha_p = \tau_p = 0.9$, $\dot{m}_f C_f = 0.035 \times 4190 = 146.65 \frac{W}{°C}$, $A_c = 2 \, m^2$ and $I(t) = 500 \, W/m^2$.

**Solution**

**For N = 10 FPC:** $T_{fo5} = 45.26 \,°C$, $T_{fi} = 15 \,°C$.
   From Eq. 8.1a, we have an expression for optical efficiency as

$$\eta_0 = \frac{\dot{Q}_{u,th}}{\gamma (\alpha_p \tau_p) I(t) N A_c} = \frac{\dot{m}_f C_f (T_{fo5} - T_{fi})}{1 \times 0.81 \times 500 \times 2 \times 10}$$
$$= \frac{146.65 \times 30.26}{8100} = \frac{4437.629}{8100} = 0.5478$$

From Eq. 8.2a, we have an expression for thermal efficiency as

$$\eta_{th} = \frac{\dot{Q}_{u,th}}{\gamma I(t) N A_c} = \frac{146.65 \times 30.26}{1 \times 500 \times 2 \times 10} = \frac{4437.629}{10,000} = 0.4437$$

This shows that the optical efficiency as well as thermal efficiency of liquid FPC is reduced as number of FPC increases from 5 to 10 due to high operating temperature of the collector as per expectation.

***Example 8.3*** Evaluate the optical and an instantaneous thermal efficiency of single glazed $N = 5$ solar air collector connected in series for Example 5.34 for air flowing below absorber plate.

**Given parameters:** $\alpha_p = \tau_p = 0.9$, $\dot{m}_f C_f = 0.029 \times 1005 = 29.145 \frac{W}{°C}$, $A_c = 2 \, m^2$ and $I(t) = 500 \, W/m^2$.

**Solution**

**For $N = 5$ solar air heater:** $T_{fo5} = 37.98 \,°C$, $T_{fi} = 17 \,°C$.
   From Eq. 8.1a, we have an expression for optical efficiency as

$$\eta_0 = \frac{\dot{Q}_{u,th5}}{\gamma\left(\alpha_p \tau_p\right)I(t)NA_c} = \frac{\dot{m}_f C_f (T_{fo5} - T_{fi})}{1 \times 0.81 \times 500 \times 2 \times 5}$$

$$= \frac{29.145 \times 20.98}{4050} = \frac{611.4621}{4050} = 0.1509$$

From Eq. 8.2a, we have an expression for thermal efficiency as

$$\eta_{th} = \frac{\dot{Q}_{u,th5}}{\gamma I(t)NA_c} = \frac{29.145 \times 20.98}{1 \times 500 \times 2 \times 5} = \frac{611.4621}{5000} = 0.12229$$

From Examples 8.1 to 8.3, one can conclude that both solar air collector efficiencies, namely optical and thermal, are significantly reduced due to least heat capacity. This means that air has no storing capacity as water.

### 8.3.2   Liquid OPVT Collector

**Example 8.4** Evaluate the optical, thermal and overall thermal instantaneous efficiencies of opaque liquid PVT (OPVT) collector for Examples 6.7 and 6.9.

**Solution**

**Known Parameters for N = 1:**

$\left(\alpha_c \tau_g\right)_{m,\text{eff}} = 0.66375$, $T_{fo} = 1\ 5.41\,°C$, $T_{fo} = 15\,°C$, $\dot{m}_f C_f = 146.65\ \frac{W}{°C}$, (Example 6.7), $E_{el} = 98.18$ W (Example 6.9).

From Eq. 8.1a, we have an expression for optical efficiency as

$$\eta_0 = \frac{\dot{Q}_{u,th}}{\gamma\left(\alpha_p \tau_p\right)I(t)NA_c} = \frac{\dot{m}_f C_f (T_{fo5} - T_{fi})}{1 \times 0.675 \times 500 \times 2}$$

$$= \frac{146.65 \times 0.41}{810} = \frac{60.1265}{663.75} = 0.0905$$

From Eq. 8.2a, we have an expression for thermal efficiency as

$$\eta_{th} = \frac{\dot{Q}_{u,th}}{\gamma I(t)NA_c} = \frac{146.65 \times 0.41}{1 \times 500 \times 2 \times 1} = \frac{60.1265}{1000} = 0.06012$$

From Eq. 8.4, we have an expression for overall instantaneous thermal efficiency as

$$\eta_{i,ov-th} = \frac{\left[\dot{Q}_{u,th} + \frac{\dot{E}_{el}}{\gamma_c}\right]}{\text{Solar radiation in } W \times A_m}$$

$$= \frac{60.1265 + \frac{98.18}{0.35}}{500 \times 2} = \frac{60.1265 + 280.51}{1000} = \frac{340.64}{1000} = 0.34$$

**Example 8.5** Evaluate the optical, thermal and overall thermal instantaneous efficiencies of semi-transparent liquid PVT (SPVT) collector for Examples 6.16 and 6.19.

**Solution**

**Known Parameters for N = 1:**

$(\alpha_c \tau_g)_{m,\text{eff}} = 0.6428($ Example 6.13$)$, $T_{\text{fo}} = 16.43\,°C$, $T_{fi} = 15\,°C$, $m_f C_f = 146.65 \frac{W}{°C}$, (Example 6.7), $E_{\text{el}} = 98.612$ W (Example 6.19).

From Eq. 8.1a, we have an expression for optical efficiency as

$$\eta_0 = \frac{\dot{Q}_{u,\text{th}}}{\gamma(\alpha_p \tau_p) I(t) N A_c} = \frac{\dot{m}_f C_f (T_{\text{fo5}} - T_{\text{fi}})}{1 \times 0.6428 \times 500 \times 2 \times 1}$$

$$= \frac{146.65 \times 1.43}{642.8} = \frac{209.71}{642.8} = 0.3264$$

From Eq. 8.2a, we have an expression for thermal efficiency as

$$\eta_{\text{th}} = \frac{\dot{Q}_{u,\text{th}}}{\gamma I(t) N A_c} = \frac{146.65 \times 1.43}{1 \times 500 \times 2 \times 1} = \frac{209.71}{1000} = 0.2097$$

From Eq. 8.4, we have an expression for overall instantaneous thermal efficiency as

$$\eta_{i,\text{ov-th}} = \frac{\left[\dot{Q}_{u,\text{th}} + \frac{\acute{E}_{\text{el}}}{\gamma_c}\right]}{\text{Solar radiation in } W \times A_m} = \frac{209.71 + \frac{98.612}{0.35}}{500 \times 2}$$

$$= \frac{209.71 + 281.18}{1000} = \frac{491.46}{1000} = 0.4915$$

This shows that semi-transparent has improved the thermal efficiency due to increased direct gain to absorber, and hence, an overall thermal efficiency has been improved from 0.34 (34%) to 0.4915 (49.15%).

**Example 8.6** Evaluate the optical, thermal and overall thermal instantaneous efficiencies of N(5)-semitransparent liquid PVT (SPVT) collectors connected in series for Examples 6.21 and 6.25.

**Solution**

**Known Parameters for N = 5:**

$(\alpha_c \tau_g)_{m,\text{eff}} = 0.6428$ (Example 6.13), $T_{\text{fo5}} = 21.72\,°C$, $T_{\text{fo}} = 15\,°C$, $\dot{m}_f C_f = 146.65 \frac{W}{°C}$, (Example 6.7), $E_{\text{elT}} = 989.89$ W (Example 6.25).

From Eq. 8.1a, we have an expression for optical efficiency as

$$\eta_0 = \frac{\dot{Q}_{u,\text{th}}}{\gamma(\alpha_p \tau_p) I(t) N A_c} = \frac{\dot{m}_f C_f (T_{\text{fo5}} - T_{\text{fi}})}{1 \times 0.6428 \times 500 \times 2 \times 5}$$

$$= \frac{146.65 \times 6.72}{3214} = \frac{985.488}{3214} = 0.3066$$

From Eq. 8.2a, we have an expression for thermal efficiency as

$$\eta_{th} = \frac{\dot{Q}_{u,th}}{\gamma I(t)NA_c} = \frac{146.65 \times 6.72}{1 \times 500 \times 2 \times 5} = \frac{985.488}{5000} = 0.1971$$

From Eq. 8.4, we have an expression for overall instantaneous thermal efficiency as

$$\eta_{i,ov-th} = \frac{\left[\dot{Q}_{u,th} + \frac{\acute{E}_{el}}{\gamma_c}\right]}{\text{Solar radiation in } W \times A_m} = \frac{985.488 + \frac{989.89}{0.35}}{500 \times 2 \times 5}$$
$$= \frac{985.488 + 2828.26}{5000} = \frac{3813.75}{5000} = 0.7627$$

From this example, one can have the following conclusions:

(i)  For given design parameters, the optimum number of semi-transparent liquid PVT collector is five (5) due to equal thermal energy and electrical power.

(ii)  An overall thermal energy is maximum [0.7627 in fraction] for a given design and climatic parameters.

(iii)  Further, this shows that 5-semitransparent PVT collector has improved the overall thermal efficiency from 0.4915 (49.15%) to 0.7627 (76.27%).

### 8.3.3  Liquid Flexible OPVT-TEC Collector

***Example 8.7*** Evaluate the optical, thermal and overall thermal instantaneous efficiencies of single flexible (Al-based) liquid OPVT-TEC collector for Examples 6.37 and 6.40.

**Solution**

**Known parameters**: $\dot{m}_f C_f = 146.65 \frac{W}{m^2}$, $(\alpha\tau)_{eff} = 0.735$ (Example 6.36), $T_{fo} = 16.91\,°C$, $T_{fi} = 15\,°C$ (Example 3.37), $\dot{E}_{el} = 106\,W$, $\dot{E}_{TEC} = 30.18\,W$ (Example 6.40).

From Eq. 8.1a, we have an expression for optical efficiency as

$$\eta_0 = \frac{\dot{Q}_{u,th}}{\gamma(\alpha_p\tau_p)I(t)NA_c}$$
$$= \frac{\dot{m}_f C_f(T_{fo5} - T_{fi})}{1 \times 0.735 \times 500 \times 2 \times 1} = \frac{146.65 \times 1.91}{735} = \frac{280.10}{735} = 0.3211$$

From Eq. 8.2a, we have an expression for thermal efficiency as

$$\eta_{th} = \frac{\dot{Q}_{u,th}}{\gamma I(t)NA_c} = \frac{146.65 \times 1.91}{1 \times 500 \times 2 \times 1} = \frac{280.10}{1000} = 0.2801$$

From Eq. 8.6, we have an expression for an overall thermal efficiency as

$$\eta_{i,ov-th} = \frac{\left[\dot{Q}_{u,th} + \frac{\dot{E}_{el}+\dot{E}_{TEC}}{\gamma_c}\right]}{\text{Solar radiation in } W \times A_m}$$

$$= \frac{280.10 + \frac{106+30.18}{0.38}}{1000} = \frac{280 + 358.37}{1000} = \frac{638.47}{1000} = 0.6385$$

**Example 8.8** Evaluate the optical, thermal and overall thermal instantaneous efficiencies of two flexible (Al-based) liquid OPVT-TEC collectors connected in series for Examples 6.42 and 6.40.

**Solution**

**Known parameters**: $\dot{m}_f c_f = 146.65$ W/°C, $(\alpha\tau)_{eff} = 0.735$ (Example 6.36), $T_{fo2} = 18.73\,°C$, $T_{fi} = 15\,°C$, $(t) = 500$ W/m², $\dot{E}_{total1} = \dot{E}_{el} + E_{TEC1} = 89.62 + 30.06 = 119.62$ W and $\dot{E}_{total2} = \dot{E}_{el} + E_{TEC} = 87.68 + 27.79 = 115.47$ W.

From Eq. 8.1a, we have an expression for optical efficiency as

$$\eta_0 = \frac{\dot{Q}_{u,th}}{\gamma(\alpha_p\tau_p)I(t)NA_c} = \frac{\dot{m}_f C_f(T_{fo5} - T_{fi})}{1 \times 0.735 \times 500 \times 2 \times 2} = \frac{146.65 \times 3.73}{1470}$$

$$= \frac{547.00}{1470} = 0.372$$

From Eq. 8.2a, we have an expression for thermal efficiency as

$$\eta_{th} = \frac{\dot{Q}_{u,th}}{\gamma I(t)NA_c} = \frac{146.65 \times 3.73}{1 \times 500 \times 2 \times 2} = \frac{547.00}{2000} = 0.2735$$

From Eq. 8.6, we have an expression for an overall thermal efficiency as

$$\eta_{i,ov-th} = \frac{\left[\dot{Q}_{u,th} + \frac{\dot{E}_{el2}+\dot{E}_{TEC2}}{\gamma_c}\right]}{\text{Solar radiation in } W \times A_m \times 2} = \frac{547.00 + \frac{119.62+115.47}{0.38}}{2000}$$

$$= \frac{547.00 + 635.38}{2000} = \frac{1182.23}{2000} = 0.5912$$

**Example 8.9** Repeat Example 8.8 for an overall thermal efficiency of two flexible (Al-based) PVT-TEC liquid collectors for Example 6.48.

**Solution**

From Example 6.48, the total electrical power $(\dot{E}_{total})$ which is the sum of electrical power of PV and TEC from two flexible PVT-TEC collectors and the same is obtained as

$$\dot{E}_{total} = \dot{E}_{el} + E_{TEC} = 177.36 + 57.78 = 235.14 \text{ W}$$

Now, an overall thermal efficiency can be calculated from Eq. 8.6 as

$$\eta_{i,ov-th} = \frac{\left[\dot{Q}_{u,th} + \frac{\dot{E}_{el2} + \dot{E}_{TEC2}}{\gamma_c}\right]}{\text{Solar radiation in } W \times A_m \times 2} = \frac{547.00 + \frac{235.14}{0.38}}{2000}$$

$$= \frac{547.00 + 671.83}{2000} = \frac{1218.29}{2000} = 0.6094$$

The value obtained for an overall thermal efficiency (0.6094) is very close to (0.5912) in Example 8.8.

### 8.3.4  Liquid SPVT-CPC Collector

**Example 8.10** Evaluate the optical, thermal and overall thermal instantaneous efficiencies of single semi-transparent liquid SPVT-CPC collector for Examples 7.7 and 7.11 for concentration ratio of two $\left[C = \frac{A_{am}}{A_{rm}} = 2\right]$.

**Solution**

**From Examples 7.7 and 7.11:** $\dot{m}_f C_f = 146.65 \frac{W}{°C}$, $(\alpha\tau)_{eff} = 0.5887$ for PVT (Example 7.2), $I_u = 475 \text{ W/m}^2$, $A_{am} = 4 \text{ m}^2$ and $A_m = 2 \text{ m}^2$, $T_{fo} = 17.42 °C$, $T_{fi} = 15 °C$ (Example 7.1), $\dot{E}_{el} = 83.09 \text{ W}$.

From Eq. 8.1a, we have an expression for optical efficiency as

$$\eta_0 = \frac{\dot{Q}_{u,th}}{\gamma(\alpha_p\tau_p)I_u NA_{am}} = \frac{\dot{m}_f C_f(T_{fo5} - T_{fi})}{1 \times 0.5887 \times 475 \times 1 \times 4}$$

$$= \frac{146.65 \times 2.42}{1118.53} = \frac{354.893}{1118.53} = 0.3173$$

From Eq. 8.2a, we have an expression for thermal efficiency as

$$\eta_{th} = \frac{\dot{Q}_{u,th}}{\gamma I(t)NA_{am}} = \frac{146.65 \times 2.42}{1 \times 475 \times 4 \times 1} = \frac{354.893}{1900} = 0.1867$$

From Eq. 8.4, we have an expression for overall instantaneous thermal efficiency as

$$\eta_{i,ov-th} = \frac{\left[\dot{Q}_{u,th} + \frac{\dot{E}_{el}}{\gamma_c}\right]}{\text{Solar radiation in } W \times A_{an}} = \frac{354.893 + \frac{83.09}{0.35}}{475 \times 4 \times 1}$$

$$= \frac{354.893 + 237.4}{1900} = \frac{592.293}{1900} = 0.3117$$

There is a decrease in an overall instantaneous thermal efficiency from 0.4915 (Example 8.5) to 0.3117 due to integration of compound parabolic concentrator (CPC) [high operating temperature]. Further, one can notice that the optical instantaneous thermal efficiency is approximately equal to an overall thermal efficiency.

**Example 8.11** Evaluate the optical, thermal and overall thermal instantaneous efficiencies of single semi-transparent liquid SPVT-CPC collector for Examples 7.8 and 7.13 for concentration ratio of two $\left[ C = \frac{A_{am}}{A_{rm}} = 1 \right]$.

**Solution**

From Example 7.8 and Example 7.13: For $C = 1$, $A_{am} = A_{rm} = A_m = 2\,\text{m}^2$ $I_u = I(t) = 500$ W/m$^2$, $(\alpha\tau)_{\text{eff}} = 0.5887$, $T_{fo} = 16.19\,^\circ$C, $T_{fi} = 15\,^\circ$C, $\dot{E}_{el} = 97.77$ W.
From Eq. 8.1a, we have an expression for optical efficiency as

$$\eta_0 = \frac{\dot{Q}_{u,\text{th}}}{\gamma\left(\alpha_p\tau_p\right)I_u NA_m} = \frac{\dot{m}_f C_f(T_{fo5} - T_{fi})}{1 \times 0.5887 \times 500 \times 1 \times 2}$$
$$= \frac{146.65 \times 1.19}{588.7} = \frac{174.51}{588.7} = 0.29$$

From Eq. 8.2a, we have an expression for thermal efficiency as

$$\eta_{\text{th}} = \frac{\dot{Q}_{u,\text{th}}}{\gamma I(t)NA_c} = \frac{146.65 \times 1.19}{1 \times 500 \times 2 \times 1} = \frac{174.51}{1000} = 0.1745$$

From Eq. 8.4, we have an expression for overall instantaneous thermal efficiency as

$$\eta_{i,\text{ov-th}} = \frac{\left[\dot{Q}_{u,\text{th}} + \frac{\dot{E}_{el}}{\gamma_c}\right]}{\text{Solar radiation in } W \times A_m} = \frac{174.51 + \frac{97.77}{0.35}}{500 \times 2 \times 1}$$
$$= \frac{174.51 + 279.34}{1000} = \frac{453.85}{1000} = 0.4538$$

For concentration ratio of one ($C = 1$), the overall instantaneous thermal efficiency [0.4538, 45.38%] is more than the overall instantaneous thermal efficiency for $C = 2$ [0.3117] due to low operating temperature.

**Example 8.12** Evaluate the optical, thermal and overall thermal instantaneous efficiencies of six ($N = 6$) semi-transparent liquid SPVT-CPC collectors connected in series for Examples 7.15 and 7.16 for concentration ratio of two $\left[ C = \frac{A_{am}}{A_{rm}} = 2 \right]$, $A_{am} = 4\,\text{m}^2$ and $A_m = 2\,\text{m}^2$.

**Solution**

**From Examples** 7.15 **and** 7.16: $\dot{m}_f C_f = 146.65 \frac{W}{°C}$, $(\alpha\tau)_{\text{eff}} = 0.5887$, $T_{\text{fo6}} = 28.49\,°C$, $T_{\text{fi}} = 15\,°C$, $\dot{E}_{\text{el6}} = 495.12\,W$.

From Eq. 8.1a, we have an expression for optical efficiency as

$$\eta_0 = \frac{\dot{Q}_{u,\text{th}}}{\gamma(\alpha_p\tau_p)I_u NA_m} = \frac{\dot{m}_f C_f(T_{\text{fo5}} - T_{\text{fi}})}{1 \times 0.5887 \times 475 \times 6 \times 4}$$

$$= \frac{146.65 \times 3.49}{6711.18} = \frac{511.81}{6711.18} = 0.0763$$

From Eq. 8.2a, we have an expression for thermal efficiency as

$$\eta_{\text{th}} = \frac{\dot{Q}_{u,\text{th}}}{\gamma I(t) NA_c} = \frac{146.65 \times 3.49}{1 \times 475 \times 6 \times 4} = \frac{511.81}{11,400} = 0.04489$$

From Eq. 8.4, we have an expression for overall instantaneous thermal efficiency as

$$\eta_{i,\,\text{ov-th}} = \frac{\left[\dot{Q}_{u,\text{th}} + \frac{\dot{E}_{\text{el}}}{\gamma_c}\right]}{\text{Solar radiation in } W \times A_m} = \frac{511.81 + \frac{495.12}{0.35}}{475 \times 6 \times 4}$$

$$= \frac{511.81 + 1414.63}{11,400} = \frac{1926,44}{11,400} = 0.17$$

This example shows that the optical, thermal and overall thermal instantaneous efficiencies of six ($N = 6$) semi-transparent liquid PVT-CPC collectors connected in series are drastically reduced in comparison with single semi-transparent liquid PVT-CPC collector due to high operating temperature range [$T_{\text{fo6}} = 28.49\,°C$].

### 8.3.5  FPC-CPC Liquid Collector

***Example 8.13*** Evaluate the optical, thermal and overall thermal instantaneous efficiencies for six ($N = 6$)-FPC-CPC liquid collectors connected in series for $C = 2$ for Example 7.20.

**Solution**

**For $N = 6$ and from Example** 7.20: $T_{\text{fo6}} = 59.14\,°C$, $T_{\text{fi}} = 15\,°C$, and $\dot{m}_f C_f = 146.65\,W/°C$, $\rho = 0.9$, $\alpha_p = 0.9$ and $\tau_g = 0.9$, $I_u = 475\,W/m^2$.

From Eq. 8.1a, we have an expression for optical efficiency as

$$\eta_0 = \frac{\dot{Q}_{u,\text{th}}}{\gamma\rho(\alpha_p\tau_p)I(t)NA_{\text{am}}} = \frac{\dot{m}_f C_f(T_{\text{fo5}} - T_{\text{fi}})}{1 \times 0.9 \times 0.81 \times 475 \times 6 \times 4}$$

$$= \frac{146.65 \times 44.14}{8310.6} = \frac{6473.13}{8310.6} = 0.7789$$

From Eq. 8.2a, we have an expression for thermal efficiency as

$$\eta_{th} = \frac{\dot{Q}_{u,th}}{\gamma I(t)NA_{am}} = \frac{146.65 \times 44.14}{1 \times 475 \times 6 \times 4} = \frac{6473.13}{11,400} = 0.5678$$

This shows that the optical efficiency is more than thermal efficiency as expected earlier.

## 8.3.6 SPVT-CPC Air Collector

**Example 8.14** Evaluate the optical, thermal and overall thermal instantaneous efficiencies for single ($N = 1$) and six ($N = 6$)-SPVT-CPC air collectors connected in series for $C = 2$ for Examples 7.25 and 7.29.

**Solution**

From Examples 7.2 and 7.29:

**For N = 1:**

$(\alpha\tau)_{m,\text{eff}} = 0.2561$ (Example 7.24), $\dot{m}_f C_f = 29.145\frac{W}{°C}$, $T_{fo1} = 21.76\,°C$, $T_{fo1} = 15\,°C$ and $\dot{E}_{el1} = 80.07\,W$.

From Eq. 8.1a, we have an expression for optical efficiency as

$$\eta_0 = \frac{\dot{Q}_{u,th}}{\gamma(\alpha_p\tau_p)I_uNA_m} = \frac{\dot{m}_f C_f(T_{fo5} - T_{fi})}{1 \times 0.2561 \times 475 \times 1 \times 4}$$
$$= \frac{29.145 \times 6.76}{486.59} = \frac{197.02}{486.59} = 0.4049$$

From Eq. 8.2a, we have an expression for thermal efficiency as

$$\eta_{th} = \frac{\dot{Q}_{u,th}}{\gamma I(t)NA_c}$$
$$= \frac{29.145 \times 6.76}{1 \times 475 \times 1 \times 4} = \frac{197.02}{1900} = 0.1037$$

From Eq. 8.4, we have an expression for overall instantaneous thermal efficiency as

$$\eta_{i,ov-th} = \frac{\left[\dot{Q}_{u,th} + \frac{\dot{E}_{el}}{\gamma_c}\right]}{\text{Solar radiation in } W \times A_m}$$

$$= \frac{197.02 + \frac{80.07}{0.35}}{475 \times 1 \times 4} = \frac{197.02 + 228.77}{1900} = \frac{425.79}{1900} = 0.22$$

**For N = 6:**

$(\alpha\tau)_{m,\mathrm{eff}} = 0.2561, \dot{m}_f C_f = 29.145 \frac{W}{°C}, T_{\mathrm{fo6}} = 47.61\,°C, T_{\mathrm{fi}} = 15\,°C, \dot{E}_{\mathrm{el6}} = 475.52\,\mathrm{W}.$

From Eq. 8.1a, we have an expression for optical efficiency as

$$\eta_0 = \frac{\dot{Q}_{u,\mathrm{th}}}{\gamma(\alpha_p\tau_p)I_u NA_m} = \frac{\dot{m}_f C_f(T_{\mathrm{fo5}} - T_{\mathrm{fi}})}{1 \times 0.2561 \times 475 \times 6 \times 4}$$

$$= \frac{29.145 \times 32.61}{2919.54} = \frac{950.42}{2919.54} = 0.3255$$

From Eq. 8.2a, we have an expression for thermal efficiency as

$$\eta_{\mathrm{th}} = \frac{\dot{Q}_{u,\mathrm{th}}}{\gamma I(t)NA_c} = \frac{29.145 \times 32.61}{1 \times 475 \times 6 \times 4} = \frac{950.42}{11{,}400} = 0.0833$$

From Eq. 8.4, we have an expression for overall instantaneous thermal efficiency as

$$\eta_{i,\mathrm{ov-th}} = \frac{\left[\dot{Q}_{u,\mathrm{th}} + \frac{\dot{E}_{\mathrm{el}}}{\gamma_c}\right]}{\text{Solar radiation in } W \times A_m} = \frac{950.42 + \frac{475.52}{0.35}}{475 \times 6 \times 4}$$

$$= \frac{950.42 + 1.358.63}{11{,}400} = \frac{2309.05}{11{,}400} = 0.2025$$

We can see that there is a drop in the optical, thermal and overall thermal instantaneous efficiencies from $N = 1$ to $N = 6$ due to high operating temperature at $N = 6$ $[T_{\mathrm{Fo6}} = 47.61\,°C]$ as expected.

**Example 8.15**  Evaluate the optical, thermal and overall thermal instantaneous efficiencies for single $(N = 1)$ and six $(N = 6)$-PVT-CPC air collectors connected in series for $C = 2$ by considering an average value of fluid and solar cell temperatures for Examples 7.26 and 7.29.

**Solution**

From Examples 7.2 and 7.29:

**For N = 1:**

$(\alpha\tau)_{m,\mathrm{eff}} = 0.2561$ (Example 7.24), $\dot{m}_f C_f = 29.145 \frac{W}{°C}, \overline{T}_{f1} = 18.41\,°C, T_{\mathrm{fo1}} = 15\,°C$ and $\dot{E}_{\mathrm{el1}} = 80.07\,\mathrm{W}, \overline{T}_{c1} = 74.61\,°C.$

From Eq. 8.1a, we have an expression for optical efficiency as

$$\eta_0 = \frac{\dot{Q}_{u,\mathrm{th}}}{\gamma(\alpha_p\tau_p)I_u NA_m} = \frac{\dot{m}_f C_f(\overline{T}_{f1} - T_{\mathrm{fi}}) \times 2}{1 \times 0.2561 \times 475 \times 1 \times 4}$$

$$= \frac{29.145 \times 3.41 \times 2}{486.59} = \frac{198.77}{486.59} = 0.4085$$

From Eq. 8.2a, we have an expression for thermal efficiency as

$$\eta_{th} = \frac{\dot{Q}_{u,th}}{\gamma I(t) N A_c} = \frac{29.145 \times 3.41 \times 2}{1 \times 475 \times 1 \times 4} = \frac{198.77}{1900} = 0.1046$$

From Eq. 8.4, we have an expression for overall instantaneous thermal efficiency as

$$\eta_{i,ov-th} = \frac{\left[\dot{Q}_{u,th} + \frac{\dot{E}_{el}}{\gamma_c}\right]}{\text{Solar radiation in} W \times A_m} = \frac{198.77 + \frac{80.07}{0.35}}{475 \times 1 \times 4}$$

$$= \frac{198.77 + 228.77}{1900} = \frac{427.54}{1900} = 0.225$$

**For N = 6:**

$(\alpha\tau)_{m,\text{eff}} = 0.2561$, $\dot{m}_f C_f = 29.145 \frac{W}{\circ C}$, $\overline{T}_{f6} = 32.79\,^\circ C$, $T_{fi} = 15\,^\circ C$, $\dot{E}_{el6} = 475.52$ W, $\overline{T}_{c6} = 76.50\,^\circ C$.

From Eq. 8.1a, we have an expression for optical efficiency as

$$\eta_0 = \frac{\dot{Q}_{u,th}}{\gamma(\alpha_p\tau_p) I_u N A_m} = \frac{\dot{m}_f C_f(\overline{T}_{f6} - T_{fi}) \times 2}{1 \times 0.2561 \times 475 \times 6 \times 4}$$

$$= \frac{29.145 \times 17.79 \times 2}{2919.54} = \frac{1036.98}{2919.54} = 0.3552$$

From Eq. 8.2a, we have an expression for thermal efficiency as

$$\eta_{th} = \frac{\dot{Q}_{u,th}}{\gamma I(t) N A_c} = \frac{29.145 \times 17.79 \times 2}{1 \times 475 \times 6 \times 4} = \frac{1036.98}{11,400} = 0.091$$

From Eq. 8.4, we have an expression for overall instantaneous thermal efficiency as

$$\eta_{i,ov-th} = \frac{\left[\dot{Q}_{u,th} + \frac{\dot{E}_{el}}{\gamma_c}\right]}{\text{Solar radiation in } W \times A_m} = \frac{1036.98 + \frac{475.52}{0.35}}{475 \times 6 \times 4}$$

$$= \frac{1036.98 + 1.358.63}{11,400} = \frac{2309.05}{11,400} = 0.21$$

We can see from Examples 8.14 and 8.15 and that there is not much difference in the results by both methods to evaluate the optical, thermal and overall thermal instantaneous efficiencies from $N = 1$ to $N = 6$ as expected.

### 8.3.7  OPVT-TEC-CPC Liquid Collector

**Example 8.16** Find out the rate of thermal energy carried out by water in W for single OPVT-TEC-CPC liquid collector for $N = 1$ and for $C = \frac{A_{am}}{A_{rm}} = 2$ for Example 7.35. Further calculate the optical and thermal instantaneous efficiencies.

**Solution**

**Given for OPVT-TEC-CPC liquid collector:**$(\alpha\tau)_{meff} = 0.7975$, $I_u = 475$ W/m$^2$, $T_a = 16\,°C$ and $T_{fi} = 15\,°C$, $T_{fo1} = 18.98\,°C$, $\dot{m}_f c_f = 146.65$W/$°C$.

The rate of thermal energy is calculated as

$$\dot{Q}_{u,\text{th}} = \dot{m}_f C_f (T_{fo1} - T_{fi}) = 146.65(18.98 - 15) = 583.667\,\text{W}$$

From Eq. 8.1a, we have an expression for optical efficiency as

$$\eta_0 = \frac{\dot{Q}_{u,\text{th}}}{\gamma(\alpha_p\tau_p)I_u NA_m} = \frac{\dot{m}_f C_f (T_{fo1} - T_{fi})}{1 \times 0.7975 \times 475 \times 1 \times 4} = \frac{583.667}{1515.25} = \frac{583.667}{1515.25} = 0.3840$$

From Eq. 8.2a, we have an expression for thermal efficiency as

$$\eta_{\text{th}} = \frac{\dot{Q}_{u,\text{th}}}{\gamma I(t) NA_c} = \frac{583.667}{1 \times 475 \times 1 \times 4} = \frac{583.667}{1900} = 0.3072$$

**Example 8.17** Find out the electrical power and equivalent thermal power obtained from TEC of W for single OPVT-TEC-CPC liquid collector for $N = 1$ and for $C = \frac{A_{am}}{A_{rm}} = 2$ for Example 7.47.

**Solution**

Given: $\overline{T}_{\text{tec,top1}} = 18.9\,°C$, $\overline{T}_{f1} = \overline{T}_{\text{tec,bottom1}} = 16.92\,°C$ (Example 7.37).

From Eq. 7.49 and Example 7.47, we have

$$\dot{E}_{\text{TEC1}} = \eta_{\text{tec}} U_{\text{tec}} \left(\overline{T}_{\text{tec,top1}} - \overline{T}_{f1}\right)\beta_{\text{tec}} A_m \times N$$
$$= 0.15 \times 450 \times (18.9 - 16.92) \times 2 \times 0.4324 \times 1 = 115.58\,\text{W}$$

Equivalent thermal power obtained from TEC of W for single PVT-TEC-CPC liquid collector is obtained as follows:

$$\dot{E}_{\text{TEC1,th}} = \frac{\dot{E}_{\text{TEC1}}}{\gamma_c} = \frac{115.58}{0.35} = 330.23\,\text{W}$$

**Example 8.18** Find out the electrical power and equivalent thermal power obtained from OPV in W for single OPVT-TEC-CPC liquid collector for $N = 1$ and for $C = \frac{A_{am}}{A_{rm}} = 2$ for Example 7.43.

**Solution**

For $N = 1$, we have an electrical power from Example 7.43 as

$$\dot{E}_{\text{ell}} = 48.15\,\text{W}$$

Now, an equivalent thermal power from PV module of single OPVT-TEC-CPC liquid collector can be obtained as

$$\dot{E}_{\text{ell,th}} = \frac{\dot{E}_{\text{ell}}}{\gamma_c} = \frac{48.15}{0.35} = 137.57\,\text{W}$$

From Examples 8.17 and 8.18 one can see that TEC produces more electrical power than PV of OPVT-TEC-CPC liquid collector.

***Example 8.19*** Find out an overall thermal efficiency of single OPVT-TEC-CPC liquid collector for $N = 1$ and for $C = \frac{A_{\text{am}}}{A_{\text{rm}}} = 2$ for Examples 8.16–8.18.

**Solution**

From Examples 8.16–8.18, we have the followings:

$$\dot{Q}_{u,\text{th}} = 583.667\,\text{W}, \quad \dot{E}_{\text{TEC1}} = 115.58\,\text{W} \text{ and } \dot{E}_{\text{ell}} = 48.15\,\text{W}.$$

From Eq. 8.4, we have an expression for overall instantaneous thermal efficiency of single OPVT-TEC-CPC liquid collector as

$$
\begin{aligned}
\eta_{i,\text{ov}-\text{th}} &= \frac{\left[\dot{Q}_{u,\text{th}} + \frac{\dot{E}_{\text{TEC1}}+\dot{E}_{\text{el}}}{\gamma_c}\right]}{\text{Solar radiation in } W \times A_m} = \frac{583.667 + \frac{115.58+48.15}{0.35}}{475 \times 1 \times 4} \\
&= \frac{583.667 + 467.8}{1900} = \frac{1051}{1900} = 0.5534
\end{aligned}
$$

From Examples 8.16 to 8.19, it is concluded that an overall instantaneous thermal efficiency [0.5534] of single OPVT-TEC-CPC liquid collector is always higher than thermal instantaneous efficiency [0.3072] of single OPVT-TEC-CPC liquid collector as expected earlier from all examples.

***Example 8.20*** Find out the rate of thermal energy carried out by water in W for six ($N = 6$)- SPVT-TEC-CPC liquid collectors connected in series and for $C = \frac{A_{\text{am}}}{A_{\text{rm}}} = 2$ for Example 7.36. Further calculate the optical and thermal instantaneous efficiencies.

**Solution**

**From Example** 7.36: $T_{\text{fo6}} = 37.11\,°C$, $T_{\text{fi}} = 15\,°C$, $(\alpha\tau)_{\text{meff}} = 0.797$, $I_u = 475\,\text{W/m}^2$, $\dot{m}_f c_f = 146.65\,\text{W/}°C$.

The rate of thermal energy of six OPVT-TEC-CPC liquid collectors is calculated as

$$\dot{Q}_{u,\text{th}} = \dot{m}_f C_f (T_{\text{fo6}} - T_{\text{fi}}) = 146.65(37.11 - 15) = 3242.43\,\text{W} = 3.242\,\text{kW}$$

From Eq. 8.1a, we have an expression for optical efficiency of six OPVT-TEC-CPC liquid collectors as

$$\eta_0 = \frac{\dot{Q}_{u,\text{th}}}{\gamma(\alpha_p\tau_p)I_u\text{NA}_m} = \frac{\dot{m}_f C_f(T_{\text{fo6}} - T_{\text{fi}})}{1 \times 0.7975 \times 475 \times 6 \times 4} = \frac{3242.43}{9091.5} = 0.3566$$

From Eq. 8.2a, we have an expression for thermal efficiency of six OPVT-TEC-CPC liquid collectors as

$$\eta_{\text{th}} = \frac{\dot{Q}_{u,\text{th}}}{\gamma I(t)\text{NA}_c} = \frac{3242.43}{1 \times 475 \times 6 \times 4} = \frac{3242.43}{11,400} = 0.2844$$

**Example 8.21** Find out the electrical power and equivalent thermal power obtained from TEC of W for six OPVT-TEC-CPC liquid collectors connected in series and for $C = \frac{A_{\text{am}}}{A_{\text{rm}}} = 2$ for Example 7.40.

**Solution**

Given: $\overline{T}_{\text{tec, top 6}} = 28.21\,^{\circ}\text{C}$, $\overline{T}_{f6} = \overline{T}_{\text{tec, Bottom6}} = 26.48\,^{\circ}\text{C}$ (Example 7.40).
    From Eq. 7.49 and Example 7.47, we have

$$\dot{E}_{\text{TEC1}} = \eta_{\text{tec}}U_{\text{tec}}(\overline{T}_{\text{tec,top1}} - \overline{T}_{f1})\beta_{\text{tec}}A_m \times N$$
$$= 0.15 \times 450 \times (28.21 - 26.48) \times 2 \times 0.4324 \times 6 = 605.92\,\text{W}$$

Equivalent thermal power obtained from TEC of W for single OPVT-TEC-CPC liquid collector is obtained as follows:

$$\dot{E}_{\text{TEC1,th}} = \frac{\dot{E}_{\text{TEC1}}}{\gamma_c} = \frac{605.92}{0.35} = 1731.21\,\text{W}$$

**Example 8.22** Find out the electrical power and equivalent thermal power obtained from OPV in W for six $(N = 6)$ OPVT-TEC-CPC liquid collectors connected in series and for $C = \frac{A_{\text{am}}}{A_{\text{rm}}} = 2$ for Example 7.44.

**Solution**

For $N = 6$, we have an electrical power from Example 7.44 as

$$\dot{E}_{\text{ell}} = 277.03\,\text{W}$$

Now, an equivalent thermal power from PV module of single OPVT-TEC-CPC liquid collector can be obtained as

$$\dot{E}_{\text{ell,th}} = \frac{\dot{E}_{\text{ell}}}{\gamma_c} = \frac{277.03}{0.35} = 791.51\,\text{W}$$

From Examples 8.21 and 8.22, one can see that TEC produces more electrical power [1731.21 W] than PV [791.51 W]  of PVT-TEC-CPC liquid collector.

***Example 8.23*** Find out an overall thermal efficiency of six OPVT-TEC-CPC liquid collectors connected in series and for $C = \frac{A_{am}}{A_{rm}} = 2$ for Examples 8.20–8.22.

**Solution**

From Examples 8.16–8.18, we have the followings:

$\dot{Q}_{u,th} = 3242.43$, $\dot{E}_{TEC1} = 115.58\,\text{W}$ and $\dot{E}_{ell} = 48.15\,\text{W}$.

From Eq. 8.4, we have an expression for overall instantaneous thermal efficiency as

$$
\eta_{i,ov-th} = \frac{\left[ \dot{Q}_{u,th} + \frac{\dot{E}_{TEC1} + \dot{E}_{el}}{\gamma_c} \right]}{\text{Solar radiation in W} \times NA_m} = \frac{3242.43 + \frac{605.92 + 791.51}{0.35}}{475 \times 6 \times 4}
$$

$$
= \frac{3242.43 + 3992.66}{11{,}400} = \frac{7235.09}{11{,}400} = 0.6346
$$

In the case of six OPVT-TEC-CPC collectors connected in series, it gives more overall thermal efficiency ($\frac{7235.09}{6} = 1205.84\,\text{W}$, Example 8.23) per single OPVT-TEC-CPC than single OPVT-TEC-CPC collector (1051 W, Example 8.19) due to increase in electrical power by TEC.

## 8.4  The Fresnel Lens

The solar concentrator is generally used to increase the solar radiation from higher surface area, known as aperture area ($A_{am}$), to lower surface area of absorber, known as receiver area ($A_{rm}$) to increase the operating temperature. The increase in solar radiation on receiver area of absorber depends on concentration ratio (C) which is ratio of aperture area to receiver area. The solar concentrator is broadly based on either principle of reflection or refraction. Both reflection and concentration work for solar radiation having direction known as beam radiation. However, there is minor improvement in incoming solar radiation to absorber due to diffuse radiation as given in Example 7.1. However, the Fresnel lens only work with beam radiation falling on the surface normally to top surface of Fresnel lens hence required tracking unlike compound parabolic concentrator (CPC) (Chap. 7). The solar concentrator based on reflection is known as compound parabolic concentrator as discussed in Chap. 7. There is limitation in concentration ratio in CPC which is three ($C = 3$) for operating point of view due to larger absorber area of 2 m$^2$. The concentration ratio (C) of two has been considered in Chap. 7 to analyze the various types of PVT-CPC collectors, Table 7.1. In order to increase the concentration ratio, the Fresnel lens has been considered in the book which works only on the basis of beam radiation with tracking system. It is also fact that any system has some advantage as well disadvantage. In the similar way, the Fresnel lens has too. These are as follows:

(i)  It is easier to make FL in plastic material due to cutting of groves of different concentration ratios (C) which has lower life and transmittivity at top flat surface reduces due to scratching during cleaning.

(ii)  Tracking system is electronic-based device and hence needs more maintenance.

(iii)  In Fresnel lens (FL), concentration ratio (C) may be increased by reducing the absorber area/receiver area. It may go up beyond three of CPC as per requirement. For example in point focus FL concentrator, it can be 1500 with effective length of 0.67–1.2 m with optical efficiency of 0.68 in fraction. However, we are going to discuss about integration of FL concentrator with evacuated tubular collector (ETC) and PVT collectors, and hence, we may consider concentration ratio of up to ten (10) (Table 8.1).

So, there are two types of Fresnel lens concentrators, namely point focus, Fig. 8.1, and line focus, Fig. 8.2. In this chapter, we will be discussing only line focus.

## 8.5   Fresnel Lens (FL) Concentrator Integrated with Evacuated Tubular Collector (ETC) [FLiETC]

There are many types of evacuated tubular collector. However, we will consider the design proposed by Mishra et al. [1] as shown in Fig. 8.3. In this case, there is outer cylindrical double glazed tube of length 1 m and diameter 'D' and U-shape copper tube of 2 m length is inserted with fitting inside glass tube. The whole glass tube is evacuated so that there is no air left inside to avoid inner convection loss. The fluid is allowed to flow at inlet temperature $(T_{fi})$ which is heated while passing through inside U-shape copper tube as shown in Fig. 8.3. The assembly is known as evacuated tubular collector (ETC). Zhai et al. [2] have proposed integration of FL (Fig. 8.2) with ETC (Fig. 8.3) which will be referred as **FLiETC** as shown in Fig. 8.4. The outer diameter of evacuated tubular collector is 0.25 m. The other design parameters of ETC are given in Table 8.2.

### 8.5.1   Energy Balances of Single FLiETC

Following Mishra et al. [1] and referring to Fig. 8.4, the basic energy balance for elemental area of 'Ddx' for different components will be written as

**(a) Cylindrical blackened and glazed copper absorber tube**

The energy balance in $W$ for cylindrical absorber of FLiETC is as follows:

$$\tau_g^2 \alpha_c I_b \frac{A_{am}}{A_{rm}}[(2r_2).dx] = U_{t,pa}(T_p - T_a)[(2r_2).dx] + h_{pf}(T_p - T_f)[(2r_2).dx]$$

$$(8.8)$$

**Table 8.1** Various solar system efficiency in fraction

| S.No | Solar system [Example] | Number of collector (N) | Optical instantaneous efficiency ($\eta_0$) | Thermal instantaneous efficiency ($\eta_{th}$) | PV electrical efficiency | TEC electrical efficiency | An overall thermal efficiency |
|---|---|---|---|---|---|---|---|
| 1 | Liquid FPC [8.1, 8.2] | 5 10 | 0.6825 0.5478 | 0.5529 0.4437 | – | – | – |
| 2 | Solar air collector [8.3] | 5 | 0.1509 | 0.12229 | – | – | – |
| 3 | Liquid OPVT –FPC[8.4] | 1 | 0.0905 | 0.06012 | 0.1103 | – | 0.34 |
| 4 | Liquid SPVT- FPC [8.5, 8.6] | 1 5 | 0.3264 0.3066 | 0.2097 0.1971 | 0.1108 0.1179 | – | 0.4915 0.7627 |
| 5 | OPVT-TEC[8.8] | 2 | 0.372 | 0.2735 | 0.09995 | 0.15 | 0.5912 |
| 6 | SPVT-CPC [8.10. 8.11] | 1 (C = 2) 1 (C = 1) | 0.3173 0.29 | 0.1867 0.1745 | 0.0983 0.10980 | – | 0.3117 0.4538 |
| 7 | SPVT-CPC [8.12] | 6 | 0.0763 | 0.04489 | 0.0976 | – | 0.17 |
| 8 | FPC-CPC [8.13] | 6 | 0.7789 | 0.5678 | – | – | – |
| 9 | OPVT-TEC-CPC [8.19, 8.23] | 1 (C = 2) 6 (C = 2) | 0.3840 0.3566 | 0.3072 0.2844 | 0.1172 0.1124 | 0.15 0.15 | 0.5534 0.6346 |
| 10 | FL:ETC [8.35] | 1(C = 5) | 0.9766 | 0.8287 | – – | – – | – |

(continued)

**Table 8.1** (continued)

| S.No | Solar system [Example] | Number of collector (N) | Optical instantaneous efficiency ($\eta_0$) | Thermal instantaneous efficiency ($\eta_{th}$) | PV electrical efficiency | TEC electrical efficiency | An overall thermal efficiency |
|---|---|---|---|---|---|---|---|
| 11 | FLiETC [8.37] | 5(C = 9) | 0.9295 | 0.7549 | – | – | – |
| 12 | FLiSPVT [case (a),8..42] [case (b),8,42] | N = 1 [C = 4, $\beta_c = 0.89$] | 0.2418 | 0.2068 | 0.0609 | – | 0.4920 |
| 13 | FLiSPVT [case (b),8..42] | N = 1 [C = 4, $\beta_c = 0.50$] | 0.4717 | 0.3995 | 0.08778 | | 0.654 |
| 14 | FLiSPVT [8.45] | N = 5 [C = 4, $\beta_c = 0.50$] | 0.4679 | 0.40 | 0.08716 | | 0.8433 |
| 15 | FLiOPVT-TEC [8.62] | N = 1, [C = 4, $\beta_{tec} = 1$] N = 5, [C = 4, $\beta_{tec} = 1$] | 0.6105 0.5904 | 0.5221 0.5047 | 0.1164 0.1131 | 0.15 0.15 | 0.9917 0.8636 |

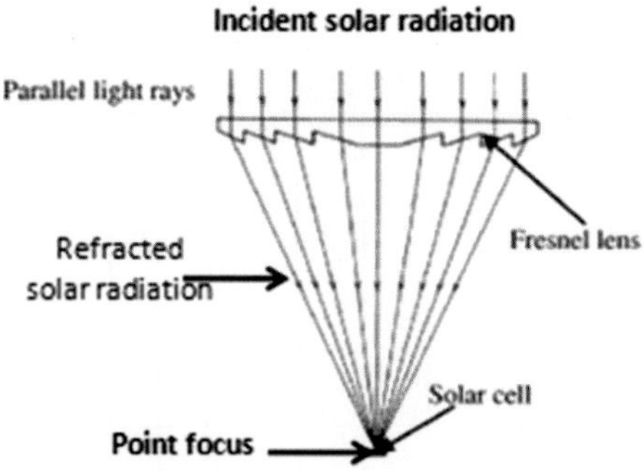

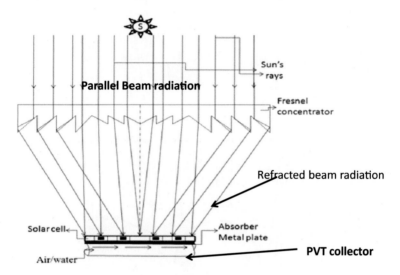

**Fig. 8.1** Cross-sectional view of point focus Fresnel lens at small piece of solar cell

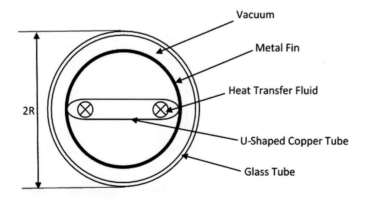

**Fig. 8.2** Cross-sectional view of line focus Fresnel lens concentrator at PVT collector

**Fig. 8.3** Cross-section view of evacuated tubular collector

**(a)**

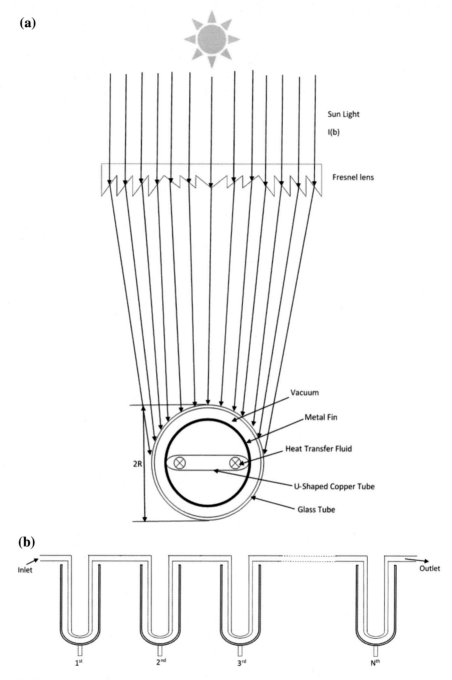

**(b)**

Fig. 8.4 **a** Integration of Fresnel lens with evacuated tubular collector, **b** Series connection of ETC

**Table 8.2**  Various design parameters of evacuated tubular collector (ETC)

| S. No. | Name of design parameters | Symbol of design parameters | Numerical values |
|---|---|---|---|
| 1 | Diameter of copper U-shape tube inside ETC | $D\,(r)$ | 0.024 m (0.012) |
| 2 | Outer radius of cylindrical glass tube ($r_2$) | $r_2$ | 0.0275 m |
| 3 | Inner radius of cylindrical glass tube | $r_1$ | 0.027 |
| 4 | Air gap between two coaxial glass tubes | | 0.0005 m |
| 5 | Outer radius of inner cylindrical glass tube ($r_2$) | $r_1'$ | 0.0265 |
| 6 | Inner radius of inner cylindrical glass tube | $r_0$ | 0.026 |
| 7 | Length of ETC | $L_0$ | 1 m |
| 8 | Specific heat of water | $C_f$ | 4190 j/kg °C |
| 9 | Copper cylindrical tube efficiency factor | $F'$ | 0.985 |
| 10 | Heat transfer coefficient from inner copper tube to working fluid | $h_{pf}$ | 633.95/363.23 W/m² °C |
| 11 | Inner heat transfer coefficient | $h_i$ | 5.7 W/m² °C |
| 12 | Outside heat transfer coefficient from glass tube to an ambient | $h_o$ | 9.5 W/m² °C |
| 13 | Penalty factor | $PF_1$ | 0.9999 |
| 14 | Thickness of glass tube | $L_g$ | 0.0005 m |
| 15 | Thermal conductivity of glass tube | $K_g$ | 1.09 W/m °C |
| 16 | An overall heat transfer coefficient from absorber cylindrical tube to outside ambient through cylindrical out glass tube | $U_{t,pa}$ | 9.73 W/m² °C |
| 17 | Length of copper tube | $L = 2L_o$ | 2 m |
| 18 | Absorptivity of absorber tube | $\alpha_c$ | 0.9 |
| 19 | Transmittivity of glass tube | $\tau_g$ | 0.95 |

where $C = \frac{A_{am}}{A_{rm}}$ with $A_{rm} = DL = 2r_2L$.

Term on LHS $= \tau_g^2 \alpha_c I_b \frac{A_{am}}{A_{rm}} D.dx$ = the rate of concentrated beam radiation absorbed by elemental area of '$D.dx$'.

First term in RHS $= U_{t,pa}(T_p - T_a)D.dx$ = the rate of heat lost from cylindrical absorber to ambient through double glazed cylindrical outer tube.

Second term in RHS $= h_{pf}(T_p - T_f)D.dx =$ the rate of heat transferred from cylindrical absorber to flowing fluid.

**Example 8.24** Evaluate concentration ratio of FL-ETC for design parameters of Table 8.2 with effective area of Fresnel lens of $1 \times 0.5 \, m^2$.

**Solution**

From Eq. 8.8, we have an expression for concentration ratio (C) as

$$C = \frac{A_{am}}{A_{rm}} = \frac{A_{am}}{DL} = \frac{0.5 \, m \times 1 \, m}{0.055 \, m \times 1 \, m} = 9$$

If area of aperture of FL is reduced to $0.275 \, m \times 1 \, m$, **then** concentration ratio (C) becomes as

$$C = \frac{A_{am}}{A_{rm}} = \frac{A_{am}}{DL} = \frac{0.275 \, m \times 1 \, m}{0.055 \, m \times 1 \, m} = 5$$

**Example 8.25** Evaluate an overall heat transfer coefficient $(U_{t,pa})$ from cylindrical absorber to ambient through cylindrical double glazed tube.

**Solution**

Following Tiwari [3], an expression **for** an overall heat transfer coefficient $(U_{t,pa})^3$ from cylindrical absorber to ambient through cylindrical double glazed tube can be written as

$$U_{t,pa} = \left[ \frac{1}{h_o} \times \frac{r_0}{r_2} + \frac{r_0}{K_g} \ln \frac{r_2}{r_1} + \frac{1}{C_0} + \frac{r_0}{K_g} \ln \frac{r_1}{r_0} \right]^{-1}$$

Since air gap between two axial glass tubes is 0.0005 m and air conductance for 0.0005 m air gap is very large [2], hence $\frac{1}{C_0} \approx 0$, then above equation reduces to

$$U_{t,pa} = \left[ \frac{1}{h_o} \times \frac{r_0}{r_2} + \frac{r_0}{K_g} \ln \frac{r_2}{r_1} + \frac{r_0}{K_g} \ln \frac{r_1}{r_0} \right]^{-1}$$

After substitution of appropriate value from Table 8.2, we have

$$U_{t,pa} = \left[ \frac{1}{9.5} \times \frac{0.026}{0.027} + \frac{0.026}{1.09} \ln \frac{0.0275}{0.027} + \frac{0.026}{1.09} \ln \frac{0.027}{0.026} \right]^{-1}$$

$$= [0.1014 + 0.0004545 + 0.0009]^{-1} = \frac{1}{0.1028} = 9.73 \, W/m^2 {}^\circ C$$

**Example 8.26** Evaluate the convective heat transfer coefficient $(h_{pf})$ from inner surface of cylindrical copper absorber for parameters given in Table 8.2 for fully developed water velocity at 0.0.0783 m/s and 0.0392 m/s with an average temperature of 50 °C at angle of 45°.

**Solution**

**Case (i): For flow rate of 0.0.0783 m/s**

The value of $\nu = \frac{\mu}{\rho} = \frac{5.62 \times 10^{-4}}{989.8} = 5.684 \times 10^{-7}$ m$^2$/s (Appendix E2) at 50 °C.
From Example 2.31, we have

$$Re = \frac{Du_0}{\nu} = \frac{(0.024)\,(0.0783)}{5.684 \times 10^{-7}} = 3.31 \times 10^3$$

In this case, Reynolds number, Re, will be replaced by Re.cos $\theta$ in Eq. 2.12b as

$$Nu = 0.023[Re \times \cos \theta]^{0.8} \times Pr^{0.4}$$

From Example 2.27, Pr = 7.88,
This is turbulent flow condition and water heating, and hence, the Nusselt number from Eq. 2.12b and Table 2.5 is given by.

$$Nu = (0.023)\,(3.31 \times 10^3 \cos 45°)^{0.8} \times (7.88)^{0.4} = 0.023 \times 495.96 \times 2.28 = 26.00.$$

The convective heat transfer coefficient from tube surface to working fluid (water) will be determined as

$$h_{pf} = \frac{K_w}{D} \times Nu = \frac{0.585}{0.024} \times 26.00 = 633.95 \text{ W/m}^2 \text{ °C.}$$

This means that there is 100% heat transfer from absorber to working fluid (water).

**Case (ii): For flow rate of 0.0392 m/s**

The value of $\nu = \frac{\mu}{\rho} = \frac{5.62 \times 10^{-4}}{989.8} = 5.684 \times 10^{-7}$ m$^2$/s (Appendix E2) at 50 °C
From Example 2.31, we have

$$Re = \frac{D\,u_0}{\nu} = \frac{(0.024)\,(0.0392)}{5.684 \times 10^{-7}} = 1.65 \times 10^3.$$

In this case, Reynolds number, Re, will be replaced by Re.cos $\theta$ in Eq. 2.12b as

$$Nu = 0.023[Re \times \cos \theta]^{0.8} \times Pr^{0.4}$$

From Example 2.27, Pr = 7.88.
This is turbulent flow condition and water heating, and hence, the Nusselt number from Eq. 2.12b and Table 2.5 is given by

$$\text{Nu} = (0.023)(1.65 \times 10^3 \cos 45°)^{0.8} \times (7.88)^{0.4} = 0.023 \times 284.17 \times 2.28 = 14.90.$$

The convective heat transfer coefficient from tube surface to working fluid (water) will be determined as

$$h_{pf} = \frac{K_w}{D} \times \text{Nu} = \frac{0.585}{0.024} \times 14.90 = 363.23 \ \text{W/m}^2 \ °\text{C}.$$

This means that there is 100% heat transfer from absorber to working fluid (water).

**(b) Cylindrical absorber with flowing fluid**: In this case, there is no bottom loss coefficient like PVGT collector [$h_b = 0$].

The energy balance of flowing fluid inside cylindrical absorber can be written as

$$\dot{m}_f C_f \frac{dT_f}{dx} dx = h_{pf}(T_p - T_f)[(2r_2).dx] \tag{8.9}$$

**Example 8.27**  Evaluate mass flow rate at 0.0783 m/s and 0.0391 m/s (Example 8.26) and $(\dot{m}_f C_f)$ water velocity flowing inside copper absorber tube.

**Solution**

The mass flow rate $(\dot{m}_f) = \rho \times$ cross sectional area of tube $\times$ velocity $= 989.8 \times 3.14 \times (0.012)^2 \times 0.0783 = 0.035$ kg/s (Example 8.26).

Now,

$$(\dot{m}_f C_f) = 0.035 \times 4174 = 146.09 \ \text{W/}°\text{C}$$

and,

$$(\dot{m}_f C_f) = 0.0175 \times 4174 = 73.04 \ \text{W/}°\text{C}$$

From Eq. 8.8, we have an expression for absorber plate temperature $(T_p)$ as

$$T_p = \frac{\tau_g^2 \alpha_c I_b C + U_{t,pa} T_a + h_f T_f}{U_{t,pa} + F'h_f} \tag{8.10}$$

Further,

$$h_f(T_p - T_f) = F'[\tau_g^2 \alpha_c I_b C - U_{t,pa}(T_f - T_a)] \tag{8.11}$$

$$\text{Here,} \quad F' = \frac{h_f}{U_{t,pa} + h_f} = \frac{633.95}{9.73 + 633.95} = 0.985$$

## 8.5.2  Analytical Expression for the Outlet Temperature of Fluid of Single FLiETC

With the help of Eq. 8.11, one can rewrite Eq. 8.9 as

$$\dot{m}_f C_f \frac{dT_f}{dx} dx = F'\left[\tau_g^2 \alpha_c I_b C - U_{t,\mathrm{pa}}(T_f - T_a)\right][(2r_2).dx] \tag{8.12}$$

Equation 8.12 can be rearranged as

$$\frac{dT_f}{dx} + aT_f = f(t) \tag{8.13}$$

where, $a = \frac{F'U_{t,\mathrm{pa}}}{\dot{m}_f C_f}(2r_2)$, $f(t) = \frac{F'\left[\tau_g^2 \alpha_c I_b C + U_{t,\mathrm{pa}} T_a\right](2r_2)}{\dot{m}_f C_f}$ and $\frac{f(t)}{a} = \frac{\tau_g^2 \alpha_c I_b C + U_{t,\mathrm{pa}} T_a}{U_{t,\mathrm{pa}}}$

The solution of Eq. 8.13 with initial condition, namely $T_f = T_{\mathrm{fi}}$ at $x = 0$, becomes as

$$T_f = \frac{f(t)}{a}\left[1 - \exp\left(-\frac{F'U_{t,\mathrm{pa}}}{\dot{m}_f C_f}(2r_2 x)\right)\right] + T_f \exp\left(-\frac{F'U_{t,\mathrm{pa}}}{\dot{m}_f C_f}(2r_2 x)\right) \tag{8.14}$$

The outlet fluid (water) temperature at $x = L_0$ becomes

$$T_{\mathrm{fo}} = \frac{f(t)}{a}\left[1 - \exp\left(-\frac{F'U_{t,\mathrm{pa}}}{\dot{m}_f C_f}(2r_2 \times L_0)\right)\right] + T_f \exp\left(-\frac{F'U_{t,\mathrm{pa}}}{\dot{m}_f C_f}(2r_2 \times L_0)\right)$$

or,

$$T_{\mathrm{fo}} = \frac{f(t)}{a}\left[1 - \exp\left(-\frac{F'U_{t,\mathrm{pa}}}{\dot{m}_f C_f}(2r_2 L_0)\right)\right] + T_f \exp\left(-\frac{F'U_{t,\mathrm{pa}}}{\dot{m}_f C_f}(2r_2 L_0)\right) \tag{8.15}$$

**Example 8.28** Evaluate the outlet temperature for beam radiation $(I_b)$ of 500 W/m$^2$ and inlet temperature of 15 °C and ambient temperature of 16 °C by using the design parameters of Table 8.2 by using Eq. 8.15 for $C = 9$.

**Solution**

Given $I_b = 500$ W/m$^2$, $T_{\mathrm{fi}} = 15$ °C, $T_a = 16$ °C, $U_{t,\mathrm{pa}} = 9.73$ W/m$^2$ °C (Example 8.25) $\tau_g = 0.95$, $\alpha_c = 0.9$, $C = 9$ (Example 8.24), $r_2 = 0.0275$ m, $F' = 0.985$ (Eq. 8.11), $\dot{m}_f C_f = 146.09$ W/°C (Example 4.27) and $L_0 = 1$ m (Table 8.2).
From Eq. 8.13, we have

**Step (i):**

$$\frac{f(t)}{a} = \frac{\tau_g^2 \alpha_c I_b C + U_{t,\mathrm{pa}} T_a}{U_{t,\mathrm{pa}}} = \frac{(0.95)^2 \times 0.9 \times 500 \times 9 + 9.73 \times 16}{9.73}$$

$$= \frac{3.655.125 + 155.68}{9.73} = \frac{3810.805}{9.73} = 391.66$$

**Step (ii):**

$$\frac{F'U_{t,\text{pa}}}{\dot{m}_f C_f}(4r_2 L_0) = \frac{0.985 \times 9.37}{146.09} \times (2 \times 0.0275 \times 1) = 0.00347$$

**Step (iii):**

$$\exp\left[-\frac{F'U_{t,\text{pa}}}{\dot{m}_f C_f}(2r_2 L_0)\right] = \exp(-0.00347) = 0.9965.$$

Substitute the appropriate value in Eq. 8.15, we get

$$T_{\text{fo}} = 391.66[1 - 0.9965] + 15 \times 0.9965 = 16.30\,^{\circ}\text{C}$$

***Example 8.29*** Repeat Example 8.26 with flow velocity of 0.0175 m/s.

**Solution**

Given: $I_b = 500$ W/m$^2$, $T_{\text{fi}} = 15\,^{\circ}$C, $T_a = 16\,^{\circ}$C, $U_{t,\text{pa}} = 9.73$W/m$^2$ $^{\circ}$C (Example 8.25) $\tau_g = 0.95$, $\alpha_c = 0.9$, $C = 9$ (Example 8.24), $r_2 = 0.0275$ m, $\dot{m}_f C_f = 73.04$ W/$^{\circ}$C (Example 4.27) and $L_0 = 1$ m (Table 8.2).
$F' = \frac{h_f}{U_{t,\text{pa}}+h_f} = \frac{363.23}{9.73+363.23} = 0.9748$ (Eq. 8.11 and Example 8.26).
From Eq. 8.13, we have

**Step (i):**

$$\frac{f(t)}{a} = \frac{\tau_g^2 \alpha_c I_b C + U_{t,\text{pa}} T_a}{U_{t,\text{pa}}} = \frac{(0.95)^2 \times 0.5 \times 500 \times 9 + 9.73 \times 16}{9.73}$$

$$= \frac{3.655.125 + 155.68}{9.73} = \frac{3810.805}{9.73} = 391.66$$

**Step (ii):**

$$\frac{F'U_{t,\text{pa}}}{\dot{m}_f C_f}(42L_0) = \frac{0.9748 \times 9.37}{73.04} \times (2 \times 0.0275 \times 1) = 0.006878$$

**Step (iii):**

$$\exp\left[-\frac{F'U_{t,\text{pa}}}{\dot{m}_f C_f}(2r_2 L_0)\right] = \exp(-0.0069) = 0.9931$$

Substitute the appropriate value in Eq. 8.15, we get

$$T_{\text{fo}} = 391.66[1 - 0.9931] + 15 \times 0.9931 = 17.60\,^{\circ}\text{C}$$

Examples 8.28 and 8.29 show that mass flow rate has significant role in the outlet fluid temperature, $T_{\text{fo}}$.

***Example 8.30*** Determine the instantaneous optical and thermal efficiencies of FLiETC for Example 8.28 for $C = 9$.

**Solution**

Here, $A_{\text{am}} = 0.5\,\text{m} \times 1\,\text{m}$ (Example 8.24), $C = 9$, $\dot{m}_f C_f = 146.09\,\text{W}/^{\circ}\text{C}$, $T_{\text{fo}} = 16.30\,^{\circ}\text{C}$ and $T_{\text{fi}} = 15\,^{\circ}\text{C}$.

With the help of Eq. 8.1a, we can define the optical efficiency as

$$\eta_0 = \frac{\dot{Q}_{u,\text{th}}}{\tau_g^2 \alpha_c I_b A_{\text{am}}} = \frac{\dot{m}_f C_f (T_{\text{fo}} - T_{\text{fi}})}{(0.95)^2 \times 0.9 \times 500 \times 0.5} = \frac{146.09 \times 1.30}{203.06} = \frac{189.92}{203.06} = 0.935$$

From Eq. 8.2a, we have an expression for thermal efficiency as

$$\eta_{ith} = \frac{\dot{Q}_{u,\text{th}}}{I_b A_{\text{am}}} = \frac{146.09 \times 1.30}{500 \times 0.5} = \frac{189.92}{250} = 0.7596$$

***Example 8.31*** Determine the instantaneous optical and thermal efficiencies of FLiETC for Example 8.29 for $C = 9$.

**Solution**

Here, $A_{\text{am}} = 0.5\,\text{m} \times 1\,\text{m}$ (Example 8.24), $C = 9$, $\dot{m}_f C_f = 73.04\,\text{W}/^{\circ}\text{C}$, $T_{\text{fo}} = 17.59\,^{\circ}\text{C}$ and $T_{\text{fi}} = 15\,^{\circ}\text{C}$.

With the help of Eq. 8.1a, we can define the optical efficiency as

$$\eta_0 = \frac{\dot{Q}_{u,\text{th}}}{\tau_g^2 \alpha_c I_b A_{\text{am}}} = \frac{\dot{m}_f C_f (T_{\text{fo}} - T_{\text{fi}})}{(0.95)^2 \times 0.9 \times 500 \times 0.5} = \frac{73.04 \times 2.60}{203.06} = \frac{189.90}{203.06} = 0.9324$$

From Eq. 8.2a, we have an expression for thermal efficiency

$$\eta_{ith} = \frac{\dot{Q}_{u,\text{th}}}{I_b A_{\text{am}}} = \frac{73.04 \times 2.60}{500 \times 0.5} = \frac{189.90}{250} = 0.7596$$

### 8.5.3   Characteristic Equation and Mass Flow Rate Factor for FLiETC

The rate of useful thermal energy for FLiETC can be derived as follows:

$$\dot{Q}_{u,\text{th}} = \dot{m}_f C_f (T_{\text{fo}} - T_{\text{fi}}) \tag{8.16a}$$

From Eqs. 8.13 and 8.15, we have an expression for $T_{\text{fo}}$ as

$$T_{\text{fo}} = \frac{\tau_g^2 \alpha_c I_b C + U_{t,\text{pa}} T_a}{U_{t,\text{pa}}} \left[ 1 - \exp\left( -\frac{F' U_{t,\text{pa}}}{\dot{m}_f C_f} (4 r_2 L_0) \right) \right]$$
$$+ T_f \exp\left( -\frac{F' U_{t,\text{pa}}}{\dot{m}_f C_f} (4 r_2 L_0) \right)$$

Substitute above equation into Eq. 8.16a, one gets after simplification

$$\dot{Q}_{u,\text{th}} = F_R A_{\text{am}} \left[ \tau_g^2 \alpha_c I_b C - U_{t,\text{pa}} (T_{\text{fi}} - T_a) \right] \tag{8.16b}$$

where, $A_{\text{am}} = C \times A_{\text{rm}}$

$F_R = \frac{\dot{m}_f C_f}{A_{\text{am}} U_{t,\text{pa}}} \left[ 1 - \exp\left( -\frac{F' U_{t,\text{pa}}}{\dot{m}_f C_f} (2 r_2 L_0) \right) \right]$ is mass flow rate factor of FLiETC.

Further, an instantaneous thermal efficiency is obtained as

$$\eta_{ith} = \frac{\dot{Q}_{u,\text{th}}}{I_b A_{\text{am}}} = F_R \left[ \tau_g^2 \alpha_c C - U_{t,\text{pa}} \frac{(T_f - T_a)}{I_b} \right]$$
$$= \left[ F_R \tau_g^2 \alpha_c C - F_R U_{t,\text{pa}} \frac{(T_{\text{fi}} - T_a)}{I_b} \right] \tag{8.17}$$

Equation 8.17 will be known as characteristic equation for FLiETC similar to Hottel–Whillier–Bliss (HWB)[3] equation for flat plate collector. This is only valid for $\frac{(T_{\text{fi}} - T_a)}{I_b} > 0$.

**Example 8.32**   Evaluate mass flow rate factor for Example 8.28.

**Solution**

From Example 8.28, we have the following:

**Step (i):** $\frac{F' U_{t,\text{pa}}}{\dot{m}_f C_f} (4 r_2 L_0) = \frac{0.985 \times 9.37}{146.09} \times (2 \times 0.0275 \times 1) = 0.00347$.

**Step (ii):** $\exp\left[ -\frac{F' U_{t,\text{pa}}}{\dot{m}_f C_f} (2 r_2 L_0) \right] = \exp(-0.00347) = 0.9965$.

From Eq. 8.16b,

Mass flow rate factor of FLiETC $= F_R = \frac{\dot{m}_f C_f}{A_{\text{am}} U_{t,\text{pa}}} \left[ 1 - \exp\left( -\frac{F' U_{t,\text{pa}}}{\dot{m}_f C_f} (2 r_2 L_0) \right) \right]$.

or,

$$F_R = \frac{146.09}{0.50 \times 9.37}[1 - 0.9965] = 0.1091$$

**Example 8.33** Evaluate an instantaneous thermal efficiency by using Eq. 8.17 by using the data of Example 8.28.

**Solution**

From Example 8.28: Given $I_b = 500$ W/m², $T_{fi} = 15\,°C$, $T_a = 16\,°C$, $U_{t,pa} = 9.73$ W/m² °C (Example 8.25) $\tau_g = 0.95$, $\alpha_c = 0.9$, $C = 9$ (Example 8.24), $F_R = 0.1091$ (Example 8.31).

From Eq. 8.28, we have the following:

$$\eta_{ith} = \left[ F_R \tau_g^2 \alpha_c C - F_R U_{t,pa} \frac{(T_f - T_a)}{I_b} \right]$$

$$= \left[ 0.1091 \times 0.9025 \times 0.9 \times 9 - 0.1091 \times 9.37 \times \frac{-1}{500} \right] = 0.7975.$$

The results for instantaneous thermal efficiency (0.7975) of Example 8.33 for FL-ETC are very close to the result (0.7596) of Example 8.30.

**Example 8.34** Evaluate the outlet temperature for beam radiation ($I_b$) of 500 W/m² and inlet temperature of 15 °C and ambient temperature of 16 °C by using the design parameters of Table 8.2 by using Eq. 8.15 for $C = 5$.

**Solution**

Given $I_b = 500$ W/m², $T_{fi} = 15\,°C$, $T_a = 16\,°C$, $U_{t,pa} = 9.73$ W/m² °C (Example 8.25) $\tau_g = 0.95$, $\alpha_c = 0.9$, $C = 5$ (Example 8.24), $r_2 = 0.0275$ m, $F' = 0.985$ (Eq. 8.11), $\dot{m}_f C_f = 146.09$ W/°C (Example 4.27) and $L_0 = 1$ m (Table 8.2).

From Eq. 8.13, we have

**Step (i):**

$$\frac{f(t)}{a} = \frac{\tau_g^2 \alpha_c I_b C + U_{t,pa} T_a}{U_{t,pa}} = \frac{(0.95)^2 \times 0.9 \times 500 \times 5 + 9.73 \times 16}{9.73}$$

$$= \frac{2186.305 + 155.68}{9.73} = \frac{2341.985}{9.73} = 240.697$$

**Step (ii):**

$$\frac{F' U_{t,pa}}{\dot{m}_f C_f}(4 r_2 L_0) = \frac{0.985 \times 9.37}{146.09} \times (2 \times 0.0275 \times 1) = 0.00347$$

**Step (iii):**

$$\mathbf{exp}\left[-\frac{F'U_{t,\mathrm{pa}}}{\dot{m}_f C_f}(2r_2 L_0)\right] = \exp(-0.00347) = 0.9965.$$

Substitute the appropriate value in Eq. 8.15, we get

$$T_{\mathrm{fo}} = 240.697[1 - 0.9965] + 15 \times 0.9965 = 15.78 \; {}^\circ\mathrm{C}$$

***Example 8.35*** Repeat Example 8.30 for $C = 5$ by using the data of Example 8.34.

**Solution**

Here, $A_{\mathrm{am}} = 0.0.275 \,\mathrm{m} \times 1 \,\mathrm{m}$ (Example 8.24), $C = 5$, $\dot{m}_f C_f = 146.09 \; \mathrm{W}/{}^\circ\mathrm{C}$, $T_{\mathrm{fo}} = 15.78 \,{}^\circ\mathrm{C}$ and $T_{\mathrm{fi}} = 15 \,{}^\circ\mathrm{C}$.

With the help of Eq. 8.1a, we can define the optical efficiency as

$$\eta_0 = \frac{\dot{Q}_{u,\mathrm{th}}}{\tau_g^2 \alpha_c I_b A_{\mathrm{am}}} = \frac{\dot{m}_f C_f (T_{\mathrm{fo}} - T_{\mathrm{fi}})}{(0.95)^2 \times 0.9 \times 500 \times 0.275}$$

$$= \frac{146.09 \times 0.78}{116.68} = \frac{113.95}{116.68} = 0.9766$$

From Eq. 8.2a, we have an expression for thermal efficiency as

$$\eta_{ith} = \frac{\dot{Q}_{u,\mathrm{th}}}{I_b A_{\mathrm{am}}} = \frac{146.09 \times 0.78}{500 \times 0.275} = \frac{113.95}{137.5} = 0.8287$$

Here, the optical as well as thermal instantaneous efficiency is reduced with decrease of concentration ratio from 9 to 5.

### 8.5.4  N-Fresnel Lens Integrated Evacuated Tubular Collector (N-FLiETC) Connected in Series

In this case, the outlet of first FLiETC is connected to inlet of second FLiETC and the outlet of second FLiETC is connected to inlet of third FLiETC, and it continues till $N^{\mathrm{th}}$ FLiETC. The series connection is similar to those as shown in Fig. 8.4b.

Following Sect. 4.5 and by using Eq. 8.15, one can derive an expression for the outlet temperature for N-FLiETC.

$$T_{\mathrm{foN}} = \frac{f(t)}{a}\left[1 - \exp\left(-\frac{NF'U_{t,\mathrm{pa}}}{\dot{m}_f C_f}(2r_2 L_0)\right)\right] + T_f \exp\left(-\frac{NF'U_{t,\mathrm{pa}}}{\dot{m}_f C_f}(2r_2 L_0)\right)$$

$$(8.18)$$

The rate of thermal energy from N-FLiETC will be obtained as

$$\dot{Q}_{u,thN} = \dot{m}_f C_f (T_{foN} - T_{fi}) \tag{8.19}$$

Substitute the expression for $T_{foN}$ from Eq. 8.18 into Eq. 8.19, one gets after simplification

$$\dot{Q}_{u,thN} = F_{RN} N A_{am}\left[\tau_g^2 \alpha_c I_b C - U_{t,pa}(T_{fi} - T_a)\right] \tag{8.20}$$

where, $A_{am} = C \times A_{rm}$

$F_{RN} = \frac{\dot{m}_f C_f}{N A_{am} U_{t,pa}}\left[1 - \exp\left(-\frac{N F' U_{t,pa}}{\dot{m}_f C_f}(2r_2 L_0)\right)\right]$ is mass flow rate factor of N-FLiETC connected in series.

Further, an instantaneous thermal efficiency is obtained as

$$\eta_{ith} = \frac{\dot{Q}_{u,th}}{N I_b A_{am}} = F_{RN}\left[\tau_g^2 \alpha_c C - U_{t,pa}\frac{(T_f - T_a)}{I_b}\right]$$
$$= \left[F_{RN}\tau_g^2 \alpha_c C - F_{RN} U_{t,pa}\frac{(T_{fia} - T_a)}{I_b}\right] \tag{8.21}$$

Equation 8.21 will be known as characteristic equation for N-FLiETC similar to Hottel–Whillier–Bliss (HWB) [3] equation for single flat plate collector. This is only valid for $\frac{(T_{fi} - T_a)}{I_b} > 0$.

***Example 8.36*** Evaluate the outlet temperature for beam radiation ($I_b$) of 500 W/m$^2$ and inlet temperature of 15 °C and ambient temperature of 16 °C by using the design parameters of Table 8.2 by using Eq. 8.21 for $C = 9$ and $N = 5$.

**Solution**

Given $I_b = 500$ W/m$^2$, $T_{fi} = 15$ °C, $T_a = 16$ °C, $U_{t,pa} = 9.73$ W/m$^2$ °C (Example 8.25) $\tau_g = 0.95, \alpha_c = 0.9, C = 9$ (Example 8.24), $N = 5, r_2 = 0.0275$ m, $F' = 0.985$ (Eq. 8.11), $\dot{m}_f C_f = 146.09$ W/ °C (Example 4.27) and $L_0 = 1$ m (Table 8.2).
From Eq. 8.13, we have

**Step (i):**

$$\frac{f(t)}{a} = \frac{\tau_g^2 \alpha_c I_b C + U_{t,pa} T_a}{U_{t,pa}} = \frac{(0.95)^2 \times 0.9 \times 500 \times 9 + 9.73 \times 16}{9.73}$$
$$= \frac{3.655.125 + 155.68}{9.73} = \frac{3810.805}{9.73} = 391.66$$

**Step (ii):**

$$\frac{N F' U_{t,pa}}{\dot{m}_f C_f}(4r_2 L_0) = \frac{5 \times 0.985 \times 9.37}{146.09} \times (2 \times 0.0275 \times 1) = 0.01735$$

**Step (iii):**

$$\exp\left[-\frac{F'U_{t,\text{pa}}}{\dot{m}_f C_f}(2r_2 L_0)\right] = \exp(-0.01735) = 0.9828.$$

Substitute the appropriate value in Eq. 8.15, we get

$$T_{\text{fo}} = 391.66[1 - 0.9828] + 15 \times 0.9828 = 21.46\,^{\circ}\text{C}$$

***Example 8.37*** Determine the instantaneous optical and thermal efficiencies of FLiETC for Example 8.28 for $C = 9$ and $N = 5$.

**Solution**

Here, $A_{\text{am}} = 0.5\,\text{m} \times 1\,\text{m}$ (Example 8.24), $C = 9$ and $N = 5$, $\dot{m}_f C_f = 146.09\,\text{W}/\,^{\circ}\text{C}$, $T_{\text{fo}} = 21.46\,^{\circ}\text{C}$ and $T_{\text{fi}} = 15\,^{\circ}\text{C}$.

With the help of Eq. 8.1a, we can define the optical efficiency as

$$\eta_0 = \frac{\dot{Q}_{u,\text{th}}}{\tau_g^2 \alpha_c I_b N A_{\text{am}}} = \frac{\dot{m}_f C_f (T_{\text{fo}} - T_{\text{fi}})}{(0.95)^2 \times 0.9 \times 500 \times 5 \times 0.5}$$
$$= \frac{146.09 \times 6.46}{1015.3} = \frac{943.74}{1015.3} = 0.9295$$

From Eq. 8.2a, we have an expression for thermal efficiency

$$\eta_{i\text{th}} = \frac{\dot{Q}_{u,\text{th}}}{I_b A_{\text{am}}} = \frac{146.09 \times 6.46}{500 \times 5 \times 0.5} = \frac{943.74}{1250} = 0.7549$$

## 8.6  The Fresnel Lens (FL) Integrated with Liquid Semi-Transparent Photo-Voltaic Thermal Collector (SPVT) Collector [FLiSPVT]

When Fresnel lens (FL) is integrated with semi-transparent photo-voltaic thermal collector (SPVT), Chap. 5, it is referred as liquid FLiSPVT similar to figure shown in Fig. 8.5 for Al/opaque PV module with airflow due to limitation of concentration ratio of compound parabolic concentrator (CPC), Chap. 7. The analysis of Fig. 8.5 can be also carried out. In this case, beam radiation ($I_b$) is concentrated on the top surface of SPVT collector and then transmitted to absorber plate. The liquid FLiSPVT works on the principle of line focus by Fresnel lens. So, in this case, the concentration ratio (C) will be more than CPC and less than FLiETC. The design parameters of liquid FLiSPVT are given in Table 8.3.

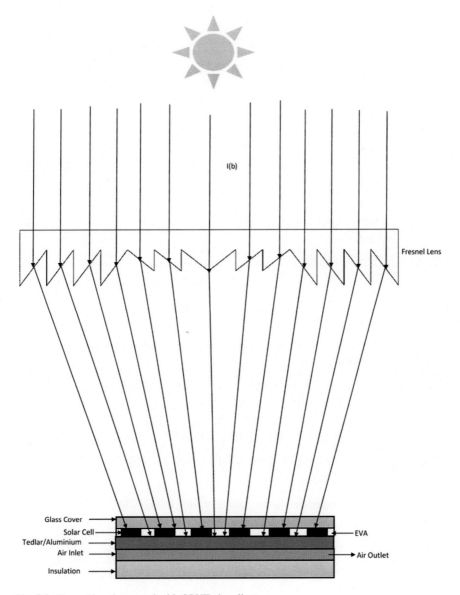

**Fig. 8.5**  Fresnel lens integrated with OPVT air collector

## *8.6.1  Energy Balance of Single Liquid FLiSPVT*

Basically, the energy balances, derivations for all variable parameters will be exactly same as written and derivation for CPC-PVT collector except $I_u$ will be replaced by

**Table 8.3** Design parameters of FLiSPVT collector

| S. No. | Name of design parameters | Symbol of design parameters | Numerical values |
|---|---|---|---|
| 1 | Aperture area of FLiSPVT | $A_{am}$ | $2\,m^2$ |
| 2 | Receiver area of FLiSPVT | $A_{rm}$ | $1\,m^2, 0.5\,m^2, 0.25\,m^2$ |
| 3 | Concentration ratio | C | 2, 4 and 8 |
| 4 | Example 7.3 | $F_0$ | 0.233 |
| 5 | Table 7.1 | $F'$ | 0.82 |
| 6 | Example 7.4 | $U_{b,cp}$ | $2.7663\,W/m^2\,°C$ |
| 7 | Do | $U_{pa}$ | $2.1225\,W/m^2\,°C$ |
| 8 | Do | $U_{t,ca}$ | $9.12\ \,W/m^2\,°C$ |
| | **Greek letters** | | |
| 9 | Absorptivity of absorber tube | $\alpha_c$ | 0.9 |
| 10 | Packing factor | $\beta_c$; | 0.89 and 0.50 |
| 11 | Transmittivity of glass tube | $\tau_g$ | 0.95 |
| 12 | Electrical efficiency of SPVT | $\eta_m$ | 0.12 |

$I_b$. For convenience of reader, the energy balances for different components of liquid FLiSPVT have been rewritten:

**(a) Solar Cell of SPVT**

Following Eq. 7.1, we have the energy balance of solar cell as

$$\alpha_c \tau_g \beta_c I_b \frac{A_{am}}{A_{rm}} = U_{t,ca}(T_c - T_a) + U_{b,cp}(T_c - T_p) + \rho I_u \eta_m \beta_c \qquad (8.22)$$

where, $A_{am}$ and $A_{rm}$ are surface areas of aperture of FL and PVT collector and the concentration ratio $(C) = \frac{A_{am}}{A_{rm}} > 1$.

**Example 8.38** Determine the various concentration ratios (C) of FLiSPVT as shown in Fig. 8.6.

**Solution**

From Eq. 8.22, we have an expression for concentration ratios (C) as

$$C = \frac{A_{am}}{A_{rm}}.$$

The value of an aperture area $(A_{am})$ should be larger than receiver area $(A_{rm})$. Since, the area of receiver is equal to area of SPVT collector. We will discuss the concentration ratio (C) for three cases, namely.

**Case (i)**: For FL area $(A_{am})$ of 2 m × 1m $= 2\,m^2$ and receiver area $(A_{rm})$ of 1 m × 1 m $=\,m^2$, in this case, the concentration ratio (C) will be as

$$C = \frac{A_{\text{am}}}{A_{\text{rm}}} = \frac{2}{1} = 2.$$

**Case (ii):** For FL area ($A_{\text{am}}$) of $2\,\text{m} \times 1\text{m} = 2\,\text{m}^2$ and receiver area ($A_{\text{rm}}$) of $0.5\,\text{m} \times 1\text{m} = \text{m}^2$, in this case, the concentration ratio (C) will be as

$$C = \frac{A_{\text{am}}}{A_{\text{rm}}} = \frac{2}{0.5} = 4.$$

**Case (iii):** For FL area ($A_{\text{am}}$) of $2\,\text{m} \times 1\,\text{m} = 2\,\text{m}^2$ and receiver area ($A_{\text{rm}}$) of $0.25\,\text{m} \times 1\,\text{m} = \text{m}^2$, in this case, the concentration ratio (C) will be as

$$C = \frac{A_{\text{am}}}{A_{\text{rm}}} = \frac{2}{0.25} = 8.$$

Following Eq. 7.2 and using Eq. 8.22, one can get an expression for solar cell temperature ($T_c$) of FLiPVT as follows:

$$T_c = \frac{(\alpha\tau)_{\text{eff}} I_b + U_{t,\text{ca}} T_a + U_{b,\text{cp}} T_p}{U_{t,\text{ca}} + U_{t,\text{cp}}} \tag{8.23}$$

where,

$(\alpha\tau)_{\text{eff}} = \left(\alpha_c \tau_g \frac{A_{\text{am}}}{A_{\text{rm}}} - \eta_m\right)\beta_c$, an effective product of absorptivity and transmittivity of semi-transparent SPV module of FLiSPVT collector.

***Example 8.39*** Determine the effective product of absorptivity and transmittivity of semi-transparent SPV module of FLiSPVT collector for $C = 4$ and $8$ for different packing factors.

**Solution**

From Eq. 8.23, we have an effective product of absorptivity and transmittivity of semi-transparent SPV module of FLiSPVT collector (($\alpha\tau)_{\text{eff}}$) as

$$(\alpha\tau)_{\text{eff}} = \left(\alpha_c \tau_g \frac{A_{\text{am}}}{A_{\text{rm}}} - \eta_m\right)\beta_c = (\alpha_c \tau_g C - \eta_m)\beta_c$$

**Case (i):** For $C = 4$ and $\beta_c = 0.89$, then substitute an appropriate value from Table 8.3 in the above equation

$$(\alpha\tau)_{\text{eff}} = (\alpha_c \tau_g C - \eta_m)\beta_c = [0.9 \times 0.9 \times 4 - 0.12] \times 0.89 = 2.937$$

**Case (ii):** For $C = 4$ and $\beta_c = 0.5$, then substitute an appropriate value from Table 8.3 in the following equation:

$$(\alpha\tau)_{\text{eff}} = (\alpha_c\tau_g C - \eta_m)\beta_c = [0.9 \times 0.9 \times 4 - 0.12] \times 0.50 = 1.65$$

**Case (iii):** For $C = 8$ and $\beta_c = 0.89$, then substitute an appropriate value from Table 8.3 in the following equation:

$$(\alpha\tau)_{\text{eff}} = (\alpha_c\tau_g C - \eta_m)\beta_c = [0.9 \times 0.9 \times 8 - 0.12] \times 0.0.89 = 5.981$$

**Case (iv):** For $C = 8$ and $\beta_c = 0.50$, then substitute an appropriate value from Table 8.3 in the following equation:

$$(\alpha\tau)_{\text{eff}} = (\alpha_c\tau_g C - \eta_m)\beta_c = [0.9 \times 0.9 \times 8 - 0.12] \times 0.0.50 = 3.36$$

For comparison purpose, we will consider cases (i), (ii) and (iii) only.

### 8.6.2   Energy Balance Equation for Absorber Plate of SPVT

The energy balance of the absorber in W can be written as

$$\alpha_p\tau_g^2(1 - \beta_c)I_b\frac{A_{\text{am}}}{A_{\text{rm}}} + U_{b,\text{cp}}(T_c - T_p) = F'h_{\text{pf}}(T_p - T_f) + U_b(T_p - T_a) \quad (8.24)$$

Following the procedure given in Sect. 7.2.2 and Examples 7.3–7.5, one can get a similar expression as Eq. 7.4 for absorber plate temperature as

$$T_p = \frac{\left[\alpha_p\tau_g^2(1 - \beta_c)C + F_0(\alpha\tau)_{\text{eff}}\right]I_u + (U_{\text{pa}} + U_b)T_a + F'h_{\text{pf}}T_f}{(U_{\text{pa}} + F'h_{\text{pf}} + U_b)} \quad (8.25)$$

where, $F_0 = \frac{U_{\text{bc},a}}{U_{\text{tc},p}+U_{\text{bc},a}} = 0.233$, $(\alpha\tau)_{\text{eff}} = (\alpha_c\tau_g C - \eta_m)\beta_c$, Eq. 8.23.

From Eq. 7.5, the rate of thermal energy transferred from absorber plate to working fluid can be obtained as

$$F'h_{\text{pf}}(T_p - T_f) = F'\left[(\alpha\tau)_{m,\text{eff}}I_b - U_{\text{Lm}}(T_f - T_a)\right] \quad (8.26)$$

where,

$(\alpha\tau)_{m,\text{eff}} = F_{01}\left[\alpha_p\tau_g^2(1 - \beta_c)C + F_0(\alpha\tau)_{\text{eff}}\right]$, an effective product of absorptivity and transmittivity of FLiSPVT collector. The values of C and $(\alpha\tau)_{\text{eff}}$ are given in Examples 8.38 and 8.39.

$$F_{01} = \frac{h_{\text{pf}}}{(U_{\text{pa}}+F'h_{\text{pf}}+U_b)} = 0.97 \text{ and } U_{\text{Lm}} = \frac{h_{\text{pf}}(U_{\text{pa}}+U_b)}{(U_{\text{pa}}+F'h_{\text{pf}}+U_b)} = 2.76 \text{ W/m}^2 \text{ }^\circ\text{C}$$

(Example 7.6).

**Example 8.40** Evaluate an effective product of absorptivity and transmittivity of FLiSPVT collector $[(\alpha\tau)_{m,\text{eff}}]$ by using Eq. 8.26.

**Solution**

**Case (a): For $C = 4$, $\beta_c = 0.89$ and $(\alpha\tau)_{\text{eff}} = 2.937$.**
From Table 8.3: $\tau_g = 0.95$, $\alpha_c = 0.9$, $\beta_c = 0.89$, $A_{\text{am}} = 2m^2$, $\alpha_p = 0.9$, $C = 4$ and $A_{\text{rm}} = 0.5\ m^2$.
(case (ii) in Example 8.38), $(\alpha\tau)_{\text{eff}} = 2.937$ (case (i) in Example 8.39).
An effective product of absorptivity and transmittivity of FLiSPVT collector is obtained as

$$
\begin{aligned}
(\alpha\tau)_{m,\text{eff}} &= F_{01}\big[\alpha_p\tau_g^2(1 - \beta_c)C + F_0(\alpha\tau)_{\text{eff}}\big] \\
&= 0.97\big[0.9 \times (0.95)^2(1 - 0.89) \times 4 + 0.233 \times 2.937\big] \\
&= 0.97 \times [0.3574 + 0.6843] = 1.0106
\end{aligned}
$$

**Case (b): For $C = 4$, $\beta_c = 0.5$ and $(\alpha\tau)_{\text{eff}} = 1.65$**
From Table 8.3: $\tau_g = 0.95$, $\alpha_c = 0.9$, $\beta_c = 0.50$, $A_{\text{am}} = 2m^2$, $\alpha_p = 0.9$, $C = 4$ and $A_{\text{rm}} = 0.5\ m^2$.
(case (ii) in Example 8.38), $(\alpha\tau)_{\text{eff}} = 1.65$ case (ii) in (Example 8.39).
An effective product of absorptivity and transmittivity of FLiSPVT collector is obtained as

$$
\begin{aligned}
(\alpha\tau)_{m,\text{eff}} &= F_{01}\big[\alpha_p\tau_g^2(1 - \beta_c)C + F_0(\alpha\tau)_{\text{eff}}\big] \\
&= 0.97\big[0.9 \times (0.95)^2(1 - 0.50) \times 4 + 0.233 \times 1.65\big] \\
&= 0.97 \times [1.6245 + 0.3845] = 1.9487
\end{aligned}
$$

**Case (c): For $C = 8$, $\beta_c = mathbf{0.89}$ and $(\alpha\tau)_{\text{eff}} = 5.981 (\alpha\tau)_{\text{eff}} = 5.981$**
From Table 8.3: $\tau_g = 0.95$, $\alpha_c = 0.9$, $\beta_c = 0.89$, $A_{\text{am}} = 2m^2$, $\alpha_p = 0.9$, $C = 8$ and $A_{\text{rm}} = 0.25\ m^2$.
(case (iii) in Example 8.38), $(\alpha\tau)_{\text{eff}} = 5.981$ case (iii) in (Example 8.39).
An effective product of absorptivity and transmittivity of FLiSPVT collector is obtained as

$$
\begin{aligned}
(\alpha\tau)_{m,\text{eff}} &= F_{01}\big[\alpha_p\tau_g^2(1 - \beta_c)C + F_0(\alpha\tau)_{\text{eff}}\big] \\
&= 0.97\big[0.9 \times (0.95)^2(1 - 0.89) \times 8 + 0.233 \times 5.981\big] \\
&= 0.97 \times [0.7148 + 1.3933] = 2.0449
\end{aligned}
$$

### 8.6.3  Energy Balance for Flowing Fluid (Water) in Tube Below the Absorber Plate of Single FLiSPVT Collector

Following Sect. 7.2.3 and Eq. 7.7a, the rate of thermal energy carried away through elemental area of 'bdx' will be equal to the rate of thermal energy available in the copper blackened absorber plate which will be written as

$$\dot{m}_f C_f \frac{dT_f}{dx} dx = F' \left[ (\alpha\tau)_{m,\text{eff}} I_b - U_{\text{Lm}} (T_f - T_a) \right] b dx \qquad (8.27)$$

### 8.6.4  The Outlet Fluid Temperature of FLiSPVT

Following Sect. 7.2.4 and after rearrangement of Eq. 8.27, the solution of Eq. 8.27 can be obtained as follows:

$$T_f = \left[ \frac{\left[ (\alpha\tau)_{m,\text{eff}} I_b + U_{\text{Lm}} T_a \right]}{U_{\text{Lm}}} \right] \left[ 1 - \exp\left\{ \frac{-F' U_{\text{Lm}} bx}{\dot{m}_f C_f} \right\} \right] + T_{\text{fi}} \exp\left\{ \frac{-F' U_{\text{Lm}} bx}{\dot{m}_f C_f} \right\}$$

$$(8.28)$$

where, $T_f|_{x=0} = T_{\text{fi}}$

Further, the outlet temperature of fluid (water) at the end of the FLiSPVT collector can be obtained as:

$$T_{\text{fo}} = T_f|_{x=L_0} = \left[ \frac{\left[ (\alpha\tau)_{m,\text{eff}} I_b + U_{\text{Lm}} T_a \right]}{U_{\text{Lm}}} \right] \left[ 1 - \exp\left\{ \frac{-F' U_{\text{Lm}} b L_0}{\dot{m}_f C_f} \right\} \right]$$

$$+ T_{\text{fi}} \exp\left\{ \frac{-F' U_{\text{Lm}} b L_0}{\dot{m}_f C_f} \right\}$$

or,

$$T_{\text{fo}} = \left[ \frac{\left[ (\alpha\tau)_{m,\text{eff}} I_b + U_{\text{Lm}} T_a \right]}{U_{\text{Lm}}} \right] \left[ 1 - \exp\left\{ \frac{-F' U_{\text{Lm}} A_{\text{rm}}}{\dot{m}_f C_f} \right\} \right]$$

$$+ T_{\text{fi}} \exp\left\{ \frac{-F' U_{\text{Lm}} A_{\text{rm}}}{\dot{m}_f C_f} \right\} \qquad (8.29)$$

where, $A_{\text{rm}} = b L_0$.

***Example 8.41*** Determine the outlet temperature ($T_{\text{fo}}$) by using Eq. 8.29 for single liquid FLiSPVT for Example 8.40 for all cases for the climatic data of $I_b = 500 \text{ W/m}^2$, $T_a = 16\,°\text{C}$ and $T_{\text{fi}} = 15\,°\text{C}$. Also determine the average fluid temperature for all cases.

**Solution**

**Known parameters**: $U_{\text{Lm}} = 2.76 \text{ W/m}^2 \, ^\circ\text{C}$ (Example 7.6), $\dot{m}_f C_f = 146.65 \text{ W/}^\circ\text{C}$, $F' = 0.82$ (Table 8.3).

From Example 8.40, we have the following cases:

**Case (a)**: For $C = 4$, $\beta_c = 0.89$ and $(\alpha\tau)_{m,\text{eff}} = 1.0106$, $A_{\text{rm}} = 0.5 \text{ m}^2$.

**Step 1**: $\left[ \dfrac{(\alpha\tau)_{m,\text{eff}} I_b + U_{\text{Lm}} T_a}{U_{\text{Lm}}} \right] = \dfrac{1.0106 \times 500 + 2.76 \times 16}{2.76} = \dfrac{549.46}{2.76} = 199.08$

**Step 2**:

$$\frac{F' U_{\text{Lm}} A_{\text{rm}}}{\dot{m}_f C_f} = \frac{0.82 \times 2.76 \times 0.5}{146.65} = 0.0077716$$

**Step 3**:

$$\exp\left\{ \frac{-F' U_{\text{Lm}} A_{\text{rm}}}{\dot{m}_f C_f} \right\} = \exp(-0.0077716) = 0.9923$$

Substitute the above value in Eq. 8.29, one gets the outlet temperature of single FLiSPVT collector as

$$T_{\text{fo}} = 199.08[1 - 0.9923] + 15 \times 0.9923 = 16.41 \, ^\circ\text{C}$$

An average fluid temperature $(\overline{T}_f) = \frac{T_{\text{fo}} + T_{\text{fi}}}{2} = \frac{16.41 + 15}{2} = 15.705 \, ^\circ\text{C}$

**Case (b)**: For $C = 4$, $\beta_c = 0.5$ and $(\alpha\tau)_{m,\text{eff}} = 1.9487$, $A_{\text{rm}} = 0.5 \text{ m}^2$.

**Step 1**:

$$\left[ \frac{(\alpha\tau)_{m,\text{eff}} I_b + U_{\text{Lm}} T_a}{U_{\text{Lm}}} \right] = \frac{1.9487 \times 500 + 2.76 \times 16}{2.76} = \frac{1018.51}{2.76} = 372.90$$

**Step 2**:

$$\frac{F' U_{\text{Lm}} A_{\text{rm}}}{\dot{m}_f C} = \frac{0.82 \times 2.76 \times 0.5}{146.65} = 0.0077716$$

**Step 3**:

$$\exp\left\{ \frac{-F' U_{\text{Lm}} A_{\text{rm}}}{\dot{m}_f C_f} \right\} = \exp(-0.0077716) = 0.9923$$

Substitute the above value in Eq. 8.29, one gets the outlet temperature of single FLiSPVT collector as

$$T_{\text{fo}} = 372.90[1 - 0.9923] + 15 \times 0.9923 = 17.75\,^\circ\text{C}$$

An average fluid temperature $\left(\overline{T}_f\right) = \frac{T_{\text{fo}} + T_{\text{fi}}}{2} = \frac{17.75 + 15}{2} = 16.375\,^\circ\text{C}$

**Case (c): For C = 8, $\beta_c = 0.89$ and $(\alpha\tau)_{m,\text{eff}} = 2.0449$, $A_{\text{rm}} = 0.25$ m$^2$.**

**Step 1:**

$$\left[\frac{\left[(\alpha\tau)_{m,\text{eff}} I_b + U_{\text{Lm}} T_a\right]}{U_{\text{Lm}}}\right] = \frac{2.0449 \times 500 + 2.76 \times 16}{2.76} = \frac{1066.61}{2.76} = 386.45$$

**Step 2:**

$$\frac{F' U_{\text{Lm}} A_{\text{rm}}}{\dot{m}_f C_f} = \frac{0.82 \times 2.76 \times 0.25}{146.65} = 0.003858$$

**Step 3:**

$$\exp\left\{\frac{-F' U_{\text{Lm}} A_{\text{rm}}}{\dot{m}_f C_f}\right\} = \exp(-0.003858) = 0.99615$$

Substitute the above value in Eq. 8.29, one gets the outlet temperature of single FLiSPVT collector as

$$T_{\text{fo}} = 386.45[1 - 0.99615] + 15 \times 0.99615 = 16.43\,^\circ\text{C}$$

An average fluid temperature $\left(\overline{T}_f\right) = \frac{T_{\text{fo}} + T_{\text{fi}}}{2} = \frac{16.43 + 15}{2} = 15.15\,^\circ\text{C}$

This example shows that importance of packing factor of SPVT determines the direct gain to absorber plate, and hence now we will consider only case (a) and case (b) for next examples.

### 8.6.5  The Rate of Useful Thermal Energy and Mass Flow Rate Factor for Single FLiSPVT

With the help of Eq. 8.29, the rate of thermal energy $\left(\dot{Q}_{u.\text{th}}\right)$ can be obtained as

$$\dot{Q}_{u.\text{th}} = \dot{m}_f C_f (T_{\text{fo}} - T_{\text{fi}}) = F_R A_{\text{rm}}\left[\left[(\alpha\tau)_{m,\text{eff}} I_b - U_{\text{Lm}}(T_{\text{fi}} - T_a)\right]\right] \qquad (8.30)$$

$$F_R = \frac{\dot{m}_f C_f}{U_{Lm} A_{rm}}\left[1 - \exp\left\{\frac{-F' U_{Lm} A_{rm}}{\dot{m}_f C_f}\right\}\right] \quad \text{is a mass flow rate factor.} \quad (8.31)$$

Further, the thermal characteristic equation can be obtained as

$$\eta_{ith} = \frac{\dot{Q}_{u.th}}{I_b A_{am}} = F_R \frac{A_{rm}}{A_{am}}\left[(\alpha\tau)_{m,\text{eff}} - U_{Lm}\frac{(T_{fi} - T_a)}{I_b}\right] \quad (8.32)$$

**Example 8.42** Determine the rate of thermal energy, mass flow rate factor, an instantaneous optical and thermal efficiency for cases (a) and (b) for Example 8.41 for single FLiSPVT collector.

**Solution**

**Given:** $\dot{m}_f C_f = 146.65$ W/ $°C$, $I_b = 500$ W/m$^2$, $\tau_g = 0.95$, $\alpha_c = 0.9$, $T_{fi} = 15\,°C$, $T_a = 16\,°C$, $U_{Lm} = 2.76$ W/m$^2$ $°C$.

**Case (a):** For $C = 4$, $\beta_c = 0.89$ and $(\alpha\tau)_{m,\text{eff}} = 1.0106$, $\alpha\tau = \alpha_p\tau_g$, $A_{rm} = 0.5$ m$^2$, $A_{am} = 2.0$ m$^2$, $\exp\left\{\frac{-F'U_{Lm}A_{rm}}{\dot{m}_f C_f}\right\} = 0.9923$, $T_{fo} = 16.41\,°C$, $\dot{m}_f C_f = 146.65$ W/ $°C$.

The rate of thermal energy $= \dot{Q}_{u.th} = \dot{m}_f C_f (T_{fo} - T_{fi}) = 146.65(16.41 - 15) = 206.78$ W.

The mass flow rate $= F_R = \frac{\dot{m}_f C_f}{U_{Lm} A_{rm}}\left[1 - \exp\left\{\frac{-F'U_{Lm}A_{rm}}{\dot{m}_f C_f}\right\}\right] = \frac{146.65}{2.76\times0.5}(1 - 0.9923) = 0.82$ (Eq. 8.31).

With the help of Eq. 8.1a, we can define the optical efficiency as

$$\eta_0 = \frac{\dot{Q}_{u.th}}{\alpha_p\tau_g I_b A_{am}} = \frac{\dot{m}_f C_f (T_{fo} - T_{fi})}{0.95 \times 0.9 \times 500 \times 2} = \frac{146.09 \times 1.41}{855} = \frac{206.78}{855} = 0.2418$$

From Eq. 8.2a, we have an expression for thermal efficiency as

$$\eta_{ith} = \frac{\dot{Q}_{u.th}}{I_b A_{am}} = \frac{146.09 \times 1.41}{500 \times 2} = \frac{206.78}{1000} = 0.2068$$

From Eq. 8.22, we also have.

$$\eta_{ith} = \frac{\dot{Q}_{u.th}}{I_b A_{am}} = F_R\left[\frac{A_{rm}}{A_{am}}\right]\left[(\alpha\tau)_{m,\text{eff}} - U_{Lm}\frac{(T_{fi}-T_a)}{I_b}\right] = 0.82 \times \frac{0.5}{2} \times 1.0106 = 0.2072.$$

Since, $U_{Lm}\frac{(T_{fi}-T_a)}{I_b}$ is negative and hence, it is considered as zero.

**Case (b):** For $C = 4$, $\beta_c = 0.50$ and $(\alpha\tau)_{m,\text{eff}} = 1.9487$, $\alpha\tau = \alpha_p\tau_g$, $A_{rm} = 0.5$ m$^2$, $A_{am} = 2.0$ m$^2$, $\exp\left\{\frac{-F'U_{Lm}A_{rm}}{\dot{m}_f C_f}\right\} = 0.9923$, $T_{fo} = 17.75\,°C$, $\dot{m}_f C_f = 146.65$ W/ $°C$.

The rate of thermal energy $= \dot{Q}_{u.th} = \dot{m}_f C_f (T_{fo} - T_{fi}) = 146.65(17.75 - 15) = 403.28$ W.

The   mass   flow   rate   $= F_R = \frac{\dot{m}_f C_f}{U_{\text{Lm}} A_{\text{rm}}}\left[1 - exp\left\{\frac{-F' U_{\text{Lm}} A_{\text{rm}}}{\dot{m}_f C_f}\right\}\right] =$
$\frac{146.65}{2.76 \times 0.5}(1 - 0.9923) = 0.82.$

With the help of Eq. 8.1a, we can define the optical efficiency as

$$\eta_0 = \frac{\dot{Q}_{u,\text{th}}}{\alpha_p \tau_g I_b A_{\text{am}}} = \frac{\dot{m}_f C_f (T_{\text{fo}} - T_{\text{fi}})}{0.95 \times 0.9 \times 500 \times 2} = \frac{146.09 \times 2.75}{855} = \frac{403.28}{855} = 0.4717$$

From Eq. 8.2a, we have an expression for thermal efficiency as

$$\eta_{ith} = \frac{\dot{Q}_{u,\text{th}}}{I_b A_{\text{am}}} = \frac{146.09 \times 2.75}{500 \times 2} = \frac{403.28}{1000} = 0.4033$$

From Eq. 8.22, we also have.

$\eta_{ith} = \frac{\dot{Q}_{u,th}}{I_b A_{am}} = F_R \frac{A_{rm}}{A_{am}}\left[(\alpha\tau)_{m,\text{eff}} - U_{\text{Lm}}\frac{(T_{fi}-T_a)}{I_b}\right] = 0.82 \times \frac{0.5}{2} \times 1.9487 = 0.3995.$

Since, $U_{\text{Lm}}\frac{(T_{fi}-T_a)}{I_b}$ is negative and hence it is considered as zero.

**Example 8.43** Evaluate the average absorber plate and solar cell temperature of single FLiSPVT liquid collector for cases (a) and (b) of Example 8.41 by using Eqs. 8.23 and 8.24, respectively.

**Solution**

**Given:** $U_{t,\text{ca}} = 9.12$ W/ m$^2$ °C, $U_{b,\text{cp}} = 2.7663$ W/m$^2$ °C, $F_0 = 0.233$, $I_b = 500$ W/m$^2$, $\tau_g = 0.95$, $\alpha_c = \alpha_p = 0.9$, $U_b = 0.7$ W/m$^2$ °C, $F' = 0.82$, $F' h_{\text{pf}} = 0.82 \times 100 = 82$ W/m$^2$ °C, $U_{\text{pa}} = 2.1225$ W/m$^2$ °C (Table 8.3).

**Case (a):** For C = 4, $\beta_c = 0.89$ and $(\alpha\tau)_{m,\text{eff}} = 1.0106$, $(\alpha\tau)_{\text{eff}} = 2.937$, $A_{\text{rm}} = 0.5$ m$^2$, $A_{\text{am}} = 2.0$ m$^2$, $\left[\alpha_p \tau_g^2(1 - \beta_c)C + F_0(\alpha\tau)_{,\text{eff}}\right] = [0.3574 + 0.6843] = 1.0417$, $F_{01} = 0.97.$

From Eq. 8.24, one can first evaluate an average absorber plate temperature $(\overline{T}_p)$ by using the average fluid temperature from Example 8.41 [$\overline{T}_f == 15.705$ °C].

$$
\begin{aligned}
\overline{T}_p &= \frac{\left[\alpha_p \tau_g^2(1 - \beta_c)C + F_0(\alpha\tau)_{,\text{eff}}\right]I_b + (U_{\text{pa}} + U_b)T_a + F' h_{\text{pf}}\overline{T}_f}{(U_{\text{pa}} + F' h_{\text{pf}} + U_b)} \\
&= \frac{1.0417 \times 500 + (2.1225 + 0.7) \times 16 + 82 \times 15.705}{(2.1225 + 82 + 0.7)} \\
&= \frac{520.85 + 45.16 + 1287.81}{84.82} = \frac{1853}{84.82} = 21.85 \text{ °C}
\end{aligned}
$$

Now, from Eq. 8.23, one gets

$$
\begin{aligned}
\overline{T}_c &= \frac{(\alpha\tau)_{\text{eff}}I_b + U_{t,\text{ca}}T_a + U_{b,\text{cp}}\overline{T}_p}{U_{t,\text{ca}} + U_{t,\text{cp}}} \\
&= \frac{2.937 \times 500 + 9,12 \times 16 + 2.7663 \times 21.85}{9.12 + 2.7663} = \frac{1674.86}{11.8863} = 140.90 \text{ °C}
\end{aligned}
$$

**Case (b):** For $C = 4$, $\beta_c = 0.50$ and $(\alpha\tau)_{m,\mathrm{eff}} = 1.9487$, $(\alpha\tau)_{\mathrm{eff}} = 1.65$, $A_{\mathrm{rm}} = 0.5\,\mathrm{m}^2$, $A_{\mathrm{am}} = 2.0\,\mathrm{m}^2$, $\left[\alpha_p\tau_g^2(1 - \beta_c)C + F_0(\alpha\tau)_{,\mathrm{eff}}\right] = [1.6245 + 0.3845] = 2.009$, $F_{01} = 0.97$.

From Eq. 8.24, one can first evaluate an average absorber plate temperature $(\overline{T}_p)$ by using the average fluid temperature from Example 8.41 $\left[\overline{T}_f == 16.375\,°\mathrm{C}\right]$.

$$
\begin{aligned}
\overline{T}_p &= \frac{\left[\alpha_p\tau_g^2(1 - \beta_c)C + F_0(\alpha\tau)_{,\mathrm{eff}}\right]I_b + \left(U_{\mathrm{pa}} + U_b\right)T_a + F'h_{\mathrm{pf}}\overline{T}_f}{\left(U_{\mathrm{pa}} + F'h_{\mathrm{pf}} + U_b\right)} \\[4pt]
&= \frac{1.9487 \times 500 + (2.1225 + 0.7) \times 16 + 82 \times 16.375}{(2.1225 + 82 + 0.7)} \\[4pt]
&= \frac{974.35 + 45.16 + 1342.76}{84.82} = \frac{2362.26}{84.82} = 27.85\,°\mathrm{C}
\end{aligned}
$$

Now, from Eq. 8.23, one gets

$$
\begin{aligned}
\overline{T}_c &= \frac{(\alpha\tau)_{\mathrm{eff}}I_b + U_{t,\mathrm{ca}}T_a + U_{b,\mathrm{cp}}\overline{T}_p}{U_{t,\mathrm{ca}} + U_{t,\mathrm{cp}}} \\[4pt]
&= \frac{1.65 \times 500 + 9{,}12 \times 16 + 2.7663 \times 27.85}{9{,}12 + 2.7663} = \frac{1047.96}{11.8863} = 88.17\,°\mathrm{C}
\end{aligned}
$$

***Example 8.44*** Evaluate the electrical efficiency, electrical power, equivalent thermal power and an overall thermal efficiency of SPVT in single FLiSPVT for both cases of Example 8.43.

**Solution**

**Case (a):** $\overline{T}_c = 140.90\,°\mathrm{C}$.

The electrical efficiency of SPVT $= \eta_m = 0.12[1 - 0.00425(140.90 - 25)] = 0.0609$, The electrical power of SPVT $\left(\acute{E}_{\mathrm{el}}\right) = \eta_m \times I_b \times \beta_c \times C == 0.0609 \times 500 \times 0.89 \times 4 = 108.38\,\mathrm{W}$.

Equivalent rate of thermal energy $= \frac{108.38}{0.38} = 285.23$.

So, an overall thermal efficiency of FLiSPVT $= \frac{206.78 + 285.23}{500 \times 2} = 0.4920$.

**Case (b):** $\overline{T}_c = 88.17\,°\mathrm{C}$.

The electrical efficiency of SPVT $= \eta_m = 0.12[1 - 0.00425(88.17 - 25)] = 0.08778$, The electrical power of SPVT $\left(\acute{E}_{\mathrm{el}}\right) = \eta_m \times I_b \times \beta_c \times C = 0.08778 \times 500 \times 0.50 \times 4 = 87.78\,\mathrm{W}$.

An equivalent rate of thermal energy $= \frac{87.78}{0.38} = 250.8$.

So, an overall thermal efficiency of FLiSPVT $= \frac{403.28 + 250.8}{500 \times 2} = 0.654$.

### 8.6.6  The Outlet Fluid Temperature at $N^{th}$ of N-FLiSPVT Liquid Collectors Connected in Series

In this case too, the outlet of first FLiSPVT collector is connected to inlet of second FLiSPVT collector and outlet of third FLiSPVT collector is connected to fourth FLiSPVT till $N^{th}$ FLiSPVT as shown in Fig. 8.7.

Following Sect. 4.5 and Eq. 8.29, the outlet fluid temperature at $N^{th}$ of N-FLiSPVT liquid collectors connected in series can be derived as follows:

$$
T_{foN} = \left[ \frac{\left[(\alpha\tau)_{m,\text{eff}} I_b + U_{Lm} T_a\right]}{U_{Lm}} \right]\left[ 1 - \exp\left\{ \frac{-N F' U_{Lm} A_{rm}}{\dot{m}_f C_f} \right\} \right]
$$
$$
+ T_{fi} \exp\left\{ \frac{-N F' U_{Lm} A_{rm}}{\dot{m}_f C_f} \right\} \tag{8.33a}
$$

$$
\text{Average fluid temperature } \left(\overline{T}_{fN}\right) = \frac{T_{foN} + T_{fi}}{N}. \tag{8.33b}
$$

**Example 8.45** Find out the outlet fluid (Eq. 8.33a) and average (Eq. 8.33b) temperature at the end of five FLiSPVT collectors connected in series for case (b) in Example 8.41.

**Solution**

**Known parameters**: $U_{Lm} = 2.76$ W/m$^2$ °C (Example 7.6), $\dot{m}_f C_f = 146.65$ W/°C, $F' = 0.82$ (Table 8.3).

From Example 8.40, we have the following cases:

**Case (b) in Example 8.41: For C = 4, $\beta_c = 0.5$ and $(\alpha\tau)_{meff} = 1.9487$, $A_{rm} = 0.5$ m$^2$.**

**Step 1:**

$$
\left[ \frac{\left[(\alpha\tau)_{m,\text{eff}} I_b + U_{Lm} T_a\right]}{U_{Lm}} \right] = 372.90
$$

**Step 2:**

$$
\frac{N F' U_{Lm} A_{rm}}{\dot{m}_f C_f} = \frac{5 \times 0.82 \times 2.76 \times 0.5}{146.65} = 0.03886
$$

**Step 3:**

$$
\exp\left\{ \frac{-N F' U_{Lm} A_{rm}}{\dot{m}_f C_f} \right\} = \exp(-0.03886) = 0.9619
$$

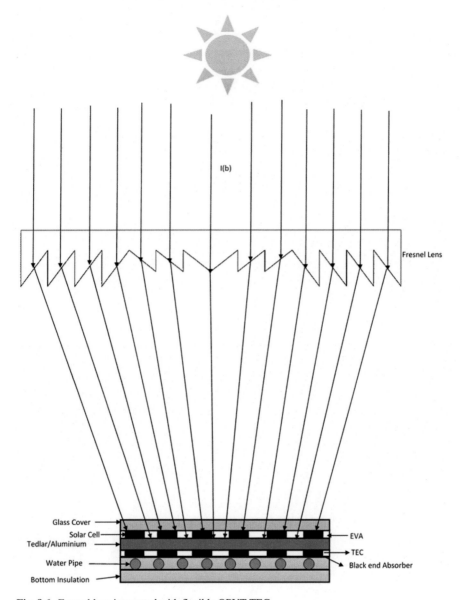

**Fig. 8.6**  Fresnel lens integrated with flexible OPVT-TEC

Substitute the above value in Eq. 8.33a, one gets the outlet temperature of five FLiSPVT collectors connected in series as

$$T_{\text{fo5}} = 372.90[1 - 0.9619] + 15 \times 0.9619 = 28.64\,°\text{C}$$

An average fluid temperature $(\overline{T}_{f5}) = \frac{T_{fo}+T_{fi}}{2} = \frac{28.64+15}{2} = 21.82\,°C$

### 8.6.6.1  The Rate of Useful Energy, Mass Flow Rate Factor and Instantaneous Thermal Efficiency for Five FLiSPVT Collectors Connected in Series

By using Eq. 8.33, the rate of thermal energy $(\dot{Q}_{uN.th})$ from N-FLiSPVT liquid collectors connected in series can be obtained as

$$\dot{Q}_{u.th} = \dot{m}_f C_f (T_{foN} - T_{fi}) = F_{RN}[A_{rm}][[(\alpha\tau)_{m,\text{eff}}I_b - U_{Lm}(T_{fi} - T_a)]] \quad (8.34)$$

where,

$$F_{RN} = \frac{\dot{m}_f C_f}{U_{Lm}A_{rm}}\left[1 - \exp\left\{\frac{-NF'U_{Lm}A_{rm}}{\dot{m}_f C_f}\right\}\right]$$

is a mass flow rate factor for N−FLiSPVT collectors.        (8.35)

Further, the thermal characteristic equation can be obtained as

$$\eta_{ith} = \frac{\dot{Q}_{uN.th}}{NI_b A_{am}} = F_{RN}\left[\frac{A_{rm}}{A_{am}}\right]\left[(\alpha\tau)_{m,\text{eff}} - U_{Lm}\frac{(T_{fi} - T_a)}{I_b}\right] \quad (8.36)$$

***Example 8.46*** Determine the rate of thermal energy, mass flow rate factor, an instantaneous optical and thermal efficiency for case (b) in Example 8.41 for five FLiSPVT collectors connected in series for $C = 4$.

**Solution**

**Given:** $\dot{m}_f C_f = 146.65\,W/°C$, $I_b = 500\,W/m^2$, $\tau_g = 0.95$, $\alpha_c = 0.9$, $T_{fi} = 15\,°C$, $T_a = 16\,°C$, $U_{Lm} = 2.76\,W/m^2\,°C$.

**Case (b) in Example 8.41:** For $C = 4$, $\beta_c = 0.50$ and $(\alpha\tau)_{m,\text{eff}} = 1.9487$, $\alpha\tau = \alpha_p\tau_g$, $A_{rm} = 0.5\,m^2$, $A_{am} = 2.0\,m^2$, $exp\left\{\frac{-F'U_{Lm}A_{rm}}{\dot{m}_f C_f}\right\} = 0.9619$, $T_{fo5} = 28.64\,°C$ $\dot{m}_f C_f = 146.65\,W/°C$.

The rate of thermal energy $= \dot{Q}_{u5.th} = \dot{m}_f C_f(T_{fo5} - T_{fi}) = 146.65(28.64\,°C - 15) = 2000.31\,W$.

The mass flow rate $= F_{RN} = \frac{\dot{m}_f C_f}{NU_{Lm}A_{rm}}\left[1 - \exp\left\{\frac{-NF'U_{Lm}A_{rm}}{\dot{m}_f C_f}\right\}\right] = \frac{146.65}{5\times2.76\times0.5}(1 - 0.9619) = 0.8097$.

With the help of Eq. 8.1a, we can define the optical efficiency as

$$\eta_0 = \frac{\dot{Q}_{u,th}}{\alpha_p\tau_g I_b N A_{am}} = \frac{\dot{m}_f C_f(T_{fo5} - T_{fi})}{0.95 \times 0.9 \times 500 \times 5 \times 2}$$
$$= \frac{146.09 \times 13.64}{4275} = \frac{2000.31}{4275} = 0.4679$$

From Eq. 8.2a, we have an expression for thermal efficiency as

$$\eta_{ith} = \frac{\dot{Q}_{u,th}}{N I_b A_{am}} = \frac{146.09 \times 13.645}{500 \times 5 \times 2} = \frac{2000.31}{5000} = 0.4000$$

From Eq. 8.22, we also have

$$\eta_{ith} = \frac{\dot{Q}_{u.th}}{I_b A_{am}} = F_{RN} \left[ \frac{A_{rm}}{A_{am}} \right] \left[ (\alpha\tau)_{m,eff} - U_{Lm} \frac{(T_{fi} - T_a)}{I_b} \right]$$

$$= 0.8097 \times \frac{0.5}{2} \times 1.9487 = 0.3945.$$

Since, $U_{Lm} \frac{(T_{fi} - T_a)}{I_b}$ is negative and hence, it is considered as zero.

### 8.6.6.2 The Average Plate and Solar Cell Temperature for N-FLiSPVT Collectors Connected in Series

From Eq. 8.24, one can first evaluate an average absorber plate temperature $(\overline{T}_{pN})$ by using the average fluid temperature from Example 8.41 $[\overline{T}_f == 16.375\,^\circ C]$

$$\overline{T}_{pN} = \frac{\left[ \alpha_p \tau_g^2 (1 - \beta_c) C + F_0 (\alpha\tau)_{,eff} \right] I_b + (U_{pa} + U_b) T_a + F' h_{pf} \overline{T}_{fN}}{(U_{pa} + F' h_{pf} + U_b)} \quad (8.37)$$

Now, from Eq. 8.23, one gets

$$\overline{T}_{cN} = \frac{(\alpha\tau)_{eff} I_b + U_{t,ca} T_a + U_{b,cp} \overline{T}_{pN}}{U_{t,ca} + U_{t,cp}} \quad (8.38)$$

***Example 8.47*** Evaluate the average absorber plate $(\overline{T}_{pN})$ and solar cell temperature $(\overline{T}_{cN})$ of five FLiSPVT liquid collectors connected in series for case (b) in Example 8.41 by using Eqs. 8.37 and 8.38, respectively.

**Solution**

**Given:** $U_{t,ca} = 9.12\,\text{W/m}^2\,^\circ C$, $U_{b,cp} = 2.7663\,\text{W/m}^2\,^\circ C$, $F_0 = 0.233$, $I_b = 500\,\text{W/m}^2$, $\tau_g = 0.95$, $\alpha_c = \alpha_p = 0.9$, $U_b = 0.7\,\text{W/m}^2\,^\circ C$, $F' = 0.82$, $F' h_{pf} = 0.82 \times 100 = 82\,\text{W/m}^2\,^\circ C$, , $U_{pa} = 2.1225\,\text{W/m}^2\,^\circ C$ (Table 8.3) and $N = 5$.

**Case (b) in Example 8.41:** For $C = 4$, $\beta_c = 0.50$ and $(\alpha\tau)_{m,eff} = 1.9487$, $(\alpha\tau)_{eff} = 1.65$, $A_{rm} = 0.5\,\text{m}^2$, $A_{am} = 2.0\,\text{m}^2$, $\left[ \alpha_p \tau_g^2 (1 - \beta_c) C + F_0 (\alpha\tau)_{,eff} \right] = [1.6245 + 0.3845] = 2.009$, $F_{01} = 0.97$.

From Eq. 8.37, one can first evaluate an average absorber plate temperature $(\overline{T}_p)$ by using the average fluid temperature from Example 8.41 $[\overline{T}_f = 21.82\,^\circ C]$.

$$\overline{T}_p = \frac{\left[\alpha_p \tau_g^2 (1 - \beta_c)C + F_0(\alpha\tau)_{,\text{eff}}\right]I_b + \left(U_{pa} + U_b\right)T_a + F'h_{pf}\overline{T}_f}{\left(U_{pa} + F'h_{pf} + U_b\right)}$$

$$= \frac{1.9487 \times 500 + (2.1225 + 0.7) \times 16 + 82 \times 21.82}{(2.1225 + 82 + 0.7)}$$

$$= \frac{974.35 + 45.16 + 1789.24}{84.82} = \frac{2808.75}{84.82} = 33.11\,°\text{C}$$

Now, from Eq. 8.38, one gets

$$\overline{T}_c = \frac{(\alpha\tau)_{\text{eff}}I_b + U_{t,\text{ca}}T_a + U_{b,\text{cp}}\overline{T}_p}{U_{t,\text{ca}} + U_{t,\text{cp}}}$$

$$= \frac{1.65 \times 500 + 9,12 \times 16 + 2.7663 \times 33.11}{9,12 + 2.7663}$$

$$= \frac{1062.51}{11.8863} = 89.39\,°\text{C}$$

***Example 8.48*** Evaluate the electrical efficiency, electrical power, equivalent thermal power and an overall thermal efficiency of five FLiSPVT collectors connected in series.

**Solution**

**Case (b):** $\overline{T}_c = 89.39\,°\text{C}$.

The electrical efficiency of SPVT $= \eta_m = 0.12[1 - 0.00425(89.39 - 25)] = 0.08716$

The electrical power of SPVT $\left(\acute{E}_{el}\right) = \eta_m \times I_b \times \beta_c \times A_{\text{am}} \times N = 0.08716 \times 500 \times 0.89 \times 2 \times 5 = 387.862\,\text{W}$.

Equivalent rate of thermal energy $= \frac{775.74}{0.38} = 2216.38$.

So, an overall thermal efficiency of FLiSPVT $= \frac{2000.31 + 2,216.38}{500 \times 2 \times 5} = 0.8433$.

### 8.6.6.3  Analytical Expression for an Average Fluid Temperature for N-FLiOPVT Collectors Connected in Series

Following Eq. 6.14 and using Eq. 8.33a, one can derive an analytical expression for $\overline{T}_{\text{fN}}$ as

$$\overline{T}_{\text{fN}} = \frac{1}{L_0}\int_0^{L_0} T_{\text{fN}}dx = \left[\frac{\left[(\alpha\tau)_{m,\text{eff}}I_b + U_{\text{Lm}}T_a\right]}{U_{\text{Lm}}}\right]\left[1 - \frac{1 - \exp\left\{\frac{-NF'U_{\text{Lm}}A_{\text{rm}}}{\dot{m}_f C_f}\right\}}{\frac{NF'U_{\text{Lm}}A_{\text{rm}}}{\dot{m}_f C_f}}\right]$$

$$+ T_{\text{fi}}\frac{1 - \exp\left\{\frac{-NF'U_{\text{Lm}}A_{\text{rm}}}{\dot{m}_f C_f}\right\}}{\frac{NF'U_{\text{Lm}}A_{\text{rm}}}{\dot{m}_f C_f}} \tag{8.39}$$

where,

$$T_{fN}(x) = \left[ \frac{\left[ (\alpha\tau)_{m,\text{eff}} I_b + U_{Lm} T_a \right]}{U_{Lm}} \right] \left[ 1 - \exp\left\{ \frac{-N F' U_{Lm} bx}{\dot{m}_f C_f} \right\} \right]$$
$$+ T_{fi} \exp\left\{ \frac{-N F' U_{Lm} bx}{\dot{m}_f C_f} \right\}$$

***Example 8.49*** Find out the average (Eq. 8.39) temperature of five FLiSPVT collectors connected in series for case (b) in Examples 8.41/8.45 and compare the results obtained in the same example.

**Solution**

**Known parameters**: $U_{Lm} = 2.76$ W/m$^2$ °C (Example 7.6), $\dot{m}_f C_f = 146.65$ W/°C, $F' = 0.82$ (Table 8.3).

From Example 8.40, we have the following cases:

**Case (b) in Example 8.41: For C = 4, $\beta_c$ = 0.5 and $(\alpha\tau)_{m\text{eff}}$ = 1.9487, $A_{rm}$ = 0.5 m$^2$.**

**Step 1**:

$$\left[ \frac{\left[ (\alpha\tau)_{m,\text{eff}} I_b + U_{Lm} T_a \right]}{U_{Lm}} \right] = 372.90$$

**Step 2**:

$$\frac{N F' U_{Lm} A_{rm}}{\dot{m}_f C_f} = \frac{5 \times 0.82 \times 2.76 \times 0.5}{146.65} = 0.03886$$

**Step 3**:

$$\exp\left\{ \frac{-N F' U_{Lm} A_{rm}}{\dot{m}_f C_f} \right\} = \exp(-0.03886) = 0.9619$$

Substitute the above value in Eq. 8.39, one gets the average temperature of five FLiSPVT collectors connected in series as

$$\overline{T}_{f5} = 372.90 \times \left[ 1 - \frac{1 - 0.9619}{0.03886} \right] + 15 \times \frac{1 - 0.9619}{0.03886}$$
$$= 372.90 \times (1 - 0.9804) + 15 \times 0.9804 = 22.01\,°C$$

which is nearly equal to the value obtained in Example 8.45 as

An average fluid temperature $\left(\overline{T}_{f5}\right) = \dfrac{T_{fo} + T_{fi}}{2} = \dfrac{28.64 + 15}{2} = 21.82\,°\text{C}.$

## 8.7　The Fresnel Lens (FL) Integrated with Flexible OPVT-TEC Liquid (FLiOPVT-TEC) Collector

If Fresnel lens (FL) is integrated with flexible opaque photo-voltaic thermal collector (OPVT), Sect. 6.6 of Chap. 6, it is referred as flexible FLiOPVT-TEC as shown in Fig. 8.6 due to limitation of concentration ratio of compound parabolic concentrator (CPC), Sect. 7.3 of Chap. 7. In the present case, only beam radiation ($I_b$) is concentrated on the top surface of flexible OPVT collector and then transmitted due to refraction toward absorber plate of flexible FLiOPVT-TEC. In this case, there is no reflection ($\rho = 1$). The flexible FLiOPVT-TEC works on the principle of line focus by Fresnel lens as in the case of FLiSPVT collector. So, in this case, the concentration ratio (C) will be more than CPC and less than FLiETC. Further, it is important to mention about an assumption that the bottom of TEC is equal to absorber plate due to proper contact and both have very high thermal conductivity without any thermal resistance. It can be expressed as

$$T_{\text{tec,bottom}} = T_p \tag{8.40}$$

Further, we will consider $C = \frac{A_{\text{am}}}{A_{\text{rm}}} = \frac{2}{0.5} = 4$, $A_{\text{am}} = 2\ \text{m}^2$, $A_{\text{rm}} = 0.5\ \text{m}^2$ and $\beta_c = 1$(packing factor of flexible OPVT) and $\beta_{\text{tec}} = 1$(packing factor of TEC) in the analysis of FLiOPVT-TEC. The design parameters of flexible FLiSPVT are given in Table 8.4.

Referring Sect. 7.4.1 of Chap. 7, the energy balance for each component of FLiOPVT-TEC liquid collector can be easily written. Here, all analytical equations of flexible OPVT-TEC, Sect. 7.4 of Chap. 7, will be exactly same with following modifications:

(i)　The incident solar radiation in flexible OPVT after reflection ($I_u$) will be replaced by beam radiation ($I_b$) in the flexible FLiOPVT-TEC collector.

(ii)　There is no reflection ($\rho = 1$).

(iii)　Numerical value of concentration ratio will be different than flexible OPVT-CPC (Table 8.4).

For convenience of readers, we have rewritten the all equations with new examples for packing factor of one ($\beta_c = 1$).

**Table 8.4**   Design parameters of FLiOPVT-TEC

| S. No. | Name of design parameters | Symbol of design parameters | Numerical values |
|---|---|---|---|
| 1 | Aperture of Fresnel lens | $A_{am}$ | 2 m$^2$ |
| 2 | Receiver area of OPVT-TEC | $A_{rm}$ | 0.5 m$^2$ |
| 3 | Breadth of OPVT-TEC | b | 1 m |
| 4 | Concentration ratio | C | 2, 4, 8 (Example 8.38) |
| 5 | | $h_t$ | |
| 6 | Incident beam radiation | $I_b$ | 500 W/m$^2$ |
| 7 | Length of OPVT-TEC | $L_0$ | 0.5 m |
| 8 | Heat transfer coefficient from solar cell to ambient air | $U_{t,ca}$ | 9.12  W/m$^2$ °C |
| 9 | Heat transfer coefficient from back of solar cell to absorber plate | $U_{b,cp}$ | 2.8  W/m$^2$ °C |
| 10 | Absorptivity of the solar cell | $\alpha_{sc} = \alpha_c$ | 0.90 |
| 11 | Transmittivity of glass cover | $\tau_g$ | 0.95 |
| 12 | Packing factor of flexible OPVT-TEC | $\beta_c$ | 1 |
| 13 | Packing factor of TEC | $\beta_{tec}$ | 1 |
| 14 | Electrical efficiency at STP | $\eta_m$ | 0.12 |

## 8.7.1   Energy Balance of Solar Cell of Single Flexible FLiOPVT-TEC Collector

The energy balance of PV module in W of single flexible FLiOPVT-TEC is as follows:

$$\tau_g \alpha_{sc} I_b C = U_{t,ca}(T_c - T_a) + h_t\left(T_c - T_{tec,top}\right)\beta_{tec}$$
$$+ U_{b,cp}\left(T_c - T_p\right)(1 - \beta_{tec}) + \eta_m I_b \qquad (8.41)$$

where, $C = \frac{A_{am}}{A_{rm}}$ is concentration ratio given in Example 8.38 (Table 8.4).
    From Eq. 8.41, we have

$$T_c = \frac{\left(\tau_g \alpha_{sc} C - \eta_m\right)I_b + U_{t,ca}T_a + h_t T_{tec,top}\beta_{tec} + U_{b,cp}(1 - \beta_{tec})T_p}{U_{t,ca} + h_t \beta_{tec} + U_{b,cp}(1 - \beta_{tec})} \qquad (8.42)$$

## 8.7.2  Energy Balance for Top of TEC

$$h_t\left(T_c - T_{\text{tec,top}}\right)\beta_{\text{tec}}bdx = U_{\text{tec}}\left(T_{\text{tec,top}} - T_{\text{tec,bottom}}\right)\beta_{\text{tec}}bdx \qquad (8.43)$$

With the help of Eq. 8.42, one can rewrite Eq. 8.43 as

$$h_t\left(T_c - T_{\text{tec,top}}\right)\beta_{\text{tec}} = F_0(\alpha\tau)_{\text{eff}}I_u - (U)_{\text{tec,top}-a}\left(T_{\text{tec,top}} - T_a\right)$$
$$- (U)_{\text{tec,top}-p}\left(T_{\text{tec,top}} - T_p\right) \qquad (8.44)$$

where,

The penalty factor, $F_0 = \frac{(h_t\beta_{\text{tec}})}{U_{t,\text{ca}}+h_t\beta_{\text{tec}}+U_{b,\text{cp}}(1-\beta_{\text{tec}})}$.

The product of absorptivity and transmittivity, $(\alpha\tau)_{\text{eff}} = \left(\tau_g\alpha_{\text{sc}}\frac{A_{\text{am}}}{A_{\text{rm}}} - \eta_m\right) = (\tau_g\alpha_{\text{sc}}C - \eta_m)$.

An overall heat transfer coefficient from top of TEC to ambient air, $(U)_{\text{tec,top}-a} = \frac{(h_t\beta_{\text{tec}})U_{t,\text{ca}}}{U_{t,\text{ca}}+h_t\beta_{\text{tec}}+U_{b,\text{cp}}(1-\beta_{\text{tec}})}$.

**Example 8.50** Evaluate (a) penalty factor, $F_0$, (b) product of absorptivity and transmittivity, $(\alpha\tau)_{\text{eff}}$, an overall heat transfer coefficient from top of TEC to ambient air $(U)_{\text{tec,top}-a}$ and an overall heat transfer coefficient from top of TEC to absorber plate $(U)_{\text{tec,top}-p}$ in Eq. 8.44 for concentration ratio of $C = \frac{A_{\text{am}}}{A_{\text{rm}}} = \frac{2}{0.5} = 4$ and $\beta_{\text{tec}} = 0.4324$.

**Solution**

From Example 3.48, we have the following parameters:

$U_{t,\text{ca}} = 9.30$ W/m$^2$ °C with wind velocity of 1 m/s; $h_t = 120$ W/m$^2$ °C and $U_{b,\text{cp}} = 2.80$ W/m$^2$ °C without wind velocity, $\beta_{\text{tec}} = 1$ packing factor of TEC, $\tau_g = 0.95$, $\alpha_{\text{sc}} = 0.9$ (Eq. 3.47a, Example 3.54).

From Eq. 8.44, we have the following expression for various constants:

$$F_0 = \frac{(h_t\beta_{\text{tec}})}{U_{t,\text{ca}} + h_t\beta_{\text{tec}} + U_{b,\text{cp}}(1 - \beta_{\text{tec}})} = \frac{120 \times 1}{9.30 + 120 \times 1 + 2.80 \times (1-1)} = \frac{120}{129.3} = 0.928$$

$$(\alpha\tau)_{\text{eff}} = \left[\tau_g\alpha_{\text{sc}}\left(\frac{A_{\text{am}}}{A_{\text{rm}}}\right) - \eta_m\right] = 0.95 \times 0.9 \times 4 - 0.12 = 3.3$$

$$(U)_{\text{tec,top}-a} = \frac{(h_t\beta_{\text{tec}})U_{t,\text{ca}}}{U_{t,\text{ca}} + h_t\beta_{\text{tec}} + U_{b,\text{cp}}(1 - \beta_{\text{tec}})}$$
$$= \frac{(120 \times 1) \times 9.30}{9.30 + 120 \times 1 + 2.80 \times (1-1)} = \frac{1116}{129.3} = 8.63 \text{ W/m}^2 \text{ °C}$$

$$(U)_{\text{tec,top}-p} = \frac{(h_t\beta_{\text{tec}})\{U_{b,\text{cp}}(1 - \beta_{\text{tec}})\}}{U_{t,\text{ca}} + h_t\beta_{\text{tec}} + U_{b,\text{cp}}(1 - \beta_{\text{tec}})}$$
$$= \frac{(120 \times 1) \times 2.80 \times (1-1)}{9.30 + 120 \times 0.4324 + 2.80 \times (1 - 0.4324)} = \frac{0}{91.765} = 0 \text{ W/m}^2 \text{ °C}$$

### 8.7.3 Absorber Plate of Flexible FLiOPVT-TEC Collector

The rate of thermal energy available in the absorber plate will be exactly same as Eq. 6.48 with change if the value of concentration ratio $(C = 4)$ and packing factor $(\beta_{tec} = 1)$ of TEC can be written as follows:

$$\dot{Q}_u = \left[(\alpha\tau)_{meff} I(t) - U_{Lm}(T_p - T_a) + F_0'' F_0' U_p T_p + U_a T_a\right] bdx \qquad (8.45)$$

where, $(\alpha\tau)_{meff} = [(1 - \eta_{tec}) F_{01} F_0 (\alpha\tau)_{eff} + F_0'' (1 - \beta_{tec})\{(\tau_g \alpha_{sc} - \eta_m) + F_0' F_0 (\alpha\tau)_{eff}\}]$; $U_{Lm} = \{(1 - \eta_{tec})(U)_{tec,top-pa} + U_b\}$, $U_p = \{(U)_{tec,top-p} + U_{tec}\beta_{tec} - (U_{t,ca} + h_t\beta_{tec})\}$ [Eq. 6.43] and $U_a = [F_0''(U_{t,ca} + F_0'(U)_{tec,top-a})]$.

$$F_{01} = \frac{U_{tec}\beta_{tec}}{\left[(UA)_{tec,top-a} + (UA)_{tec,top-p} + U_{tec}\beta_{tec}\right]},$$

$$F_0' = \frac{(h_t\beta_{tec})}{\left\{(U)_{tec,top-a} + (U)_{tec,top-p} + U_{tec}\beta_{tec}\right\}},$$

$$F_0'' = \frac{U_{b,cp}}{U_{t,ca} + h_t\beta_{tec} + U_{b,cp}(1 - \beta_{tec})} \quad \text{and}$$

$$(U)_{tec,top-pa} = \frac{U_{tec}\beta_{tec}(U)_{tec,top-a}}{\left[(U)_{tec,top-a} + (U)_{tec,top-p} + U_{tec}\beta_{tec}\right]}$$

**Example 8.51** Evaluate the penalty factor, $F_{01}$, and an overall heat transfer coefficient from plate to ambient through top of TEC, $(U)_{tec,top-pa}$, for Eq. 8.45.

**Solution**

We have the following known parameters:
$(U)_{tec,top-a} = 8.63 \text{ W/m}^2 \,^\circ\text{C}$ and $(U)_{tec,top-p} = 0$ and $\beta_{tec} = 1$ (Example 8.50); $U_{tec} = 450 \text{ W/m}^2 \,^\circ\text{C}$ (Eq. 3.48a).

From Eq. 8.45, we have

The penalty factor, $F_{01} = \frac{U_{tec}\beta_{tec}}{\left[(UA)_{tec,top-a} + (UA)_{tec,top-p} + U_{tec}\beta_{tec}\right]} = \frac{450 \times 1}{8.63 + 0 + 450 \times 1} = \frac{194.58}{200.7334} = 0.9812$.

An overall heat transfer coefficient from plate to ambient through top of TEC,

$(U)_{tec,top-pa} = \frac{U_{tec}\beta_{tec}(U)_{tec,top-a}}{\left[(U)_{tec,top-a} + (U)_{tec,top-p} + U_{tec}\beta_{tec}\right]} = \frac{(450 \times 1) \times 8.63}{8.63 + 0 + 450 \times 1} = 8.46 \text{ W/m}^2 \,^\circ\text{C}$

**Example 8.52** Evaluate the penalty factors, $F_0'$, $F_0''$ and $U_p$, for Eq. 8.45 by using the parameters of Table 8.4 and Example 8.50.

**Solution**

The penalty factor, $F_0' = \frac{(h_t\beta_{tec})}{\left\{(U)_{tec,top-a} + (U)_{tec,top-p} + U_{tec}\beta_{tec}\right\}} = \frac{120 \times 1}{8.63 + 450 \times 1} = \frac{120}{200.7334} = 0.2616$.

The penalty factor, $F_0'' = \dfrac{U_{b,cp}}{U_{t,ca}+h_t\beta_{tec}+U_{b,cp}(1-\beta_{tec})} = \dfrac{2.80}{9.30+120\times1+2.80\times(1-1)} = \dfrac{2.80}{91.765} = 0.02165.$

An overall net heat transfer coefficient, $U_p = \{(U)_{tec,top-p} + U_{tec}\beta_{tec} - (U_{t,ca}+h_t\beta_{tec})\} = 8.63+450\times1-(9.30+120\times1) = 8.63+450-129.3 = 329.33$ W/m$^2$ °C

**Example 8.53** Evaluate the product of absorptivity and transmittivity, $(\alpha\tau)_{meff}$, for $C = 4$ and an overall heat transfer coefficient $U_{Lm}$ and $U_a$ of OPVT-TEC-CPC liquid collector for Eq. 8.45 by using the data of Examples 8.50–8.52 and Table 8.4. The $\rho = 1$ is assumed.

**Solution**

**Known parameters**: $F_0 = 0.928$, $F_{01} = 0.9812$, $F_0' = 0.2616$, $F_0'' = 0.02165$, $(\alpha\tau)_{eff} = \left[\rho\tau_g\alpha_{sc}\left(\dfrac{A_{am}}{A_{rm}}\right) - \eta_m\right] = 3.3$ (Example 8.50), $\beta_{tec} = 1$, $U_b = 0.7$ W/m$^2$ °C and $\eta_{tec} = 0.15$ (Eq. 3.45).

$(U)_{ec,top-pa} = 8.46$ W/m$^2$ °C W/m$^2$ °C , $(U)_{tec,top-a} = 0$ (Example 8.50), $U_{t,ca} = 9.30$ W/m$^2$ °C .

From Eq. 6.44, we have the following:

$$(\alpha\tau)_{meff} = \left[(1-\eta_{tec})F_{01}F_0(\alpha\tau)_{eff} + F_0''(1-\beta_{tec})\left\{\left(\rho\tau_g\alpha_{sc}\dfrac{A_{am}}{A_{rm}} - \eta_m\right) + F_0'F_0(\alpha\tau)_{eff}\right\}\right]$$

$$= (1-0.15)\times0.9812\times0.928\times3.3 + 0.02165\times(1-1)$$
$$\times\{3.3 + 0.2616\times0.928\times3.3\} = 2.5541 + 0\times4.101 = 2.5541$$

$$\mathbf{U_{Lm}} = \{(1-\eta_{tec})(U)_{tec,top-pa} + U_b\} = (1-0.15)\times8.46 + 0.7 = 7.891 \text{ W/m}^2\,°C$$

$$U_a = \left[F_0''(U_{t,ca} + F_0'(U)_{tec,top-a})\right]$$
$$= 0.02165\times(9.30 + 0.2616\times0) = 0.2015 \text{ W/m}^2\,°C$$

### 8.7.4  The Outlet Fluid Temperature at $N^{th}$ Flexible FLiOPVT-TEC Liquid Collectors

Following Sect. 7.4.2, the outlet fluid (water) temperature at $N$th flexible FLiOPVT-TEC liquid collectors connected in series can be derived as

$$T_{foN} = \dfrac{\left[(\alpha\tau)_{meff}I_b + (U_{Lm}+U_a)T_a\right]}{U_{Lm-eff}}\left[1 - \exp\left\{-\dfrac{NF'U_{Lm-eff}b}{\dot{m}_f c_f}L_0\right\}\right]$$
$$+ T_{fi}\exp\left\{-\dfrac{NF'U_{Lm-eff}b}{\dot{m}_f c_f}L_0\right\} \qquad (8.46)$$

where, $U_{\text{Lm-eff}} = \left(U_{\text{Lm}} - F_0'' F_0' U_p\right)$

Further, an average temperature of N-FLiOPVT-TEC collector can also be obtained as

$$\overline{T}_{\text{fN}} = \frac{\left[(\alpha\tau)_{m\text{eff}} I_b + (U_{\text{Lm}} + U_a)T_a\right]}{U_{\text{Lm-eff}}}\left[1 - \frac{1 - \exp\left\{-\frac{NF'U_{\text{Lm-eff}}b}{\dot{m}_f c_f}L_0\right\}}{\frac{NF'U_{\text{Lm-eff}}b}{\dot{m}_f c_f}L_0}\right]$$

$$+ T_{\text{fi}}\frac{1 - \exp\left\{-\frac{NF'U_{\text{Lm-eff}}b}{\dot{m}_f c_f}L_0\right\}}{\frac{NF'U_{\text{Lm-eff}}b}{\dot{m}_f c_f}L_0} \qquad (8.47)$$

Further, with the help of Eq. 8.47 for $\overline{T}_{\text{fN}} = \overline{T}_p$, one can get an expression for $\overline{T}_{\text{tec,topN}}$ and $\overline{T}_{\text{cN}}$ from Eqs. 7.46 and 7.47, respectively, with the help of Eq. 8.47 for $\overline{T}_{\text{fN}} = \overline{T}_{pN}$ as

$$\overline{T}_{\text{tec,topN}} = \frac{F_0(\alpha\tau)_{\text{eff}} I_b + \left\{(U)_{\text{tec,top}-p} + U_{\text{tec}}\beta_{\text{tec}}\right\}\overline{T}_p + (U)_{\text{tec,top}-a}\overline{T}_a}{(U)_{\text{tec,top}-a} + (U)_{\text{tec,top}-p} + U_{\text{tec}}\beta_{\text{tec}}} \qquad (8.48)$$

and,

$$\overline{T}_{\text{cN}} = \frac{\left(\tau_g\alpha_{\text{sc}}\frac{A_{\text{am}}}{A_{\text{rm}}} - \eta_m\right)I_b + U_{t,\text{ca}}T_a + h_t\overline{T}_{\text{tec,topN}}\beta_{\text{tec}} + U_{b,\text{cp}}(1 - \beta_{\text{tec}})\overline{T}_{pN}}{U_{t,\text{ca}} + h_t\beta_{\text{tec}} + U_{b,\text{cp}}(1 - \beta_{\text{tec}})}$$

$$\left[\overline{T}_{\text{fN}} = \overline{T}_{pN}\right] \qquad (8.49)$$

***Example 8.54*** Compute an effective overall heat loss coefficient ($U_{\text{Lm-eff}}$) by using Eq. 8.46 for data of Examples 8.53 and 8.54 for $C = 4$ and $\beta_{\text{tec}} = 1$.

**Solution**

From Examples 8.53 and Table 8.4, we have the followings:
$U_{Lw} = 7.891$ W/m °C, $U_p = 329.33$ W/m² °C, $F_0' = 0.2616$, $F_0'' = 0.02165$.
From Eq. 8.46, one has the following:

$$U_{\text{Lm-eff}} = \left(U_{\text{Lm}} - F_0'' F_0' U_p\right) = 7.891 - 0.02165 \times 0.2616$$
$$\times 329.33 = 6.026 \text{ W/m}^2 \text{ }^\circ\text{C}$$

***Example 8.55*** Compute the outlet and average fluid temperatures for $N = 1$ and 5 for flexible FLiOPVT-TEC collector for beam radiation of $I_b = 500$ W/m², ambient air temperature ($T_a$) of 16 °C and inlet temperature ($T_{\text{fi}}$) = 15 °C for Examples 8.53 and 8.54 for $C = 4$.

**Solution**

From Examples 8.53 and 8.54 : $U_{\text{Lm-eff}} = 6.026$ W/m$^2\,^{\circ}$C, $U_{Lw} = 7.891\,\frac{W}{m^2}\,^{\circ}$C, $U_a = 0.2015$ W/m$^2\,^{\circ}$C, $(\alpha\tau)_{meff} = 2.5541$, $\dot{m}_f c_f = 146.65$ W/$^{\circ}$C and $F' = 0.82$.

From Eq. 8.43, we have

$$
T_{\text{fo}N} = \frac{\left[(\alpha\tau)_{\textit{meff}} I_b + (U_{\text{Lm}} + U_a)T_a\right]}{U_{\text{Lm-eff}}}\left[1 - \exp\left\{-\frac{NF'U_{\text{Lm-eff}}b}{\dot{m}_f c_f}L_0\right\}\right]
$$
$$
+ T_{\text{fi}}\exp\left\{-\frac{NF'U_{\text{Lm-eff}}b}{\dot{m}_f c_f}L_0\right\}
$$

Now,

$$
\frac{\left[(\alpha\tau)_{\textit{meff}} I_b + (U_{\text{Lm}} + U_a)T_a\right]}{U_{\text{Lm-eff}}} = \frac{2.5541 \times 500 + (6.026 + 0.2015) \times 16}{6.026}
$$
$$
= \frac{1376.69}{6.026} = 227.18
$$

For $N = 1$:

**Step 1**: $\frac{NF'U_{\text{Lm-eff}}b}{\dot{m}_f c_f}L_0 = \frac{1\times0.82\times6.06\times1\times0.5}{146.65} = \frac{2.4846}{146.65} = 0.0169$.

**Step 2**: $\exp(-0.0169) = 0.9832$.

Substitute the appropriate value in the above equation for $N = 1$, one gets

$$
T_{\text{fo}1} = 227.18[1 - 0.9832] + 15 \times 0.9832 = 18.56\,^{\circ}\text{C}
$$

Average of fluid temperature $\left(\overline{T}_{f1}\right) = \frac{18.5+15}{2} = 16.75\,^{\circ}$C.

For $N = 5$:

**Step 1**: $\frac{NF'U_{\text{Lm-eff}}b}{\dot{m}_f c_f}L_0 = \frac{5\times0.82\times6.06\times1\times0.5}{146.65} = \frac{12.423}{146.65} = 0.0845$.

**Step 2**: $\exp(-0.0845) = 0.9189$.

Substitute the appropriate value in the above equation for $N = 1$, one gets

$$
T_{\text{fo}1} = 227.18[1 - 0.9189] + 15 \times 0.9189 = 32.21\,^{\circ}\text{C}
$$

Average of fluid temperature $\left(\overline{T}_{f5}\right) = \frac{32.21+15}{2} = 23.60\,^{\circ}$

***Example 8.56*** Compute the average fluid temperature for $N = 1$ and 5 for flexible FLiOPVT-TEC collector for beam radiation of $I_b = 500$ W/m$^2$, ambient air temperature $(T_a)$ of 16 $^{\circ}$C and inlet temperature $(T_{\text{fi}}) = 15\,^{\circ}$C for Example 8.55 by using Eq. 8.47.

**Solution**

From Eq. 8.47, We Have

$$\overline{T}_{\text{fN}} = \frac{\left[(\alpha\tau)_{\text{meff}} I_b + (U_{\text{Lm}} + U_a)T_a\right]}{U_{\text{Lm-eff}}} \left[1 - \frac{1 - exp\left\{-\frac{NF'U_{\text{Lm-eff}}b}{\dot{m}_f c_f}L_0\right\}}{\frac{NF'U_{\text{Lm-eff}}b}{\dot{m}_f c_f}L_0}\right]$$

$$+ T_{\text{fi}}\frac{1 - exp\left\{-\frac{NF'U_{\text{Lm-eff}}b}{\dot{m}_f c_f}L_0\right\}}{\frac{NF'U_{\text{Lm-eff}}b}{\dot{m}_f c_f}L_0} \tag{8.47}$$

Now,

$$\frac{\left[(\alpha\tau)_{\text{meff}} I_b + (U_{\text{Lm}} + U_a)T_a\right]}{U_{\text{Lm-eff}}} = \frac{2.5541 \times 500 + (6.026 + 0.2015) \times 16}{6.026}$$

$$= \frac{1376.69}{6.026} = 227.18$$

For $N = 1$:

**Step 1**: $\frac{NF'U_{\text{Lm-eff}}b}{\dot{m}_f c_f}L_0 = \frac{1 \times 0.82 \times 6.06 \times 1 \times 0.5}{146.65} = \frac{2.4846}{146.65} = 0.0169.$

**Step 2**: $\exp(-0.0169) = 0.9832.$

Substitute the appropriate value in the above equation for $N = 1$, one gets

$$\overline{T}_{f1} = 227.18\left[1 - \frac{1 - 0.9832}{0.0169}\right] + 15 \times \frac{1 - 0.9832}{0.0169}$$

$$= 227.18 \times (1 - 0.99408) + 15 \times 0.99408 = 16.23\,°C$$

Average of fluid temperature $\left(\overline{T}_{f1}\right) = \frac{18.5+15}{2} = 16.75\,°C$

There is difference of about 3% by both methods which is acceptable within limit (Example 8.55).

For $N = 5$:

**Step 1**: $\frac{NF'U_{\text{Lm-eff}}b}{\dot{m}_f c_f}L_0 = \frac{5 \times 0.82 \times 6.06 \times 1 \times 0.5}{146.65} = \frac{12.423}{146.65} = 0.0845.$

**Step 2**: $\exp(-0.0845) = 0.9189.$

Substitute the appropriate value in the above equation for $N = 1$, one gets

$$\overline{T}_{f5} = 227.18\left[1 - \frac{1 - 0.9189}{0.0845}\right] + 15 \times \frac{1 - 0.9189}{0.0845}$$

$$= 227.18(1 - 0.9609) + 15 \times 0.9609 = 23.28\,°C$$

Average of fluid temperature $\left(\overline{T}_{f5}\right) = \frac{32.21+15}{2} = 23.60\,°C$

There is difference of about 1.3% by both methods which is acceptable within limit (Example 8.55).

**Example 8.57** Find out an average top surface of TEC of FLiOPVT-TEC liquid collectors for $N = 1$ and 5 for Example 8.56 by using Eq. 8.48 for $I_b = 500\ \text{W/m}^2$ and $\overline{T}_a = 15\,°C$.

**Solution**

From Eq. 8.48, we have

$$\overline{T}_{\text{tec,topN}} = \frac{F_0(\alpha\tau)_{eff} I_b + \left\{(U)_{\text{tec,top}-p} + U_{\text{tec}}\beta_{\text{tec}}\right\}\overline{T}_p + (U)_{\text{tec,top}-a}\overline{T}_a}{(U)_{\text{tec,top}-a} + (U)_{\text{tec,top}-p} + U_{\text{tec}}\beta_{\text{tec}}}$$

For $N = 1$:

**Known constants:** $\overline{T}_{f1} = \overline{T}_{p1} = 16.23\,°C$ (Example 8.56), $F_0 = 0.928$, $(\alpha\tau)_{eff} = 3.3$, $\beta_{\text{tec}} = 1$, $(U)_{\text{tec,top}-p} = 0$, $(U)_{\text{tec,top}-a} = 8.63\,\text{W/m}^2\,°C$ (Table 8.4), $U_{\text{tec}} = 450\,\text{W/m}^2\,°C$ (Eq. 3.48a).

Substitute the above values in the above equation (Eq. 8.48) to get $\overline{T}_{\text{tec,top1}}$ as

$$\overline{T}_{\text{tec,top1}} = \frac{0.928 \times 3.3 \times 500 + \{0 + 450 \times 1\} \times 16.23 + 8.63 \times 16}{8.63 + 0 + 450 \times 1}$$

$$= \frac{1531.2 + 7303.5 + 138.08}{458.63} = \frac{8972.78}{458.63} = 19.56\,°C$$

For $N = 5$:

**Known constants**: $\overline{T}_{f6} = \overline{T}_{p6} = 23.28$ [Example 8.56], $F_0 = 0.928$, $(\alpha\tau)_{eff} = 3.3$, $\beta_{\text{tec}} = 1$, $(U)_{\text{tec,top}-p} = 0$, $(U)_{\text{tec,top}-a} = 8.63\,\text{W/m}^2\,°C$ (Table 8.4), $U_{\text{tec}} = 450\,\text{W/m}^2\,°C$ (Eq. 3.48a).

Substitute the above values in Eq. 8.46 to get $\overline{T}_{\text{tec,top6}}$, one gets

$$\overline{T}_{\text{tec,top5}} = \frac{0.928 \times 3.3 \times 500 + \{0 + 450 \times 1\} \times 23.28 + 8.63 \times 16}{8.63 + 0 + 450 \times 1}$$

$$= \frac{1531.2 + 10,476.6 + 138.08}{458.63} = \frac{12,145.28}{458.63} = 26.48\,°C$$

***Example 8.58*** Evaluate an average solar cell temperature for $N = 1$ and 5 for flexible FLiOPVT-TEC collectors connected in series by using Eq. 8.49 for design parameters of Example 8.57 and Table 8.4.

**Solution**

From Eq. 8.49, we have an expression for $\overline{T}_{\text{cN}}$ as

$$\overline{T}_{\text{cN}} = \frac{\left(\tau_g\alpha_{\text{sc}}\frac{A_{\text{am}}}{A_{\text{rm}}} - \eta_m\right)I_b + U_{t,\text{ca}}T_a + h_t\overline{T}_{\text{tec,topN}}\beta_{\text{tec}} + U_{b,\text{cp}}(1 - \beta_{\text{tec}})\overline{T}_{pN}}{U_{t,\text{ca}} + h_t\beta_{\text{tec}} + U_{b,\text{cp}}(1 - \beta_{\text{tec}})} \quad [\overline{T}_{\text{fN}} = \overline{T}_{pN}]$$

**Given and known parameters:** $\overline{T}_{f1} = \overline{T}_{p1} = 16.23\,°C$ (Example 8.56)$= \left[\tau_g\alpha_{\text{sc}}\left(\frac{A_{\text{am}}}{A_{\text{rm}}}\right) - \eta_m\right] = 3.3$, $U_{t,\text{ca}} = 9.30\,\text{W/m}^2\,°C$ with wind velocity of 1 m/s; $h_t = 120\,\text{W/m}^2\,°C$ and $U_{b,\text{cp}} = 2.80\,\text{W/m}^2\,°C$ without wind velocity, $\beta_{\text{tec}} = 1$, packing factor of TEC, $\tau_g = 0.95$, $\alpha_{sc} =$, $\alpha_{sc} = 0.9$, $I_b = 500$ W/m² (Eq. 3.47a, Example 3.54) (Example 7.31).

**For $N = 1$:** $\overline{T}_{tec,top1} = 19.56\,°C$ (Example 8.57).

Substitute the above values in above equation (from Eq. 8.49), value for $\overline{T}_{c1}$ as

$$\overline{T}_{c1} = \frac{3.3 \times 500 + 9.12 \times 16 + 120 \times 19.56 \times 1 + 2.80 \times (1-1) \times 16.92}{9.12 + 120 \times \times 1 + 2.80(1-1)}$$

$$= \frac{1650 + 145.92 + 23472 + 0}{129.12} = \frac{4143.12}{129.12} = 32.09\,°C$$

**For $N = 5$:** $\overline{T}_{toc,top\,5} = 26.48\,°C$ (Example 8.57).

Substitute the above values in above equation (from Eq. 8.49), value for $\overline{T}_{c1}$ as

$$\overline{T}_{c5} = \frac{3.3 \times 500 + 9.12 \times 16 + 120 \times 26.48 \times 1 + 2.80 \times (1-1) \times 16.92}{9.12 + 120 \times \times 1 + 2.80(1-1)}$$

$$= \frac{1650 + 145.92 + 3277.6 + 0}{129.12} = \frac{4973.52}{129.12} = 38.52\,°C$$

***Example 8.59*** Evaluate an electrical efficiency and electrical power for single ($N = 1$) and five ($N = 5$) flexible FLiOPVT-TEC liquid collectors connected in series for Example 8.58 and data of Table 8.4.

**Solution**

For $N = 1$:

**Given parameters**: $N = 1$, $\overline{T}_{c1} = 32.09\,°C$, $I_b = 500$ W/m² (Example 8.58), $\eta_{m0} = 0.12$, $A_{rm} = 0.5$ m × 1 m = 0.5 m².

From Eq. 7.48a, one gets the values of $\overline{\eta}_{m1}$ for $N = 1$ for given above parameters as

$$\overline{\eta}_{m1} = \eta_{m0}\big[1 - \beta_0(\overline{T}_{c1} - 25)\big] = 0.12[1 - 0.00425(32.09\,°C - 25)] = 0.1164(11.64\%) \text{ and}$$

$$\dot{E}_{ell} = 0.1164 \times 500 \times 1 \times 0.5 \times 1 = 58.2\,W.$$

For $N = 5$:

**Given parameters**: $\overline{T}_{c1} = 32.09\,°C$, $I_b = 500$W/m² (Example 8.58), $\eta_{m0} = 0.12$, $A_{rm} = 0.5$ m × 1 m = 0.5 m².

From Eq. 7.48a, one gets the values of $\overline{\eta}_{m1}$ for $N = 1$ for given above parameters as

$$\overline{\eta}_{m5} = \eta_{m0}\big[1 - \beta_0(\overline{T}_{c1} - 25)\big] = 0.12[1 - 0.00425(38.52 - 25)] = 0.1131(11.31\%)$$

and,

$$\dot{E}_{el5} = 0.1131 \times 500 \times 1 \times 0.5 \times 5 = 141.75 \text{ W}$$

**Example 8.60** Evaluate electrical power from TEC of N-FLiOPVT-TEC liquid collectors connected in series for $N = 1$ and 5, respectively, for Example 8.57.

**Solution**

**For $N = 1$, known:** $\overline{T}_{\text{tec,top1}} = 19.56\,°\text{C}$ (Example 8.57), $\overline{T}_{f1} = \overline{T}_{p1} = \overline{T}_{\text{tec,bottom1}}$

$= 16.23\,°\text{C}$ ( Example 8. 56), $A_{rn} = 0.5$ m $\times$ 1m $= 0.5$ m$^2$
From Eq. 7.49, we have

$$\dot{E}_{\text{TEC1}} = \eta_{\text{tec}} U_{\text{tec}} \left( \overline{T}_{\text{tec,top1}} - \overline{T}_{f1} \right) \beta_{\text{tec}} A_{rm} \times N = 0.15 \times 450 \times (19.56 - 16.23)$$
$$\times 0.5 \times 1 \times 5 = 115.58 \text{ W}$$

**For $N = 5$, known:** $\overline{T}_{\text{tec,top5}} = 26.48\,°\text{C}$ (Example 8.57), $\overline{T}_{f5} = \overline{T}_{p5} = \overline{T}_{\text{tec, bottom5}} = 23.28\,°\text{C}$ Example 8.56), $A_{rn} = 0.5$ m $\times$ 1 m $= 0.5$ m$^2$
From Eq. 7.49, we have

$$\dot{E}_{\text{TEC5}} = \eta_{\text{tec}} U_{\text{tec}} \left( \overline{T}_{\text{tec,top5}} - \overline{T}_{f5} \right) \beta_{\text{tec}} A_{rm} \times N$$
$$= 0.15 \times 450 \times (26.48 - 23.28\,°\text{C}) \times 0.5 \times 1 \times 5 = 540 \text{ W}$$

**Example 8.61** Find out the rate of thermal energy and instantaneous optical and thermal efficiencies for $N = 1$ and $N = 5$ from flexible FLiOPVT-TEC collector for Example 8.55.

**Solution**

**For N = 1:** $T_{\text{fol}} = 18.56\,°\text{C}$, $T_{\text{fi}} = 15\,°\text{C}$ and $\dot{m}_f c_f = 146.65 \text{ W/}°\text{C}$.
   Now, the rate of thermal energy $= \dot{Q}_{uth1} = \dot{m}_f c_f (T_{\text{fol}} - T_{\text{fi}}) = 146.65 \times (18.56 - 15) = 522.07 \text{ W}$.
   The optical instantaneous efficiency $(\eta_0) = \dfrac{\dot{Q}_{uth1}}{\alpha_c \tau_g I_b A_{\text{am}}} = \dfrac{522.07}{0.9 \times 0.95 \times 500 \times 2} = \dfrac{522.07}{855} = 0.6105$.
   The thermal instantaneous efficiency $(\eta_{i1}) = \dfrac{\dot{Q}_{uth1}}{I_b A_{\text{am}}} = \dfrac{522.07}{500 \times 2} = \dfrac{522.07}{1000} = 0.5221$.

**For $N = 5$:** $T_{\text{fol}} = 32.21\,°\text{C}$, $T_{\text{fi}} = 15\,°\text{C}$ and $\dot{m}_f c_f = 146.65 \text{W/}°\text{C}$
   Now, the rate of thermal energy $= \dot{Q}_{uth5} = \dot{m}_f c_f (T_{\text{fo5}} - T_{\text{fi}}) = 146.65 \times (32.21 - 15) = 2523.85 \text{ W}$.
   The optical instantaneous efficiency $(\eta_0) = \dfrac{\dot{Q}_{uth5}}{\alpha_c \tau_g I_b N A_{\text{am}}} = \dfrac{2,523.85}{0.9 \times 0.95 \times 500 \times 5 \times 2} = \dfrac{2,523.85}{4275} = 0.5904$.
   The thermal instantaneous efficiency $(\eta_{i5}) = \dfrac{\dot{Q}_{uth5}}{I_b N A_{\text{am}}} = \dfrac{2523.85}{500 \times 5 \times 2} = \dfrac{2523.85}{5000} = 0.5047$.

**Example 8.62** Find out the rate of an overall total thermal energy and efficiency for $N = 1$ and $N = 5$ from flexible FLiOPVT-TEC collector for Examples 8.59–8.61.

## Solution

**For $N = 1$:** $\dot{E}_{el1} = 58.2$ W, $\dot{E}_{TEC1} = 115.58$ W and $\dot{Q}_{uth1} = 522.07$ W.

Hence,

The rate of an overall total thermal energy $\left(\dot{Q}_{u1,\text{total}}\right) = 522.07 + \frac{58.2+115.58}{0.38} = 991.74$ W.

and,

An overall thermal efficiency $\left(\eta_{i1,\text{th-ov}}\right) = \frac{\dot{Q}_{u1,total}}{I_b \times A_{am}} = \frac{991.74}{500 \times 2} = 0.9917$.

**For $N = 5$:** $\dot{E}_{el5} = 141.75$ W, $\dot{E}_{TEC1} = 540$ W and $\dot{Q}_{uth5} = 2,523.85$ W.

Hence,

The rate of an overall total thermal energy $\left(\dot{Q}_{u5,\text{total}}\right) = 2523.85 + \frac{141.75+5408}{0.38} = 4317.93$ W.

and,

An overall thermal efficiency $\left(\eta_{i5,\text{th-ov}}\right) = \frac{\dot{Q}_{u5,total}}{I_b \times A_{am} \times 5} = \frac{4317.93}{500 \times 2 \times 5} = \frac{4317.93}{5000} = 0.8636$.

# References

1. Mishra, R. K., Garg, V., & Tiwari, G. N. (2015). Thermal modeling and development of characteristic equations of evacuated tubular collector (ETC). *Solar Energy, 116*, 165–176. https://doi.org/10.1016/j.solener.2015.04.00
2. Zhai, H., Dai, Y. J., Wu, J. Y., Wang, R. Z., & Zhang, L. Y. (2010). *Experimental investigation and analysis on a concentrating solar collector using linear Fresnel lens., 51*(1), 48–55. https://doi.org/10.1016/j.enconman.2009.08.018
3. Tiwari, G. N. (2004). *Solar energy: Fundamental, design, modelling and applications*, Narosa Publishing House, New Delhi and CRC Press, Eq. 2.34b and Fig. 2.5c, New York.

# Chapter 9
# Thermal Energy and Exergy Matrices of Solar Energy Systems

## 9.1 Introduction

Basically, we have discussed so far two type energy based on two thermodynamics law.

Namely

(a) **First law of thermodynamics (Section 2.2)**: On the first law of thermodynamics, we use energy conservation to write energy balance equation of solar energy systems (Chap. 3) for their thermal modeling (Chaps. 4 and 5) for energy analysis.
(b) **Second law of thermodynamics (Section 2.9)**: On the second law thermodynamics, the solar energy system should be analyzed from exergy point of view.

However, we have seen in Chaps. 6–8 that one gets two type of energy which are thermal energy and electrical energy. Such solar energy system is referred as photovoltaic thermal (PVT) collector. In order to obtain their instantaneous efficiency, the hourly, daily, monthly or yearly yield cannot be obtained just by adding their thermal and electrical output. This should be done by

(i) Convert electrical power into equivalent rate of thermal energy by dividing electrical power by conversion factor of fossil fuel into electrical energy and then add it to thermal energy $\left(\dot{Q}_{u,th}\right)$. It is referred as an overall thermal energy output of PVT solar collector $\left(\dot{Q}_{u,thT}\right)$.

Mathematically, it is expressed as follows:

$$\dot{Q}_{u,thT} = \dot{Q}_{u,th} + \frac{\acute{E}_{el}}{\gamma_c} + \frac{\dot{E}_{TEC}}{\gamma_c} \tag{9.1}$$

© Bag Energy Research Society 2024
G. N. Tiwari, *Fundamental of Mathematical Tools for Thermal Modeling of Solar Thermal and Photo-voltaic Systems-Part-I*,
https://doi.org/10.1007/978-981-99-7085-8_9

where, $\dot{E}_{el}$ and $\dot{E}_{TEC}$ are electrical power from PV and TEC and $\gamma_c$ is conversion factor of fossil fuel for thermal power plant as discussed in Chap. 8.

The instantaneous optical, thermal and an overall thermal efficiency of various solar energy systems have already discussed in Chap. 8, and the results are given in Table 8.1.

(ii)  Convert the rate of thermal energy into rate of thermal exergy (Eq. 2.35a) and add it into electrical power which is equivalent to high grade energy. It is referred as an overall rate of exergy of PVT solar systems $(\dot{E}_{ex-ov})$ based on second law of thermodynamics.

Mathematically, it is expressed as follows:

$$\dot{E}x_{u,ov} = \dot{Q}_{u,thex} + \dot{E}_{el} + \dot{E}_{TEC} \tag{9.2a}$$

where, $\dot{E}_{el}$ and $\dot{E}_{TEC}$ are electrical power from PV and TEC and $\gamma_c$ is conversion factor of fossil fuel for thermal power plant as discussed in Chap. 8 and

$$\dot{Q}_{u,thex} = \dot{m}_f C_f \left[ (T_{fo} - T_{fi}) - (T_a + 273) \ln \frac{T_{fo} + 273}{T_{fi} + 273} \right] \text{ [From Eq.2.35a]} \tag{9.2b}$$

An instantaneous overall exergy efficiency of solar energy $(\eta_{i,ex-ov})$ is defined as follows:

$$\eta_{i,ex-ov} = \frac{\dot{E}x_{u-ov}}{[I(t)]_{ex} \times \text{area of solar energy system}} \tag{9.3}$$

where, $[I(t)]_{ex}$ is the exergy of solar radiation $= I(t) \times U_{ee} = I(t) \times 0.96$ (Example 2.75).

**Example 9.1** Evaluate exergy of solar radiation for CPC and Fresnel lens integrated collectors.

**Solution**

From Example 7.1, $I_u = 475$ W for CPC.
   So,

$$[I_u]_{ex} = 475 \times 0.96 = 456 \text{ W/m}^2$$

For Fresnel lens, $I_b = 500$ W, so

$$[I_b]_{ex} = 500 \times 0.96 = 480 \text{ W/m}^2$$

In this chapter, we will address the overall exergy of solar energy systems and its exergy efficiency of all solar systems already discussed in Chaps. 4–8 (Fig. 9.1).

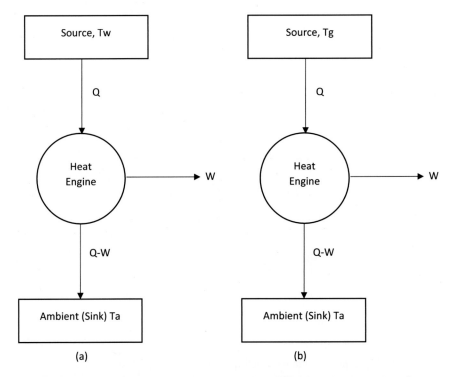

**Fig. 9.1** Block diagram for determining the maximum work done by using the concept of entropy

## 9.2 Exergy of Solar Thermal and Photo-Voltaic Thermal (PVT)

In this case, we will only consider the collectors connected in series to higher outlet temperature for positive thermal exergy as experienced earlier (Chaps. 4–8).

### 9.2.1 The Rate of Thermal Exergy and Exergy Efficiency for Flat Plate Collector (FPC)

**Example 9.2** Evaluate the rate of thermal exergy and instantaneous exergy efficiency for Example 4.44 for N = 5 and 10 of liquid flat plate collector.

**Solution**

Given: $\dot{m}_f C_f = 146.65 \, \text{W/°C}$, $I(t) = 500 \, \text{W/m}^2$, $A_c = 2 \, \text{m}^2$.

For **N = 5**: $T_{fo5} = 34.85 \, °\text{C}$, $T_{fi} = 15 \, °\text{C}$ and $T_a = 16 \, °\text{C}$.

From Eq. 2.35a and Eq. 9.2b, we have an expression for the rate of thermal exergy as

$$\dot{Q}_{u,thex} = \dot{m}_f C_f \left[ (T_{fo5} - T_{fi}) - (T_a + 273) \ln \frac{T_{fo5} + 273}{T_{fi} + 273} \right]$$

$$= 146.65 \left[ 19.89 - 289 \ln \frac{307.85}{288} \right] = 146.65[19.89 - 19.26] = 92.39 \text{ W}$$

From Eq. 9.3, the instantaneous exergy efficiency will be as

$$\eta_{i,ex} = \frac{\dot{Q}_{u,thex}}{[I(t)]_{ex} \times \text{area of solar energy system}}$$

$$= \frac{92.39}{500 \times \times 0.96 \times 5 \times 2} = \frac{92.39}{4800} = 0.0193$$

**For N = 10:** $T_{fo10} = 45.26$ °C, $T_{fi} = 15$ °C and $T_a = 16$ °C.

From Eq. 2.35a and Eq. 9.2b, we have an expression for the rate of thermal exergy as

$$\dot{Q}_{u,thex} = \dot{m}_f C_f \left[ (T_{fo10} - T_{fi}) - (T_a + 273) \ln \frac{T_{fo10} + 273}{T_{fi} + 273} \right]$$

$$= 146.65 \left[ 30.26 - 289 \ln \frac{318.26}{288} \right] = 146.65[30.26 - 28.87] = 203.34 \text{ W}$$

From Eq. 9.3, the instantaneous exergy efficiency will be as

$$\eta_{i,ex} = \frac{\dot{Q}_{u,thex}}{[I(t)]_{ex} \times \text{area of solar energy system}}$$

$$= \frac{203.34}{500 \times \times 0.96 \times 10 \times 2} = \frac{203.34}{9600} = 0.021$$

One can see that the rate of exergy as well as instantaneous exergy efficiency is increased due to increase in the outlet temperature of fluid from FPC.

**Example 9.3** Evaluate the rate of thermal exergy and instantaneous exergy efficiency for Example 5.34 (solar air heaters) for N = 5.

**Solution**

**For N = 5:** $T_{fo5} = 37.98$ °C, $T_{fi} = 17$ °C and $T_a = 16$ °C, $\dot{m}_f C_f = 0.029 \times 1005 = 29.116$ W/°C and $F' = 0.22$.

From Eq. 2.35a and Eq. 9.2b, we have an expression for the rate of thermal exergy as

$$\dot{Q}_{u,thex} = \dot{m}_f C_f \left[ (T_{fo5} - T_{fi}) - (T_a + 273) \ln \frac{T_{fo5} + 273}{T_{fi} + 273} \right]$$

$$= 29.116 \left[ 20.98 - 289 \ln \frac{310.98}{290} \right] = 29.116[20.98 - 20.18] = 49.68 \text{ W}$$

From Eq. 9.3, the instantaneous exergy efficiency will be as

$$\eta_{i,ex} = \frac{\dot{Q}_{u,thex}}{[I(t)]_{ex} \times \text{area of solar energy system}}$$
$$= \frac{49.68}{500 \times 0.96 \times 5 \times 2} = \frac{49.68}{4800} = 0.01$$

## 9.2.2 The Rate of Thermal Exergy and Overall Exergy Efficiency for SPVT Liquid Collectors Connected in Series

**Example 9.4** Evaluate the rate of thermal exergy and instantaneous overall exergy efficiency for Examples 6.21 and 6.25 (SPVT liquid collector) for N = 5.

**Solution**

**For N = 5:** $T_{fo5} = 21.72\,°C, T_{fi} = 15\,°C$ and $T_a = 16\,°C, \dot{m}_f C_f = 146.65\,W/°C$.
From Eq. 2.35a and Eq. 9.2b, we have an expression for the rate of thermal exergy as

$$\dot{Q}_{u,thex5} = \dot{m}_f C_f \left[ (T_{fo5} - T_{fi}) - (T_a + 273) \ln \frac{T_{fo5} + 273}{T_{fi} + 273} \right]$$
$$= 146.65 \left[ 6.72 - 289 \ln \frac{294.72}{288} \right] = 146.65[6.72 - 6.66] = 7.939\,W$$

The total electrical power $(\dot{E}_{elT}) = \dot{E}_{el1} + \dot{E}_{el2} + \dot{E}_{el3} + \dot{E}_{el4} + \dot{E}_{el5} = 98.26 + 98.12 + 97.97 + 97.81 + 97.68 = 489.84\,W$ (Example 6.25).
From Eq. 9.2a, the overall exergy of 5-SPVT liquid collectors connected in series will be determined as

$$\dot{E}x_{u,ov} = \dot{Q}_{u,thex5} + \dot{E}_{elT} = 7.939 + 489.84 = 497.779\,W$$

Further, an instantaneous overall exergy efficiency of $(\eta_{i,ex-ov})$ of 5-SPVT liquid collectors from Eq. 9.3 is as follows:

$$\eta_{i,ex-ov} = \frac{\dot{E}x_{u-ov}}{[I(t)]_{ex} \times \text{area of solar energy system}} = \frac{997.829}{500 \times \times 0.96 \times 5 \times 2} = \frac{497.779}{4800} = 0.1037$$

**Example 9.5** Evaluate the rate of thermal exergy for individual SPVT liquid collectors [N from 1 to 5] connected in series and instantaneous overall exergy efficiency for Examples 6.21 and 6.25 (SPVT liquid collector) and compare the results of Example 9.4 for N = 5.

**Solution**

(a) **For N = 1:** $T_{fo1} = 16.43\ °C$ and $T_{fi} = 15\ °C$; $\dot{m}_f C_f = 146.65\ W/°C$.

From Eq. 9.2b, we have an expression for the rate of thermal exergy for 1st SPVT liquid collector as

$$\dot{Q}_{u,thex1} = \dot{m}_f C_f \left[ (T_{fo1} - T_{fi}) - (T_a + 273) \ln \frac{T_{fo1} + 273}{T_{fi} + 273} \right]$$

$$= 146.65 \left[ 1.43 - 289 \ln \frac{289.43}{288} \right] = 146.65[1.43 - 1.43] = 0\ W$$

(b) **For N = 2:** $T_{fo2} = 17.68\ °C$ and $T_{fo1} = T_{fi2} = 16.43\ °C$; $\dot{m}_f C_f = 146.65\ W/°C$.

From Eq. 9.2b, we have an expression for the rate of thermal exergy for 2nd SPVT liquid collector as

$$\dot{Q}_{u,thex2} = \dot{m}_f C_f \left[ (T_{fo2} - T_{fi2}) - (T_a + 273) \ln \frac{T_{fo2} + 273}{T_{fi2} + 273} \right]$$

$$= 146.65 \left[ 1.43 - 289 \ln \frac{290.68}{289.43} \right] = 146.65[1.25 - 1.2455] = 0.66\ W$$

(c) **For N = 3:** $T_{fo3} = 19.17\ °C$ and $T_{fo2} = T_{fi3} = 17.68\ °C$; $\dot{m}_f C_f = 146.65\ W/°C$.

From Eq. 9.2b, we have an expression for the rate of thermal exergy for 3rd SPVT liquid collector as

$$\dot{Q}_{u,thex3} = \dot{m}_f C_f \left[ (T_{fo3} - T_{fi3}) - (T_a + 273) \ln \frac{T_{fo3} + 273}{T_{fi3} + 273} \right]$$

$$= 146.65 \left[ 1.49 - 289 \ln \frac{292.17}{290.68} \right] = 146.65[1.49 - 1.4776] = 1.8185\ W$$

(d) **For N = 4:** $T_{fo4} = 20.47\ °C$ and $T_{fo3} = T_{fi4} = 19.17\ °C$ ; $\dot{m}_f C_f = 146.65\ W/°C$.

From Eq. 9.2b, we have an expression for the rate of thermal exergy for 3rd SPVT liquid collector as

$$\dot{Q}_{u,thex4} = \dot{m}_f C_f \left[ (T_{fo4} - T_{fi4}) - (T_a + 273) \ln \frac{T_{fo4} + 273}{T_{fi4} + 273} \right]$$

$$= 146.65 \left[ 1.3 - 289 \ln \frac{293.47}{292.17} \right] = 146.65[1.3 - 1.283] = 2.49\ W$$

(e)  **For N = 5:** $T_{fo5} = 21.72$ °C and $T_{fo4} = T_{fi5} = 20.47$ °C ; $\dot{m}_f C_f = 146.65$ W/°C.

From Eq. 9.2b, we have an expression for the rate of thermal exergy for 3rd SPVT liquid collector as

$$\dot{Q}_{u,thex5} = \dot{m}_f C_f \left[ (T_{fo5} - T_{fi5}) - (T_a + 273) \ln \frac{T_{fo5} + 273}{T_{fi5} + 273} \right]$$

$$= 146.65 \left[ 1.25 - 289 \ln \frac{294.72}{293.47} \right] = 146.65[1.25 - 1.2283] = 3.17 \text{ W}$$

The total rate of thermal exergy will be sum of all five exergy evaluated above as

$$\dot{Q}_{u,thex5} = \dot{Q}_{u,thex1} + \dot{Q}_{u,thex2} + \dot{Q}_{u,thex3} + \dot{Q}_{u,thex4} + \dot{Q}_{u,thex5} = 0 + 0.66$$
$$+ 1.8185 + 2.49 + 3.17 = 8.1385 \text{ W}$$

There is an improvement of $\frac{8.1385 - 7.939}{7.939} = 2.5\%$ in the rate of thermal exergy obtained in Example 9.4.

### 9.2.3  The Rate of Thermal Exergy and Overall Exergy Efficiency for Flexible OPVT-TEC Liquid Collectors Connected in Series

**Example 9.6** Find out the rate of thermal exergy and instantaneous overall exergy efficiency for flexible OPVT-TEC liquid collectors connected in series for $N = 1$ and 2 respectively for Examples 6.42–6.45.

**Solution**

**For N = 1st:** $T_{fo1} = 16.91$ °C, $T_{fi} = 15$ °C, $\dot{m}_f c_f = 146.65$ W/°C, $\overline{T}_{tec,top1} = 16.51$ °C, $\overline{T}_{f1} = 15.995$ °C.
From Eq. 9.2b, we have an expression for the rate of thermal exergy as
The rate of thermal exergy $\left( \dot{Q}_{u,thex1} \right) = 146.65 \left[ 1.91 - 289 \ln \frac{289.91}{288} \right] = 146.65[1.91 - 1.91] = 0$.
Electrical power from 1st TEC is

$$\dot{E}_{TEC1} = 0.15 \times 450 \times (16.51 - 15.995) \times 0.4324 \times 2 = 30.06 \text{ W}$$

Following Eq. 6.31a, an electrical power can be obtained as

$$\dot{E}_{ell} = 0.12[1 - 0.00425(62.82 - 25)] \times 0.89 \times 500 \times 2 = 0.1010$$
$$\times 0.89 \times 500 \times 2 = 89.62 \text{ W}$$

The total electrical power is the sum of electrical power from PV and TEC, and hence

$$\dot{E}_{\text{total1}} = \dot{E}_{el1} + E_{TEC1} = 89.62 + 30.06 = 119.62 \text{ W}$$

Further, an instantaneous overall exergy efficiency of $\left(\eta_{i,ex-ov}\right)$ of single $(N = 1)$ OPVT-TEC liquid collector from Eq. 9.3 is as follows:

$$\eta_{i,ex-ov} = \frac{\cdot Ex_{u-ov}}{[I(t)]_{ex} \times \text{area of solar energy system}}$$

$$= \frac{0 + 119.62}{500 \times \times 0.96 \times 1 \times 2} = \frac{119.62}{960} = 0.1246$$

**For N = 2nd**: $T_{fo2} = 18.73$ °C, $T_{fi} = T_{fo1} = 16.91$ °C, $\dot{m}_f c_f = 146.65$ W/°C, $\overline{T}_{tec,top2} = 18.296$ °C (Example 6.43), $\overline{T}_{f2} = 17.82$ °C (Example 6.42).

The rate of thermal exergy $\left(\dot{Q}_{u,thex2}\right) = 146.65\left[1.82 - 289\ln\frac{291.73}{289.91}\right] = 146.65[1.82 - 1.79] = 4.3995$ W.

Electrical power from 2nd TEC can be evaluated as

$$\dot{E}_{TEC2} = 0.15 \times 450 \times (18.296 - 17.82) \times 0.4324 \times 2 = 27.79 \text{ W}$$

Following Eq. 6.31a, an electrical power from 2nd PV can be obtained as

$$\dot{E}_{el2} = 0.12[1 - 0.00425(67.12 - 25)] \times 0.89 \times 500 \times 2$$
$$= 0.0985 \times 0.89 \times 500 \times 2 = 87.68 \text{ W}$$

The total electrical power is the sum of electrical power from PV and TEC, and hence

$$\dot{E}_{total2} = \dot{E}_{el} + \dot{E}_{TEC} = 87.68 + 27.79 = 115.47 \text{ W}$$

**For N = 2 (total)**: $T_{fo2} = 18.73$ °C, $T_{fi} = 15$ °C, $\dot{m}_f c_f = 146.65$ W/°C, $\overline{T}_{tec,top2} = 18.296$ °C (Example 6.43), $\overline{T}_{f2} = 17.82$ °C (Example 6.42).

From Eq. 9.2b, we have an expression for the rate of thermal exergy as

The rate of thermal exergy $\left(\dot{Q}_{u,thex1+2}\right) = 146.65\left[3.73 - 289ln\frac{291.73}{288}\right] = 146.65[3.73 - 3.718] = 1.761.63$ W.

Total electrical power from both TEC of OPVT-TEC can be evaluated as

$$\dot{E}_{total1+2} = \dot{E}_{TEC1} + \dot{E}_{TEC2} = 30.06 + 27.79 = 57.85 \text{ W}$$

Total electrical power from both PV of OPVT-TEC collector can be obtained as

$$\dot{E}_{el1+2} = \dot{E}_{el1} + \dot{E}_{el2} = 89.62 + 87.68 = 177.3 \text{ W}$$

The total electrical power from both TEC and PV of OPVT-TEC is the sum of electrical power from PV and TEC, and hence

$$\dot{E}_{elT} = \dot{E}_{el1+2} + \dot{E}_{total1+2} = 177.3 + 57.85 = 235.15 \text{ W}$$

The total rate of thermal energy from OPVT-TEC is the sum of rate of thermal from both OPVT-TEC, and hence

$$\dot{Q}_{u,thexT} = \dot{Q}_{u,thex1} + \dot{Q}_{u,thex2} = 0 + 4.3995 = 4.3995 \text{ W}$$

The, an overall thermal energy from both OPVT-TEC collectors connected in series is

$$\dot{Ex}_{u,ov} = \dot{Q}_{u,thexT} + \dot{E}_{elT} = 4.3995 + 235.15 = 239.55 \text{ W}$$

From this example, the following points to be noted are as follows:

(i) The rate of thermal exergy should be calculated individually then added to get total rate of thermal energy for all solar energy collectors connected in series.
(ii) The electrical power should also be calculated individually similar to rate of thermal energy and then added to get total electrical power.

Now, an overall exergy efficiency is obtained as

$$\eta_{i,ex-ov} = \frac{\dot{Ex}_{u-ov}}{[I(t)]_{ex} \times \text{area of solar energy system}}$$
$$= \frac{239.55}{500 \times 0.96 \times 2 \times 2} = \frac{239.55}{1920} = 0.1247$$

### 9.2.4 The Rate of Thermal Exergy and Overall Exergy Efficiency for SPVT-CPC Liquid Collectors Connected in Series

**Example 9.7** Calculate an overall exergy efficiency of six SPVT-CPC liquid collectors connected in series for Example 7.16 for $C = 2$.

**Solution**

From Example 7.16, we have
Total exergy of six PVT-CPC liquid collectors connected in series as

$$\left(\dot{Ex}_{u-ov6}\right) = \text{rate of thermal exergy} + \text{electrical power} = 99.97 + 495.12 = 595.10 \text{ W}$$

Now, an overall exergy efficiency is obtained as

$$\eta_{i,ex-ov} = \frac{\dot{E}x_{u-ov}}{[I(t)]_{ex} \times \text{area of solar energy system}} = \frac{595.10}{500 \times \times 0.96 \times 6 \times 2} = \frac{595.10}{5760} = 0.1033$$

**Example 9.8** Repeat the Example 9.7 for Example 7.19 for six SPVT-CPC for C = 1.

**Solution**

From Example 7.20, we have

The rate of thermal exergy of six SPVT-CPC liquid collectors for C=1 connected in series as

$$\dot{E}x_{u,th6} = 0.035 \times 4190\left[(24.20 - 15) - (16 + 273)\ln\frac{24.20 + 273}{15 + 273}\right]$$
$$= 146.65[12.452 - 289\ln 1.0319] = 146.65 \times [9.2 - 9.0875] = 16.498 \text{ W}$$

and,

$$\dot{E}_{el6} = \overline{\eta}_m \times I(t) \times \beta_c A_m \times N = 0.1118 \times 500 \times 0.89 \times 2 \times 6 = 597.17 \text{ W}$$

Now, an overall exergy efficiency is obtained as

$$\eta_{i,ex-ov} = \frac{\dot{E}x_{u,th6} + \dot{E}_{el6}}{[I(t)]_{ex} \times \text{area of solar energy system}}$$
$$= \frac{16.498 + 597.17}{500 \times \times 0.96 \times 6 \times 2} = \frac{613.67}{5760} = 0.1065$$

**Example 9.9** Evaluate overall exergy efficiency for Example 7.20 of six FPC-CPC air collectors connected in series for C = 2.

**Solution**

Now, an overall exergy efficiency of six FPC-CPC air collectors connected in series for C = 2 is obtained as

$$\eta_{i,ex-ov} = \frac{\dot{E}x_{u,th6}}{[I(t)]_{ex} \times \text{area of solar energy system}}$$
$$= \frac{429.68}{500 \times \times 0.96 \times 6 \times 4} = \frac{429.68}{11.540} = 0.0372$$

**Example 9.10** Evaluate overall exergy efficiency of PVT-CPC air collectors connected in series for $N = 1$ and 6 for $C = 2$ for Example 7.29.

**Solution**

**For N = 1**: The rate of thermal exergy $= 0$.

Overall exergy efficiency of single PVT-CPC air collector for $C = 2$ is obtained as

$$\eta_{i,ex-ov} = \frac{\dot{Ex}_{u,th1}}{[I(t)]_{ex} \times \text{area of solar energy system}}$$

$$= \frac{0 + 80.07}{500 \times \times 0.96 \times 1 \times 4} = \frac{80.07}{1920} = 0.0417$$

**For N = 6**: The rate of thermal exergy $= 0$.

Overall exergy efficiency of single PVT-CPC air collector for $C = 2$ is obtained as

$$\eta_{i,ex-ov} = \frac{\dot{Ex}_{u,th6}}{[I(t)]_{ex} \times \text{area of solar energy system}}$$

$$= \frac{0 + 475.52}{500 \times \times 0.96 \times 6 \times 4} = \frac{475.52}{11,540} = 0.0412$$

**Example 9.11** Evaluate an overall exergy (total) efficiency for single and six flexible OPVT-TEC-CPC collectors connected in series for $N = 1$ and 6 for $C = 2$ for Example 7.48.

**Solution**

**For N = 1, Known**: $\dot{Ex}_{u1,th} = 1.98$ W (Example 7.45), $\dot{E}_{el1} = 48.15$ W (Example 7.43) and $\dot{E}_{TEC1} = 115.58$ (Example 7.46).

Overall exergy efficiency of single flexible PVT-TEC-CPC collector for $C = 2$ is obtained as

$$\eta_{i,ex-ov} = \frac{\dot{Ex}_{u,0v}}{[I(t)]_{ex} \times \text{areaof solar energy system}} = \frac{1.98 + 48.15 + 115.58}{500 \times \times 0.96 \times 1 \times 4}$$

$$= \frac{165.63}{1920} = 0.0863$$

**For N = 6, Known**: $\dot{Ex}_{u6,th} = 107.61$ W (Example 7.46), $\dot{E}_{el6} = 277.03$ W (Example 7.44) and $\dot{E}_{TEC6} = 605$ W (Example 7.46).

Overall exergy efficiency of six flexible PVT-TEC-CPC collectors connected in series for $C = 2$ is obtained as

$$\eta_{i,ex-ov} = \frac{\dot{Ex}_{u,0v}}{[I(t)]_{ex} \times \text{area of solar energy system}}$$

$$= \frac{107.61 + 277.03 \text{ W} + 605 \text{ W}}{500 \times \times 0.96 \times 6 \times 4} = \frac{989.64}{11,520} = 0.0858$$

### 9.2.5  The Fresnel Lens (FL) Integrated Evacuated Tubular Collector (ETC) [FLiETC]

**Example 9.12** Compute an instantaneous overall thermal exergy efficiency of FLiETC for Example 8.37 for $C = 9$ and mass flow rate of 0.035 kg/s.

**Solution**

Given: $T_{fo5} = 21.46\,°C$, $T_{fi} = 15\,°C$ and $\dot{m}_f C_f = 0.035 \times 4190 = 146.65$ W/ °C.
The rate of thermal exergy of FLiETC for

$$\dot{Ex}_{u,th6} = 0.035 \times 4190 \left[ (21.46 - 15) - (16 + 273) \ln \frac{21.46 + 273}{15 + 273} \right]$$

$$= 146.65[6.46 - 289 ln\,1.0224] = 146.65 \times [6.46 - 6.41] = 7.33 \text{ W}$$

From Example 8.37, we have the following instantaneous overall exergy efficiency of FLiETC as

$$\eta_{i,ex-ov} = \frac{7.33}{500 \times 0.96 \times 5 \times 0.5} = \frac{7.33}{1200} = 0.0061$$

### 9.2.6  The Fresnel Lens (FL) Integrated SPVT Liquid Collector [FLiSPVT]

**Example 9.13** Compute an instantaneous overall exergy efficiency of FLiSPVT for Example 8.46 and 8.48 for $C = 4$ and mass flow rate of 0.035 kg/s.

**Solution**

Given: $T_{fo5} = 28.64\,°C$, $T_{fi} = 15\,°C$ and $\dot{m}_f C_f = 0.035 \times 4190 = 146.65$ W/°C.
The rate of thermal exergy of five FLiSPVT collectors connected in series as

$$\dot{Ex}_{u,th6} = 0.035 \times 4190 \left[ (28.64 - 15) - (16 + 273) \ln \frac{28.64 + 273}{15 + 273} \right]$$

$$= 146.65[13.64 - 289 \ln 1.0474] = 146.65 \times [13.64 - 13.38] = 38.129 \text{ W}$$

The electrical power of SPVT $\left(\acute{E}_{el5}\right) = \eta_m \times I_b \times \beta_c \times A_{am} \times N == 0.08716 \times 500 \times 0.89 \times 2 \times 5 = 387.862$ W.

From Eq. 9.3, we have the following instantaneous overall exergy efficiency of FLiSPVT as

$$\eta_{i,ex-ov} = \frac{38.129 + 387.862}{500 \times 0.96 \times 5 \times 2} = \frac{425.991}{4800} = 0.0887$$

### 9.2.7 The Fresnel Lens (FL) Integrated Flexible OPVT-TEC Liquid Collector [FLiOPVT-TEC]

**Example 9.14** Compute an instantaneous overall exergy efficiency of FLiOPVT-TEC for Examples 8.55 to 8.61 for $C = 4$ and mass flow rate of 0.035 kg/s.

**Solution**

**For N = 1:** $T_{fo1} = 18.56$ °C, $T_{fi} = 15$ °C and $\dot{m}_f c_f = 146.65$ W/°C.

The rate of thermal exergy of single FLiOPVT-TEC collector as

$$\dot{Ex}_{u,th1} = 0.035 \times 4190 \left[ (18.56 - 15) - (16 + 273) \ln \frac{18.56 + 273}{15 + 273} \right]$$
$$= 146.65[3.56 - 289 \ln 1.01236] = 146.65 \times [3.56 - 3.55] = 1.4665 \text{ W}$$

Electrical power from TEC of FLiOPVT-TEC (Example 8.59) as

$$\dot{E}_{el1} = 0.1164 \times 500 \times 1 \times 0.5 \times 1 = 58.2 \text{ W}$$

Electrical power from TEC of FLiOPVT-TEC (Example 8.60) as

$$\dot{E}_{TEC1} = \eta_{tec} U_{tec} \left(\overline{T}_{tec,top1} - \overline{T}_{f1}\right) \beta_{tec} A_{rm} \times N = 0.15 \times 450 \times (19.56 - 16.23)$$
$$\times 0.5 \times 1 \times 5 = 115.58 \text{ W}$$

From Eq. 9.3, we have the following instantaneous overall exergy efficiency of FLiOPVT-TEC as

$$\eta_{i,ex-ov1} = \frac{\dot{Ex}_{u-ov}}{[I(t)]_{ex} \times \text{areaof solar energy system}}$$
$$= \frac{1.4665 + 58.2 + 115.58}{500 \times 0.96 \times 1 \times 2} = \frac{175.25}{960} = 0.183$$

**For N = 5:** $T_{fo1} = 32.21$°C, $T_{fi} = 15$°C and $\dot{m}_f c_f = 146.65$ W/°C.

The rate of thermal exergy of five FLiOPVT-TEC collectors connected in series as

$$\dot{E}x_{u,th5} = 0.035 \times 4190\left[(32.21 - 15) - (16 + 273)\ln\frac{32.21 + 273}{15 + 273}\right]$$
$$= 146.65[17.21 - 289\ln 1.05975] = 146.65 \times [17.21 - 16.77] = 64.02\ \text{W}$$

Electrical power from TEC of FLiOPVT-TEC (Example 8.59) as

$$\dot{E}_{el5} = 0.1131 \times 500 \times 1 \times 0.5 \times 5 = 141.75\ \text{W}$$

Electrical power from TEC of FLiOPVT-TEC (Example 8.60) as

$$\dot{E}_{TEC1} = \eta_{tec}U_{tec}\left(\overline{T}_{tec,top1} - \overline{T}_{f1}\right)\beta_{tec}A_{rm} \times N = 0.15 \times 450 \times (19.56 - 16.23)$$
$$\times 0.5 \times 1 \times 5 = 540\ \text{W}$$

From Eq. 9.3, we have the following instantaneous overall exergy efficiency of FLiOPVT-TEC as

$$\eta_{i,ex-ov5} = \frac{\dot{E}x_{u-ov5}}{[I(t)]_{ex} \times \text{areaof solar energy system}}$$
$$= \frac{64.02 + 141.75 + 540}{500 \times 0.96 \times 5 \times 2} = \frac{745.77}{4.800} = 0.1553$$

All results of an overall exergy efficiency obtained in Examples 9.2–9.14 have been summarized in Table 9.1.

## 9.3  Energy Matrices

The energy matrices mainly consist of (i) energy payback time (EPBT), (ii) energy production factor (EPF) and (iii) life cycle conversion efficiency (LCEC). These energy matrices mainly depend on

(a) **Embodied energy** $(E_{in})$: It should be in kWh of solar energy system. The embodied energy consists of embodied e Embodied energy coefficient nergy coefficient $(e_i)$ of each material (Appendix F) used to manufactured solar energy system and its mass of each components $(m_i)$. Mathematically, it can be expressed as follows:

$$E_{in} = \sum_{i=1}^{n} e_i m_i \tag{9.4}$$

**Table 9.1** Overall thermal exergy and its instantaneous overall exergy efficiency in fraction of solar energy system

| S. No. | Solar energy system | Example | An overall rate of thermal exergy | An overall rate of exergy (electrical) | An instantaneous exergy efficiency in fraction |
|---|---|---|---|---|---|
| 1 | Liquid FPC (N = 5)<br>Liquid FPC (N = 10) | 9.2 | 92.39 W<br>203.34 W | | 0.0193<br>0.021 |
| 2 | Solar air collector (N = 5) | 9.3 | 23.2928 W | | 0.01 |
| 3 | SPVT collector (N = 5) | 9.4 | 7.939 W | 497.779 W | 0.1037 |
| 4 | Flexible OPVT-TEC (N = 1)<br>Flexible OPVT-TEC (N = 2) | 9.6 | 0<br>4. 3995 W | 119.62 W<br>235.15 W | 0.1246<br>0.1247 |
| 5 | SPVT-CPC [N = 6, C = 2]<br>SPVT-CPC [N = 6, C = 1] | 9.7<br>9.8 | 99.97 W<br>16.498 W | 495.12 W<br>597.17 W | 0.1033<br>0.1065 |
| 6 | Air FPC-CPC [N = 6, C = 2] | 9.9 | 429.68 W | | 0.0372 |
| 7 | Air PVT-CPC [N = 1, C = 2]<br>Air PVT-CPC [N = 6, C = 2] | 9.10 | 0<br>0 | 80.07 W<br>475.52 W | 0.0417<br>0.0412 |
| 8 | Flexible OPVT-TEC-CPC [N = 1, C = 2] | 9.11 | 1.98 W | 163.73W | 0.0863 |
| 9 | Flexible PVT-TEC-CPC [N = 6, C = 2] | 9.11 | 107.61 W | 832.03 W | 0.0858 |
| 10 | Liquid FLiETC [N = 5, C = 9] | 9.12 | 7.33 | | 0.0061 |
| 11 | Liquid FLiSPVT [N = 5, C = 4] | 9.13 | 38.129 W | 387 W | 0.0887 |
| 12 | FLiOPVT-TEC[N = 1, C = 4]<br>FLiOPVT-TEC[N = 5, C = 4] | 9.14<br>9.14 | 1.4665 W<br>64.02 W | 173.78<br>681.75 | 0.183<br>0.1553 |

and,

(b) **Annual energy**: It should be either an annual overall thermal or an annual overall exergy. It can be obtained by summing either first hourly (Eq. 9.1) to make it daily, then monthly and finally yearly or evaluating monthly performance of solar energy system and summing monthly performance to get yearly performance in terms of kWh. Similarly, an overall annual exergy can be determined.

Mathematically, these can be expressed as follows:

$$Q_{th,\mathrm{annul}} = \sum_{j=1} \dot{Q}_{u,thTj} \tag{9.5a}$$

or,

$$Ex_{u,\mathrm{annual}} = \sum_{j=1} Ex_{u,ovj} \tag{9.5b}$$

For evaluating energy matrices, the following assumptions have been considered for clear understanding to readers:

(i)   The annual average solar radiation as considered in the various chapters has been assumed to be 500 W/m$^2$.
(ii)  The number of clear days has been considered as 300 for convenience in computing energy matrices.
(iii) The number of bright/blue sky sunshine hours has been considered as 5 h.

The energy matrices [1] can be expressed as follows:

### 9.3.1  Energy Payback Time (EPBT)

It is a period over which the embodied energy $(E_{in})$ of solar energy system is recovered. It can be expressed as

$$\mathrm{EPBT} = \frac{\mathrm{Embodied\ energy}(E_{in}),\ \mathrm{Eq.9.4}}{\mathrm{An\ annual\ thermal\ or\ exergy\ of\ solar\ energy\ system,\ Eq.9.5}} \tag{9.6}$$

The EPBT should be always less than life of the solar system $(T)$.

### 9.3.2  Energy Production Factor (EPF)

It is the ratio of an annual thermal oe exergy of solar energy system, Eq. 9.5 to embodied energy, Eq. 9.4. Basically, it is reciprocal of EPBT, i.e.,

$$\text{EPF} = \frac{T}{\text{EPBT}} \qquad (9.7)$$

The energy production factor should be always more than one.

### 9.3.3 Life Cycle Conversion Efficiency (LCEC)

The LCCE is the net energy productivity of the system with respect to the solar input (radiation) over the life time of the system (T years). In other words, it is the ratio of net energy saved over the life time to total solar energy utilized over the life time. The expression for LCCE is given by

$$\eta_{\text{life}}$$
$$= \frac{Q_{th,\text{annul}} \times T - E_{\text{in}}}{[I(t)]_{\text{annual}} \times T} \text{ on the basis of thermal energy of solar energy system (8.7)a}$$
$$(9.8)$$

and,

$$\eta_{\text{life}} = \frac{Ex_{u,\text{annual}} \times T - E_{\text{in}}}{[I(t)]_{\text{annualex}} \times T} \text{ on the basis of exergy of solar energy system (9.7b}$$
$$(9.9)$$

where, $T$ is the life of solar energy system. The $\eta_{life}$ should be less than one in fraction.

### 9.3.4 Energy Matrices and Embodied Energy of Liquid and Air Flat Plate Collectors (FPC)

Generally, liquid and air flat plate collectors are 2 m². However, an effective area of glazed surface is 1.875 m². However, the remaining area of 0.125 m² is occupied by frame. Hence, we will determine the embodied energy of 1.875 m² of liquid and air FPC.

**Example 9.15** Calculate embodied energy of two liquid flat plate collectors (FPC) with each effective area of 1.875 m² by using the data of Table 9.2 and Appendix F and Eq. 9.4.

**Solution**

From Eq. 9.4, we have an expression for total embodied energy of solar system as

**Table 9.2** Embodied energy coefficient [2] and embodied energy of two flat plate collector (FPC) with each effective area of 1.875 m$^2$

| S. No. | Components of FPC | Quantity | Total weight ($m_i$) | Embodied energy coefficient ($e_i$) (MJ/kg) | Embodied energy (MJ) ($e_i m_i$) | Embodied energy (kWh) ($e_i m_i$) |
|---|---|---|---|---|---|---|
| 1 | Copper riser 1/2″ | 20 × 1.8 = 36 m | 8.2 kg | 81.00 | 664.20 | 184.50 |
| 2 | Header 1″ | 4 × 1.15 = 4.6 m | 3.8 kg | 81.00 | 307.80 | 85.50 |
| 3 | Al box | 2 | 10 kg | 199.00 | 1990.00 | 552.00 |
| 4 | Cu sheet | 2 | 11 kg | 70.6 | 776.6 | 215.72 |
| 5 | Glass cover toughened 4 mm | 2 (3.75m$^2$) | 0.01464 m$^3$ | 66,020.00 MJ/m$^3$ | 966.53 | 268.48 |
| 6 | Glass wool | 13 m$^2$ | 0.064 m$^3$ | 139.00 MJ/m$^3$ | 8.896 | 2.50 |
| 7 | Nuts / bolts/ screws | 32 | 1 kg | 31.06 | 31.06 | 8.60 |
| 8 | Union /elbow | 8 | 1.5 kg | 46.80 | 70.20 | 19.50 |
| 9 | Nozzle/ flange | 8 | 1 kg | 62.10 | 62.10 | 17.30 |
| 10 | Mild steel stand | 1 | 40 kg | 34.20 | 1368.00 | 380.00 |
| 11 | Paint | 1L | 1L | 90.40 | 90.40 | 25.10 |
| 12 | Rubber gasket | 18 m | 4.2 kg | 11.83 | 49.686 | 13.80 |
| 13 | G I pipes 1/2″ | | 9.5 kg | 44.10 | 418.95 | 116.10 |
| 14 | Al frame 1″ | 12 m | 2.5 kg | 170.00 | 425.00 | 118.00 |
| 15 | Al sheet 24 gauge | | 2.5 kg | 170.00 | 425.00 | 118.00 |
| | Total | | | | 7654.42 | 2126.23 |

$$E_{in} = \sum_{i=1}^{n} e_i m_i$$

Substitute the appropriate value of $e_i$ and $m_i$ from Table 9.2 for two FPC, one gets

$$\begin{aligned}
E_{in2} = {} & 81.0 \times 8.2 + 81.0 \times 3.8 + 199.0 \times 10 + 70.6 \times 11 \\
& + 66020 \times 0.01464 + 139 \times 0.064 + 31.06 \times 1 + 46.8 \\
& \times 1.5 + 62.1 \times 1 + 34.2 \times 40 + 90.4 \times 1 + 11.83 \times 4.2
\end{aligned}$$

$$+ 170 \times 2.5 + 170 \times 2.5$$
$$= [664.2 + 307.8 + 1990 + 766.60 + 966.53 + 8.896 + 31.06 + 70.20 + 62.10]$$
$$+ [1,368 + 90.4 + 49.686 + 418.95 + 425 + 425]$$
$$= 5560.486 + 2777.036 = 7654.422 \text{ MJ} = 2126.23 \text{ kWh}$$

The embodied energy of single liquid flat plate collector with 1.875 m² area $(E_{in1}) = \frac{2126.23}{2} = 1159.49$ kWh.

**Example 9.16** Evaluate thermal energy matrices for liquid flat plate collectors connected in series for Example 8.1 for $N = 5$ by using Eqs. 9.6–9.8.

**Solution**

From Example 8.1: $\dot{Q}_{u,th} = 2764.35$ W $= 2.764$ kW and $T = 10$ years (Life of FPC).

As per assumption mentioned in Sect. 9.3 and following Eq. 9.5a, we have

$$Q_{th,\text{annul}}/\text{year} = \sum_{j=1} \dot{Q}_{u,thTj} = 2.764 \times 5 \times 300 \text{ kWh} = 4146 \text{ kWh}$$

The annual solar energy per year $[I(t)]_{\text{annual}} = 500 \frac{W}{m^2} \times 5 \text{ h} \times 300 \text{ days} = 750{,}000$ Wh/m² $= 750$ kWh/m².

The annual solar energy on five FPC with 2 m² area $[I(t)]_{\text{annual5}} = 750 \times 5 \times 2$ kWh $= 7500$ kWh.

Further, the embodied energy for five FPC will be evaluated as follows:
$E_{in5} = E_{in1} \times 5 = 1159.49 \times 5$ kWh $= 5797.45$ kWh (Example 9.15).

From Eq. 9.6, one gets energy payback time (EPBT) as

$$\text{EPBT} = \frac{\text{Embodied energy}(E_{in5}),}{\text{An annual thermal energy of five liqid FPC}} = \frac{5797.45}{4146} = 1.4 \text{ years}$$

From Eq. 9.7, one has an expression for energy production factor (EPF) as

$$\text{EPF} = \frac{T}{\text{EPBT}} = \frac{10}{1.4} = 7.14$$

From Eq. 9.8, we have thermal life cycle conversion efficiency (LCCE) as

$$\eta_{\text{life}} = \frac{Q_{th,\text{annul}} \times T - E_{in}}{[I(t)]_{\text{annual}} \times T} = \frac{4146 \times 10 - 5797.45}{7500 \times 10} = \frac{35{,}662.55}{75{,}000} = 0.47$$

**Example 9.17** Evaluate thermal exergy matrices for five liquid flat plate collectors connected in series for Example 9.2 by using Eqs. 9.6–9.9.

**Solution**

From Example 8.1: $\dot{Q}_{u,thex} = 92.39$ W $= 0.0924$ kW and T $= 10$ years (Life of FPC).

As per assumption mentioned in Sect. 9.3 and following Eq. 9.5b, we have

$$\frac{Ex_{u,\text{annual}}}{\text{year}} = \sum_{j=1} \dot{Ex}_{\cdot u,ovj} = 0.0924 \times 5 \times 300 \text{ kWh} = 138.6 \text{ kWh}$$

The annual solar energy $[I(t)]_{\text{annualex}} = 500 \times 0.96 \times 5 \times 300$ W/m$^2$ = 720,000 W/m$^2$ = 720 kWh/m$^2$ (Example 9.1).

The annual solar energy exergy on five FPC with 2 m$^2$ area $[I(t)]_{\text{annual5}} = 720 \times 5 \times 2$ kWh = 7200 kWh.

Further, the embodied energy for five FPC will be evaluated as follows:

$E_{in5} = E_{in1} \times 5 = 1159.49 \times 5 = 5797.45$ kWh (Example 9.15).

From Eq. 9.6, one gets energy payback time (EPBT) as

$$\text{EPBT} = \frac{\text{Embodied energy}(E_{in5}),}{\text{An annual thermal exergy of five liqid FPC}} = \frac{5797.45}{138.6} = 41.82 \text{ years}$$

From Eq. 9.7, one has an expression for energy production factor (EPF) as

$$\text{EPF} = \frac{T}{\text{EPBT}} = \frac{10}{41.82} = 0.24$$

From Eq. 9.8, we have thermal life cycle conversion efficiency (LCCE) as

$$\eta_{life} - \frac{Ex_{u,\text{annual}} \times T - E_{in}}{[I(t)]_{\text{annualex}} \times T} = \frac{138.6 \times 10 - 5797.45}{7200 \times 10} = \frac{-4411.45}{72,000} = -\text{ve}$$

So, the liquid flat plate collector is unacceptable from thermal exergy point of view because energy matrices are not satisfying criteria led down in section Eqs. 9.6–9.9.

**Example 9.18** Calculate embodied energy of two flat plate solar air collectors (SAC) with each effective area of 1.875 m$^2$ by using the data of Table 9.3 and Appendix F and Eq. 9.4.

**Solution**

From Eq. 9.4, we have an expression for total embodied energy of solar system as

$$E_{in} = \sum_{i=1}^{n} e_i m_i$$

Substitute the appropriate value of $e_i$ and $m_i$ from Table 9.2, one gets

$$E_{in} = 199.0 \times 10 + 70.6 \times 11 + 66020.00 \times 0.01464$$
$$+ 139 \times 0.064 + 31.06 \times 1 + 46.8 \times 1.5 + 62.1 \times 1$$
$$+ 34.2 \times 40 + 90.4 \times 1 + 11.83 \times 4.2 + 170 \times 2.5$$

**Table 9.3** Embodied energy coefficient and embodied energy of two flat plate solar air collector (Solar air collector, SAC) with each effective area of 1.875 m$^2$

| S. No. | Components of FPC | Quantity | Total weight ($m_i$) | Embodied energy coefficient ($e_i$) (MJ/kg) | Embodied energy (MJ) ($e_i m_i$) | Embodied energy (kWh) ($e_i m_i$) |
|--------|------------------|----------|----------------------|---------------------------------------------|----------------------------------|-----------------------------------|
| 1 | Al box | 2 | 10 kg | 199.00 | 1990.00 | 552.00 |
| 2 | Cu sheet | 2 | 11 kg | 70.6 | 776.6 | 215.72 |
| 3 | Glass cover toughened 4 mm | 2 (3.75m$^2$) | 0.01464 m$^3$ | 66,020.00 MJ/m$^3$ | 966.53 | 268.48 |
| 4 | Glass wool | 13 m$^2$ | 0.064 m$^3$ | 139.00 MJ/m$^3$ | 8.896 | 2.50 |
| 5 | Nuts / bolts/ screws | 32 | 1 kg | 31.06 | 31.06 | 8.60 |
| 6 | Union / elbow | 8 | 1.5 kg | 46.80 | 70.20 | 19.50 |
| 7 | Nozzle/ flange | 8 | 1 kg | 62.10 | 62.10 | 17.30 |
| 8 | Mild steel stand | 1 | 40 kg | 34.20 | 1368.00 | 380.00 |
| 9 | Paint | 1L | 1L | 90.40 | 90.40 | 25.10 |
| 10 | Rubber gasket | 18 m | 4.2 kg | 11.83 | 49.686 | 13.80 |
| 11 | Al frame 1/2″ | 12 m | 2.5 kg | 170.00 | 425.00 | 118.00 |
| 12 | Al sheet 24 gauge | | 2.5 kg | 170.00 | 425.00 | 118.00 |
| | Total | | | | 6263.472 | 1739.85 |

$$+ 170 \times 2.5 = [1990 + 776.6 + 966.53 + 8.896$$
$$+ 31.06 + 70.20 + 62.10] + [1368 + 90.4 + 49.686$$
$$+ 425 + 425] = 5560.486 + 2777.036$$
$$= 6263.472 \text{ MJ} = 1739.85 \text{ kWh}$$

The embodied energy of single flat plate air collector with copper absorber plate $(E_{in1}) = \frac{1739.85}{2} = 869.92$ kWh.

**Example 9.19** Evaluate thermal energy matrices for flat plate solar air collectors (SAC) connected in series for Example 8.3 for $N = 5$ by using Eqs. 9.6–9.8.

From Example 8.3: $\dot{Q}_{u,th} = 611.4621$ W $= 0.6115$ kW and T $= 10$ years (Life of FPC).

As per assumption mentioned in Sect. 9.3 and following Eq. 9.5a, we have

$$Q_{th,\text{annul}}/\text{year} = \sum_{j=1} \dot{Q}_{u,thTj} = 0.6115 \times 5 \times 300 \text{ kWh} = 917.25 \text{ kWh}$$

The annual solar energy $[I(t)]_{\text{annual}} = 500 \times 5 \times 300 = 750,000 \text{ W} = 750 \text{ kWh/m}^2$.

The annual solar energy on five solar air (SAC) with 2 m$^2$ area $[I(t)]_{\text{annual5}} = 500 \times 5 \times 300 \times 5 \times 2 = 750 \times 5 \times 2 = 7500 \text{ kWh}$.

Further, the embodied energy for five solar air (SAC) will be evaluated as follows:
$E_{in5} = E_{in1} \times 5 = 869.92 \times 5 = 4349.6 \text{ kWh}$ (Example 9.18).

From Eq. 9.6, one gets energy payback time (EPBT) as

$$\text{EPBT} = \frac{\text{Embodied energy}(E_{in5}),}{\text{An annualthermal energy of five liqid FPC}} = \frac{4349.6}{917.25} = 4.74 \text{ years}$$

$$(9.6)$$

From Eq. 9.7, one has an expression for energy production factor (EPF) as

$$\text{EPF} = \frac{T}{\text{EPBT}} = \frac{10}{4.7} = 2.13$$

From Eq. 9.8, we have thermal life cycle 1 conversion efficiency (LCCE) as

$$\eta_{\text{life}} = \frac{Q_{th,\text{annul}} \times T - E_{in}}{[I(t)]_{\text{annual}} \times T} = \frac{917.25 \times 10 - 4349.6}{7500 \times 10} = \frac{4822.9}{75,00} = 0.06$$

**Example 9.20** Evaluate thermal exergy matrices for five solar air collectors connected in series for Example 9.3 by using Eqs. 9.6–9.9.

**Solution**

From Example 9.3: $\dot{Q}_{u,thex} = 49.68 \text{ W} = 0.04968 \text{ kW}$ due to very small value of solar air collector efficiency of $F' = 0.22$ and T = 10 years (Life of FPC).

As per assumption mentioned in Sect. 9.3 and following Eq. 9.5b, we have

$$Ex_{u,\text{annual}} = \sum_{j=1} Ex_{u,ovj} = 0.04968 \times 5 \times 300 \text{ kWh} = 74.52 \text{ kWh per year}$$

The annual solar energy $[I(t)]_{\text{annualex}}$ on 5 FPC with 2 m$^2$= $500 \times 0.96 \times 5 \times 300 \times 5 \times 2$= 7,200,000 W/m$^2$ = 7200 kWh/m$^2$ (Example 9.1).

Further, the embodied energy for five solar air collectors will be evaluated as follows:
$E_{in5} = E_{in1} \times 5 = 869.92 \times 5 \text{ kWh} = 4349.6 \text{ kWh}$ (Example 9.18).

From Eq. 9.6, one gets energy payback time (EPBT) as

$$\text{EPBT} = \frac{\text{Embodied energy}(E_{in5}),}{\text{An annual thermal energy of five liqid FPC}} = \frac{4349.65}{74.52} = 57 \text{ years}$$

$$(9.6)$$

From Eq. 9.7, one has an expression for energy production factor (EPF) as

$$\text{EPF} = \frac{T}{\text{EPBT}} = \frac{10}{3257} = 0.175$$

From Eq. 9.8, we have thermal life cycle conversion efficiency (LCCE) as

$$\eta_{\text{life}} = \frac{Ex_{u,\text{annual}} \times T - E_{in}}{[I(t)]_{\text{annualex}} \times T} = \frac{74.52 \times 10 - 4349.65}{72,000} = \frac{-3603.8}{72,000} = -\text{ve value}$$

So, the solar air collector with $F' = 0.22$ is not acceptable from thermal exergy point of view because energy matrices are not satisfying criteria led down in section Eqs. 9.6–9.9.

## 9.3.5 Energy Matrices and Embodied Energy of Photo-Voltaic (PV) Module

The high grade energy, based on 2nd law of thermodynamics, is required for different processes for production of PV module. These energy requirements for 1 m² of PV module are given in Table 9.4 for three different cases

**Case (i):** It is the actual data obtained from various literatures available for different processes.

**Case (ii):** It is 10% of case (i) due to advanced technology as well as reduction in mass of solar cell in the future.

**Case (iii):** It is the average value of case (i) and case (ii), respectively.

Here, it assumed that embodied energy of opaque PV (OPV) and semi-transparent PV (SPV) module is approximately same and expected life ($T$) of 30 years.

**Table 9.4** Embodied energy of PV module for 1 m² for each processes

| S. No. | Processes | MG-Si (kWh) | EG-Si (kWh) | Eg-Si for Cz-Si (kWh) | Total Processed energy (kWh) | Fabrication (kWh) | Total (kWh) |
|--------|-----------|-------------|-------------|----------------------|------------------------------|-------------------|-------------|
| 1 | Case (i) | 48 | 230 | 483 | 761 | 310 | 1071 |
| 2 | Case (ii) | 4.8 | 23.0 | 48.3 | 76.1 | 184.9 | 261 |
| 3 | Case (iii) | 26.40 | 126,5 | 265.65 | 418.55 | 188.45 | 607 |

The embodied energy of PV module with 2 m$^2$ = 2 × embodied energy of 1 m$^2$ [case (i) of Table 9.4] = 2 × 1071 = 2142 kWh.

The embodied energy of PV module with 2 m$^2$ = 2 × embodied energy of 1 m$^2$ [case (iii) of Table 9.4] = 2 × 607 = 1214 kWh.

**Example 9.21** Evaluate the exergy matrices of PV module with 2 m$^2$ area and packing factor of 0.89 for Example 3.30 with wind velocity of 1 m/s [case (i) of Table 9.4].

**Solution**

**From Example 3.30**: $T$ = 30 years (assumed).

Electrical power from 1 m$^2$ $\left(\dot{E}_{pv}\right)$ = 54.735 W and electrical power from 2 m$^2$ $\left(\dot{E}_{pv}\right)$ = 2 × 54.735 W = 109.49 W = 0.10949 kw.

The embodied energy of PV module with 2 m$^2$ $(E_{in})$ = 2142 kWh [case (i) of Table 9.4].

Annual electrical power $\left(E_{el,\text{annual}}\right)$ = 0.10949 × 5 × 300 kWh = 164 kWh = 0.10949.

The annual solar energy $[I(t)]_{\text{annualex}}$ = 500 × 0.96 × 5 × 2 × 300 = 1,500,000 Wh = 1500 kWh (Example 9.1).

From Eq. 9.6, one gets energy payback time (EPBT) as

$$EPBT = \frac{\text{Embodied energy } (E_{in5}),}{\text{An annual thermal energy of five liqid FPC}} = \frac{2142}{164} = 13 \text{ years} \quad (9.6)$$

From Eq. 9.7, one has an expression for energy production factor (EPF) as

$$\text{EPF} = \frac{T}{\text{EPBT}} = \frac{30}{13} = 2.31$$

From Eq. 9.9, we have exergy life cycle conversion efficiency (LCCE) as

$$\eta_{life} = \frac{Ex_{u,\text{annual}} \times T - E_{in}}{[I(t)]_{\text{annualex}} \times T} = \frac{164 \times 30 - 2142}{1500 \times 30} = \frac{2778}{45,000} = 0.017$$

So, the PV module is acceptable for production of electrical power. If embodied energy of cases (ii) and (ii) are considered, the exergy matrices will give very good results.

### 9.3.6 Energy Matrices and Embodied Energy of Opaque Photo-Voltaic Thermal (OPVT) Collector

**Example 9.22** Find out embodied energy of 2 m$^2$ single liquid OPVT collector by using the data of Example 9.15 and 9.21.

**Solution**

From Example 9.15:

The embodied energy of single liquid flat plate collector with 2 m² area $(E_{in1}) =$ 1159.49 kWh.

In the above value, embodied energy of toughen glass which is 268.48 kWh (Table 9.2) should be replaced by embodied energy of PV module with same area (2142 kWh) from Example 9.20.

**For case (i):**

So, the embodied energy of 2 m² single liquid OPVT collector $(E_{inopvt}) =$ $1159.49 - 268.48 + 2142 = 3033.01$ kWh [case (i) of Table 9.4].

**For case (ii):**

So, the embodied energy of 2 m² single liquid OPVT collector $(E_{inopvt}) =$ $1159.49 - 268.48 + 522 = 1413.01$ kWh [case (ii) of Table 9.4].

**For case (iii):**

So, the embodied energy of 2 m² single liquid OPVT collector $(E_{inopvt}) =$ $1159.49 - 268.48 + 1214 = 2105.01$ kWh [case (iiii) of Table 9.4 and S. No. 3 (nc-Si) Table 3.2].

**Example 9.23** Evaluate the thermal energy of single liquid OPVT collector for Example 8.4 with expected life $(T)$ of 25 years.

**Solution**

From Example 8.4, the rate of overall thermal energy is given by

$$\dot{Q}_{u,thT} = \dot{Q}_{u,th} + \frac{E'_{el}}{\gamma_c} = 60.1265 + \frac{98.18}{0.35} = 340.64 \text{ W} = 0.34064 \text{ kW (Eq. 9.1)}$$

Annual thermal energy $(Q_{th,annul}) = 0.34064 \times 5 \times 300$ kWh $= 510.96$ kWh.

The embodied energy of 2 m² single liquid OPVT collector $(E_{inopvt}) =$ 3033.01 kWh [case (i) of Example 9.22].

The annual solar energy $[I(t)]_{annualex}$ on single liquid OPVT collector with 2 m² $= 500 \times 5 \times 300 \times 2 = 1,500,000$ W $= 1500$ kWh (Example 9.1).

From Eq. 9.6, one gets energy payback time (EPBT) as

$$\text{EPBT} = \frac{\text{Embodied energy}(E_{inopvt}),}{\text{An annual thermal energy of single liqid OPVT collector}} = \frac{3033.01}{510.96}$$

$$= 5.94 \text{ years}$$

From Eq. 9.7, one has an expression for energy production factor (EPF) as

$$\text{EPF} = \frac{T}{\text{EPBT}} = \frac{25}{5.94} = 4.2$$

From Eq. 9.8, we have thermal life cycle conversion efficiency (LCCE) as

$$\eta_{\text{life}} = \frac{Q_{th,\text{annul}} \times T - E_{in}}{[I(t)]_{\text{annual}} \times T} = \frac{510.96 \times 25 - 3033.01}{1500 \times 25} = \frac{9740.99}{37,500} = 0.2598$$

**Example 9.24** Evaluate the overall thermal energy matrices of single liquid SPVT collector for Example 8.5 with expected life ($T$) of 25 years.

**Solution**

From Example 8.4, the rate of overall thermal energy is given by

$$\dot{Q}_{u,thT} = \dot{Q}_{u,th} + \frac{\dot{E}_{el}}{\gamma_c} = 209.71 + \frac{98.612}{0.35} = 209.71 + 281.18 = 491.46 \text{ W} =$$
0.491 kW (Eq. 9.1).

Annual thermal energy $(Q_{th,\text{annul}}) = 0.491 \times 5 \times 300 \text{ kWh} = 736.50 \text{ kWh}$.

The embodied energy of 2 m$^2$ single liquid OPVT/SPVT collector $(E_{\text{inopvt}}) = 3033.01$ kWh (assumed).

The annual solar energy $[I(t)]_{\text{annualex}}$ on single liquid OPVT collector with 2 m$^2$ $= 500 \times 5 \times 300 \times 2 = 1,500,000 \text{ W} = 1500 \text{ kWh}$ (Example 9.1).

From Eq. 9.6, one gets energy payback time (EPBT) as

$$\text{EPBT} = \frac{\text{Embodied energy}(E_{\text{inopvt}}),}{\text{An annualthermal energy of single liqid OPVT collector}} = \frac{3033.01}{736.50} = 4.12 \text{ years}$$

From Eq. 9.7, one has an expression for energy production factor (EPF) as

$$\text{EPF} = \frac{T}{\text{EPBT}} = \frac{25}{4.12} = 6.06$$

From Eq. 9.8, we have thermal life cycle conversion efficiency (LCCE) as

$$\eta_{\text{life}} = \frac{Q_{th,\text{annul}} \times T - E_{in}}{[I(t)]_{\text{annual}} \times T} = \frac{736.50 \times 25 - 3033.01}{1500 \times 25}$$
$$= \frac{15,379.49}{37,500} = 0.4101$$

One can observe that liquid SPVT collector (Example 9.24) gives best performance in comparison with OPVT collector (Example 9.23), and hence now onward we will consider only SPVT collector if required.

**Example 9.25** Evaluate the exergy matrices for Examples 9.4 and 9.33 (SPVT liquid collector) for $N = 5$.

**Solution**

**From Example 9.4 for N = 5:**

From Eq. 9.2a, the overall exergy of 5-SPVT liquid collectors connected in series will be as follows:

$$\dot{Ex}_{u,ov} = \dot{Q}_{u,thex5} + \dot{E}_{elT} = 7.939 + 489.84 = 497.779 \text{ W} = 0.4978 \text{ kWh}$$

The overall annual exergy of 5-SPVT liquid collectors connected in series will be

$$Ex_{u,\text{annual}} = \dot{E}x_{u,ov} \times 5 \times 300 = 0.4978 \times 5 \times 300 = 749.67 \text{ kWh}$$

The embodied energy of $2\,m^2$ single liquid OPVT/SPVT collector $\left(E_{\text{inopvt}}\right) = 3033.01$ kWh (assumed).

The embodied energy of five liquid OPVT/SPVT collectors is $\left(E_{\text{inopvt5}}\right) = 3033.01 \times 5 = 15,165.05$ kWh.

The annual solar energy $[I(t)]_{\text{annualex}}$ on five liquid OPVT collector with 2 m$^2$ = $500 \times 0.96 \times 5 \times 300 \times 2 \times 5 = 7,200,000$ W $= 7200$ kWh (Example 9.1).

From Eq. 9.6, one gets energy payback time (EPBT) as

$$\text{EPBT} = \frac{\text{Embodied energy} \left(E_{\text{inopvt5}}\right),}{\text{An annual thermal exergy of five liqid OPVT collector}}$$

$$= \frac{15,165.05}{749.67} = 20.25 \text{ years}$$

From Eq. 9.7, one has an expression for energy production factor (EPF) as

$$\text{EPF} = \frac{T}{\text{EPBT}} = \frac{25}{20.25} = 1.23$$

From Eq. 9.9, we have thermal life cycle conversion efficiency (LCCE) as

$$\eta_{\text{life}} = \frac{Ex_{u,\text{annual}} \times T - E_{in}}{[I(t)]_{\text{annualex}} \times T} = \frac{749.67 \times 25 - 15,165.05}{7200 \times 25}$$

$$= \frac{3576.67}{180,000} = 0.02$$

**Example 9.26**  Repeat the Example 9.25 for case (ii) of Table 9.4.

**Solution**

In this case, the embodied energy of 2 m$^2$ single liquid OPVT/SPVT collector $\left(E_{\text{inopvt}}\right) = 1413.01$ kWh.

The embodied energy of five liquid OPVT/SPVT collectors is $\left(E_{\text{inopvt5}}\right) = 1413.01 \times 5 = 7065.05$ kWh.

From Eq. 9.6, one gets energy payback time (EPBT) as

$$\text{EPBT} = \frac{\text{Embodied energy} \left(E_{\text{inopvt5}}\right),}{\text{An annual thermal exergy of five liqid OPVT collector}}$$

$$= \frac{7065.05}{749.67} = 9.42 \text{ years}$$

From Eq. 9.7, one has an expression for energy production factor (EPF) as

$$EPF = \frac{T}{EPBT} = \frac{25}{9.42} = 2.65$$

From Eq. 9.9, we have thermal life cycle conversion efficiency (LCCE) as

$$\eta_{life} = \frac{Ex_{u,\text{annual}} \times T - E_{in}}{[I(t)]_{\text{annualex}} \times T} = \frac{749.67 \times 25 - 7065.05}{7200 \times 25} = \frac{11,677}{180,000} = 0.065$$

**Example 9.27** Repeat the Example 9.25 for case (iii) of Table 9.4.

**Solution**

From Example 9.22 for case (iii), we have

The embodied energy of 2 m² single liquid OPVT collector $(E_{\text{inopvt}}) = 1159.49 - 268.48 + 1214 = 2105.01$ kWh (case (iiii) of Table 9.4).

So, embodied energy of five liquid OPVT collector $(E_{\text{inopvt}}) = 2105.01 \times 5 = 10,525.05$ kWh.

From Eq. 9.6, one gets energy payback time (EPBT) as

$$EPBT = \frac{\text{Embodied energy}(E_{\text{inopvt5}}),}{\text{An annual thermal exergy of five liqid OPVT collector}} = \frac{10,525.05}{749.67} = 14.05 \text{ years}$$

From Eq. 9.7, one has an expression for energy production factor (EPF) as

$$EPF = \frac{T}{EPBT} = \frac{25}{14.05} = 1.77$$

From Eq. 9.9, we have thermal life cycle conversion efficiency (LCCE) as

$$\eta_{life} = \frac{Ex_{u,\text{annual}} \times T - E_{in}}{[I(t)]_{\text{annualex}} \times T} = \frac{749.67 \times 25 - 10,525.05}{7200 \times 25} = \frac{3576.67}{180,000} = 0.045$$

From Examples 9.25–9.27, one can conclude the importance of embodied energy in deciding exergy matrices of any PVT systems (Tables 9.5 and 9.6).

### 9.3.7 Thermal Energy Matrices, an Overall Exergy Matrices and Embodied Energy of Single Flexible (Al-Based) Liquid OPVT-TEC Collector

**Example 9.28** Evaluate the thermal energy matrices of single flexible (Al-based) liquid OPVT-TEC collector for Example 8.7 and Table 9.7 with life time $(T)$ of 25 years.

**Table 9.5** Thermal energy matrices of various solar energy system

| S. No. | Name of solar energy system (example) | Annual thermal energy, (kWh) | Annual solar energy (kWh) | Embodied energy (kWh) | Energy matrices EPBT EPF LCCE (Years) (fraction) |
|---|---|---|---|---|---|
| 1 | Liquid FPC (N = 5) [9.16] | 4146 | 7500 | 5797.45 | 1.4 7.14 0.47 |
| 2 | Air FPC (N = 5) [9.19] | 917.25 | 7500 | 4349.6 | 4.71 2.13 0.06 |
| 3 | Liquid OPVT collector (N = 1) [5.23] | 510.96 | 1500 | 3033.1 | 5.9 4.2 0.2598 |
| 4 | Liquid SPVT (N = 1) [5.24] | 736.50 | 1500 | 3033.01 | 4.12 6.06 0.4101 |
| 5 | Flexible liquid OPVT-TEC (N = 1) [5.28] | 957.75 | 1500 | 2349.02 | 2.45 10.20 0.5758 |
| 6 | Liquid SPVT-CPC (N = 1, C = 2) [9.30] | 888.45 | 1500 | 2,388.36 | 2.69 9.29 0.5564 |
| 7 | Liquid FPC-CPC (N = 6, C = 2) [9.31] | 9709.5 | 1425 | 8657.04 | 0.89 11.23 1.03 |
| 8 | SPVT-CPC air collector (N = 6, C = 2) [9.32] | 3463.5 | 1425 | 6919.62 | 3.45 7.25 0.3726 |
| 9 | Flexible OPVT-TEC-CPC liquid collector [9.33] | 8646.75 | 1425 | 10.448.16 | 1.21 20.66 0.96 |
| 10 | FLiETC (N = 5) [9.35] | 1415.61 | 1875 | 592.87 | 0.42 23.80 0.72 |
| 11 | Liquid FLiSPVT (N = 1, C = 4) [9.36] | 981.12 | 1500 | 1109.72 | 1.13 22.12 0.82 |

## Solution

With help of Example 9.22, the embodied energy of 2 m$^2$ single liquid OPVT-TEC collector $\left(E_{inopvt-tec}\right) = 1159.49 - 268.48 + 1458.01 = 2349.02$ kWh (Table 9.7).

The rate of overall thermal energy of single flexible (Al-based) liquid OPVT-TEC collector is obtained from Example 8.7 as

**Table 9.6** Thermal exergy matrices of various solar energy systems

| S. No. | Name of solar energy system (Example) | Annual thermal exergy (kWh) | Annual solar energy (kWh) | Embodied energy (kWh) | Energy matrices EPBT EPF LCCE (Years) (Fraction) |
|---|---|---|---|---|---|
| 1 | Liquid FPC (N = 5) (9.17) | 4146 | 7500 | 5797.45 | 1.40 7.15 0.4953 |
| 2 | Air FPC (N = 5) (9.20) | 74.52 $(F^/ = 0.22)$ | 7200 | 4349.6 | 57 0.175 $-ve$ |
| 3 | PV module [case (ii) of Table 9.4] (9.21) | 164 | 1500 | 2142 | 13 2.31 0.017 |
| 4 | Liquid SPVT [N = 5, case (i) of Table 9.4] (5.25) | 749.67 | 7200 | 15,165.05 | 20.25 1.23 0.02 |
| 5 | Liquid SPVT [N = 5, case (ii) of Table 9.4] [9.26] | 749.67 | 7200 | 7065.05 | 9.42 2.65 0.065 |
| 6 | Liquid SPVT [N = 5, case (iii) of Table 9.4] (9.27) | 749.67 | 7200 | 2105.01 | 14.05 1.77 0.045 |
| 7 | Flexible OPVT-TEC (N = 1) (9.29) | 179.43 | 1500 | 2349.02 | 13 1.92 0.057 |
| 8 | Flexible OPVT-TEC-CPC (N = 6, C = 2) (9.34) | 1484.46 | 1425 | 10,448.16 | 7.18 3.48 0.1247 |
| 9 | Liquid FLiSPVT (N = 5, C = 5, $\beta_c = 0.50$) (5.37) | 638.985 | 1500 | 5548.60 | 8.6 2.9 0.53 |

$$\dot{Q}_{u,thT} = \left[\dot{Q}_{u,th} + \frac{\dot{E}_{el} + \dot{E}_{TEC}}{\gamma_c}\right] = 280.10 + \frac{106 + 30.18}{0.38}$$
$$= 280 + 358.3 = 638.47 \text{ W} = 0.6385 \text{ kW}$$

Annual overall thermal energy is calculated as

$$Q_{th,annul} = \dot{Q}_{u,thT} \times 5 \times 300 = 0.6385 \times 5 \times 300 \text{ kWh} = 957.75 \text{ kWh}$$

The annual solar energy received by single flexible (Al-based) liquid OPVT-TEC collector is

**Table 9.7** Embodied energy coefficient [2] and embodied energy of single flexible OPVT-TEC collector with each effective area of 1.875 m²

| S. No | Components of FPC | Quantity | Total weight ($m_i$) | Embodied energy coefficient ($e_i$) (MJ/kg) | Embodied energy (MJ) ($e_i m_i$) | Embodied energy (kWh) ($e_i m_i$) |
|---|---|---|---|---|---|---|
| 1 | Al virgin sheet | 1 | 10 kg | 191 | 1910 | 530 |
| 2 | Transparent flexible plastic low density sheet | 1 (2 m²) | 2 kg | 103 | 206 | 57.22 |
| 3 | Solar cells | Case (iii) (Table 9.4) | 2m² | – | – | 836 |
| 4 | Copper wire | | 0.5 kg | 70.6 | 35.25 | 9.792 |
| 5 | Others including TEC material | | | | | 25 (assumed) |
| | Total | | | | | 1458.01 |

$[I(t)]_{\text{annualex}}$ on flexible(Al based)liquid OPVT − TEC collector with 2 m² = $500 \times 5 \times 300 \times 2 = 1,500,000$ Wh $= 1500$ kWh (Example 9.1).

From Eq. 9.6, one gets energy payback time (EPBT) in years as

$$\text{EPBT} = \frac{\text{Embodied energy}(E_{\text{inopvt}}),}{\text{An annual thermal energy of single liqid OPVT collector}} = \frac{2349.02}{957.75} = 2.45 \text{ years}$$

From Eq. 9.7, one has an expression for energy production factor (EPF) as

$$\text{EPF} = \frac{T}{\text{EPBT}} = \frac{25}{2.45} = 10.20$$

From Eq. 9.8, we have thermal life cycle conversion efficiency (LCCE) as

$$\eta_{\text{life}} = \frac{Q_{th,\text{annul}} \times T - E_{\text{in}}}{[I(t)]_{\text{annual}} \times T} = \frac{957.75 \times 25 - 2349.02}{1500 \times 25} = \frac{21,594.75}{37,500} = 0.5758$$

Due to integration of TEC in flexible OPVT collector energy matrices have improved significantly as obtained in Example 9.24.

**Example 9.29** Find out the thermal exergy matrices of single flexible OPVT-TEC liquid collector for Example 9.6 and 9.28.

**Solution**

Since, the rate of thermal exergy $\left(\dot{Q}_{u,thex1}\right) = 146.65\left[1.91 - 289\ln\frac{289.91}{288}\right] = 146.65[1.91 - 1.91] = 0$ (Example 9.6).

The total electrical power is the sum of electrical power from PV and TEC, and hence

$$\dot{E}_{total1} = \dot{E}_{el1} + \dot{E}_{TEC1} = 89.62 + 30.06 = 119.62 \text{ W} = 0.11962 \text{ kW}$$

The annual electrical energy is obtained as

$$Ex_{u,\text{annual}} = \dot{E}_{total1} \times 5 \times 300 = 0.11962 \times 5 \times 300 = 179.43 \text{ kWh}$$

The embodied energy of 2 m$^2$ single liquid OPVT-TEC collector $\left(E_{\text{inopvt-tec}}\right) = 2349.02$ kWh (Example 9.28 and Table 9.7).

The annual solar energy received by single flexible (Al-based) liquid OPVT-TEC collector is given by

$[I(t)]_{\text{annualex}} = 500 \times 5 \times 300 \times 2 = 1,500,000$ Wh $= 1500$ kWh (Example 9.1).

From Eq. 9.6, one gets energy payback time (EPBT) as

$$\text{EPBT} = \frac{\text{Embodied energy}(E_{\text{inopvt5}}),}{\text{An annualthermal exergy of five liqid OPVT collector}}$$

$$= \frac{2349.02}{179.43} = 13 \text{ years}$$

From Eq. 9.7, one has an expression for energy production factor (EPF) as

$$\text{EPF} = \frac{T}{\text{EPBT}} = \frac{25}{13} = 1.92$$

From Eq. 9.9, we have thermal life cycle conversion efficiency (LCCE) as

$$\eta_{\text{life}} = \frac{Ex_{u,\text{annual}} \times T - E_{\text{in}}}{[I(t)]_{\text{annualex}} \times T} = \frac{179.43 \times 25 - 2349.02}{1500 \times 25} = \frac{2136.73}{37,500} = 0.057$$

### 9.3.8   Thermal Energy and Thermal Exergy Matrices of Liquid PVT-CPC Collector

**Example 9.30** Evaluate thermal energy matrices for single semi-transparent liquid SPVT-CPC collector for Example 8.10 for $C = 2$.

**Solution**

Embodied energy of single SPVT-CPC = Embodied energy of single SPVT/OPVT + embodied energy of single CPC (mass × energy coefficient of Al foil) = 2105.01 kWh (Example 9.27) + 5 kg × 56.67 kWh/kg = 2105.01 + 283.35 = 2388.36 kWh.

The rate of thermal energy of single semi-transparent liquid SPVT-CPC collector for $C = 2$ is given by Example 8.10 as

$$\dot{Q}_{u,thT} = \left[\dot{Q}_{u,th} + \frac{E'_{el}}{\gamma_c}\right] = 354.893 + \frac{83.09}{0.35} = 354.893 + 237.4 = 592.293 \text{ W} = 0.5923 \text{ kW}$$

So, annual thermal energy can be calculated as

$$Q_{th,\text{annul}} = \dot{Q}_{u,thT} \times 5 \times 300 = 0.5923 \times 5 \times 300 \text{ kWh} = 888.45 \text{ kWh}$$

and, an annual solar energy is

$[I(t)]_{\text{annualex}} = 475 \times 5 \times 300 \times 2 = 1,425,000 \text{ W} = 1425 \text{ kWh}$ (Since for CPC, $I_u = 475 \text{ W/m}^2$, Example 9.1).

From Eq. 9.6, one gets energy payback time (EPBT) as

$$\text{EPBT} = \frac{\text{Embodied energy}(E_{\text{inopvt}}),}{\text{An annualthermal energy of single liqid OPVT collector}}$$

$$= \frac{2388.36}{888.45} = 2.69 \text{ years}$$

From Eq. 9.7, one has an expression for energy production factor (EPF) as

$$\text{EPF} = \frac{T}{\text{EPBT}} = \frac{25}{2.69} = 9.29$$

From Eq. 9.8, we have thermal life cycle conversion efficiency (LCCE) as

$$\eta_{\text{life}} = \frac{Q_{th,\text{annul}} \times T - E_{\text{in}}}{[I(t)]_{\text{annual}} \times T} = \frac{888.45 \times 25 - 2388.36}{1425 \times 25} = \frac{19,822.89}{35.625} = 0.5564$$

Here, one can notice that energy matrices for OPVT-TEC (Example 9.28) is approximately same as energy matrices of SPVT-CPC (Example 9.30).

### 9.3.9  Thermal Energy Matrices of FPC-CPC Collectors

**Example 9.31** Compute the thermal energy matrices for six ($N = 6$)-FPC-CPC liquid collectors connected in series for $C = 2$ for Example 8.13.

**Solution**

Embodied energy of single FPC-CPC liquid collectors connected in series [$C = 2$] = Embodied energy of single liquid FPC (Example 9.15) + embodied energy of CPC (Example 9.30) = 1159.49 + 283.35 = 1442.84 $kWh$.

So, embodied energy of six FPC-CPC liquid collectors ($E_{in6}$) = 1442.84 × 6 = 8657.04 kWh.

The rate of thermal energy $(\dot{Q}_{u,th6})$ = 6473.13 $W$ = 6.473 kW.

So, an annual thermal energy can be calculated as

$$Q_{th,annul6} = \dot{Q}_{u,th6} \times 5 \times 300 = 6.473 \times 5 \times 300 \, \text{kWh} = 9709.5 \, \text{kWh}$$

Annual solar energy on single FPC-CPC liquid collectors is

$[I(t)]_{annual6} = 475 \times 5 \times 300 \times 2 = 1425,000 \, \text{Wh} = 1425 \, \text{kWh}$ (Since for CPC, $I_u = 475 \, \text{W/m}^2$, Example 9.1).

From Eq. 9.6, one gets energy payback time (EPBT) as

$$\text{EPBT} = \frac{\text{Embodied energy}(E_{\text{inopvt}}),}{\text{An annual thermal energy of single liqid OPVT collector}}$$

$$= \frac{8657.04}{9709.5} = 0.89 \text{ years}$$

From Eq. 9.7, one has an expression for energy production factor (EPF) as

$$\text{EPF} = \frac{T}{\text{EPBT}} = \frac{10}{0.89} = 11.23$$

From Eq. 9.8, we have thermal life cycle conversion efficiency (LCCE) as

$$\eta_{\text{life}} = \frac{Q_{\text{th,annul}} \times T - E_{in}}{[I(t)]_{\text{annual}} \times T} = \frac{9709.5 \times 10 - 8657.04}{1425 \times 6 \times 10} = \frac{88,437.96}{85,500} = 1.03$$

### 9.3.10  Thermal Energy Matrices of SPVT-CPC Air Collectors

**Example 9.32**  Compute the thermal energy matrices for six (N = 6)-SPVT-CPC air collectors connected in series for $C = 2$ for Example 8.14 with life (T) of 25 years.

**Solution**

Embodied energy of single SPVT-CPC air collectors connected in series $[C = 2] =$ Embodied energy of single SPVT air collector (Example 9.18)] + embodied energy of CPC (Example 9.30) = 869.92 kWh + 283.35 = 1153.27 kWh.

So, an embodied energy of six FPC-CPC liquid collectors $(E_{in6}) = 1153.27 \times 6 = 6919.62$ kWh.

The rate of thermal energy $\left(\dot{Q}_{u,th6}\right) = \left[\dot{Q}_{u,th} + \frac{\dot{E}_{el}}{\gamma_c}\right] = 950.42 + \frac{475.52}{0.35} = 2309.05 \, \text{W} = 2.309 \, \text{kW}$.

So, an annual thermal energy can be calculated as

$$Q_{th,annul6} = \dot{Q}_{u,th6} \times 5 \times 300 = 2.309 \times 5 \times 300 \, \text{kWh/year} = 3463.5 \, \text{kWh/year}$$

Annual solar energy on single SPVT-CPC air collectors is

$[I(t)]_{\text{annual}} = 475 \times 5 \times 300 \times 2 = 1,425,000 \, \text{Wh} = 1425 \, \text{kWh}$ (Since for CPC, $I_u = 475 \, \text{W/m}^2$, Example 9.1).

From Eq. 9.6, one gets energy payback time (EPBT) as

$$\text{EPBT} = \frac{\text{Embodied energy } \left(E_{\text{inopvt}}\right),}{\text{An annualthermal energy of single liqid OPVT collector}}$$

$$= \frac{6919.62}{3463.5} = 3.45 \text{ years}$$

From Eq. 9.7, one has an expression for energy production factor (EPF) as

$$\text{EPF} = \frac{T}{\text{EPBT}} = \frac{25}{3.45} = 7.25$$

From Eq. 9.8, we have thermal life cycle conversion efficiency (LCCE) as

$$\eta_{\text{life}} = \frac{Q_{\text{th,annul}} \times T - E_{\text{in}}}{[I(t)]_{\text{annual}} \times T} = \frac{3463.5 \times 25 - 6919.62}{1425 \times 6 \times 25} = \frac{79,643.5}{213,750} = 0.3726$$

## 9.3.11 Thermal Energy Matrices and an Overall Thermal Exergy Matrices of SPVT-CPC Air Collectors

**Example 9.33** Find out thermal energy matrices for six (N = 6) flexible OPVT-TEC-CPC liquid collectors connected in series and for $C = \frac{A_{\text{am}}}{A_{\text{rm}}} = 2$ from Examples 8.20–8.22.

**Solution**

The embodied energy of single OPVT-TEC-CPC liquid collector = [Embodied energy of OPVT + Embodied energy of TEC] (Table 9.7) + embodied energy of CPC (Example 9.30) = 1458.01 + 283.35 = 1741.36 kWh.

So, the embodied energy of six flexible OPVT-TEC-CPC liquid collectors will be as

$$E_{in6} = 1741.36 \times 6 = 10,448.16 \text{ kWh}$$

The rate of thermal energy $\left(\dot{Q}_{u,th6}\right) = \left[\dot{Q}_{u,th} + \frac{\dot{E}_{TEC6}}{\gamma_c} + \frac{\dot{E}'_{el6}}{\gamma_c}\right] = 3.242 + 1.731 +$ 791.51 == 5.7645 kW (Examples 8.20–8.22).

The annual thermal energy $\left(Q_{\text{th,annul}}\right) = 5.7645 \times 5 \times 300 = 8646.75 \text{ kWh}$.

Annual solar energy on single OPVT-TEC-CPC air collectors is

$[I(t)]_{\text{annual}} = 475 \times 5 \times 300 \times 2 = 1,425,000 \text{ Wh} = 1425 \text{ kWh}$ (Since for CPC, $I_u = 475 \text{ W/m}^2$, Example 9.1).

From Eq. 9.6, one gets energy payback time (EPBT) as

$$EPBT = \frac{\text{Embodied energy}\left(E_{\text{inopvt}}\right),}{\text{An annualthermal energy of single liqid OPVT collector}}$$

$$= \frac{10,448.16}{8646.75} = 1.21 \text{ years}$$

From Eq. 9.7, one has an expression for energy production factor (EPF) as

$$EPF = \frac{T}{EPBT} = \frac{25}{1.21} = 20.66$$

From Eq. 9.8, we have thermal life cycle conversion efficiency (LCCE) as

$$\eta_{\text{life}} = \frac{Q_{\text{th,annul}} \times T - E_{in}}{[I(t)]_{\text{annual}} \times T} = \frac{8646.75 \times 25 - 10,448.16}{1425 \times 6 \times 25} = \frac{205,720.59}{213,750} = 0.96$$

**Example 9.34** Evaluate an overall exergy matrices for six ($N = 6$) flexible OPVT-TEC-CPC liquid collectors connected in series and for $C = \frac{A_{\text{am}}}{A_{\text{rm}}} = 2$ from Example 9.11.

**Solution**

The embodied energy of six flexible OPVT-TEC-CPC liquid collectors (Example 9.33) will be as

$$E_{in6} = 1741.36 \times 6 = 10,448.16 \text{ kWh}$$

Annual solar energy on single OPVT-TEC-CPC air collectors is
$[I(t)]_{\text{annual}} = 1425 \text{ kWh}$ (Example 9.33)
The an overall electrical power from six flexible OPVT-TEC-CPC liquid collectors is given as
$\dot{E}x_{u,0v} = 989.64 \text{ W}$ (Example 9.11)
The overall annual electrical energy can be obtained as

$$Ex_{u,annual} = 989.64 \times 5 \times 300 = 1,484,460 \ Wh = 1484.46 \text{ kWh}$$

From Eq. 9.6, one gets energy payback time (EPBT) as

$$EPBT = \frac{\text{Embodied energy}\left(E_{\text{inopvt5}}\right),}{\text{An annualthermal exergy of five liqid OPVT collector}}$$

$$= \frac{10,448.16}{1484.46} = 7.18 \text{ years}$$

From Eq. 9.7, one has an expression for energy production factor (EPF) as

$$EPF = \frac{T}{EPBT} = \frac{25}{7.18} = 3.48$$

From Eq. 9.9, we have thermal life cycle conversion efficiency (LCCE) as

$$\eta_{\text{life}} = \frac{Ex_{u,\text{annual}} \times T - E_{\text{in}}}{[I(t)]_{\text{annualex}} \times T} = \frac{1484.46 \times 25 - 10{,}448.16}{1425 \times 6 \times 25} = \frac{26{,}663.34}{213{,}750} = 0.1247$$

### 9.3.12 Thermal Energy Matrices of Fresnel Lens Integrated Evacuated Tubular Collectors [FLiETC] Connected in Series

**Example 9.35** Find out thermal energy matrices of Fresnel lens integrated evacuated tubular collectors [FLiETC] connected in series for Example 8.37 for $C = 9$ and $N = 5$ with life time of 10 years.

**Solution**

The embodied energy of five Fresnel lens integrated evacuated tubular collectors [FLiETC] is calculated by using the data of Table 9.8 as

$$E_{in5} = 118.574 \times 5 = 592.87 \text{ kWh}$$

**Table 9.8** Energy coefficient and embodied energy of one evacuated tubular collector [3] (Appendix F)

| S. No. | Components of FPC | Quantity | Total weight $(m_i)$ (kg) | Embodied energy coefficient $(e_i)$ (MJ/kg) | Embodied energy (MJ) $(e_i m_i)$ | Embodied energy (kWh) $(e_i m_i)$ |
|--------|------------------|----------|---------------------------|---------------------------------------------|----------------------------------|-----------------------------------|
| 1 | Double glass tube | 1 | 2.67 | 14.9 | 39.783 | 11.051 |
| 2 | Copper U-tube and fin material | 1 | 1.33 | 70.60 | 93.898 | 26.083 |
| 3 | Vacuum creation | | | | 71 | 19.722 |
| 4 | Wielding | | | | 8.85 | 2.458 |
| 5 | Brazing | | | | 7.33 | 2.04 |
| | Total (ETC) | | | | 223.861 | 61.354 |
| 1 | Fresnel lens (plastic, low density) | 1 | 2 | 103 | 206 | 57.22 |
| | Total (FLiETC) | | | | 429.861 | 118.574 |

The rate of thermal energy from five FLiETC is given by
$\dot{Q}_{u,th5} = 943.74$ W (Example 8.37).
The annual thermal energy from five FLiETC will be obtained as

$$Q_{th,annul} = 943.74 \times 5 \times 300 = 1,415,6106 \text{ Wh} = 1415.61 \text{ kWh}$$

Annual solar energy incident on five FLiETC is calculated as follows:
$[I(t)]_{annual} = 500 \times 5 \times 300 \times 5 \times 0.5 = 1,875,000$ W $= 1875$ kWh (Since
$A_{am} = 2$ m$^2$, Example 8.24)
From Eq. 9.6, one gets energy payback time (EPBT) as

$$EPBT = \frac{\text{Embodied energy } (E_{inopvt}),}{\text{An annual thermal energy of single liqid OPVT collector}}$$
$$= \frac{592.87}{1415.61} = 0.42 \text{ years}$$

From Eq. 9.7, one has an expression for energy production factor (EPF) as

$$EPF = \frac{T}{EPBT} = \frac{10}{0.42} = 23.80$$

From Eq. 9.8, we have thermal life cycle conversion efficiency (LCCE) as

$$\eta_{life} = \frac{Q_{th,annul} \times T - E_{in}}{[I(t)]_{annual} \times T} = \frac{1415.61 \times 10 - 592.87}{1875 \times 10} = \frac{13,563.23}{18750} = 0.72$$

Because of thermal energy matrices obtained in Example 9.34, the evacuated tubular collector is most preferable in comparison with other solar thermal devices.

### 9.3.13   Fresnel Lens Integrated with Liquid Semi-transparent Photo-Voltaic Thermal (SPVT) [FLiSPVT] Collector

**Example 9.36** Evaluate the thermal energy matrices for liquid FLiSPVT with $C = 4$, $\beta_c = 50$ for case (b) of Examples 8.42 and 8.44.

**Solution**

**For N = 1**
    For $C = 4$, area of Fresnel lens (FL) $= 2$ m$^2$ and area of liquid SPVT collector $= 0.5$ m$^2$ [case (ii) of Example 8.38].
    The embodied energy of 2 m$^2$ single liquid OPVT collector $(E_{inopvt})$ from Example 9.22 $= 2105.01$ kWh (case (iiii) of Table 9.4 and S. No. 3 (nc-Si) Table 3.2).

So, the embodied energy of single FLiSPVT = Embodied energy of FLiSPVT with effective area of 0.5 m² SPVT + embodied energy of FL (Table 9.8) = 0.5 × 2105.01 + 57.22 = 1109.72 kWh.

The rate of thermal energy from Example 8.42 for liquid FLiSPVT is given by $\dot{Q}_{u,th} = \dot{m}_f C_f (T_{fo} - T_{fi}) = 146.09 \times 2.75 = 403.28$ W (Example 8.42).

An equivalent rate of thermal energy of electrical power from PV $= \frac{87.78}{0.38} = 250.8$ W (Example 8.44).

So, the total rate of thermal energy $(\dot{Q}_{u,thT}) = 403.28 + 250.8 = 654.08$ W $= 0.65408$ kW.

Now, an annual overall thermal energy $(Q_{th,annual}) = 0.65408 \times 5 \times 300$ kWh $= 981.12$ kWh.

Annual solar energy incident on liquid single FLiSPVT is calculated as follows: $[I(t)]_{annual} = 500 \times 5 \times 300 \times 2 = 1,500,000$ W $= 1500$ kWh (Since $A_{am} = 2$ m², Example 8.24)

From Eq. 9.6, one gets energy payback time (EPBT) as

$$\text{EPBT} = \frac{\text{Embodied energy } (E_{inopvt}),}{\text{An annual thermal energy of single liqid OPVT collector}}$$

$$= \frac{1109.72}{981.12} = 1.13 \text{ years}$$

From Eq. 9.7, one has an expression for energy production factor (EPF) as

$$\text{EPF} = \frac{T}{\text{EPBT}} = \frac{25}{1.13} = 22.12$$

From Eq. 9.8, we have thermal life cycle conversion efficiency (LCCE) as

$$\eta_{life} = \frac{Q_{th,annul} \times T - E_{in}}{[I(t)]_{annual} \times T} = \frac{1274.53 \times 25 - 1109.72}{1500 \times 25} = \frac{23,418.28}{37,500} = 0.62$$

**Example 9.37** Evaluate the thermal exergy matrices for liquid FLiSPVT with $C = 4$, $\beta_c = 0.50$ for Example 9.13 for $N = 5$.

**Solution**

**For N = 5**

So, the embodied energy of five FLiSPVT $(E_{in5}) = 1109.72 \times 5 = 5548.60$ kWh (Example 9.36).

The rate of thermal exergy of five FLiSPVT collectors connected in series as $\dot{E}x_{uth6} = 38.129$ W (Example 9.13)

The electrical power of 5-SPVT $(\dot{E}_{el5}) = 387.862$ W (Example 9.13).

The rate of overall exergy is the sum of rate of thermal exergy and electrical power $(\dot{E}e_{T5}) = 38.129 + 387.862 = 425.99$ W $= 0.42599$ kW

An overall exergy for 5-FLiSPVT becomes

$$Ex_{u,\text{annual}} = 0.42599 \times 5 \times 300 = 638.985 \text{ kWh}$$

Annual solar energy incident on liquid single FLiSPVT is calculated as follows: $[I(t)]_{annual} = 500 \times 5 \times 300 \times 2 = 1{,}500{,}000\ W = 1500\ \text{kWh}$ (Since $A_{am} = 2\ \text{m}^2$, Example 8.24).

From Eq. 9.6, one gets energy payback time (EPBT) as

$$\text{EPBT} = \frac{\text{Embodied energy}\left(E_{\text{inopvt5}}\right),}{\text{An annual thermal exergy of five liqid OPVT collector}}$$
$$= \frac{5548.60}{638.985} = 8.6 \text{ years}$$

From Eq. 9.7, one has an expression for energy production factor (EPF) as

$$EPF = \frac{T}{EPBT} = \frac{25}{8.6} = 2.9$$

From Eq. 9.9, we have thermal life cycle conversion efficiency (LCCE) as

$$\eta_{life} = \frac{Ex_{u,annual} \times T - E_{in}}{[I(t)]_{annualex} \times T} = \frac{638.985 \times 25 - 5548.60}{1500 \times 5 \times 25} = \frac{10{,}426.025}{19.375} = 0.53$$

## 9.4  Carbon Credit Earned by Solar Energy System

Most of thermal power plant based on fossil fuel emits thermal energy and $CO_2$ to atmosphere. This is responsible to increase the global air temperature and level of $CO_2$ in atmosphere. The increase of global temperature is responsible for melting the iceberg which raises the sea level. The business of carbon credits was created at international level to reduce the effect of greenhouse gases including $CO_2$ emitted by thermal power plant based on fossil fuel in early twenty-first century. Carbon credits are defined as 'a key component of national and international emissions trading schemes that have been implemented to mitigate global warming'. This encourages to the individual, business community as well as industrial sector to adopt renewable energy sources as maximum as possible to meet their energy demand. The renewable energy sources are less polluting in comparison with fossil fuel to sustain the climate which includes atmospheric temperature as well $CO_2$ level.

### *9.4.1 Formulation to Evaluate Carbon ($CO_2$) Credit*

If one kWh (one unit) power is used by a consumer and the losses due to poor domestic appliances, namely fan, electric bulb, fridge and connecting wires, etc., is La, then the transmitted power should be $\frac{1}{1-L_a}$ kWh (units). If the average transmission and distribution losses from the thermal power plants to users end are $L_{td}$, then the power is further reduced by $\frac{1}{1-L_a} \times \frac{1}{1-L_{td}}$ units from one kWh generated at thermal power plant.

The average $CO_2$ equivalent intensity for electricity generation from various fossil fuel including hard coal is given in Table 9.9. For coal, it is approximately 1.001 kg of $CO_2$ per kWh at the source. Thus, for unit (kWh) power consumption, Table 9.9 by the consumer, the amount of $CO_2$ emission into atmosphere is as follows:

**For hard coal:**

$$CO_2 \text{ emission, } [CO_2]_{emission}/kWh = \frac{1}{1-L_a} \times \frac{1}{1-L_{td}} \times 1.001 \text{ kg} \qquad (9.10a)$$

**For Petroleum (Diesel):**

$$CO_2 \text{ emission, } [CO_2]_{emission}/kWh = \frac{1}{1-L_a} \times \frac{1}{1-L_{td}} \times 0.865 \text{ kg} \qquad (9.10b)$$

**For natural gas:**

$$CO_2 \text{ emission, } [CO_2]_{emission}/kWh = \frac{1}{1-L_a} \times \frac{1}{1-L_{td}} \times 0.443 \text{ kg} \qquad (9.10c)$$

This shows that among fossil fuel the natural gas is most preferable fuel to be used on priority to get less emission of $CO_2$ in the atmosphere.

If $E_{in}$ is the embodied energy (Eq. 9.4) and T is the life of the solar energy system, then the annual $CO_2$ emission due to embodied energy can be expressed as

$$\text{The } CO_2 \text{ emission per year} = \frac{E_{in}}{T} \times [CO_2]_{emission} \text{ kg } CO_2/kWh \qquad (9.11a)$$

Also,

**Table 9.9** Specific $CO_2$ emissions from fossil fuels in kg/kWh$_e$ [4]

| S. No. | Name of fuel | Mean electrical efficiency (%) | Kg $CO_2$/kWh$_{PE}$ |
|---|---|---|---|
| 1 | Lignite | 38 | 1.093 |
| 2 | Hard coal | 39 | 1.001 |
| 3 | Natural gas | 56.1 | 0.443 |
| 4 | Petroleum (Diesel) | ≈ 40 | 0.8–0.93 |

The $CO_2$ emission over the lifetime of the system

$$= E_{in} \times \frac{1}{1 - L_a} \times \frac{1}{1 - L_{td}} \times Kg\ CO_2/kWh \qquad (9.11b)$$

The net $CO_2$ mitigation over the lifetime of the system is

(a) **Based on thermal energy**

Total $CO_2$ mitigation–$CO_2$ emission due to embodied energy

$$Net\ [CO_2]_{mitigation} = (Q_{th,annul} \times T - E_{in}) \times \frac{1}{1 - L_a} \times \frac{1}{1 - L_{td}} \times Kg\ CO_2/kWh$$
$$(9.12a)$$

The above Eq. 9.12a is only applicable for $(Q_{th,annul} \times T - E_{in}) > 0$.

(b) **Based on thermal exergy**

Total $CO_2$ mitigation–$CO_2$ emission due to embodied energy

$$Net\ [CO_2]_{mitigation} = (Ex_{u,annual} \times T - E_{in}) \times \frac{1}{1 - L_a} \times \frac{1}{1 - L_{td}} \times Kg\ CO_2/kWh$$
$$(9.12b)$$

where, $Q_{th,annul}$, Eq. 9.51 and $Ex_{u,annual}$, Eq. 9.5b are annual thermal energy and thermal exergy gain of solar energy system.

The above Eq. 9.12b is also only applicable for $(Ex_{u,annual} \times T - E_{in}) > 0$.

The net $CO_2$ mitigation over the lifetime in tones of $CO_2$ is given by

$$[CO_2]_{tones} = Net\ [CO_2]_{mitigation} \times 10^{-3} tones \qquad (9.13)$$

If $CO_2$ emission is being traded at US $C$ per tones of $CO_2$ mitigation, then the carbon credit earned by the system is evaluated as

$$[CC]_{earned}\ in(\$) = C \times [CO_2]_{tones} \qquad (9.14)$$

**Example 9.38** Evaluate $[CO_2]_{emission}/kWh$ for hard coal, petroleum and natural gas if domestic loss $(L_a)$ and transmission and distributions $(L_{td})$ losses are 10% and 40%, respectively.

**Solution**

From Eq. 9.10, we have the following expression for $[CO_2]_{emission}/kWh$ as
     **For hard coal:**

$$CO_2\ emission,\ [CO_2]_{emission}/kWh = \frac{1}{1 - L_a} \times \frac{1}{1 - L_{td}} \times 1.001\ kg/kWh$$

$$= \frac{1}{0.9} \times \frac{1}{0.6} \times 1.001 = 1.8537 \text{ kg/kWh}$$

**For Petroleum (Diesel):**

$$CO_2 \text{ emission}, [CO_2]_{\text{emission}}/\text{kWh} = \frac{1}{1 - L_a} \times \frac{1}{1 - L_{td}} \times 0.865 \text{ kg/kWh}$$

$$= \frac{1}{0.9} \times \frac{1}{0.6} \times 0.865 = 1.6019 \text{ kg/kWh}$$

**For natural gas:**

$$CO_2 \text{ emission}, [CO_2]_{\text{emission}}/\text{kWh} = \frac{1}{1 - L_a} \times \frac{1}{1 - L_{td}} \times 0.443 \text{ kg/kWh}$$

$$= \frac{1}{0.9} \times \frac{1}{0.6} \times 0.443 = 0.8204 \text{ kg/kWh}$$

**This example further shows that natural gas emits about 55.74% less $CO_2$.**

### 9.4.2 Carbon Credit Earned by Liquid Flat Plate Collector (FPC)

**Example 9.39** Evaluate carbon credit earned by five liquid flat plate collectors by using Eq. 9.14 based on thermal energy for Example 9.16 at rate of $5 per tones for different fossil fuel (Table 9.9).

**Solution**

From Example 9.16, based on thermal energy, the net thermal energy gain with respect to embodied energy of five liquid flat plate collectors (FPC) can be determined as

$$Q_{\text{th,annul-gain}} = Q_{\text{th,annul}} \times T - E_{in} = 4146 \times 10 - 5797.45 = 35{,}662.55 \text{ kWh}$$

(Example 9.16).

Now, the $\text{Net}[CO_2]_{\text{mitigation}}$ can be determined from Eq. 9.12a as

$$\text{Net}[CO_2]_{\text{mitigation}} = \left(Q_{\text{th,annul}} \times T - E_{in}\right) \times \frac{1}{1 - L_a} \times \frac{1}{1 - L_{td}} \times \text{Kg } CO_2/\text{kWh}$$

Substitute the value of $Q_{\text{th,annul-gain}}$ from above obtained value and $[CO_2]_{\text{emission}}/\text{kWh}$ from Example 9.38 for different fossil fuel, one gets

## (a) **For hard coal**

$\text{Net}[CO_2]_{\text{mitigation}} = \left(Q_{\text{th,annul}} \times T - E_{\text{in}}\right) \times \frac{1}{1-L_a} \times \frac{1}{1-L_{td}} \times \text{Kg}\frac{CO_2}{kWh} = 35,662.55 \times$
$1.8537 = 66,107.67$ kg $= 66.108$ tones (Eq. 9.13 and Example 9.38)

From Eq. 9.14, one gets

$$[CC]_{\text{earned}}\text{in } (\$) = C \times [CO_2]_{\text{tones}} = 5 \times 66.108 = \$ \, 330.54$$

## (b) **For petroleum**

$\text{Net}[CO_2]_{\text{mitigation}} = \left(Q_{\text{th,annul}} \times T - E_{\text{in}}\right) \times \frac{1}{1-L_a} \times \frac{1}{1-L_{td}} \times \text{Kg}\frac{CO_2}{kWh} = 35,662.55 \times$
$1.6019 = 57,127.94$ kg $= 57.128$ tones (Eq. 9.13 and Example 9.38).

From Eq. 9.14, one gets

$$[CC]_{\text{earned}}\text{in } (\$) = C \times [CO_2]_{\text{tones}} = 5 \times 57.128 = \$285.64$$

## (c) **For natural gas**

$\text{Net}[CO_2]_{\text{mitigation}} = \left(Q_{\text{th,annul}} \times T - E_{\text{in}}\right) \times \frac{1}{1-L_a} \times \frac{1}{1-L_{td}} \times \text{Kg}\frac{CO_2}{kWh} = 35,662.55 \times$
$0.8204 = 29,441.33$ kg $= 29.441$ tones (Eq. 9.13 and Example 9.38).

From Eq. 9.14, one gets

$$[CC]_{\text{earned}}\text{in } (\$) = C \times [CO_2]_{\text{tones}} = 5 \times 29.441 = \$147.21$$

**Example 9.40** Evaluate carbon credit earned by six ($N = 6$)-FPC-CPC liquid collectors connected in series for $C = 2$ by using Eq. 9.14 based on thermal energy for Example 9.31 at rate of $5 *per tones* for different fossil fuel (Table 9.9).

**Solution**

From Example 9.31, based on thermal energy, the net thermal energy gain with respect to embodied energy of six ($N = 6$)-FPC-CPC liquid collectors can be determined as

$Q_{\text{th,annul - gain}} = Q_{\text{th,annul}} \times T - E_{\text{in}} = 9709.5 \times 10 - 8657.04 = 88,437.96$ kWh
(Example 9.16)

Now, the Net $[CO_2]_{\text{mitigation}}$ can be determined from Eq. 9.12a as

$$\text{Net } [CO_2]_{\text{mitigation}} = \left(Q_{\text{th,annul}} \times T - E_{\text{in}}\right) \times \frac{1}{1 - L_a} \times \frac{1}{1 - L_{td}} \times \text{Kg } CO_2/kWh$$

Substitute the value o f $Q_{\text{th,annul - gain}}$ from above obtained value and $[CO_2]_{\text{emission}}/kWh$ from Example 9.38 for different fossil fuel, one gets

(a) **For hard coal**

$\text{Net}[CO_2]_{\text{mitigation}} = \left(Q_{\text{th,annul}} \times T - E_{\text{in}}\right) \times \frac{1}{1-L_a} \times \frac{1}{1-L_{td}} \times \text{Kg}\frac{CO_2}{\text{kWh}} = 88{,}437.96 \times 1.8537 = 163{,}937.45 \text{ kg} = 163{,}937 \text{ tones (Eq. 9.13 and Example 9.38)}.$

From Eq. 9.14, one gets

$$[CC]_{\text{earned}}\text{in}(\$) = C \times [CO_2]_{\text{tones}} = 5 \times 163.937 = \$819.68$$

(b) **For petroleum**

$Net \ [CO_2]_{\text{mitigation}} = \left(Q_{th,annul} \times T - E_{in}\right) \times \frac{1}{1-L_a} \times \frac{1}{1-L_{td}} \times \text{Kg}\frac{CO_2}{\text{kWh}} = 88{,}437.96 \times 1.6019 = 141{,}668.79 \text{ kg} = 141.669 \text{ tones} = 88{,}437.96 \times 1.6019 = 141{,}668.79 \text{ kg} = 141.669 \text{ tones (Eq. 9.13 and Example 9.38)}.$

From Eq. 9.14, one gets

$$[CC]_{\text{earned}}\text{in}(\$) = C \times [CO_2]_{\text{tones}} = 5 \times 141.669 = \$708.345$$

(c) **For natural gas**

$\text{Net}[CO_2]_{\text{mitigation}} = \left(Q_{\text{th,annul}} \times T - E_{\text{in}}\right) \times \frac{1}{1-L_a} \times \frac{1}{1-L_{td}} \times \text{Kg}\frac{CO_2}{\text{kWh}} = 88{,}437.96 \times 0.8204 = 72{,}554.50 \text{ kg} = 72.555 \text{ tones (Eq. 9.13 and Example 9.38)}.$

From Eq. 9.14, one gets

$$[CC]_{\text{earned}}\text{in}(\$) = C \times [CO_2]_{\text{tones}} = 5 \times 72.555 = \$362.773$$

From Examples 9.39 and 9.40, one can conclude that an integration of CPC with FPC plays an important role in improving the carbon credit due to significant improvement in net thermal energy gain $\left[Q_{th,annul-gain}\right]$.

### 9.4.3 Carbon Credit Earned by PV and PVT Module

**Example 9.41** Evaluate carbon credit earned by photo-voltaic (PV) module [case (i) of Table 9.4] by using Eq. 9.14 based on thermal exergy for Example 9.21 at rate of \$5pertones for different fossil fuel (Table 9.9).

**Solution**

From Example 9.21, based on thermal exergy, the net thermal exergy gain with respect to embodied energy of photo-voltaic (PV) module [case (i) of Table 9.4] can be obtained from Example 9.21 as

$Ex_{u,\text{annual-gain}} = Ex_{u,\text{annual}} \times T - E_{\text{in}} = 164 \times 30 - 2142 = 2778 \text{ kWh (Example 9.21)}.$

Now, the Net $[CO_2]_{\text{mitigation}}$ can be determined from Eq. 9.12b as

$$\text{Net } [CO_2]_{\text{mitigation}} = \left(Q_{\text{th,annul}} \times T - E_{\text{in}}\right) \times \frac{1}{1 - L_a} \times \frac{1}{1 - L_{td}} \times \text{Kg } CO_2/\text{kWh}$$

Substitute the value of $Ex_{u,annual-gain}$ from above obtained value and $[CO_2]_{\text{emission}}/\text{kWh}$ from Example 9.38 for different fossil fuel, one gets

(a) **For hard coal**

$\text{Net}[CO_2]_{\text{mitigation}} = \left(Ex_{u,\text{annual}} \times T - E_{in}\right) \times \frac{1}{1-L_a} \times \frac{1}{1-L_{td}} \times \text{Kg}\frac{CO_2}{\text{kWh}} = 2778 \times 1.8537 = 5149.786 \text{ kg} = 5.15 \text{ tones (Eq. 9.13 and Example 9.38)}.$
From Eq. 9.14, one gets

$$[CC]_{\text{earned}}\text{in}(\$) = C \times [CO_2]_{\text{tones}} = 5 \times 5.15 = \$25.75$$

(b) **For petroleum**

$\text{Net}[CO_2]_{\text{mitigation}} = \left(Ex_{u,\text{annual}} \times T - E_{in}\right) \times \frac{1}{1-L_a} \times \frac{1}{1-L_{td}} \times \text{Kg}\frac{CO_2}{\text{kWh}} = 2778 \times 1.6019 = 4450.08 \text{ kg} = 4.45 \text{ tones (Eq. 9.13 and Example 9.38)}.$
From Eq. 9.14, one gets

$$[CC]_{\text{earned}}\text{in}(\$) = C \times [CO_2]_{\text{tones}} = 5 \times 4.45 = \$22.25$$

(c) **For natural gas**

$\text{Net}[CO_2]_{\text{mitigation}} = \left(Ex_{u,\text{annual}} \times T - E_{in}\right) \times \frac{1}{1-L_a} \times \frac{1}{1-L_{td}} \times \text{Kg}\frac{CO_2}{\text{kWh}} = 2778 \times 0.8204 = 2279.07 \text{ kg} = 2.279 \text{ tones (Eq. 9.13 and Example 9.38)}.$
From Eq. 9.14, one gets

$$[CC]_{\text{earned}}\text{in}(\$) = C \times [CO_2]_{\text{tones}} = 5 \times 2.279 = \$11.395$$

**Example 9.42** Evaluate carbon credit earned by single liquid OPVT collector by using Eq. 9.14 based on thermal energy for Example 9.23 at rate of \$5 per tones for different fossil fuel (Table 9.9).

**Solution**

From Example 9.23, based on thermal energy, the net thermal energy gain with respect to embodied energy of single liquid OPVT collector can be determined as
$Q_{\text{th,annul-gain}} = Q_{\text{th,annul}} \times T - E_{in} = 510.96 \times 25 - 3033.01 = 9740.99 \text{ kWh}$
(Example 9.23)
Now, the Net$[CO_2]_{\text{mitigation}}$ can be determined from Eq. 9.12a as

$$\text{Net } [CO_2]_{\text{mitigation}} = \left(Q_{\text{th,annul}} \times T - E_{\text{in}}\right) \times \frac{1}{1 - L_a} \times \frac{1}{1 - L_{td}} \times \text{Kg } CO_2/\text{kWh}$$

Substitute the value of $Q_{th,\text{annul-gain}}$ from above obtained value and $[CO_2]_{\text{emission}}/kWh$ from Example 9.38 for different fossil fuel, one gets

### (a) For hard coal

$Net[CO_2]_{\text{mitigation}} = (Q_{th,\text{annul}} \times T - E_{in}) \times \frac{1}{1-L_a} \times \frac{1}{1-L_{td}} \times Kg\frac{CO_2}{kWh} = 9740.99 \times$
$1.8537 = 18,047.13$ kg $= 18.047$ tones (Eq. 9.13 and Example 9.38).
From Eq. 9.14, one gets

$$[CC]_{\text{earned}} \text{ in}(\$) = C \times [CO_2]_{\text{tones}} = 5 \times 18.047 = \$90.24$$

### (b) For petroleum

$Net[CO_2]_{\text{mitigation}} = (Q_{th,\text{annul}} \times T - E_{in}) \times \frac{1}{1-L_a} \times \frac{1}{1-L_{td}} \times Kg\frac{CO_2}{kWh} = 9740.99 \times$
$1.6019 = 15,604.09$ kg $= 15.604$ tones (Eq. 9.13 and Example 9.38).
From Eq. 9.14, one gets

$$[CC]_{\text{earned}} \text{in}(\$) = C \times [CO_2]_{\text{tones}} = 5 \times 57.128 = \$78.02$$

### (c) For natural gas

$Net[CO_2]_{mitigation} = (Q_{th,\text{annul}} \times T - E_{in}) \times \frac{1}{1-L_a} \times \frac{1}{1-L_{td}} \times Kg\frac{CO_2}{kWh} = 9740.99 \times$
$0.8204 = 7991.51$ kg $= 7.99$ tones (Eq. 9.13 and Example 9.38).
From Eq. 9.14, one gets

$$[CC]_{\text{earned}} \text{in}(\$) = C \times [CO_2]_{\text{tones}} = 5 \times 7.99 = \$39.95$$

**Example 9.43** Evaluate carbon credit earned by single liquid SPVT collector by using Eq. 9.14 based on thermal energy for Examples 9.24 at rate of $5 per tones for different fossil fuel (Table 9.9).

**Solution**

From Example 9.24, based on thermal energy, the net thermal energy gain with respect to embodied energy of single liquid SPVT collector can be determined as
$Q_{th,\text{annul-gain}} = Q_{th,\text{annul}} \times T - E_{in} = 736.50 \times 25 - 3033.01 = 15,379.49$ kWh
(Example 9.24).
Now, the $Net[CO_2]_{\text{mitigation}}$ can be determined from Eq. 9.12a as

$$Net[CO_2]_{\text{mitigation}} = (Q_{th,\text{annul}} \times T - E_{in}) \times \frac{1}{1 - L_a} \times \frac{1}{1 - L_{td}} \times Kg\ CO_2/kWh$$

Substitute the value o f $Q_{th,\text{annul - gain}}$ from above obtained value and $[CO_2]_{\text{emission}}/kWh$ from Example 9.38 for different fossil fuel, one gets

## (a) For hard coal

$\text{Net[CO}_2]_{\text{mitigation}} = \left(Q_{\text{th,annul}} \times T - E_{\text{in}}\right) \times \frac{1}{1-L_a} \times \frac{1}{1-L_{td}} \times \text{Kg}\frac{CO_2}{\text{kWh}} = 15{,}379.49 \times$
$1.8537 = 28{,}508.96 \text{ kg} = 28.509 \text{ tones (Eq. 9.13 and Example 9.38).}$
From Eq. 9.14, one gets

$$[CC]_{\text{earned}}\text{in}(\$) = C \times [CO_2]_{\text{tones}} = 5 \times 28.509 = \$142.54$$

## (b) For petroleum

$\text{Net[CO}_2]_{\text{mitigation}} = \left(Q_{\text{th,annul}} \times T - E_{\text{in}}\right) \times \frac{1}{1-L_a} \times \frac{1}{1-L_{td}} \times \text{Kg}\frac{CO_2}{\text{kWh}} = 15{,}379.49 \times$
$1.6019 = 24{,}636.41 \text{ kg} = 24.636 \text{ tones (Eq. 9.13 and Example 9.38).}$
From Eq. 9.14, one gets

$$[CC]_{\text{earned}}\text{in}(\$) = C \times [CO_2]_{\text{tones}} = 5 \times 24.636 = \$123.18$$

## (c) For natural gas

$\text{Net[CO}_2]_{\text{mitigation}} = \left(Q_{\text{th,annul}} \times T - E_{\text{in}}\right) \times \frac{1}{1-L_a} \times \frac{1}{1-L_{td}} \times \text{Kg}\frac{CO_2}{\text{kWh}} = 5379.49 \times$
$0.8204 = 12{,}617.33 \text{ kg} = 12.617 \text{ tones (Eq. 9.13 and Example 9.38)}$
From Eq. 9.14, one gets

$$[CC]_{\text{earned}}\text{in}(\$) = C \times [CO_2]_{\text{tones}} = 5 \times 12.617 = \$63.09$$

On the basis of Examples 9.40–9.42, the following conclusions can be drawn:

$$[CC]_{\text{earned}} \, by \, PV\left[\text{Example } 9.41\right] < [CC]_{\text{earned}} by \, \text{OPVT}\left[\text{Example } 9.42\right]$$
$$< [CC]_{\text{earned}} by \, SPVT[9.43]$$

**Example 9.44** Evaluate carbon credit earned by single semi-transparent liquid SPVT-CPC collector by using Eq. 9.14 based on thermal energy for Example 9.30 at rate of \$5 per tones for different fossil fuel (Table 9.9).

**Solution**

From Example 9.30, based on thermal energy, the net thermal energy gain with respect to embodied energy of single semi-transparent liquid SPVT-CPC collector can be determined as

$Q_{\text{th,annul - gain}} = Q_{\text{th,annul}} \times T - E_{\text{in}} = 888.45 \times 25 - 2388.36 = 19{,}822.89 \text{ kWh}$
(Example 9.30)
Now, the Net $[CO_2]_{\text{mitigation}}$ can be determined from Eq. 9.12a as

$$\text{Net } [CO_2]_{\text{mitigation}} = \left(Q_{\text{th,annul}} \times T - E_{in}\right) \times \frac{1}{1 - L_a} \times \frac{1}{1 - L_{td}} \times \text{Kg } CO_2/\text{kWh}$$

Substitute the value of $Q_{th,annul\text{-}gain}$ from above obtained value and $[CO_2]_{emission}/kWh$ from Example 9.38 for different fossil fuel, one gets

### (a) For hard coal

$Net[CO_2]_{mitigation} = \left(Q_{th,annul} \times T - E_{in}\right) \times \frac{1}{1-L_a} \times \frac{1}{1-L_{td}} \times Kg\frac{CO_2}{kWh} = 19{,}822.89 \times 1.8537 = 36{,}745.69 \text{ kg} = 36.746 \text{ tones (Eq. 9.13 and Example 9.38)}.$
From Eq. 9.14, one gets

$$[CC]_{earned}in(\$) = C \times [CO_2]_{tones} = 5 \times 36.746 = \$183.729$$

### (b) For petroleum

$Net[CO_2]_{mitigation} = \left(Q_{th,annul} \times T - E_{in}\right) \times \frac{1}{1-L_a} \times \frac{1}{1-L_{td}} \times Kg\frac{CO_2}{kWh} = 19{,}822.89 \times 1.6019 = 31{,}754.28 \text{ kg} = 31.754 \text{ tones (Eq. 9.13 and Example 9.38)}$
From Eq. 9.14, one gets

$$[CC]_{earned}in(\$) = C \times [CO_2]_{tones} = 5 \times 31.754 = \$158.77$$

### (c) For natural gas

$Net[CO_2]_{mitigation} = \left(Q_{th,annul} \times T - E_{in}\right) \times \frac{1}{1-L_a} \times \frac{1}{1-L_{td}} \times Kg\frac{CO_2}{kWh} = 19{,}822.89 \times 0.8204 = 16{,}262.698 \text{ kg} = 16.263 \text{ tones (Eq. 9.13 and Example 9.38)}.$
From Eq. 9.14, one gets

$$[CC]_{earned}in(\$) = C \times [CO_2]_{tones} = 5 \times 16.263 = \$81.31$$

From Examples 9.43 and 9.44, one can see improvement in carbon credit due to integration of CPC with SVT due to increased solar radiation.

**Example 9.45** Evaluate carbon credit earned by for six ($N = 6$)- SPVT-CPC air collectors connected in series for $C = 2$ by using Eq. 9.14 based on thermal energy for Example 9.32 at rate of $5 *per tones* for different fossil fuel (Table 9.9).

**Solution**

From Example 9.32, based on thermal energy, the net thermal energy gain with respect to embodied energy of single semi-transparent SPVT-CPC air collector can be determined as
$Q_{th,annul\text{-}gain} = Q_{th,annul} \times T - E_{in} = 3463.5 \times 25 - 6919.62 = 9643.5 \text{ kWh}$
(Example 9.32)
Now, the $Net[CO_2]_{mitigation}$ can be determined from Eq. 9.12a as

$$\text{Net}[CO_2]_{\text{mitigation}} = \left(Q_{\text{th,annul}} \times T - E_{\text{in}}\right) \times \frac{1}{1 - L_a} \times \frac{1}{1 - L_{td}} \times \text{Kg } CO_2/\text{kWh}$$

Substitute the value of $Q_{\text{th,annul - gain}}$ from above obtained value and $[CO_2]_{\text{emission}}/\text{kWh}$ from Example 9.38 for different fossil fuel, one gets

### (a)  For hard coal

$$\text{Net}[CO_2]_{\text{mitigation}} = \left(Q_{\text{th,annul}} \times T - E_{\text{in}}\right) \times \frac{1}{1-L_a} \times \frac{1}{1-L_{td}} \times \text{Kg}\frac{CO_2}{\text{kWh}} = 79{,}643.5 \times$$
$1.8537 = 146{,}635.156 \text{ kg} = 147.635 \text{ tones (Eq. 9.13 and Example 9.38)}.$
From Eq. 9.14, one gets

$$[CC]_{\text{earned}}\text{in}(\$) = C \times [CO_2]_{\text{tones}} = 5 \times 147.635 = \$738.18$$

### (b)  For petroleum

$$\text{Net}[CO_2]_{\text{mitigation}} = \left(Q_{\text{th,annul}} \times T - E_{\text{in}}\right) \times \frac{1}{1-L_a} \times \frac{1}{1-L_{td}} \times \text{Kg}\frac{CO_2}{\text{kWh}} = 79{,}643.5 \times$$
$1.6019 = 127{,}580.92 \text{ kg} = 127.581 \text{ tones (Eq. 9.13 and Example 9.38)}$
From Eq. 9.14, one gets

$$[CC]_{\text{earned}}\text{in}(\$) = C \times [CO_2]_{\text{tones}} = 5 \times 127.581 = \$637.91$$

### (c)  For natural gas

$$\text{Net}[CO_2]_{\text{mitigation}} = \left(Q_{\text{th,annul}} \times T - E_{in}\right) \times \frac{1}{1-L_a} \times \frac{1}{1-L_{td}} \times \text{Kg}\frac{CO_2}{\text{kWh}} = 79{,}643.5 \times$$
$0.8204 = 65{,}339.52 \text{ kg} = 65.339 \text{ tones (Eq. 9.13 and Example 9.38)}.$
From Eq. 9.14, one gets

$$[CC]_{earned}in(\$) = C \times [CO_2]_{tones} = 5 \times 65.339 = \$326.69$$

## 9.4.4  Carbon Credit Earned by Single Flexible (Al-Based) Liquid OPVT-TEC Collector

**Example 9.46** Evaluate carbon credit earned by single flexible (Al-based) liquid OPVT-TEC collector by using Eq. 9.14 based on thermal energy for Example 9.28 at rate of $5 *per tones* for different fossil fuel (Table 9.9).

**Solution**

From Example 9.28, based on thermal energy, the net thermal energy gain with respect to embodied energy of single flexible (Al-based) liquid OPVT-TEC collector can be determined as

$Q_{\text{th,annul - gain}} = Q_{\text{th,annul}} \times T - E_{\text{in}} = 957.75 \times 25 - 2349.02 = 21{,}594.75 \text{ kWh}$
(Example 9.28)

Now, the Net $[CO_2]_{mitigation}$ can be determined from Eq. 9.12a as

$$\text{Net } [CO_2]_{mitigation} = \left(Q_{th,annul} \times T - E_{in}\right) \times \frac{1}{1-L_a} \times \frac{1}{1-L_{td}} \times \text{Kg } CO_2/kWh$$

Substitute the value o f $Q_{th,annul\text{-}gain}$ from above obtained value and $[CO_2]_{emission}/kWh$ from Example 9.38 for different fossil fuel, one gets

(a)  **For hard coal**

$\text{Net}[CO_2]_{mitigation} = \left(Q_{th,annul} \times T - E_{in}\right) \times \frac{1}{1-L_a} \times \frac{1}{1-L_{td}} \times \text{Kg}\frac{CO_2}{kWh} = 21,594.75 \times$
$1.8537 = 40,030.19 \text{ kg} = 40.030 \text{ tones (Eq. 9.13 and Example 9.38)}.$
From Eq. 9.14, one gets

$$[CC]_{earned} \text{in}(\$) = C \times [CO_2]_{tones} = 5 \times 40.030 = \$200.15$$

(b)  **For petroleum**

$\text{Net}[CO_2]_{mitigation} = \left(Q_{th,annul} \times T - E_{in}\right) \times \frac{1}{1-L_a} \times \frac{1}{1-L_{td}} \times \text{Kg}\frac{CO_2}{kWh} = 21,594.75 \times$
$1.6019 = 34,592.63 \text{ kg} = 34.593 \text{ tones (Eq. 9.13 and Example 9.38)}$
From Eq. 9.14, one gets

$$[CC]_{earned} \text{in}(\$) = C \times [CO_2]_{tones} = 5 \times 34.593 = \$172.96$$

(c)  **For natural gas**

$\text{Net}[CO_2]_{mitigation} = \left(Q_{th,annul} \times T - E_{in}\right) \times \frac{1}{1-L_a} \times \frac{1}{1-L_{td}} \times \text{Kg}\frac{CO_2}{kWh} = 21,594.75 \times$
$0.8204 = 17,715.51 \text{ kg} = 17.716 \text{ tones (Eq. 9.13 and Example 9.38)}$
From Eq. 9.14, one gets

$$[CC]_{earned} \text{in } (\$) = C \times [CO_2]_{tones} = 5 \times 17.716 = \$88.58$$

**Example 9.47** Evaluate carbon credit earned by single flexible (Al-based) liquid OPVT-TEC collector by using Eq. 9.14 based on thermal exergy for Example 9.29 at rate of \$5 per tones for different fossil fuel (Table 9.9).

**Solution**

From Example 9.29, based on thermal exergy, the net thermal exergy gain with respect to embodied energy of photo-voltaic (PV) module [case (i) of Table 9.4] can be obtained as

$Ex_{u,annual\text{-}gain} = Ex_{u,annual} \times T - E_{in} = 179.43 \times 25 - 2349.02 = 2136.73 \text{ kWh}$
(Example 9.29)

Now, the Net $[CO_2]_{mitigation}$ can be determined from Eq. 9.12b as

$$\text{Net}[CO_2]_{mitigation} = \left(Q_{th,annul} \times T - E_{in}\right) \times \frac{1}{1-L_a} \times \frac{1}{1-L_{td}} \times \text{Kg } CO_2/kWh$$

Substitute the value of $Ex_{u,\text{annual - gain}}$ from above obtained value and $[CO_2]_{\text{emission}}/\text{kWh}$ from Example 9.38 for different fossil fuel, one gets

### (a)  For hard coal

$\text{Net}[CO_2]_{\text{mitigation}} = \left(Ex_{u,\text{annual}} \times T - E_{\text{in}}\right) \times \frac{1}{1-L_a} \times \frac{1}{1-L_{td}} \times \text{Kg}\frac{CO_2}{\text{kWh}} = 2136.73 \times 1.8537 = 3960.86 \text{ kg} = 3.961 \text{ tones}$ (Eq. 9.13 and Example 9.38).

From Eq. 9.14, one gets

$$[CC]_{earned}\text{in}(\$) = C \times [CO_2]_{tones} = 5 \times 3.961 = \$19.80$$

### (b)  For petroleum

$\text{Net}[CO_2]_{\text{mitigation}} = \left(Ex_{u,\text{annual}} \times T - E_{\text{in}}\right) \times \frac{1}{1-L_a} \times \frac{1}{1-L_{td}} \times \text{Kg}\frac{CO_2}{\text{kWh}} = 2136.73 \times 1.6019 = 3422.83 \text{ kg} = 3.423 \text{ tones}$ (Eq. 9.13 and Example 9.38).

From Eq. 9.14, one gets

$$[CC]_{\text{earned}}\text{in}(\$) = C \times [CO_2]_{\text{tones}} = 5 \times 3.423 = \$17.114$$

### (c)  For natural gas

$\text{Net}[CO_2]_{\text{mitigation}} = \left(Ex_{u,\text{annual}} \times T - E_{\text{in}}\right) \times \frac{1}{1-L_a} \times \frac{1}{1-L_{td}} \times \text{Kg}\frac{CO_2}{\text{kWh}} = 2136.73 \times 0.8204 = 1752.97 \text{ kg} = 1.753 \text{ tones}$ (Eq. 9.13 and Example 9.38).

From Eq. 9.14, one gets

$$[CC]_{\text{earned}}\text{in}(\$) = C \times [CO_2]_{\text{tones}} = 5 \times 1.753 = \$8.765$$

From Examples 9.46 and 9.47, one can also observe that carbon credit earned by thermal energy basis is significantly higher than carbon credit earned by thermal exergy basis due to low grade energy based on first law of thermodynamics.

## 9.4.5  Carbon Credit Earned by Flexible OPVT-TEC-CPC Liquid Collector

**Example 9.48** Evaluate carbon credit earned by six ($N = 6$) flexible OPVT-TEC-CPC liquid collectors connected in series and for $C = \frac{A_{\text{am}}}{A_{\text{rm}}} = 2$ by using Eq. 9.14 based on thermal energy for Example 9.33 at rate of \$5 per tones for different fossil fuel (Table 9.9).

## Solution

From Example 9.33, based on thermal energy, the net thermal energy gain with respect to embodied energy of flexible OPVT-TEC-CPC liquid collector can be determined as

$$Q_{th,annul\text{-}gain} = Q_{th,annul} \times T - E_{in} = 8646.75 \times 25 - 10{,}448.16 = 205{,}720.59 \text{ kWh (Example 9.33)}$$

Now, the $Net[CO_2]_{mitigation}$ can be determined from Eq. 9.12a as

$$Net\ [CO_2]_{mitigation} = \left(Q_{th,annul} \times T - E_{in}\right) \times \frac{1}{1-L_a} \times \frac{1}{1-L_{td}} \times Kg\ CO_2/kWh$$

Substitute the value o f $Q_{th,annul\text{-}gain}$ from above obtained value and $[CO_2]_{emission}/kWh$ from Example 9.38 for different fossil fuel, one gets

### (a) For hard coal

$$Net[CO_2]_{mitigation} = \left(Q_{th,annul} \times T - E_{in}\right) \times \frac{1}{1-L_a} \times \frac{1}{1-L_{td}} \times Kg\frac{CO_2}{kWh} = 205{,}720.59 \times$$
$$1.8537 = 381{,}344.26 \text{ kg} = 381.344 \text{ tones (Eq. 9.13 and Example 9.38)}.$$

From Eq. 9.14, one gets

$$[CC]_{earned}in(\$) = C \times [CO_2]_{tones} = 5 \times 381.344 = \$1{,}906.72$$

### (b) For petroleum

$$Net[CO_2]_{mitigation} = \left(Q_{th,annul} \times T - E_{in}\right) \times \frac{1}{1-L_a} \times \frac{1}{1-L_{td}} \times Kg\frac{CO_2}{kWh} = 205{,}720.59 \times$$
$$1.6019 = 329{,}543.81 \text{ kg} = 329.54 \text{ tones (Eq. 9.13 and Example 9.38)}$$

From Eq. 9.14, one gets

$$[CC]_{earned}in\ (\$) = C \times [CO_2]_{tones} = 5 \times 329.54 = \$1{,}647.72$$

### (c) For natural gas

$$Net[CO_2]_{mitigation} = \left(Q_{th,annul} \times T - E_{in}\right) \times \frac{1}{1-L_a} \times \frac{1}{1-L_{td}} \times Kg\frac{CO_2}{kWh} = 205{,}720.59 \times$$
$$0.8204 = 168773.32 \text{ kg} = 168.773 \text{ tones (Eq. 9.13 and Example 9.38)}$$

From Eq. 9.14, one gets

$$[CC]_{earned}in(\$) = C \times [CO_2]_{tones} = 5 \times 168.773 = \$843.87$$

**Example 9.49** Evaluate carbon credit earned by six (N = 6) flexible OPVT-TEC-CPC liquid collectors connected in series and for $C = \frac{A_{am}}{A_{rm}} = 2$ by using Eq. 9.14 based on thermal exergy for Examples 9.34 at rate of $5 per tones for different fossil fuel (Table 9.9).

**Solution**

From Example 9.34, based on thermal exergy, the net thermal exergy gain with respect to embodied energy of flexible OPVT-TEC-CPC liquid collectors can be obtained as

$Ex_{u,\text{annual - gain}} = Ex_{u,\text{annual}} \times T - E_{in} = 1484.46 \times 25 - 10{,}448.16 = 26{,}663.34$ kWh (Example 9.29).

Now, the Net $[CO_2]_{\text{mitigation}}$ can be determined from Eq. 9.12b as

$$\text{Net } [CO_2]_{\text{mitigation}} = \left( Q_{\text{th,annul}} \times T - E_{in} \right) \times \frac{1}{1 - L_a} \times \frac{1}{1 - L_{td}} \times \text{Kg } CO_2/\text{kWh}$$

Substitute the value of $Ex_{u,\text{annual - gain}}$ from above obtained value and $[CO_2]_{\text{emission}}/\text{kWh}$ from Example 9.38 for different fossil fuel, one gets

(a) **For hard coal**

$\text{Net}[CO_2]_{\text{mitigation}} = \left( Ex_{u,\text{annual}} \times T - E_{in} \right) \times \frac{1}{1-L_a} \times \frac{1}{1-L_{td}} \times \text{Kg} \frac{CO_2}{\text{kWh}} = 26{,}663.34 \times 1.8537 = 49{,}425.83$ kg $= 40.426$ tones (Eq. 9.13 and Example 9.38)

From Eq. 9.14, one gets

$$[CC]_{\text{earned}}\text{in}(\$) = C \times [CO_2]_{\text{tones}} = 5 \times 40.426 = \$247.13$$

(b) **For petroleum**

$Net \ [CO_2]_{mitigation} = \left( Ex_{u,annual} \times T - E_{in} \right) \times \frac{1}{1-L_a} \times \frac{1}{1-L_{td}} \times \text{Kg} \frac{CO_2}{\text{kWh}} = 26{,}663.34 \times 1.6019 = 42712.00$ kg $= 42.712$ tones (Eq. 9.13 and Example 9.38)

From Eq. 9.14, one gets

$$[CC]_{\text{earned}}\text{in}(\$) = C \times [CO_2]_{\text{tones}} = 5 \times 42.712 = \$213.56$$

(c) **For natural gas**

$\text{Net}[CO_2]_{\text{mitigation}} = \left( Ex_{u,\text{annual}} \times T - E_{in} \right) \times \frac{1}{1-L_a} \times \frac{1}{1-L_{td}} \times \text{Kg} \frac{CO_2}{\text{kWh}} = 26{,}663.34 \times 0.8204 = 21{,}874.60$ kg $= 21.875$ tones (Eq. 9.13 and Example 9.38)

From Eq. 9.14, one gets

$$[CC]_{\text{earned}}\text{in}(\$) = C \times [CO_2]_{\text{tones}} = 5 \times 21.875 = \$109.37$$

The same results have been obtained as reported in Examples 9.46 and 9.47, respectively.

## 9.4.6  Carbon Credit Earned by Fresnel Lens (FL) Integrated SPVT Liquid Collector [FLiSPVT]

**Example 9.50** Evaluate carbon credit earned by single Fresnel lens (FL) integrated SPVT liquid collector [FLiSPVT] by using Eq. 9.14 based on thermal energy for Example 9.36 at rate of $5 per tones for different fossil fuel (Table 9.9).

**Solution**

**For N = 1**

From Example 9.36, based on thermal energy, the net thermal energy gain with respect to embodied energy of Fresnel lens (FL) integrated SPVT liquid collector [FLiSPVT] can be determined as

$$Q_{th,annul\text{-}gain} = Q_{th,annul} \times T - E_{in} = 1274.53 \times 25 - 1109.72 = 23,418.28 \text{ kWh}$$

(Example 9.36)

Now, the $Net[CO_2]_{mitigation}$ can be determined from Eq. 9.12a as

$$Net[CO_2]_{mitigation} = \left(Q_{th,annul} \times T - E_{in}\right) \times \frac{1}{1 - L_a} \times \frac{1}{1 - L_{td}} \times \text{Kg } CO_2/\text{kWh}$$

Substitute the value of $Q_{th,annul\text{-}gain}$ from above obtained value and $[CO_2]_{emission}/\text{kWh}$ from Example 9.38 for different fossil fuel, one gets

(a) **For hard coal**

$$Net[CO_2]_{mitigation} = \left(Q_{th,annul} \times T - E_{in}\right) \times \frac{1}{1-L_a} \times \frac{1}{1-L_{td}} \times \text{Kg}\tfrac{CO_2}{\text{kWh}} = 23,418.28 \times$$
$$1.8537 = 43,410.46 \text{ kg} = 43.410 \text{ tones (Eq. 9.13 and Example 9.38)}$$

From Eq. 9.14, one gets $[CC]_{earned}\text{in}(\$) = C \times [CO_2]_{tones} = 5 \times 43.410 = \$217.05$

(b) **For petroleum**

$$Net[CO_2]_{mitigation} = \left(Q_{th,annul} \times T - E_{in}\right) \times \frac{1}{1-L_a} \times \frac{1}{1-L_{td}} \times \text{Kg}\tfrac{CO_2}{\text{kWh}} = 23,418.28 \times$$
$$1.6019 = 37,513.74 \text{ kg} = 37.514 \text{ tones (Eq. 9.13 and Example 9.38)}$$

From Eq. 9.14, one gets

$$[CC]_{earned}\text{in}(\$) = C \times [CO_2]_{tones} = 5 \times 37.514 = \$187.57$$

(c) **For natural gas**

$$Net[CO_2]_{mitigation} = \left(Q_{th,annul} \times T - E_{in}\right) \times \frac{1}{1-L_a} \times \frac{1}{1-L_{td}} \times \text{Kg}\tfrac{CO_2}{\text{kWh}} = 23,418.28 \times$$
$$0.8204 = 19,212.36 \text{ kg} = 19.212 \text{ tones (Eq. 9.13 and Example 9.38)}$$

From Eq. 9.14, one gets

$$[CC]_{earned}\text{in}(\$) = C \times [CO_2]_{tones} = 5 \times 19.212 = \$96.06$$

**Example 9.51** Evaluate carbon credit earned by five liquid FLiSPVT collectors connected in series with $C = 4$, $\beta_c = 0.50$ by using Eq. 9.14 based on thermal exergy for Example 9.37 at rate of \$5 per tones for different fossil fuel (Table 9.9).

## Solution

From Example 9.37, based on thermal exergy, the net thermal exergy gain with respect to embodied energy of five liquid FLiSPVT collectors can be obtained as

$Ex_{u,annual\,-\,gain} = Ex_{u,annual} \times T - E_{in} = 638.985 \times 25 - 5548.60 = 10,426.025$ kWh (Example 9.37)

Now, the Net$[CO_2]_{mitigation}$ can be determined from Eq. 9.12b as

$$\text{Net } [CO_2]_{mitigation} = \left(Q_{th,annul} \times T - E_{in}\right) \times \frac{1}{1 - L_a} \times \frac{1}{1 - L_{td}} \times \text{Kg } CO_2/kWh$$

Substitute the value of $Ex_{u,annual\,-\,gain}$ from above obtained value and $[CO_2]_{emission}/kWh$ from Example 9.38 for different fossil fuel, one gets

### (a)  For hard coal

$Net[CO_2]_{mitigation} = \left(Ex_{u,annual} \times T - E_{in}\right) \times \frac{1}{1-L_a} \times \frac{1}{1-L_{td}} \times Kg\frac{CO_2}{kWh} = 10,426.025 \times$
$1.853 = 19,326.72$ kg $= 19.327$ tones (Eq. 9.13 and Example 9.38)
From Eq. 9.14, one gets

$$[CC]_{earned}in(\$) = C \times [CO_2]_{tones} = 5 \times 19.327 = \$96.63$$

### (b)  For petroleum

$Net[CO_2]_{mitigation} = \left(Ex_{u,annual} \times T - E_{in}\right) \times \frac{1}{1-L_a} \times \frac{1}{1-L_{td}} \times Kg\frac{CO_2}{kWh} = 10,426.025 \times$
$1.6019 = 16,701.45$ kg $= 16.701$ tones (Eq. 9.13 and Example 9.38)
From Eq. 9.14, one gets

$$[CC]_{earned}in\,(\$) = C \times [CO_2]_{tones} = 5 \times 16.701 = \$83.51$$

### (c)  For natural gas

$Net[CO_2]_{mitigation} = \left(Ex_{u,annual} \times T - E_{in}\right) \times \frac{1}{1-L_a} \times \frac{1}{1-L_{td}} \times Kg\frac{CO_2}{kWh} = 10,426.025 \times$
$0.8204 = 8553.51$ kg $= 8.554$ tones (Eq. 9.13 and Example 9.38)
From Eq. 9.14, one gets

$$[CC]_{earned}in(\$) = C \times [CO_2]_{tones} = 5 \times 8.554 = \$42.77$$

The same results have been obtained as reported in Examples 9.46 and 9.27, respectively.

# References

1. Tiwari, G. N., & Mishra, R. K. (2012) Advanced renewable energy sources. *Royal Society of Chemistry (RSC)*
2. A.1 Embodied Energy Coefficients—Alphabetical. https://www.wgtn.ac.nz/architecture/centres/cbpr/resources/pdfs/ee-coefficients.pdf
3. Mishra, R. K., Garg, V., & Tiwari, G. N. (2017). Energy matrices of U-shaped evacuated tubular collector (ETC) integrated with compound parabolic concentrator (CPC). *Solar Energy, 153,* 531–539. https://doi.org/10.1016/j.solener.2017.06.00410.1016/j
4. Specific carbon dioxide emissions of various fuels, https://www.volker-quaschning.de › $CO_2$-spez › index_e

# Appendix A

## Conversion of Units

### (i) Length, m

| |
|---|
| 1 yd (yard) = 3 ft = 36 in (inches) = 0.9144 m |
| 1 m = 39.3701 in = 3.280839 ft = 1.093613 yd = 1,650,763.73 wavelength |
| 1 ft = 12 in = 0.3048 m |
| 1 in = 2.54 cm = 25.4 mm |
| 1 mil = 2.54 × $10^{-3}$ cm |
| 1 μm = $10^{-6}$ m |
| 1 nm = $10^{-9}$ m = $10^{-3}$ μm |

### (ii) Area, m²

| |
|---|
| 1 ft$^2$ = 0.0929 m$^2$ |
| 1 in$^2$ = 6.452 cm$^2$ = 0.00064516 m$^2$ |
| 1 cm$^2$ = $10^{-4}$ m$^2$ = 10.764 × $10^{-4}$ ft$^2$ = 0.1550 in$^2$ |
| 1 ha = 10,000 m$^2$ |

### (iii) Volume, m³

| |
|---|
| 1 ft$^3$ = 0.02832 m$^3$ = 28.3168 l (liter) |
| 1 in$^3$ = 16.39 cm$^3$ = 1.639 ×$10^2$ l |
| 1 yd$^3$ = 0.764555 m$^3$ = 7.646 × $10^2$ l |
| 1 UK gallon = 4.54609 l |
| 1 US gallon = 3.785 l = 0.1337 ft$^3$ |
| 1 m$^3$ = 1.000 × $10^6$ cm$^3$ = 2.642 × $10^2$ US gallons = 109 l |
| 1 l = $10^{-3}$ m$^3$ |
| 1 fluid ounce = 28.41 cm$^3$ |

### (iv) Mass, kg

| |
|---|
| 1 kg = 2.20462 lb = 0.068522 slug |
| 1 ton (short) = 2000 lb (pounds) = 907.184 kg |

(continued)

© Bag Energy Research Society 2024
G. N. Tiwari, *Fundamental of Mathematical Tools for Thermal Modeling of Solar Thermal and Photo-voltaic Systems-Part-I*, https://doi.org/10.1007/978-981-99-7085-8

(continued)

| |
|---|
| 1 ton (long) = 1016.05 kg |
| 1 lb = 16 oz (ounces) = 0.4536 kg |
| 1 oz = 28.3495 g |
| 1 quintal = 100 kg |
| 1 kg = 1000 g = 10,000 mg |
| 1 $\mu$g = $10^{-6}$ g |
| 1 ng = $10^{-9}$ g |
| **(v) Density and specific volumes, kg/m$^3$, m$^3$/kg** |
| 1 lb/ft$^3$ = 16.0185 kg/m$^3$ = 5.787 $\times$ $10^{-4}$ lb/in$^3$ |
| 1 g/cm$^3$ = $10^3$ kg/m$^3$ = 62.43 lb/ft$^3$ |
| 1 lb/ft$^3$ = 0.016 g/cm$^3$ = 16 kg/m$^3$ |
| 1 ft$^3$ (air) = 0.08009 lb = 36.5 g at N.T.P |
| 1 gallon/lb = 0.010 cm$^3$/kg |
| 1 $\mu$g/m$^3$ = $10^{-6}$ g/m$^3$ |
| **(vi) Pressure, Pa (Pascal)** |
| 1 lb/ft$^2$ = 4.88 kg/m$^2$ = 47.88 Pa |
| 1 lb/in$^2$ = 702.7 kg/m$^2$ = 51.71 mm Hg = 6.894757 $\times$ $10^3$ Pa = 6.894757 $\times$ $10^3$ N/m$^2$ |
| 1 atm = 1.013 $\times$ $10^5$ N/m$^2$ = 760 mm Hg = 101.325 kPa |
| 1 in $H_2O$ = 2.491 $\times$ $10^2$ N/m$^2$ = 248.8 Pa = 0.036 lb/in$^2$ |
| 1 bar = 0.987 atm = 1.000 $\times$ $10^6$dynes/cm$^2$ = 1.020 kgf/cm$^2$ = 14.50 lbf/in$^2$ = $10^5$ N/m$^2$ = 100 kPa |
| 1 torr (mm Hg 0 °C) = 133 Pa |
| 1 Pa (Pa) = 1 N/m$^2$ = 1.89476 kg |
| 1 inch of Hg = 3.377 kPa = 0.489 lb/in$^2$ |
| **(vii) Velocity, m/s** |
| 1 ft/s = 0.3041 m/s |
| 1 mile/h = 0.447 m/s = 1.4667 ft/s = 0.8690 knots |
| 1 km/h = 0.2778 m/s |
| 1 ft/min = 0.00508 m/s |
| **(viii) Force, N** |
| 1 N (Newton) = $10^5$ dynes = 0.22481 lb wt = 0.224 lb f |
| 1 pdl (poundal) = 0.138255 N (Newton) = 13.83 dynes = 14.10 gf |
| 1 lbf (i.e. wt of 1 lb mass) = 4.448222 N = 444.8222 dynes |
| 1 ton = 9.964 $\times$ $10^3$ N |
| 1 bar = $10^5$ Pa (Pascal) |
| 1 ft of $H_2O$ = 2.950 $\times$ $10^{-2}$ atm = 9.807 $\times$ $10^3$ N/m$^2$ |
| 1 in $H_2O$ = 249.089 Pa |
| 1 mm $H_2O$ = 9.80665 Pa |

(continued)

(continued)

| |
|---|
| 1 dyne $= 1.020 \times 10^{-6}$ kg f $= 2.2481 \times 10^{-6}$ lb f $= 7.2330 \times 10^{-5}$ pdl $= 10^{-5}$ N |
| 1 mm of Hg $= 133.3$ Pa |
| 1 atm $= 1$ kg f/cm$^2 = 98.0665$ k Pa |
| 1 Pa (Pascal) $= 1$ N/m$^2$ |
| **(ix) Mass flow rate and discharge, kg/s, m$^3$/s** |
| 1 lb/s $= 0.4536$ kg/s |
| 1 ft$^3$/min $= 0.4720$ 1/s $= 4.179 \times 10^{-4}$ m$^3$/s |
| 1 m$^3$/s $= 3.6 \times 10^6$ l/h |
| 1 g/cm$^3 = 10^3$ kg/m$^3$ |
| 1 lb/h ft$^2 = 0.001356$ kg/s m$^2$ |
| 1 lb/ft$^3 = 16.2$ kg/m$^2$ |
| 1 L/s (l/s) $= 10^{-3}$ m$^3$/s |
| **(x) Energy, J** |
| 1 cal $= 4.187$ J (Joules) |
| 1 kcal $= 3.97$ Btu $= 12 \times 10^{-4}$ kWh $= 4.187 \times 10^3$ J |
| 1 W $= 1.0$ J/s |
| 1 Btu $= 0.252$ kcal $= 2.93 \times 10^{-4}$ kWh $= 1.022 \times 10^3$ J |
| 1 hp $= 632.34$ kcal $= 0.736$ kWh |
| 1 kWh $= 3.6 \times 10^6$ J $= 1$ unit |
| 1 J $= 2.390 \times 10^{-4}$ kcal $= 2.778 \times 10^{-4}$ Wh |
| 1 kWh $= 860$ kcal $= 3413$ Btu |
| 1 erg $= 1.0 \times 10^{-7}$ J $= 1.0 \times 10^{-7}$ Nm $= 1.0$ dyne cm |
| 1 J $= 1$ Ws $= 1$ Nm |
| 1 eV $= 1.602 \times 10^{-19}$ J |
| 1 GJ $= 10^9$ J |
| 1 MJ $= 10^6$ J |
| 1 TJ (Terajoules) $= 10^{12}$ J |
| 1EJ (Exajoules) $= 10^{18}$ J |
| **(xi) Power, Watt (J/s)** |
| 1 Btu/h $= 0.293071$ W $= 0.252$ kcal/h |
| 1 Btu/h $= 1.163$ W $= 3.97$ Btu/h |
| 1 W $= 1.0$ J/s $= 1.341 \times 10^{-3}$ hp $= 0.0569$ Btu/min $= 0.01433$ kcal/min |
| 1 hp (F.P.S.) $= 550$ ft lb f/s $= 746$ W $= 596$ kcal/h $= 1.015$ hp (M.K.S.) |
| 1 hp (M.K.S.) $= 75$ mm kg f/s $= 0.17569$ kcal/s $= 735.3$ W |
| 1 W/ft$^2 = 10.76$ W/m$^2$ |
| 1 ton (Refrigeration) $= 3.5$ kW |
| 1 kW $= 1000$ W |
| 1 GW $= 10^9$ W |

(continued)

(continued)

| |
|---|
| 1 W/m$^2$ = 100 lux |

**(xii) Specific Heat, J/kg °C**

| |
|---|
| 1 Btu/lb °F = 1.0 kcal/kg °C = 4.187 × 10$^3$ J/kg °C |
| 1 Btu/lb = 2.326 kJ/kg |

**(xiii) Temperature, °C and K used in SI**

| |
|---|
| $T_{(Celcius, °C)} = (5/9) [T_{(Fahrenheit, °F)} + 40] - 40$ |
| $T_{(°F)} = (9/5) [T (°C) + 40] - 40$ |
| $T_{(Rankine, °R)} = 460 + T (°F)$ |
| $T_{(Kelvin,K)} = (5/9) T (°R)$ |
| $T_{(Kelvin,K)} = 273.15 + T (°C)$ |
| $T_{(°C)} = T (°F)/1.8 = (5/9) T (°F)$ |

**(xiv) Rate of heat flow per unit area or heat flux, W/m$^2$**

| |
|---|
| 1 Btu/ft$^2$ h = 2.713 kcal/m$^2$ h = 3.1552 W/m$^2$ |
| 1 kcal/m$^2$ h = 0.3690 Btu/ft$^2$ h = 1.163 W/m$^2$ = 27.78 × 10$^{-6}$ cal/s cm$^2$ |
| 1 cal/cm$^2$ min = 221.4 Btu/ft$^2$ h |
| 1 W/ft$^2$ = 10.76 W/m$^2$ |
| 1 W/m$^2$ = 0.86 kcal/hm$^2$ = 0.23901 × 10$^{-4}$ cal/s cm$^2$ = 0.137 Btu/h ft$^2$ |
| 1 Btu/h ft = 0.96128 W/m |

**(xv) Heat transfer coefficient, W/m$^2$ °C**

| |
|---|
| 1 Btu/ft$^2$h°F = 4.882 kcal/m$^2$h °C = 1.3571 × 10$^{-4}$ cal/cm$^2$ s °C |
| 1 Btu/ft$^2$h °F = 5.678 W/m$^2$ °C |
| 1 kcal/m$^2$h °C = 0.2048 Btu/ft$^2$ h °F = 1.163 W/m$^2$ °C |
| 1 W/m$^2$K = 2.3901 × 10$^{-5}$ cal/cm$^2$sK = 1.7611 × 10$^{-1}$ Btu/ft$^2$ °F = 0.86 kcal/m$^2$h °C |

**(xvi) Thermal Conductivity, W/m °C**

| |
|---|
| 1 Btu/ft h °F = 1.488 kcal/m h °C = 1.73073 W/m °C |
| 1 kcal/m h °C = 0.6720 Btu/ft h °F = 1.1631 W/m °C |
| 1 Btu in/ft$^2$ h °F = 0.124 kcal/m h °C = 0.144228 W/m °C |
| 1 Btu/in h °F = 17.88 kcal/mh °C |
| 1 cal/cm s °F = 4.187 × 10$^2$ W/m °C = 242 Btu/h ft °F |
| 1 W/cm °C = 57.79 Btu/h ft °F |

**(xvii) Angle, rad**

| |
|---|
| 2π rad (radian) = 360° (degree) |
| 1° (degree) = 0.0174533 rad = 60′ (minutes) |
| 1′ = 0.290888 × 10$^{-3}$ rad = 60″ (seconds) |
| 1″ = 4.84814 × 10$^{-6}$ rad |
| 1° (hour angle) = 4 min (time) |

**(xviii) Illumination**

| |
|---|
| 1 lx (lux) = 1.0 lm (lumen)/m$^2$ |

(continued)

(continued)

| |
|---|
| 1 lm/ft$^2$ = 1.0 foot candle |
| 1 foot candle = 10.7639 lx |
| 100 lux = 1 W/m$^2$ |

**(xix) Time, h**

| |
|---|
| 1 week = 7 days = 168 h = 10,080 min = 6,04,800 s |
| 1 mean solar day = 1440 min = 86,400 s |
| 1 calendar year = 365 days = 8760 h = 5.256 × 10$^5$ min |
| 1 tropical mean solar year = 365.2422 days |
| 1 sidereal year = 365.2564 days (mean solar) |
| 1 s (second) = 9.192631770 × 10$^9$ Hz (Hz) |
| 1 day = 24 h = 360° (hour angle) |

**(xx) Concentration, kg/m$^3$ and g/m$^3$**

| |
|---|
| 1 g/l = 1 kg/m$^3$ |
| 1 lb/ft$^3$ = 6.236 kg/m$^3$ |

**(xxi) Diffusivity, m$^2$/s**

| |
|---|
| 1 ft$^2$/h = 25.81 × 10$^{-6}$ m$^2$/s |

# Appendix B

The value of $f_{0-\lambda T}$ for different $\lambda T, \mu mK$, for even increment of $\lambda T$

| $\lambda T, \mu mK$ | $f_{0-\lambda T}$ | $\lambda T, \mu mK$ | $f_{0-\lambda T}$ | $\lambda T, \mu mK$ | $f_{0-\lambda T}$ |
|---|---|---|---|---|---|
| 1000 | 0.0003 | | | 8000 | 0.8562 |
| 1200 | 0.0021 | 4600 | 0.5793 | 8200 | 0.8639 |
| 1400 | 0.0077 | 4800 | 0.6075 | 8400 | 0.8711 |
| 1600 | 0.0197 | 5000 | 0.6337 | 8600 | 0.8778 |
| 1800 | 0.0393 | 5200 | 0.6579 | 8800 | 0.8841 |
| 2000 | 0.0667 | 5400 | 0.6803 | 9000 | 0.8899 |
| 2200 | 0.1009 | 5600 | 0.7010 | 9200 | 0.8954 |
| 2400 | 0.1402 | 5800 | 0.7201 | 9400 | 0.9005 |
| 2600 | 0.1831 | 6000 | 0.7378 | 9600 | 0.9054 |
| 2800 | 0.2279 | 6200 | 0.7451 | 9800 | 0.9099 |
| 3000 | 0.2730 | 6400 | 0.7692 | 10,000 | 0.9141 |
| 3200 | 0.3181 | 6600 | 0.7831 | 12,000 | 0.9450 |
| 3400 | 0.3617 | 6800 | 0.7961 | 14,000 | 0.9628 |
| 3600 | 0.4036 | 7000 | 0.8080 | 16,000 | 0.9737 |
| 3800 | 0.4434 | 7200 | 0.8191 | 18,000 | 0.9807 |
| 4000 | 0.4829 | 7400 | 0.8295 | 20,000 | 0.9855 |
| 4200 | 0.5160 | 7600 | 0.8390 | 50,000 | 0.9988 |
| 4400 | 0.5488 | 7800 | 0.8479 | $\infty$ | 1 |

© Bag Energy Research Society 2024
G. N. Tiwari, *Fundamental of Mathematical Tools for Thermal Modeling of Solar Thermal and Photo-voltaic Systems-Part-I*, https://doi.org/10.1007/978-981-99-7085-8

# Appendix
# C1

Parameters on horizontal surface for sunshine hours = 10 for all four weather type of days for different Indian climates [1]

© Bag Energy Research Society 2024

G. N. Tiwari, *Fundamental of Mathematical Tools for Thermal Modeling of Solar Thermal and Photo-voltaic Systems-Part-I*, https://doi.org/10.1007/978-981-99-7085-8

## (a): New Delhi

| Type of day | Month ▶<br>Parameters ▼ | January | February | March | April | May | June | July | August | September | October | November | December |
|---|---|---|---|---|---|---|---|---|---|---|---|---|---|
| a | $T_R$ | 2.25 | 2.79 | 2.85 | 2.72 | 3.54 | 2.47 | 2.73 | 2.58 | 2.53 | 1.38 | 0.62 | 0.72 |
|   | $\alpha$ | 0.07 | 0.10 | 0.17 | 0.23 | 0.16 | 0.28 | 0.37 | 0.41 | 0.29 | 0.47 | 0.59 | 0.54 |
|   | $K_1$ | 0.47 | 0.39 | 0.33 | 0.28 | 0.20 | 0.27 | 0.41 | 0.40 | 0.23 | 0.21 | 0.21 | 0.28 |
|   | $K_2$ | -13.17 | -6.25 | 5.61 | 38.32 | 65.04 | 31.86 | -40.57 | -55.08 | 39.92 | 32.77 | 30.62 | 9.73 |
| b | $T_R$ | 2.28 | 2.78 | 2.89 | 3.15 | 5.44 | 4.72 | 5.58 | 5.43 | 3.23 | 4.56 | 0.19 | 1.83 |
|   | $\alpha$ | 0.15 | 0.13 | 0.14 | 0.17 | 0.16 | 0.20 | 0.24 | 0.18 | 0.31 | 0.22 | 1.14 | 0.42 |
|   | $K_1$ | 0.51 | 0.54 | 0.49 | 0.46 | 0.45 | 0.45 | 0.53 | 0.39 | 0.37 | 0.42 | 0.35 | 0.40 |
|   | $K_2$ | -21.77 | -28.26 | -9.22 | -11.55 | 1.54 | 23.99 | -51.61 | 9.46 | 14.07 | -9.50 | 17.47 | -0.07 |
| c | $T_R$ | 5.88 | 6.36 | 6.11 | 7.77 | 9.20 | 10.54 | 7.13 | 7.97 | 5.51 | 5.01 | 4.93 | 3.23 |
|   | $\alpha$ | 0.27 | 0.37 | 0.37 | 0.31 | 0.07 | 0.06 | 0.41 | 0.51 | 0.49 | 1.26 | 1.06 | 0.64 |
|   | $K_1$ | 0.39 | 0.36 | 0.33 | 0.35 | 0.56 | 0.48 | 0.47 | 0.35 | 0.39 | 0.36 | 0.31 | 0.43 |
|   | $K_2$ | -14.73 | -7.97 | 10.87 | 20.45 | -56.00 | -0.37 | -52.27 | 47.70 | 35.64 | -0.68 | 13.06 | -7.04 |
| d | $T_R$ | 7.47 | 8.97 | 10.77 | 11.18 | 13.69 | 12.47 | 8.21 | 8.58 | 9.40 | 7.24 | 4.30 | 4.02 |
|   | $\alpha$ | 0.96 | 1.04 | 0.24 | 0.07 | 0.07 | 0.61 | 1.26 | 1.10 | 0.84 | 1.29 | 1.43 | 1.70 |
|   | $K_1$ | 0.35 | 0.30 | 0.43 | 0.49 | 0.48 | 0.46 | 0.43 | 0.43 | 0.41 | 0.36 | 0.31 | 0.38 |
|   | $K_2$ | -25.89 | -6.48 | -36.46 | -44.07 | -42.58 | -62.66 | -56.75 | -61.08 | -27.09 | 3.90 | 20.10 | -11.78 |

**(b): Bangalore**

| Type of day | Parameters ▼ | January | February | March | April | May | June | July | August | September | October | November | December |
|---|---|---|---|---|---|---|---|---|---|---|---|---|---|
| a | $T_R$ | 3.36 | 3.27 | 3.63 | 5.05 | 4.24 | 4.32 | 5.18 | 4.75 | 4.10 | 2.28 | 1.66 | 1.65 |
| | $\alpha$ | 0.07 | 0.13 | 0.06 | −0.06 | 0.10 | 0.19 | 0.10 | 0.18 | 0.13 | 0.33 | 0.35 | 0.36 |
| | $K_1$ | 0.33 | 0.35 | 0.33 | 0.29 | 0.21 | 0.25 | 0.32 | 0.23 | 0.20 | 0.05 | 0.03 | 0.12 |
| | $K_2$ | −18.05 | −22.11 | −5.44 | 14.54 | 47.81 | 22.40 | −26.04 | 10.14 | 38.54 | 107.04 | 103.64 | 47.70 |
| b | $T_R$ | 3.24 | 5.25 | 6.21 | 5.72 | 5.90 | 7.35 | 4.12 | 5.27 | 4.83 | 2.43 | 1.89 | 3.68 |
| | $\alpha$ | 0.31 | 0.24 | 0.21 | 0.19 | 0.25 | 0.17 | 0.51 | 0.44 | 0.62 | 0.56 | 0.78 | 0.39 |
| | $K_1$ | 0.50 | 0.45 | 0.48 | 0.50 | 0.41 | 0.50 | 0.46 | 0.50 | 0.33 | 0.26 | 0.37 | 0.41 |
| | $K_2$ | −60.12 | −60.50 | −80.04 | −75.59 | −28.55 | −103.35 | −90.54 | −115.27 | 13.80 | 69.14 | 9.08 | −33.76 |
| c | $T_R$ | 3.70 | 4.51 | 7.74 | 5.83 | 4.95 | 4.39 | 5.68 | 2.67 | 6.64 | 4.71 | 5.68 | 2.02 |
| | $\alpha$ | 0.96 | 0.94 | 0.63 | 0.98 | 0.96 | 1.12 | 1.07 | 1.35 | 0.78 | 1.03 | 0.93 | 1.44 |
| | $K_1$ | 0.46 | 0.57 | 0.36 | 0.50 | 0.53 | 0.58 | 0.50 | 0.55 | 0.48 | 0.43 | 0.36 | 0.43 |
| | $K_2$ | −63.02 | −129.68 | −20.76 | −61.13 | −103.14 | −156.14 | −108.34 | −161.61 | −52.93 | −26.53 | −15.95 | −47.21 |
| d | $T_R$ | 6.13 | 7.49 | 7.35 | 6.86 | 6.33 | 4.84 | 4.45 | 6.68 | 3.94 | 3.91 | 3.84 | 2.80 |
| | $\alpha$ | 1.61 | 1.31 | 1.41 | 1.48 | 1.59 | 2.00 | 2.32 | 1.69 | 2.16 | 2.00 | 2.04 | 2.58 |
| | $K_1$ | 0.29 | 0.30 | 0.40 | 0.45 | 0.53 | 0.61 | 0.41 | 0.50 | 0.38 | 0.42 | 0.55 | 0.27 |
| | $K_2$ | 36.80 | 83.73 | −39.85 | −72.22 | −99.52 | −213.29 | −79.79 | −146.94 | −88.62 | −125.35 | −177.28 | −12.29 |

**(c): Jodhpur**

| Type of day | Parameters | January | February | March | April | May | June | July | August | September | October | November | December |
|---|---|---|---|---|---|---|---|---|---|---|---|---|---|
| a | $T_R$ | 1.26 | 1.33 | 1.59 | 2.82 | 3.72 | 3.87 | 3.25 | 3.39 | 3.20 | 2.26 | 1.56 | 1.54 |
|   | $\alpha$ | 0.37 | 0.38 | 0.37 | 0.27 | 0.21 | 0.21 | 0.27 | 0.28 | 0.27 | 0.33 | 0.39 | 0.31 |
|   | $K_1$ | 0.22 | 0.14 | 0.18 | 0.21 | 0.20 | 0.13 | 0.10 | 0.17 | 0.26 | 0.24 | 0.23 | 0.26 |
|   | $K_2$ | 30.67 | 63.90 | 56.40 | 47.66 | 50.84 | 87.88 | 105.23 | 59.41 | 14.42 | 27.40 | 22.71 | 9.48 |
| b | $T_R$ | 2.34 | 2.03 | 3.00 | 4.07 | 5.21 | 5.50 | 5.07 | 4.73 | 3.81 | 2.90 | 2.28 | 3.43 |
|   | $\alpha$ | 0.46 | 0.55 | 0.42 | 0.31 | 0.23 | 0.28 | 0.37 | 0.40 | 0.35 | 0.38 | 0.46 | 0.24 |
|   | $K_1$ | 0.33 | 0.29 | 0.31 | 0.34 | 0.33 | 0.33 | 0.34 | 0.33 | 0.34 | 0.30 | 0.33 | 0.40 |
|   | $K_2$ | 12.89 | 43.13 | 42.22 | 23.50 | 31.22 | 33.40 | 35.81 | 29.57 | 8.71 | 24.12 | 12.35 | −11.64 |
| c | $T_R$ | 3.81 | 4.78 | 4.04 | 4.97 | 6.87 | 5.58 | 4.90 | 5.10 | 3.40 | 3.71 | 3.28 | 4.23 |
|   | $\alpha$ | 0.93 | 1.32 | 0.98 | 0.64 | 0.61 | 0.67 | 1.02 | 0.88 | 0.97 | 2.05 | 1.31 | 1.06 |
|   | $K_1$ | 0.43 | 0.40 | 0.42 | 0.47 | 0.47 | 0.46 | 0.41 | 0.50 | 0.48 | 0.53 | 0.44 | 0.44 |
|   | $K_2$ | −33.72 | 12.44 | −19.11 | −26.93 | −44.76 | −35.15 | 2.06 | −60.42 | −26.96 | −62.06 | −35.85 | −32.84 |
| d | $T_R$ | 2.25 | 5.20 | 7.09 | 9.33 | 8.01 | 3.52 | 9.62 | 3.17 | 1.63 | 7.67 | 1.71 | 1.94 |
|   | $\alpha$ | 1.89 | 1.64 | 2.03 | 1.59 | 1.66 | 2.37 | 2.37 | 2.77 | 3.24 | 0.86 | 2.89 | 2.03 |
|   | $K_1$ | 0.44 | 0.46 | 0.42 | 0.44 | 0.43 | 0.28 | 0.52 | 0.44 | 0.44 | 0.52 | 0.36 | 0.39 |
|   | $K_2$ | −19.31 | −45.44 | −89.92 | −149.27 | −117.01 | 60.69 | −221.29 | −87.34 | −77.55 | −26.47 | −15.46 | −14.88 |

**(d): Mumbai**

| Type of day | Parameters | January | February | March | April | May | June | July | August | September | October | November | December |
|---|---|---|---|---|---|---|---|---|---|---|---|---|---|
| a | $T_R$ | 1.95 | 1.80 | 2.88 | 3.95 | 5.40 | 3.20 | 3.31 | 4.25 | 4.22 | 3.16 | 2.97 | 3.27 |
|  | $\alpha$ | 0.34 | 0.37 | 0.23 | 0.14 | −0.02 | 0.16 | 0.61 | 0.33 | 0.15 | 0.30 | 0.23 | 0.18 |
|  | $K_1$ | 0.26 | 0.19 | 0.28 | 0.34 | 0.28 | 0.25 | 0.09 | 0.12 | 0.24 | 0.24 | 0.26 | 0.30 |
|  | $K_2$ | 19.77 | 53.96 | 27.13 | −0.75 | 30.06 | 4.55 | 27.28 | 47.27 | 30.02 | 15.87 | 9.11 | −4.81 |
| b | $T_R$ | 2.96 | 2.68 | 3.57 | 4.98 | 6.25 | 6.08 | 7.74 | 6.70 | 4.78 | 3.93 | 3.40 | 4.21 |
|  | $\alpha$ | 0.43 | 0.49 | 0.37 | 0.25 | 0.15 | 0.19 | 0.20 | 0.37 | 0.47 | 0.47 | 0.45 | 0.24 |
|  | $K_1$ | 0.35 | 0.31 | 0.35 | 0.40 | 0.42 | 0.44 | 0.31 | 0.39 | 0.41 | 0.36 | 0.34 | 0.37 |
|  | $K_2$ | −0.14 | 24.17 | 11.73 | −13.57 | −13.69 | −19.52 | 61.35 | 22.16 | −14.71 | 5.99 | 0.60 | −14.17 |
| c | $T_R$ | 3.06 | 2.26 | 3.24 | 4.39 | 5.91 | 5.97 | 8.17 | 4.24 | 5.36 | 3.16 | 2.97 | 3.75 |
|  | $\alpha$ | 1.14 | 1.18 | 1.10 | 1.00 | 0.79 | 0.86 | 0.62 | 1.26 | 0.98 | 1.13 | 1.10 | 0.91 |
|  | $K_1$ | 0.59 | 0.58 | 0.52 | 0.54 | 0.60 | 0.52 | 0.54 | 0.43 | 0.44 | 0.47 | 0.57 | 0.54 |
|  | $K_2$ | −59.86 | −47.12 | −58.09 | −78.37 | −111.97 | −81.79 | −95.21 | −34.40 | −39.31 | −28.02 | −48.41 | −52.45 |
| d | $T_R$ | 3.38 | 7.42 | 4.45 | 2.30 | 4.71 | 4.71 | 6.41 | 7.40 | 7.46 | 3.22 | 5.13 | 3.05 |
|  | $\alpha$ | 1.71 | 1.73 | 2.29 | 2.08 | 2.95 | 2.66 | 2.68 | 1.81 | 2.14 | 2.15 | 1.53 | 1.51 |
|  | $K_1$ | 0.52 | 0.56 | 0.50 | 0.35 | 0.41 | 0.38 | 0.32 | 0.47 | 0.34 | 0.42 | 0.57 | 0.53 |
|  | $K_2$ | −59.78 | −26.16 | −82.34 | 63.52 | −101.81 | −87.19 | −61.50 | −108.37 | −38.68 | −25.89 | −78.03 | −40.51 |

**(e): Srinagar**

| Type of day | Month▶ Parameters▼ | January | February | March | April | May | June | July | August | September | October | November | December |
|---|---|---|---|---|---|---|---|---|---|---|---|---|---|
| a | $T_R$ | 1.45 | 5.37 | 3.31 | 4.25 | 5.41 | 3.63 | 5.77 | 6.45 | 4.06 | 2.61 | 4.03 | 0.72 |
|  | $\alpha$ | 0.33 | −0.36 | −0.03 | −0.03 | −0.12 | 0.08 | −0.09 | −0.23 | 0.03 | 0.20 | −0.37 | 0.53 |
|  | $K_1$ | 0.37 | 0.63 | 0.69 | 0.37 | 0.51 | 0.33 | 0.17 | 0.37 | 0.46 | 0.43 | 0.66 | 0.33 |
|  | $K_2$ | −6.14 | −82.86 | −94.01 | −10.95 | −79.57 | −13.73 | 68.06 | −42.79 | −60.27 | −47.83 | −37.00 | −6.60 |
| b | $T_R$ | 3.09 | 6.98 | 4.65 | 6.92 | 5.86 | 6.82 | 7.40 | 7.58 | 6.41 | 4.04 | 0.04 | 0.35 |
|  | $\alpha$ | 0.38 | −0.48 | 0.23 | 0.06 | 0.29 | 0.11 | 0.00 | −0.13 | −0.04 | 0.19 | 1.16 | 1.00 |
|  | $K_1$ | 0.39 | 0.83 | 0.59 | 0.42 | 0.32 | 0.63 | 0.48 | 0.38 | 0.48 | 0.52 | 0.37 | 0.41 |
|  | $K_2$ | −23.08 | −110.23 | −107.74 | −49.61 | 0.26 | −167.86 | −80.06 | −13.91 | −66.64 | −62.52 | −14.63 | −12.20 |
| c | $T_R$ | 2.35 | 6.59 | 6.31 | 7.57 | 8.69 | 8.00 | 9.72 | 8.23 | 7.36 | 5.02 | 1.86 | 0.76 |
|  | $\alpha$ | 1.64 | 0.86 | 1.35 | 0.57 | 0.61 | 0.81 | 0.69 | 0.90 | 0.99 | 1.49 | 1.47 | 1.98 |
|  | $K_1$ | 0.41 | 0.42 | 0.48 | 0.54 | 0.50 | 0.39 | 0.56 | 0.49 | 0.44 | 0.52 | 0.41 | 0.31 |
|  | $K_2$ | −37.87 | −85.68 | −180.45 | −120.38 | −146.97 | −87.44 | −228.91 | −147.96 | −62.10 | −93.64 | −40.07 | −12.15 |
| d | $T_R$ | 1.69 | 1.36 | 7.52 | 9.09 | 9.48 | 10.79 | 10.93 | 8.54 | 8.16 | 7.75 | 3.78 | 2.44 |
|  | $\alpha$ | 2.63 | 2.97 | 1.87 | 1.35 | 1.13 | 1.56 | 3.08 | 1.71 | 3.15 | 1.70 | 1.74 | 2.04 |
|  | $K_1$ | 0.43 | 0.36 | 0.35 | 0.62 | 0.92 | 0.80 | 0.45 | 0.75 | 0.67 | 0.55 | 0.48 | 0.63 |
|  | $K_2$ | −41.27 | −44.68 | −65.17 | −254.24 | −467.30 | −421.63 | −129.49 | −356.92 | −261.85 | −119.53 | −49.16 | −64.02 |

# Appendix
## C2

The turbidity factor $(T_R)$ for different months [1].

| Month→Region↓ | 1 | 2 | 3 | 4 | 5 | 6 | 7 | 8 | 9 | 10 | 11 | 12 |
|---|---|---|---|---|---|---|---|---|---|---|---|---|
| Mountain | 1.8 | 1.9 | 2.1 | 2.2 | 2.4 | 2.7 | 2.7 | 2.7 | 2.5 | 2.1 | 1.9 | 1.8 |
| Flat land | 2.2 | 2.2 | 2.5 | 2.9 | 3.2 | 3.4 | 3.5 | 3.3 | 2.9 | 2.6 | 2.3 | 2.2 |
| City | 3.1 | 3.2 | 3.5 | 3.9 | 4.1 | 4.2 | 4.3 | 4.2 | 3.9 | 3.6 | 3.3 | 3.1 |

For cloudy condition the value of $T_R$ will be more than 10.0

© Bag Energy Research Society 2024
G. N. Tiwari, *Fundamental of Mathematical Tools for Thermal Modeling of Solar Thermal and Photo-voltaic Systems-Part-I*, https://doi.org/10.1007/978-981-99-7085-8

# Appendix
# C3

G. N. Tiwari, *Fundamental of Mathematical Tools for Thermal Modeling of Solar Thermal and Photo-voltaic Systems-Part-I*, https://doi.org/10.1007/978-981-99-7085-8

| Model Type | Model Name | Model correlations | Remarks-I | Remarks-II |
|---|---|---|---|---|
| Solar radiation model for calculation of hourly radiation | Hottel model (1958) [2] | $I_N = I_{ON}[a_0 + a_1 \exp(-k/\cos\theta_z)]$ <br> $a_0 = 0.4237 - 0.00821(6 - A)^2$ <br> $a_1 = 0.5055 - 0.00595(6.5 - A)^2$ <br> $k = 0.2711 - 0.01858(2.5 - A)^2$ | The constants $a_0$, $a_1$ and $k$ are functions of the altitude of the location, and $A$ is the altitude in kilometers | Relation between Terrestrial, $I_N$ and extra-terrestrial, $I_{ON}$ radiation |
| | Kasten and Young model (1989) [3] | $I_N = I_{ON}\exp(-m.\varepsilon.T_R)$ <br> $m =$ <br> $\left[\cos\theta_z + 0.15 \times (93.885 - \theta_z)^{-1.253}\right]^{-1}$ <br> $\varepsilon = 4.529 \times 10^{-4}\mathrm{m}^2 - 9.66865 \times$ <br> $10^{-3}\mathrm{m} + 0.108014$ <br> $\theta_z = \cos^{-1}[\cos\varphi\cos\delta\cos\omega + \sin\delta\sin\varphi]$ | $m$ is air mass, $\varepsilon$ is optical thickness of atmosphere and $T_R$ Linke turbidity factor, $\varphi$ (Table 1.2), $\delta$ (Table 1.6, Eq. 1.9) and $\omega$ (Table 1.4) | Relation between Terrestrial, $I_N$ and extra-terrestrial, $I_{ON}$ radiation |
| | Perez et al. model (1990) [4] | $I_N = I_{ON}\exp\left[-T_R/(0.9 + 9.4\cos\theta_z)\right]$ <br> $\theta_z = \cos^{-1}[\cos\varphi\cos\delta\cos\omega + \sin\delta\sin\varphi]$ | $I_{ON}$ is solar radiation in extra-terrestrial region and $T_R$ is Linke turbidity factor,, $\varphi$ (Table 1.2), $\delta$ (Table 1.6, Eq. 1.9) and $\omega$ (Table 1.4) | Relation between Terrestrial, $I_N$ and extra-terrestrial, $I_{ON}$ radiation |
| | ASHRAE model (1999) [5] | $I = I_N\cos\theta_z + I_d$ (on horizontal surface in terrestrial region) <br> $I_N = A\exp(-B/\cos\theta_z)$ <br> $I_d = CI_N$ | $A$, $B$ and $C$ are constants depends on locations. These constants were also found by G. V. Parishwad et al. [29] (Appendix C3) | Valid in terrestrial region between horizontal, $I$ and normal irradiance, $I_N$ |

(continued)

(continued)

| Model Type | Model Name | Model correlations | Remarks-I | Remarks-II |
|---|---|---|---|---|
| | Singh and Tiwari model (2005) [6] | $I_N = I_{ON}\exp\{-(m.\varepsilon.T_R + \alpha)\}$ <br> $I_{HD} = K_1(I_{ON} - I_N)\cos\theta_z + K_2$ | $\alpha$ is atmospheric transmittance for beam radiation $I_{HD}$ is diffuse radiation on horizontal surface $K_1$ and $K_2$ are atmospheric transmittances for diffuse radiation. Applicable to different weather conditions, Appendix C1 | Relation between Terrestrial, $I_N$ and extra-terrestrial, $I_{ON}$ radiation |
| | Jamil and Tiwari model (2008) [7] | $I_N =$ $I_{ON}\exp\{-((m.\varepsilon)^2 T_{RO} + (m.\varepsilon).T_R + \alpha)\}$ <br> $I_{HD} = K_0\{(I_{ON} - I_N)\cos\theta_z\}^2 +$ $K_1(I_{ON} - I_N)\cos\theta_z + K_2$ | Applicable to different weather conditions. Validated for composite climatic condition of New Delhi, India | Relation between Terrestrial, $I_N$ and extra-terrestrial, $I_{ON}$ radiation |
| Radiation on horizontal surface (decomposition model) | Liu and Jordan model (1960) [8] | $k_D = \frac{I_d}{I_0} = 0.384 - 0.416k_t$ <br> $k_t = \frac{I}{I_0}, k_d = \frac{I_d}{I}, k_D = \frac{I_d}{I_0}, k_b = \frac{I_b}{I_0}$ | $k_t$ is clearness index, $k_d$ is diffuse fraction, $k_D$ is diffuse coefficient and $k_b$ is direct transmittance $I, I_b, I_d$ and $I_0$ being the global, direct, diffuse and extra-terrestrial irradiances, respectively, on a horizontal surface (all in $MJm^{-2}$) | Relation between Terrestrial,$(I, I_b, I_d$ and $I_0)$ and extra-terrestrial,$I_0$ radiation |

(continued)

(continued)

| Model Type | Model Name | Model correlations | Remarks-I | Remarks-II |
|---|---|---|---|---|
| | Orgill and Hollands model (1977) [9] | $k_d = \frac{I_d}{I} = 1.557 - 1.846 k_t$ for $0.35 \leq k_t \leq 0.75$, $k_d = 1.0 - 0.249 k_t$ for $k_t < 0.35$, $k_d = 0.177$ for $k_t > 0.75$. | The model was based on the global and diffuse irradiance values registered in Toronto (Canada, 42.81N) during the years 1967–1971 | Relation between diffuse, $I_d$ and total radiation, I in terrestrial region for as given range of $k_t = \frac{I}{I_0}$ between terrestrial and extra-terrestrial |
| | Erbs et al. model (1982) [10] | $k_d = \frac{I_d}{I} = 0.951 - 0.1604 k_t + 4.388 k_t^2 - 16.638 k_t^3 + 12.336 k_t^4$ for $0.22 \leq k_t \leq 0.80$ $k_d = 1.0 - 0.09 k_t$ for $k_t \leq 0.22$ $k_d = 0.165$ for $k_t > 0.80$ | Correlations developed using data from five stations in the USA with latitudes between 31 and 42° | Relation between diffuse, $I_d$ and total radiation, I in terrestrial region for as given range of $k_t = \frac{I}{I_0}$ between terrestrial and extra-terrestrial |
| | Spencer model (1982) [11] | $k_d = \frac{I_d}{I} = a_3 - b_3 k_t$ for $0.35 \leq k_t \leq 0.75$ $a_3 = 0.94 + 0.0118|\phi|$ $b_3 = 1.185 + 0.0135|\phi|$ | Correlations developed from five stations in Australia (20–45 °S latitude) $\phi$ (degrees) is the latitude | Relation between diffuse, $I_d$ and total radiation, I in terrestrial region for as given range of $k_t = \frac{I}{I_0}$ between terrestrial and extra-terrestrial |
| | Muneer et al. model (1984) [12] | $k_d = \frac{I_d}{I} = 0.9698 + 0.4353 k_t - 3.4499 k_t^2 + 2.1888 k_t^3$ for $0.175 \leq k_t \leq 0.775$ $k_d = 0.95$ for $k_t < 0.175$ $k_d = 0.26$ for $k_t > 0.775$ | Correlation developed using data from New Delhi, India | Relation between diffuse, $I_d$ and total radiation, I in terrestrial region for as given range of $k_t = \frac{I}{I_0}$ between terrestrial and extra-terrestrial |

(continued)

(continued)

| Model Type | Model Name | Model correlations | Remarks-I | Remarks-II |
|---|---|---|---|---|
| | Hawlader model (1984) [13] | $k_d = \frac{I_d}{I} = 1.135 - 0.9422k_t - 0.3878k_t^2$ for $0.225 < k_t < 0.775$<br>$k_d = 0.915$ for $k_t \leq 0.225$<br>$k_d = 0.215$ for $k_t \geq 0.775$ | Correlation developed using data from a tropical site in Singapore | Relation between diffuse, $I_d$ and total radiation, I in terrestrial region for as given range of $k_t = \frac{I}{I_0}$ between terrestrial and extra-terrestrial |
| | Reindl et al. model (1990) [14] | First correlation:<br>$k_d = \frac{I_d}{I} = 1.02 - 0.248 k_t$ for $k_t \leq 0.30$<br>$k_d = 1.45 - 1.67k_t$ for $0.3 < k_t < 0.78$<br>$k_d = 0.147$ for $k_t \geq 0.78$<br>Second correlation:<br>$k_d = 1.02 - 0.254k_t + 0.0123\sin\alpha$ for $k_t \leq 0.30$<br>$k_d = 1.4 - 1.749k_t + 0.177\sin\alpha$ for $0.3 < k_t < 0.78k_d = 0.486k_t - 0.182\sin\alpha$ for $k_t \geq 0.78$ | Correlations developed from five locations in the USA and Europe (28–60 °N latitude) | Relation between diffuse, $I_d$ and total radiation, I in terrestrial region for as given range of $k_t = \frac{I}{I_0}$ between terrestrial and extra-terrestrial |
| | Chandrasekaran and Kumar model (1994) [15] | $k_d = \frac{I_d}{I} = 0.9686 + 0.1325k_t + 1.4183k_t^2 - 10.1862k_t^3 + 8.3733k_t^4$ for $0.24 < k_t \leq 0.80$<br>$k_d = 1.0086 - 0.178k_t$ for $k_t \leq 0.24$<br>$k_d = 0.197$ for $k_t > 0.80$ | Correlation developed using data from a tropical environment in Chennai, India | Relation between diffuse, $I_d$ and total radiation, I in terrestrial region for as given range of $k_t = \frac{I}{I_0}$ between terrestrial and extra-terrestrial |
| | Lam and Li model (1996) [16] | $k_d = 0.977$ for $k_t \leq 0.15$<br>$k_d = 1.237 - 1.361k_t$ for $0.15 < k_t \leq 0.7$<br>$k_d = 0.273$ for $k_t > 0.7$ | Correlations developed for Hong Kong (22.31N latitude) with the measured data in 1991–1994 | Relation between diffuse, $I_d$ and total radiation, I in terrestrial region for as given range of $k_t = \frac{I}{I_0}$ between terrestrial and extra-terrestrial |

(continued)

(continued)

| Model Type | Model Name | Model correlations | Remarks-I | Remarks-II |
|---|---|---|---|---|
| | Boland et al. model (2001) [17] | $k_d = \frac{1}{1+e^{7.997(k_t-0.586)}}$ for all values of $k_t$ | Correlation developed using data from one location in Victoria, Australia | Relation between diffuse, $I_d$ and total radiation, I in terrestrial region for as given range of $k_t = \frac{I}{I_0}$ between terrestrial and extra-terrestrial |
| | Miguel et al. model (2001) [18] | $k_d = 0.724 + 2.738k_t - 8.32k_t^2 + 4.967k_t^3 + 12.336k_t^4$ for $0.21 < k_t \le 0.76$<br>$k_d = 0.995 - 0.081k_t$ for $k_t \le 0.21$<br>$k_d = 0.18$ for $k_t \ge 0.76$ | Correlation developed using data from several countries in the North Mediterranean Belt area | do |
| | Oliveira et al. model (2002) [19] | $k_d = 0.97 + 0.8k_t - 3.0k_t^2 - 3.1k_t^3 + 5.2k_t^4$ for $0.17 < k_t < 0.75$<br>$k_d = 1.0$ for $k_t \le 0.17$<br>$k_d = 0.17$ for $k_t > 0.75$ | Correlation developed using data from Sao Paolo site, Brazil | do |
| | Karatasou et al. model (2003) [20] | $k_d = 0.9995 - 0.05k_t - 2.4156k_t^2 + 1.4926k_t^3$ for $0 < k_t \le 0.78$<br>$k_d = 0.20$ for $k_t > 0.78$ | Correlation developed using data from Athens, Greece | do |
| | Soares et al. model (2004) [21] | $k_d = 0.90 + 1.1k_t - 4.5k_t^2 + 0.01k_t^3 + 3.14k_t^4$ for $0.17 < k_t < 0.75$<br>$k_d = 1.0$ for $k_t \le 0.17$<br>$k_d = 0.17$ for $k_t > 0.75$ | Correlation developed with neural network technique using data from Sao Paolo site, Brazil | do |

(continued)

(continued)

| Model Type | Model Name | Model correlations | Remarks-I | Remarks-II |
|---|---|---|---|---|
| Models for predicting the mean hourly global radiation from daily summations | Whillier/Liu and Jordan model (1956) [22] | $r_0 = \frac{I(t)}{H_0} = (\cos\omega - \cos\omega_0)/kA(\omega_0)$ <br> $A(\omega_0) = \sin\omega_0 - \omega_0\cos\omega_0$ <br> $\omega_0$ is the sunrise hour angle (in radians) <br> $\omega$ is the hour angle | It is assumed that global radiation follow the same hourly distribution as if there were no atmosphere $r_0$ is the extra-terrestrial hourly/daily ratio | Only valid for extra-terrestrial region for horizontal surface: $H_0$ is daily solar radiation in extra-terrestrial region (MJ/m$^2$) |
| | Collares-Pereira and Rabl model (1979) [23] | $r_{CPR} = \frac{I(t)}{H_0} = (a + b\cos\omega)r_0$ <br> $r_0 = (\cos\omega - \cos\omega_0)/kA(\omega_0), k = 24/\pi$ <br> $A(\omega_0) = \sin\omega_0 - \omega_0\cos\omega_0$ <br> $\omega_0 = \frac{2\pi\omega_0}{360}$ is the sunrise hour angle (in radians) <br> $a = 0.4090 + 0.5016\sin(\omega_0 - 1.047)$ <br> $b = 0.6609 + 0.4767\sin(\omega_0 - 1.047)$ | $r_{CPR}$ is the extra-terrestrial hourly/daily ratio | Only valid for extra-terrestrial region for horizontal surface: $H_0$ is daily solar radiation in extra-terrestrial region (MJ/m$^2$) |
| | Newell model (1983) [24] | $r_N = \frac{I(t)}{H_0} = (1.5/S_0)\left[1 - 4(t - 12)^2/S_0^2\right]$ <br> $S_0 = k\omega_0, k = 24/\pi$ <br> $\omega_0$ is the sunrise hour angle (in radians) <br> $\cos\omega_0 = -\tan\phi\tan\delta$ <br> Here t varies from 6 to 18 h | $\phi$ is the site's latitude, $\delta$ is solar declination $r_N$ is the extra-terrestrial hourly/daily ratio | Only valid for extra-terrestrial region for horizontal surface: $H_0$ is daily solar radiation in extra-terrestrial region (MJ/m$^2$) |

(continued)

(continued)

| Model Type | Model Name | Model correlations | Remarks-I | Remarks-II |
|---|---|---|---|---|
| | Jain model (1984) [25] | $r_J = \frac{I(t)}{H_0} = \frac{1}{\sigma_J\sqrt{2\pi}}\exp\left[-\frac{(t-12)^2}{2\sigma_J^2}\right]$<br><br>$\sigma_J = 0.461 + 0.192 S_0,$<br>$S_0 = k\omega_0, k = 24/\pi$<br>$\omega_0 = \frac{2\pi\omega_s}{360}$ is the sunrise hour angle (in radians)<br>$\cos\omega_0 = -\tan\phi\tan\delta$<br>Here t varies from 6 to 18 h | $\phi$ is the site's latitude, $\delta$ is solar declination $r_J$ is the extra-terrestrial hourly/daily ratio | Only valid for extra-terrestrial region for horizontal surface: $H_0$ is daily solar radiation in extra-terrestrial region (MJ/m²) |
| | Gueymard model (1986) [26] | $r_{CPRG} = \frac{I(t)}{H_0} = (a + b\cos\omega)r_0/f$<br>$r_0 = (\cos\omega - \cos\omega_0)/kA(\omega_0)$<br>$f = a + 0.5b(\omega_0 - \sin\omega_0\cos\omega_0)/A(\omega_0)$<br>$A(\omega_0) = \sin\omega_0 - \omega_0\cos\omega_0$<br>$a = 0.4090 + 0.5016\sin(\omega_0 - 1.047)$<br>$b = 0.6609 + 0.4767\sin(\omega_0 - 1.047)$ | Modified Collares-Pereira and Rabl model $r_{CPRG}$ is the extra-terrestrial hourly/daily ratio | Only valid for extra-terrestrial region for horizontal surface: $H_0$ is daily solar radiation in extra-terrestrial region (MJ/m²) |
| | Garg and Garg Model (1987) [27] | $r_G = \frac{I(t)}{H_0} = r_0 - 0.008\sin3(\omega - 0.65)$<br>$r_0 = (\cos\omega - \cos\omega_0)/kA(\omega_0)$<br>$A(\omega_0) = \sin\omega_0 - \omega_0\cos\omega_0$<br>$k = 24/\pi$ | Corrected Whillier/Liu and Jordan model for Indian climatic condition $r_G$ is the extra-terrestrial hourly/daily ratio | do |

(continued)

(continued)

| Model Type | Model Name | Model correlations | Remarks-I | Remarks-II |
|---|---|---|---|---|
| | Baig et al. model (1991) [28] | $r_B = \frac{I(t)}{H_0} = \frac{1}{2\sigma_B\sqrt{2\pi}}\exp\left[-\frac{(t-12)^2}{2\sigma_B^2}\right]$ $\sigma_B = 0.26 + 0.21S_0,\ S_0 = k\omega_0, k = 24/\pi$ $\omega_0 = \frac{2\pi\omega_s}{360}$ is the sunrise hour angle (in radians) $\cos\omega_0 = -\tan\phi\tan\delta$ Here t varies from 6 to 18 h | $\phi$ is the site's latitude, $\delta$ is solar declination It is corrected Jain model for better accuracy for values of solar radiation during sunrise and sunset $r_B$ is the extra-terrestrial hourly/daily ratio | do |

# Appendix

# C4

Values of constants A, B and C obtained for predicting hourly solar radiation in India

| S. No. | Day | A (W/m$^2$) | B (W/m$^2$) | C (W/m$^2$) |
|---|---|---|---|---|
| 1 | January, 21 | 708.00 | 0.000 | 0.192 |
| 2 | February, 21 | 732.20 | 0.010 | 0.209 |
| 3 | March, 21 | 767.86 | 0.046 | 0.229 |
| 4 | April, 21 | 713.35 | 0.131 | 0.385 |
| 5 | May, 21 | 798.39 | 0.150 | 0.250 |
| 6 | June, 21 | 440.71 | 0.398 | 1.108 |
| 7 | July, 21 | 222.87 | 0.171 | 1.721 |
| 8 | August, 21 | 240.80 | 0.148 | 1.624 |
| 9 | September, 21 | 396.21 | 0.074 | 0.748 |
| 10 | October, 21 | 644.73 | 0.020 | 0.256 |
| 11 | November, 21 | 666.60 | 0.008 | 0.213 |
| 12 | December, 21 | 692.52 | 0.000 | 0.193 |

Further, the weather classifications for a given climatic condition are defined according to sunshine hours ($N$) and ratios of daily diffuse to daily global radiation. These are briefly described as follows:

(a) **Clear day (blue sky)**: For clear days the ratio of daily diffuse radiation in J/m$^2$ to daily total (global) radiation in J/m$^2$ less than or equal to 0.25 and sunshine hours (N) greater than or equal to 9 h.
(b) **Hazy day (fully)**: For hazy days the ratio of daily diffuse radiation in J/m$^2$ to daily total (global) radiation in J/m$^2$ between 0.25 and 0.5 and sunshine hours (N) between 7 and 9 h.
(c) **Hazy and cloudy (partially)**: For hazy and cloudy days the ratio of daily diffuse radiation in J/m$^2$ to daily total (global) radiation in J/m$^2$ between 0.5 and 0.75 and sunshine hours (N) between 5 and 7 h.

© Bag Energy Research Society 2024
G. N. Tiwari, *Fundamental of Mathematical Tools for Thermal Modeling of Solar Thermal and Photo-voltaic Systems-Part-I*, https://doi.org/10.1007/978-981-99-7085-8

(d)  **Cloudy day (fully)**: For cloudy days the ratio of daily diffuse radiation in $J/m^2$ to daily total (global) in $J/m^2$ radiation greater than or equal to 0.75 and sunshine hours less than or equal to 5 h.

# References

1.  Tiwari, G. N., & Mishra, R. K. (2012). *Advance renewable energy sources*. RSC publishing.
2.  Hottel H. C., & Whiller A. (1958). *Transactions of the conference on use of solar energy, The Scientific Basis, vol. II(1), Section A* (Vol. 74). University of Arizona Press.
3.  Kasten, F., & Young, A. T. (1989). *Applied optics, 28*, 4735.
4.  Perez, R., Ineichen, P., Maxwell, E., Seals, R., & Zelenka, A. (1990). *ASHRAE Transactions, 98*, 354.
5.  American Society of Heating. (1999). *Refrigeration and air-conditioning engineers*. ASHRAE Applications Handbook (SI).
6.  Singh, H. N., & Tiwari, G. N. (2005). *Energy, 30*, 1589.
7.  Ahmad, M. J., & Tiwari, G. N. (2008). *CIGR Ejournal, 10*, 1.
8.  Liu, B. Y. H., & Jordan, R. C. (1960). *Solar Energy, 4*, 1–19.
9.  Orgill, J. F., &Hollands, K. (1977). *G. T. Solar Energy, 19*, 357.
10. Erbs, D. G., Klein, S. A., & Duffie, J. A. (1982). *Solar Energy, 28*, 293.
11. Spencer, J. W. (1982). *Solar Energy, 29*(1), 19.
12. Muneer, T., Hawas, M. M., & Sahili, K. (1984). *Energy Conversion and Management, 24*(4), 265.
13. Hawlader, M. N. A. (1984). *International Journal of Ambient Energy, 5*, 31.
14. Reindl, D. T., Beckman, W. A., & Duffie, J. A. (1990). *Solar Energy, 45*, 1.
15. Chandrasekaran, J., & Kumar, S. (1994). *Solar Energy, 53*, 505.
16. Lam, J. C., & Li, D. H. W. (1996). *Building and Environment, 31*(6), 527.
17. Boland, J., Scott, L., & Luther, M. (2001). *Environmetrics, 12*, 103.
18. Miguel, A., Bilbao, J., Aguiar, R., Kambezidis, H., & Negro, E. (2001). *Solar Energy, 70*, 143.
19. Oliveira, A. P., Escobedo, J. F., Machado, A. J., & Soares, J. (2002). *Applied Energy, 71*, 59.
20. Karatasou, S., Santamouris, M., & Geros, V. (2003). *International Journal of Sustainable Energy, 23*, 1.
21. Soares, J., Oliveira, A. P., Boznar, M. Z., Mlakar, P., Escobedo, J. F., & Machado, A. (2004). *J. Applied Energy, 79*, 201.
22. Whillier, A. (1956). *Archives for Meteorology Geophysics and Bioclimatology, B8*, 197.
23. Collares-Pereira, M., & Rabl, A. (1979). *Solar Energy, 22*, 155.
24. Newell, T. A. (1983). *Solar Energy, 31*, 339.
25. Jain, P. C. (1984). *Solar Wind Technology, 1*, 123.
26. Gueymard, C. (1986). Journal of Solar Energy Engineering. *Transactions on ASME, 108*, 320.
27. Garg, H. P., & Garg, S. N. (1987). *Solar and Wind Technology, 4*, 113.
28. Baig, A., Akhter, P., & Mufti, A. (1991). *Renewable Energy, 1*, 119.
29. Parishwad, G. V., Bhardwaj, R. K., & Nema, V. K. (1997). *Renewable Energy, 12*(3), 303.

# Appendix D

**Specifications of solar cell material (at solar intensity 1000 W/m² and cell temperature 25 °C) and cost (from Tiwari and Mishra, 2012)**

| Cell technology | Efficiency (%) | Fill Factor (FF) | Aperture area ($10^{-4} \times m^2$) | Life time* (years) | Manufacturing cost ($/kWp in 2007) | Selling price ($/kWp in 2007) |
|---|---|---|---|---|---|---|
| Monocrystalline silicon | $24.7 \pm 0.5$ | 0.828 | 4.0 | 30 | 2.5 | 3.7 |
| Multicrystalline silicon | $19.8 \pm 0.5$ | 0.795 | 1.09 | 30 | 2.4 | 3.5 |
| Copper indium diselenide (CIS/ CIGS) | $18.4 \pm 0.5$ | 0.77 | 1.04 | 5 | 1.5 | 2.5 |
| Thin silicon cell | $16.6 \pm 0.4$ | 0.782 | 4.02 | 25 | 2.0 | 3.3 |
| Cadmium telluride (CdTe) | $16.5 \pm 0.5$ | 0.755 | 1.03 | 15 | 1.5 | 2.5 |
| Amorphous silicon (a-si) | $10.1 \pm 0.2$ | 0.766 | 1.2 | 20 | 1.5 | 2.5 |

*Based on experience.
*Source: B. Agarwal, G.N. Tiwari, Development in Environmental Durability for Photovoltaics, Pira International Ltd., UK, 2008.*

© Bag Energy Research Society 2024
G. N. Tiwari, *Fundamental of Mathematical Tools for Thermal Modeling of Solar Thermal and Photo-voltaic Systems-Part-I*, https://doi.org/10.1007/978-981-99-7085-8

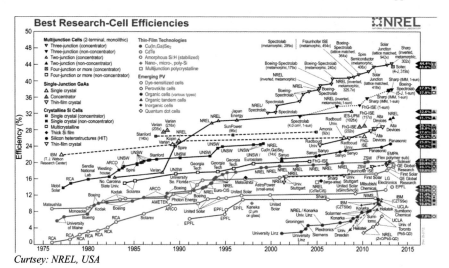

*Curtsey: NREL, USA*

# Appendix E

$\phi_P$ = Volume fraction of nanoparticles, $k_{nf}$ = *thermal* Conductivity of nanofluids, $k_{bf}$ = Thermal conductivity of base fluid, $k_p$ = Thermal conductivity of nanoparticles, $k_H = 6.2$, Huggins coefficient, n = Empirical shape factor, $r_p$ = Particle radius, h = Inter-particle spacing, t = thickness of the nanolayer, T = Temperature, $\phi_{p,\max}$ = maximum volume fraction of nanoparticles, $k_{layer}$ = thermal conductivity of the nanolayer.

**Sub-script** f = fluid, bf = base fluid, p = nanoparticle, nf = nanofluid

$(P_e)_p$ = Particle Peclet number of the nanoparticle, $Re_{nf}$ = Reynolds number of nanofluids, $Pr_{nf}$ = Prandtl number of nanofluids, $\alpha_{nf}$ = Thermal diffusivity of nanofluids, $K_{nf}$ = Thermal conductivity of nanofluids, D- Tube diameter, $d_p$- Particle diameter, u = fluid velocity, A = cross-section area of the tube, D = diameter of the tube, L = tube length, $T_{b1}$ = inlet bulk temperature (K), $T_{b2}$ = exit

**Table E1** Properties of air at atmospheric pressure. The values of $\mu$, $K$, $C_p$ and $Pr$ are not strongly pressure-dependent and may be used over a fairly wide range of pressures

| $T(K)$ | $\rho(kg/m^3)$ | $C_p(kJ/kg\ K)$ | $\mu$ (kg/m-s) $\times 10^{-5}$ | $v$ (m2/s) $\times 10^{-6}$ | $K$ (W/m2K) $\times 10^{-3}$ | $\alpha$ (m2/s) $\times 10^{-5}$ | $Pr$ |
|---|---|---|---|---|---|---|---|
| 100 | 3.6010 | 1.0259 | 0.6924 | 1.923 | 9.239 | 0.2501 | 0.770 |
| 150 | 2.3675 | 1.0092 | 1.0283 | 4.343 | 13.726 | 0.5745 | 0.753 |
| 200 | 1.7684 | 1.0054 | 1.3289 | 7.490 | 18.074 | 1.017 | 0.739 |
| 250 | 1.4128 | 1.0046 | 1.488 | 9.49 | 22.26 | 1.3161 | 0.722 |
| 300 | 1.1774 | 1.0050 | 1.983 | 15.68 | 26.22 | 2.216 | 0.708 |
| 350 | 0.9980 | 1.0083 | 2.075 | 20.76 | 30.00 | 2.983 | 0.697 |
| 400 | 0.8826 | 1.0134 | 2.286 | 25.90 | 33.62 | 3.760 | 0.689 |

Note: If one desire to calculate the thermal conductivity of air at 313 K, then

$$K \text{ at } 313 \text{ K} = K \text{ at } 300 + \frac{K\ at\ 350\ K - K\ at\ 300}{K\ at\ 300\ K} = 0.002622 +$$

$\frac{0.0030 - 0.002622}{0.002622} = 0.002622 + 0.000144 = 0.002762$ .

**Similarly, other properties at various temperature can be evaluated.**

G. N. Tiwari, *Fundamental of Mathematical Tools for Thermal Modeling of Solar Thermal and Photo-voltaic Systems-Part-I*, https://doi.org/10.1007/978-981-99-7085-8

**Table E.2** Properties of Water (saturated liquid)s

| Temperature | | $C_p$(kJ/kg K) | $\rho$ (kg/m$^3$) | $\mu_k$(kg/m s) | $K$ (W/m K) | $Pr$ | $\frac{g\,\beta\,\rho^2\,C_p}{\mu_k}$ (I/m$^3$ K) |
|---|---|---|---|---|---|---|---|
| °F | °C | | | | | | |
| 32 | 0.00 | 4.225 | 999.8 | $1.79 \times 10^{-3}$ | 0.566 | 13.25 | $1.91 \times 10^9$ |
| 40 | 4.44 | 4.208 | 999.8 | 1.55 | 0.575 | 11.35 | $6.34 \times 10^9$ |
| 50 | 10.00 | 4.195 | 999.2 | 1.31 | 0.585 | 9.40 | $1.08 \times 10^{10}$ |
| 60 | 15.56 | 4.186 | 998.6 | 1.12 | 0.595 | 7.88 | $1.46 \times 10^{10}$ |
| 70 | 21.11 | 4.179 | 997.4 | $9.8 \times 10^{-4}$ | 0.604 | 6.78 | $1.46 \times 10^{10}$ |
| 80 | 26.67 | 4.179 | 995.8 | 8.6 | 0.614 | 5.85 | $1.91 \times 10^{10}$ |
| 90 | 32.22 | 4.174 | 994.9 | 7.65 | 0.623 | 5.12 | $2.48 \times 10^{10}$ |
| 100 | 37.78 | 4.174 | 993.0 | 6.82 | 0.630 | 4.53 | $3.3 \times 10^{10}$ |
| 110 | 43.33 | 4.174 | 990.6 | 6.16 | 0.637 | 4.04 | $4.19 \times 10^{10}$ |
| 120 | 48.89 | 4.174 | 988.8 | 5.62 | 0.644 | 3.64 | $4.89 \times 10^{10}$ |
| 130 | 54.44 | 4.179 | 985.7 | 5.13 | 0.649 | 3.30 | $5.66 \times 10^{10}$ |
| 140 | 60.00 | 4.179 | 983.3 | 4.71 | 0.654 | 3.01 | $6.48 \times 10^{10}$ |
| 150 | 65.55 | 4.183 | 980.3 | 4.3 | 0.659 | 2.73 | $7.62 \times 10^{10}$ |
| 160 | 71.11 | 4.186 | 977.3 | 4.01 | 0.665 | 2.53 | $8.84 \times 10^{10}$ |
| 170 | 76.67 | 4.191 | 973.7 | 3.72 | 0.668 | 2.33 | $9.85 \times 10^{10}$ |
| 180 | 82.22 | 4.195 | 970.2 | 3.47 | 0.673 | 2.16 | $1.09 \times 10^{10}$ |
| 190 | 87.78 | 4.199 | 966.7 | 3.27 | 0.675 | 2.03 | |
| 200 | 93.33 | 4.204 | 963.2 | 3.06 | 0.678 | 1.90 | |
| 210 | 104.40 | 4.216 | 955.1 | 2.67 | 0.684 | 1.66 | |

bulk temperature (K), $T_w$ = wall temperature of the tube (K), $(T_w - T_b)_{LM}$ = logarithmic mean temperature difference in which $T_w$ is the wall temperature that is the average of ten measured temperatures on tube wall at different positions. $P_{nf}$ = Pressure drop of the pressure drop test section, L = Length of the pressure drop test section, g = Acceleration gravity, f = friction factor, volume concentration is $\varphi_v = \left[1/(100/\varphi_m)\left((\rho_p/\rho_w) + 1\right)\right] \times 100\%$

**Table E.3** Properties of conducting metals

| Metal | | Properties at 20 °C | | | |
|---|---|---|---|---|---|
| | | $\rho$(kg/ m$^3$) | $C_p$ (kJ/ kg K) | $K$ (W/m K) | A (m$^2$/s $\times 10^{-5}$) |
| Aluminum | Pure | 2707 | 0.896 | 204 | 8.418 |
| | Al-Si (Silumin, copper bearing) 86% Al, 1% Cu | 2659 | 0.867 | 137 | 5.933 |
| Lead | Pure | 11,400 | 0.1298 | 34.87 | 7.311 |
| Iron | Pure | 7897 | 0.452 | 73 | 2.034 |
| | Steel (Carbon steel) | 7753 | 0.486 | 63 | 0.970 |
| Copper | Pure | 8954 | 0.3831 | 386 | 11.234 |
| | Aluminum bronze (95% Cu, 5% Al) | 8666 | 0.410 | 383 | 2.330 |
| Bronze | 75% Cu, 25% Sn | 8666 | 0.343 | 326 | 0.859 |
| Red Brass | 85% Cu, 9% Sn 6% Zn | 8714 | 0.385 | 61 | 1.804 |
| Brass | 70% Cu, 30% Zn | 8600 | 0.877 | 85 | 3.412 |
| German Silver | 62% Cu, 15% Ni, 22% Zn | 8618 | 0.394 | 24.9 | 0.733 |
| Constantan | 60% Cu, 40% Ni | 8922 | 0.410 | 22.7 | 0.612 |
| Magnesium | Pure | 1746 | 1.013 | 171 | 9.708 |
| Nickel | Pure | 8906 | 0.4459 | 90 | 2.266 |
| Silver | Purest | 10,524 | 0.2340 | 419 | 17.004 |
| | Pure (99.9%) | 10,524 | 0.2340 | 407 | 16.563 |
| Tin | Pure | 7304 | 0.2265 | 64 | 3.884 |
| Tungsten | Pure | 19,350 | 0.1344 | 163 | 6.271 |
| Zinc | Pure | 7144 | 0.3843 | 112.2 | 4.106 |

**Table E.4** Properties of non-metals

| Material | Temperature (°C) | K (W/m k) | $\rho$ (kg/m$^3$) | C (kJ/kg K) | A (m$^2$/s) $\times$ 10$^{-7}$ | Remarks |
|---|---|---|---|---|---|---|
| Asbestos | 50 | 0.08 | 470 | – | – | Insulating |
| Building brick | 20 | 0.69 | 1600 | 0.84 | 5.2 | Conductive |
| Common face | – | 1.32 | 2000 | – | – | Conductive |
| Concrete, Cinder | 23 | 0.76 | – | – | – | Conductive |
| Stone 1-2-4 mix | 20 | 1.37 | 1900–2300 | 0.88 | 8.2–6.8 | Conductive |
| Glass, window | 20 | 0.78 (avg) | 2700 | 0.84 | 3.4 | Conductive |
| Toughened glass | 25 | 0.668 | 2500 | 0.80 | 3.4 | |
| Borosilicate | 30–75 | 1.09 | 2200 | – | – | Conductive |
| Plaster, Gypsum | 20 | 0.48 | 1440 | 0.84 | 4.0 | Conductive |
| Granite | – | 1.73–3.98 | 2640 | 0.82 | 8–18 | Conductive |
| Limestone | 100–300 | 1.26–1.33 | 2500 | 0.90 | 5.6–5.9 | Conductive |
| Marble | – | 2.07–2.94 | 2500–2700 | 0.80 | 10–13.6 | Conductive |
| Sandstone | 40 | 1.83 | 2160–2300 | 0.71 | 11.2–11.9 | Conductive |
| Fir | 23 | 0.11 | 420 | 2.72 | 0.96 | Conductive |
| Maple or Oak | 30 | 0.166 | 540 | 2.4 | 1.28 | Conductive |
| Solar cell | 25 | 130 | 2330 | 6.77 | | Conductive |
| Tedlar | 25 | 0.033 | 1200 | 1.25 | | Insulating |
| Yellow Pine | 23 | 0.147 | 640 | 2.8 | 0.82 | Conductive |
| Cord board | 30 | 0.043 | 160 | 1.88 | 2–5.3 | Insulating |
| Cork, regranulated | 32 | 0.045 | 45–120 | 1.88 | 2–5.3 | Insulating |
| Ground | 32 | 0.043 | 150 | – | – | Insulating |
| Sawdust | 23 | 0.059 | – | – | – | Insulating |
| Wood shaving | 23 | 0.059 | – | – | – | Insulating |

**Table E.5** Physical properties of some other materials

| S. No. | Material | Density (kg/m$^3$) | Thermal conductivity (W/m K) | Specific Heat (J/kg K) |
|---|---|---|---|---|
| 1 | Air | 1.117 | 40.026 | 1006 |
| 2 | Alumina | 3800 | 29.0 | 800 |
| 3 | Aluminum | 41–45 | 211 | 0.946 |
| 4 | Asphalt | 1700 | 0.50 | 1000 |
| 5 | Brick | 1700 | 0.84 | 800 |
| 6 | Carbon dioxide | 1.979 | 0.145 | 871 |
| 7 | Cement | 1700 | 0.80 | 670 |
| 8 | Clay | 1458 | 11.28 | 879 |
| 9 | Concrete | 2400 | 1.279 | 1130 |
| 10 | Copper | 8795 | 385 | – |
| 11 | Cork board | 240 | 0.04 | 2050 |
| 12 | Cotton Wool | 1522 | – | 1335 |
| 13 | Fiber board | 300 | 0.057 | 1000 |
| 14 | Glass-Crown | 2600 | 1.0 | 670 |
| 15 | Glass window | 2350 | 0.816 | 712 |
| 16 | Glass wool | 50 | 0.042 | 670 |
| 17 | Transparent flexible polymer | 1290 | 0.30 | – |
| 18 | Ice | 920 | 2.21 | 1930 |
| 19 | Iron | 7870 | 80 | 106 |
| 20 | Lime stone | 2180 | 1.5 | – |
| 21 | Mudphuska | – | – | – |
| 22 | Oxygen | 1.301 | 0.027 | 920 |
| 23 | Plaster-board | 950 | 0.16 | 840 |
| 24 | Polyesterene-expanded | 25 | 0.033 | 1380 |
| 25 | P.V.C.—rigid foam | 25–80 | 0.035–0.041 | – |
| 26 | P.V.C.—rigid sheet | 1350 | 0.16 | – |
| 27 | Saw dust | 188 | 0.57 | – |
| 28 | Thermocole | 22 | 0.03 | – |
| 29 | Timber | 600 | 0.14 | 1210 |
| 30 | Turpentine | 870 | 0.136 | 1760 |
| 31 | Water ($H_2O$) | 998 | 0.591 | 4190 |
| 32 | Seawater | 1025 | – | 3900 |
| 33 | Water vapor | 0.586 | 0.025 | 2060 |
| 34 | Wood wool | 500 | 0.10 | 1000 |

**Table E.6** Absorptivity of various surfaces for sun's ray

| Surface | Absorptivity | Surface | Absorptivity |
|---|---|---|---|
| White paint | 0.12–0.26 | **Walls** | |
| Whitewash/glossy white | 0.21 | White/yellow brick tiles | 0.30 |
| Bright aluminum | 0.30 | White stone | 0.40 |
| Flat white | 0.25 | Cream brick tile | 0.50 |
| Yellow | 0.48 | Burl brick tile | 0.60 |
| Bronze | 0.50 | Concrete/red brick tile | 0.70 |
| Silver | 0.52 | Red sand line brick | 0.72 |
| Dark aluminum | 0.63 | White sand stone | 0.76 |
| Bright red | 0.65 | Stone rubble | 0.80 |
| Brown | 0.70 | Blue brick tile | 0.88 |
| Light green | 0.73 | **Surroundings** | |
| Medium red | 0.74 | Sea/lake water | 0.29 |
| Medium green | 0.85 | Snow | 0.30 |
| Dark green | 0.95 | Grass | 0.80 |
| Blue/black | 0.97 | Light-colored grass | 0.55 |
| **Roof** | | Sand Gray | 0.82 |
| Asphalt | 0.89 | Rock | 0.84 |
| White asbestos cement | 0.59 | Green leaf | 0.85 |
| Cooper sheeting | 0.64 | Earth (black plowed field) | 0.92 |
| Uncolored roofing tile | 0.67 | White leaves | 0.20 |
| Red roofing tiles | 0.72 | Yellow leaves | 0.58 |
| Galvanized iron, clean | 0.77 | Aluminum foil | 0.39 |
| Brown roofing tile | 0.87 | Unpainted wood | 0.60 |
| Galvanized iron, dirty | 0.89 | | |
| Black roofing tile | 0.92 | | |
| **Metals** | | | |
| Polished aluminum/copper | 0.26 | | |
| New galvanized iron | 0.66 | | |
| Old galvanized iron | 0.89 | | |
| Polished iron | 0.45 | | |
| Oxidized rusty iron | 0.38 | | |

# Appendix F

## List of Embodied Energy Coefficients

| Material | MJ/kg | MJ/m3 |
|---|---|---|
| Aggregate, general | 0.10 | 150 |
| Virgin rock | 0.04 | 63 |
| River | 0.02 | 36 |
| Aluminum, virgin | 191 | 515,700 |
| Extruded | 201 | 542,700 |
| Extruded, anodized | 227 | 612,900 |
| Extruded, factory painted | 218 | 588,600 |
| Foil | 204 | 550,800 |
| Sheet | 199 | 537,300 |
| Aluminum, recycled | 8.1 | 21,870 |
| Extruded | 17.3 | 46,710 |
| Extruded, anodized | 42.9 | 115,830 |
| Extruded, factory painted | 34.3 | 92,610 |
| Foil | 20.1 | 54,270 |
| Sheet | 14.8 | 39,960 |
| Asphalt (paving) | 3.4 | 7140 |
| Bitumen | 44.1 | 45,420 |
| Brass | 62.0 | 519,560 |
| Carpet | 72.4 | – |
| Felt underlay | 18.6 | – |
| Nylon | 148 | – |
| Polyester | 53.7 | – |
| Polyethylterepthalate (PET) | 107 | – |
| Polypropylene | 95.4 | – |

(continued)

© Bag Energy Research Society 2024

639

G. N. Tiwari, *Fundamental of Mathematical Tools for Thermal Modeling of Solar Thermal and Photo-voltaic Systems-Part-I*, https://doi.org/10.1007/978-981-99-7085-8

(continued)

| Material | MJ/kg | MJ/m3 |
|---|---|---|
| Wool | 106 | – |
| Cement | 7.8 | 15,210 |
| Cement mortar | 2.0 | 3200 |
| Fiber cement board | 9.5 | 13550 |
| Soil–cement | 0.42 | 819 |
| Ceramic | | – |
| Brick | 2.5 | 5170 |
| Brick, glazed | 7.2 | 14760 |
| Pipe | 6.3 | – |
| Tile | 2.5 | 5250 |
| Concrete | | – |
| Block | 0.94 | – |
| Brick | 0.97 | – |
| GRC | 7.6 | 14820 |
| Paver | 1.2 | – |
| Precast | 2 .0 | – |
| Ready mix, 17.5 MPa | 1.0 | 2350 |
| 30 MPa | 1.3 | 3180 |
| 40 MPa | 1.6 | 3890 |
| Roofing tile | 0.81 | – |
| Copper | 70.6 | 631160 |
| Earth, raw | | – |
| Adobe block, straw stabilized | 0.47 | 750 |
| Adobe, bitumen stabilized | 0.29 | – |
| Adobe, cement stabilized | 0.42 | – |
| Rammed soil–cement | 0.80 | – |
| Pressed block | 0.42 | – |
| Fabric | | – |
| Cotton | 143 | – |
| Polyester | 53.7 | – |
| Glass | 66.2 | – |
| Float | 15.9 | 40060 |
| Toughened | 26.2 | 66020 |
| Laminated | 16.3 | 41080 |
| Tinted | 14.9 | 375450 |
| Insulation | | – |
| Cellulose | 3.3 | 112 |

(continued)

(continued)

| Material | MJ/kg | MJ/m3 |
|---|---|---|
| Fiberglass | 30.3 | 970 |
| Polyester | 53.7 | 430 |
| Polystyrene | 117 | 2340 |
| Wool (recycled) | 14.6 | 139 |
| Lead | 35.1 | 398,030 |
| Linoleum | 116 | 150,930 |
| Paint | 90.4 | 118 per liter |
| Solvent based | 98.1 | 128 per liter |
| Water based | 88.5 | 115 per liter |
| Paper | 36.4 | 33670 |
| Building | 25.5 | – |
| Kraft | 12.6 | – |
| Recycled | 23.4 | – |
| Wall | 36.4 | – |
| Plaster, gypsum | 4.5 | 6460 |
| Plaster board | 6.1 | 5890 |
| Plastics | | – |
| ABS | 111 | – |
| High-density polyethylene (HDPE) | 103 | 97,340 |
| Low-density polyethylene (LDPE) | 103 | 91,800 |
| Polyester | 53.7 | 7710 |
| Polypropylene | 64.0 | 57,600 |
| Polystyrene, expanded | 117 | 2340 |
| Polyurethane | 74.0 | 44,400 |
| PVC | 70.0 | 93,620 |
| Rubber | | – |
| Natural latex | 67.5 | 62,100 |
| Synthetic | 110 | – |
| Sand | 0.10 | 232 |
| Sealants and adhesives | | – |
| Phenol formaldehyde | 87.0 | – |
| Urea formaldehyde | 78.2 | – |
| Steel, recycled | 10.1 | 37,210 |
| Reinforcing, sections | 8.9 | – |
| Wire rod | 12.5 | – |
| Steel, virgin, general | 32.0 | 251,200 |
| Galvanized | 34.8 | 273,180 |

(continued)

(continued)

| Material | MJ/kg | MJ/m3 |
|---|---|---|
| Imported, structural | 35.0 | 274,570 |
| Stone, dimension | | – |
| Local | 0.79 | 1890 |
| Imported | 6.8 | 1890 |
| Straw, baled | 0.24 | 30.5 |
| Timber, softwood | | – |
| Air dried, rough sawn | 0.3 | 165 |
| Kiln dried, rough sawn | 1.6 | 880 |
| Air dried, dressed | 1.16 | 638 |
| Kiln dried, dressed | 2.5 | 1380 |
| Moldings, etc | 3.1 | 1710 |
| Hardboard | 24.2 | 13,310 |
| MDF | 11.9 | 8330 |
| Glulam | 4.6 | 2530 |
| Particle bd | 8.0 | – |
| Plywood | 10.4 | – |
| Shingles | 9.0 | – |
| Timber, hardwood | | – |
| Air dried, rough sawn | 0.50 | 388 |
| Kiln dried, rough sawn | 2.0 | 1550 |
| Vinyl flooring | 79.1 | 105,990 |
| Zinc | 51.0 | 364,140 |
| Galvanizing, per kg steel | 2.8 | – |

# Appendix G

**Heating Values of Various Combustibles and their Conversion Efficiencies**

| Fuel | Heating value (kJ/kg) | Efficiency of device |
|---|---|---|
| Coal coke | 29,000 | 70 |
| Wood | 15,000 | 60 |
| Straw | 14,000–16,000 | 60 |
| Gasoline | 43,000 | 80 |
| Kerosene | 42,000 | 80 |
| Methane (Natural gas) | 50,000 | 80 |
| Biogas (60% methane) | 20,000 | 80 |
| Electricity | – | 95 |

© Bag Energy Research Society 2024
G. N. Tiwari, *Fundamental of Mathematical Tools for Thermal Modeling of Solar Thermal and Photo-voltaic Systems-Part-I*, https://doi.org/10.1007/978-981-99-7085-8

# Appendix H

**Steam table for saturation vapor pressure**

| Temp. (K) | $P\ (N/m^2)$ | Temp. (K) | $P\ (N/m^2)$ | Temp. (K) | $P\ (N/m^2)$ |
|---|---|---|---|---|---|
| 273 | 610.8 | 304 | 4491.0 | 334 | 20,860.0 |
| 274 | 656.6 | 305 | 4743.0 | 335 | 21,840.0 |
| 275 | 705.5 | 306 | 5029.0 | 336 | 22,860.0 |
| 276 | 757.6 | 307 | 5318.0 | 337 | 23,710.0 |
| 277 | 812.0 | 308 | 5622.0 | 338 | 25,010.0 |
| 278 | 871.8 | 309 | 5940.0 | 339 | 26,150.0 |
| 279 | 934.5 | 310 | 6274.0 | 340 | 27,330.0 |
| 280 | 1001.2 | 311 | 6624.0 | 341 | 28,560.0 |
| 281 | 1072.0 | 312 | 6991.0 | 342 | 29,840.0 |
| 282 | 1147.2 | 313 | 7375.0 | 343 | 31,160.0 |
| 283 | 1227.0 | 314 | 7777.0 | 344 | 32,530.0 |
| 284 | 1311.6 | 315 | 8198.0 | 345 | 33,960.0 |
| 285 | 1401.4 | 316 | 8639.0 | 346 | 35,430.0 |
| 286 | 1496.5 | 317 | 9100.0 | 347 | 36,960.0 |
| 287 | 1597.3 | 318 | 9583.0 | 348 | 38,550.0 |
| 288 | 1703.9 | 319 | 10,086.0 | 349 | 40,190.0 |
| 289 | 1816.8 | 320 | 10,612.0 | 350 | 41,890.0 |
| 290 | 1936.2 | 321 | 11,162.0 | 351 | 43,650.0 |
| 291 | 2062.0 | 322 | 11,736.0 | 352 | 45,470.0 |
| 292 | 2190.0 | 323 | 12,335.0 | 353 | 47,360.0 |
| 293 | 2337.0 | 324 | 12,961.0 | 354 | 49,310.0 |
| 294 | 2485.0 | 325 | 13,613.0 | 355 | 51,350.0 |
| 295 | 2642.0 | 326 | 14,340.0 | 356 | 53,420.0 |
| 296 | 2808.0 | 327 | 15,002.0 | 357 | 55,570.0 |

(continued)

© Bag Energy Research Society 2024
G. N. Tiwari, *Fundamental of Mathematical Tools for Thermal Modeling of Solar Thermal and Photo-voltaic Systems-Part-I*, https://doi.org/10.1007/978-981-99-7085-8

(continued)

| Temp. (K) | P (N/m$^2$) | Temp. (K) | P (N/m$^2$) | Temp. (K) | P (N/m$^2$) |
|---|---|---|---|---|---|
| 297 | 2982.0 | 328 | 15,641.0 | 358 | 57,800.0 |
| 298 | 3166.0 | 329 | 16,511.0 | 359 | 60,110.0 |
| 299 | 3360.0 | 330 | 17,313.0 | 360 | 62,490.0 |
| 300 | 3564.0 | 331 | 18,147.0 | 361 | 64,950.0 |
| 301 | 3778.0 | 332 | 19,016.0 | 362 | 67,490.0 |
| 302 | 4004.0 | 333 | 19,920.0 | 363 | 70,110.0 |
| 303 | 4241.0 | | | | |

# Appendix I

A1. Derivation of exergy of thermal energy based on entropy concept.

Referring to Fig. A.1a, the entropy of water heat engine and ambient can be written as follows:

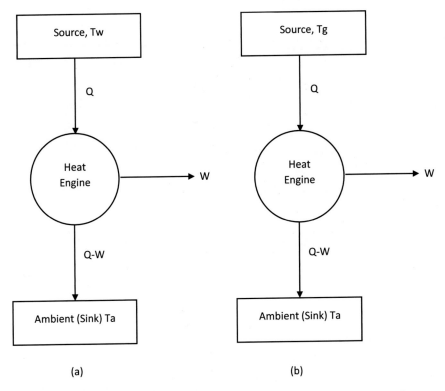

(a)                                       (b)

**Fig. A.1** Block diagram for determining the maximum work done by using the concept of entropy

© Bag Energy Research Society 2024
G. N. Tiwari, *Fundamental of Mathematical Tools for Thermal Modeling of Solar Thermal and Photo-voltaic Systems-Part-I*, https://doi.org/10.1007/978-981-99-7085-8

$$(\Delta S)_{water} = \int_{T_w}^{T_a} \frac{C_p dT}{T_w} = C_p ln \frac{T_a}{T_w} \tag{A1}$$

$$(\Delta S)_{heat\ engine} = 0 \tag{A2}$$

$$(\Delta S)_{ambient} = \frac{Q - W}{T_a} \tag{A3}$$

The total entropy of a system, Fig. A.1a can be written as follows:

$$(\Delta S)_{total} = (\Delta S)_{water} + (\Delta S)_{heat\ engine} + (\Delta S)_{ambient} = C_p ln \frac{T_a}{T_w} + \frac{Q - W}{T_a} \tag{A4}$$

Eq. (A4) can be referred to as universe entropy as:

$$(\Delta S)_{universe} = C_p ln \frac{T_a}{T_w} + \frac{Q - W}{T_a} \tag{A5}$$

By entropy principle, $(\Delta S)_{universe} \geq 0$; hence Eq. (A5) becomes as:

$$C_p ln \frac{T_a}{T_w} + \frac{Q - W}{T_a} \geq 0$$

Or

$$W \leq Q + T_a C_p ln \frac{T_a}{T_w} \tag{A6}$$

From the above equation, maximum work between water and ambient becomes as:

$$W_{max,1} = C_p \left[ (T_w - T_a) + T_a ln \frac{T_a}{T_w} \right] \tag{A7}$$

Similarly, in Fig. A1b, $W_{max}$ between glass and ambient can be written as:

$$W_{max,2} = C_p \left[ (T_g - T_a) + T_a ln \frac{T_a}{T_g} \right] \tag{A8}$$

By using Eqs. (A7) and (A8), maximum work between water and glass (condensing cover) can be written as:

$$W_{max} = W_{max,1} - W_{max,2} = C_p \left[ (T_w - T_g) + T_a ln \frac{T_w}{T_g} \right] \tag{A9}$$

The above maximum work is nothing but exergy; hence exergy of a system is given by:

$$E_{x,wg} = C_p \left[ (T_w - T_g) + T_a ln \frac{T_w}{T_g} \right]$$  (A10)

Since $\dot{m}C_p = hA$, (W/°C) hence exergy in terms of heat transfer coefficient (h) becomes:

$$E_{x,wg} = h \left[ (T_w - T_g) + T_a ln \frac{T_w}{T_g} \right]$$  (A11)

It is important to mention that all temperature values are considered in Kelvin. Further, $C_p = C_w$ is the same for both Fig. A.1a, b due to condensed water at the inner surface of condensing cover

# Glossary

| S. No. | Symbol | Explanation |
|---|---|---|
| 1 | C | Doller per tone |
| 2 | CCE | Carbon credit earned |
| 3 | CCT | Constant collection temperature |
| 4 | $CO_2$ | Carbon dioxide |
| 5 | CPC | Compound parabolic concentrator |
| 6 | DC | Direct current |
| 7 | ETC | Evacuated tubular collector |
| 8 | FPC | Flat plate collector |
| 9 | FL | Fresnel lens |
| 10 | FLiETC | Fresnel lens integrated with evacuated tubular collector |
| 11 | FLiOPVT-TEC | Fresnel lens integrated with flexible opaque PVT-TEC module |
| 12 | FLiSPVT | Fresnel lens integrated with SPVT module |
| 13 | FPC-CPC | Flat plate collector integrated with Compound parabolic concentrator |
| 14 | $m - FPC$ | m-FPC connected in parallel |
| 15 | m-SAC | m solar air collector connected in parallel |
| 16 | N-FPC | N FPC connected in series |
| 17 | PV | Photo-voltaic module |
| 18 | PVT | Photo-voltaic thermal module |
| 19 | PVT-CPC | Photo-voltaic thermal module integrated with compound parabolic concentrator |
| 20 | N-PVT | N-PVT collectors connected in series |
| 21 | N-SAC | N-solar air collector connected in series |
| 22 | OPVT | Opaque PVT module |

(continued)

© Bag Energy Research Society 2024
G. N. Tiwari, *Fundamental of Mathematical Tools for Thermal Modeling of Solar Thermal and Photo-voltaic Systems-Part-I*, https://doi.org/10.1007/978-981-99-7085-8

(continued)

| S. No. | Symbol | Explanation |
| --- | --- | --- |
| 23 | OPVT-TEC | Opaque PV module integrated with TEC |
| 24 | OPVT-TEC-CPC | OPVT-TEC module integrated with CPC |
| 25 | PV-TEC | Photo-voltaic module integrated with thermoelectric cooler |
| 26 | PVT-TEC-CPC | PV-TEC module integrated with CPC |
| 27 | TEC | Thermoelectric cooler |
| 28 | SAC | Solar air collector |
| 29 | SPVT | Semi-transparent PVT module |
| 30 | T | Life of solar energy system |
| 31 | N-SAC | N solar air collector connected in series |